Universal Algebra

Second Edition

George Grätzer

Universal Algebra

Second Edition

Springer

George Grätzer
Department of Mathematics
University of Manitoba
Winnipeg, Manitoba R3T 2N2
Canada
gratzer@ms.umanitoba.ca

ISBN: 978-0-387-77486-2 e-ISBN: 978-0-387-77487-9
DOI: 10.1007/978-0-387-77487-9

Library of Congress Control Number: 2008922052

Mathematics Subject Classification (2000): 08-01, 08-02

Originally published in the University Series in Higher Mathematics (D. Van Nostrand Company); edited by M. H. Stone, L. Nirenberg, and S. S. Chern, 1968

Printed on acid-free paper.

9 8 7 6 5 4 3 2 1

springer.com

INTRODUCTION TO THE SECOND EDITION

About a year ago I was approached by the publisher of the present volume about a new edition of the book, Universal Algebra. This was a good opportunity to review the book I wrote in 1964–1965, about thirteen years ago, to find out whether I can still subscribe to the presentation of the book or a complete revision is necessary.

It is my opinion that the definitions and results presented in this book form a foundation of universal algebra as much today as they did a decade ago. Some concepts became more important and some new ones appeared, but the foundation has not changed much.

On the other hand, my point of view changed rather substantially in a number of areas. Compare the elementary approach to the congruence lattice characterization theorem of a decade ago with the axiomatic approach of today (see Appendix 7).

The obstacles in the way of a complete revision appeared just as formidable. An initial appraisal put the number of papers written since the bibliography of Universal Algebra was closed (around 1967) near to 1000, making it very difficult for someone to pretend to be an expert on all the major developments in universal algebra. At twenty-seven I thought nothing of establishing as my goal "to give a systematic treatment of the most important results"; at forty (with a thousand more papers to contend with and in the middle of proofreading my General Lattice Theory) I was not so sure of being able to undertake the same.

So I decided to obtain the help of a number of experts to review various aspects of recent developments. B. Jónsson agreed to survey congruence varieties, a fast evolving chapter of universal algebra, based on his 1974 lecture at the Vancouver meeting of the International Mathematical Union (Appendix 3). Walter Taylor consented to have an abbreviated version of his survey on equational theories included (Appendix 4). R. W. Quackenbush undertook to present primal algebras and their generalizations, a vast field containing many important results (Appendix 5). Finally, G. H. Wenzel agreed to survey equational compactness (Appendix 6).

In addition, Appendix 1 surveys the developments of the last decade: in §55, the survey follows the sections of the book; §56 surveys related structures and §57 outlines some important new topics. Appendix 2 reviews the problems given in the first edition. Finally, Appendix 7 contains a proof of the independence of congruence lattices, automorphism groups, and subalgebra lattices of infinitary algebras and the characterization of type-2 congruence lattices by modularity; these have not previously appeared in print.

Referencing to the new bibliography is by year of publication, e.g., [1975], [1975 a]; items not in print at the time of the original compilation are listed as [a], [b], etc.

All the appendices and the new bibliography have been widely circulated. I would like to thank all those who sent in corrections and additions, especially H. Andréka, J. Berman, C. C. Chen, A. P. Huhn, L. Márki, I. Németi, B. M. Schein, W. Taylor, and A. Waterman.

In the compilation of the new bibliography I was greatly assisted by M. E. Adams. The typing and clerical work was done by L. Gushulak, M. McTavish, and S. Padmanabhan. In the proofreading I was helped by M. E. Adams, W. J. Blok, and P. Köhler.

INTRODUCTION TO THE FIRST EDITION

In A. N. Whitehead's book on Universal Algebra,† published in 1898, the term universal algebra had very much the same meaning that it has today.

Universal algebra started to evolve when mathematics departed from the study of operations on real numbers only. Hamilton's quaternions, Boole's symbolic logic, and so forth, brought to light operations on objects other than real numbers and operations which are very different from the traditional ones.

"Such algebras have an intrinsic value for separate detailed study; also they are worthy of a comparative study, for the sake of the light thereby thrown on the general theory of symbolic reasoning, and on algebraic symbolism in particular. The comparative study necessarily presupposes some previous separate study, comparison being impossible without knowledge"; so wrote Whitehead in 1898 and his point of view is still shared by many.

Thus universal algebra is the study of finitary operations on a set, and the purpose of research is to find and develop the properties which such diverse algebras as rings, fields, Boolean algebras, lattices, and groups may have in common.

Although Whitehead recognized the need for universal algebra, he had no results. The first results were published by G. Birkhoff in the thirties. Some thirty years elapsed between Whitehead's book and Birkhoff's first paper, despite the fact that the goal of research was so beautifully stated in Whitehead's book. However, to generalize, one needs experience, and before the thirties most of the branches of modern algebra were not developed sufficiently to give impetus to the development of universal algebras.

In the period from 1935 to 1950 most papers were written along the lines suggested by Birkhoff's papers: free algebras, the homomorphism theorem and the isomorphism theorems, congruence lattices, and subalgebra lattices were discussed. Many of the results of this period can be found in Birkhoff's book [6].

† According to A. N. Whitehead, the subject matter originated with W. R. Hamilton and A. DeMorgan, and the name for it was coined by J. J. Sylvester.

In the meantime, mathematical logic developed to the point where it could be applied to algebra. K. Gödel's completeness theorem (Gödel [1]), A. Tarski's definition of satisfiability, and so on, made mathematicians realize the possibility of applications. Such applications came about slowly. A. I. Mal'cev's 1941 paper [2] was the first one, but it went unnoticed because of the war. After the war, A. Tarski, L. A. Henkin, and A. Robinson began working in this field and they started publishing their results about 1950.

A. Tarski's lecture at the International Congress of Mathematicians (Cambridge, Massachusetts, 1950) may be considered as the beginning of the new period.

The model-theoretic aspect of universal algebras was mostly developed by Tarski himself and by C. C. Chang, L. A. Henkin, B. Jónsson, H. J. Keisler, R. C. Lyndon, M. Morley, D. Scott, R. L. Vaught, and others, and to a certain extent by A. I. Mal'cev.

In the late fifties E. Marczewski [2] emphasized the importance of bases of free algebras; he called them independent sets. As a result Marczewski, J. Mycielski, W. Narkiewicz, W. Nitka, J. Płonka, S. Świerczkowski, K. Urbanik, and others were responsible for more than 50 papers on the algebraic theory of free algebras.

There are a number of individuals who have not been mentioned yet and who have made significant contributions to universal algebra. It is hoped that the references in the text will give everyone his due credit.

Because of the way in which universal algebras developed, many elementary results have never been published but have been used without any reference in the papers, sometimes only in the form of a "therefore". It is hoped that this book will give an adequate background for the explanation of the "therefore's".

The purpose of this book is to give a systematic treatment of the most important results in the field of universal algebras. We will consider generalizations of universal algebras only to the extent that they are necessary for the development of the theory of universal algebras themselves. Therefore, the particular problems of partial algebras and structures are not discussed. Infinitary algebras will be touched upon only in the exercises. Multi-algebras are scarcely mentioned at all. This limitation is quite natural. First of all, to keep the length of a book within reasonable bounds, some limitations are necessary. Secondly, it so happens that most of the results on universal algebras can be extended in each of the directions mentioned, at the expense of more involved notations. Since the purpose of a book should be, in the author's opinion, to present ideas and methods, the framework of universal algebras is sufficiently wide enough to accomplish this. However, each of these directions has problems of its own. For instance, infinitary partial algebras contain topological spaces as special cases, and the nature of these investigations does not have much to do with the topic of this book. Topological universal algebras and partially ordered universal algebras have not been included because of lack of material.

Category theory is excluded from this book because a superficial treatment seems to present pedagogical difficulties and it would be mathematically not too effective; moreover, those topics that can be treated in depth in a categorical framework (in particular, parts of Chapters 4 and 6) are to be discussed by S. Eilenberg in a book (universal algebra and automata theory) and by F. W. Lawvere in a book (on elementary theories), in lecture notes (on algebraic theories), and in an expository article (on the category of sets). However, there are a number of exercises originating in category theory.

Since a short description of the content is given at the beginning of each chapter, we will include here only a brief outline of the book.

In Chapter 0 the set-theoretic notations together with some basic facts are given, of course, without proof. The last section is on a special type of lattices that are useful in algebraic applications. One can hardly expect everyone to agree with the presentation of Chapter 0. Some will find it too short, some too long. However, it is hoped that the reader without set-theoretic knowledge will find sufficient background material there for an understanding of the remainder of the book, and, if he wants to delve deeper into set theory, at least he will know what to look for.

Chapters 1–3 develop the basic results. In Chapter 1, polynomials, polynomial symbols, homomorphisms, congruence relations, and subalgebras are discussed and the standard results, the isomorphism theorems, and the like are given. The same results for partial algebras are presented in Chapter 2, but only from the point of view of applications to algebras. To show the usefulness of partial algebras, the last two sections of Chapter 2 give the characterization theorem of congruence lattices of algebras, due to E. T. Schmidt and the author. Constructions of new algebras from given ones play a central role in universal algebras. Direct products, subdirect products, direct and inverse limits, and many related constructions are given in Chapter 3.

In Chapters 4 and 5 one of the most important concepts of universal algebras, namely that of free algebras, is discussed. The constructions, basic properties, and several applications of free algebras are given in Chapter 4, and in Chapter 5 we consider the bases of free algebras, a concept identical with E. Marczewski's notion of independence.

A short introduction to model theory is given in Chapter 6. The basic tool is J. Łoś' concept of prime product.

In Chapter 7 these results are applied to determine the properties that are preserved under certain algebraic constructions using generalized atomic formulas of H. J. Keisler; for direct products, the method of S. Feferman and R. L. Vaught is used.

It is hoped that most experts will agree that in these chapters the author has selected the most important topics not biased by his own research, but this obviously does not apply to Chapter 8, which is the author's theory of free structures over first order axiom systems. However, this topic seems to be as good as any to yield further applications of the methods developed in Chapter 6 to the purely algebraic problems of free algebras.

Each chapter is followed by exercises and problems. There are more than 650 exercises and over 100 research problems in the book. Many of the exercises are simple illustrations of new concepts, some ask for (or give) counterexamples, and some review additional results in the field. The problems list some open questions which the author thought interesting.

The numbering system of theorems, lemmas, corollaries, definitions, exercises, and problems is self-explanatory. Within each section, theorems and lemmas are numbered consecutively. A single corollary to a theorem or lemma is not numbered; however, if more than one corollary follows a lemma or theorem, they are numbered from one in each case. Theorem 2 refers to Theorem 2 of the section in which it occurs; Theorem 38.2 refers to Theorem 2 of §38; Exercise 3.92 refers to Exercise 92 of Chapter 3.

The present book is intended for the mathematician who wants to use the methods and results of universal algebra in his own field and also for those who want to specialize in universal algebra. For applications of universal algebra to groups, rings, Lie algebras, and so on, the reader should consult P. M. Cohn [1] and §6 of the author's report [14].

The first version of the Bibliography was sent out to about 50 experts. Numerous suggestions were received, for which the author wants to thank each contributor. In the compilation of the original bibliography, and also of the revised form, the author was helped by Catherine M. Gratzer.

This book is based on the notes of the lectures delivered at the Pennsylvania State University between October 1, 1964, and November 1, 1965. Professor Leo F. Boron took notes of the lectures, and after his notes were reviewed (many times, rewritten), he typed them up and had them duplicated. He worked endless hours on this. The author finds it hard to find the words which would express his gratitude for Professor Boron's unselfish help. These lectures notes were completely rewritten by the author and mimeographed. Thanks are due to the Mathematics Department of the Pennsylvania State University for providing partial funds for this project and to Mrs. L. Moyer who did all the typing of this second version.

The author cannot be too grateful to the large number of mathematicians who took the time and trouble to read the second mimeographed version and to send him detailed lists of suggestions and corrections, major and minor. The author wants to thank all of them for their help, especially P. M. Cohn, K. H. Diener, B. Jónsson, H. F. J. Lowig, D. Monk, M. Novotny, H. Ribeiro, B. M. Schein, J. Schmidt, and A. G. Waterman; their generous interest was invaluable in writing the third, final version.

The author's students, especially M. I. Gould, G. H. Wenzel, and also R. M. Vancko and C. R. Platt, contributed many suggestions, simplifications of proofs, and corrections at all stages of the work. They also helped in checking the Bibliography and in presenting papers in the seminar. The task of the final revision of the manuscript, including a final check of the Bibliography, was undertaken by C. C. Chen. E. C. Johnston, W. A. Lampe, H. Pesotan, C. R. Platt, R. M. Vancko, and G. H. Wenzel aided the author in the proofreading.

Thanks are also due to the Mathematics Department of McMaster University, Hamilton, Ontario, Canada, and especially to Professors B. Banaschewski and G. Bruns, for making it possible for the author to give three series of lectures (December 1964, June 1965, and December 1965) on parts of this book, and for their several suggestions.

CONTENTS

CHAPTER 3. CONSTRUCTIONS OF ALGEBRAS

CHAPTER 4. FREE ALGEBRAS

CHAPTER 5. INDEPENDENCE

CHAPTER 6. ELEMENTS OF MODEL THEORY

CHAPTER 7. ELEMENTARY PROPERTIES OF ALGEBRAIC CONSTRUCTIONS

APPENDIX 5. PRIMALITY: THE INFLUENCE OF BOOLEAN ALGEBRAS IN UNIVERSAL ALGEBRA

by Robert W. Quackenbush

APPENDIX 6. EQUATIONAL COMPACTNESS

by Günter H. Wenzel

APPENDIX 7. THE INDEPENDENCE PROOF

by G. Grätzer and W. A. Lampe

TABLE OF NOTATION

Following a symbol the page number of the first occurrence is given in parentheses.

Algebra, structure	$\mathfrak{A}$	$\mathfrak{B}$	$\mathfrak{C}$	$\mathfrak{D}$	$\mathfrak{E}$	$\mathfrak{F}$	$\mathfrak{G}$	$\mathfrak{H}$	$\mathfrak{I}$	$\mathfrak{K}$	$\mathfrak{L}$	$\mathfrak{M}$	$\mathfrak{N}$	$\mathfrak{O}$	$\mathfrak{P}$	$\mathfrak{Q}$	$\mathfrak{R}$	$\mathfrak{S}$	$\mathfrak{T}$	$\mathfrak{U}$	$\mathfrak{V}$	$\mathfrak{W}$	$\mathfrak{X}$	$\mathfrak{Y}$	$\mathfrak{Z}$
Base set	A	B	C	D	E	F	G	H	I	K	L	M	N	O	P	Q	R	S	T	U	V	W	X	Y	Z

Algebras, structures. $\langle A; F\rangle$ (8, 33, 224), $\langle A; F, R\rangle$ (223), $\langle A, R\rangle$ (8, 224), $\mathfrak{A}/\Theta$ (36, 82), $\mathfrak{P}^{(n)}(\mathfrak{A})$ (38), $\mathfrak{P}^{(n)}(\tau)$ (40, 84) $\mathfrak{P}^{(\alpha)}(\tau)$ (41), $\mathfrak{P}^{(\alpha)}(K)$ (43), $\mathfrak{C}(\mathfrak{A})$ (51), $\mathfrak{E}(\mathfrak{A})$ (67, 98), $\mathfrak{G}(\mathfrak{A})$ (68), $\mathfrak{E}_F(\mathfrak{A})$, $\mathfrak{E}_S(\mathfrak{A})$ (98), $\mathfrak{L}(\mathfrak{A})$ (72), $\mathfrak{L}(\tau)$ (172), $\mathfrak{F}_K(\mathfrak{m})$, $\mathfrak{F}_K(\alpha)$ (163), $\mathfrak{F}_K(\mathfrak{A})$ (180), $\mathfrak{F}_K(\alpha, \Omega)$ (183), $\mathfrak{F}_\Sigma(\alpha)$ (307), $\langle\mathfrak{A}, \mathsf{a}\rangle$ (239).

Classes of algebras, structures. $K(\tau)$ (34, 223), $K(\mathfrak{O}, X)$ (165), $\hat{K}$ (278), $Sp(K)$ (159), $Id(K)$ (170), $v(K)$, $f(K)$ (191), $A(K)$, $\hat{A}(K)$ (327).

Algebraic constructions. $\mathfrak{A}/\Theta$ (36, 82), $\prod(\mathfrak{A}_i \mid i \in I)$ (118), $\mathfrak{A}^I$ (119), $\prod_{\mathscr{D}}(\mathfrak{A}_i \mid i \in I)$ (144, 145, 240), $\mathfrak{A}_{\mathscr{D}}{}^I$ (144, 145, 246), $\lim\limits_{\rightarrow} \mathscr{A}$ (129, 130), $\lim\limits_{\leftarrow} \mathscr{A}$ (131), $\mathscr{A}/P$ (135, 136), $\mathfrak{A}[\mathfrak{B}]$ (147), $\mathbf{I}$, $\mathbf{S}$, $\mathbf{H}$, $\mathbf{P}$, $\mathbf{P}^*$, $\mathbf{P}_S$, $\mathbf{P}_S{}^*$, $\underset{\rightarrow}{\mathbf{L}}$, $\underset{\leftarrow}{\mathbf{L}}$, (152), $\mathbf{C}$ (158), $\mathbf{P}_P$ (244).

Sets. $\in$, $\notin$, $\subseteq$, $\subset$, $\cup$, $\cap$, $'$, $-$, $\bigcup$, $\bigcap$, $\varnothing$, (1, 4), $P(A)$ (1), $(a_i \mid i \in I)$, $\{a_i \mid i \in I\}$ (4) Part(A) (1), ι, ι_A, ω, ω_A (2), $A \times B$, $\prod(A_i \mid i \in I)$, A^n (2, 4, 5), $E(A)$ (3), $|A|$ (13), A/ε (6), $[x]\varepsilon$ (6), ε_φ (7).

Mappings. φ: $A \rightarrow B$, $a\varphi$, $\varphi(a)$, $A\varphi$, $D(\varphi)$ (3), $D(f, \mathfrak{A})$ (80), φ_A (4), A^B (4), A^α (16), $e_i{}^I$, $e_i{}^n$ (5), $\varphi\psi$ (6), ε (17), $M(A)$, $M_O(A)$, $M_I(A)$ (17), $\mathbf{p}_{\mathfrak{A}}$, $\mathbf{p}^{\mathfrak{A}}$ (40, 304), $p(\bar{a})$ (42) a, a_k, $\mathsf{a}(k/b)$ (227).

Partially ordered sets, lattices, Boolean algebras. $\leqq$, $<$ (8), l.u.b.(H), g.l.b. (H) (10), $\vee$, $\wedge$, $\bigvee$, $\bigwedge$, 0, 1, $'$ (8, 9, 10, 11), $\mathfrak{P}(A)$ (9), $(H]$, $(a]$ (20), $[H)$, $[a)$ (26), $I(\mathfrak{L})$, $\mathfrak{I}(\mathfrak{L})$ (20).

Cardinals, ordinals. $|A|$ (13) $\aleph_0$, $\mathfrak{c}$ (13), $\aleph_\alpha$ (16), $\mathfrak{m} \leqq \mathfrak{n}$, $\mathfrak{m}\cdot\mathfrak{n}$, $\mathfrak{m}+\mathfrak{n}$, $\mathfrak{m}^\mathfrak{n}$, $\Sigma(\mathfrak{m}_i \mid i \in I)$, $\prod(\mathfrak{m}_i \mid i \in I)$ (13), $\bar{\alpha}$ (15), ω (14), $\omega_\mathfrak{m}$ (15), ω_α (16), $\alpha+\beta$, $\alpha\cdot\beta$, $\Sigma(\alpha_i \mid i \in I)$, $\lim(\alpha_i \mid i \in I)$ (15).

Closures. $[X]_\mathscr{E}$ (6), $[H]_\mathscr{A}$ (24), $[H]$ (24, 35), $[H]_\Sigma$ (303), $\mathscr{S}(\mathfrak{A})$ (45, 96), $\mathscr{S}^+(\mathfrak{A})$ (49), $\mathscr{S}^0(\mathfrak{A})$ (72), $[a_0, \cdots, a_{n-1}]$ (202).

Congruence relations. $\Theta_{\bar{a}}$ (42, 84), Θ_K, $\Theta_\mathfrak{A}$ (42), $\Theta(H)$, $\Theta(a, b)$ (52), ι, ω (2), Φ/Θ (59), Θ^σ (105), Θ_L (144), $\mathscr{O}_K(a)$ (214), $\Sigma\Theta_i$ (215).

Logic. $L(\tau)$ (225), $\mathbf{L}(\tau)$ (226), $L_\mathfrak{A}(\tau)$ (239), $L_{\mathfrak{A},\mathfrak{B}}(\tau)$ (271), $L(\tau \oplus \xi)$ (253), $L^{(\alpha)}(\tau)$ (249), $\wedge$, $\vee$, $\neg$, $\rightarrow$, $\leftrightarrow$, $\equiv$, $\bigwedge$, $\bigvee$, $\exists$, $\forall$ (225, 226), $\models$, $\Leftrightarrow$, $_\mathfrak{A}\!\models$ (232), $\mathfrak{A} \equiv \mathfrak{B}$ (234), $\langle\mathfrak{A}, \mathsf{a}\rangle$ (239), K^* (255), Σ^* (170, 255), $S(\Phi, \mathsf{f})$ (241), $\mathscr{S}(\Phi)$ (291), $e(\Phi)$ (301).

Properties and conditions. U_N (294), B_n, C_n (315), B, C (317), P (318), IP (322), P_n, P_ω (327), SP (35, 44), EIS (217).

Miscellaneous. $r_0 \cdot r_1$, r^{-1} (2), r_A (3), $\cong$ (34), $o(\tau)$ (33), $P^{(n)}(\mathfrak{A})$, $P^{(n,k)}(\mathfrak{A})$ (38), $\mathbf{P}^{(n)}(\tau)$ (40, 84), $T(\mu)$ (73), g^*, g, i, i_*, p (205, 206), τ_a (239), f^v (240), $\mathbf{P}_n(\Sigma)$ (304), $E(\Sigma)$ (311).

Universal Algebra

CHAPTER 0
BASIC CONCEPTS

In this chapter we will review briefly the basic concepts of set theory. The results that can be found in any standard book on set theory or algebra will be stated without proof. Those who are familiar with the basic concepts of set theory should only check the notations. Ideal theory of semilattices with complete proofs is presented in §6. This chapter, including the exercises, gives an adequate set theoretical background for the book.

§1. SETS AND RELATIONS

We accept the intuitive concept of a *set* as a collection of objects, called *elements* or *members* of the set. (See also the remark in §4 concerning classes.) The notation $a \in A$ means that a is an element of the set A. If a is not an element of A, we write $a \notin A$. If A and B are sets, $A \subseteq B$ denotes *inclusion*, that is, that A is a *subset* of B, or, all the elements of A are also in B. *Equality* of the sets A and B, in symbols $A = B$, holds if and only if $A \subseteq B$ and $B \subseteq A$. If $A = B$ does not hold, we write $A \neq B$. *Proper inclusion* is denoted by $A \subset B$; by definition $A \subset B$ means $A \subseteq B$ and $A \neq B$.

The *void set* is denoted by $\varnothing$; note that $\varnothing \subseteq A$ for every set A.

The *set theoretic operations* $\cup$, $\cap$, $-$ (they are called *union*, *intersection* and *difference*, respectively) have their usual meaning. If a set A is fixed, then for subsets B of A the *complement* B' of B is defined by $A - B$; by definition, $B \cup (A - B) = A$ and $B \cap (A - B) = \varnothing$. Note that $B \subseteq A$ is equivalent to $B = B \cap A$, which, in turn, is equivalent to $A = B \cup A$. If $A \cap B = \varnothing$, we say that A and B are *disjoint*.

If A is a set, then $P(A)$ (called the *power set* of A) denotes the set of all subsets of A.

A subset of $P(A)$ will be called a *system*, or more precisely, a system over A. A *partition* π of A is a system (over A) not containing $\varnothing$, satisfying the following property: every $a \in A$ is an element of exactly one $B \in \pi$. The members of π are called *blocks* of the partition π. We use $\mathrm{Part}(A)$ for the set of all partitions of A. Note that $\mathrm{Part}(A) \neq \varnothing$; indeed, if $A = \varnothing$, then $\mathrm{Part}(A) = P(A) = \{\varnothing\}$, and if $A \neq \varnothing$, then $\mathrm{Part}(A)$ contains the partition which has one block, namely A. If π_0 and π_1 are partitions of A, we will write $\pi_0 \leqq \pi_1$ if for every block B of π_0, there exists a block C of π_1, with $B \subseteq C$; in this case π_0 is a *refinement* of π_1.

If A and B are sets, the *Cartesian product* $A \times B$ of A and B is defined as the set of all ordered pairs $\langle a, b\rangle$, with $a \in A$ and $b \in B$. In symbols,

$$A \times B = \{\langle a, b\rangle \mid a \in A \text{ and } b \in B\},$$

where $\mid$ reads "for those which satisfy". In general, if $A_0, \cdots, A_{n-1}$ are sets, then

$$A_0 \times A_1 \times \cdots \times A_{n-1} = \{\langle a_0, a_1, \cdots, a_{n-1}\rangle \mid a_0 \in A_0, a_1 \in A_1, \cdots, a_{n-1} \in A_{n-1}\}.$$

If $A_0 = \cdots = A_{n-1} = A$, then we set

$$A^n = A_0 \times \cdots \times A_{n-1}.$$

We define A^0 to be $\{\varnothing\}$.

* * *

For a positive integer n and for a set A, we define an *n-ary relation* r on A as a subset of A^n. n is called the *type* of r. If r is an n-ary relation on A and $a_0, \cdots, a_{n-1} \in A$, we say that $a_0, \cdots, a_{n-1}$ are r-related, in notation $r(a_0, \cdots, a_{n-1})$, if and only if $\langle a_0, \cdots, a_{n-1}\rangle \in r$.

If r_0 and r_1 are n-ary relations on A, then so are $r_0 \cup r_1$, $r_0 \cap r_1$ and $\neg r_0 = A^n - r_0$ (read: not r_0).

If r is a relation on A and $A \subseteq B$, then we can consider r as a relation on B, since $r \subseteq A^n \subseteq B^n$.

We shall be particularly interested in binary relations. For a binary relation r, $\langle a, b\rangle \in r$ will also be denoted by one of the three equivalent notations: $r(a, b)$, arb, and $a \equiv b(r)$. For binary relations we also define the product and inverse. If r_0 and r_1 are binary relations on A, then the *product* $r_0 \cdot r_1$ (or simply $r_0 r_1$) is defined by the rule: for $a, b \in A$, $a(r_0 r_1)b$ if and only if there exists a $c \in A$ with ar_0c and cr_1b. Note that in general $r_0 r_1 \neq r_1 r_0$. If r is a binary relation on A, then the *inverse* r^{-1} of r is defined by the rule: $ar^{-1}b$ if and only if bra.

Two binary relations ι_A and ω_A on the set A are frequently used: $a \equiv b(\iota_A)$ for all $a, b \in A$; $a \equiv b(\omega_A)$ if and only if $a = b$. ι_A is called the *complete relation* on A and ω_A the *equality relation* on A. If there is no danger of confusion, we will omit the index A, i.e., we write ι and ω for ι_A and ω_A, respectively. By definition, $\iota = A \times A$ and $\omega = \{\langle a, a\rangle \mid a \in A\}$ (this set is sometimes called the *diagonal of* A^2).

ω and ι are examples of an important class of binary relations, called equivalence relations. A binary relation Θ on A is defined to be an *equivalence relation* if the following three conditions hold for all $a, b, c \in A$:

(i) $a \Theta a$ (Θ is *reflexive*);
(ii) $a \Theta b$ implies $b \Theta a$ (Θ is *symmetric*);
(iii) $a \Theta b$ and $b \Theta c$ imply $a \Theta c$ (Θ is *transitive*).

The set of all equivalence relations on A will be denoted by $E(A)$. For $\varepsilon_0, \varepsilon_1 \in E(A)$ we agree to write $\varepsilon_0 \leqq \varepsilon_1$ for $\varepsilon_0 \subseteq \varepsilon_1$.

We shall now state a theorem which relates partitions and equivalence relations.

Theorem 1. 1. *Let π be a partition of the set A and define a binary relation ε_π on A by $a\varepsilon_\pi b$ if and only if a and b are in the same block of the partition π. Then ε_π is an equivalence relation on A.*

2. *Let ε be an equivalence relation on A. For $a \in A$ set*

$$A_a = \{b \mid b \in A \text{ and } a\varepsilon b\}.$$

Let π_ε be the system of all $B \subseteq A$ which are of the form A_a. Then π_ε is a partition of A.

3. *If $\pi_0 \leqq \pi_1$, then $\varepsilon_{\pi_0} \leqq \varepsilon_{\pi_1}$. If $\varepsilon_0 \leqq \varepsilon_1$, then $\pi_{\varepsilon_0} \leqq \pi_{\varepsilon_1}$.*

4. *$\pi = \pi_{(\varepsilon_\pi)}$ and $\varepsilon = \varepsilon_{(\pi_\varepsilon)}$.*

If r is an n-ary relation on A and $B \subseteq A$, then $r_B = r \cap B^n$ is an n-ary relation on B (the notation $r|B$ for r_B is very common in the literature but will not be used in this book). The relation r_B is called the *restriction* of r to B. If there is no danger of confusion, we shall omit the subscript B. For instance, if A is the set of all real numbers, $\leqq$ is the usual ordering of real numbers, and B is the set of all rational numbers, then we shall write $\leqq$ instead of $\leqq_B$ for the usual ordering restricted to the rationals.

Note that a restriction of an equivalence relation is always an equivalence relation.

§2. MAPPINGS AND OPERATIONS

Given two sets A and B and a binary relation φ on $A \cup B$, we call φ a *mapping* (or a *function*) of A into B if: $\langle a, b\rangle \in \varphi$ only if $a \in A$, $b \in B$ and, for every $a \in A$, there exists exactly one $b \in B$ satisfying $\langle a, b\rangle \in \varphi$; this element b is called the *image* of the element a under the mapping φ and a is called an *inverse-image* of b under φ. For a mapping φ we introduce the notations $\varphi\colon a \to b$ and $a\varphi = b$ for $\langle a, b\rangle \in \varphi$ (and the functional notation $\varphi(a) = b$), and we write $\varphi\colon A \to B$ to indicate that φ is a mapping of A into B. A is called the *domain* of φ, in notation $D(\varphi) = A$. If the inverse relation is also a mapping, we will denote it by φ^{-1}. We set

$$A\varphi = \{b \mid b \in B, \text{ and there exists an } a \in A \text{ with } a\varphi = b\},$$

and we call $A\varphi$ the *image* of A under φ.

It is easily seen that if φ is a mapping of A into B and $C \subseteq A$, then

$\varphi \cap (C \times B)$ is a mapping of C into B, denoted by φ_C, and called the *restriction* of φ to C.

The mapping $\varphi: A \to B$ is *onto* if $A\varphi = B$; φ is 1–1 (*one-to-one*) if every element of B has at most one inverse-image, that is, if $a_0\varphi = a_1\varphi$ ($a_0, a_1 \in A$) implies $a_0 = a_1$.

It should be emphasized that a mapping $\varphi: A \to B$ is by definition a subset of $(A \cup B)^2$, thus we can form unions, intersections, and so on, of mappings; whatever we get this way will again be binary relations on $A \cup B$, but they will very seldom be mappings.

If A and B are sets, the set of all mappings of A into B will be denoted by B^A. Note that if A has n elements and B has m elements, then B^A has m^n elements, and $B^\varnothing = \{\varnothing\}$.

A *family* $(a_i \mid i \in I)$ of elements of A is a mapping φ from the set I into the set A, where $a_i = i\varphi$. This notation will be used when the emphasis is on listing the elements of A (probably with repetitions) rather than on the set I. I is called the *index set* of the family $(a_i \mid i \in I)$. The image of I under φ will be denoted by

$$\{a_i \mid i \in I\}.$$

Let us remark that every set A gives rise to a family, whose index set is A with φ as the identity map: $x \to x$. Thus every set can be written as

$$\{a \mid a \in A\}.$$

If $(A_i \mid i \in I)$ is a family of subsets of a certain set A, the union and intersection of these sets are denoted by

$$\bigcup (A_i \mid i \in I)$$

and

$$\bigcap (A_i \mid i \in I)$$

respectively.

Thus if the r_i, $i \in I$ are n-ary relations on A, then $\bigcup (r_i \mid i \in I)$ and $\bigcap (r_i \mid i \in I)$ are also n-ary relations on A.

As usual, if $I = \varnothing$, we set $\bigcup (A_i \mid i \in I) = \varnothing$ and $\bigcap (A_i \mid i \in I) = A$.

Let $(A_i \mid i \in I)$ be a family of sets. The *Cartesian product* (or *direct product*)

$$\prod (A_i \mid i \in I)$$

is defined as the subset of $(\bigcup (A_i \mid i \in I))^I$ of all functions f for which $f(i) \in A_i$ for all $i \in I$. To relate this to the definition of $A_0 \times \cdots \times A_{n-1}$ in §1, let us agree that from now on we use an n-tuple $\langle a_0, \cdots, a_{n-1}\rangle$ of elements of A as a notation for $f \in A^{\{0,1,\ldots,n-1\}}$, for which $a_0 = f(0), \cdots, a_{n-1} = f(n-1)$. In other words, n-tuples are used as notations for special types of functions. Then we have indeed that

$$A_0 \times \cdots \times A_{n-1} = \prod (A_i \mid i \in \{0, \cdots, n-1\}).$$

Let us also agree that we will write A^n for $A^{\{0,\cdots,n-1\}}$ and then we see that the definitions of §1 are special cases of the general definition of direct product. (Note that in many developments of set theory n is defined as $\{0, 1, \cdots, n-1\}$.)

For $i \in I$, we can define a mapping e_i^I of $\prod (A_i \mid i \in I)$ into A_i by

$$e_i^I : f \to f(i).$$

e_i^I is called a *projection*. In particular, the projections of A^n are $e_0^n, \cdots, e_{n-1}^n$. Since these are mappings of A^n into A, we will consider them as functions on n variables. Thus $e_i^n(a_0, \cdots, a_{n-1}) = a_i$.

Let A be a set and n a nonnegative integer. An *n-ary operation* on the set A is a mapping f of A^n into A; n is called the *type* of f. Thus an n-ary operation assigns to every n-tuple $\langle a_0, \cdots, a_{n-1}\rangle$ of elements of A a unique element of A, which will be denoted by $f(a_0, \cdots, a_{n-1})$. Hence

$$f(a_0, \cdots, a_{n-1}) = a$$

means $f\colon \langle a_0, \cdots, a_{n-1}\rangle \to a$. Since an operation is a mapping of A^n into A, we can also say that an n-ary operation is an element of $A^{(A^n)}$.

We observe that a 0-ary (nullary) operation is a mapping $f\colon \{\varnothing\} \to A$, which is determined by the single image element of $\varnothing$, $f(\varnothing) \in A$. Examples of nullary operations will be given in §3. One can think of a nullary operation as a constant unary operation, where the variable was omitted since the operation does not depend on it. Of course, one cannot identify the constant unary operation f with the nullary operation g which arises from f by "omitting the variable". Among other things, f and g are not of the same type!

An n-ary operation f on A can also be described by an $(n+1)$-ary relation r defined by

$$r(a_0, \cdots, a_{n-1}, a) \quad \text{if and only if} \quad f(a_0, \cdots, a_{n-1}) = a.$$

(See Exercise 36.)

Unary and binary operations will sometimes be given by means of *Cayley tables*. For instance if $A = \{a, b\}$, then

f	a	b
	b	a

represents the unary operation f defined by $f(a) = b$, $f(b) = a$, and

f	a	b
a	a	b
b	a	a

represents the binary operation f defined by $f(a, a)=a$, $f(a, b)=b$, $f(b, a)=a$, $f(b, b)=a$.

Let $B\subseteq A^n$. Then $f\colon B \to A$ is called a *partial operation* on A of *type n*. A partial operation assigns elements of A to certain n-tuples and is undefined for others. (The operation a^{-1} for real numbers is an example of a partial operation, since it is defined only for $a\neq 0$.)

Although our primary interest is in operations, we consider partial operations because they provide an important tool for the study of operations themselves. Partial operations can also be described by relations (see Exercise 36).

If f is an n-ary operation on A and $B\subseteq A^n$, then $f_B\colon B \to A$ is a partial operation on A.

Binary operations (and partial operations) play an important role; sometimes we use *infix* notation for them (as for binary relations), e.g., $a+b$, $a\cdot b$ (rather than $+(a, b)$ and $\cdot(a, b)$). For some unary operations we use *exponent* notation, e.g., B' for complement (rather than $'(B)$). Infix notations were used for the three binary operations on binary relations.

The product operation on binary relations can be applied to define the *product of mappings*. If $\varphi\colon A \to B$ and $\psi\colon B \to C$, then $\varphi\psi$ will be the consecutive application of φ and ψ (Prove it.). Thus $\varphi\psi$ is a mapping of A into C and for $a \in A$

$$a(\varphi\psi) = (a\varphi)\psi.$$

If $\varphi\colon A \to B$, $\psi\colon B_1 \to C$, then $\varphi\psi$ will be a mapping of A into C if and only if $A\varphi\subseteq B_1$. This condition is always satisfied if φ and ψ are mappings of a set A into itself. The properties of this operation on A^A will be discussed in §5.

* * *

In §1 we set up a 1–1 correspondence between $E(A)$ and $\mathrm{Part}(A)$. There is also an interesting relationship between $E(A)$ and the mappings of A.

Let ε be an equivalence relation on A and let π_ε be the corresponding partition (see Theorem 1.1). For $H\subseteq A$ set

$$[H]\varepsilon = \{a \mid a \in A \text{ and } h\varepsilon a \text{ for some } h \in H\};$$

this set is called the *closure* of H under ε. If $H=\{x\}$, we will write $[x]\varepsilon$ for $[\{x\}]\varepsilon$. By Theorem 1.1, for every $x \in A$, $[x]\varepsilon \in \pi_\varepsilon$; the block $[x]\varepsilon$ is called the *equivalence class* containing x. (In this expression, class is a synonym for set.) Thus $[H]\varepsilon$ is the union of all blocks of π_ε which contain at least one element of H. The mapping

$$\varphi_\varepsilon\colon x \to [x]\varepsilon$$

is called the *natural mapping* of A onto A/ε, the set of all equivalence classes under ε. (Of course, π_ε and A/ε are identical.) A/ε is called the *quotient set* of A modulo ε.

Just as an equivalence relation ε defines a natural mapping of A onto its quotient set A/ε, a mapping $\varphi: A \to B$ defines a natural equivalence relation ε_φ on A under which two elements are related if and only if they have the same image in B under φ. We will call ε_φ the *equivalence relation induced by* φ. (Note that $\varepsilon_\varphi = \varphi\varphi^{-1}$.)

Theorem 1. *Any mapping* $\chi: A \to B$ *can be represented as a product of two mappings* φ *and* ψ, $\chi = \varphi\psi$, *where* φ *is onto and* ψ *is 1–1; if* ε *is the equivalence relation induced by* χ, *then we can set* $\varphi = \varphi_\varepsilon: A \to A/\varepsilon$ *and* $\psi: A/\varepsilon \to B$, *defined by* $\psi: [x]\varepsilon \to x\chi \ (x \in A)$.

Remark. Theorem 1 can be visualized using Fig. 1, where A, B, A/ε are the sets of Theorem 1, and an arrow indicates a mapping; the arrows are

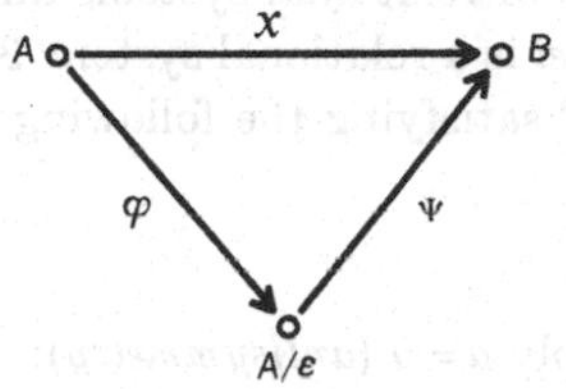

Fig. 1

labeled by the symbol of the mapping. The diagram is *commutative*, that is, if we can get from a set to another one by different sequences of arrows, the product of the corresponding mappings is always the same.

Proof. First we prove that ψ is well defined, that is, if $[x]\varepsilon = [y]\varepsilon$ $(x, y \in A)$, then $x\chi = y\chi$. Indeed, $[x]\varepsilon = [y]\varepsilon$ implies that $y \in [x]\varepsilon$, and so $x\varepsilon y$. By the definition of ε this means $x\chi = y\chi$.

$\varphi\psi$ is a mapping from A into B since $D(\varphi)\varphi = A/\varepsilon$ and $D(\psi) = A/\varepsilon$. Finally we prove that $\varphi\psi = \chi$. Indeed, for $x \in A$ we have

$$x(\varphi\psi) = (x\varphi)\psi = ([x]\varepsilon)\psi = x\chi,$$

completing the proof of Theorem 1.

§3. ALGEBRAS AND RELATIONAL SYSTEMS

We will study the basic properties of universal algebras in Chapter 1, and of relational systems in Chapter 6. However, we want to discuss some results concerning special types of algebras and relational systems, e.g., semigroups, partially ordered sets, and so on. We give at this point the general definitions only to see that these specific systems are special cases of the general definitions.

A *universal algebra* or, briefly, *algebra* $\mathfrak{A}$ is a pair $\langle A; F\rangle$, where A is a nonvoid set and F is a family of *finitary* operations on A. F is not necessarily finite, and it may be void. When F is finite, $F=\{f_0, \cdots, f_{n-1}\}$, we denote the algebra $\langle A; F\rangle$ by $\langle A; f_0, \cdots, f_{n-1}\rangle$.

A *relational system* $\mathfrak{A}$ is a pair $\langle A; R\rangle$, where A is a nonvoid set and R is a family of (finitary) relations on A. Again, if $R=\{r_0, \cdots, r_{n-1}\}$ we write $\langle A; r_0, \cdots, r_{n-1}\rangle$ for $\mathfrak{A}$.

In both cases, A is called the *base set* of $\mathfrak{A}$. Algebras and relational systems will be denoted by German capital letters: $\mathfrak{A}, \mathfrak{B}, \mathfrak{C}, \cdots, \mathfrak{L}, \cdots, \mathfrak{P}, \cdots$ and the base sets by the corresponding italic capital letters $A, B, C, \cdots, L, \cdots, P, \cdots$. Thus if we say that $\mathfrak{A}$ is an algebra and $a \in A$, it is understood that A is the base set of $\mathfrak{A}$. In Chapter 1 and Chapter 6 the definitions of algebras and relational systems will be slightly modified.

Now we give examples of relational systems and algebras:

1. A *partially ordered set* is a relational system $\mathfrak{P}=\langle P; \leqq\rangle$, where "$\leqq$" is a binary relation on P satisfying the following three conditions for all $a, b, c \in P$:

(i) $a \leqq a$ (*reflexivity*);
(ii) $a \leqq b$ and $b \leqq a$ imply $a=b$ (*antisymmetry*);
(iii) $a \leqq b$ and $b \leqq c$ imply $a \leqq c$ (*transitivity*).

"$\leqq$" is called a *partial ordering relation.* $a<b$ will stand for $a \leqq b$ and $a \neq b$, and $a \geqq b$ for $b \leqq a$. Examples of partially ordered sets are $\langle \text{Part}(A); \leqq\rangle$, $\mathfrak{E}(A)=\langle E(A); \leqq\rangle$, and $\langle P(A); \subseteq\rangle$.

If $\mathfrak{P}$ is a partially ordered set and $H \neq \varnothing$, $H \subseteq P$, then $\langle H; \leqq_H\rangle$ is also a partially ordered set. As we agreed, we will write $\langle H; \leqq\rangle$ for $\langle H; \leqq_H\rangle$.

2. A *chain* $\mathfrak{C}=\langle C; \leqq\rangle$ is a partially ordered set satisfying the additional condition

(iv) $a \leqq b$ or $b \leqq a$ for all $a, b \in C$.

(A chain is also called a *linearly ordered set* or a *totally ordered set.*)

If $\mathfrak{P}$ is a partially ordered set, $\varnothing \neq C \subseteq P$ and $\mathfrak{C}=\langle C; \leqq\rangle$ is a chain, where $\leqq$ is the restriction of the partial ordering of $\mathfrak{P}$ to C, then $\mathfrak{C}$ is called a *subchain of* $\mathfrak{P}$ or a *chain in* $\mathfrak{P}$.

Example: $\langle\{\varnothing, \{a\}, A\}; \subseteq\rangle$ is a chain in $\langle P(A); \subseteq\rangle$ if $a \in A$.

3. A *lattice* is an algebra $\langle A; \vee, \wedge\rangle$, where $\vee$ and $\wedge$ are binary operations on A, called *join* and *meet*, respectively, satisfying the following laws for all $a, b, c \in A$:

(i) $a \vee a=a$,
$a \wedge a=a$; (*idempotent laws*)

(ii) $a \vee b = b \vee a$, $a \wedge b = b \wedge a$; (*commutative laws*)

(iii) $a \vee (b \vee c) = (a \vee b) \vee c$, $a \wedge (b \wedge c) = (a \wedge b) \wedge c$; (*associative laws*)

(iv) $a \wedge (a \vee b) = a$, $a \vee (a \wedge b) = a$. (*absorption laws*)

Example: $\langle P(A); \cup, \cap \rangle$ is a lattice.

4. A *distributive lattice* $\langle A; \vee, \wedge \rangle$ is a lattice in which for all $a, b, c \in A$ we have $a \vee (b \wedge c) = (a \vee b) \wedge (a \vee c)$.

Example: $\langle P(A); \cup, \cap \rangle$ is distributive.

5. A *Boolean algebra* is an algebra $\mathfrak{B} = \langle B; \vee, \wedge, ', 0, 1 \rangle$ with two binary operations $\vee$, $\wedge$, one unary operation $'$, and two nullary operations 0, 1, such that the following conditions are satisfied:

(i) $\langle B; \vee, \wedge \rangle$ is a distributive lattice;

(ii) $0 \vee a = a$, $a \wedge 1 = a$ for all $a \in B$;

(iii) $a \vee a' = 1$ and $a \wedge a' = 0$ for all $a \in B$.

Example: $\langle P(A); \cup, \cap, ', \varnothing, A \rangle$ is a Boolean algebra. This Boolean algebra will be called a *Boolean set algebra* and will be denoted by $\mathfrak{P}(A)$. If $\langle B; \vee, \wedge, ', 0, 1 \rangle$ is a Boolean algebra, then $\langle B; \vee, \wedge \rangle$ will be called a *Boolean lattice*.

6. A *semigroup* $\langle A; \cdot \rangle$ is an algebra with one binary operation such that

$$(a \cdot b) \cdot c = a \cdot (b \cdot c) \text{ for all } a, b, c \in A.$$

An example of a semigroup is the algebra of all mappings of A to A, $\langle A^A; \cdot \rangle$ (see §5). In semigroups (and in general, whenever an operation is associative) we will write $a_0 \cdot a_1 \cdots a_{n-1}$ for $(\cdots (a_0 \cdot a_1) \cdots) \cdot a_{n-1}$, and in semigroups we will write ab for $a \cdot b$.

7. A *group* is an algebra $\mathfrak{G} = \langle G; \cdot, 1 \rangle$, with one binary operation $\cdot$ and one nullary operation 1 such that the following conditions are satisfied:

(i) $\langle G; \cdot \rangle$ is a semigroup;

(ii) $1 \cdot a = a \cdot 1 = a$ for all $a \in G$;

(iii) For all $a \in G$, there exists a $b \in G$ such that $a \cdot b = b \cdot a = 1$.

Sometimes, by a group we mean an algebra $\langle G; \cdot, ^{-1}, 1 \rangle$ where $^{-1}$ is a unary operation such that (i) and (ii) hold, and

(iii′) $aa^{-1} = a^{-1}a = 1$.

It will always be made clear which definition is used. If $ab = ba$ for all $a, b \in G$, the group will be called *commutative* or *abelian*.

8. A *ring* $\mathfrak{R} = \langle R; +, \cdot, 0 \rangle$ is an algebra with two binary operations $+$, $\cdot$ and a nullary operation 0 such that

(i) $\langle R; +, 0 \rangle$ is a commutative group;

(ii) $\langle R; \cdot\rangle$ is a semigroup;
(iii) $a\cdot(b+c)=a\cdot b+a\cdot c$ and $(b+c)\cdot a=b\cdot a+c\cdot a$ for all $a, b, c \in R$.

9. A *division ring* $\langle R; +, \cdot, ^{-1}, 0, 1\rangle$ is a system with two binary operations $+$, $\cdot$, one unary partial operation $^{-1}$, and two nullary operations 0, 1 such that

(i) $\langle R; +, \cdot, 0\rangle$ is a ring and for all $a \in R$, $a\cdot 1=1\cdot a=a$;
(ii) a^{-1} is defined for all $a \neq 0$ and $a\cdot a^{-1} = a^{-1}\cdot a = 1$.

A system like a division ring, that is, a system in which we have partial operations as well as operations, is called a *partial algebra*. To fit partial algebras into the framework of relational systems we must replace every partial operation by the relation which describes it; in Chapter 2, however, we do not use this way of dealing with partial algebras.

§4. PARTIALLY ORDERED SETS

Let $\mathfrak{P}=\langle P, \leqq\rangle$ be a partially ordered set. The relation $\geqq$ (the inverse of $\leqq$) is also reflexive, antisymmetric, and transitive, hence $\langle P; \geqq\rangle$ is also a partially ordered set, called the *dual* of $\mathfrak{P}$. (This is an ambiguous notation, used for want of a better one.) The *duality principle* utilizes this simple observation; it states that if a statement on partially ordered sets is dualized, that is, all $\leqq$ are replaced by $\geqq$, then if the statement is true, so is its dual.

Given $H \subseteq P$, $a \in P$ is an *upper bound* of H if $b \leqq a$ for all $b \in H$; a is called the *least upper bound* of H, in symbols, l.u.b. (H), if:

(i) a is an upper bound of H;
(ii) if b is any upper bound of H, then $a \leqq b$.

If the l.u.b. (H) exists, then it is unique. Consider $\langle I, \leqq\rangle$, where I is the set of all integers and $\leqq$ is the usual ordering of integers. Take $H=I$. Here, no upper bounds for H exist; hence, there does not exist a l.u.b. for H.

Consider $\langle R; \leqq\rangle$, where R is the set of rationals with the usual partial ordering $\leqq$. Let $H=\{r \mid r \in R \text{ and } r^2 \leqq 2\}$. H has upper bounds, but the l.u.b. of H does not exist because $\sqrt{2}$ is not rational.

In general, we write

$$\text{l.u.b. } (H) = \bigvee (h \mid h \in H),$$

which becomes $a \vee b$ in the case when H has two elements, a and b.

Lower bounds and the g.l.b. (*greatest lower bound*) are defined dually.

Also, g.l.b. $(H) = \bigwedge (h \mid h \in H)$, which becomes $a \wedge b$ in the case H has two elements, a and b.

A partially ordered set $\mathfrak{P}$ is called *directed* if any two element subset of P has an upper bound in P.

A partially ordered set $\mathfrak{L}$ is a *lattice* if $a \vee b$, $a \wedge b$ exist for all $a, b \in L$. Note that the dual of a lattice is a lattice again. Hence the duality principle applies also to lattices. Now we have two definitions of lattice: lattice as an algebra and lattice as a partially ordered set. The two definitions are equivalent in the following sense:

Theorem 1. 1. *Let $\mathfrak{L} = \langle L; \vee, \wedge \rangle$ be a lattice. Define a binary relation $\leqq$ on L by $a \leqq b$ if and only if $a \vee b = b$. Then $\mathfrak{L}^0 = \langle L; \leqq \rangle$ is a partially ordered set, and as a partially ordered set it is a lattice; furthermore* l.u.b. $(\{a, b\}) = a \vee b$ *and* g.l.b. $(\{a, b\}) = a \wedge b$.

2. *Let $\mathfrak{L} = \langle L; \leqq \rangle$ be a partially ordered set which is a lattice. Set*

$$a \vee b = \text{l.u.b.}\ (\{a, b\})$$

and

$$a \wedge b = \text{g.l.b.}\ (\{a, b\}).$$

Then $\mathfrak{L}^+ = \langle L; \vee, \wedge \rangle$ is a lattice, and $a \leqq b$ if and only if $a \vee b = b$.

3. *Let $\mathfrak{L} = \langle L; \vee, \wedge \rangle$ be a lattice. Then $(\mathfrak{L}^0)^+ = \mathfrak{L}$.*

4. *Let $\mathfrak{L} = \langle L; \leqq \rangle$ be a lattice. Then $(\mathfrak{L}^+)^0 = \mathfrak{L}$.*

A lattice $\mathfrak{L}$ is called *complete* if l.u.b. (H) and g.l.b. (H) exist for all H, $H \subseteq L$. The dual of a complete lattice is a complete lattice.

If $\mathfrak{L}$ is a complete lattice then it always has a least element 0 and greatest element 1. Then for all $H \subseteq L$

$$\bigvee (a \mid a \in H), \qquad \bigwedge (a \mid a \in H)$$

exist and if $H = \varnothing$, then

$$\bigvee (a \mid a \in H) = 0, \qquad \bigwedge (a \mid a \in H) = 1.$$

The Boolean algebra $\langle B; \vee, \wedge, ', 0, 1 \rangle$ is called *complete*, if $\langle B; \vee, \wedge \rangle$ is complete.

Examples: Every finite lattice is complete. The set of all reals in $[0, 1]$ with the usual ordering is a complete lattice. The lattice of all closed subspaces of a topological space is complete. $\langle P(A); \subseteq \rangle$ is a complete lattice. The partially ordered set of all rationals in $[0, 1]$ is not complete. $\mathfrak{P}(A)$ is a complete Boolean algebra.

As we remarked in §1, our entire discussion is based on intuitive set theory. That is, we consider the basic concepts as intuitively clear notions and the facts we use from set theory are also intuitively clear. There is, however, one exception, which arises in the following way: If we are given a nonempty set A, then, by definition, we can select an element $a \in A$. Similarly, if we are given n nonempty sets, $A_0, \cdots, A_{n-1}$, then we can select one element from each, $a_0 \in A_0, \cdots, a_{n-1} \in A_{n-1}$. In other words, we can select one element from each set simultaneously, that is, there exists a function f from $\{0, \cdots, n-1\}$ into $\bigcup (A_i \mid 0 \leqq i < n)$ such that $f(i) \in A_i$ for all $0 \leqq i < n$. The controversial question is the following: If we are given an infinite number of nonvoid sets, is it still possible to select simultaneously one element from each set? The axiom which states that this can be done is called the Axiom of Choice, two equivalent formulations of which follow.

Axiom of Choice.

(AC_1) *Given any set A, there exists a function f: $P(A) \to A$ such that if $\varnothing \neq B \in P(A)$, then $f(B) \in B$.* (f is called a *choice function* on A.)

(AC_2) *Let $(B_i \mid i \in I)$ be a family of nonvoid sets. Then* $\prod (B_i \mid i \in I) \neq \varnothing$.

(AC_1) Implies (AC_2). Set $A = \bigcup (B_i \mid i \in I)$. By ($AC_1$), there exists a mapping f: $P(A) \to A$, with $f(B_i) \in B_i$. An element g in $\prod (B_i \mid i \in I)$ can be defined by $g(i) = f(B_i)$, for $i \in I$. Thus $\prod (B_i \mid i \in I) \neq \varnothing$.

(AC_2) Implies (AC_1). Take a set A. By (AC_2),

$$\prod (B \mid B \in P(A) \text{ and } B \neq \varnothing)$$

is nonvoid. Let $g \in \prod (B \mid B \in P(A) \text{ and } B \neq \varnothing)$. Then $g(B) \in B$. We define g in any way at $\varnothing$. Thus we get a choice function g.

We will give without proof four nontrivially equivalent formulations of the Axiom of Choice.

Let $\mathfrak{P} = \langle P; \leqq \rangle$ be a partially ordered set and let $\mathfrak{C}$ be a chain in $\mathfrak{P}$. The chain $\mathfrak{C}$ is called *maximal* if $C \subset D \subseteq P$ implies that $\mathfrak{D}$ is not a chain in $\mathfrak{P}$. It is easy to see that $\mathfrak{C}$ is maximal if and only if for every $a \in P$ which is not in C, there exists a $b \in C$ such that neither $a \geqq b$ nor $a \leqq b$ holds.

(1) *Maximal Chain Principle.* Every chain in a partially ordered set is included in a maximal chain.

(2) *Zorn's Lemma.* Let A be a set, and let $P \subseteq P(A)$. Assume that if $\mathfrak{C}$ is a chain in $\langle P; \subseteq \rangle$, then $\bigcup (X \mid X \in C) \in P$. Then P has a *maximal* element M (i.e., $M \in P$ and if $M \subseteq X \in P$, then $M = X$).

A system P over a set A is of *finite character* when $B \in P$ if and only if every finite subset of B is in P.

(3) *Teichmüller-Tukey Lemma.* Let P be a nonvoid system of finite character of subsets of A. Then there exists a maximal subset of A which belongs to P.

The partially ordered set $\langle P; \leqq \rangle$ is called a *well-ordered* set if any nonvoid subset H of P has a least element. It is easy to see that every well-ordered set is a chain.

(4) *Well-Ordering Principle.* Given a set $A \neq \varnothing$, we can define a binary relation $\leqq$ on A so that $\langle A; \leqq \rangle$ is a well-ordered set.

In other words, every nonvoid set can be well ordered.

It will be convenient to consider the void-set $\varnothing$ as a well-ordered set; then (4) can be rephrased to state that *every* set can be well ordered.

The well-ordering principle (and its equivalence to the Axiom of Choice) is due to E. Zermelo, who was the first to recognize the importance of the Axiom of Choice. The maximal chain principle is due to F. Hausdorff. For historical notes and proofs of the equivalence, see, e.g., G. Birkhoff [6].

To prove that all the statements listed above are equivalent is not very easy; however, the only difficult step is to prove that the Axiom of Choice implies any one of the others.

It is well known that intuitive set theory is contradictory. The contradictions arise not from sets which we use in everyday mathematics, but from considering "very large" sets, as, for instance, the set of all sets. The contradictions can be resolved (or so we hope) by, for instance, the introduction of the concept of *class* for "very large" sets; classes cannot be elements of sets or classes.

For a very clear discussion of Axiomatic Set Theory and the definition of classes, see E. Mendelson, Introduction to Mathematical Logic, D. Van Nostrand Co., Princeton, N.J., 1964, Chapter 4.

* * *

The sets A, B are *equipotent*, provided there exists a mapping $\varphi: A \to B$ which is 1–1 and onto. It is easy to see that the equipotency of sets is reflexive, symmetric, and transitive. The equivalence classes (these are really classes) are called *cardinal numbers* or *cardinals*.

If A is any set, let $|A|$ denote the equivalence class containing A, i.e., the cardinal number of A. $|A|$ is also called the *power* of A.

The equivalence class containing n-element sets will be denoted by n; the equivalence classes containing the integers and the reals are denoted by $\aleph_0$ and $\mathfrak{c}$, respectively.

Operations on cardinal numbers are:

(i) $\mathfrak{m}+\mathfrak{n}$ (*addition*);

(ii) $\mathfrak{m}\cdot\mathfrak{n}$ (*multiplication*);

(iii) $\mathfrak{m}^{\mathfrak{n}}$ (*exponentiation*).

To define the operations, take $\mathfrak{m}=|A|$, $\mathfrak{n}=|B|$, $A \cap B=\varnothing$. Then $|A \cup B|=\mathfrak{m}+\mathfrak{n}$, $|A \times B|=\mathfrak{m}\cdot\mathfrak{n}$, and $|A^B|=\mathfrak{m}^{\mathfrak{n}}$. Of course, it has to be proved that the results of the operations depend only on $\mathfrak{m}$ and $\mathfrak{n}$, and not on any particular choice of the sets A and B.

If $(\mathfrak{m}_i \mid i \in I)$ is a family of cardinals, $|A_i|=\mathfrak{m}_i$, and $i \neq j$ implies that $A_i \cap A_j=\varnothing$, then $|\bigcup (A_i \mid i \in I)|=\sum (\mathfrak{m}_i \mid i \in I)$, and $|\prod (A_i \mid i \in I)|=\prod (\mathfrak{m}_i \mid i \in I)$. (The Axiom of Choice is used!)

We write $\mathfrak{m} \leqq \mathfrak{n}$ if there exists a mapping $\varphi: A \to B$ which is 1–1. Again we must prove that the definition is independent of A, B.

If $\mathfrak{n} < \aleph_0$, then $\mathfrak{n}$ is called a *finite cardinal*; otherwise it is an *infinite cardinal*.

Theorem 1. (i) *Let A be a set of cardinals. Then $\langle A; \leqq \rangle$ is a well-ordered set.*

(ii) *If $\mathfrak{m}$ or $\mathfrak{n}$ is infinite, then*

$$\mathfrak{m} + \mathfrak{n} = \max(\mathfrak{m}, \mathfrak{n});$$

if in addition $\mathfrak{m} \neq 0$, $\mathfrak{n} \neq 0$, then

$$\mathfrak{m} \cdot \mathfrak{n} = \max(\mathfrak{m}, \mathfrak{n}).$$

(iii) $2^{\mathfrak{m}} > \mathfrak{m}$ *and* $2^{\mathfrak{m}} = |P(A)|$, *where* $|A| = \mathfrak{m}$.

A cardinal $\mathfrak{m}$ is *regular* if for any family of cardinals $(\mathfrak{m}_i \mid i \in I)$, $|I| < \mathfrak{m}$ and $\mathfrak{m}_i < \mathfrak{m}$ for $i \in I$ imply that $\sum(\mathfrak{m}_i \mid i \in I) < \mathfrak{m}$.

For example, $\aleph_0$ is regular.

We now discuss the concept of ordinal.

Take two partially ordered sets $\langle A; \leqq \rangle$, $\langle B; \leqq \rangle$. We define the concept of isomorphism of $\mathfrak{A} = \langle A; \leqq \rangle$ and $\mathfrak{B} = \langle B; \leqq \rangle$: $\mathfrak{A}$ and $\mathfrak{B}$ are said to be *isomorphic* if there exists a mapping $\varphi \colon A \to B$ which is 1–1 and onto and for which

$$a_0 \leqq a_1 \quad \text{if and only if} \quad a_0\varphi \leqq a_1\varphi \qquad (a_0, a_1 \in A).$$

Such a mapping φ is called an *isomorphism*.

Two well-ordered sets $\mathfrak{A}$ and $\mathfrak{B}$ have the same *order type* if they are isomorphic. The equivalence classes obtained this way are called *ordinals*. Consider the set $\{0, 1, \cdots, n-1\}$ with the usual ordering $0 < 1 < \cdots < n-1$. The equivalence class containing this chain is the ordinal denoted by n.

The equivalence class containing $\varnothing$ consists of $\varnothing$; it will be denoted by 0.

Assume that the order type of $\langle A; \leqq \rangle$ is α and that the order type of $\langle B; \leqq \rangle$ is β. If there exists a mapping $\varphi \colon A \to B$ which is 1–1 and which preserves ordering (that is, $a_0 \leqq a_1$ implies $a_0\varphi \leqq a_1\varphi$) then we write $\alpha \leqq \beta$. By definition, $0 \leqq \alpha$, for all ordinals α.

Consider the chain N of nonnegative integers; the corresponding ordinal is denoted by ω. An ordinal α is called *infinite* if $\omega \leqq \alpha$; otherwise it is *finite*.

Theorem 2. (i) *Let A be a set of ordinals; then $\langle A; \leqq \rangle$ is a well-ordered set.*

(ii) *The order type of $\langle \{\gamma \mid \gamma < \alpha\}; \leqq \rangle$ is α.*

Theorem 2 implies that every well-ordered set is isomorphic to the well-ordered set of all ordinals less than a given ordinal. (Note that in many axiomatic set theories, an ordinal equals the set of smaller ordinals.) Thus if $\langle I; \leqq \rangle$ is a well-ordered set, then there exists an ordinal α such that we can write I in the form $I = \{x_\gamma \mid \gamma < \alpha\}$ and $x_\gamma \leqq x_\delta$ is equivalent to $\gamma \leqq \delta$.

We now define the *sum* of two ordinals. Let α, β, $\langle A; \leqq \rangle$ and $\langle B; \leqq \rangle$ be as above and in addition assume that $A \cap B = \varnothing$. Define $\leqq$ on $A \cup B$ as follows:

(i) if $x, y \in A$, $x \leqq y$ has its original meaning;
(ii) same as (i) for $x, y \in B$;
(iii) $x \in A$ and $y \in B$, then $x < y$.

Now we take $\langle A \cup B; \leqq \rangle$. It can be shown to be a well-ordered set. The order type of $\langle A \cup B; \leqq \rangle$ is defined to be $\alpha + \beta$.

This definition can be extended to the case of an infinite number of ordinals as follows: Let $(\alpha_i \mid i \in I)$ be a family of ordinals and let $\langle I; \leqq \rangle$ be a well-ordered set of order type β. Take a well-ordered set $\langle A_i; \leqq \rangle$ of order type α_i for each $i \in I$ and assume that $A_i \cap A_j = \varnothing$ if $i \neq j$. Set $A = \bigcup (A_i \mid i \in I)$ and define $\leqq$ on A as follows: if $x, y \in A_i$, then $x \leqq y$ keeps its original meaning; if $x \in A_i$, $y \in A_j$, $i \neq j$, then $x < y$ if and only if $i < j$. Then $\langle A; \leqq \rangle$ will be a well-ordered set and its order type will be defined to be $\sum (\alpha_i \mid i \in I)$. If $\alpha_i = \alpha$ for all $i \in I$, then $\beta \cdot \alpha$ will denote the ordinal $\sum (\alpha_i \mid i \in I)$.

An alternative way of defining multiplication of ordinals is as follows: Let α, β, $\langle A; \leqq \rangle$ and $\langle B; \leqq \rangle$ be given as above. Set $\langle a_0, b_0 \rangle < \langle a_1, b_1 \rangle$ if and only if $a_0 < a_1$, or $a_0 = a_1$ and $b_0 < b_1$. Then $\langle A \times B; \leqq \rangle$ is a well-ordered set, the order type of which will be denoted by $\alpha \cdot \beta$. This partial ordering of $A \times B$ is called the *lexicographic ordering*.

Note that neither the addition nor the multiplication of ordinals is commutative.

Given an ordinal α, we denote by $\bar{\alpha}$ the *power* of α, defined as follows: Let α be the order type of $\langle A; \leqq \rangle$; then $\bar{\alpha} = |A|$.

Let $\mathfrak{m}$ be a given infinite cardinal and $A = \{\alpha \mid \bar{\alpha} = \mathfrak{m}\}$. It follows from the well-ordering principle that A is not void. Thus by Theorem 2, it has a smallest element. It is called the *initial ordinal* of power $\mathfrak{m}$, and it will be denoted by $\omega_{\mathfrak{m}}$.

Let $\langle I; \leqq \rangle$ be well ordered, let α_i be an ordinal for all $i \in I$ and let $\alpha_i \leqq \alpha_j$ if $i < j$. Then $\lim (\alpha_i \mid i \in I)$ will denote the smallest ordinal α such that $\alpha_i \leqq \alpha$ for all $i \in I$. *A limit ordinal* α is an ordinal which can be expressed as $\lim (\alpha_i \mid i \in I) = \alpha$ with $\alpha_i < \alpha$ for all $i \in I$. Thus 0 and ω are limit ordinals. Either α is a limit ordinal or $\alpha = \beta + 1$ for some ordinal β.

If we take any subset of the class of all infinite cardinals, then the usual ordering makes it a well-ordered set. This implies that we can index the infinite cardinals with the class of all ordinals; let $\aleph_\alpha$ denote the infinite cardinal which is indexed by the ordinal α. Then

$$\aleph_\alpha < \aleph_\beta \quad \text{if and only if} \quad \alpha < \beta$$

and thus all the infinite cardinals are members of the sequence

$$\aleph_0, \aleph_1, \cdots, \aleph_\omega, \aleph_{\omega+1}, \cdots.$$

If $\mathfrak{m} = \aleph_\alpha$, we will write ω_α for $\omega_{\mathfrak{m}}$. Thus $\omega_0 = \omega$.

The *Generalized Continuum Hypothesis* is the assumption that $2^{\aleph_\alpha}$ is the cardinal which follows $\aleph_\alpha$, that is

$$2^{\aleph_\alpha} = \aleph_{\alpha+1}.$$

This can be expressed without the $\aleph_\alpha$ notation as follows: if $\mathfrak{m}$ is an infinite cardinal, then there is no cardinal $\mathfrak{n}$ with $\mathfrak{m} < \mathfrak{n} < 2^{\mathfrak{m}}$. Although we shall use the Axiom of Choice without any special reference to it, whenever we use the Generalized Continuum Hypothesis in proving a theorem we shall mention this fact explicitly as an assumption in the theorem.

Using ordinals, we can generalize finite induction as follows:

Transfinite Induction. *Let the statement $\Phi(\alpha)$ be defined for all ordinals α. Assume that*

(*) *if $\Phi(\beta)$ holds for all $\beta < \alpha$, then $\Phi(\alpha)$ holds.*

Then $\Phi(\alpha)$ holds for all ordinals α.

Proof. Assume that there exists a δ for which $\Phi(\delta)$ does not hold. Consider the set $I = \{\gamma \mid \gamma \leqq \delta \text{ and } \Phi(\gamma) \text{ does not hold}\}$; by Theorem 2, I has a smallest member α. Obviously, $\alpha \neq 0$. If we apply (*), we get that $\Phi(\alpha)$ holds, which is a contradiction.

In most cases it is useful to replace (*) by the following three conditions:

(i) $\Phi(0)$ holds;

(ii) If $\Phi(\beta)$ holds, then $\Phi(\beta+1)$ holds;

(iii) if α_λ are ordinals for $\lambda < \beta$, with $\alpha_\lambda \leqq \alpha_\delta$ for $\lambda \leqq \delta < \beta$, and $\Phi(\alpha_\lambda)$ holds for all $\lambda < \beta$, then $\Phi(\lim (\alpha_\lambda \mid \lambda < \beta))$ holds.

Using ordinals, we can introduce transfinite sequences: let α be an ordinal; then the elements of $A^{\{\gamma \mid \gamma < \alpha\}}$ are called *α-termed sequences*, or, simply, *α-sequences*. We also agree to write A^α for $A^{\{\gamma \mid \gamma < \alpha\}}$. If $f \in A^\alpha$, we will sometimes write f as

$$\langle f(0), f(1), \cdots, f(\gamma), \cdots \rangle_{\gamma < \alpha}.$$

A mapping $f: A^\alpha \to A$ is an α-ary operation; if $\omega \leqq \alpha$, we will call f an *infinitary operation* of type α. An *algebra* $\langle A; F \rangle$ *with infinitary operations* will mean that every $f \in F$ is a finitary or infinitary operation.

We can define similarly α-ary *partial operations* as mappings from some

$B \subseteq A^\alpha$ into A, *α-ary relations* as subsets of A^α, *and relational systems with infinitary relations.*

It is not the purpose of this book to consider algebras with infinitary operations. However, many results of the book can be proved for algebras with infinitary operations. Some results along this line will be given as exercises.

§5. STRUCTURE OF MAPPINGS AND EQUIVALENCE RELATIONS

The following theorem describes some important properties of mappings.

Theorem 1. *The algebra $\langle M(A); \cdot \rangle$, that is, the set of mappings of a set A into itself under product of mappings, is a semigroup. Consider the subsets of $M(A)$, denoted as follows: $M_O(A)$, the set of all mappings of A onto A; and $M_I(A)$, the set of all 1–1 mappings of A into A. Then $\langle M_O(A); \cdot \rangle$ and $\langle M_I(A); \cdot \rangle$ are both semigroups. If $\gamma \in M_O(A)$ and $\alpha, \beta \in M(A)$, then $\gamma\alpha = \gamma\beta$ implies that $\alpha = \beta$. If $\gamma \in M_I(A)$ and $\alpha, \beta \in M(A)$, then $\alpha\gamma = \beta\gamma$ implies that $\alpha = \beta$.*

Let ε denote the identity mapping, i.e., $x\varepsilon = x$ for all $x \in A$. Then

$$\varepsilon \in M_O(A) \cap M_I(A).$$

Every $\gamma \in M_O(A)$ ($\gamma \in M_I(A)$) has a left inverse (right inverse) δ, i.e., $\delta\gamma = \varepsilon$ ($\gamma\delta = \varepsilon$). Furthermore, $\langle M_O(A) \cap M_I(A); \cdot, \varepsilon \rangle$ is a group.

The last part of the theorem can be generalized to A^B (see Exercises 61–63).

We will now establish the main properties of equivalence relations.

Let A be a set and let $E(A)$ be the set of all equivalence relations on A. We have already introduced a partial ordering on $E(A)$. Let us recall that for $\varepsilon_0, \varepsilon_1 \in E(A)$,

$$\varepsilon_0 \leqq \varepsilon_1 \quad \text{if and only if} \quad x\varepsilon_0 y \text{ implies } x\varepsilon_1 y.$$

Theorem 2. $\mathfrak{E}(A) = \langle E(A); \leqq \rangle$ *is a complete lattice.*

Proof. $E(A)$ has a least element, namely, ω. (Recall that $x \equiv y(\omega)$ if and only if $x = y$.) $E(A)$ has a greatest element, ι. (Recall that $x \equiv y(\iota)$ for all $x, y \in A$.) (This is why the letters ω and ι were chosen to denote these equivalence relations: ω is the greek o and ω is the zero of $\mathfrak{E}(A)$; ι is the greek i and ι is the 1 of $\mathfrak{E}(A)$.)

Let $(\varepsilon_i \mid i \in I)$ be a family of equivalence relations $I \neq \varnothing$. Define a new equivalence relation Φ in the following way:

$$x \equiv y(\Phi) \quad \text{if and only if} \quad x \equiv y(\varepsilon_i) \text{ for all } i \in I.$$

Then $\Phi = \bigwedge (\varepsilon_i \mid i \in I)$.

We will verify this statement in three steps.

(i) $\Phi \in E(A)$. This means that Φ is an equivalence relation, i.e., it is reflexive, symmetric, and transitive.

(a) $a \equiv a(\Phi)$ is equivalent to $a \equiv a(\varepsilon_i)$ for all $i \in I$, which is true since all the ε_i are reflexive.

(b) $a \equiv b(\Phi)$ is equivalent to $a \equiv b(\varepsilon_i)$ for all $i \in I$; thus $b \equiv a(\varepsilon_i)$ for all $i \in I$ and so $b \equiv a(\Phi)$, since all the ε_i are symmetric. Therefore, Φ is symmetric.

(c) A similar argument shows the transitivity of Φ.

Hence $\Phi \in E(A)$.

(ii) $\Phi \leqq \varepsilon_i$ for all $i \in I$. Indeed, $x \equiv y(\Phi)$ implies $x \equiv y(\varepsilon_i)$ for all $i \in I$.

(iii) Let $\Theta \leqq \varepsilon_i$ for all $i \in I$. This implies that $\Theta \leqq \Phi$. Indeed, take $x, y \in A$ such that $x \equiv y(\Theta)$; then $x \equiv y(\varepsilon_i)$ for all $i \in I$, which is equivalent to $x \equiv y(\Phi)$. Hence, $\Theta \leqq \Phi$.

Hence, by (i), (ii), and (iii), $\bigwedge (\varepsilon_i \mid i \in I)$ exists and equals Φ.

By Exercise 31, the existence of arbitrary meets implies that $\mathfrak{E}(A)$ is a complete lattice. However, this only guarantees the existence of infinite joins without explicitly describing them. Such a description follows.

Let

$$(\varepsilon_i \mid i \in I)$$

be a family of equivalence relations on A. We define a relation Ψ: $x \equiv y(\Psi)$ if and only if there exists a finite sequence of elements $z_0, z_1, \cdots, z_n$, $x = z_0$, $y = z_n$, such that for all $1 \leqq j \leqq n$ there exists an $i_j \in I$ with the property:

$$z_j \equiv z_{j-1}(\varepsilon_{i_j}) \qquad (j = 1, 2, \cdots, n).$$

Then $\Psi = \bigvee (\varepsilon_i \mid i \in I)$. Again, we verify this statement in three steps.

(i) $\Psi \in E(A)$.

(a) (Reflexivity) $x \equiv x(\Psi)$ because the sequence x, x satisfies the requirement.

(b) (Symmetry) If $x \equiv y(\Psi)$, let $z_0, z_1, \cdots, z_n$ be the corresponding sequence. Then the sequence $z_n, z_{n-1}, \cdots, z_0$ will guarantee $y \equiv x(\Psi)$.

(c) (Transitivity) Let $x \equiv y(\Psi)$ and $y \equiv z(\Psi)$. If we put together the two sequences that correspond to the two relations, then we get a sequence which guarantees $x \equiv z(\Psi)$.

(ii) $\varepsilon_i \leqq \Psi$ for all $i \in I$. If $x \equiv y(\varepsilon_i)$, then $x \equiv y(\Psi)$, since the sequence x, y guarantees this.

(iii) If $\Theta \in E(A)$ and $\varepsilon_i \leqq \Theta$ for all $i \in I$, then $\Psi \leqq \Theta$. Let $x \equiv y(\Psi)$. By definition, this means that there exists a sequence $z_0, \cdots, z_n$ $(z_j \in A)$, $x = z_0$, $y = z_n$, with $z_j \equiv z_{j-1}(\varepsilon_{i_j})$ $(j = 1, \cdots, n)$. Then also $z_j \equiv z_{j-1}(\Theta)$ for $j = 1, 2, \cdots, n$. But Θ is an equivalence relation, and so it is transitive. Hence, $x \equiv y(\Theta)$.

This completes the proof of our theorem.

We will give an alternative description of $\Theta_0 \vee \Theta_1$. Consider the relations

$$\begin{aligned}
\varepsilon_0 &= \Theta_0,\\
\varepsilon_1 &= \Theta_0\Theta_1,\\
\varepsilon_2 &= \Theta_0\Theta_1\Theta_0,\\
\varepsilon_3 &= \Theta_0\Theta_1\Theta_0\Theta_1,\\
&\vdots
\end{aligned}$$

We have immediately that $\varepsilon_0 \leqq \varepsilon_1 \leqq \varepsilon_2 \leqq \cdots$ and $\varepsilon_i \leqq \Theta_0 \vee \Theta_1$. Then

$$\bigcup (\varepsilon_i \mid 0 \leqq i < \omega) = \Theta_0 \vee \Theta_1.$$

The proof is immediate.

§6. IDEALS AND SEMILATTICES

Consider the algebra $\mathfrak{F} = \langle F; \vee \rangle$ with one binary operation. $\mathfrak{F}$ is a *semilattice* if for all $a, b, c \in F$:

(i) $a \vee a = a$,
(ii) $a \vee b = b \vee a$,
(iii) $a \vee (b \vee c) = (a \vee b) \vee c$.

If we have a partially ordered set $\langle F; \leqq \rangle$, it is called a *semilattice* if l.u.b. $(\{a, b\})$ always exists.

The "equivalence" of the two notions can be formulated in the same way as was done for lattices in §4.

In a semilattice, *zero* is an element 0 satisfying $0 \vee a = a$ for every a.

An *ideal* of a semilattice $\mathfrak{F}$ is a nonvoid subset I of F such that, for all $a, b \in F$

$$a \vee b \in I \quad \text{if and only if} \quad a, b \in I.$$

An ideal can also be characterized by:

(i) $a, b \in I$ implies that $a \vee b \in I$;
(ii) $a \in I$ and $c \leqq a$ imply that $c \in I$.

Indeed, assume that I is an ideal. Then (i) is trivial; further, if $a \in I$, $c \leqq a$, then $c \vee a = a \in I$; thus $c, a \in I$, and so $c \in I$, proving (ii).

Now assume (i) and (ii) hold for a nonvoid $I \subseteq F$. We will show that I is an ideal. By (i), $a, b \in I$ implies that $a \vee b \in I$. Let us now assume that $a \vee b \in I$; since $a \leqq a \vee b$, $b \leqq a \vee b$, we have, by (ii), that $a, b \in I$, which was to be proved.

Theorem 1. *Let $\mathfrak{F}$ be a semilattice with* 0. *Let $I(\mathfrak{F})$ be the set of all ideals in $\mathfrak{F}$. Then $\langle I(\mathfrak{F}); \subseteq \rangle = \mathfrak{I}(\mathfrak{F})$ is a complete lattice, called the lattice of ideals of $\mathfrak{F}$.*

Proof. The zero element in $\mathfrak{I}(\mathfrak{F})$ is the ideal $\{0\}$ consisting of the zero element 0 only. The greatest element in $\mathfrak{I}(\mathfrak{F})$ is F.

Let $(I_j \mid j \in J)$ be a family of ideals in F, $J \neq \varnothing$. Then $\bigcap (I_j \mid j \in J)$ is an ideal, i.e.,

$$\bigcap (I_j \mid j \in J) \in I(\mathfrak{F}).$$

Since $\{0\} \subseteq \bigcap (I_j \mid j \in J)$, $\bigcap (I_j \mid j \in J)$ is nonvoid. $\bigcap (I_j \mid j \in J)$ is an ideal because if $a \vee b$ is in this intersection then $a \vee b \in I_j$ for all $j \in J$; since all the I_j are ideals, $a, b \in I_j$ for all $j \in J$; therefore, a and b are in the intersection of all the I_j. By a similar argument, if a and b are in the intersection, then so is $a \vee b$.

Since $\mathfrak{I}(\mathfrak{F})$ has a unit element and infinite meets always exist, we get from Exercise 31 that $\mathfrak{I}(\mathfrak{F})$ is a complete lattice. This completes the proof of Theorem 1.

Let H be any nonvoid subset of F, and let $(H]$ denote the smallest ideal containing H. $(H]$ is called the *ideal generated by* H and can be constructed as the intersection of all ideals containing H.

If $H = \{a, b, \cdots\}$, we shall write $(a, b, \cdots]$ instead of $(\{a, b, \cdots\}]$. $(a]$ is called the *principal ideal* generated by a; for instance, $(0] = \{0\}$.

It is obvious that

$$(a] = \{x \mid x \leqq a\}.$$

This implies that if $(a] = (b]$, then $a = b$. A general description of $(H]$ is the following:

$$(H] = \{t \mid t \leqq h_0 \vee \cdots \vee h_{n-1} \text{ for some } h_i \in H\}.$$

Let K denote the righthand side of the preceding equality. Obviously, if K is an ideal, then it is the smallest ideal containing H. Property (ii) is trivial for K. To prove (i), let $t \leqq h_0 \vee \cdots \vee h_{n-1}$ and $s \leqq h'_0 \vee \cdots \vee h'_{m-1}$. Then

$$s \vee t \leqq h_0 \vee \cdots \vee h_{n-1} \vee h'_0 \vee \cdots \vee h'_{m-1};$$

hence, s and $t \in K$ imply $s \vee t \in K$, verifying property (i).

Note that we have used Exercise 67 in the proof.

A simple application of this result is the following: Since

$$\bigvee (I_j \mid j \in J) = (\bigcup (I_j \mid j \in J)]$$

therefore we get:

Corollary. *$t \in \bigvee (I_j \mid j \in J)$ $(J \neq \varnothing)$ if and only if there exist $j_0, \cdots, j_{n-1} \in J$ and $x_{j_i} \in I_{j_i}$ such that*

$$t \leqq x_{j_0} \vee x_{j_1} \vee \cdots \vee x_{j_{n-1}}.$$

To characterize the lattice of ideals of a given semilattice, we need two concepts.

Let $\mathfrak{L} = \langle L; \leqq \rangle$ be a complete lattice. Let $a \in L$. The element a is called *compact* if the following condition is satisfied:

If $a \leqq \bigvee (x_i \mid i \in I)$, where $x_i \in L$ for each $i \in I$, then there exists $I_1 \subseteq I$ such that I_1 is finite and $a \leqq \bigvee (x_i \mid i \in I_1)$, i.e., if a is contained in an infinite join, it is already contained in a finite join.

(The adjective "compact" is used because of the analogy with the concept of compact subspaces in topology.)

A lattice is called *algebraic* if:

(i) it is complete;

(ii) every a in the lattice can be written as $a = \bigvee (x_i \mid i \in I)$, where all the x_i are compact.

Examples: $\langle \omega + 1; \leqq \rangle$ is an algebraic lattice, where every element, except the greatest one, is compact. Every finite lattice is algebraic.

Lemma 1. *Let $I \in I(\mathfrak{F})$. I is compact in $\mathfrak{I}(\mathfrak{F})$ if and only if I is a principal ideal.*

Proof. Let I be a principal ideal, $I = (a]$. Suppose that

$$(a] = I \subseteq \bigvee (I_j \mid j \in J), J \neq \varnothing.$$

Then $a \in \bigvee (I_j \mid j \in J)$ which implies that $a \leqq x_{j_0} \vee x_{j_1} \vee \cdots \vee x_{j_{n-1}}$, where $x_{j_i} \in I_{j_i}$, $j_i \in J$. Let $J' = \{j_0, \cdots, j_{n-1}\}$. This implies that

$$a \in \bigvee (I_j \mid j \in J')$$

and therefore $(a] \subseteq \bigvee (I_j \mid j \in J')$, which means that I is compact.

For every ideal I, we have the equality

$$I = \bigvee ((a] \mid a \in I).$$

Assume now that I is compact. Since $I \subseteq \bigvee ((a] \mid a \in I)$, there exists a finite $J \subseteq I$ such that $I \subseteq \bigvee ((a] \mid a \in J)$. Set $J = \{a_0, \cdots, a_{n-1}\}$. Then $I = (a_0] \vee (a_1] \vee \cdots \vee (a_{n-1}]$, i.e., I is a finite join of principal ideals. It follows that $I = (a_0 \vee a_1 \vee \cdots \vee a_{n-1}]$, i.e., I is a principal ideal. This completes the proof of the lemma.

The formula $I = \bigvee ((a] \mid a \in I)$ means that every ideal I is the join of principal ideals, which by Lemma 1 means that in the complete lattice $\mathfrak{I}(\mathfrak{F})$, every element is a join of compact elements.

Theorem 2. *Let $\mathfrak{F}$ be a semilattice with zero. Then $\mathfrak{I}(\mathfrak{F})$ is an algebraic lattice.*

Remark. As we will prove later on, many lattices constructed from universal algebras are algebraic. The lattice $\mathfrak{I}(\mathfrak{F})$ was first characterized by A. Komatu, Proc. Imp. Acad. Tokyo, 19 (1943), 119–124, in the special case when $\mathfrak{F}$ is a lattice. A general characterization theorem is in G. Birkhoff and O. Frink [1]. Compact elements were introduced by L. Nachbin, Fund. Math. 36 (1949), 137–142; see also J. R. Büchi [1]. For Theorem 3, see also G. Grätzer and E. T. Schmidt [2].

Next we prove that $\mathfrak{I}(\mathfrak{F})$ is a typical example of an algebraic lattice.

Theorem 3. *Let $\mathfrak{L}$ be an algebraic lattice. Then there exists a semilattice $\mathfrak{F}$ with zero such that $\mathfrak{L}$ is isomorphic to $\mathfrak{I}(\mathfrak{F})$.*

Remark. $\mathfrak{L}$ and $\mathfrak{I}(\mathfrak{F})$ are partially ordered sets. Isomorphism of partially ordered sets was defined in §4.

Proof. Let F be the set of compact elements of L.

(i) If $a, b \in F$, then also $a \vee b \in F$.

Let $a \vee b \leqq \bigvee (x_i \mid i \in J)$, where $x_i \in L$ for $i \in J$. Since $a \leqq a \vee b$, $a \leqq \bigvee (x_i \mid i \in J)$, which by the compactness of a implies that there exists a finite $J' \subseteq J$ such that

$$a \leqq \bigvee (x_i \mid i \in J').$$

Similarly, $b \leqq \bigvee (x_j \mid j \in J'')$, where $J'' \subseteq J$ and J'' is finite. Then

$$a \vee b \leqq \bigvee (x_i \mid i \in J' \cup J''),$$

where $J' \cup J''$ is finite. Hence, $a \vee b$ is compact, that is, $a \vee b \in F$.

(ii) $\langle F; \vee \rangle$ is a semilattice with the zero 0.

Note that the zero element is always compact.

For $a \in L$ set

$$I_a = \{x \mid x \in F \text{ and } x \leqq a\}.$$

(iii) $I_a \in I(\mathfrak{F})$.

This is immediate from (i). Note that $I_a = (a] \cap F$.

(iv) If $a \neq b$, then $I_a \neq I_b$. If $a > b$, then $I_a \supset I_b$.

We first verify the following formula:

$$a = \bigvee (x \mid x \in I_a) \quad \text{for every} \quad a \in L.$$

Since $\mathfrak{L}$ is an algebraic lattice, there exists a set H of compact elements such that

$$a = \bigvee (x \mid x \in H).$$

Since $x \in H$ implies $x \leqq a$, therefore $H \subseteq I_a$. Thus,

$$a = \bigvee (x \mid x \in H) \leqq \bigvee (x \mid x \in I_a) \leqq a,$$

which implies the required formula.

This formula implies that if $I_a = I_b$, then

$$a = \bigvee (x \mid x \in I_a) = \bigvee (x \mid x \in I_b) = b,$$

which proves the first part of (iv).

Now assume that $a > b$. By the definition of I_a, I_b it is obvious that $I_a \supseteq I_b$, and since $I_a \neq I_b$, we obtain $I_a \supset I_b$.

(v) Let $I \in I(\mathfrak{F})$; then there exists an $a \in L$ such that $I = I_a$; in fact, $a = \bigvee (y \mid y \in I)$.

Let $a = \bigvee (y \mid y \in I)$. Then we have an I_a, which consists of all compact elements $\leqq a$. Hence $I_a \supseteq I$. To show that $I_a = I$, we must prove that if $x \in I_a$, then $x \in I$. Let $x \in I_a$; this implies that x is a compact element and

$$x \leqq a = \bigvee (y \mid y \in I).$$

Hence, $x \leqq \bigvee (y \mid y \in I')$ for some finite $I' \subseteq I$. Set $I' = \{y_0, \cdots, y_{n-1}\}$ and $y = y_0 \vee \cdots \vee y_{n-1}$. Then $x \leqq y$ and $y \in I$ because I is an ideal; thus, $x \in I$, which was to be proved.

(vi) The correspondence $\varphi\colon a \to I_a$ sets up an isomorphism between $\mathfrak{L}$ and $\mathfrak{I}(\mathfrak{F})$.

φ is 1–1 by the first part of (iv). φ is onto by (v). φ preserves the order by the second part of (iv), and $a\varphi \geqq b\varphi$ implies $a \geqq b$ by (v).

Hence, φ is an isomorphism. This completes the proof of Theorem 3.

Another useful representation of algebraic lattices can be given in terms of closure systems.

Let A be a set and let $\mathscr{A}$ be a system over A, that is $\mathscr{A} \subseteq P(A)$. If $\mathscr{A}$ is closed under arbitrary intersection, then $\mathscr{A}$ is called a *closure system*. Note that $A \in \mathscr{A}$, by the definition of the intersection of a void family of sets.

Take $X \subseteq A$. Set

$$[X] = \bigcap (B \mid B \in \mathscr{A},\ B \supseteq X).$$

$[X]$ is called the member of $\mathcal{A}$ *generated by* X and if more than one closure system is under consideration, we will write $[X]_{\mathcal{A}}$. Then:

(a) $[X]$ always exists for all $X \subseteq A$;
(b) $[X] \in \mathcal{A}$;
(c) $[X]$ is the smallest member of $\mathcal{A}$ containing X.

It is easy to prove that $X \subseteq Y$ implies $[X] \subseteq [Y]$ and that $[[X]] = [X]$.

If $\mathcal{A}$ is a closure system, then by Exercise 31, $\langle \mathcal{A}; \subseteq \rangle$ is a complete lattice. We want to impose a further condition on $\mathcal{A}$ which will make this lattice algebraic.

Let $\mathcal{B} \subseteq P(A)$. $\mathcal{B} \neq \varnothing$ is a *directed system* if $\langle \mathcal{B}; \subseteq \rangle$ is a directed partially ordered set. Now we define the notion of an algebraic closure system.

An *algebraic closure system* $\mathcal{A}$ satisfies:

(i) $\mathcal{A}$ is a closure system;
(ii) if $\varnothing \neq \mathcal{B} \subseteq \mathcal{A}$ and $\mathcal{B}$ is a directed system, then

$$\bigcup (X \mid X \in \mathcal{B}) \in \mathcal{A}.$$

It should be remarked that it is enough to formulate condition (ii) for chains rather than directed systems. This would not affect the notion of an algebraic closure system.

We give an example of an algebraic closure system. Let $\mathfrak{F}$ be a semilattice with 0. Set $\mathcal{A} = I(\mathfrak{F}) \subseteq P(F)$.

Lemma 2. *$\mathcal{A}$ is an algebraic closure system.*

Proof. (i) We already know that $\mathcal{A}$ is a closure system.

(ii) Given a directed system $\mathcal{B} \neq \varnothing$ of ideals, we must show that the union of the members of $\mathcal{B}$ is an ideal of $\mathfrak{F}$. Set $B = \bigcup (X \mid X \in \mathcal{B})$. To show that B is an ideal, let $x, y \in B$ and $z \leqq x$. Since $x, y \in B$, there exist X, $Y \in \mathcal{B}$ such that $x \in X$, $y \in Y$. Since $\mathcal{B}$ is directed, there exists a $Z \in \mathcal{B}$ such that $X \subseteq Z$ and $Y \subseteq Z$; then $x, y \in Z$ and $z \leqq x$. Inasmuch as Z is an ideal, this implies that $x \vee y$, $z \in Z \subseteq B$. Thus, B is an ideal.

Let $\mathcal{A} \subseteq P(A)$ be an algebraic closure system which is kept fixed in the next three lemmas and Theorem 4. Since $\mathcal{A}$ is a closure system, we again have the notion of the member $[X]$ of $\mathcal{A}$ generated by X for all $X \subseteq A$.

Lemma 3. *Let $X \subseteq A$. Then $[X] = \bigcup ([Y] \mid Y \subseteq X$ and Y is finite$)$.*

Proof. Set $\bar{X} = \bigcup ([Y] \mid Y \subseteq X$ and Y is finite$)$.

(i) Obviously, $X \subseteq \bar{X}$. For, $x \in X$ implies $[\{x\}] \subseteq \bar{X}$. But we always have $x \in \{x\} \subseteq [\{x\}]$, so $x \in \bar{X}$.

(ii) $\bar{X} \in \mathcal{A}$. The system $\mathcal{B} = \{[Y] \mid Y \subseteq X$ and Y finite$\}$ is directed since

$[Y_0], [Y_1] \subseteq [Y_0 \cup Y_1] \in \mathscr{B}$ if Y_0 and Y_1 are finite subsets of A. Therefore, $\bar{X} = \bigcup (Z \mid Z \in \mathscr{B}) \in \mathscr{A}$.

(iii) $\bar{X} \subseteq [X]$, since $Y \subseteq X$ implies $[Y] \subseteq [X]$.

(i), (ii) and (iii) imply, by the definition of $[X]$, that $\bar{X} = [X]$.

Lemma 4. *In the lattice* $\langle \mathscr{A}; \subseteq \rangle$,

$$\bigvee (A_i \mid i \in I) = [\bigcup (A_i \mid i \in I)]$$

if $A_i \in \mathscr{A}$ *for all* $i \in I$, $I \neq \varnothing$.

Proof. See Exercise 31.

Lemma 5. *Let* $B \in \mathscr{A}$. *Then* B *is compact in* $\langle \mathscr{A}; \subseteq \rangle$ *if and only if* $B = [X]$, *for some finite* $X \subseteq A$.

Proof. Assume that $B = [X]$, X finite. Suppose that $B \subseteq \bigvee (A_i \mid i \in I)$; then

$$X \subseteq [X] = B \subseteq \bigvee (A_i \mid i \in I) = [\bigcup (A_i \mid i \in I)],$$

where the last equality holds by virtue of Lemma 4.

Let $X = \{x_0, \cdots, x_{n-1}\}$. Then $x_j \in [\bigcup (A_i \mid i \in I)]$ and so by Lemma 3 there exists $H_j \subseteq \bigcup (A_i \mid i \in I)$ such that H_j is finite and $x_j \in [H_j]$.

Thus, we can find $I_j \subseteq I$, I_j finite, such that $H_j \subseteq \bigcup (A_i \mid i \in I_j)$. Set $J = \bigcup (I_j \mid 0 \leqq j < n)$. Then $X \subseteq [\bigcup (A_i \mid i \in J)]$ so that

$$B = [X] \subseteq [[\bigcup (A_i \mid i \in J)]] = [\bigcup (A_i \mid i \in J)] = \bigvee (A_i \mid i \in J).$$

This proves that B is compact.

Conversely, assume that $B \in \mathscr{A}$ is compact. Now using Lemmas 3 and 4,

$$\begin{aligned} B = [B] &= \bigcup ([Y] \mid Y \subseteq B,\ Y \text{ finite}) \\ &= \bigvee ([Y] \mid Y \subseteq B,\ Y \text{ finite}). \end{aligned}$$

Therefore, $B = [Y_0] \vee \cdots \vee [Y_{k-1}]$, where $Y_0, \cdots, Y_{k-1}$ are finite and $\subseteq B$. Set $Y = Y_0 \cup \cdots \cup Y_{k-1}$. Then $Y \subseteq B$, Y is finite and $B = [Y]$.

Lemma 3 shows that every element of $\langle \mathscr{A}; \subseteq \rangle$ is a union of compact elements. Thus:

Theorem 4. $\langle \mathscr{A}; \subseteq \rangle$ *is an algebraic lattice.*

Summarizing, we get the following result.

Theorem 5. *Given a lattice* $\mathfrak{L}$ *the following conditions are equivalent:*

(i) $\mathfrak{L}$ *is an algebraic lattice;*

(ii) $\mathfrak{L}$ *is isomorphic to some* $\mathfrak{I}(\mathfrak{F})$, *where* $\mathfrak{F}$ *is a semilattice with* 0;

(iii) *there exists an algebraic closure system* $\mathscr{A}$ *such that* $\mathfrak{L}$ *is isomorphic to* $\langle \mathscr{A}; \subseteq \rangle$.

Proof. (i) implies (ii) by Theorem 3. (ii) implies (iii) by Lemma 2, and (iii) implies (i) by Theorem 4.

* * *

Let $\langle L; \vee, \wedge \rangle$ be a lattice; then $\langle L; \vee \rangle$ and $\langle L; \wedge \rangle$ are semilattices. An ideal of $\langle L; \vee \rangle$ is called an *ideal* of $\langle L; \vee, \wedge \rangle$, while an ideal of $\langle L; \wedge \rangle$ is called a *dual ideal* of $\langle L; \vee, \wedge \rangle$. We keep the notations $(H]$, $(a]$ for ideals, while we use $[H)$ and $[a)$ for dual ideals.

An ideal P of $\mathfrak{L}$ is called *proper* if $P \neq L$. A proper ideal P is called *prime* if $a, b \notin P$ implies that $a \wedge b \notin P$.

Theorem 6 (*G. Birkhoff and M. H. Stone*). *Let* $\mathfrak{L} = \langle L; \vee, \wedge \rangle$ *be a distributive lattice,* I *an ideal of* L, $a \in L$, *and* $a \notin I$. *Then there exists a prime ideal* P *with* $a \notin P \supseteq I$.

Proof. Let $\mathscr{P}$ denote the system of all ideals J of $\mathfrak{L}$ with $a \notin J \supseteq I$. We will show that $\mathscr{P}$ satisfies the assumption of Zorn's Lemma: Let $\mathscr{C}$ be a chain in $\langle \mathscr{P}; \subseteq \rangle$ and let $K = \bigcup (X \mid X \in \mathscr{C})$. Since $a \notin X$ for all $X \in \mathscr{C}$, $a \notin K$; further, $X \supseteq I$ for all $X \in \mathscr{C}$; therefore, $K \supseteq I$. Then by Lemma 2, K is an ideal and $a \notin K \supseteq I$; thus $K \in \mathscr{P}$. By Zorn's Lemma, $\mathscr{P}$ has a maximal element P. Assume that P is not a prime ideal; then there exist elements $x, y \in L$ such that $x, y \notin P$ and $x \wedge y \in P$. Let $I_0 = (P \cup \{x\}]$ and $I_1 = (P \cup \{y\}]$. Then $I_0 \supset P$ and $I_1 \supset P$ and so $a \in I_0$ and $a \in I_1$. By the corollary to Theorem 1, this implies the existence of $p_0, p_1 \in P$ with $a \leqq p_0 \vee x$ and $a \leqq p_1 \vee y$; thus

$$a \leqq (p_0 \vee x) \wedge (p_1 \vee y) = (p_0 \wedge p_1) \vee (p_0 \wedge y) \vee (x \wedge p_1) \vee (x \wedge y).$$

Since the element on the right-hand side is in P, $a \in P$, which is a contradiction.

A dual ideal D is called *proper* if $D \neq L$. A proper dual ideal D for which $x, y \notin D$ implies that $x \vee y \notin D$ is called a *prime dual ideal.*

Theorem 7. *Every proper ideal (dual ideal) of a distributive lattice* $\mathfrak{L}$ *is contained in a prime ideal (prime dual ideal).*

A system $\mathscr{A}$ over A is said to have the *finite intersection property* if the intersection of finitely many members of $\mathscr{A}$ is never void. It is obvious, e.g., from Lemma 3, that $\mathscr{A}$ has the finite intersection property if and only if there exists a proper dual ideal $\mathscr{D}$ of $\langle P(A); \subseteq \rangle$ containing all members of $\mathscr{A}$. Thus we have the following result.

Corollary. *Every system of subsets of A with the finite intersection property can be extended to a prime dual ideal of* $\langle P(A); \subseteq \rangle$.

Using the concept of a closure system one can state a very important technique of proof.

Principle of $\mathscr{A}$-induction. Let $\mathscr{A}$ be a closure system over the set A. In order to prove that a proposition P holds for all elements of $B = [M] \subseteq A$ it is sufficient to prove that the set of all elements for which P holds

(1) includes the generating set M,

and

(2) belongs to $\mathscr{A}$.

We can also define $\mathscr{A}$-independence: $M \subseteq A$ will be called *$\mathscr{A}$-independent*, if for all $x \in M$,

$$x \notin [M - \{x\}].$$

Otherwise, M is *$\mathscr{A}$-dependent*.

Special cases of these concepts will be used later.

EXERCISES

1. Let I, J be finite sets and let $(A_i \mid i \in I)$, $(B_j \mid j \in J)$ be families of sets. Then
$$\bigcup (A_i \mid i \in I) \cap \bigcup (B_j \mid j \in J) = \bigcup (A_i \cap B_j \mid i \in I \text{ and } j \in J).$$
2. Prove that in a distributive lattice $a \wedge (b \vee c) = (a \wedge b) \vee (a \wedge c)$.
3. Formulate and prove the statement of Ex. 1 for distributive lattices.
4. Does Ex. 1 hold if we drop the assumption that I and J are finite?
5. Let $(A_{i,j} \mid i \in I \text{ and } j \in J)$ be a family of sets. Then
$$\bigcap (\bigcup (A_{i,j} \mid j \in J) \mid i \in I) = \bigcup (\bigcap (A_{i,i\varphi} \mid i \in I) \mid \varphi \in J^I).$$
6. Find and prove the analogue of Ex. 5, for expressions of the form
$$\bigcup (\bigcap (A_{i,j} \mid j \in J) \mid i \in I).$$
7. Define $A + B = (A - B) \cup (B - A)$ and $A \cdot B = A \cap B$. Then
$$\langle P(X);\ +, \cdot, \varnothing, X \rangle$$
is a ring with $\varnothing$ as null and X as unit element. In this ring, $B + B = \varnothing$ and $B \cdot B = B$ for every $B \subseteq X$.
8. A ring $\langle R;\ +, \cdot, 0 \rangle$ is called *Boolean* if $x \cdot x = x$ for every $x \in R$. Prove that every Boolean ring is commutative, and $x + x = 0$ for every $x \in R$.

9. Prove that a binary relation r on A is an equivalence relation if and only if $r = r^{-1}$, $r \cdot r \subseteq r$ and $r \cap \omega = \omega$.
10. $\langle A; r \rangle$ is a partially ordered set if and only if $r \cdot r \subseteq r$, $r \cap r^{-1} \subseteq \omega$, and $\omega \subseteq r$.
11. $\langle A; r \rangle$ is a chain if and only if it is a partially ordered set and $r \cup r^{-1} = \iota$.
12. Let $T(A)$ be the set of all partial orderings on A. Then $\langle T(A); \subseteq \rangle$ is a partially ordered set. Let $C \subseteq T(A)$ such that $\langle C; \subseteq \rangle$ is a chain and $r = \bigcup (s \mid s \in C)$. Prove that $r \in T(A)$.
13. Combine Ex. 12 and Zorn's Lemma to prove the following: Let $\langle A; r \rangle$ be a partially ordered set; then there exists a chain $\langle A; s \rangle$ such that $r \subseteq s$.
14. Let r and s be binary relations on A. Is it possible to express $r \cdot s$ and r^{-1} in terms of $\cup$, $\cap$, and $\neg$?
15. Let Θ and Φ be equivalence relations on A. Prove that $\Theta \cdot \Phi$ is an equivalence relation on A if and only if $\Theta \cdot \Phi = \Phi \cdot \Theta$. Give an example of $\Theta \cdot \Phi = \Phi \cdot \Theta$ and of $\Theta \cdot \Phi \neq \Phi \cdot \Theta$.
16. Let $A = \{1, 2, 3, 4\}$, $B = \{1, 2\}$, $s = (B \times B) - \{\langle 1, 1 \rangle\}$. What is the number of binary relations r on A satisfying $r_B = s$. Can such an r be an equivalence relation?
17. Show that the binary relation r is an equivalence relation if and only if $r = (\neg((\neg(r^{-1})) \cdot r)) \cap (\neg(r^{-1} \cdot (\neg r)))$.
18. Let $A = \{1, 2, 3, 4\}$, $\varepsilon = \omega \cup \{\langle 1, 2 \rangle, \langle 2, 1 \rangle\}$. Find a set B and a mapping $\varphi \colon A \to B$ such that $\varepsilon = \varepsilon_\varphi$.
19. Let $A_0, \cdots, A_{n-1}$ be subsets of A. Let π be the system of all nonvoid subsets B of A which can be represented in the form

$$\bigcap (A_i \mid i \in t) \cap \bigcap (A_i' \mid i \notin t)$$

for some $t \subseteq \{0, 1, \cdots, n-1\}$. Prove that π is a partition of A.
20. Using the notation of Ex. 19, prove that if $A_i \neq \varnothing$, then it is a union of blocks of π.
21. Using the notation of Ex. 19 and 20, prove that if π_1 is a partition such that every $A_i \neq \varnothing$ is a union of blocks of π_1, then $\pi_1 \leqq \pi$.
22. Let $A, B \subseteq X$ and let Θ be an equivalence relation on X. Prove that $[A]\Theta \cap [B]\Theta \supseteq [A \cap B]\Theta$, and equality does not hold in general, but $[A]\Theta \cup [B]\Theta = [A \cup B]\Theta$ always.
23. Prove that $[A]\Theta \cap [B]\Theta = [A \cap B]\Theta$ if $[A]\Theta = A$.
24. Let Θ_0 and Θ_1 be binary relations. Define $\Theta_0 \circ \Theta_1$ as follows: $x \equiv y(\Theta_0 \circ \Theta_1)$ if and only if there exists a sequence $z_0 = x, z_1, \cdots, z_n = y$ such that $z_i \equiv z_{i-1}(\Theta_{j_i})$ for $1 \leqq i \leqq n$, where $j_i = 0$ or 1. Set $\varepsilon_0 = \Theta_0$, $\varepsilon_1 = \Theta_0 \Theta_1$, $\varepsilon_2 = \Theta_0 \Theta_1 \Theta_0$, and so on. Prove that $\Theta_0 \circ \Theta_1 = \bigcup (\varepsilon_i \mid 0 \leqq i < \omega)$ if Θ_0 and Θ_1 are reflexive.
25. Using the notations of Ex. 24, show that if Θ_0 and Θ_1 are reflexive and transitive, then $\Theta_0 \circ \Theta_1 = \varepsilon_1$ if and only if $\Theta_0 \Theta_1 = \Theta_1 \Theta_0$.
26. Under what conditions is $\Theta_0 \circ \Theta_1$ (for the notations, see Ex. 24) an equivalence relation?
27. Let π be a partition of the set A and let $B \subseteq A$. Define the restriction π_B of π to B.

28. What is the number of different partitions on a four-element set? Let $\pi(n)$ be the number of partitions on an n-element set. Prove that

$$\sum_{n=0}^{\infty} \frac{\pi(n)}{n!} x^n = e^{e^x - 1}.$$

29. Prove that every chain is a distributive lattice.
30. The l.u.b. and g.l.b. in a partially ordered set are unique if they exist.
31. Let $\mathfrak{P} = \langle P; \leqq \rangle$ be a partially ordered set with a 0 ($0 \leqq x$ for all $x \in P$) and assume that any subset of P has a least upper bound. Prove that $\mathfrak{P}$ is a complete lattice. (Hint: The greatest lower bound of a subset H of P is the least upper bound of the set $\{x \mid x \text{ is a lower bound of } H\}$.)
32. What is $\prod (A_i \mid i \in I)$ if $I = \varnothing$? What is A^I if $I = \varnothing$?
33. Let A be a set, R a transitive relation on A, and $\langle a_0, a_1, \cdots, a_\gamma, \cdots \rangle_{\gamma \leq \alpha}$ a sequence of elements of A. Assume that $a_\gamma R a_{\gamma+1}$ holds for all $\gamma < \alpha$. Does this imply that $a_0 R a_\alpha$?
34. Let $\langle P; \leqq \rangle$ be a partially ordered set. Prove that $\langle P; \geqq \rangle$ is also a partially ordered set, where $\geqq$ is the inverse of $\leqq$.
35. Is it true that every mapping φ can be factored into the product $\varphi = \psi \cdot \chi$, where ψ is 1–1 and χ is onto?
36. The $(n+1)$-ary relation R is associated with an n-ary partial operation if and only if $R(a_0, \cdots, a_{n-1}, a)$ and $R(a_0, \cdots, a_{n-1}, a')$ imply that $a = a'$. If, in addition, R satisfies the condition that for every $a_0, \cdots, a_{n-1} \in A$ there exists an $a \in A$ such that $R(a_0, \cdots, a_{n-1}, a)$, then R is associated with an n-ary operation.
37. In the axiom system of a lattice as an algebra (see §3), the first axiom can be deleted.
38. Describe the lattice of all equivalence relations of a four-element set.
39. (A. Tarski) Let $\mathfrak{L} = \langle L; \leqq \rangle$ be a complete lattice and let φ be a mapping of L into itself. If $x \leqq y$ implies that $x\varphi \leqq y\varphi$, then there exists an $a \in L$ with $a\varphi = a$.
40. (S. Banach) Let A and B be sets, let $\varphi\colon A \to B$ and $\psi\colon B \to A$. Then there exist $A_0, A_1 \subseteq A$ with $A = A_0 \cup A_1$ and $A_0 \cap A_1 = \varnothing$ and $B_0, B_1 \subseteq B$, with $B_0 \cup B_1 = B$ and $B_0 \cap B_1 = \varnothing$ such that φ_{A_0} maps A_0 onto B_0 and ψ_{B_1} maps B_1 onto A_1. (Hint: apply Ex. 39 to $\mathfrak{L} = \langle P(A); \subseteq \rangle$ and the mapping $X \to A - (B - X\varphi)\psi$.)
41. Use Ex. 40 to prove that if $\mathfrak{m}$ and $\mathfrak{n}$ are cardinals, $\mathfrak{m} \leqq \mathfrak{n}$ and $\mathfrak{n} \leqq \mathfrak{m}$, then $\mathfrak{m} = \mathfrak{n}$.
42. Let r_i be a mapping of A_i into B_i, for $i \in I$. Under what conditions is $\bigcup (r_i \mid i \in I)$ a mapping of $\bigcup (A_i \mid i \in I)$ into $\bigcup (B_i \mid i \in I)$?
43. Let $\mathfrak{P}$ be a directed partially ordered set of power $\aleph_0$. ($|P| = \aleph_0$). Prove that there is a chain $\mathfrak{C}$ of order type $\leqq \omega$ in $\mathfrak{P}$, which is *cofinal* with $\mathfrak{P}$ (i.e., for every $x \in P$ there exists a $y \in C$ with $x \leqq y$).
44. Prove that the condition that we get from (AC_2) by adding "$A_i \cap A_j = \varnothing$ if $i \neq j$" is equivalent to (AC_2).
45. Prove that the well-ordering principle implies the axiom of choice. (Hint: define the choice function in terms of a well ordering.)

46. Prove that the maximal chain principle implies Zorn's Lemma (find the maximal element as an upper bound of a maximal chain).
47. Prove the following equivalent form of Zorn's Lemma: Let $\langle P; \leqq \rangle$ be a partially ordered set in which every chain has an upper bound; then there exists a maximal element in P.
48. Prove that the following statement is equivalent to the Axiom of Choice: For every binary relation r on a set A, there exists a unary partial operation $f(x)$ on A such that $r(a, b)$ implies that $f(a)$ is defined and $r(a, f(a))$.
49. Let r be a binary relation on A. For $X \subseteq A$ set $X\varphi = \{y \mid y \in A$ and there exists an $x \in X$ with $r(x, y)\}$. Then φ is a mapping of $P(A)$ into itself and $(\bigcup (X_i \mid i \in I))\varphi = \bigcup (X_i\varphi \mid i \in I)$ for any family $(X_i \mid i \in I)$ of subsets of A.
50. Prove the converse of Ex. 49.
51. Let φ be a mapping of $P(A)$ into $P(A)$. Set $X\varphi^{-1} = \bigcup (Y \mid Y \subseteq A$ and $Y\varphi \subseteq X\}$. Using the notations of Ex. 49, prove that r determines a unary partial operation (in the sense of Ex. 36) if and only if $X\varphi^{-1}\varphi \subseteq X$ for all $X \subseteq A$.
52. Combine Ex. 48–51 to give an equivalent form of the Axiom of Choice in terms of mappings of $P(A)$ into itself.
53. Let f be an n-ary partial operation on A. Does there exist an n-ary operation g on A such that $g_B = f$, where $B = D(f)$?
54. Prove that the addition and multiplication of cardinals are commutative. Are the addition and multiplication of ordinals also commutative?
55. Consider the set of all ordinals α with $\bar{\alpha} = \aleph_0$. What is the cardinality of this set?
56. Get a contradiction from the assumption that all cardinals (ordinals) form a set.
57. Prove that if α is an infinite ordinal then $n + \alpha = \alpha$ for all finite ordinals n. When is $\omega + \alpha = \alpha$ true?
58. Prove that every ordinal has a unique representation of the form $\lambda + n$, where λ is a limit ordinal and $n < \omega$.
59. Condition (iii) of transfinite induction can be replaced by the following: if α is a limit ordinal and $\Phi(\beta)$ holds for $\beta < \alpha$, then $\Phi(\alpha)$ holds.
60. Prove that in the formulation of Zorn's Lemma and in Ex. 47, "every chain" can be replaced by "every well-ordered chain".
61. Let φ be a mapping of A onto B; then there exists a mapping ψ of B into A such that $b = b\psi\varphi$ for every $b \in B$.
62. Prove that the statement of Ex. 61 is equivalent to the Axiom of Choice. Prove that Ex. 61 characterizes the onto mappings.
63. Let φ be a mapping of A into B, and $A \neq \varnothing$. Then φ is 1–1 if and only if there exists a mapping ψ of B into A with $a\varphi\psi = a$ for all $a \in A$.
64. An element $\varphi \in M(A)$ is called a *right-annihilator* if $\psi\varphi = \varphi$ for all $\psi \in M(A)$. Describe all the right-annihilators of $M(A)$. Are there any left-annihilators?
65. Describe the join of equivalence relations in terms of partitions.
66. (B. Jónsson) Let $L \subseteq E(A)$ with the property that if ε_0 and $\varepsilon_1 \in L$ then $\varepsilon_0 \vee \varepsilon_1$ and $\varepsilon_0 \wedge \varepsilon_1 \in L$. Then $\mathfrak{L} = \langle L; \leqq \rangle$ is a lattice. Assume that for all ε_0,

$e_1 \in L$ we have that $e_0 \vee e_1 = e_0 e_1 e_0$. Prove that $\mathfrak{L}$ is *modular*, that is $e_0 \geqq e_2$ implies $e_0 \wedge (e_1 \vee e_2) = (e_0 \wedge e_1) \vee e_2$ for all $e_0, e_1, e_2 \in L$.

67. In any semilattice $x_i \leqq y_i$, $i = 0, 1, \cdots, n-1$ imply

$$\bigvee (x_i \mid 0 \leqq i < n) \leqq \bigvee (y_i \mid 0 \leqq i < n).$$

68. Which of the following lattices are algebraic?

(i) $\langle P(A); \cup, \cap \rangle$;

(ii) $\langle \mathrm{Part}(A); \vee, \wedge \rangle$;

(iii) $\langle [0, 1]; \leqq \rangle$, where $[0, 1]$ is the set of all reals (rationals) satisfying $0 \leqq x \leqq 1$.

69. Let I and J be ideals of a distributive lattice. Prove that $x \in I \vee J$ if and only if $x = i \vee j$ for some $i \in I$ and $j \in J$.
70. Prove that if $\langle L; \vee, \wedge \rangle$ is a lattice, then $(a] \vee (b] = (a \vee b]$ and $(a] \wedge (b] = (a \wedge b]$.
71. Let $\mathfrak{L} = \langle L; \vee, \wedge \rangle$ be a distributive lattice. Then $\langle I(\mathfrak{L}); \subseteq \rangle$ is also distributive.
72. Let $\mathfrak{F}_0 = \langle F_0; \leqq \rangle$ and $\mathfrak{F}_1 = \langle F_1; \leqq \rangle$ be semilattices with 0. Prove that $\langle F_0; \leqq \rangle$ is isomorphic to $\langle F_1; \leqq \rangle$ if and only if $\langle I(\mathfrak{F}_0); \subseteq \rangle$ is isomorphic to $\langle I(\mathfrak{F}_1); \subseteq \rangle$.
73. An element p in a lattice with 1 is called a *dual atom* if $p < 1$ and $p < x \leqq 1$ imply that $x = 1$. Prove that if the lattice is distributive then $(p]$ is a prime ideal, whenever p is a dual atom.
74. Let H be an infinite set and let I be the system of finite subsets of H. Then I is an ideal in $\langle P(H); \cup, \cap \rangle$. Let P be a prime ideal containing I. Prove that P is not a principal ideal generated by a dual atom.
75. Let $\langle B; \vee, \wedge, ', 0, 1 \rangle$ be a Boolean algebra. Every ideal of $\langle B; \vee, \wedge \rangle$ is principal if and only if every prime ideal of $\langle B; \vee, \wedge \rangle$ is principal, which is equivalent to B being finite.
76. P is a prime ideal if and only if $L - P$ is a dual prime ideal.
77. Let $\langle L; \leqq \rangle$ be a complete lattice. Then there exists a set A and a closure system $\mathscr{A}$ of subsets of A such that $\langle L; \leqq \rangle$ is isomorphic to $\langle \mathscr{A}; \subseteq \rangle$.
78. Let $X \to \overline{X}$ be a mapping of $P(A)$ into itself such that

(i) $X \subseteq \overline{X}$;

(ii) if $X \subseteq Y$, then $\overline{X} \subseteq \overline{Y}$;

(iii) $\overline{\overline{X}} = \overline{X}$.

Let $\mathscr{A} = \{\overline{X} \mid X \subseteq A\}$. Then $\mathscr{A}$ is a closure system, and $\overline{X} = [X]_{\mathscr{A}}$.

79. For a Boolean algebra $\mathfrak{B} = \langle B; \vee, \wedge, ', 0, 1 \rangle$, the lattice $\langle B; \vee, \wedge \rangle$ is algebraic if and only if $\mathfrak{B}$ is isomorphic to some $\mathfrak{P}(I)$.
80. Let $\mathfrak{L} = \langle L; \leqq \rangle$ be a complete lattice and $H \subseteq L$, $H \neq \varnothing$. $\langle H; \leqq \rangle$ is called a *complete sublattice* of $\mathfrak{L}$ if for every $K \subseteq H$, l.u.b. $(K) \in H$ and g.l.b. $(K) \in H$. Prove that a complete sublattice of an algebraic lattice is always an algebraic lattice.
81. Every well-ordered set with a largest element is an algebraic lattice.

82. (G. Grätzer [8]) Let $\mathfrak{L}$ be a complete lattice and let $\mathfrak{m}$ be a regular cardinal. $a \in L$ is called *$\mathfrak{m}$-compact* if $a \leqq \bigvee (x_i \mid i \in I)$ implies that $a \leqq \bigvee (x_i \mid i \in I')$ for some $I' \subseteq I$ with $|I'| < \mathfrak{m}$. $\mathfrak{L}$ is called *$\mathfrak{m}$-algebraic* if every element is the join of $\mathfrak{m}$-compact elements. Find the analogue of Theorem 6.5 for $\mathfrak{m}$-algebraic lattices.
83. Let $\mathfrak{L} = \langle L; \vee, \wedge \rangle$ be a distributive lattice with 1. An ideal I of $\mathfrak{L}$ is *maximal* ($I \neq L$ and if $I \subseteq J \neq L$, then $I = J$) if and only if for $a \in L$, the condition that $a \vee b \neq 1$ for all $b \in I$, implies that $a \in I$.
84. The condition of Ex. 83 characterizes prime ideals of Boolean algebras.

CHAPTER 1

SUBALGEBRAS AND HOMOMORPHISMS

The basic results on subalgebras and homomorphisms are presented in this chapter. Many of these results belong to the "folklore" of the theory, and therefore some results are given without references. The systematic treatment of polynomial symbols, which will be continued in Chapter 2, turns out to be one of the most useful topics of this chapter and, surprisingly, one of the topics most neglected in the literature.

§7. BASIC CONCEPTS

The concept of an algebra $\langle A; F\rangle$ as introduced is quite adequate in dealing with such problems as the structure of subalgebras (§9) or endomorphisms (§12). However, to introduce the concept of a similarity class of algebras one has to consider F as a family of operations with a fixed index set. For the sake of convenience we choose this index set to be a set of ordinals, but this is not very important.

Remark. Sometimes we will say, let $\langle A; F\rangle$ be an algebra, where F is a family of operations on A, and we will not well order F. This always will mean that the well ordering of F does not matter. We can always do this if we consider only a single algebra, or if, for some other reason, we already have names for the operations.

A *type* of algebras τ is a sequence $\langle n_0, n_1, \cdots, n_\gamma, \cdots\rangle$ of nonnegative integers, $\gamma < o(\tau)$, where $o(\tau)$ is an ordinal, called the *order of* τ. For every $\gamma < o(\tau)$ we have a *symbol* $\mathbf{f}_\gamma$ of an n_γ-ary operation.

An *algebra* $\mathfrak{A} = \langle A; F\rangle$ *of type* τ is a pair, where A is a nonvoid set (the *base set* of $\mathfrak{A}$), and for every $\gamma < o(\tau)$, we realize $\mathbf{f}_\gamma$ as an n_γ-ary operation on A: $(\mathbf{f}_\gamma)_\mathfrak{A}$, and $F = \langle(\mathbf{f}_0)_\mathfrak{A}, (\mathbf{f}_1)_\mathfrak{A}, \cdots, (\mathbf{f}_\gamma)_\mathfrak{A}, \cdots\rangle$.

$(\mathbf{f}_\gamma)_\mathfrak{A}$ is the *realization* of $\mathbf{f}_\gamma$ and if there is no danger of confusion, we will write f_γ for $(\mathbf{f}_\gamma)_\mathfrak{A}$ and $F = \langle f_0, \cdots, f_\gamma, \cdots\rangle$. Thus if $\mathfrak{A}$ and $\mathfrak{B}$ are both algebras of the same type τ, f_γ will denote an operation on A as well as on B. In general there is no danger of confusion since, if we write $f_\gamma(a_0, \cdots, a_{n_\gamma - 1})$, $a_0, \cdots, a_{n_\gamma - 1} \in A$, then f_γ obviously means an operation on A.

Let us remark that this usage is generally accepted in algebra, e.g., $+$ is used to denote an operation in every abelian group.

If $F = \langle f_0, \cdots, f_{n-1} \rangle$ we will write $\langle A; f_0, \cdots, f_{n-1} \rangle$ for $\langle A; F \rangle$.

The class of all algebras of type τ will be denoted by $K(\tau)$; it will be called a ***similarity class*** of algebras (also called a ***species*** of algebras).

Let $\mathfrak{A}, \mathfrak{B} \in K(\tau)$. A mapping $\varphi: A \to B$ is called an ***isomorphism*** between the algebras $\mathfrak{A}$ and $\mathfrak{B}$ if it is 1–1, onto, and for every $\gamma < o(\tau)$, $a_0, \cdots a_{n_\gamma - 1} \in A$ we have

$$f_\gamma(a_0, \cdots, a_{n_\gamma - 1})\varphi = f_\gamma(a_0\varphi, \cdots, a_{n_\gamma - 1}\varphi).$$

If there is an isomorphism between $\mathfrak{A}$ and $\mathfrak{B}$, then $\mathfrak{A}$ and $\mathfrak{B}$ are called ***isomorphic***. If $\mathfrak{A}$ and $\mathfrak{B}$ are algebras, we write $\mathfrak{A} \cong \mathfrak{B}$ for "$\mathfrak{A}$ and $\mathfrak{B}$ are isomorphic".

The purpose of the theory of universal algebras is to find and examine those properties of universal algebras which are invariant under isomorphism. Therefore, in general, we will not distinguish between isomorphic algebras (one notable exception is, if they are both subalgebras of the same algebra, see below).

Being of the same type means very little. If, unlike in §3, we define a ring as $\langle R; +, \cdot \rangle$, then rings and lattices are of type $\langle 2, 2 \rangle$. However, to develop algebraic constructions (Chapters 1 and 3), it is enough in most cases to assume that the algebras belong to the same similarity class.

An algebra $\mathfrak{A}$ is called ***unary*** if it is of type $\langle 1, 1, \cdots, 1, \cdots \rangle$.

Next we define the three most important algebraic concepts; namely those of subalgebra, homomorphism, and congruence.

Let $\mathfrak{A}$ be an algebra of type τ and B a nonvoid subset of A. $\mathfrak{B} = \langle B; F \rangle$ is called a ***subalgebra*** of $\mathfrak{A}$ (and $\mathfrak{A}$ an ***extension*** of $\mathfrak{B}$) if and only if

$$b_0, \cdots, b_{n_\gamma - 1} \in B \text{ implies that } (f_\gamma)_\mathfrak{A}(b_0, \cdots, b_{n_\gamma - 1}) = (f_\gamma)_\mathfrak{B}(b_0, \cdots, b_{n_\gamma - 1}) \in B$$

for all $\gamma < o(\tau)$, that is, if and only if B is closed under all the operations of $\mathfrak{A}$ and $(\mathbf{f}_\gamma)_\mathfrak{B}$ is the restriction of $(\mathbf{f}_\gamma)_\mathfrak{A}$ to B (or more precisely, to B^{n_γ}).

If $\mathfrak{A}$ is an algebra, $B \subseteq A$, $B \neq \varnothing$, and for all $\gamma < o(\tau)$, $b_0, \cdots, b_{n_\gamma - 1} \in B$, $f_\gamma(b_0, \cdots, b_{n_\gamma - 1}) \in B$, then there is exactly one subalgebra of $\mathfrak{A}$ on the base set B. Thus if we write "so $\langle B; F \rangle$ is a subalgebra" or "let $\langle B; F \rangle$ be a subalgebra", this always means that B has the property described above.

Lemma 1. *Let $\langle B_i; F \rangle$, $i \in I$ be subalgebras of $\mathfrak{A}$, $B = \bigcap (B_i \mid i \in I)$. If $B \neq \varnothing$, then $\langle B; F \rangle$ is a subalgebra of $\mathfrak{A}$.*

Proof. Let $b_0, \cdots, b_{n_\gamma - 1} \in B$. Then $b_0, \cdots, b_{n_\gamma - 1} \in B_i$ and so

$$f_\gamma(b_0, \cdots, b_{n_\gamma - 1}) \in B_i,$$

for all $i \in I$. Thus $f_\gamma(b_0, \cdots, b_{n_\gamma - 1}) \in B$.

Lemma 1 implies that if $\mathfrak{A}$ is an algebra, $H \subseteq A$, $H \neq \varnothing$, then there is a smallest subset B containing H such that $\langle B; F\rangle$ is a subalgebra. We will denote this B by $[H]$ and $\langle [H]; F\rangle$ will be called the *subalgebra generated by H* and H is a *generating set* of $\langle [H]; F\rangle$.

We extend the notation $[H]$ to the void set:

$[\varnothing] = \varnothing$ if there are no nullary operations;

$[\varnothing]$ is the subalgebra generated by the values of nullary operations, if there are nullary operations.

$\langle A; F\rangle$ is *finitely generated* if $A = [H]$ for some finite $H \subseteq A$.

Note that if $\langle G; \cdot, 1\rangle$ is a group, then $\langle H; \cdot, 1\rangle$ is a subalgebra if and only if $1 \in H$, H is closed under multiplication, and the multiplication of elements of H is the same in $\langle H; \cdot, 1\rangle$ and $\langle G; \cdot, 1\rangle$ (which does not mean that $\langle H; \cdot, 1\rangle$ is a subgroup).

Let us also note that the elements picked out by the nullary operations are contained in every subalgebra of the given algebra.

Let $\mathfrak{A}$ and $\mathfrak{B}$ be two algebras belonging to the same similarity class $K(\tau)$. A mapping $\varphi: A \to B$ such that

$$f_\gamma(a_0, \cdots, a_{n_\gamma - 1})\varphi = f_\gamma(a_0\varphi, \cdots, a_{n_\gamma - 1}\varphi)$$

for all $\gamma < o(\tau)$ is called a *homomorphism* of $\mathfrak{A}$ into $\mathfrak{B}$.

Let us note that if a nullary operation f_γ picks out a from $\mathfrak{A}$ and b from $\mathfrak{B}$, then $a\varphi = b$, more precisely, $((\mathbf{f}_\gamma)_{\mathfrak{A}})\varphi = (\mathbf{f}_\gamma)_{\mathfrak{B}}$.

We will call the homomorphism 1–1 (onto) if the mapping φ is 1–1 (onto). 1–1 homomorphisms are also called *embeddings*. If φ is 1–1 and onto, it is thus an isomorphism. If φ is onto, then $\mathfrak{B}$ is called a *homomorphic image* of $\mathfrak{A}$. Some authors use special names for 1–1 and onto homomorphisms. 1–1 homomorphisms are called *injections* and *monomorphisms*, onto homomorphisms are called *surjections* and *epimorphisms*. Isomorphisms are also called *bijections*.

Let $\mathfrak{A}$ be an algebra and Θ a binary relation defined on A. Θ is called a *congruence relation* if it is an equivalence relation satisfying the *substitution property* (SP):

(SP) If $\gamma < o(\tau)$, $a_i \equiv b_i(\Theta)$, $a_i, b_i \in A$, $0 \leqq i < n_\gamma$, then

$$f_\gamma(a_0, \cdots, a_{n_\gamma - 1}) \equiv f_\gamma(b_0, \cdots, b_{n_\gamma - 1})(\Theta).$$

If we are given an algebra $\mathfrak{A}$ and a congruence relation Θ on $\mathfrak{A}$, we can construct a new algebra called the *quotient algebra* as follows: the new algebra is defined on the quotient set

$$A/\Theta = \{[a]\Theta \mid a \in A\}$$

(for the notation, see §2), with operations defined as

$$f_\gamma([a_0]\Theta, \cdots, [a_{n_\gamma - 1}]\Theta) = [f_\gamma(a_0, \cdots, a_{n_\gamma - 1})]\Theta.$$

The new algebra is denoted by

$$\mathfrak{A}/\Theta = \langle A/\Theta; F \rangle.$$

Since the operations are defined in terms of the representatives $a_0, \cdots, a_{n_\gamma - 1}$ of the equivalence classes, we have to prove that the operations are well defined, that is, the result of the operation does not depend on the representatives chosen. Indeed, if $b_0, \cdots, b_{n_\gamma - 1}$ is another set of representatives, that is, if

$$a_i \equiv b_i(\Theta), \qquad 0 \leqq i < n_\gamma,$$

then, by (SP),

$$f_\gamma(a_0, \cdots, a_{n_\gamma - 1}) \equiv f_\gamma(b_0, \cdots, b_{n_\gamma - 1})(\Theta),$$

and thus,

$$[f_\gamma(a_0, \cdots, a_{n_\gamma - 1})]\Theta = [f_\gamma(b_0, \cdots, b_{n_\gamma - 1})]\Theta,$$

which was to be proved.

Now consider the mapping

$$\varphi_\Theta \colon a \to [a]\Theta,$$

which is the natural mapping of A onto the quotient set A/Θ (see §2).

Lemma 2. *φ_Θ is a homomorphism of $\mathfrak{A}$ onto $\mathfrak{A}/\Theta$.*

That is, every quotient algebra is a homomorphic image of the algebra.

Proof. Let $a_0, \cdots, a_{n_\gamma - 1} \in A$. Then

$$\begin{aligned} f_\gamma(a_0, \cdots, a_{n_\gamma - 1})\varphi_\Theta &= [f_\gamma(a_0, \cdots, a_{n_\gamma - 1})]\Theta \\ &= f_\gamma([a_0]\Theta, \cdots, [a_{n_\gamma - 1}]\Theta) \\ &= f_\gamma(a_0\varphi_\Theta, \cdots, a_{n_\gamma - 1}\varphi_\Theta), \end{aligned}$$

which was to be proved.

In conclusion, we state four other elementary facts.

Lemma 3. *Suppose $\varphi \colon A \to B$ is a homomorphism of $\mathfrak{A}$ into $\mathfrak{B}$. Then $\langle A\varphi; F \rangle$ is a subalgebra of $\mathfrak{B}$.*

Proof. Let $b_0, \cdots, b_{n_\gamma - 1}$ be elements in $A\varphi$. Then there exist $a_0, \cdots, a_{n_\gamma - 1} \in A$ such that $b_i = a_i\varphi$, $0 \leqq i < n_\gamma$. Since

$$f_\gamma(b_0, \cdots, b_{n_\gamma - 1}) = f_\gamma(a_0\varphi, \cdots, a_{n_\gamma - 1}\varphi) = f_\gamma(a_0, \cdots, a_{n_\gamma - 1})\varphi \in A\varphi,$$

we see that $A\varphi$ is closed under the operations.

Lemma 4. *Let $\mathfrak{A}$ be an algebra and $\mathfrak{B}$ a subalgebra of $\mathfrak{A}$. Let Θ be a congruence relation of $\mathfrak{A}$ and φ a homomorphism of $\mathfrak{A}$ into $\mathfrak{C}$. Then Θ_B, that is, the restriction of Θ to B, is a congruence relation of $\mathfrak{B}$ and φ_B, that is, the restriction of φ to B, is a homomorphism of $\mathfrak{B}$ into $\mathfrak{C}$.*

Lemma 5. *Let $\mathfrak{A}$, $\mathfrak{B}$ and $\mathfrak{C}$ be algebras, φ a homomorphism of $\mathfrak{A}$ into $\mathfrak{B}$, and ψ a homomorphism of $\mathfrak{B}$ into $\mathfrak{C}$. Then $\varphi\psi$ is a homomorphism of $\mathfrak{A}$ into $\mathfrak{C}$.*

Lemma 6. *Let φ be a homomorphism of $\mathfrak{A}$ into $\mathfrak{B}$. Then ε_φ, the equivalence relation on A induced by φ, is a congruence relation of $\mathfrak{A}$.*

The proofs of Lemmas 4–6 are left to the reader.

§8 POLYNOMIAL SYMBOLS AND POLYNOMIAL ALGEBRAS

Definition 1. *Let $\mathfrak{A}$ be an algebra; the* n-ary polynomials of $\mathfrak{A}$ *are certain mappings from A^n into A, defined as follows:*

(i) *The projections (see §2) e_i^n are n-ary polynomials* $(0 \leq i < n)$;

(ii) *If $p_0, \cdots, p_{n_\gamma - 1}$ are n-ary polynomials, then so is $f_\gamma(p_0, \cdots, p_{n_\gamma-1})$, defined by*

$$\begin{aligned} f_\gamma(p_0, \cdots, p_{n_\gamma-1})&(x_0, \cdots, x_{n-1}) \\ &= f_\gamma(p_0(x_0, \cdots, x_{n-1}), \cdots, p_{n_\gamma-1}(x_0, \cdots, x_{n-1})); \end{aligned}$$

(iii) *n-ary polynomials are those and only those which we get from* (i) *and* (ii) *in a finite number of steps.*

Since the n-ary polynomials of $\langle A; F\rangle$ are functions, their equality is defined as the equality of functions.

Note that in (ii) $n_\gamma = 0$ is not excluded. Therefore every nullary operation is an n-ary polynomial, for every n. Moreover, $n=0$ is also allowed, in which case there is no e_i^n, and therefore we get nullary polynomials if and only if there is at least one nullary operation.

Examples: Let $\langle L; \vee, \wedge\rangle$ be a lattice; then examples of unary polynomials are $e_0^1(x_0) = x_0$, $(e_0^1 \vee e_0^1)(x_0) = e_0^1(x_0) \vee e_0^1(x_0) = x_0$, and so on. It is easy to see that there is only one unary polynomial. $e_0^2(x_0, x_1) = x_0$, $(e_1^2 \vee e_0^2)(x_0, x_1) = x_1 \vee x_0$, and so on, are examples of binary polynomials.

Let $\langle R; +, \cdot, 0, 1\rangle$ be a ring with a unit element. It is easy to prove that in this case every unary polynomial p is of the form

$$p(x_0) = n_0 + n_1x_0 + n_2x_0^2 + \cdots + n_{m-1}x_0^{m-1},$$

where the n_i are elements of the form $1+1+\cdots+1$ or 0, and conversely.

Let $\langle G; \cdot\rangle$ be a semigroup; then all unary polynomials are of the form x_0^n.

Let $\mathfrak{A}$ be an algebra. Then $P^{(n)}(\mathfrak{A})$ will denote the set of n-ary polynomials. (ii) can be considered as a definition of the operations on the set of n-ary polynomials; the resultant algebra $\mathfrak{P}^{(n)}(\mathfrak{A}) = \langle P^{(n)}(\mathfrak{A}); F\rangle$ is the *algebra of n-ary polynomials*. Note that $P^{(0)}(\mathfrak{A})$ consists of the nullary polynomials, that is, of those polynomials which we build up from the nullary operations (if there are any); therefore $\mathfrak{P}^{(0)}(\mathfrak{A})$ is defined only if there are nullary operations.

Lemma 1. *Let $p \in P^{(n)}(\mathfrak{A})$ and $n > 1$. Then there exists a $q \in P^{(n-1)}(\mathfrak{A})$ such that for all $x_0, \cdots, x_{n-2} \in A$*

$$p(x_0, \cdots, x_{n-2}, x_{n-2}) = q(x_0, \cdots, x_{n-2}).$$

Proof. If $p = e_i^n$ and $i \neq n-1$, then set

$$q = e_i^{n-1};$$

if $i = n-1$, then set

$$q = e_{n-2}^{n-1}.$$

If the statement has already been proved for $p_0, \cdots, p_{n_\gamma - 1}$, and the corresponding polynomials are $q_0, \cdots, q_{n_\gamma - 1}$, then the polynomial

$$f_\gamma(p_0, \cdots, p_{n_\gamma - 1})$$

will correspond to

$$f_\gamma(q_0, \cdots, q_{n_\gamma - 1}).$$

Lemma 2. *Let $p \in P^{(n)}(\mathfrak{A})$ and let σ be a permutation of $0, \cdots, n-1$. Define $p^\sigma(x_0, \cdots, x_{n-1}) = p(x_{0\sigma}, \cdots, x_{(n-1)\sigma})$. Then $p^\sigma \in P^{(n)}(\mathfrak{A})$.*

Proof. $(e_i^n)^\sigma = e_{i\sigma}^n \in P^{(n)}(\mathfrak{A})$. The induction step is the same as in Lemma 1.

Corollary 1. *Let φ be a mapping of $\{0, \cdots, n-1\}$ into $\{0, \cdots, m-1\}$, $n \geqq m$, and let $p \in P^{(n)}(\mathfrak{A})$. Then there exists a $q \in P^{(m)}(\mathfrak{A})$ such that*

$$p(x_{0\varphi}, \cdots, x_{(n-1)\varphi}) = q(x_0, \cdots, x_{m-1}).$$

Proof. By Lemmas 1 and 2.

Let $p \in P^{(n)}(\mathfrak{A})$. We say that p *depends on* x_i if there exist $a_0, \cdots, a_{n-1}$ and $a_i' \in A$ such that

$$p(a_0, \cdots, a_i, \cdots, a_{n-1}) \neq p(a_0, \cdots, a_i', \cdots, a_{n-1}).$$

$P^{(n,k)}(\mathfrak{A})$ denotes those n-ary polynomials which depend on at most k variables. The polynomials in $P^{(n,0)}(\mathfrak{A})$ are constant. If for $a \in A$ there exists a $p \in P^{(n,0)}(\mathfrak{A})$ such that $a = p(x_0, \cdots, x_{n-1})$, then a is called a *constant of* $\mathfrak{A}$.

Lemma 3. *If a is a constant of $\mathfrak{A}$, then there exists a $p \in P^{(1,0)}(\mathfrak{A})$ such that $a = p(x_0)$.*

Proof. Identify all variables and use Lemma 1, if $a = p(x_0, \cdots, x_{n-1})$ with $n \geqq 1$, and recall that a nullary polynomial is also a unary polynomial.

Lemma 4. *Let a be a constant of $\mathfrak{A}$. There exists a $p \in P^{(0)}(\mathfrak{A})$ such that $a = p$, if and only if there is at least one nullary operation.*

Proof. Trivial, by the definition of nullary polynomials.

Let $\mathfrak{A}$ and $\mathfrak{B}$ be algebras of type τ. If we consider the n-ary polynomials over $\mathfrak{A}$ and the n-ary polynomials over $\mathfrak{B}$, we observe that they are built up the same way, and we use the same symbols in both cases. For instance, $f_\gamma(e_0{}^n, \cdots, e_i{}^n)$ denotes an n-ary polynomial for $\mathfrak{A}$ and also for $\mathfrak{B}$. This is similar to the situation, that f_γ denotes an operation on both, and suggests that just as we have an operation symbol $\mathbf{f}_\gamma$, it would be useful to have polynomial symbols as well.

Definition 2. *The* n-ary polynomial symbols *of type τ are defined as follows:*

(i) $\mathbf{x}_0, \cdots, \mathbf{x}_{n-1}$ *are n-ary polynomial symbols;*

(ii) *if* $\mathbf{p}_0, \cdots, \mathbf{p}_{n_\gamma - 1}$ *are n-ary polynomial symbols and* $\gamma < o(\tau)$, *then* $\mathbf{f}_\gamma(\mathbf{p}_0, \cdots, \mathbf{p}_{n_\gamma - 1})$ *is an n-ary polynomial symbol;*

(iii) *n-ary polynomial symbols are those and only those which we get from* (i) *and* (ii) *in a finite number of steps.*

Remark. If $\mathbf{f}_\gamma$ is the symbol of a nullary operation, then $\mathbf{f}_\gamma$ is an n-ary polynomial symbol for any n. Nullary polynomial symbols exist if and only if there are nullary operation symbols. We consider polynomial symbols as sequences of symbols; thus equality means formal equality.

Now we will show that a polynomial symbol is indeed a symbol for a polynomial.

Definition 3. *The n-ary polynomial p over the algebra $\mathfrak{A}$, associated with (or induced by) the n-ary symbol* $\mathbf{p}$ *is defined as follows:*

(i) $\mathbf{x}_i$ *induces* $e_i{}^n$;

(ii) *if* $\mathbf{p} = \mathbf{f}_\gamma(\mathbf{p}_0, \cdots, \mathbf{p}_{n_\gamma - 1})$ *and* $\mathbf{p}_i$ *induces* p_i *for* $0 \leqq i < n_\gamma$, *then* $\mathbf{p}$ *induces* $f_\gamma(p_0, \cdots, p_{n_\gamma - 1})$.

Corollary 1. *Every n-ary polynomial p over $\mathfrak{A}$ is induced by some n-ary polynomial symbol* $\mathbf{p}$.

If $\mathbf{p}$ is a polynomial symbol, $(\mathbf{p})_{\mathfrak{A}}$ or $\mathbf{p}^{\mathfrak{A}}$ will denote the polynomial of $\mathfrak{A}$ induced by $\mathbf{p}$.

Corollary 2. *For every n-ary polynomial p over $\mathfrak{A}$ and for $m>n$, there exists an m-ary polynomial q over $\mathfrak{A}$ such that*

$$p(a_0, \cdots, a_{n-1}) = q(a_0, \cdots, a_{m-1})$$

for every $a_0, \cdots, a_{m-1} \in A$.

Proof. By Definition 2, if $n < m$, then every n-ary polynomial symbol is also an m-ary polynomial symbol. Thus Corollary 2 is trivial from Corollary 1.

Let $\mathfrak{A}$ be an algebra and $a \in A$. Then a is called an *algebraic constant* if $a = (\mathbf{p})_{\mathfrak{A}}$ for some nullary polynomial symbol $\mathbf{p}$.

Corollary 3. *Let $\mathfrak{A}$ be an algebra and $a \in A$. If a is an algebraic constant, then a is a constant in $\mathfrak{A}$. Conversely, if a is a constant in $\mathfrak{A}$ and there are nullary operations, then a is an algebraic constant.*

Proof. By Corollary 1 and by Lemma 4.

Corollary 1 suggests a natural way of defining the equality of polynomial symbols: we want two polynomial symbols to be equal if in any algebra of $K(\tau)$ they induce the same polynomial. We will prove that formal equality is, in fact, equivalent to this condition. To this end, we construct the *n-ary polynomial algebra* $\mathfrak{P}^{(n)}(\tau) = \langle \mathbf{P}^{(n)}(\tau); F \rangle$ as follows:

$\mathbf{P}^{(n)}(\tau)$ is the set of all n-ary polynomial symbols; the operations on $\mathbf{P}^{(n)}(\tau)$ are defined by

$$f_\gamma(\mathbf{p}_0, \cdots, \mathbf{p}_{n_\gamma - 1}) = \mathbf{f}_\gamma(\mathbf{p}_0, \cdots, \mathbf{p}_{n_\gamma - 1}).$$

Then $\mathfrak{P}^{(n)}(\tau) \in K(\tau)$.

Remark. We will prove propositions concerning polynomial symbols in the same way that we did for polynomials, using the following scheme:

(i) The statement is true for $\mathbf{x}_i$;
(ii) if it is true for $\mathbf{p}_0, \cdots, \mathbf{p}_{n_\gamma - 1}$ then it is true for

$$\mathbf{f}_\gamma(\mathbf{p}_0, \cdots, \mathbf{p}_{n_\gamma - 1}).$$

This scheme is simply proof by induction on the "rank" of a polynomial symbol. Rank can be defined in any way, only it has to be a positive integer, and the rank of $\mathbf{f}_\gamma(\mathbf{p}_0, \cdots, \mathbf{p}_{n_\gamma-1})$ must be greater than the rank of any $\mathbf{p}_i$. For instance, we can define the rank of $\mathbf{p}$ as the number of symbols which occur in $\mathbf{p}$; thus, the rank of $\mathbf{x}_i$ is 2 and the rank of $\mathbf{f}_3(\mathbf{x}_0, \mathbf{f}_1(\mathbf{x}_1))$ is 13.

Lemma 5. *Let* $\mathbf{p}$ *be an n-ary polynomial symbol and* p *the polynomial of* $\mathfrak{P}^{(n)}(\tau)$ *associated with* $\mathbf{p}$. *Then*

$$p(\mathbf{x}_0, \cdots, \mathbf{x}_{n-1}) = \mathbf{p}.$$

Proof. (i) If $\mathbf{p} = \mathbf{x}_i$, then

$$e_i^n(\mathbf{x}_0, \cdots, \mathbf{x}_{n-1}) = \mathbf{x}_i;$$

(ii) if the statement is true for $\mathbf{p}_0, \cdots, \mathbf{p}_{n_\gamma-1}$ and $\mathbf{p} = \mathbf{f}_\gamma(\mathbf{p}_0, \cdots, \mathbf{p}_{n_\gamma-1})$, then

$$\begin{aligned} p(\mathbf{x}_0, \cdots, \mathbf{x}_{n-1}) &= f_\gamma(p_0(\mathbf{x}_0, \cdots, \mathbf{x}_{n-1}), \cdots, p_{n_\gamma-1}(\mathbf{x}_0, \cdots, \mathbf{x}_{n-1})) \\ &= f_\gamma(\mathbf{p}_0, \cdots, \mathbf{p}_{n_\gamma-1}) = \mathbf{f}_\gamma(\mathbf{p}_0, \cdots, \mathbf{p}_{n_\gamma-1}) = \mathbf{p}. \end{aligned}$$

This completes the proof of Lemma 5.

Theorem 1. *Let* $\mathbf{p}, \mathbf{q} \in \mathbf{P}^{(n)}(\tau)$. *If* $\mathbf{p}$ *and* $\mathbf{q}$ *induce equal polynomials in every algebra of type* τ, *then* $\mathbf{p} = \mathbf{q}$.

Proof. Let p and q be the polynomials associated with $\mathbf{p}$ and $\mathbf{q}$, respectively, in the algebra $\mathfrak{P}^{(n)}(\tau)$. By assumption, $p(a_0, \cdots, a_{n-1}) = q(a_0, \cdots, a_{n-1})$ for any $a_0, \cdots, a_{n-1} \in \mathbf{P}^{(n)}(\tau)$. Put $a_i = \mathbf{x}_i$. Then applying Lemma 5 twice:

$$\mathbf{p} = p(\mathbf{x}_0, \cdots, \mathbf{x}_{n-1}) = q(\mathbf{x}_0, \cdots, \mathbf{x}_{n-1}) = \mathbf{q},$$

which was to be proved.

In most applications it is useful to consider *α-ary polynomial symbols* where α is an arbitrary ordinal. It is easy to modify Definitions 1–3; we only have to replace the $e_0^n, \cdots, e_{n-1}^n$ by

$$e_0^\alpha, \cdots, e_\delta^\alpha, \cdots, \delta < \alpha$$

in Definitions 1 and 3, and in Definition 2 we have to replace $\mathbf{x}_0, \cdots, \mathbf{x}_{n-1}$ by $\mathbf{x}_0, \cdots, \mathbf{x}_\gamma, \cdots$ for $\gamma < \alpha$.

The corresponding algebras will be denoted by

$$\mathfrak{P}^{(\alpha)}(\tau) = \langle \mathbf{P}^{(\alpha)}(\tau); F \rangle \quad \text{and} \quad \mathfrak{P}^{(\alpha)}(\mathfrak{A}), \text{ respectively.}$$

Lemma 5′. *Lemma 5 holds for α-ary polynomial symbols for arbitrary α.*

Theorem 1′. *Theorem 1 holds for α-ary polynomial symbols for arbitrary α.*

Let $\mathfrak{A} \in K(\tau)$, and $\bar{a} = \langle a_0, \cdots, a_\gamma, \cdots \rangle_{\gamma<\alpha}$, where $a_\gamma \in A$ for $\gamma < \alpha$. If $\mathbf{p} \in \mathbf{P}^{(\alpha)}(\tau)$, we will sometimes denote $p(a_0, \cdots, a_\gamma, \cdots)$ by $p(\bar{a})$.

Theorem 2. *We define a binary relation $\Theta_{\bar{a}}$ on $\mathbf{P}^{(\alpha)}(\tau)$ as follows:*

$$\mathbf{p} \equiv \mathbf{q}(\Theta_{\bar{a}})$$

if and only if

$$p(a_0, \cdots, a_\gamma, \cdots) = q(a_0, \cdots, a_\gamma, \cdots).$$

Then $\Theta_{\bar{a}}$ is a congruence relation of $\mathfrak{P}^{(\alpha)}(\tau)$.

Proof. That $\Theta_{\bar{a}}$ is an equivalence relation follows simply from the fact that "$=$" on A is an equivalence relation.

To prove the substitution property, let $\mathbf{p}_i \equiv \mathbf{q}_i(\Theta_{\bar{a}})$, $0 \leqq i < n_\gamma$, and consider $\mathbf{p} = \mathbf{f}_\gamma(\mathbf{p}_0, \cdots, \mathbf{p}_{n_\gamma - 1})$, $\mathbf{q} = \mathbf{f}_\gamma(\mathbf{q}_0, \cdots, \mathbf{q}_{n_\gamma - 1})$.

Then

$$\begin{aligned} p(a_0, \cdots, a_\gamma, \cdots) &= f_\gamma(p_0(a_0, \cdots, a_\delta, \cdots), \cdots, p_{n_\gamma - 1}(a_0, \cdots, a_\delta, \cdots)) \\ &= f_\gamma(q_0(a_0, \cdots, a_\delta, \cdots), \cdots, q_{n_\gamma - 1}(a_0, \cdots, a_\delta, \cdots)) \\ &= q(a_0, \cdots, a_\delta, \cdots); \end{aligned}$$

thus $\mathbf{p} \equiv \mathbf{q}(\Theta_{\bar{a}})$, that is,

$$f_\gamma(\mathbf{p}_0, \cdots, \mathbf{p}_{n_\gamma - 1}) \equiv f_\gamma(\mathbf{q}_0, \cdots, \mathbf{q}_{n_\gamma - 1})(\Theta_{\bar{a}}),$$

which was to be proved.

Corollary. *The mapping*

$$\varphi: [p(\mathbf{x}_0, \cdots, \mathbf{x}_\gamma, \cdots)]\Theta_{\bar{a}} \rightarrow p(a_0, \cdots, a_\gamma, \cdots)$$

is a 1–1 homomorphism of $\mathfrak{P}^{(\alpha)}(\tau)/\Theta_{\bar{a}}$ into $\mathfrak{A}$ and $[\mathbf{x}_\gamma]\Theta_{\bar{a}} \rightarrow a_\gamma$ under this mapping.

Another interesting congruence relation is defined on $\mathfrak{P}^{(\alpha)}(\tau)$ as follows: Let $K \subseteq K(\tau)$; let $\mathbf{p} \equiv \mathbf{q}(\Theta_K)$ if and only if $p(a_0, \cdots, a_\gamma, \cdots) = q(a_0, \cdots, a_\gamma, \cdots)$ holds for any $a_0, \cdots, a_\gamma, \cdots \in A$, $\mathfrak{A} \in K$.

Theorem 3. *Θ_K is a congruence relation of $\mathfrak{P}^{(\alpha)}(\tau)$.*

Proof. Similar to that of Theorem 2.

Let us note that Theorem 1′ states that $\Theta_{K(\tau)} = \omega$. If $K = \{\mathfrak{A}\}$, let us write $\Theta_{\mathfrak{A}}$ for Θ_K.

Corollary. *$\mathfrak{P}^{(\alpha)}(\tau)/\Theta_{\mathfrak{A}}$ is isomorphic to $\mathfrak{P}^{(\alpha)}(\mathfrak{A})$. An isomorphism is given by*

$$\psi: [\mathbf{p}]\Theta_{\mathfrak{A}} \rightarrow p,$$

where p is the polynomial over $\mathfrak{A}$ induced by $\mathbf{p}$.

Proof. ψ is onto by Corollary 1 to Definition 3. That ψ is well defined and 1–1 follows from the fact that $[\mathbf{p}]\Theta_{\mathfrak{A}} = [\mathbf{q}]\Theta_{\mathfrak{A}}$ means that $\mathbf{p}$ and $\mathbf{q}$ induce the same mapping in A. Since ψ obviously preserves the operations, the proof is complete.

As suggested by this corollary, we will use the notation $\mathfrak{P}^{(\alpha)}(K)$ for $\mathfrak{P}^{(\alpha)}(\tau)/\Theta_K$.

In the next lemma, we will show that every α-ary polynomial depends on only a finite number of its variables, and, conversely, every n-ary polynomial can be enlarged to an α-ary polynomial.

Lemma 6. *Let* $\mathbf{p}$ *be an* α*-ary polynomial symbol; then for some* $n<\omega$ *there exists an* n*-ary polynomial symbol* $\mathbf{p}'$ *and* $\gamma_0, \cdots, \gamma_{n-1}$ *with* $0 \leqq \gamma_0 < \cdots < \gamma_{n-1} < \alpha$ *such that for every algebra* $\mathfrak{A} \in K(\tau)$, *if* p *and* p' *denote the polynomials over* $\mathfrak{A}$ *induced by* $\mathbf{p}$ *and* $\mathbf{p}'$, *respectively, then*

$$p(a_0, \cdots, a_\gamma, \cdots) = p'(a_{\gamma_0}, \cdots, a_{\gamma_{n-1}})$$

for all $a_0, \cdots, a_\gamma, \cdots \in A$.

Conversely, if $\mathbf{p}'$ *is an* n*-ary polynomial symbol and* $n < \omega$ *and the ordinals* $\gamma_0, \cdots, \gamma_{n-1}$ *with* $0 \leqq \gamma_0 < \gamma_1 < \cdots < \gamma_{n-1} < \alpha$ *are fixed, then there exists an* α*-ary polynomial symbol* $\mathbf{p}$ *such that for the induced polynomials the above equality holds.*

Proof. The first statement can be proved by the usual inductive argument. The second statement can be proved as Corollary 2 to Definition 3 was.

This lemma shows that the equality of the α-ary polynomial symbols $\mathbf{p}$ and $\mathbf{q}$, both of which are associated with the same sequence, $\gamma_0, \cdots, \gamma_{n-1}$, and with the n-ary polynomial symbols $\mathbf{p}'$, $\mathbf{q}'$, is equivalent to the equality of $\mathbf{p}'$ and $\mathbf{q}'$. This establishes the following result:

Theorem 4. *Let* $\mathbf{P}^{(\alpha)}(\gamma_0, \cdots, \gamma_{n-1})$ *denote those* α*-ary polynomial symbols which are built up from*

$$\mathbf{x}_{\gamma_0}, \cdots, \mathbf{x}_{\gamma_{n-1}}$$

where

$$0 \leqq \gamma_0 < \cdots < \gamma_{n-1} < \alpha.$$

Then $\langle \mathbf{P}^{(\alpha)}(\gamma_0, \cdots, \gamma_{n-1}); F\rangle$ *is a subalgebra of* $\mathfrak{P}^{(\alpha)}(\tau)$ *and*

$$\langle \mathbf{P}^{(\alpha)}(\gamma_0, \cdots, \gamma_{n-1}); F\rangle \cong \mathfrak{P}^{(n)}(\tau).$$

These results show that, in fact, ω-ary polynomials are simply a common notation for all n-ary polynomials.

At this point we agree, if there is no danger of confusion, that if $\mathbf{p} \in \mathbf{P}^{(n)}(\tau)$ and $\mathfrak{A} \in K(\tau)$, $a_0, \cdots, a_{n-1} \in A$, then in the expression $p(a_0, \cdots, a_{n-1})$, p denotes the polynomial over $\mathfrak{A}$ induced by $\mathbf{p}$ (and e_i^n denotes the polynomial induced by $\mathbf{x}_i \in \mathbf{P}^{(n)}(\tau)$).

The next three lemmas show that from the point of view of congruence relations, homomorphisms, and subalgebras, polynomials behave the same way as operations.

Lemma 7. *Let $\mathfrak{A}$ be an algebra and Θ be a congruence relation on $\mathfrak{A}$. Let* $\mathbf{p}$ *be an n-ary polynomial symbol. Then $a_i \equiv b_i(\Theta)$ for $0 \leqq i < n$ implies that*

$$p(a_0, \cdots, a_{n-1}) \equiv p(b_0, \cdots, b_{n-1})(\Theta).$$

Proof. First step. Proof for $\mathbf{p} = \mathbf{x}_i$.

$$p(a_0, \cdots, a_{n-1}) = e_i^n(a_0, \cdots, a_{n-1}) = a_i,$$

and

$$p(b_0, \cdots, b_{n-1}) = e_i^n(b_0, \cdots, b_{n-1}) = b_i.$$

Indeed, $a_i \equiv b_i(\Theta)$.

Second step. Suppose the statement has been proved for the polynomial symbols $\mathbf{p}_0, \cdots, \mathbf{p}_{n_\gamma - 1}$ and that

$$\mathbf{p} = \mathbf{f}_\gamma(\mathbf{p}_0, \cdots, \mathbf{p}_{n_\gamma - 1}).$$

Then $p_i(a_0, \cdots, a_{n-1}) \equiv p_i(b_1, \cdots, b_{n-1})(\Theta)$, for $0 \leqq i < n_\gamma$. Indeed,

$$\begin{aligned} p(a_0, \cdots, a_{n-1}) &= f_\gamma(p_0(a_0, \cdots, a_{n-1}), \cdots, p_{n_\gamma - 1}(a_0, \cdots, a_{n-1})) \\ &\equiv f_\gamma(p_0(b_0, \cdots, b_{n-1}), \cdots, p_{n_\gamma - 1}(b_0, \cdots, b_{n-1})) \\ &= p(b_0, \cdots, b_{n-1}), \end{aligned}$$

which was to be proved.

We have a similar statement for homomorphisms:

Lemma 8. *Let $\mathfrak{A}$ and $\mathfrak{B}$ be algebras and let $\varphi: A \to B$ be a homomorphism of $\mathfrak{A}$ into $\mathfrak{B}$; let $\mathbf{p} \in \mathbf{P}^{(n)}(\tau)$. Then*

$$p(a_0, \cdots, a_{n-1})\varphi = p(a_0\varphi, \cdots, a_{n-1}\varphi).$$

Lemma 9. *Let $\mathfrak{A}$ be an algebra and $\mathfrak{B}$ a subalgebra of $\mathfrak{A}$. If p is an n-ary polynomial over $\mathfrak{A}$ and $b_0, \cdots, b_{n-1} \in B$, then $p(b_0, \cdots, b_{n-1}) \in B$.*

The proofs of Lemmas 8 and 9 are left to the reader.

Let p be a mapping of A^n into A with the property that if $a_i \equiv b_i(\Theta)$ for $0 \leqq i < n$, then $p(a_0, \cdots, a_{n-1}) \equiv p(b_0, \cdots, b_{n-1})(\Theta)$ for any congruence relation Θ; such a function is said to have the *substitution property* (SP).

Let p be an n-ary polynomial over $\mathfrak{A}$ and let us substitute fixed elements of A for certain variables. If k variables have been substituted, then p induces a mapping of A^{n-k} into A, that is, it induces a function of $n-k$ variables; such functions are called *algebraic* functions.

It is easy to see that Lemma 7 holds for algebraic functions as well. Let us adjoin to F every element of A as a nullary operation. It is obvious that in the algebra $\langle A; F \cup A\rangle$ the function constructed above is an n-ary polynomial over $\langle A; F \cup A\rangle$.

In many investigations (see e.g., the entire Chapter 5) it is irrelevant what are the basic operations F of the algebra $\mathfrak{A} = \langle A; F\rangle$; we are only interested in the polynomials over $\mathfrak{A}$. In such cases we are only interested in algebras up to equivalence in the following sense:

The algebras $\mathfrak{A}_0 = \langle A; F_0\rangle$ and $\mathfrak{A}_1 = \langle A; F_1\rangle$ are *equivalent*, if for $n = 1, 2, \cdots$ we have $P^{(n)}(\mathfrak{A}_0) = P^{(n)}(\mathfrak{A}_1)$.

The algebra $\mathfrak{A} = \langle A; F\rangle$ is called *trivial* if it is equivalent to $\langle A; \varnothing\rangle$. This means that $P^{(n)}(\mathfrak{A}) = \{e_i^n \mid 0 \leqq i < n\}$ for all $n = 1, 2, \cdots$ and $P^{(0)}(\mathfrak{A}) = \varnothing$.

In investigating algebras up to equivalence the most natural device to use is P. Hall's clones. The *clone* of the algebra $\mathfrak{A}$ is a family of sets $(P^{(n)}(\mathfrak{A}) \mid n = 1, 2, \cdots)$ along with a "partial operation" which assigns to an element p of $P^{(n)}(\mathfrak{A})$, and n elements $q_0, \cdots, q_{n-1}$ of $P^{(k)}(\mathfrak{A})$ an element $p(q_0, \cdots, q_{n-1})$ of $P^{(k)}(\mathfrak{A})$ defined by

$$p(q_0, \cdots, q_{n-1})(a_0, \cdots, a_{k-1}) = p(q_0(a_0, \cdots, a_{k-1}), \cdots, q_{n-1}(a_0, \cdots, a_{k-1})).$$

The most elegant treatment of clones is given by F. W. Lawvere [1], [2] and J. Benabou [1] using categories (including an application of clones to the characterization problem of equational classes as categories). See also Ja. V. Hion [1].

§9. STRUCTURE OF SUBALGEBRAS

We will now establish some of the most important properties of subalgebras.

Lemma 1. *If there is at least one nullary operation symbol* $\mathbf{f}_\gamma$, $\gamma < o(\tau)$, *then every algebra* $\mathfrak{A}$ *of type* τ *has a smallest subalgebra* $\mathfrak{B}$ *and* $b \in B$ *if and only if* b *is an algebraic constant.*

Proof. This follows from the fact that if $\mathbf{p}_0, \cdots, \mathbf{p}_{n_\delta - 1}$ are nullary polynomial symbols, then so is $\mathbf{f}_\delta(\mathbf{p}_0, \cdots, \mathbf{p}_{n_\delta - 1})$ for all $\delta < o(\tau)$.

For an algebra $\mathfrak{A} = \langle A; F\rangle$, let $\mathscr{S}(\mathfrak{A})$ denote the system of subsets B of

A of the form $[H]$ for some $H \subseteq A$. $\mathscr{S}(\mathfrak{A})$ will be called the *subalgebra system* of $\mathfrak{A}$ since if $B \in \mathscr{S}(\mathfrak{A})$ and $B \neq \varnothing$, then $\langle B; F\rangle$ is a subalgebra of $\mathfrak{A}$.

Lemma 2. *$\mathscr{S}(\mathfrak{A})$ is a closure system.*

Proof. By Lemma 7.1 a nonvoid intersection of elements of $\mathscr{S}(\mathfrak{A})$ always belongs to $\mathscr{S}(\mathfrak{A})$. If the intersection $\bigcap (B_i \mid i \in I)$ is $\varnothing$, then either $B_i = \varnothing$ for some $i \in I$, and thus the intersection belongs to $\mathscr{S}(\mathfrak{A})$, or $B_i \neq \varnothing$ for all $i \in I$, in which case there are no nullary operations, hence $[\varnothing] = \varnothing \in \mathscr{S}(\mathfrak{A})$, by the convention adopted in §7.

Thus by the comments at the end of §6 we get from $\mathscr{S}(\mathfrak{A})$ a principle of induction, and a concept of independence. These will be called *$\mathscr{S}$-induction* and *$\mathscr{S}$-independence*, and a set $X \subseteq A$ will be called *$\mathscr{S}$-independent*, and *$\mathscr{S}$-dependent*, as the case may be.

Lemma 3. *Let $\mathfrak{A}$ be an algebra and $H \subseteq A$.*

(i) *If $H = \{h_0, \cdots, h_{n-1}\}$, $n < \omega$, then $a \in [H]$ if and only if there exists an n-ary polynomial p over $\mathfrak{A}$ such that*

$$a = p(h_0, \cdots, h_{n-1}).$$

(ii) *In general, $a \in [H]$ if and only if there exist an $n < \omega$, an n-ary polynomial p over $\mathfrak{A}$, and $h_0, \cdots, h_{n-1} \in H$ such that $a = p(h_0, \cdots, h_{n-1})$.*

Proof. If $H = \varnothing$, then (i) follows from Lemma 2 and the definition of $[\varnothing]$. It is now obvious from Lemma 8.6 that (i) and (ii) follow from the following proposition:

(*) If $H \neq \varnothing$ and $H = \{a_\gamma \mid \gamma < \alpha\}$, where α is an ordinal, then $a \in [H]$ if and only if there exists an α-ary polynomial p over $\mathfrak{A}$ such that

$$a = p(a_0, \cdots, a_\gamma, \cdots).$$

To prove (*), set

$$K = \{a \mid a = p(a_0, \cdots, a_\gamma, \cdots) \quad \text{for some} \quad p \in P^{(\alpha)}(\mathfrak{A})\}.$$

By Lemmas 8.6 and 8.9, $K \subseteq [H]$. K is closed under the operations since if $b_0, \cdots, b_{n_\gamma - 1} \in K$, then

$$b_i = p_i(a_0, \cdots, a_\gamma, \cdots), \quad \text{for} \quad 0 \leq i < n_\gamma$$

and so with $p = f_\gamma(p_0, \cdots, p_{n_\gamma - 1})$, we have that

$$f_\gamma(b_0, \cdots, b_{n_\gamma - 1}) = p(a_0, \cdots, a_\gamma, \cdots) \in K.$$

Furthermore, $H \subseteq K$, since

$$a_\gamma = e_\gamma^{\alpha}(a_0, \cdots, a_\gamma, \cdots).$$

Thus $K \in \mathscr{S}(\mathfrak{A})$, $H \subseteq K$, and $K \subseteq [H]$, which imply $K = [H]$. This completes the proof of Lemma 3.

Corollary 1. $|[H]| \leqq (|H| + \overline{o(\tau)})\aleph_0$.

Proof. If $|H| = \overline{o(\tau)} = 0$, then the statement is trivial. If this is not the case, then every $a \in [H]$ can be associated with a finite sequence of elements of H and with a polynomial symbol, which can be regarded as a finite sequence of symbols from the set $\{\mathbf{x}_i \mid i < \omega\} \cup \{\mathbf{f}_\gamma \mid \gamma < o(\tau)\}$, which is of power $\aleph_0 + \overline{o(\tau)}$. There are at most $(|H| + \aleph_0 + \overline{o(\tau)}) \cdot \aleph_0$ such sequences. Thus

$$|[H]| \leqq (|H| + \aleph_0 + \overline{o(\tau)})\aleph_0 = (|H| + \overline{o(\tau)})\aleph_0.$$

Corollary 2. *Let* $\overline{o(\tau)} \leqq \aleph_0$. *Then*

$$|[H]| \leqq |H| + \aleph_0.$$

Lemma 4. *Let* $\mathfrak{A}$ *be an algebra and let*

$$\mathscr{B} = \{B_i \mid i \in I\},$$

where each $\langle B_i; F\rangle$ *is a subalgebra of* $\mathfrak{A}$. *If* $\mathscr{B}$ *is directed, then*

$$\langle \bigcup (B_i \mid i \in I); F\rangle$$

is a subalgebra of $\langle A; F\rangle$.

Proof. Set $B = \bigcup (B_i \mid i \in I)$. Let $a_0, \cdots, a_{n_\gamma - 1} \in B$. Then

$$a_i \in B_{j_i} \quad \text{for some} \quad j_i \in I.$$

Let $\bar{B}$ denote a common upper bound for $B_{j_0}, \cdots, B_{j_{n_\gamma - 1}}$, that is,

$$\bar{B} \supseteq B_{j_0}, \cdots, B_{j_{n_\gamma - 1}} \text{ and } \bar{B} \in \mathscr{B}.$$

Such a $\bar{B}$ exists because $\mathscr{B}$ is directed.

Then $a_0, \cdots, a_{n_\gamma - 1} \in \bar{B} \in \mathscr{B}$ and so $f_\gamma(a_0, \cdots, a_{n_\gamma - 1}) \in \bar{B} \subseteq B$, which was to be proved.

Corollary. *For every algebra* $\mathfrak{A}$, $\mathscr{S}(\mathfrak{A})$ *is closed under directed unions.*

The previous statements can be summarized as follows.

Theorem 1. *Let* $\mathfrak{A}$ *be an algebra. Then* $\mathscr{S}(\mathfrak{A})$ *is an algebraic closure system.*

This was proved in G. Birkhoff and O. Frink [1], in which it was also proved that the following converse of Theorem 1 holds.

Theorem 2. *Let $\mathscr{S}$ be an algebraic closure system over A. Then we can define an algebra $\mathfrak{A}=\langle A; F\rangle$ such that $\mathscr{S}(\mathfrak{A})=\mathscr{S}$.*

Proof. We wish to define operations on A.

(i) For every $a \in [\varnothing]$ define a nullary operation f_a whose value is a.

(ii) Let $0<n<\omega$, $\bar{a}=\langle a_0, \cdots, a_{n-1}\rangle \in A^n$ and $a \in [\{a_0, \cdots, a_{n-1}\}]$; we define an n-ary operation $f_a{}^{\bar{a}}$ by the rule:

$$f_a{}^{\bar{a}}(b_0, \cdots, b_{n-1}) = \begin{cases} a, \text{ if } \langle b_0, \cdots, b_{n-1}\rangle = \langle a_0, \cdots, a_{n-1}\rangle, \\ b_0 \text{ otherwise.} \end{cases}$$

Let F denote the collection of all operations defined in (i) and (ii). We claim that $\mathscr{S}(\mathfrak{A})=\mathscr{S}$, where $\mathfrak{A}=\langle A; F\rangle$.

Let $B \in \mathscr{S}$. If $B=\varnothing$, then under (i) no nullary operation was defined; thus $B \in \mathscr{S}(\mathfrak{A})$. If $B \neq \varnothing$, then $[\varnothing]_{\mathscr{S}} \subseteq B$; therefore B is closed under all nullary operations (if any). Let $f \in F$ be an n-ary operation, $0<n<\omega$, $b_0, \cdots, b_{n-1} \in B$, $b=f(b_0, \cdots, b_{n-1})$. Then $f=f_a{}^{\bar{a}}$ for some $\bar{a} \in A^n$. If $\langle b_0, \cdots, b_{n-1}\rangle \neq \bar{a}$, then $b=b_0 \in B$. If $\langle b_0, \cdots, b_{n-1}\rangle=\bar{a}$, then

$$\{a_0, \cdots, a_{n-1}\} \subseteq B,$$

hence

$$b = a \in [\{a_0, \cdots, a_{n-1}\}]_{\mathscr{S}} \subsetneq [B]_{\mathscr{S}} = B.$$

Thus B is closed under all the operations, and so $B \in \mathscr{S}(\mathfrak{A})$.

Conversely, let $B \in \mathscr{S}(\mathfrak{A})$. If $B=\varnothing$, then there are no nullary operations in $\mathfrak{A}$, which by rule (i) means that $[\varnothing]_{\mathscr{S}}=\varnothing$, and so $B \in \mathscr{S}$.

Let us assume that $B \neq \varnothing$. Let $H \subseteq B$, where H is finite

$$H = \{a_0, \cdots, a_{n-1}\}.$$

First we verify that $[H]_{\mathscr{S}} \subseteq B$. Let $a \in [H]_{\mathscr{S}}$, and $\bar{a}=\langle a_0, \cdots, a_{n-1}\rangle$. Then

$$a = f_a{}^{\bar{a}}(a_0, \cdots, a_{n-1}) \in B.$$

Thus $[H]_{\mathscr{S}} \subseteq B$.

Now we form the family

$$\mathscr{A} = \{[H]_{\mathscr{S}} \mid H \subseteq B \text{ and } H \text{ finite}\}.$$

Note that $\mathscr{A} \subseteq \mathscr{S}$ and $\mathscr{A}$ is directed. Since $\mathscr{S}$ is an algebraic closure system we get that

$$B = \bigcup (X \mid X \in \mathscr{A}) \in \mathscr{S}$$

completing the proof of Theorem 2.

The concept of subalgebra is invariant under equivalence of algebras; however, the closure system $\mathscr{S}(\mathfrak{A})$ and $\mathscr{S}$-independence are not invariant. For instance, the group $\mathfrak{G}=\langle G; \cdot, ^{-1}, 1\rangle$ is equivalent to $\langle G; \cdot, ^{-1}\rangle=\mathfrak{G}_0$, but $\mathscr{S}(\mathfrak{G})\neq\mathscr{S}(\mathfrak{G}_0)$, since $\varnothing\notin\mathscr{S}(\mathfrak{G})$, but $\varnothing\in\mathscr{S}(\mathfrak{G}_0)$. Also, $\{1\}$ is $\mathscr{S}$-dependent in $\mathfrak{G}$ but $\mathscr{S}$-independent in $\mathfrak{G}_0$.

This situation can be rectified by introducing the closure system $\mathscr{S}^+(\mathfrak{A})$:

For $B\subseteq A$, $B\neq\varnothing$, $B\in\mathscr{S}^+(\mathfrak{A})$ if and only if $\langle B; F\rangle$ is a subalgebra of $\mathfrak{A}$; $\varnothing\in\mathscr{S}^+(\mathfrak{A})$ if and only if there is no element of A which is constant in $\mathfrak{A}$.

Theorems 1 and 2 can be proved with $\mathscr{S}^+(\mathfrak{A})$ in place of $\mathscr{S}(\mathfrak{A})$. The construction of Theorem 2 can still be used; however, one has to verify that if $\varnothing\in\mathscr{S}$, then there is no constant polynomial in $P^{(1)}(\mathfrak{A})$. The details are left to the reader.

$\mathscr{S}^+$-independence, defined in terms of $\mathscr{S}^+(\mathfrak{A})$, will be used in Chapter 5. Note that $\mathscr{S}^+$-independence implies $\mathscr{S}$-independence and $\mathscr{S}$-dependence implies $\mathscr{S}^+$-dependence. Furthermore, if $|X|>1$, then X is $\mathscr{S}$-independent if and only if X is $\mathscr{S}^+$-independent.

If $\overline{o(\tau)}\leqq\aleph_0$, then by Corollary 2 to Lemma 3, $\mathscr{S}(\mathfrak{A})$ has the following property:

$$(*)\ \text{If } H\subseteq A,\ \text{then } |[H]|\ \leqq\ |H|+\aleph_0.$$

Theorem 3. *Let $\mathscr{A}$ be a closure system with property (*). Then every $B\in\mathscr{A}$ with $|B|>\aleph_0$ can be represented as*

$$B=\bigcup(X\mid X\in\mathscr{C}),$$

where $\mathscr{C}$ is a well-ordered chain in $\mathscr{A}$, and for every $D\in\mathscr{C}$, $|D|<|B|$.

Proof. Let α be the initial ordinal of power $|B|$. Then $B=\{b_\gamma\mid\gamma<\alpha\}$. Set $B_\gamma=\{b_\delta\mid\delta<\gamma\}$ for all $\gamma<\alpha$. Since $\bar{\gamma}<\bar{\alpha}$, $|B_\gamma|<|B|$. Now set $C_\gamma=[B_\gamma]$ for $\gamma<\alpha$ and $\mathscr{C}=\{C_\gamma\mid\gamma<\alpha\}$. Then $\mathscr{C}$ is a well-ordered chain of sets and obviously $\bigcup(X\mid X\in\mathscr{C})=B$. Furthermore, by (*),

$$|C_\gamma|\ \leqq\ |B_\gamma|+\aleph_0\ <\ \bar{\alpha}+\aleph_0\ =\ |B|,$$

which completes the proof of Theorem 3.

Corollary. *Let $\mathfrak{A}$ be an algebra of type τ and $\overline{o(\tau)}\leqq\aleph_0$. If $|A|>\aleph_0$, then $\mathfrak{A}$ is the union of a well-ordered chain of subalgebras, each of which is of cardinality $<|A|$.*

For an interesting application, see Exercise 44.

§10. STRUCTURE OF CONGRUENCE RELATIONS

In the following lemmas we will establish some of the most important properties of congruence relations.

Lemma 1. *Let* Θ_i, $i \in I$, *be congruence relations of the algebra* $\mathfrak{A}$. *Then* $\bigcap (\Theta_i \mid i \in I)$ *is also a congruence relation.*

Proof. Recall that $\bigcap (\Theta_i \mid i \in I)$ stands for the set-theoretical intersection and that in §5 we proved that this intersection is again an equivalence relation. So all we have to prove is that it satisfies the substitution property. To this end, let $a_j \equiv b_j(\bigcap (\Theta_i \mid i \in I))$, $0 \leqq j < n_\gamma$. Then $a_j \equiv b_j(\Theta_i)$ holds for all $i \in I$ and so $f_\gamma(a_0, \cdots, a_{n_\gamma - 1}) \equiv f_\gamma(b_0, \cdots, b_{n_\gamma - 1})(\Theta_i)$ for all $i \in I$ which means that

$$f_\gamma(a_0, \cdots, a_{n_\gamma - 1}) \equiv f_\gamma(b_0, \cdots, b_{n_\gamma - 1})(\bigcap (\Theta_i \mid i \in I)),$$

which was to be proved.

Lemma 2. *Let* Θ_i, $i \in I$, *be congruence relations of the algebra* $\mathfrak{A}$. *Then* $\bigvee (\Theta_i \mid i \in I)$ *is again a congruence relation.*

Proof. Recall that $\bigvee (\Theta_i \mid i \in I)$ stands for the join of the Θ_i as equivalence relations and thus Lemma 2 states that if we take the join of equivalence relations which are congruence relations, then the join is again a congruence relation. Again, we only have to prove that $\bigvee (\Theta_i \mid i \in I)$ satisfies the substitution property.

Let $a_j \equiv b_j(\bigvee (\Theta_i \mid i \in I))$, $j = 0, \cdots, n_\gamma - 1$. Then there exists a sequence $a_j = z_0{}^j, z_1{}^j, \cdots, z_{n_j}^j = b_j$ for every j such that for every i with $0 \leqq i < n_j$ we have $z_i{}^j \equiv z_{i+1}^j(\Theta_i{}^j)$ for some $\Theta_i{}^j \in \{\Theta_i \mid i \in I\}$, $j = 0, \cdots, n_\gamma - 1$. We say that these sequences are *uniform* if $n_0 = n_1 = \cdots = n_{n_\gamma - 1} = n$ and $\Theta_i{}^j = \Theta_i^{j'}$ for every $0 \leqq j, j' \leqq n_\gamma - 1$ and $0 \leqq i < n$. We will prove that we can choose the sequences such that they are uniform.

Assume that we have j sequences and we will use induction on j. If $j = 1$, we have nothing to prove. If the first j sequences are already uniform, then extend them by adding at the end of each its last term n_j-times and extend the sequences of congruences by $\Theta_0{}^j, \cdots, \Theta_{n_j - 1}^j$, and extend the j-th sequence of elements by adding n-times (where n stands for the number $n_0 = n_1 = \cdots = n_{j-1}$) the first term at the beginning of this sequence and by adding $\Theta_0, \cdots, \Theta_{n-1}$ at the beginning of the sequence of congruences, where Θ_i stands for $\Theta_i{}^k$ for any $k \leqq j - 1$. It is obvious that this makes the first j sequences uniform.

For example, if we have the two sequences a_1, z_1, b_1 and a_2, z_1', z_2', b_2, where the corresponding sequences of congruences are $\Theta_1{}^1$, $\Theta_2{}^1$ and $\Theta_1{}^2$, $\Theta_2{}^2$, $\Theta_3{}^2$, then the uniform sequences are $a_1, z_1, b_1, b_1, b_1, b_1$ and $a_2, a_2, a_2, z_1', z_2', b_2$ and the associated sequence of congruences is $\Theta_1{}^1$, $\Theta_2{}^1$, $\Theta_1{}^2$, $\Theta_2{}^2$, $\Theta_3{}^2$.

So we can assume that the sequences are uniform. Let n denote the common length of these sequences and $\Theta_0, \cdots, \Theta_{n-1}$ the common associated sequence of congruences. To verify that

$$f_\gamma(a_0, \cdots, a_{n_\gamma-1}) \equiv f_\gamma(b_0, \cdots, b_{n_\gamma-1})(\bigvee(\Theta_i \mid i \in I)),$$

we construct the sequence:

$$\begin{aligned} &f_\gamma(z_0{}^0, \cdots, z_0^{n_\gamma-1}) = f_\gamma(a_0, \cdots, a_{n_\gamma-1}),\\ &f_\gamma(z_1{}^0, \cdots, z_1^{n_\gamma-1}),\\ &\cdot\quad\cdot\quad\cdot\quad\cdot\quad\cdot\quad\cdot\quad\cdot\quad\cdot\quad\cdot\\ &f_\gamma(z_{n-1}^0, \cdots, z_{n-1}^{n_\gamma-1}),\\ &f_\gamma(z_n{}^0, \cdots, z_n^{n_\gamma-1}) = f_\gamma(b_0, \cdots, b_{n_\gamma-1}).\end{aligned}$$

Since $z_j{}^i \equiv z_{j+1}^i(\Theta_j)$, we have that

$$f_\gamma(a_0, \cdots, a_{n_\gamma-1}) \equiv f_\gamma(b_0, \cdots, b_{n_\gamma-1})(\bigvee(\Theta_i \mid i \in I)).$$

For an algebra $\mathfrak{A}$, let $C(\mathfrak{A})$ denote the set of all congruence relations of $\mathfrak{A}$, and let $\mathfrak{C}(\mathfrak{A})$ denote $\langle C(\mathfrak{A}); \leq\rangle$. $\mathfrak{C}(\mathfrak{A})$ is called the *congruence lattice* of $\mathfrak{A}$.

Corollary 1. *$\mathfrak{C}(\mathfrak{A})$ is a lattice.*

Corollary 2. *$\mathfrak{C}(\mathfrak{A})$ is a complete sublattice of $\mathfrak{E}(A)$, the lattice of all equivalence relations on A.*

Corollary 3 (*G. Birkhoff* [2]). *$\mathfrak{C}(\mathfrak{A})$ is a complete lattice.*

Lemma 3. *Let $(\Theta_i \mid i \in I)$ be a directed family of congruences. Then*

$$\bigcup(\Theta_i \mid i \in I) = \bigvee(\Theta_i \mid i \in I).$$

Note that for intersection we always have

$$\bigcap(\Theta_i \mid i \in I) = \bigwedge(\Theta_i \mid i \in I).$$

Proof. It is trivial that $\bigcup(\Theta_i \mid i \in I) \subseteq \bigvee(\Theta_i \mid i \in I)$.
To prove that $\bigcup(\Theta_i \mid i \in I) \supseteq \bigvee(\Theta_i \mid i \in I)$, assume that

$$a \equiv b(\bigvee(\Theta_i \mid i \in I)).$$

Then there exists a sequence of elements of A, $a=z_0, z_1, \cdots, z_n=b$, and congruences $\Theta_{j_0}, \cdots, \Theta_{j_{n-1}} (j_k \in I)$ such that

$$z_k \equiv z_{k+1}(\Theta_{j_k}).$$

Since $(\Theta_i \mid i \in I)$ is a directed family, it has an element Θ such that $\Theta \geqq \Theta_{j_k}$ for $0 \leqq k < n$.

Hence, $z_k \equiv z_{k+1}(\Theta)$ and so $a \equiv b(\Theta)$ which implies that

$$a \equiv b(\bigcup (\Theta_i \mid i \in I)).$$

Therefore,

$$\bigvee (\Theta_i \mid i \in I) \subseteq \bigcup (\Theta_i \mid i \in I),$$

which completes the proof.

Now we shall prove that the congruence relations of the algebra $\mathfrak{A}$ form an algebraic closure system over $A \times A$.

(i) The whole set $A \times A$ is in the system since ι is a congruence relation and $\iota = A \times A$.

(ii) This system is closed under arbitrary intersection according to Lemma 1.

(iii) This system is also closed under directed union by Lemma 3.

Thus we have proved (G. Birkhoff and O. Frink [1]):

Theorem 1. *$C(\mathfrak{A})$ is an algebraic closure system over $A \times A$.*

Theorem 1 combined with Theorem 6.5 yields the following result.

Theorem 2. *$\mathfrak{C}(\mathfrak{A})$ is an algebraic lattice.*

Now, let $\mathfrak{A}$ be an algebra and let $H \subseteq A \times A$, that is, H is a collection of ordered pairs. Let $\Theta(H)$ be the smallest congruence relation Θ such that $a \equiv b(\Theta)$ for all $\langle a, b\rangle \in H$. Obviously, $\Theta(H)$ exists and

$$\Theta(H) = \bigcap (\Theta \mid a \equiv b(\Theta) \quad \text{for all} \quad \langle a, b\rangle \in H).$$

If $H = \{\langle a, b\rangle\}$, then $\Theta(\{\langle a, b\rangle\})$ will be denoted by $\Theta(a, b)$, and $\Theta(a, b)$ is called the *principal* congruence relation induced by $a \equiv b$ (also called "minimal").

Let $\mathfrak{A}$ be an algebra and $a, b \in A$. We define a binary relation Θ on A as follows:

$x \equiv y(\Theta)$ if and only if there exists a sequence $x = z_0, z_1, \cdots, z_n = y$ of elements of A and an associated sequence $p_0, p_1, \cdots, p_{n-1}$ of unary algebraic functions such that

$$\{z_i, z_{i+1}\} = \{p_i(a), p_i(b)\}, \, i = 0, \cdots, n-1.$$

At the end of §8 we noted that algebraic functions satisfy the substitution property. This implies that if Φ is a congruence relation and $a \equiv b(\Phi)$, then $z_i \equiv z_{i+1}(\Phi)$ and so $x \equiv y(\Phi)$. In particular, $x \equiv y(\Theta(a, b))$. Hence, $x \equiv y(\Theta)$ implies $x \equiv y(\Theta(a, b))$. Thus, to prove that $\Theta = \Theta(a, b)$ it is enough to verify that $a \equiv b(\Theta)$ and that Θ is a congruence relation. However, $a \equiv b(\Theta)$ is trivial since we can choose the sequence a, b and the unary algebraic function $e_0{}^1$. Now we prove that Θ is a congruence relation.

(i) Θ is reflexive, that is, $c \equiv c(\Theta)$ for every $c \in A$; indeed, choose the sequence c, c and a unary algebraic function which is identically c (e.g., $e_0{}^2(c, x_1)$).

(ii) Θ is symmetric; if $x \equiv y(\Theta)$, then $y \equiv x(\Theta)$. This can be established by taking the reverse sequence.

(iii) Θ is transitive; if $x \equiv y(\Theta)$ and $y \equiv z(\Theta)$, then taking the composition of the two sequences establishing these congruences proves that $x \equiv z(\Theta)$.

(iv) Θ satisfies the substitution property: Let

$$a_0 \equiv b_0(\Theta), \cdots, a_{n_\gamma - 1} \equiv b_{n_\gamma - 1}(\Theta).$$

Then we have sequences

$$a_0 = z_0{}^0, \cdots, z^0_{n_0} = b_0,$$

$$\cdot \quad \cdot \quad \cdot \quad \cdot \quad \cdot \quad \cdot \quad \cdot \quad \cdot \quad \cdot \quad \cdot$$

$$a_{n_\gamma - 1} = z_0^{n_\gamma - 1}, \cdots, z^{n_\gamma - 1}_{n(n_\gamma - 1)} = b_{n_\gamma - 1}$$

and the associated sequences of unary algebraic functions

$$p_0{}^0, \cdots, p^0_{n_0 - 1},$$

$$\cdot \quad \cdot \quad \cdot \quad \cdot \quad \cdot \quad \cdot \quad \cdot \quad \cdot$$

$$p_0^{n_\gamma - 1}, \cdots, p^{n_\gamma - 1}_{n(n_\gamma - 1) - 1}.$$

We prove by induction on i that

$$f_\gamma(a_0, \cdots, a_{n_\gamma - 1}) \equiv f_\gamma(b_0, \cdots, b_{i-1}, a_i, \cdots, a_{n_\gamma - 1})(\Theta).$$

The statement is obvious for $i = 0$. Let us assume that it holds for i $(< n_\gamma)$. Since $a_i \equiv b_i(\Theta)$, there are sequences

$$a_i = z_0, \cdots, z_m = b_i,$$
$$p_0, \cdots, p_{m-1},$$

such that

$$\{z_j, z_{j+1}\} = \{p_j(a), p_j(b)\}, j = 0, \cdots, m-1.$$

Then the sequences

$$t_0 = f_\gamma(b_0, \cdots, b_{i-1}, z_0, a_{i+1}, \cdots, a_{n_\gamma - 1}),$$
$$t_1 = f_\gamma(b_0, \cdots, b_{i-1}, z_1, a_{i+1}, \cdots, a_{n_\gamma - 1}),$$
$$\cdots\cdots\cdots$$
$$t_m = f_\gamma(b_0, \cdots, b_{i-1}, z_m, a_{i+1}, \cdots, a_{n_\gamma - 1}),$$

and

$$q_0 = f_\gamma(b_0, \cdots, b_{i-1}, p_0, a_{i+1}, \cdots, a_{n_\gamma - 1})$$
$$\cdots\cdots\cdots$$
$$q_{m-1} = f_\gamma(b_0, \cdots, b_{i-1}, p_{m-1}, a_{i+1}, \cdots, a_{n_\gamma - 1})$$

establish that

$$f_\gamma(b_0, \cdots, b_{i-1}, a_i, \cdots, a_{n_\gamma - 1}) \equiv f_\gamma(b_0, \cdots, b_{i-1}, b_i, a_{i+1}, \cdots, a_{n_\gamma - 1})(\Theta),$$

and by the transitivity of Θ,

$$f_\gamma(a_0, \cdots, a_{n_\gamma - 1}) \equiv f_\gamma(b_0, \cdots, b_i, a_{i+1}, \cdots, a_{n_\gamma - 1})(\Theta).$$

Now the substitution property follows by setting $i = n_\gamma$.

Thus we have proved the characterization theorem of principal congruence relations.

Theorem 3. *$x \equiv y(\Theta(a, b))$ if and only if there exists an $n < \omega$, a sequence $x = z_0, z_1, \cdots, z_n = y$ of elements of A and a sequence $p_0, \cdots, p_{n-1}$ of unary algebraic functions such that*

$$\{p_i(a), p_i(b)\} = \{z_i, z_{i+1}\}, \qquad i = 0, 1, \cdots, n-1.$$

Theorem 3 is implicit in Mal'cev [3].

Examples. (1) Let $\langle G; \cdot, 1\rangle$ be a group and Θ a congruence relation of $\langle G; \cdot, 1\rangle$. Let N be the equivalence class containing 1. It is known that there is a 1–1 correspondence between congruences Θ and normal subgroups N. Thus a and b are congruent modulo Θ if and only if $ab^{-1} \in N$. Let $N(a, b)$ be the normal subgroup which corresponds to $\Theta(a, b)$. Then $N(a, b)$ is the normal subgroup generated by ab^{-1}. An n-ary polynomial of a group is always equivalent to a polynomial of the form

$$e^n_{i_1} \cdot e^n_{i_2} \cdots e^n_{i_k}, \quad \text{where} \quad 0 \leqq i_j < n.$$

A unary algebraic function is thus of the form

$$c_1 \cdot x_0 \cdot c_2 \cdot x_0 \cdot c_3 \cdot x_0 \cdot \cdots \cdot x_0 \cdot c_k \ (c_i \in G)$$

which is obtained by fixing all the variables except one in a given polynomial.

We will prove that in the case of groups the following simpler version of the above theorem holds:

$$c \equiv d(\Theta(a, b))$$

if and only if there exists a unary algebraic function p such that $p(a)=c$ and $p(b)=d$.

Let $u \in G$; then the normal subgroup generated by u consists of the elements of the form

$$x_1 u^{n_1} x_1^{-1} x_2 u^{n_2} x_2^{-1} \cdots x_k u^{n_k} x_k^{-1},$$

where the n_i are integers and $x_i \in G$. Since $N(a, b)$ is generated by ab^{-1}, and since $cd^{-1} \in N(a, b)$, we get that cd^{-1} can be expressed in the form

$$cd^{-1} = x_1 u^{n_1} \cdots u^{n_k} x_k^{-1}$$

or

$$c = x_1 u^{n_1} \cdots x_k^{-1} d,$$

where $u = ab^{-1}$. Set

$$p(x) = x_1(xb^{-1})^{n_1} x_1^{-1} x_2 (xb^{-1})^{n_2} x_2 \cdots x_k (xb^{-1})^{n_k} x_k^{-1} d.$$

Then p is a unary algebraic function and

$$p(a) = c,$$
$$p(b) = d,$$

which completes the proof of our assertion.

(2) Let $\langle R; +, \cdot, 0\rangle$ be a ring. Every congruence relation is also a congruence relation of the corresponding additive group $\langle R; +, 0\rangle$. The stronger version of the above theorem which was given in Example 1 can also be proved in this case.

(3) Let $\langle L; \vee, \wedge\rangle$ be a lattice. Then it is easy to give examples to show that the theorem cannot be sharpened as in Examples 1 and 2. (See Exercises.)

Now we can describe $\Theta(H)$.

Lemma 4. $\Theta(H) = \bigvee (\Theta(a, b) \mid \langle a, b\rangle \in H)$.

Proof. Trivial.

Lemma 4, combined with Theorem 3 and the description of the join of equivalence relations given in §5, gives the following explicit description of $\Theta(H)$.

Theorem 4. $c \equiv d(\Theta(H))$ *if and only if there exist* $n < \omega$, *a sequence* $c = z_0, z_1, \cdots, z_n = d$ *and pairs of elements* $\langle a_i, b_i\rangle \in H$ *and unary algebraic functions* p_i, $i = 1, \cdots, n$, *such that*

$$\{p_i(a_i), p_i(b_i)\} = \{z_{i-1}, z_i\}, \qquad i = 1, \cdots, n.$$

A useful formula is the following.

Lemma 5. $\Theta = \bigcup (\Theta(a, b) \mid a \equiv b(\Theta))$, *for all* $\Theta \in C(\mathfrak{A})$.

Proof. Let $\Theta_1 = \bigcup(\Theta(a, b) \mid a \equiv b(\Theta))$. If $a \equiv b(\Theta)$, then $\Theta(a, b) \leqq \Theta$; hence, $\Theta_1 \subseteq \Theta$.

On the other hand, if $u \equiv v(\Theta)$, then $\Theta(u, v) \subseteq \Theta_1$; thus, $u \equiv v(\Theta_1)$, i.e., $\Theta \subseteq \Theta_1$, which was to be proved.

Lemma 6. *A congruence relation Θ is compact in $\mathfrak{C}(\mathfrak{A})$ if and only if it can be represented as a finite join of principal congruences, that is,*

$$\Theta = \bigvee(\Theta(a_i, b_i) \mid 0 \leqq i < n).$$

Remark. Not every compact congruence relation is principal (see Exercise 51).

Proof. The statement is a special case of Lemma 6.5. A direct proof is the following.

First we prove that $\Theta(a, b)$ is compact. Let $\Theta(a, b) \leqq \bigvee(\Theta_i \mid i \in I)$, which means that

$$a \equiv b(\bigvee(\Theta_i \mid i \in I)).$$

This is equivalent to the existence of a sequence $a = z_0, \cdots, z_n = b$ such that

$$z_j \equiv z_{j+1}(\Theta_{i_j}),$$

$j = 0, \cdots, n-1$, $i_j \in I$.

Take $I' = \{i_0, \cdots, i_{n-1}\}$. Then

$$a \equiv b(\bigvee(\Theta_i \mid i \in I')),$$

which means that $\Theta(a, b) \leqq \bigvee(\Theta_i \mid i \in I')$, verifying that $\Theta(a, b)$ is compact.

We already know (proof of Theorem 6.3, step (i)) that a finite join of compact elements is compact.

Thus,

$$\bigvee(\Theta(a_i, b_i) \mid 0 \leqq i < n)$$

is always compact.

To prove the converse, assume that Θ is compact. Then, by Lemma 5, $\Theta = \bigvee(\Theta(a, b) \mid a \equiv b(\Theta))$; thus by the compactness

$$\Theta = \Theta(a_0, b_0) \vee \cdots \vee \Theta(a_{n-1}, b_{n-1}),$$

which proves the statement.

Let $K(\mathfrak{A})$ denote the set of congruence relations of $\mathfrak{A}$ of the form

$$\bigvee(\Theta(a_i, b_i) \mid 0 \leqq i < n),$$

where $n < \omega$.

Theorem 5 (*J. Hashimoto* [1]). *$\mathfrak{K}(\mathfrak{A}) = \langle K(\mathfrak{A}); \vee \rangle$ is a semilattice and the lattice of all ideals of this semilattice is isomorphic to $\mathfrak{C}(\mathfrak{A})$. In symbols:*

$$\mathfrak{I}(\mathfrak{K}(\mathfrak{A})) \cong \mathfrak{C}(\mathfrak{A}).$$

Proof. By Theorem 6.5.

In Theorem 3, we described the smallest congruence relation under which $a \equiv b$. Now for $a \neq b$ we will prove the existence of a maximal one under which $a \not\equiv b$.

Theorem 6. *Let* $\mathfrak{A}$ *be an algebra and* $a, b \in A, a \neq b$. *There exists a congruence relation* $\Psi(a, b)$ *such that* $a \not\equiv b(\Psi(a, b))$ *and* $\Psi(a, b)$ *is maximal with respect to this property* (*i.e., if* $\Psi(a, b) < \Theta$, *then* $a \equiv b(\Theta)$).

Proof. Let $\mathscr{P} = \{\Phi \mid a \not\equiv b(\Phi)\}$. Note that $\mathscr{P}$ is not void since $a \neq b$ implies that $\omega \in \mathscr{P}$.

Consider $\mathfrak{P} = \langle \mathscr{P}; \leqq \rangle$. Let C be a chain in $\mathfrak{P}$. Then $\Psi = \bigvee (\Phi \mid \Phi \in C)$ is a congruence relation. By Lemma 3,

$$\bigvee (\Phi \mid \Phi \in C) = \bigcup (\Phi \mid \Phi \in C);$$

thus we have that $x \equiv y(\Psi)$ if and only if $x \equiv y(\Phi)$ for some $\Phi \in C$. Therefore, $a \not\equiv b(\Psi)$ and so $\Psi \in \mathscr{P}$.

Hence, the hypothesis of Zorn's Lemma is satisfied; $\mathfrak{P}$ has a maximal element $\Psi(a, b)$.

§11. THE HOMOMORPHISM THEOREM AND SOME ISOMORPHISM THEOREMS

Theorem 1 (Homomorphism Theorem). *Let* $\mathfrak{A}$ *and* $\mathfrak{B}$ *be algebras, and* $\varphi: A \to B$ *a homomorphism of* $\mathfrak{A}$ *onto* $\mathfrak{B}$. *Let* Θ *denote the congruence relation induced by* φ (*that is,* $\Theta = \varepsilon_\varphi$). *Then we have that* $\mathfrak{A}/\Theta$ *is isomorphic to* $\mathfrak{B}$, *and an isomorphism is given by* $[a]\Theta \to a\varphi$ ($a \in A$).

Remark. In short, the homomorphism theorem asserts that every homomorphic image of an algebra $\mathfrak{A}$ is isomorphic to a quotient algebra of $\mathfrak{A}$.

Since a quotient algebra is completely determined by $\mathfrak{A}$ and a congruence relation Θ of $\mathfrak{A}$, we can say that the homomorphism theorem establishes the fact that the concept of homomorphism in a sense can be replaced by that of congruence relation.

Another aspect of this result is that while the homomorphic image is an *extrinsic* notion, it is, in a certain sense, equivalent to the concept of congruence relation and quotient algebra which are *intrinsic* notions.

Thus, all homomorphic images of an algebra $\mathfrak{A}$ can be found up to isomorphism "within" the algebra $\mathfrak{A}$.

Proof. The mapping $\psi: [a]\Theta \to a\varphi$ is well defined, 1–1 and onto by

Theorem 2.1. To prove that it is an isomorphism we still have to verify that it preserves the operations; indeed,

$f_\gamma(([a_0]\Theta)\psi, \cdots, ([a_{n_\gamma-1}]\Theta)\psi) =$ (by the definition of ψ)
$= f_\gamma(a_0\varphi, \cdots, a_{n_\gamma-1}\varphi) =$ (since φ is a homomorphism)
$= f_\gamma(a_0, \cdots, a_{n_\gamma-1})\varphi =$ (by the definition of ψ)
$= ([f_\gamma(a_0, \cdots, a_{n_\gamma-1})]\Theta)\psi =$ (by the definition of f_γ in the quotient algebra)
$= f_\gamma([a_0]\Theta, \cdots, [a_{n_\gamma-1}]\Theta)\psi$, which was to be proved.

The following lemma will be needed in the first isomorphism theorem.

Lemma 1. *Let $\mathfrak{A}$ be an algebra, let $\mathfrak{B}$ be a subalgebra of $\mathfrak{A}$, and let Θ be a congruence relation of $\mathfrak{A}$. Then $\langle [B]\Theta; F\rangle$ is also a subalgebra of $\mathfrak{A}$.*

Proof. Let $a_0, \cdots, a_{n_\gamma-1} \in [B]\Theta$. Then $a_i \equiv b_i(\Theta)$ for some $b_i \in B$. Thus

$$f_\gamma(a_0, \cdots, a_{n_\gamma-1}) \equiv f_\gamma(b_0, \cdots, b_{n_\gamma-1})(\Theta),$$

and since $f_\gamma(b_0, \cdots, b_{n_\gamma-1}) \in B$, we have that

$$f_\gamma(a_0, \cdots, a_{n_\gamma-1}) \in [B]\Theta,$$

which is what we had to prove.

Theorem 2 (First Isomorphism Theorem). *Let $\mathfrak{A}$ be an algebra, $\mathfrak{B}$ a subalgebra of $\mathfrak{A}$, and Θ a congruence relation of $\mathfrak{A}$. Then*

$$\langle [B]\Theta/\Theta_{[B]\Theta}; F\rangle \cong \langle B/\Theta_B; F\rangle;$$

an isomorphism is given by

$$\psi\colon [b]\Theta \to [b]\Theta_B \quad \text{for} \quad b \in B.$$

Corollary. *If we assume further that $[B]\Theta = A$, then*

$$\mathfrak{A}/\Theta \cong \mathfrak{B}/\Theta_B.$$

Remark. In other words, the corollary asserts that if $\mathfrak{B}$ is a subalgebra of $\mathfrak{A}$, Θ is a congruence relation of $\mathfrak{A}$, and every congruence class contains an element of B, then $\mathfrak{A}/\Theta \cong \mathfrak{B}/\Theta_B$.

Proof. By Lemma 1, $\langle [B]\Theta; F\rangle$ is a subalgebra of $\mathfrak{A}$. If we replace the algebra $\mathfrak{A}$ by the subalgebra $\langle [B]\Theta; F\rangle$ and the congruence relation Θ by $\Theta_{[B]\Theta}$, then we arrive at the special case of the corollary. Hence, it is enough to verify the corollary.

Consider the mapping $\varphi: B \to A/\Theta$ defined by $b\varphi = [b]\Theta$. Then φ is obviously onto since $[B]\Theta = A$ means that for every $a \in A$ there is a $b \in B$ with $[a]\Theta = [b]\Theta$. φ is obviously a homomorphism and φ induces the congruence relation Θ_B. Hence, by the Homomorphism Theorem (Theorem 1), $\mathfrak{B}/\Theta_B \cong \mathfrak{A}/\Theta$, completing the proof.

A direct proof, establishing that ψ is an isomorphism, would also be very simple.

The statement of the first isomorphism theorem can be visualized by means of a diagram. The whole algebra A is partitioned into subclasses

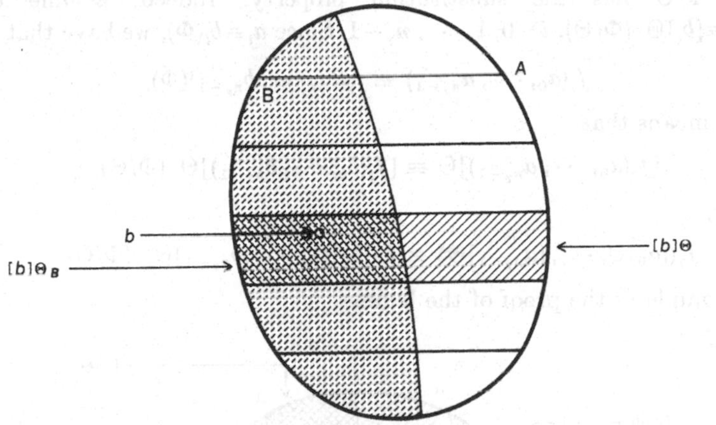

by the congruence relation Θ. In the figure, the classes are separated by the horizontal lines. The dotted area represents B. Picking out a $b \in B$, the doubly shaded area is $[b]\Theta_B$ and the doubly shaded area *together* with the simply shaded area represents $[b]\Theta$. The elements of B/Θ_B are the intersections of the equivalence classes with B; that this intersection is always nonvoid is guaranteed by the assumption that $[B]\Theta = A$. The mapping ψ makes an equivalence class correspond to its intersection with B.

Let $\mathfrak{A}$ be an algebra and let Θ be a congruence relation of $\mathfrak{A}$. Our object is to obtain a complete description of the congruence relations of the quotient algebra $\mathfrak{A}/\Theta$. To this end, we introduce the following notation. Let Φ be a congruence relation of $\mathfrak{A}$ with $\Phi \geqq \Theta$; then Φ/Θ is a binary relation on A/Θ which is defined as follows:

$$[a]\Theta \equiv [b]\Theta \ \ (\Phi/\Theta)$$

if and only if

$$a \equiv b(\Phi) \qquad (a, b \in A).$$

Lemma 2. *Φ/Θ is a congruence relation of $\mathfrak{A}/\Theta$.*

Proof. (i) Φ/Θ is well defined because if $[a]\Theta = [a_1]\Theta$ and $[b]\Theta = [b_1]\Theta$, i.e., $a \equiv a_1(\Theta)$ and $b \equiv b_1(\Theta)$, then

$$[a]\Theta \equiv [b]\Theta \ (\Phi/\Theta)$$

if and only if

$$a \equiv b(\Phi),$$

which is equivalent to

$$a_1 \equiv b_1(\Phi),$$

since $a \equiv a_1(\Phi)$ and $b \equiv b_1(\Phi)$ (in the last step $\Theta \leqq \Phi$ was used).

(ii) Φ/Θ is an equivalence relation. This is obvious.

(iii) Φ/Θ has the substitution property. Indeed, assume that $[a_i]\Theta \equiv [b_i]\Theta \ (\Phi/\Theta)$, $i = 0, 1, \cdots, n_\gamma - 1$. Since $a_i \equiv b_i(\Phi)$, we have that

$$f_\gamma(a_0, \cdots, a_{n_\gamma - 1}) \equiv f_\gamma(b_0, \cdots, b_{n_\gamma - 1})(\Phi),$$

which means that

$$[f_\gamma(a_0, \cdots, a_{n_\gamma - 1})]\Theta \equiv [f_\gamma(b_0, \cdots, b_{n_\gamma - 1})]\Theta \ (\Phi/\Theta).$$

Hence,

$$f_\gamma([a_0]\Theta, \cdots, [a_{n_\gamma - 1}]\Theta) \equiv f_\gamma([b_0]\Theta, \cdots, [b_{n_\gamma - 1}]\Theta) \ (\Phi/\Theta).$$

This completes the proof of the lemma.

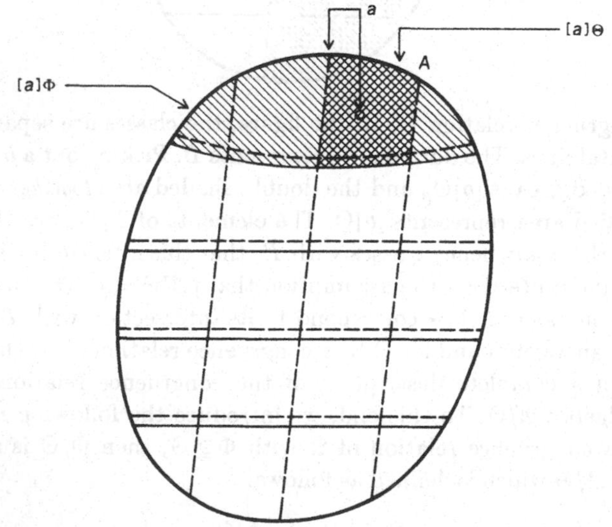

The diagram can be used to visualize the statement of the lemma. In the diagram Θ effects a partition of the elements of A as shown by the dotted lines and Φ a partition as shown by the solid lines. Φ/Θ is the natural partition of classes modulo Θ, i.e., Φ/Θ is the partition of the "blocks" modulo Θ of A.

Consider the quotient algebra $\mathfrak{A}/\Theta$ and the congruence relation Φ of this quotient algebra. We define a congruence relation $\overline{\Phi}$ on A in terms of Φ as follows:

$$a \equiv b(\overline{\Phi}) \quad \text{if and only if} \quad [a]\Theta \equiv [b]\Theta \ (\Phi).$$

Lemma 3. *$\overline{\Phi}$ is a congruence relation of $\mathfrak{A}$ and $\overline{\Phi} \geqq \Theta$.*

Proof. (i) $\overline{\Phi}$ is an equivalence relation. This is trivial, since Φ is an equivalence relation.

(ii) The substitution property holds for $\overline{\Phi}$. Assume that $a_i \equiv b_i(\overline{\Phi})$, $0 \leqq i < n_\gamma$. Then, by the definition of $\overline{\Phi}$, we have that $[a_i]\Theta \equiv [b_i]\Theta \ (\Phi)$. Hence,

$$f_\gamma([a_0]\Theta, \cdots, [a_{n_\gamma - 1}]\Theta) \equiv f_\gamma([b_0]\Theta, \cdots, [b_{n_\gamma - 1}]\Theta) \ (\Phi);$$

therefore,

$$[f_\gamma(a_0, \cdots, a_{n_\gamma - 1})]\Theta \equiv [f_\gamma(b_0, \cdots, b_{n_\gamma - 1})]\Theta \ (\Phi),$$

which implies that

$$f_\gamma(a_0, \cdots, a_{n_\gamma - 1}) \equiv f_\gamma(b_0, \cdots, b_{n_\gamma - 1})(\overline{\Phi}),$$

which was to be proved.

(iii) $\overline{\Phi} \geqq \Theta$. If $a \equiv b(\Theta)$, then $[a]\Theta = [b]\Theta$, which implies that $[a]\Theta \equiv [b]\Theta \ (\Phi)$, and so $a \equiv b(\overline{\Phi})$, which was to be proved. This completes the proof of Lemma 3.

Lemma 4. $\overline{\Phi/\Theta} = \Phi$ *if* $\Phi \geqq \Theta$.

Proof. $a \equiv b(\overline{\Phi/\Theta})$ if and only if $[a]\Theta \equiv [b]\Theta \ (\Phi/\Theta)$, which is equivalent to $a \equiv b(\Phi)$.

Lemma 5. *If $\Phi \in C(\mathfrak{A}/\Theta)$, then $\overline{\Phi}/\Theta = \Phi$.*

Proof. Trivial.

Theorem 3. *Let $\mathfrak{A}$ be an algebra and let Θ be a congruence relation of $\mathfrak{A}$; consider $\langle [\Theta); \leqq \rangle$, that is, the dual ideal $[\Theta) = \{\Psi \mid \Psi \geqq \Theta\}$ of $\mathfrak{C}(\mathfrak{A})$ generated by Θ. Then $\mathfrak{C}(\mathfrak{A}/\Theta) \cong \langle [\Theta); \leqq \rangle$. This isomorphism is effected by*

$$\Phi \to \overline{\Phi} \qquad (\Phi \in C(\mathfrak{A}/\Theta)),$$

the inverse of which is

$$\Phi \to \Phi/\Theta \qquad (\Phi \in [\Theta)).$$

In other words, the congruence relations of $\mathfrak{A}/\Theta$ behave exactly as do the congruence relations Φ of $\mathfrak{A}$ with $\Phi \geqq \Theta$.

Proof. Let $\psi\colon \Phi \to \overline{\Phi}$, $\Phi \in C(\mathfrak{A}/\Theta)$ and let $\chi\colon \Phi \to \Phi/\Theta$, $\Phi \in [\Theta)$. Then, by Lemmas 4 and 5, $\psi\chi$ and $\chi\psi$ are the identity mappings; therefore, both are 1–1 and onto.

Since $\Phi_0 \geqq \Phi_1$ (for $\Phi_0, \Phi_1 \in C(\mathfrak{A}/\Theta)$) implies that $\overline{\Phi}_0 \geqq \overline{\Phi}_1$, and conversely, we have that ψ is an isomorphism.

Theorem 4 (Second Isomorphism Theorem). *Let $\mathfrak{A}$ be an algebra, let Θ, Φ be congruence relations of $\mathfrak{A}$, and assume that $\Theta \leqq \Phi$. Then*

$$\mathfrak{A}/\Phi \cong \mathfrak{A}/\Theta/\Phi/\Theta.$$

This isomorphism is effected by

$$[a]\Phi \to [[a]\Theta](\Phi/\Theta).$$

This situation can be viewed as shown in the diagram, where Φ effects the partition shown by solid lines and Θ the partition shown by dotted *and* solid lines.

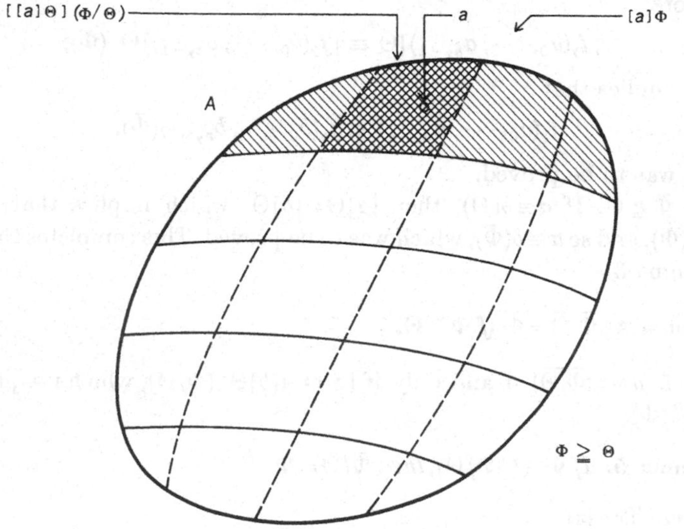

Proof. This theorem follows directly from Lemma 5.

Corollary. *Let φ be a homomorphism of $\mathfrak{A}$ into $\mathfrak{B}$. If Θ is a congruence relation of $\mathfrak{A}$ such that the equivalence relation induced by φ, $\varepsilon_\varphi \geqq \Theta$, then*

$$\psi\colon [a]\Theta \to a\varphi$$

is a homomorphism of $\mathfrak{A}/\Theta$ into $\mathfrak{B}$.

§12. HOMOMORPHISMS

Our first goal in this section is to find all algebras which can be constructed from a given one by constructing subalgebras and homomorphic images repeatedly.

Lemma 1. *Let $\mathfrak{A}$ and $\mathfrak{B}$ be algebras. Let $\varphi: A \to B$ be a homomorphism and $\mathfrak{C}$ a subalgebra of $\mathfrak{B}$. Consider*

$$C\varphi^{-1} = \{x \mid x\varphi \in C, x \in A\}.$$

If $C\varphi^{-1}$ is nonvoid, then $\langle C\varphi^{-1}; F\rangle$ is a subalgebra of $\mathfrak{A}$.

Proof. Let $c_0, \cdots, c_{n_\gamma - 1} \in C\varphi^{-1}$ and form $f_\gamma(c_0, \cdots, c_{n_\gamma - 1})$; applying φ, we obtain

$$f_\gamma(c_0, \cdots, c_{n_\gamma - 1})\varphi = f_\gamma(c_0\varphi, \cdots, c_{n_\gamma - 1}\varphi) \in C$$

because $\mathfrak{C}$ is a subalgebra and is thus closed under all operations. Hence, $f_\gamma(c_0, \cdots, c_{n_\gamma - 1}) \in C\varphi^{-1}$.

Lemma 2. *Let $\mathfrak{A}$, $\mathfrak{B}$, and $\mathfrak{C}$ be algebras.*

(i) *Let $\mathfrak{B}$ be a homomorphic image of $\mathfrak{A}$ and let $\mathfrak{C}$ be a homomorphic image of $\mathfrak{B}$. Then $\mathfrak{C}$ is a homomorphic image of $\mathfrak{A}$.*

(ii) *Let $\mathfrak{B}$ be a subalgebra of $\mathfrak{A}$ and let $\mathfrak{C}$ be a subalgebra of $\mathfrak{B}$. Then $\mathfrak{C}$ is a subalgebra of $\mathfrak{A}$.*

(iii) *Let $\mathfrak{B}$ be a homomorphic image of $\mathfrak{A}$ and let $\mathfrak{C}$ be a subalgebra of $\mathfrak{B}$; then there exists a subalgebra $\mathfrak{D}$ of $\mathfrak{A}$ such that $\mathfrak{C}$ is a homomorphic image of $\mathfrak{D}$.*

Proof.

(i) By Lemma 7.5.

(ii) It is obvious.

(iii) Let φ be a homomorphism of $\mathfrak{A}$ onto $\mathfrak{B}$. Set $D = C\varphi^{-1}$. Then $D \neq \varnothing$, so by Lemma 1, $\mathfrak{D}$ is a subalgebra of $\mathfrak{A}$. By Lemma 7.4, $\mathfrak{C}$ is a homomorphic image of $\mathfrak{D}$.

Viewed diagrammatically, we have the following: Let $\circ$ denote algebras; then let

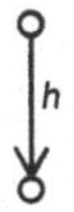

denote that the lower algebra is a homomorphic image of the upper algebra, and let

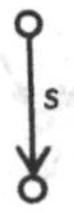

denote that the lower algebra is a subalgebra of the upper algebra. Then:

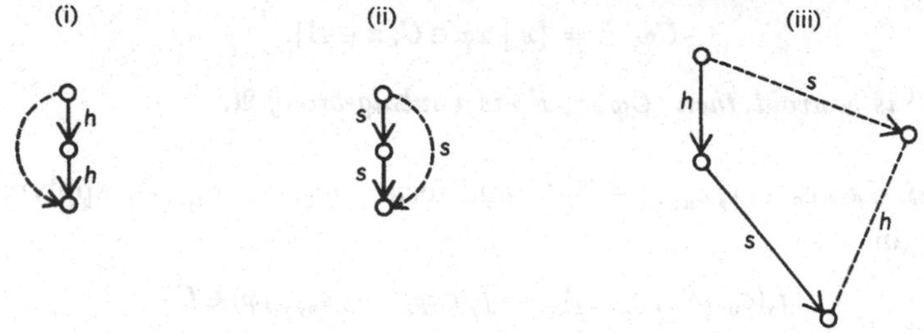

where the arrows with the solid lines denote the assumptions and the arrows with the dashed lines denote the result.

Definition 1. *Let* $\mathfrak{A}$ *and* $\mathfrak{B}$ *be algebras. We say that* $\mathfrak{B}$ *is a* derived algebra *of* $\mathfrak{A}$ *if there exists a sequence of algebras,* $\mathfrak{A} = \mathfrak{X}_0, \mathfrak{X}_1, \cdots, \mathfrak{X}_n = \mathfrak{B}$ *such that* $\mathfrak{X}_i$ *is either a subalgebra or a homomorphic image of* $\mathfrak{X}_{i-1}$, $i = 1, 2, \cdots, n$.

Theorem 1. *Let* $\mathfrak{B}$ *be a derived algebra of* $\mathfrak{A}$*. Then* $\mathfrak{B}$ *is a homomorphic image of a subalgebra of* $\mathfrak{A}$.

Remark. As above, this statement can be visualized by observing the following diagram, where the letters $a_1, \cdots, a_n$ stand for h or s.

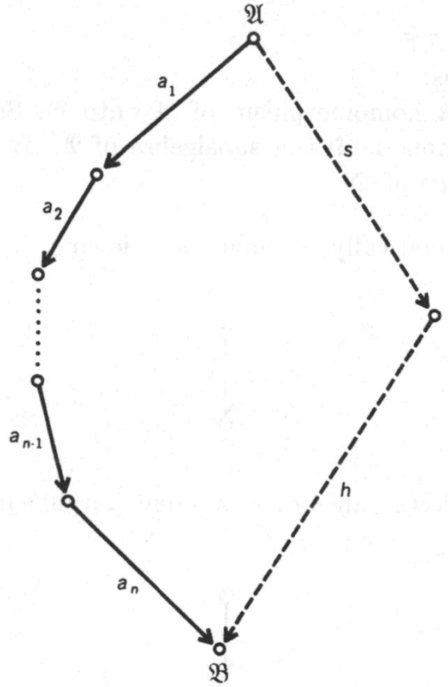

Proof (by induction on n). $n=1$. There are then two possibilities:

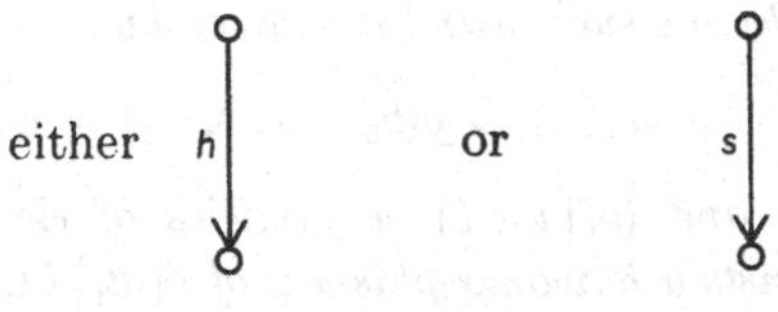

These sequences can be enlarged to:

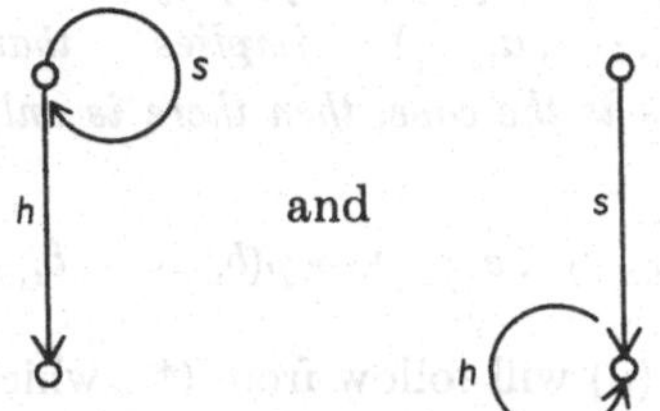

respectively, since every algebra is a subalgebra and a homomorphic image of itself. Thus the theorem holds for $n=1$. We assume that the result is true for n, and prove it for $n+1$.

1. If in the sequence there are two successive arrows having the same letter, then by (i) or (ii) of Lemma 2 they can be replaced by one arrow and we can apply the induction hypothesis. From now on, we can assume that there are no two consecutive arrows labelled by the same letter.

2. Assume the first arrow is labelled by h and that the second is labelled by s. Then by (iii) of Lemma 2 they can be replaced by two arrows the first labelled by s and the second by h. If there are no more arrows, we are through. Otherwise the next arrow is h, and we come back to Case 1.

3. If the first two labels are s and h, and there are no more, then we are through. If there is a third, it is an s and then, by (iii) of Lemma 2, the second and third arrows can be replaced by two arrows, the first labelled by s and the second by h, and the new sequence again falls into the category of Case 1. This completes the proof of the theorem.

Theorem 1 can be very neatly put in terms of "operators on classes of algebras", see §23.

The following problem will come up frequently. We are given a mapping φ of a subset H of an algebra $\mathfrak{A}$ into an algebra $\mathfrak{B}$. When can φ be extended to a homomorphism of $\mathfrak{A}$ into $\mathfrak{B}$? This question can easily be answered if $A=[H]$.

Theorem 2. *Let* $\mathfrak{A}$ *and* $\mathfrak{B}$ *be algebras.*

(i) *If* $a_0, \cdots, a_{n-1} \in A$ *and* $b_0, \cdots, b_{n-1} \in B$, *then there exists a homomorphism* φ *of* $\langle [a_0, \cdots, a_{n-1}]; F\rangle$ *into* $\mathfrak{B}$ *with* $a_0\varphi = b_0, \cdots, a_{n-1}\varphi = b_{n-1}$ *if and only if for every pair* $\mathbf{p}, \mathbf{q}$ *of n-ary polynomial symbols*, $p(a_0, \cdots, a_{n-1}) = q(a_0, \cdots, a_{n-1})$ *implies that* $p(b_0, \cdots, b_{n-1}) = q(b_0, \cdots, b_{n-1})$. *If this is the case, then there is only one such* φ *and it is given by:*

$$\varphi: p(a_0, \cdots, a_{n-1}) \rightarrow p(b_0, \cdots, b_{n-1}), \; p \in \mathbf{P}^{(n)}(\tau).$$

(ii) *Let* $(a_i \mid i \in I)$ *and* $(b_i \mid i \in I)$ *be families of elements of* A *and* B, *respectively; there exists a homomorphism* φ *of* $\langle[\{a_i \mid i \in I\}]; F\rangle$ *into* $\mathfrak{B}$ *with* $a_i\varphi = b_i$, *for* $i \in I$, *if and only if for every* $n \leqq |I|$, *for every choice of* $i_0, \cdots, i_{n-1} \in I$, *and for every pair* $\mathbf{p}, \mathbf{q}$ *of n-ary polynomial symbols*, $p(a_{i_0}, \cdots, a_{i_{n-1}}) = q(a_{i_0}, \cdots, a_{i_{n-1}})$ *implies that* $p(b_{i_0}, \cdots, b_{i_{n-1}}) = q(b_{i_0}, \cdots, b_{i_{n-1}})$. *If this is the case, then there is only one such* φ *and it is given by*

$$\varphi: p(a_{i_0}, \cdots, a_{i_{n-1}}) \rightarrow p(b_{i_0}, \cdots, b_{i_{n-1}}).$$

Proof. Both (i) and (ii) will follow from (*), which is (i) formulated for an arbitrary ordinal α and for

$$\bar{a} = \langle a_0, \cdots, a_\gamma, \cdots \rangle_{\gamma < \alpha}, \; \bar{b} = \langle b_0, \cdots, b_\gamma, \cdots \rangle_{\gamma < \alpha}.$$

If φ is a homomorphism, and $a_\gamma\varphi = b_\gamma$ then for any pair $\mathbf{p}, \mathbf{q}$ of α-ary polynomial symbols $p(a_0, \cdots, a_\gamma, \cdots) = q(a_0, \cdots, a_\gamma, \cdots)$ implies that

$$\begin{aligned} p(b_0, \cdots, b_\gamma, \cdots) &= p(a_0, \cdots, a_\gamma, \cdots)\varphi = q(a_0, \cdots, a_\gamma, \cdots)\varphi \\ &= q(b_0, \cdots, b_\gamma, \cdots). \end{aligned}$$

Thus the uniqueness and the formula for φ follow.

Now suppose that $p(a_0, \cdots, a_\gamma, \cdots) = q(a_0, \cdots, a_\gamma, \cdots)$ implies that $p(b_0, \cdots, b_\gamma, \cdots) = q(b_0, \cdots, b_\gamma, \cdots)$. Using the notation of Theorem 8.2, this means that $\Theta_{\bar{a}} \leqq \Theta_{\bar{b}}$. By the corollary to Theorem 8.2,

$$\varphi_1: p(\mathbf{x}_0, \cdots, \mathbf{x}_\gamma, \cdots) \rightarrow p(b_0, \cdots, b_\gamma, \cdots)$$

is a homomorphism of $\mathfrak{P}^{(\alpha)}(\tau)$ into $\mathfrak{B}$ with $\mathbf{x}_\gamma\varphi_1 = b_\gamma$ for $\gamma < \alpha$, and the congruence relation induced by φ_1 is $\Theta_{\bar{b}}$. Thus by the corollary to Theorem 11.4, $\varphi_2: [x]\Theta_{\bar{a}} \rightarrow x\varphi_1$ is a homomorphism of $\mathfrak{P}^{(\alpha)}(\tau)/\Theta_{\bar{a}}$ into $\mathfrak{B}$. By the corollary to Theorem 8.2, there is an isomorphism ψ between

$$\langle[a_0, \cdots, a_\gamma, \cdots]; F\rangle$$

and $\mathfrak{P}^{(\alpha)}(\tau)/\Theta_{\bar{a}}$, and $a_\gamma\psi = [\mathbf{x}_\gamma]\Theta_{\bar{a}}$, for $\gamma < \alpha$. So $\psi\varphi_2$ is a homomorphism of $\langle[a_0, \cdots, a_\gamma, \cdots]; F\rangle$ into $\mathfrak{B}$; since $a_\gamma\psi\varphi_2 = (a_\gamma\psi)\varphi_2 = ([\mathbf{x}_\gamma]\Theta_{\bar{a}})\varphi_2 = \mathbf{x}_\gamma\varphi_1 = b_\gamma$, we have completed the proof of Theorem 2.

Theorem 2 is so important that we give also a direct proof of the existence of φ. To simplify the notation, we prove it only in case (i). So let us assume that the condition of (i) is satisfied and consider the mapping

$$\varphi: p(a_0, \cdots, a_{n-1}) \rightarrow p(b_0, \cdots, b_{n-1}).$$

(1) φ is defined on the whole of $[a_0, \cdots, a_{n-1}]$ because if

$$a \in [a_0, \cdots, a_{n-1}],$$

then by Lemma 9.3, there exists a polynomial symbol $\mathbf{p} \in \mathbf{P}^{(n)}(\tau)$ such that $a = p(a_0, \cdots, a_{n-1})$, proving that every element of A can be represented in the form $p(a_0, \cdots, a_{n-1})$.

(2) φ is well defined, for, if $a \in A$ has two distinct representations

$$a = p(a_0, \cdots, a_{n-1}) \quad \text{and} \quad a = q(a_0, \cdots, a_{n-1}),$$

then $p(a_0, \cdots, a_{n-1}) = q(a_0, \cdots, a_{n-1})$ and thus, by the condition of (i), $p(b_0, \cdots, b_{n-1}) = q(b_0, \cdots, b_{n-1})$. From the first representation, $a\varphi = p(b_0, \cdots, b_{n-1})$; from the second, $a\varphi = q(b_0, \cdots, b_{n-1})$, and in both cases we indeed got the same element of B.

(3) φ is a homomorphism. Let $c_0, \cdots, c_{n_\gamma - 1} \in [a_0, \cdots, a_{n-1}]$. Then c_i can be represented as

$$c_i = p_i(a_0, \cdots, a_{n-1}).$$

Let $p = f_\gamma(p_0, \cdots, p_{n_\gamma - 1})$. Hence

$$\begin{aligned} f_\gamma(c_0, \cdots, c_{n_\gamma - 1}) &= f_\gamma(p_0(a_0, \cdots, a_{n-1}), \cdots, p_{n_\gamma - 1}(a_0, \cdots, a_{n-1})) \\ &= p(a_0, \cdots, a_{n-1}). \end{aligned}$$

Therefore,

$$\begin{aligned} f_\gamma(c_0, \cdots, c_{n_\gamma - 1})\varphi &= p(a_0, \cdots, a_{n-1})\varphi \\ &= p(b_0, \cdots, b_{n-1}) \\ &= f_\gamma(p_0(b_0, \cdots, b_{n-1}), \cdots, p_{n_\gamma - 1}(b_0, \cdots, b_{n-1})) \end{aligned}$$

[since $c_i = p_i(a_0, \cdots, a_{n-1})$, $c_i\varphi = p(b_0, \cdots, b_{n-1})$]

$$= f_\gamma(c_0\varphi, \cdots, c_{n_\gamma - 1}\varphi).$$

(4) Since $a_i = e_i^n(a_0, \cdots, a_{n-1})$ and $e_i^n\ (b_0, \cdots, b_{n-1}) = b_i$, we have that $\varphi\colon a_i \to b_i$.

This completes the second proof of the theorem.

We shall now consider homomorphisms of an algebra into itself.

Definition 2. *A homomorphism φ of an algebra $\mathfrak{A}$ into itself is called an* endomorphism *of $\mathfrak{A}$.*

Denote the set of endomorphisms of $\mathfrak{A}$ by $E(\mathfrak{A})$. Then $E(\mathfrak{A}) \subseteq M(A)$, the set of all mappings of A into itself.

Lemma 3. *$\mathfrak{E}(\mathfrak{A}) = \langle E(\mathfrak{A}); \cdot \rangle$ is a semigroup and ε, the identity mapping, is the unit element of this semigroup.*

This semigroup is called the *endomorphism semigroup* of the algebra $\mathfrak{A}$. It is a subalgebra of the algebra $\langle M(A); \cdot \rangle$.

Theorem 3. *Let $\mathfrak{S}$ be a semigroup. There exists an algebra $\mathfrak{A}$ such that $\mathfrak{S}$ is isomorphic to $\mathfrak{E}(\mathfrak{A})$ if and only if $\mathfrak{S}$ has a unit element (i.e., an element* 1 *such that* $1 \cdot x = x \cdot 1 = x$ *for all* $x \in S$*).*

Proof. The "only if" part follows from Lemma 3. To prove the "if" part, assume that $\mathfrak{S}$ has a unit element 1. Construct an algebra as follows: Let $A = S$ and for $a \in S$, define a unary operation $f_a(x) = ax$. Set

$$F = (f_a \mid a \in S)$$

and consider the algebra $\mathfrak{A} = \langle A; F\rangle$. We want to describe all the endomorphisms of $\mathfrak{A}$. To this end, for $a \in A$ define a mapping φ_a by $x\varphi_a = xa$. This is a mapping of A into itself.

(i) $\varphi_a = \varphi_b$ if and only if $a = b$.
Indeed, $1\varphi_a = a$, $1\varphi_b = b$. Hence, $\varphi_a = \varphi_b$ is equivalent to $a = b$.

(ii) $\varphi_a \in E(\mathfrak{A})$.
Since $f_b(x)\varphi_a = (bx)\cdot a = b(xa) = f_b(x\varphi_a)$.

(iii) $\varphi_a \cdot \varphi_b = \varphi_{ab}$.
Indeed, $x(\varphi_a\varphi_b) = (x\varphi_a)\varphi_b = (xa)b = x(ab) = x\varphi_{ab}$.

(iv) Let $\varphi \in E(\mathfrak{A})$. Set $a = 1\varphi$. Then $\varphi = \varphi_a$.
Compute: $x\varphi = f_x(1)\varphi = f_x(1\varphi) = f_x(a) = xa = x\varphi_a$.

(v) $a \to \varphi_a$ is an isomorphism between $\mathfrak{S}$ and $\mathfrak{E}(\mathfrak{A})$.
Consider the mapping $\psi\colon a \to \varphi_a$. ψ is 1–1 by (i). ψ is onto by (iv). It preserves multiplication by (iii). Therefore, ψ is an isomorphism. This completes the proof of Theorem 3.

Remark. This result was found and was semi-published by A. G. Waterman (in the preliminary version of the third edition of G. Birkhoff's Lattice Theory) and by G. Grätzer (Some results on universal algebras, mimeographed notes, August 1962). The first published proof is in M. Armbrust and J. Schmidt [1]. The idea of the proof comes from G. Birkhoff [5]. The result is implicit in J. R. Isbell [1]. It is also a special case of Yoneda's Lemma, well known in category theory. In their recent paper [1], Z. Hedrlin and A. Pultr prove that in Theorem 3 the algebra $\mathfrak{A}$ can be chosen to be of type $\langle 1, 1\rangle$; see Exercises to Chapter 2.

Let φ be an endomorphism of the algebra $\mathfrak{A}$. If φ is also 1–1 and onto, then φ is called an *automorphism*. Let $G(\mathfrak{A})$ denote the set of all automorphisms of A, $G(\mathfrak{A}) \subseteq E(\mathfrak{A})$.

Lemma 4. $\mathfrak{G}(\mathfrak{A}) = \langle G(\mathfrak{A}); \cdot\rangle$ *is a group and it is a subalgebra of* $\mathfrak{E}(\mathfrak{A})$ *and of* $\langle M(A); \cdot\rangle$.

Proof. Clear.

$\mathfrak{G}(\mathfrak{A})$ is called the *automorphism group* of $\mathfrak{A}$.

Corollary 1 (*to Theorem 3*) (*G. Birkhoff* [5]). *Every group is isomorphic to an automorphism group of some algebra.*

Proof. Consider a group $\mathfrak{G}$. Apply Theorem 3 and obtain the algebra $\mathfrak{A}$. Then

$$\mathfrak{G} \cong \mathfrak{E}(\mathfrak{A}).$$

By this isomorphism, every endomorphism φ has a two-sided inverse; hence by Theorem 5.1 $E(\mathfrak{A}) = G(\mathfrak{A})$ and so $\mathfrak{G} \cong \mathfrak{E}(\mathfrak{A}) = \mathfrak{G}(\mathfrak{A})$. In fact, we proved somewhat more than was required.

Corollary 1′. *Given a group $\mathfrak{G}$, there exists an algebra $\mathfrak{A}$ such that the automorphism group of $\mathfrak{A}$ is isomorphic to $\mathfrak{G}$ and every endomorphism of $\mathfrak{A}$ is an automorphism.*

For a general study of automorphism groups, see the monograph of B. I. Plotkin [1].

Suppose we are given a semigroup $\mathfrak{S}$ and a group $\mathfrak{G}$. When is it possible to find an algebra $\mathfrak{A}$ such that

$$\mathfrak{E}(\mathfrak{A}) \cong \mathfrak{S}$$

and

$$\mathfrak{G}(\mathfrak{A}) \cong \mathfrak{G}\,?$$

Let us observe (Theorem 5.1) that an endomorphism φ of the algebra $\mathfrak{A}$ is an automorphism if and only if there exists an endomorphism ψ such that $\psi\varphi = \varphi\psi = \varepsilon$.

Corollary 2. *Let $\mathfrak{S}$ be a semigroup with identity and $\mathfrak{G}$ a group. Set*

$$G' = \{x \mid x \in S,\ x \text{ has a two-sided inverse}\}.$$

Then there exists an algebra $\mathfrak{A}$ such that

$$\mathfrak{S} \cong \mathfrak{E}(\mathfrak{A})$$

and

$$\mathfrak{G} \cong \mathfrak{G}(\mathfrak{A})$$

if and only if $\mathfrak{G} \cong \mathfrak{G}'$.

Proof. The "only if" part follows from the previous remark. To verify the "if" part, let $\mathfrak{A}$ be the algebra constructed in Theorem 3. Then φ_a is an automorphism of $\mathfrak{A}$ if and only if a has a two-sided inverse in $\mathfrak{S}$.

Indeed, a has a two-sided inverse b, that is, $ab = ba = 1$, if and only if $\varphi_a \cdot \varphi_b = \varphi_b \cdot \varphi_a = \varphi_1 = \varepsilon$.

Some other semigroups can also be constructed from an algebra $\mathfrak{A}$, namely, the semigroups of all 1–1 homomorphisms and all onto homomorphisms, respectively. Their properties will be described in the exercises. For the best result in this field, see M. Makkai [1].

The algebra $\mathfrak{A}$ is called *simple* if its only congruence relations are ω, ι.

Suppose φ is an endomorphism; then ε_φ, defined by $x\varepsilon_\varphi y$ if and only if $x\varphi = y\varphi$, is a congruence relation. Suppose we consider now endomorphisms of a simple algebra $\mathfrak{A}$. Then $\varepsilon_\varphi = \iota$ or ω. $\varepsilon_\varphi = \omega$ means that φ is 1–1 and thus by Theorem 5.1, $\beta\varphi = \gamma\varphi$ implies $\beta = \gamma$ if β and γ are arbitrary endomorphisms. $\varepsilon_\varphi = \iota$ means that φ maps every element onto a single element a, $\varphi\colon x \to a \in A$. This implies that $\alpha \cdot \varphi = \varphi$ for any endomorphism α.

Lemma 5 (*G. Grätzer* [7]). *Let $\mathfrak{A}$ be a simple algebra. Then in the endomorphism semigroup of $\mathfrak{A}$ every element φ is either a right annihilator or φ satisfies the right cancellation law.*

Some further information on this topic can be found in the exercises.

Endomorphisms can be considered as unary operations; thus from the algebra $\mathfrak{A} = \langle A; F\rangle$ we can form $\langle A; F \cup E(\mathfrak{A})\rangle = \mathfrak{A}'$. The congruence relations of $\mathfrak{A}'$ are called the *fully invariant congruence relations* of $\mathfrak{A}$. Thus they form an algebraic lattice. Theorem 10.6 applied to $\mathfrak{A}'$ gives the following useful result.

Theorem 4. *Let $\mathfrak{A}$ be an algebra, $a, b \in A$ and $a \neq b$. Then there exists a maximal fully invariant congruence relation Ψ of $\mathfrak{A}$ such that $a \not\equiv b\ (\Psi)$.*

If we specialize the corollary to Theorem 11.4 to the special case $\mathfrak{A} = \mathfrak{B}$, we get the following result.

Lemma 6. *Let φ be an endomorphism and Θ a congruence relation of the algebra $\mathfrak{A}$. Then $\bar{\varphi}\colon [a]\Theta \to [a\varphi]\Theta$ is an endomorphism of $\mathfrak{A}/\Theta$ if and only if $a \equiv b(\Theta)$ implies that $a\varphi \equiv b\varphi(\Theta)$.*

Corollary. *If Θ is fully invariant, then $\varphi \to \bar{\varphi}$ is a homomorphism of $\mathfrak{E}(\mathfrak{A})$ into $\mathfrak{E}(\mathfrak{A}/\Theta)$.*

EXERCISES

1. Let A and B be sets and φ a mapping of A into B. Let Θ be an equivalence relation on A. Prove that $\bar{\varphi}\colon [x]\Theta \to x\varphi$ is a mapping of A/Θ into B if and only if $a \equiv b(\Theta)$ implies that $a\varphi = b\varphi$, i.e., $\Theta \leqq \varepsilon_\varphi$.

2. Which groups $\langle G; \cdot, 1\rangle$ have the property that every subalgebra is a group?
3. Let $\langle G; \cdot, 1\rangle$ be a group, Θ a congruence relation of $\langle G; \cdot, 1\rangle$, $N=[1]\Theta$. Then N is *normal*, i.e., $a \in N$, $x \in G$ imply $xax^{-1} \in N$. Further, $x \equiv y(\Theta)$ if and only if $xy^{-1} \in N$.
4. If $\langle N; \cdot, 1\rangle$ is a *normal subgroup* (i.e., N is normal and $\langle N; \cdot, 1\rangle$ is a subgroup) of $\langle G; \cdot, 1\rangle$, then the relation Θ defined by $x \equiv y(\Theta)$ if and only if $xy^{-1} \in N$ is a congruence relation and $[1]\Theta = N$.
5. Formulate and prove the statements of Ex. 3 and Ex. 4 for rings.
6. Let $\langle L; \vee, \wedge\rangle$ be a lattice with a zero element, 0, and Θ a congruence relation of $\langle L; \vee, \wedge\rangle$. Then $[0]\Theta$ is an ideal.
7. Is the converse of Ex. 6 true? That is, can every ideal I of a lattice $\langle L; \vee, \wedge\rangle$ with a zero be represented as $I=[0]\Theta$ for some congruence relation Θ? If I can be so represented, is Θ unique?
8. The same question as in Ex. 7 for distributive lattices.
9. The same question as in Ex. 7 for Boolean algebras.
10. Consider $\mathfrak{A} = \langle A; f_0, f_1\rangle$ of type $\langle 1, 1\rangle$, where $A = \{a, b\}$, and $f_0(a) = b$, $f_0(b) = a$, $f_1(x) = b$. Find $P^{(1)}(\mathfrak{A})$.
11. Find $P^{(n)}(\mathfrak{A})$, if $\mathfrak{A}$ is the algebra of Ex. 10.
12. Let $\mathfrak{A}$ be a unary algebra, i.e., of type $\langle 1, 1, \cdots, 1, \cdots\rangle$. Describe $P^{(n)}(\mathfrak{A})$ in terms of $P^{(1)}(\mathfrak{A})$.
13. Let $\mathfrak{B}$ be a Boolean algebra with more than one element. An *n-ary atomic Boolean polynomial* is one of the form $x_0^{i_0} \wedge \cdots \wedge x_{n-1}^{i_{n-1}}$, where $i_0, \cdots, i_{n-1}$ are zeros and ones, $x_j^{i_j} = x_j$ if $i_j = 0$, $x_j^{i_j} = x_j'$ if $i_j = 1$. Prove that every n-ary polynomial over $\mathfrak{B}$ is a join of atomic polynomials.
14. Prove that there are exactly 2^{2^n} elements of $P^{(n)}(\mathfrak{B})$. ($\mathfrak{B}$ as in Ex. 13.)
15. Let $\mathfrak{L}$ be a lattice. Describe $P^{(1)}(\mathfrak{L})$ and $P^{(2)}(\mathfrak{L})$.
16. Prove: if $\mathfrak{A}$ is finite, $o(\tau) < \omega$, then $P^{(n)}(\mathfrak{A})$ is finite.
17. Characterize $P^{(\omega)}(\mathfrak{A})$ in terms of the following composition of functions: if

$$p_0 \in P^{(n_0)}(\mathfrak{A}), \cdots, p_{n_\gamma - 1} \in P^{(n_{n_\gamma - 1})}(\mathfrak{A}),$$

then $p = f_\gamma^\infty(p_0, \cdots, p_{n_\gamma - 1})$ is a function of $n_0 + \cdots + n_{n_\gamma - 1}$ variables, and

$$\begin{aligned} p(x_0, \cdots, x_{n_0 + \cdots + n_{n_\gamma - 1} - 1}) = f_\gamma(&p_0(x_0, \cdots, x_{n_0 - 1}), \\ &p_1(x_{n_0}, \cdots, x_{n_0 + n_1 - 1}), \cdots, \\ &p_{n_\gamma - 1}(x_{n_0 + \cdots + n_{n_\gamma - 2}}, \cdots, x_{n_0 + \cdots + n_{n_\gamma - 1} - 1})). \end{aligned}$$

(See, e.g., J. Schmidt [5]).
18. For an algebra $\mathfrak{A}$ set $L(\mathfrak{A}) = \{n \mid n \geqq 2 \text{ and } P^{(n)}(\mathfrak{A}) \neq P^{(n, n-1)}(\mathfrak{A})\}$ (see K. Urbanik [4]). Let $\mathfrak{B}$ be a Boolean algebra; set $g(x_0, x_1, x_2) = x_0 + x_1 + x_2$, where $u + v = (u \wedge v') \vee (u' \wedge v)$. Find $L(\langle B; g\rangle)$. (Hint: if $|B| > 1$, then $L(\langle B; g\rangle) = \{2n+1 \mid 0 < n < \omega\}$.)
19. Describe all possible ways of constructing "ranks" of polynomial symbols.
20. Let $\mathbf{p}, \mathbf{p}_0, \cdots, \mathbf{p}_{n-1}$ be α-ary polynomial symbols, $\mathbf{r}$ an n-ary polynomial symbol, and $\mathbf{p} = r(\mathbf{p}_0, \cdots, \mathbf{p}_{n-1})$. If $\mathbf{r} = f_\gamma(\mathbf{r}_0, \cdots, \mathbf{r}_{n_\gamma - 1})$, then $\mathbf{p} = f_\gamma(\mathbf{q}_0, \cdots, \mathbf{q}_{n_\gamma - 1})$, for some $\mathbf{q}_i \in \mathbf{P}^{(\alpha)}(\tau)$.

21. Let $\mathbf{p}, \mathbf{q}$ be n-ary polynomial symbols, $\mathbf{r}_0, \cdots, \mathbf{r}_{n-1}$ be β-ary polynomial symbols, and $p(\mathbf{r}_0, \cdots, \mathbf{r}_{n-1}) = q(\mathbf{r}_0, \cdots, \mathbf{r}_{n-1})$. What conclusion can be drawn from this?
22. Let $K \subseteq K(\tau)$ be a class of algebras such that no polynomial algebra belongs to K. Is it still possible that $\Theta_K = \omega$?
23. Express Θ_K in terms of $\Theta_{\bar{\mathfrak{a}}}$.
24. Prove Lemma 8.5′ and Theorem 8.1′.
25. Formulate and prove the converses of Lemmas 8.7, 8.8, and 8.9.
26. Prove that if $f(x_0, \cdots, x_{n-1})$ is an algebraic function of $\mathfrak{A}$ and Θ is a congruence relation on $\mathfrak{A}$, and $a_i \equiv b_i(\Theta)$ for $0 \leq i < n$, then

$$f(a_0, \cdots, a_{n-1}) \equiv f(b_0, \cdots, b_{n-1})\ (\Theta).$$

27. Prove that if $f(x_0, \cdots, x_{n-1})$ is a function defined on a Boolean algebra $\mathfrak{B}$ such that f has the substitution property with respect to *any* congruence relation, then f is an algebraic function. (G. Grätzer, Revue de Math. Pure et Appliquées, 7 (1962), 693–697).
28. Find all functions on a distributive lattice $\mathfrak{L}$ with 0 and 1 which have the substitution property. (G. Grätzer, Acta Math. Acad. Sci. Hungar., 15 (1964), 195–201).
29. Let $\mathfrak{A}$ be an algebra, $\varnothing \neq B \subseteq A$ such that for any algebraic function f and $b_0, \cdots, b_{n-1} \in B$, we have $f(b_0, \cdots, b_{n-1}) \in B$. What can we say about B?
30. Let $\mathfrak{A}$ be an algebra and let $U(\mathfrak{A})$ be the set of all unary polynomials, considered as mappings of A into itself. Prove that $\langle U(\mathfrak{A}); \cdot \rangle$ is a semigroup with identity and if the algebra is finite then so is this semigroup. Prove that every finite semigroup with an identity is isomorphic to some $\langle U(\mathfrak{A}); \cdot \rangle$ where $\mathfrak{A}$ is a finite algebra.
31. If a is constant in $\mathfrak{A}$, then $\mathfrak{A}$ has a smallest subalgebra $\mathfrak{B}$ and $b \in B$ if and only if b is constant in $\mathfrak{A}$.
32. Let $\mathfrak{B}$ be a subalgebra of $\mathfrak{A}$. Then $p \to p_B$ is a homomorphism of $\mathfrak{P}^{(n)}(\mathfrak{A})$ onto $\mathfrak{P}^{(n)}(\mathfrak{B})$
33. Let K_0 and K_1 be classes of algebras, $K_0 \subseteq K_1$. Then $\mathfrak{P}^{(n)}(K_0)$ is a homomorphic image of $\mathfrak{P}^{(n)}(K_1)$.
34. Let K be the class of Boolean algebras. Prove that $P^{(n)}(K)$ has 2^{2^n} elements. (Hint: prove that $P^{(n)}(K) \cong P^{(n)}(\mathfrak{B})$ for every $\mathfrak{B} \in K$, with $|B| > 1$ and use Ex. 14.)
35. For an algebra $\mathfrak{A}$ let $\mathscr{S}^0(\mathfrak{A})$ denote the system of all subsets B of A such that $\langle B; F \rangle$ is a subalgebra of $\mathfrak{A}$. If $\mathscr{S}^0(\mathfrak{A})$ is not a closure system put $\mathscr{S}^*(\mathfrak{A}) = \mathscr{S}^0(\mathfrak{A}) \cup \{\varnothing\}$, otherwise put $\mathscr{S}^*(\mathfrak{A}) = \mathscr{S}^0(\mathfrak{A})$. Prove that $\mathscr{S}^*(\mathfrak{A})$ is an algebraic closure system. Characterize $\mathscr{S}^*(\mathfrak{A})$.
36. Let $\mathfrak{A}$ be a unary algebra. Prove that if $A_i \in \mathscr{S}(\mathfrak{A})$ for $i \in I$ and $I \neq \varnothing$, then $\bigcup (A_i \mid i \in I) \in \mathscr{S}(\mathfrak{A})$.
37. Characterize $\mathscr{S}(\mathfrak{A})$ for unary algebras.
38. Define the subalgebra lattice $\mathfrak{L}(\mathfrak{A})$ as $\langle \mathscr{S}(\mathfrak{A}); \subseteq \rangle$. Prove that $\mathfrak{L}(\mathfrak{A})$ is an algebraic lattice.
39. Let $\mathfrak{L}$ be an algebraic lattice. Prove that there exists an algebra $\mathfrak{A}$ with $\mathfrak{L}(\mathfrak{A}) \cong \mathfrak{L}$ ($\mathfrak{L}(\mathfrak{A})$ was defined in Ex. 38).
40. The *multiplicity-type* μ of an algebra $\mathfrak{A}$ is a sequence of cardinals

$\langle \mathfrak{m}_0, \mathfrak{m}_1, \cdots, \mathfrak{m}_i, \cdots \rangle_{i<\omega}$, where $\mathfrak{m}_i = |\{\gamma \mid \gamma < o(\tau), n_\gamma = i\}|$, that is, $\mathfrak{m}_i$ is the "number" of i-ary operations. Let $T(\mu)$ denote the class of all algebraic closure systems $\mathscr{S}(\mathfrak{A})$, where the multiplicity type of $\mathfrak{A}$ is μ. Find an algebra $\mathfrak{A}$ for which $\mathscr{S}(\mathfrak{A}) \notin T(\langle \mathfrak{m}, \mathfrak{n}, 1, 0, 0, \cdots, 0, \cdots \rangle)$ for any $\mathfrak{m}, \mathfrak{n}$.

41. Let $\mu = \langle \mathfrak{m}_0, \cdots, \mathfrak{m}_i, \cdots \rangle$ and $\mu' = \langle \mathfrak{m}'_0, \cdots, \mathfrak{m}'_i, \cdots \rangle$ be multiplicity-types (see Ex. 40). Is it true that if $\sum (\mathfrak{m}_i \mid i < j) \leqq \sum (\mathfrak{m}'_i \mid i < j)$ holds for infinitely many j, then $T(\mu) \subseteq T(\mu')$?
42. Find $\mu \neq \mu'$ such that $T(\mu) = T(\mu')$.
43.* Prove that in Ex. 39 $\mathfrak{A}$ can always be chosen of multiplicity-type $\langle 0, \mathfrak{m}, 1, 0, 0, \cdots, 0, \cdots \rangle$ for some $\mathfrak{m}$.
44. Let $\mathfrak{P}$ be an infinite directed partially ordered set. Prove that P can always be represented in the form $P = \bigcup (P_\gamma \mid \gamma < \alpha)$, where $\langle P_\gamma; \leqq \rangle$ is directed, $P_\gamma \subseteq P_\delta$ for $\gamma < \delta < \alpha$ and $|P_\gamma| < |P|$ for all $\gamma < \alpha$. (T. Iwamura, Zenkoku Shijo Sugaku Danwakai, 262 (1944), 107–111). (Hint: Use Theorem 9.3.)
45. Let $\mathfrak{A}$ be an algebra and Θ a binary relation on A. Then Θ is a congruence relation of $\mathfrak{A}$ if and only if it is a congruence relation of each algebra $\langle A; f_\gamma \rangle$, $\gamma < o(\tau)$.
46. Prove Lemma 10.2 without using the concept of a uniform sequence.
47. Let $F \subseteq F_1$. Prove that $\mathfrak{C}(\langle A; F_1 \rangle)$ is a complete sublattice of $\mathfrak{C}(\langle A; F \rangle)$. Derive Corollary 2 of Lemma 10.2 from this statement.
48. Prove Lemma 10.3 directly, that is, without using Lemma 10.2.
49. Let A be a set and $\mathscr{A}$ an algebraic closure system over $A \times A$. Give a necessary and sufficient condition on $\mathscr{A}$ for it to represent a set of equivalence relations on A which forms a complete sublattice of the lattice of all equivalence relations on A.
50. To every algebra $\mathfrak{A} = \langle A; F \rangle$ there corresponds a unary algebra $\mathfrak{A}' = \langle A; F_1 \rangle$ such that Θ is a congruence relation of $\mathfrak{A}$ if and only if it is a congruence relation of $\mathfrak{A}'$.
51. Let $\mathfrak{L} = \langle L; \vee, \wedge \rangle$ be the four element chain. Find in $\mathfrak{L}$ a compact congruence relation which is not principal.
52. Let $\mathfrak{L} = \langle L; \vee, \wedge \rangle$ be a distributive lattice. Prove that every compact congruence relation of $\mathfrak{L}$ is principal if and only if $\mathfrak{L}$ is *relatively complemented*, that is, $a \geqq b \geqq c$ ($a, b, c \in L$) implies that there exists a $d \in L$ with $b \vee d = a$, $b \wedge d = c$. (Use the following result of G. Grätzer and E. T. Schmidt (Publ. Math. Debrecen, 5 (1958), 275–287): if a property P of distributive lattices is preserved under the formation of homomorphic images and the three element chain does not have property P, then every distributive lattice having property P is relatively complemented.)
53. (G. Grätzer and E. T. Schmidt [1]) Prove that for every algebra $\mathfrak{A}$ there exists an algebra $\mathfrak{A}_1$ such that $\mathfrak{C}(\mathfrak{A}) \cong \mathfrak{C}(\mathfrak{A}_1)$ and every compact congruence relation of $\mathfrak{A}_1$ is principal.
54. Show by example that in Theorem 10.3 n cannot be fixed.

* Necessary and sufficient conditions for $\mathscr{S}(\mathfrak{A}) \in T(\mu)$ and $T(\mu) \subseteq T(\mu')$ were given by M. I. Gould.

55. Let $\mathfrak{A}$ be an algebra and $\mathfrak{B}$ a subalgebra of $\mathfrak{A}$, Θ a congruence relation of $\mathfrak{B}$. Assume that there exists a congruence relation Φ of $\mathfrak{A}$ such that $\Phi_B = \Theta$. Prove that there exists a smallest Φ with this property.
56. Under the conditions of Ex. 55, does there exist a maximal Φ?
57. Work out Example (2) of §10.
58. Find a lattice $\langle L; \vee, \wedge\rangle$, congruence relations Θ, Φ of $\langle L; \vee, \wedge\rangle$ and elements a, b in L such that $a \equiv b(\Theta \vee \Phi)$ for which any sequence $a = z_0, z_1, \cdots, z_n = b$ $(z_i \in L)$ satisfying

$$z_i \equiv z_{i+1}(\Theta) \quad \text{or} \quad z_i \equiv z_{i+1}(\Phi), \qquad i = 0, 1, \cdots, n-1,$$

is at least of length m for a given positive integer m.
59. Let $\mathfrak{m}$ be a fixed cardinal number. Find an algebra $\mathfrak{A}$ and $a, b \in A$ such that the set of all congruence relations Ψ which are maximal with respect to the property $a \not\equiv b(\Psi)$ is of cardinality $\mathfrak{m}$.
60. Let $\mathfrak{A}$ be an algebra, $\varnothing \neq B \subseteq A$ and assume that $\langle [B]\Theta; F\rangle$ is a subalgebra of $\mathfrak{A}$ for every congruence relation Θ. Prove that $\mathfrak{B}$ is a subalgebra of $\mathfrak{A}$.
61. Let $\mathfrak{A}$ be an algebra and $\varnothing \neq B \subseteq A$, Θ and Φ congruence relations of $\mathfrak{A}$. Define $B_0 = B$, $B_1 = [B_0]\Theta$, $B_2 = [B_1]\Phi$, $B_3 = [B_2]\Theta, \cdots$. Prove that

$$[B](\Theta \vee \Phi) = B_0 \cup B_1 \cup B_2 \cup \cdots.$$

62. Prove that if $\Theta\Phi = \Phi\Theta$, then

$$[B](\Theta \vee \Phi) = B_2.$$

63. Let Θ and Φ be congruence relations of $\mathfrak{A}$ and $B \subseteq A$. Θ and Φ are *weakly associable over* B if $[B](\Theta \vee \Phi) = [[B]\Theta]\Phi = [[B]\Phi]\Theta$. Does $[B](\Theta \vee \Phi) = [[B]\Theta]\Phi$ imply that Θ and Φ are weakly associable over B?
64. Let $\mathfrak{A}$ be an algebra, and $\mathfrak{B}$ a subalgebra of $\mathfrak{A}$. Let Θ be a congruence relation of $\mathfrak{A}$ and Φ be a congruence relation of $\mathfrak{B}$ such that $\Theta_B \leq \Phi$. We define a binary relation $\Theta(\Phi)$ on $[B]\Theta$ as follows: $a \equiv b(\Theta(\Phi))$ if and only if there exist $c, d \in B$ such that $a \equiv c(\Theta)$, $c \equiv d(\Phi)$, $d \equiv b(\Theta)$. Prove that $\Theta(\Phi)$ is a congruence relation of the algebra $\langle [B]\Theta; F\rangle$.
65. Using the notation of Ex. 64, prove the isomorphism

$$[B]\Theta/\Theta(\Phi) \cong B/\Phi.$$

66. (Zassenhaus Lemma) Let $\langle D; F\rangle$ and $\langle E; F\rangle$ be subalgebras of $\langle A; F\rangle$ and assume that $D \cap E \neq \varnothing$. Let Θ and Φ be congruence relations of $\langle D; F\rangle$ and $\langle E; F\rangle$, respectively. Set

$$\Psi = \Theta_{D\cap E} \vee \Phi_{D \cap E}.$$

Then we have the following isomorphism:

$$[D \cap E]\Theta/\Theta(\Psi) \cong [D \cap E]\Phi/\Phi(\Psi)$$

(A. W. Goldie [1]).
67. Find a mapping which sets up the isomorphism of Ex. 66.
68. (Jordan-Hölder-Schreier Theorem) A *normal series* of an algebra $\mathfrak{A}$ is a finite sequence (*) $\mathfrak{A} = \mathfrak{A}_0, \mathfrak{A}_1, \cdots, \mathfrak{A}_n$ of subalgebras such that (i)

$A_0 \supseteq A_1 \supseteq \cdots \supseteq A_n$, (ii) there exist $\Theta_i \in C(\mathfrak{A}_i)$, $i=0,\cdots,n$ with $\Theta_n=\omega$ and $A_i=[A_n]\Theta_{i-1}$, $i=1,\cdots,n$. If (**) $\mathfrak{A}=\mathfrak{B}_0,\cdots,\mathfrak{B}_m$ is another normal series, the two normal series are said to be *isomorphic* if $n=m$, $\mathfrak{A}_n=\mathfrak{B}_m$ and the $\Theta_i \in C(\mathfrak{A}_i)$, $\Phi_j \in C(\mathfrak{B}_j)$ can be chosen in such a way that $\mathfrak{A}_i/\Theta_i \cong \mathfrak{B}_{k_i}/\Phi_{k_i}$, $i=0,\cdots,n-1$, for a certain permutation $k_0,\cdots,k_{n-1}$ of $0,\cdots,n-1$. (**) is a *refinement* of (*) if every $\mathfrak{B}_i$ is an $\mathfrak{A}_j$. Prove that (*) and (**) have isomorphic refinements if $\mathfrak{A}_n=\mathfrak{B}_m$ and the Θ_i and Φ_j can be chosen in such a way that $(\Theta_i)_{A_i\cap B_j}$ is weakly associable with $(\Phi_j)_{A_i\cap B_j}$ over $A_n = B_m$. (This formulation is from Grätzer [4]. See also A. W. Goldie [1] and M. I. Gould [1].)

69. Interpret Ex. 64, 65, 66, 67, and 68 for the cases of groups and rings.
70. Let Θ and Φ be congruence relations of a group or a ring. Prove that $\Theta\Phi=\Phi\Theta$.
71. Can the statement of Ex. 70 be proved for lattices (distributive lattices)?
72. Let Θ and Φ be congruence relations of a group. Describe $[1](\Theta \vee \Phi)$ in terms of $[1]\Theta$ and $[1]\Phi$.
73. Give the group-theoretic form of Theorem 11.3.
74. Simplify the statement of Theorem 10.3 for lattices. (R. P. Dilworth).
75. Prove that in a distributive lattice $c \equiv d(\Theta(a, b))$ if and only if

$$[(a \wedge b) \vee (c \wedge d)] \wedge (c \vee d) = c \wedge d$$

and

$$[(a \vee b) \vee (c \wedge d)] \wedge (c \vee d) = c \vee d.$$

(G. Grätzer and E. T. Schmidt, Acta Math. Acad. Sci. Hungar., 9 (1958), 137–175.)

76. Give the ring-theoretic form of Theorem 11.4.
77. Let $\mathfrak{A}$ be an algebra which has a smallest subalgebra. Prove that every derived algebra of $\mathfrak{A}$ has a smallest subalgebra.
78. Prove the corollary of Theorem 8.2 from Theorem 12.2.
79. (G. Grätzer [7]) Prove the following converse of Lemma 12.5: Let $\mathfrak{E}$ be a semigroup with identity in which every element a is either a right annihilator or a satisfies the right cancellation law; then there exists a simple algebra $\mathfrak{A}$ such that $\mathfrak{E}(\mathfrak{A}) \cong \mathfrak{E}$.
80. Let $E_0(\mathfrak{A})$ denote the set of all onto endomorphisms of $\mathfrak{A}$. Prove that $\mathfrak{E}_0(\mathfrak{A})=\langle E_0(\mathfrak{A}); \cdot\rangle$ is a semigroup with identity satisfying the left cancellation law.
81. Prove the converse of Ex. 80.
82. Let $E_I(\mathfrak{A})$ denote the set of all 1–1 endomorphisms of $\mathfrak{A}$. Prove that $\mathfrak{E}_I=\langle E_I(\mathfrak{A}); \cdot\rangle$ is a semigroup satisfying the right cancellation law.
83. (E. Fried and M. Makkai) Let $\alpha \in E_I(\mathfrak{A})$, $\beta \in E_0(\mathfrak{A})$, and $\gamma, \delta \in E(\mathfrak{A})$ such that $\gamma\alpha=\beta\delta$. Then there exists a $\varphi \in E(\mathfrak{A})$ such that $\gamma=\beta\varphi$ and $\varphi\alpha=\delta$.
84. (M. Makkai [1]) Find additional properties of $\langle \mathfrak{E}(\mathfrak{A}), \mathfrak{E}_I(\mathfrak{A}), \mathfrak{E}_0(\mathfrak{A})\rangle$.
85. $\mathfrak{A}$ and $\mathfrak{B}$ are said to be *weakly isomorphic* if $\langle A, P^{(\omega)}(\mathfrak{A})\rangle$ is isomorphic to $\langle B; P^{(\omega)}(\mathfrak{B})\rangle$ ($\mathfrak{A}$ and $\mathfrak{B}$ may be of different types). Prove that if $\mathfrak{A}$ and $\mathfrak{B}$ are weakly isomorphic, then the subalgebra lattices, congruence lattices, and endomorphism semigroups are isomorphic.

86. (Goetz[1]) Let $x \to x$ be a weak isomorphism of the group $\langle G; \cdot, 1\rangle$ to the group $\langle G; \circ, 1'\rangle$. If $x^2y = yx^2$ for every $x, y \in G$ (i.e., x^2 is in the center of $\mathfrak{G}$), then $a \circ b = a \cdot b$ for every $a, b \in G$ or $a \circ b = b \cdot a$ for every $a, b \in G$, and $1 = 1'$.
87. Define the concepts of homomorphism, congruence relation, and subalgebra for infinitary algebras. Prove the results of §7.
88. The *type* τ of an infinitary algebra $\mathfrak{A}$ is a sequence $\langle \alpha_0, \cdots, \alpha_\gamma, \cdots\rangle$, $\gamma < o(\tau)$, where α_γ is an ordinal. The *characteristic* $\mathfrak{m}(\tau)$ of τ is the smallest *infinite regular* cardinal $\mathfrak{m}$ such that $\bar{\alpha}_\gamma < \mathfrak{m}$ for all $\gamma < o(\tau)$. Then $\mathfrak{A}$ is an algebra if and only if the characteristic is $\aleph_0$.
89. Define polynomials and polynomial symbols for infinitary algebras of type τ, by transfinite recursion. Prove that every polynomial (polynomial symbol) can be built up in less than $\omega_{\mathfrak{m}(\tau)}$ steps.
90. Generalize the results of §8 for infinitary algebras.
91. Prove that $|[H]| \leq (|H|^{\underline{\mathfrak{m}}} + \mathfrak{m}(\tau)) \cdot \overline{o(\tau)}$, for infinitary algebras, where $|H|^{\underline{\mathfrak{m}}} = \sum (|H|^{\underline{\mathfrak{n}}} \mid \mathfrak{n} < \mathfrak{m})$.
92. (G. Grätzer [8]) Generalize Theorems 9.1 and 9.2 to infinitary algebras using the concept of $\mathfrak{m}$-*algebraic closure systems* ($a \in [H]$ implies $a \in [H_1]$ for some $H_1 \subseteq H$ with $|H_1| < \mathfrak{m}$).
93. (G. Grätzer [8]) Generalize Ex. 38 to infinitary algebras (use Ex. 0.82).
94. Show that $\mathfrak{C}(\mathfrak{A})$ is always a complete lattice even for infinitary algebras, but Lemma 10.2 is false in general.
95. Find counterexamples for Theorems 10.3 and 10.6 among infinitary algebras.
96. Prove that the results of §11 carry over to infinitary algebras.
97. (M. Armbrust and J. Schmidt [1]) Let $\mathfrak{G}$ be a group of permutations of a set A. Prove that there exists an infinitary algebra $\mathfrak{A}$ whose automorphism group is $\mathfrak{G}$.
98. (P. Erdös and A. Hajnal [1]) Prove that for each $n < \omega$ there exists an algebra $\mathfrak{A} = \langle A; f\rangle$ of type $\langle 2\rangle$ such that $|A| = \aleph_n$ and $\mathfrak{A}$ has no proper subalgebra of power $\aleph_n$.
99. Find two algebras $\mathfrak{A}$ and $\mathfrak{B}$ such that there exist homomorphisms of $\mathfrak{A}$ onto $\mathfrak{B}$ and of $\mathfrak{B}$ onto $\mathfrak{A}$ but $\mathfrak{A}$ and $\mathfrak{B}$ are not isomorphic. Can $\mathfrak{A}$ and $\mathfrak{B}$ be chosen to be semigroups?
100. (D. Monk [1]) Let $\mathfrak{A}$ be an algebra with more than one element with the property that every homomorphic image of $\mathfrak{A}$ with more than one element is isomorphic to $\mathfrak{A}$. Prove that $\mathfrak{C}(\mathfrak{A})$ is well ordered.

PROBLEMS

1. (a) Let μ_0 and μ_1 be multiplicity types. (See Ex. 40–42.)

(1) Find necessary and sufficient conditions for $T(\mu_0) \subseteq T(\mu_1)$ and $T(\mu_0) = T(\mu_1)$.

(2) Let $\mathscr{A}$ be an algebraic closure system. Find necessary and sufficient conditions for $\mathscr{A} \in T(\mu)$, where μ is a fixed multiplicity type.

(3) Find a "normal form" theorem for multiplicity types, that is, find a set N of multiplicity types, such that (i) $\mu \in N$ can be easily determined; (ii) for every multiplicity type μ_0 there exists a $\mu_1 \in N$ with $T(\mu_0) = T(\mu_1)$; and (iii) if $\mu_0 \neq \mu_1$, $\mu_0, \mu_1 \in N$, then $T(\mu_0) \neq T(\mu_1)$.

(b) Same as (a) for infinitary algebras.

(c) Can $T(\mu_0) \subseteq T(\mu_1)$ be tested using the polynomial algebras only?

(d) Can $T(\mu_0) \subseteq T(\mu_1)$ be tested using "small" algebras only? Is small = finite for countable multiplicity types?

2. Let $\mathscr{A} \subseteq P(A \times A)$. Find necessary and sufficient conditions on $\mathscr{A}$ for the existence of an algebra $\mathfrak{A} = \langle A; F \rangle$ with $\mathscr{A} = C(\mathfrak{A})$.
3. Let $E \subseteq M(A)$. Find necessary and sufficient conditions on E for the existence of an algebra $\mathfrak{A} = \langle A; F \rangle$ with $E = E(\mathfrak{A})$.†
4. Characterize the closure system of the closed subalgebras of a topological algebra. ($\mathfrak{A}$ is a *topological algebra* if a topology T is defined on A such that every f_γ is continuous. A similar, but easier, problem was solved in O. Frink and G. Grätzer [1].)
5. Is the isomorphism of normal series (Ex. 68) transitive? Is it transitive for *composition series* (normal series with no proper refinement)?
6. Describe all algebras $\mathfrak{A}$ with the property that all functions having the substitution property with respect to any congruence relation are algebraic functions. (See Ex. 27.)
7. Let $\mathfrak{E}$ be a semigroup and $\mathfrak{L}$ an algebraic lattice. When‡ is it possible to find an algebra $\mathfrak{A}$ with $\mathfrak{E}(\mathfrak{A}) \cong \mathfrak{E}$ and $\mathfrak{C}(\mathfrak{A}) \cong \mathfrak{L}$?
8. Let $\mathfrak{A}$ be an infinitary algebra of characteristic $\mathfrak{m}$; let a topology $\mathscr{T}$ be defined on A (i.e, $\mathscr{T}$ is a closure system; an $X \in \mathscr{T}$ is called a closed set) and let us assume that if $\mathfrak{B}$ is a subalgebra of $\mathfrak{A}$ and C is the topological closure of B, then $\mathfrak{C}$ is also a subalgebra of $\mathfrak{A}$. Let $\mathscr{S}_\mathfrak{n}(\mathfrak{A})$ denote the system of all closed subalgebras of $\mathfrak{A}$ which can be generated by $<\mathfrak{n}$ elements. Characterize $\mathscr{S}_\mathfrak{n}(\mathfrak{A})$. (References: O. Frink and G. Grätzer [1], G. Grätzer§ [8].)
9. Describe the semigroups which are isomorphic to an endomorphism semigroup of an algebra with a given "small" congruence lattice. (For simple algebras, see G. Grätzer [7].)

† The case when all the $\varphi \in E$ are permutations has been completely settled by B. Jónsson, W. A. Lampe, and independently by P. Goralčik, Z. Hedrlin, and J. Sichler, who also have interesting contributions to the general case.

‡ It was conjectured by the author that the automorphism group and congruence lattice are independent (with the trivial exception mentioned in Exercise 2.34). This was claimed to have been proved by E. T. Schmidt [2], the proof, however, was incorrect (two computations went wrong, one is mentioned in Exercise 2.31). Nevertheless, the conjecture still seems to be true. References: G. Grätzer [7], Exercise 79, Exercise 2.35. W. A. Lampe has some relevant unpublished results, see Exercises 2.43 and 2.44.

§ Note that Theorem 4 of G. Grätzer [8] is incorrect as stated, but the mistake is easy to correct.

10. Let K be a class of algebras and $\mathscr{K}$ the category whose morphisms are the homomorphisms in K. Let $\mathscr{M}$ and $\mathscr{H}$ denote the 1–1 and onto homomorphisms, respectively. Characterize the triple $\langle \mathscr{K}, \mathscr{M}, \mathscr{H} \rangle$. (If K consists of a single algebra, this was solved by M. Makkai [1]. Note that all conditions of M. Makkai [1] can be formulated in the general case.)

CHAPTER 2
PARTIAL ALGEBRAS

§13 and §16 contain the elements of the theory of partial algebras. §14 and §15 are rather technical; the reader is advised to omit the proofs at the first reading. §17 and §18 give the characterization theorem of congruence lattices; the reader should omit these sections completely at first reading. Since §17 and §18 contain a long series of results, it is useful to cover them first without reading the proofs. These two sections were included to show the usefulness of partial algebras.

§13. BASIC NOTIONS

Let us recall that a *partial algebra* $\mathfrak{A}$ is a pair $\langle A; F\rangle$ where A is a nonvoid set and F is a collection of partial operations on A. We will always assume that F is well ordered, $F=\langle f_0, f_1, \cdots, f_\gamma, \cdots\rangle_{\gamma<o(\tau)}$. The *type* τ of the partial algebra $\mathfrak{A}$ is defined in the same way as for algebras.

Two partial algebras $\mathfrak{A}$, $\mathfrak{B}$ of the same type τ are *isomorphic* if there exists a 1–1 mapping φ of A onto B such that $f_\gamma(a_0, \cdots, a_{n_\gamma-1})$ exists if and only if $f_\gamma(a_0\varphi, \cdots, a_{n_\gamma-1}\varphi)$ exists and

$$f_\gamma(a_0, \cdots. a_{n_\gamma-1})\varphi = f_\gamma(a_0\varphi, \cdots, a_{n_\gamma-1}\varphi).$$

The first question that arises is why we consider partial algebras in the study of algebras. Our most important motivation is the following: Consider an algebra $\mathfrak{A}$ and a nonvoid subset B of A. Restrict all the operations to B in the following way: Let $f_\gamma \in F$, $b_0, \cdots, b_{n_\gamma-1} \in B$; if $f_\gamma(b_0, \cdots, b_{n_\gamma-1}) \in B$, then we do not change $f_\gamma(b_0, \cdots, b_{n_\gamma-1})$. However, if $f_\gamma(b_0, \cdots, b_{n_\gamma-1}) \notin B$, we will say that $f_\gamma(b_0, \cdots, b_{n_\gamma-1})$ is *not* defined. We will denote by $\mathfrak{B}=\langle B; F\rangle$ the system that arises.

In spite of the fact that we started out with an algebra, $\mathfrak{B}$ is only a partial algebra unless B is closed under *all* the operations.

Thus we can say that the language of partial algebras is the natural one if we want to talk about subsets of an algebra and the properties of operations on these subsets even if the subsets are not closed under all the operations. The question we are now going to settle is a very simple one. Is the concept of partial algebras too general from this point of view?

Theorem 1. *Let $\mathfrak{B}$ be a partial algebra. Then there exists an algebra $\mathfrak{A}$ and $A_1 \subseteq A$ such that*

$$\mathfrak{B} \cong \langle A_1; F \rangle.$$

Proof. Construct A as $B \cup \{p\}$ $(p \notin B)$. If $f_\gamma(a_0, \cdots, a_{n_\gamma - 1}) = a$ in $\mathfrak{B}$, keep it. Otherwise, let $f_\gamma(a_0, \cdots, a_{n_\gamma - 1}) = p$. Take $A_1 = B$; the rest is trivial.

For algebras, there is only one reasonable way to define the concepts of subalgebra, homomorphism, and congruence relation. For partial algebras we will define three different types of subalgebra, three types of homomorphism, and two types of congruence relation. In many papers, the authors select one of each (probably based on the assumption that if there was one good concept for algebras then there is only one good concept for partial algebras) and give the reasons for their choices. In the author's opinion, all these concepts have their merits and drawbacks, and each particular situation determines which one should be used.

First we define the three subalgebra concepts.

Let $\mathfrak{A}$ be a partial algebra and let $\varnothing \neq B \subseteq A$. We say that $\mathfrak{B}$ is a *subalgebra* of $\mathfrak{A}$ if it is closed under all operations in $\mathfrak{A}$, i.e., if $b_0, \cdots, b_{n_\gamma - 1} \in B$ and $f_\gamma(b_0, \cdots, b_{n_\gamma - 1})$ is defined in $\mathfrak{A}$, then

$$f_\gamma(b_0, \cdots, b_{n_\gamma - 1}) \in B.$$

In this case,

$$D(f_\gamma, \mathfrak{A}) \cap B^{n_\gamma} = D(f_\gamma, \mathfrak{B}) \qquad \text{for } \gamma < o(\tau),$$

where $D(f_\gamma, \mathfrak{A})$ and $D(f_\gamma, \mathfrak{B})$ denote the domain of f_γ in $\mathfrak{A}$, and in $\mathfrak{B}$, respectively.

In the case of algebras, the new notion of subalgebra is the same as the old one.

We shall now describe other ways of obtaining partial algebras from a given one.

Consider a partial algebra $\mathfrak{A}$ and let $\varnothing \neq B \subseteq A$. For every $\gamma < o(\tau)$ we define f_γ on B as follows: $f_\gamma(b_0, \cdots, b_{n_\gamma - 1})$ is defined for $b_0, \cdots, b_{n_\gamma - 1}$ and equals b if and only if $f_\gamma(b_0, \cdots, b_{n-1})$ is defined in $\mathfrak{A}$ and $f_\gamma(b_0, \cdots, b_{n_\gamma - 1}) = b \in B$ in $\mathfrak{A}$. Thus for $\mathfrak{B} = \langle B; F \rangle$ we have that

$$D(f_\gamma, \mathfrak{B}) = \{\langle b_0, \cdots, b_{n_\gamma - 1}\rangle \mid \langle b_0, \cdots, b_{n_\gamma - 1}\rangle \in D(f_\gamma, \mathfrak{A}) \cap B^{n_\gamma} \text{ and } f_\gamma(b_0, \cdots, b_{n_\gamma - 1}) \in B\}.$$

In this case, we say that $\mathfrak{B}$ is a *relative subalgebra* of $\mathfrak{A}$, and $\mathfrak{A}$ an *extension* of $\mathfrak{B}$. We will use the convention that if we write, "let $\mathfrak{A}$ be a partial algebra, $B \subseteq A$, then the partial algebra $\mathfrak{B} \cdots$", then $\mathfrak{B}$ always means the relative subalgebra determined by B. Observe that a subalgebra $\mathfrak{B}$ of a partial algebra $\mathfrak{A}$ is only a partial algebra, and that a subalgebra $\mathfrak{B}$ is a relative subalgebra of $\mathfrak{A}$ with $D(f_\gamma, \mathfrak{B}) = D(f_\gamma, \mathfrak{A}) \cap B^{n_\gamma}$, for $\gamma < o(\tau)$.

To introduce the third kind of subalgebra, we will have to be somewhat more careful about our notation. Let $\mathfrak{A}$ be a partial algebra and $\varnothing \neq B \subseteq A$. Suppose we have partial operations f_γ' defined on B such that if $f_\gamma'(b_0, \cdots, b_{n_\gamma - 1}) = b$, then $f_\gamma(b_0, \cdots, b_{n_\gamma - 1}) = b$. Let

$$F' = \langle f_0', f_1', \cdots, f_\gamma', \cdots \rangle_{\gamma < o(\tau)}.$$

Then we say that $\mathfrak{B}_1 = \langle B; F' \rangle$ is a *weak subalgebra* of $\mathfrak{A}$. In this case,

$$D(f_\gamma', \mathfrak{B}_1) \subseteq D(f_\gamma, \mathfrak{B}) \subseteq D(f_\gamma, \mathfrak{A}).$$

Note that we could not use the notation $\langle B; F \rangle$ in this case because this would suggest that the partial operations on B are the restrictions of the partial operations on A which is not at all the case.

Next we define three notions of homomorphism.

Suppose that $\mathfrak{A}$ and $\mathfrak{B}$ are partial algebras. $\varphi: A \to B$ is called a *homomorphism* of $\mathfrak{A}$ into $\mathfrak{B}$ if whenever $f_\gamma(a_0, \cdots, a_{n_\gamma - 1})$ is defined, then so is $f_\gamma(a_0\varphi, \cdots, a_{n_\gamma - 1}\varphi)$ and

$$f_\gamma(a_0, \cdots, a_{n_\gamma - 1})\varphi = f_\gamma(a_0\varphi, \cdots, a_{n_\gamma - 1}\varphi).$$

By the definition of homomorphism, if f_γ can be performed on some elements of A, then f_γ can be performed on their images. A homomorphism is called full if the only partial operations which can be performed on the image are the ones that follow from the definition of homomorphism.

Formally, the homomorphism φ of $\mathfrak{A}$ into $\mathfrak{B}$ is a *full homomorphism* if

$$f_\gamma(a_0\varphi, \cdots, a_{n_\gamma - 1}\varphi) = a\varphi, \quad a_0, \cdots, a_{n_\gamma - 1}, a \in A$$

imply that there exist $b_0, \cdots, b_{n_\gamma - 1}, b \in A$ with $b_0\varphi = a_0\varphi, \cdots, b_{n_\gamma - 1}\varphi = a_{n_\gamma - 1}\varphi$, $b\varphi = a\varphi$ and $f_\gamma(b_0, \cdots, b_{n_\gamma - 1}) = b$.

A *strong homomorphism* φ is a homomorphism such that $f_\gamma(a_0, \cdots, a_{n_\gamma - 1})$ is defined in $\mathfrak{A}$ if and only if $f_\gamma(a_0\varphi, \cdots, a_{n_\gamma - 1}\varphi)$ is defined in $\mathfrak{B}$.

Every strong homomorphism is thus a full homomorphism, but the converse is false. Every full homomorphism is a homomorphism, and the converse is again false. In the case of algebras, all three concepts are equivalent to the concept of a homomorphism of an algebra.

Let φ be a homomorphism of $\mathfrak{A}$ into $\mathfrak{B}$, $C = A\varphi$, and $\mathfrak{C}$ the corresponding relative subalgebra of $\mathfrak{B}$. If φ is an isomorphism of $\mathfrak{A}$ and $\mathfrak{C}$, then φ is called an *embedding of* $\mathfrak{A}$ into $\mathfrak{B}$.

We shall now discuss congruence relations.

Given a partial algebra $\mathfrak{A}$ and Θ, an equivalence relation, Θ is called a *congruence relation* if we have:

(SP) If $a_i \equiv b_i(\Theta)$ and if $f_\gamma(a_0, \cdots, a_{n_\gamma - 1})$ and $f_\gamma(b_0, \cdots, b_{n_\gamma - 1})$ are *both* defined, then

$$f_\gamma(a_0, \cdots, a_{n_\gamma - 1}) \equiv f_\gamma(b_0, \cdots, b_{n_\gamma - 1})(\Theta).$$

A congruence relation Θ on $\mathfrak{A}$ is called *strong* if whenever $a_i \equiv b_i(\Theta)$, $0 \leqq i < n_\gamma$, and $f_\gamma(a_0, \cdots, a_{n_\gamma - 1})$ exists, then $f_\gamma(b_0, \cdots, b_{n_\gamma - 1})$ also exists.

The following four lemmas connect up the above defined concepts.

Lemma 1. *Let $\mathfrak{A}$ and $\mathfrak{B}$ be partial algebras and let φ be a homomorphism of $\mathfrak{A}$ into $\mathfrak{B}$. Let ε_φ be the equivalence relation induced by φ. Then ε_φ is a congruence relation.*

Proof. Suppose that $f_\gamma(a_0, \cdots, a_{n_\gamma - 1})$ and $f_\gamma(b_0, \cdots, b_{n_\gamma - 1})$ are both defined and that $a_i \equiv b_i(\varepsilon_\varphi)$. Since $a_i \equiv b_i(\varepsilon_\varphi)$ is equivalent to $a_i\varphi = b_i\varphi$, we have that

$$\begin{aligned} f_\gamma(a_0, \cdots, a_{n_\gamma - 1})\varphi &= f_\gamma(a_0\varphi, \cdots, a_{n_\gamma - 1}\varphi) = f_\gamma(b_0\varphi, \cdots, b_{n_\gamma - 1}\varphi) \\ &= f_\gamma(b_0, \cdots, b_{n_\gamma - 1})\varphi, \end{aligned}$$

so that

$$f_\gamma(a_0, \cdots, a_{n_\gamma - 1}) \equiv f_\gamma(b_0, \cdots, b_{n_\gamma - 1})(\varepsilon_\varphi).$$

Lemma 2. *Let $\mathfrak{A}$ and $\mathfrak{B}$ be partial algebras and let φ be a strong homomorphism; then ε_φ is a strong congruence relation.*

Proof. It suffices to verify that if $f_\gamma(a_0, \cdots, a_{n_\gamma - 1})$ is defined and $a_i \equiv b_i(\varepsilon_\varphi)$, then $f_\gamma(b_0, \cdots, b_{n_\gamma - 1})$ is also defined.

Since $f_\gamma(a_0, \cdots, a_{n_\gamma - 1})$ is defined and φ is a homomorphism, we have that $f_\gamma(a_0\varphi, \cdots, a_{n_\gamma - 1}\varphi)$ is also defined and so

$$f_\gamma(a_0, \cdots, a_{n_\gamma - 1})\varphi = f_\gamma(a_0\varphi, \cdots, a_{n_\gamma - 1}\varphi) = f_\gamma(b_0\varphi, \cdots, b_{n_\gamma - 1}\varphi).$$

By the definition of strong homomorphism, $f_\gamma(b_0\varphi, \cdots, b_{n_\gamma - 1}\varphi)$ is defined if and only if $f_\gamma(b_0, \cdots, b_{n_\gamma - 1})$ is defined; thus $f_\gamma(b_0, \cdots, b_{n_\gamma - 1})$ is defined.

To prove the converse of Lemmas 1 and 2 we need to define a quotient partial algebra.

Let $\mathfrak{A}$ be a partial algebra and let Θ be a congruence relation of $\mathfrak{A}$. We define the *quotient partial algebra* $\mathfrak{A}/\Theta = \langle A/\Theta; F\rangle$ as follows:

If $b_0, \cdots, b_{n_\gamma - 1} \in A/\Theta$, then $f_\gamma(b_0, \cdots, b_{n_\gamma - 1})$ is defined to be equal to b

if and only if *there exist* $a_i \in A$ and $a \in A$ such that $b_i = [a_i]\Theta$, $b = [a]\Theta$ and $f_\gamma(a_0, \cdots, a_{n_\gamma - 1}) = a$.

Lemma 3. *Let $\mathfrak{A}$ be a partial algebra and Θ a congruence relation of $\mathfrak{A}$. Then the mapping $\varphi: a \to [a]\Theta$ is a full homomorphism of $\mathfrak{A}$ onto $\mathfrak{A}/\Theta = \langle A/\Theta; F\rangle$ and $\varepsilon_\varphi = \Theta$.*

Proof. The proof follows directly from the definition.

Lemma 4. *Let $\mathfrak{A}$ be a partial algebra and Θ a strong congruence relation of $\mathfrak{A}$. Then the mapping $\varphi: a \to [a]\Theta$ is a strong homomorphism of $\mathfrak{A}$ onto $\mathfrak{A}/\Theta$ and $\varepsilon_\varphi = \Theta$.*

Proof. Again, by the definitions.

Summarizing, we have the following theorem.

Theorem 2. *Under the correspondence $\varphi \to \varepsilon_\varphi$ homomorphisms correspond to congruence relations on the one hand and strong homomorphisms correspond to strong congruence relations on the other hand.*

There is no such concept as "full congruence relation", which would correspond to full homomorphism, since "φ is full" means a relationship between $\mathfrak{A}$ and $\mathfrak{B}$ and is not a property of ε_φ.

As we explained at the beginning of this section, we develop the theory of partial algebras in order to obtain a theory to use when considering the properties of an operation on a subset of an algebra. Therefore, if $\mathfrak{A}$ is an algebra, $\varnothing \neq B \subseteq A$ and Θ is a congruence relation of $\mathfrak{A}$, then it is quite natural to require that Θ_B be a congruence relation of the partial algebra $\mathfrak{B}$, and every congruence relation of $\mathfrak{B}$ can be so obtained from some algebra $\mathfrak{A}$. Our next theorem states that the notion of congruence relation as defined above does exactly this.

Theorem 3. *Let $\mathfrak{B}$ be a partial algebra and let Θ be a congruence relation of $\mathfrak{B}$. Then there exists an algebra $\mathfrak{A}$ which is an extension of $\mathfrak{B}$, and a congruence relation Φ of $\mathfrak{A}$ such that $\Phi_B = \Theta$.*

Theorem 3 will be proved in §14 and §15 in a much stronger form. It was proved in another form by G. Grätzer and E. T. Schmidt [2]. A similar characterization of strong congruence relations will be given in §16. A very simple direct proof of Theorem 3 is given in G. Grätzer and G. H. Wenzel [1].

§14. POLYNOMIAL SYMBOLS OVER A PARTIAL ALGEBRA†

Let τ be a fixed type of partial algebras. The polynomial symbols $\mathbf{P}^{(\alpha)}(\tau)$ are defined the same as they were for algebras. In this case, an α-ary polynomial symbol does not always induce a mapping of A^α into A, if $\mathfrak{A}$ is a partial algebra. However, some of them do; this will be clear from the following definition.

Definition 1. *Let $\mathfrak{A}$ be a partial algebra of type τ, $a_0, \cdots, a_\gamma, \cdots \in A$, $\gamma < \alpha$, $\mathbf{p} \in \mathbf{P}^{(\alpha)}(\tau)$. Then $p(a_0, \cdots, a_\gamma, \cdots)$ is defined and equals $a \in A$ if and only if it follows from the following rules:*

(i) *If $\mathbf{p} = \mathbf{x}_\delta$, for $\delta < \alpha$, then $p(a_0, \cdots, a_\gamma, \cdots) = a_\delta$;*

(ii) *if $p_i(a_0, \cdots)$ are defined and $p_i(a_0, \cdots) = b_i$ $(0 \leqq i < n_\gamma)$, $f_\gamma(b_0, \cdots, b_{n_\gamma - 1})$ is defined and $\mathbf{p} = \mathbf{f}_\gamma(\mathbf{p}_0, \cdots, \mathbf{p}_{n_\gamma - 1})$, then $p(a_0, \cdots)$ is defined and*

$$p(a_0, \cdots) = f_\gamma(b_0, \cdots, b_{n_\gamma - 1}).$$

The basic difficulty which arises is that if we take

$$\bar{a} = \langle a_0, \cdots, a_\gamma, \cdots \rangle_{\gamma < \alpha}$$

($a_\gamma \in A$) where $\mathfrak{A}$ is a partial algebra, then the congruence relation $\Theta_{\bar{a}}$ of $\mathfrak{P}^{(\alpha)}(\tau)$ cannot be defined as in Theorem 8.2. As a matter of fact, it can be defined that way if and only if the a_γ generate a subalgebra which is an algebra.

Our main result in this section is the following theorem.

Theorem 1. *Let $\mathfrak{A}$ be a partial algebra, $\bar{a} \in A^\alpha$, $\bar{a} = \langle a_0, \cdots, a_\gamma, \cdots \rangle$. Define a binary relation $\Theta_{\bar{a}}$ on $\mathbf{P}^{(\alpha)}(\tau)$ as follows:*

$\mathbf{p} \equiv \mathbf{q}(\Theta_{\bar{a}})$ if and only if there exist $\mathbf{r} \in \mathbf{P}^{(k)}(\tau)$, $\mathbf{p}_i$, $\mathbf{q}_i \in \mathbf{P}^{(\alpha)}(\tau)$ $(0 \leqq i < k)$ such that $p_i(a_0, \cdots, a_\gamma, \cdots)$ and $q_i(a_0, \cdots, a_\gamma, \cdots)$ exist and

$$p_i(a_0, \cdots, a_\gamma, \cdots) = q_i(a_0, \cdots, a_\gamma, \cdots)$$

and

$$\mathbf{p} = r(\mathbf{p}_0, \cdots, \mathbf{p}_{k-1}),$$
$$\mathbf{q} = r(\mathbf{q}_0, \cdots, \mathbf{q}_{k-1}).$$

Then $\Theta_{\bar{a}}$ is a congruence relation of $\mathfrak{P}^{(\alpha)}(\tau)$.

Remark. If we want to find a congruence relation Θ of $\mathfrak{P}^{(\alpha)}(\tau)$ such that $p(a_0, \cdots, a_\gamma, \cdots) = q(a_0, \cdots, a_\gamma, \cdots)$ implies $\mathbf{p} \equiv \mathbf{q}(\Theta)$, then it is obvious that our $\Theta_{\bar{a}}$ is contained in Θ. One does not expect, however, that $\Theta_{\bar{a}}$ is

† The results of this section are taken from G. Grätzer [13].

transitive. Thus the natural statement would be that the smallest such congruence relation is the transitive extension of $\Theta_{\mathfrak{a}}$.

Proof. $\Theta_{\mathfrak{a}}$ is reflexive; indeed, let $\mathbf{p} \in \mathbf{P}^{(\alpha)}(\tau)$; then, by Lemma 8.5′,

$$\mathbf{p} = p(\mathbf{x}_0, \cdots, \mathbf{x}_\gamma, \cdots).$$

By Lemma 8.6,

$$p(\mathbf{x}_0, \cdots, \mathbf{x}_\gamma, \cdots) = r(\mathbf{x}_{\gamma_0}, \cdots, \mathbf{x}_{\gamma_{k-1}}),$$

for some $\mathbf{r} \in \mathbf{P}^{(k)}(\tau)$. Thus

$$\mathbf{p} = r(\mathbf{x}_{\gamma_0}, \cdots, \mathbf{x}_{\gamma_{k-1}});$$

since $x_{\gamma_i}(a_0, \cdots, a_\gamma, \cdots)$ always exists, this verifies that $\mathbf{p} \equiv \mathbf{p}(\Theta_{\mathfrak{a}})$.

It is trivial that $\Theta_{\mathfrak{a}}$ is symmetric. To prove the substitution property, let $\mathbf{p} = \mathbf{f}_\gamma(\mathbf{p}_0, \cdots, \mathbf{p}_{n_\gamma - 1})$, $\mathbf{q} = \mathbf{f}_\gamma(\mathbf{q}_0, \cdots, \mathbf{q}_{n_\gamma - 1})$ and

$$\mathbf{p}_i \equiv \mathbf{q}_i(\Theta_{\mathfrak{a}}), \qquad 0 \leqq i < n_\gamma.$$

Then

$$\mathbf{p}_i = r_i(\mathbf{p}_0^i, \cdots, \mathbf{p}_{n_i - 1}^i),$$

$$\mathbf{q}_i = r_i(\mathbf{q}_0^i, \cdots, \mathbf{q}_{n_i - 1}^i),$$

and $p_j^i(a_0, \cdots, a_\gamma, \cdots)$, $q_j^i(a_0, \cdots, a_\gamma, \cdots)$ exist and

$$p_j^i(a_0, \cdots, a_\gamma, \cdots) = q_j^i(a_0, \cdots, a_\gamma, \cdots).$$

Set $n = n_0 + n_1 + \cdots + n_{n_\gamma - 1}$. By the second part of Lemma 8.6, for $0 \leqq i < n_\gamma$ there exists an n-ary polynomial symbol $\mathbf{r}_i'$, such that

$$\begin{aligned} r_i(b_0, \cdots, b_{n_i - 1}) \\ = r_i'(c_0, \cdots, c_{n_0 + \cdots + n_{i-1} - 1}, b_0, \cdots, b_{n_i - 1}, c_{n_0 + \cdots + n_i}, \cdots, c_{n-1}) \end{aligned}$$

for any values b_j and c_j. Thus we have that

$$\mathbf{p}_i = r_i'(\mathbf{p}_0^0, \cdots, \mathbf{p}_{n_0 - 1}^0, \cdots, \mathbf{p}_0^{n_\gamma - 1}, \cdots, \mathbf{p}_{n(n_\gamma - 1) - 1}^{n_\gamma - 1})$$

for all $0 \leqq i < n_\gamma$.

Set

$$\mathbf{r} = \mathbf{f}_\gamma(\mathbf{r}_0', \cdots, \mathbf{r}_{n_\gamma - 1}').$$

Then

$$r(\mathbf{p}_0^0, \cdots, \mathbf{p}_{n_0 - 1}^0, \mathbf{p}_0^1, \cdots, \mathbf{p}_{n_1 - 1}^1, \cdots, \mathbf{p}_0^{n_\gamma - 1}, \cdots, \mathbf{p}_{n(n_\gamma - 1) - 1}^{n_\gamma - 1}) = \mathbf{p},$$

$$r(\mathbf{q}_0^0, \cdots, \mathbf{q}_{n_0 - 1}^0, \cdots, \mathbf{q}_0^{n_\gamma - 1}, \cdots, \mathbf{q}_{n(n_\gamma - 1) - 1}^{n_\gamma - 1}) = \mathbf{q},$$

establishing that

$$\mathbf{p} \equiv \mathbf{q}(\Theta_{\mathfrak{a}}).$$

which was to be proved.

To establish the transitivity of $\Theta_{\mathfrak{a}}$, we need a lemma.

Lemma 1†. *Let* $\mathbf{p}=\mathbf{f}_\gamma(\mathbf{p}_0,\cdots,\mathbf{p}_{n_\gamma-1})$ *and* $\mathbf{q}=\mathbf{f}_\delta(\mathbf{q}_0,\cdots,\mathbf{q}_{n_\delta-1})$. *Then* $\mathbf{p}\equiv\mathbf{q}(\Theta_{\bar a})$, *if and only if either* $p(\bar a)$ *and* $q(\bar a)$ *exist and* $p(\bar a)=q(\bar a)$, *or* $\gamma=\delta$ and $\mathbf{p}_i\equiv\mathbf{q}_i(\Theta_{\bar a})$. *Moreover, if* $\mathbf{p}\equiv\mathbf{q}(\Theta_{\bar a})$ *and* $p(\bar a)$ *and* $q(\bar a)$ *exist, then* $p(\bar a)=q(\bar a)$.

Proof. Let us assume that $p(\bar a)$ does not exist. By the definition of $\Theta_{\bar a}$, $\mathbf{p}$ and $\mathbf{q}$ have representations of the form

$$\mathbf{p} = r(\mathbf{p}_0',\cdots,\mathbf{p}_{k-1}'),$$
$$\mathbf{q} = r(\mathbf{q}_0',\cdots,\mathbf{q}_{k-1}'),$$

where $p_i'(\bar a)=q_i'(\bar a)$, $0\leqq i<k$ and $\mathbf{r}\in\mathbf{P}^{(k)}(\tau)$. Since $p(\bar a)$ does not exist, $\mathbf{r}\neq\mathbf{x}_i$ for $0\leqq i<k$, and so

$$\mathbf{r} = \mathbf{f}_\nu(\mathbf{r}_0,\cdots,\mathbf{r}_{n_\nu-1}).$$

Therefore,

$$\mathbf{p} = \mathbf{f}_\gamma(\mathbf{p}_0,\cdots,\mathbf{p}_{n_\gamma-1}) = \mathbf{f}_\nu(r_0(\mathbf{p}_0',\cdots,\mathbf{p}_{k-1}'),\cdots,r_{n_\nu-1}(\mathbf{p}_0',\cdots,\mathbf{p}_{k-1}'))$$

and

$$\mathbf{q} = \mathbf{f}_\delta(\mathbf{q}_0,\cdots,\mathbf{q}_{n_\delta-1}) = \mathbf{f}_\nu(r_0(\mathbf{q}_0',\cdots,\mathbf{q}_{k-1}'),\cdots,r_{n_\nu-1}(\mathbf{q}_0',\cdots,\mathbf{q}_{k-1}')).$$

Thus $\gamma=\nu$ and $\delta=\nu$ and so $\gamma=\delta$. From the equalities given above we conclude that

$$\mathbf{p}_i = r_i(\mathbf{p}_0',\cdots,\mathbf{p}_{k-1}')$$

and

$$\mathbf{q}_i = r_i(\mathbf{q}_0',\cdots,\mathbf{q}_{k-1}')$$

for $i=0,\cdots,k-1$. Since $\mathbf{p}_i'\equiv\mathbf{q}_i'(\Theta_{\bar a})$ for $0\leqq i<k$ and $\Theta_{\bar a}$ has (SP), we conclude that

$$\mathbf{p}_i \equiv \mathbf{q}_i(\Theta_{\bar a}), \qquad i=0,\cdots,k-1,$$

which was to be proved. The other statements of Lemma 1 are trivial.

Now we return to the proof of transitivity of the $\Theta_{\bar a}$. Let $\mathbf{q}\equiv\mathbf{p}(\Theta_{\bar a})$ and $\mathbf{p}\equiv\mathbf{r}(\Theta_{\bar a})$. It follows from the definition of $\Theta_{\bar a}$, that if $q(\bar a)$ exists, then $p(\bar a)$ and $r(\bar a)$ exist and $q(\bar a)=p(\bar a)=r(\bar a)$, hence $\mathbf{q}\equiv\mathbf{r}(\Theta_{\bar a})$.

Let us assume now that $q(\bar a)$ does not exist. Then $p(\bar a)$ and $r(\bar a)$ do not exist. Let n be the maximum of the ranks of $\mathbf{p}$, $\mathbf{q}$, and $\mathbf{r}$. We prove the transitivity by induction on n. If $n=2$, we get a contradiction to the assumption that $q(\bar a)$ does not exist. Let us assume that the transitivity has been proven for maximum rank $<n$, and apply Lemma 1 to the two congruences.

† This lemma and the conclusion of the proof of Theorem 1 are due to G. H. Wenzel; the original proof was much longer.

We get that

$$\mathbf{p} = \mathbf{f}_\gamma(\mathbf{p}_0, \cdots, \mathbf{p}_{n_\gamma - 1}),$$
$$\mathbf{q} = \mathbf{f}_\gamma(\mathbf{q}_0, \cdots, \mathbf{q}_{n_\gamma - 1}),$$
$$\mathbf{r} = \mathbf{f}_\gamma(\mathbf{r}_0, \cdots, \mathbf{r}_{n_\gamma - 1}),$$

and $\mathbf{q}_i \equiv \mathbf{p}_i(\Theta_{\bar{a}})$, $\mathbf{p}_i \equiv \mathbf{r}_i(\Theta_{\bar{a}})$ for $i = 0, \cdots, n_{\gamma-1}$. Since for a fixed i, the maximum of the ranks of $\mathbf{q}_i$, $\mathbf{p}_i$, $\mathbf{r}_i$ is less than n, we get $\mathbf{q}_i \equiv \mathbf{r}_i(\Theta_{\bar{a}})$, and so by (SP), $\mathbf{q} \equiv \mathbf{r}(\Theta_{\bar{a}})$. This completes the proof of Theorem 1.

Let $\mathfrak{A}$ be a partial algebra, $\bar{a} = \langle a_0, \cdots, a_\gamma, \cdots \rangle_{\gamma < \alpha}$, and assume that each element of A occurs once and only once in this sequence. We consider the quotient algebra $\mathfrak{P}^{(\alpha)}(\tau)/\Theta_{\bar{a}}$ and we denote by A^* the set of elements of the form $[\mathbf{x}_\gamma]\Theta_{\bar{a}}$.

Theorem 2. *The relative subalgebra* $\mathfrak{A}^* = \langle A^*; F \rangle$ *of* $\mathfrak{P}^{(\alpha)}(\tau)/\Theta_{\bar{a}}$ *is isomorphic to* $\mathfrak{A}$, *and the correspondence*

$$\varphi \colon a_\gamma \to [\mathbf{x}_\gamma]\Theta_{\bar{a}}$$

is an isomorphism between $\mathfrak{A}$ *and* $\mathfrak{A}^*$.

Proof. As the first step, we prove that

$$[\mathbf{x}_\gamma]\Theta_{\bar{a}} = [\mathbf{x}_\delta]\Theta_{\bar{a}}$$

if and only if $\gamma = \delta$.

Assume that $[\mathbf{x}_\gamma]\Theta_{\bar{a}} = [\mathbf{x}_\delta]\Theta_{\bar{a}}$, that is,

$$\mathbf{x}_\gamma \equiv \mathbf{x}_\delta(\Theta_{\bar{a}}).$$

Then by Lemma 1, $x_\gamma(\bar{a}) = x_\delta(\bar{a})$, that is, $a_\gamma = a_\delta$, and so $\gamma = \delta$.

Thus, we have proved that the mapping φ is 1–1; φ is obviously onto.

To conclude the proof of Theorem 2, we must verify that

$$f_\gamma(a_{\delta_0}, \cdots, a_{\delta_{n_\gamma - 1}}) = a_\delta \tag{1}$$

if and only if

$$f_\gamma([\mathbf{x}_{\delta_0}]\Theta_{\bar{a}}, \cdots, [\mathbf{x}_{\delta_{n_\gamma - 1}}]\Theta_{\bar{a}}) = [\mathbf{x}_\delta]\Theta_{\bar{a}}. \tag{2}$$

(2) is equivalent to

$$f_\gamma(\mathbf{x}_{\delta_0}, \cdots, \mathbf{x}_{\delta_{n_\gamma - 1}}) \equiv \mathbf{x}_\delta(\Theta_{\bar{a}}). \tag{3}$$

Using the same argument as we used for the congruence

$$\mathbf{x}_\gamma \equiv \mathbf{x}_\delta(\Theta_{\bar{a}})$$

above, we can prove analogously that the two sides of (3) have only trivial representations and then the equivalence of (1) and (3) follows. This concludes the proof of Theorem 2.

Theorem 2 gives another proof of Theorem 13.1, namely, it gives an embedding of a partial algebra $\mathfrak{A}$ into an algebra. While Theorem 13.1 gives the most economical construction, Theorem 2 gives the least economical one, that is, $\mathfrak{P}^{(\alpha)}(\tau)/\Theta_{\bar{a}}$ is the largest algebra into which $\mathfrak{A}$ can be embedded, such that the image of $\mathfrak{A}$ is a generating set.

We conclude this section by describing the structure of the algebra $\mathfrak{P}^{(\alpha)}(\tau)/\Theta_{\bar{a}}$.

First we define certain subsets $A_{\langle n,\gamma\rangle}$ and $A'_{\langle n,\gamma\rangle}$ $(0 \leqq n < \omega, 0 \leqq \gamma < o(\tau))$ of this algebra as follows:

$$A'_{\langle 0,0\rangle} = A^*,$$

where A^* was defined before Theorem 2;

$$A_{\langle n,\delta\rangle} = A'_{\langle n,\delta\rangle} \cup \{f_\delta(b_0, \cdots, b_{n_\delta - 1}) \mid b_0, \cdots, b_{n_\delta - 1} \in A'_{\langle n,\delta\rangle}\};$$

$$A'_{\langle n,\delta\rangle} = \bigcup (A_{\langle m,\gamma\rangle} \mid \langle m, \gamma\rangle < \langle n, \delta\rangle), \quad \text{if} \quad \langle n, \delta\rangle \neq \langle 0, 0\rangle,$$

where $\langle m, \gamma\rangle < \langle n, \delta\rangle$ means that $m < n$ or $m = n$ and $\gamma < \delta$ (thus the $\langle m, \gamma\rangle$ form a well-ordered set of order type $\omega \cdot o(\tau)$).

Lemma 2. *The following equality holds:*

$$\mathbf{P}^{(\alpha)}(\tau)/\Theta_{\bar{a}} = \bigcup (A_{\langle n,\delta\rangle} \mid 0 \leqq n < \omega, 0 \leqq \delta < o(\tau)).$$

Proof. The following inclusions are trivial, by the definitions of $A_{\langle n,\gamma\rangle}$ and $A'_{\langle n,\gamma\rangle}$:

$$A_{\langle n,\gamma\rangle} \subseteq A'_{\langle n,\delta\rangle} \subseteq A_{\langle n,\delta\rangle} \quad \text{if} \quad \gamma < \delta,$$
$$A_{\langle n,\gamma\rangle} \subseteq A'_{\langle m,\delta\rangle} \subseteq A_{\langle m,\delta\rangle} \quad \text{if} \quad n < m.$$

Take $\mathbf{p} \in \mathbf{P}^{(\alpha)}(\tau)$. We will prove by induction on the rank of $\mathbf{p}$ that

$$[\mathbf{p}]\Theta_{\bar{a}} \in A_{\langle n,\delta\rangle} \tag{4}$$

for some $n < \omega$ and $\delta < o(\tau)$. If $\mathbf{p} = \mathbf{x}_\gamma$, then $[\mathbf{x}_\gamma]\Theta_{\bar{a}} \in A'_{\langle 0,0\rangle}$. Let

$$\mathbf{p} = \mathbf{f}_\gamma(\mathbf{p}_0, \cdots, \mathbf{p}_{n_\gamma - 1}),$$

and assume that (4) holds for each $\mathbf{p}_i$, that is,

$$[\mathbf{p}_i]\Theta_{\bar{a}} \in A_{\langle n_i,\delta_i\rangle}.$$

We set

$$n = \max(n_0, \cdots, n_{n_\gamma - 1})$$

and

$$\delta = \max(\delta_0, \cdots, \delta_{n_\gamma - 1}).$$

Then

$$A_{\langle n_i,\delta_i\rangle} \subseteq A_{\langle n,\delta\rangle} \subseteq A'_{\langle n+1,0\rangle} \subseteq A'_{\langle n+1,\gamma\rangle}.$$

Thus,

$$[\mathbf{p}]\Theta_{\bar{a}} = f_\gamma([\mathbf{p}_0]\Theta_{\bar{a}}, \cdots, [\mathbf{p}_{n_\gamma - 1}]\Theta_{\bar{a}}) \in A_{\langle n+1, \gamma\rangle},$$

which was to be proved.

To get our final result in this section, we introduce the following notation.

Definition 2. *Let $\mathfrak{B}$ be a partial algebra, $X \subseteq B$ and*

$$Y = X \cup \{f_\gamma(x_0, \cdots, x_{n_\gamma - 1}) \mid x_i \in X\}, \qquad \gamma < o(\tau).$$

We will write

$$Y = X[f_\gamma]$$

if $f_\gamma(x_0, \cdots, x_{n_\gamma - 1}) = f_\delta(x_0', \cdots, x'_{n_\delta - 1}) \in Y - X$, $\delta < o(\tau)$ imply that

$$\gamma = \delta,\, x_0 = x_0', \cdots, x_{n_\gamma - 1} = x'_{n_\gamma - 1},$$

and if whenever $\{x_0, \cdots, x_{n_\rho - 1}\} \nsubseteq X$, then $f_\rho(x_0, \cdots, x_{n_\rho - 1})$ does not exist in Y, for any $\rho < o(\tau)$. If $\mathfrak{A}$ and $\mathfrak{B}$ are partial algebras, $\mathfrak{A}$ is a relative subalgebra of $\mathfrak{B}$ and $B = A[f_\gamma]$, then we will write $\mathfrak{B} = \mathfrak{A}[f_\gamma]$.

Lemma 3. $A_{\langle n,\gamma\rangle} = A'_{\langle n,\gamma\rangle}[f_\gamma]$.

Proof. We start with the following observation which follows immediately from the definition of $\Theta_{\bar{a}}$:

(*) For any $\mathbf{p} \in \mathbf{P}^{(\alpha)}(\tau)$, $p(\bar{a})$ is defined if and only if $[\mathbf{p}]\Theta_{\bar{a}} \in A^*$.

Now to prove Lemma 3 we first observe that the first requirement of Definition 2 follows trivially from Lemma 1 and (*). Now assume that $\{a_0, \cdots, a_{n_\delta - 1}\} \nsubseteq A'_{\langle n,\gamma\rangle}$, but that $f_\delta(a_0, \cdots, a_{n_\delta - 1})$ exists in $A_{\langle n,\gamma\rangle}$. By Lemma 1 and (*) we get that $f_\delta(a_0, \cdots, a_{n_\delta - 1}) \in A'_{\langle n,\gamma\rangle}$. Let $a_i = [\mathbf{p}_i]\Theta_{\bar{a}}$, $\mathbf{p} = \mathbf{f}_\delta(\mathbf{p}_0, \cdots, \mathbf{p}_{n_\delta - 1})$. Since $[\mathbf{p}]\Theta_{\bar{a}} \in A'_{\langle n,\gamma\rangle}$, we have that

$$[\mathbf{p}]\Theta_{\bar{a}} \in A_{\langle m,\lambda\rangle}$$

for some smallest $\langle m, \lambda\rangle < \langle n, \gamma\rangle$. By (*) and the assumption that $\{a_0, \cdots, a_{n_\delta - 1}\} \nsubseteq A'_{\langle n,\gamma\rangle}$, we have that $\langle m, \lambda\rangle \neq \langle 0, 0\rangle$ so

$$[\mathbf{p}]\Theta_{\bar{a}} \in A_{\langle m,\lambda\rangle} - A'_{\langle m,\lambda\rangle}.$$

Hence $\mathbf{p} \equiv f_\lambda(\mathbf{q}_0, \cdots, \mathbf{q}_{n_\lambda - 1})$ for some $[\mathbf{q}_i]\Theta_{\bar{a}} \in A'_{\langle m,\lambda\rangle}$, which implies by Lemma 1 and (*) that $\lambda = \delta$ and $\mathbf{q}_i \equiv \mathbf{p}_i(\Theta_{\bar{a}})$. Thus $a_i = [\mathbf{p}_i]\Theta_{\bar{a}} \in A'_{\langle m,\lambda\rangle} \subseteq A'_{\langle n,\gamma\rangle}$, a contradiction. This completes the proof of Lemma 3.

We now summarize what we have proved so far concerning the structure of $\mathfrak{P}^{(\alpha)}(\tau)/\Theta_{\bar{a}}$:

Theorem 3. *$\mathfrak{P}^{(\alpha)}(\tau)/\Theta_{\bar{a}}$ contains a relative subalgebra $\mathfrak{A}^*$ isomorphic to the partial algebra $\mathfrak{A}$; if we start with A^* and we perform two kinds of constructions,*

(i) *taking the set union of the previously constructed sets,*
(ii) *constructing $X[f_\gamma]$ from X,*

then we get a transfinite sequence of increasing subsets of $\mathbf{P}^{(\alpha)}(\tau)/\Theta_{\bar{a}}$ such that the union of all these subsets is the whole set.

It is obvious from Theorem 3 that $\mathfrak{B}=\mathfrak{P}^{(\alpha)}(\tau)/\Theta_{\bar{a}}$ has the following properties:

(α) $\mathfrak{B}$ has a relative subalgebra $\mathfrak{A}^+$ isomorphic to $\mathfrak{A}$ and A^+ generates $\mathfrak{B}$;

(β) if $f_\gamma(b_0, \cdots, b_{n_\gamma-1})=f_\delta(b_0', \cdots, b'_{n_\delta-1}) \notin A^+$, then $\gamma=\delta$ and $b_0=b_0', \cdots, b_{n_\gamma-1}=b'_{n_\gamma-1}$;

(γ) if $f_\gamma(b_0, \cdots, b_{n_\gamma-1}) \in A^+$, then $b_0, \cdots, b_{n_\gamma-1} \in A^+$.

Theorem 4†. *Conditions (α)–(γ) characterize $\mathfrak{P}^{(\alpha)}(\tau)/\Theta_{\bar{a}}$ up to isomorphism.*

Proof. Let $\mathfrak{B}$ satisfy (α)–(γ). Then $B_{\langle n,\gamma\rangle}$ and $B'_{\langle n,\gamma\rangle}$ can be defined in $\mathfrak{B}$ as $A_{\langle n,\gamma\rangle}$ and $A'_{\langle n,\gamma\rangle}$ were defined in $\mathfrak{P}^{(\alpha)}(\tau)/\Theta_{\bar{a}}$, respectively.

Let $\varphi'_{\langle 0,0\rangle}$ be an isomorphism between $\mathfrak{A}^+$ and $\mathfrak{A}^*$. If $\varphi_{\langle m,\delta\rangle}$ is defined for all $\langle m, \delta\rangle < \langle n, \gamma\rangle$, set

$$\varphi'_{\langle n,\gamma\rangle} = \bigcup (\varphi_{\langle m,\delta\rangle} \mid \langle m, \delta\rangle < \langle n, \gamma\rangle).$$

Then $\varphi'_{\langle n,\gamma\rangle}$ will map $B'_{\langle n,\gamma\rangle}$ into $A'_{\langle n,\gamma\rangle}$, and it is 1–1 and onto. If $x \in B_{\langle n,\gamma\rangle} = B'_{\langle n,\gamma\rangle}[f_\gamma]$, then $x=f_\gamma(x_0, \cdots, x_{n_\gamma-1})$, where $x_0, \cdots, x_{n_\gamma-1}$ are uniquely determined elements of $B'_{\langle n,\gamma\rangle}$. Set

$$x\varphi_{\langle n,\gamma\rangle} = f_\gamma(x_0\varphi'_{\langle n,\gamma\rangle}, \cdots, x_{n_\gamma-1}\varphi'_{\langle n,\gamma\rangle}).$$

Then

$$\varphi = \bigcup (\varphi_{\langle n,\gamma\rangle} \mid n < \omega, \gamma < o(\tau))$$

will be the required isomorphism. The easy details are left to the reader.

It should be noted that if $\mathfrak{A}=\langle\{0\}; {}'\rangle$, $\mathfrak{A}$ is of type $\langle 1\rangle$, and $D(', \mathfrak{A})=\varnothing$, then ($\alpha$)–($\gamma$) is the usual Peano axiom system of natural numbers. If $\mathfrak{A}$ is arbitrary with $D(f_\gamma, \mathfrak{A})=\varnothing$ for all $\gamma<o(\tau)$, then $\Theta_{\bar{a}}=\omega$, and thus (α)–(γ)

† J. Schmidt (oral communication).

characterize $\mathfrak{P}^{(\alpha)}(\tau)$ up to isomorphism. In this special case, algebras satisfying (α)–(γ) are called *absolutely free algebras* or *Peano algebras* in the literature.

§15. EXTENSION OF CONGRUENCE RELATIONS

In this section we will prove a strong version of Theorem 13.3. Using the notation of §14, we proved that $\mathfrak{A}$ and $\mathfrak{A}^*$ are isomorphic (Theorem 14.2). Let us identify these two partial algebras; then we can say that $\mathfrak{P}^{(\alpha)}(\tau)/\Theta_{\bar{a}}$ is an algebra which contains $\mathfrak{A}$ as a relative subalgebra.

Theorem 1. *Let Θ be a congruence relation of $\mathfrak{A}$. There exists a congruence relation $\overline{\Theta}$ of $\mathfrak{P}^{(\alpha)}(\tau)/\Theta_{\bar{a}}$ such that $\overline{\Theta}_A = \Theta$.*

According to Theorem 14.3, it suffices to prove the following two lemmas.

Lemma 1. *Let $\mathfrak{A}$ be a partial algebra, $A = \bigcup (X_\gamma \mid \gamma < \alpha)$, and $X_{\gamma_0} \subseteq X_{\gamma_1}$ if $\gamma_0 < \gamma_1$.*

Let Θ^γ be a congruence relation of $\mathfrak{X}_\gamma$ such that

$$\Theta^{\gamma_1}_{X_{\gamma_0}} = \Theta^{\gamma_0}$$

if $\gamma_0 < \gamma_1$.

Then there exists a congruence relation Θ of $\mathfrak{A}$ such that

$$\Theta_{X_\gamma} = \Theta^\gamma$$

for each $\gamma < \alpha$.

Lemma 2. *Let $\mathfrak{A}$ be a partial algebra and $\mathfrak{B}$ a relative subalgebra of $\mathfrak{A}$. Assume that $\mathfrak{A} = \mathfrak{B}[f_\gamma]$ for some $\gamma < o(\tau)$. Then to every congruence relation Θ of $\mathfrak{B}$ there corresponds a congruence relation $\overline{\Theta}$ of $\mathfrak{A}$ such that $\overline{\Theta}_B = \Theta$.*

Remark. Let us note that Theorem 1 is stronger than Theorem 13.3 since we extended $\mathfrak{A}$ to an algebra such that *every* congruence relation of $\mathfrak{A}$ can be extended—not merely a given one.

Theorem 1 was first given in G. Grätzer and E. T. Schmidt [2], but in a weaker version; namely, in that paper it was proved that every partial algebra can be extended to an algebra which satisfies the requirements of Theorem 1 but it was not proved that this algebra can be represented as $\mathfrak{P}^{(\alpha)}(\tau)/\Theta_{\bar{a}}$. As a matter of fact, that version follows directly from Lemmas 1 and 2; for that we do not need the investigations of §14 at all. A minor difference is that in that paper a third construction was also needed to get the algebra (besides the constructions given by Lemmas 1 and 2), but it is easy to see that it can be eliminated.

Proof of Lemma 1. Set

$$\Theta = \bigcup (\Theta^{\gamma} \mid \gamma < \alpha).$$

It is routine to check that Θ is a congruence relation. As an illustration, we prove the transitivity of Θ.

Let $a \equiv b(\Theta)$ and $b \equiv c(\Theta)$. Then $\langle a, b\rangle, \langle b, c\rangle \in \bigcup (\Theta^{\gamma} \mid \gamma < \alpha)$. Therefore, $\langle a, b\rangle \in \Theta^{\gamma_0}$, $\langle b, c\rangle \in \Theta^{\gamma_1}$. Suppose, for instance, that $\gamma_0 \leqq \gamma_1$. Then

$$\langle a, b\rangle, \langle b, c\rangle \in \Theta^{\gamma_1}$$

and thus by the transitivity of Θ^{γ_1}, $\langle a, c\rangle \in \Theta^{\gamma_1}$. The proof of reflexivity, symmetry, and the substitution property is similar.

Finally, let us compute Θ_{X_γ};

$$\begin{aligned} \Theta_{X_\gamma} &= \Theta \cap (X_\gamma \times X_\gamma) = \bigcup (\Theta^{\delta} \mid \delta < \alpha) \cap (X_\gamma \times X_\gamma) \\ &= \bigcup (\Theta^{\delta} \cap (X_\gamma \times X_\gamma) \mid \delta < \alpha) \\ &= \bigcup (\Theta^{\delta} \cap (X_\gamma \times X_\gamma) \mid \gamma \leqq \delta < \alpha) \\ &= \bigcup (\Theta^{\delta}_{X_\gamma} \mid \gamma \leqq \delta < \alpha) \\ &= \bigcup (\Theta^{\gamma} \mid \gamma \leqq \delta < \alpha) \\ &= \Theta^{\gamma}, \end{aligned}$$

which was to be proved.

Lemma 3. *Under the conditions of Lemma 2, for a fixed* Θ, *define a relation* Φ *on* $\mathfrak{A}$ *as follows:*

(i) $a \equiv b(\Phi)$, $a, b \in B$ *if and only if* $a \equiv b(\Theta)$;

(ii) $a \equiv b(\Phi)$, $a \in B, b \notin B$ $(b = f_\gamma(x_0, \cdots, x_{n_\gamma - 1}))$ *if and only if there exists a* $u = f_\gamma(y_0, \cdots, y_{n_\gamma - 1}) \in B$ *such that* $a \equiv u(\Theta)$, $x_i \equiv y_i(\Theta)$, $0 \leqq i < n_\gamma$; *and the symmetric condition holds for* $a \notin B, b \in B$;

(iii) $a \equiv b(\Phi)$, $a, b \notin B$ $(a = f_\gamma(x_0, \cdots, x_{n_\gamma - 1}), b = f_\gamma(y_0, \cdots, y_{n_\gamma - 1}))$ *if and only if*

(iii$_1$) $x_i \equiv y_i(\Theta)$, $0 \leqq i < n_\gamma$, *or*

(iii$_2$) *there exist* $u = f_\gamma(u_0, \cdots, u_{n_\gamma - 1}) \in B$, $v = f_\gamma(v_0, \cdots, v_{n_\gamma - 1}) \in B$ *such that* $x_i \equiv u_i(\Theta)$, $v_i \equiv y_i(\Theta)$, $0 \leqq i < n_\gamma$, *and* $u \equiv v(\Theta)$.

Then Φ *is a congruence relation of* $\mathfrak{A}$.

Let us note that Lemma 3 implies Lemma 2 since $\Phi_B = \Theta$ is equivalent to (i).

The following diagrams illustrate rules (i)–(iii), in case $f_\gamma = f$ is binary. Dotted lines denote congruence modulo Φ and solid lines denote congruence modulo Θ.

Rule (i)

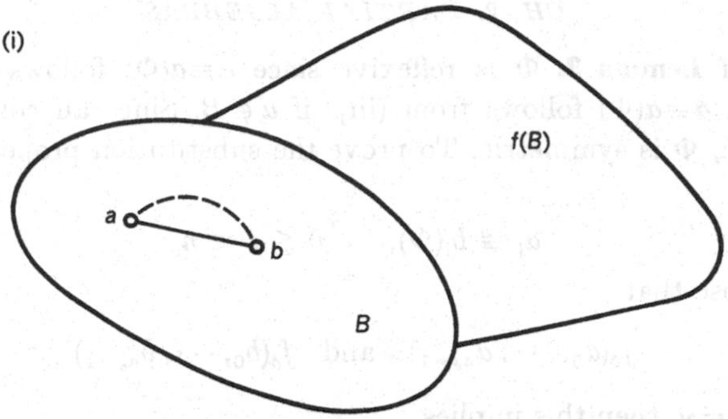

Rule (ii)

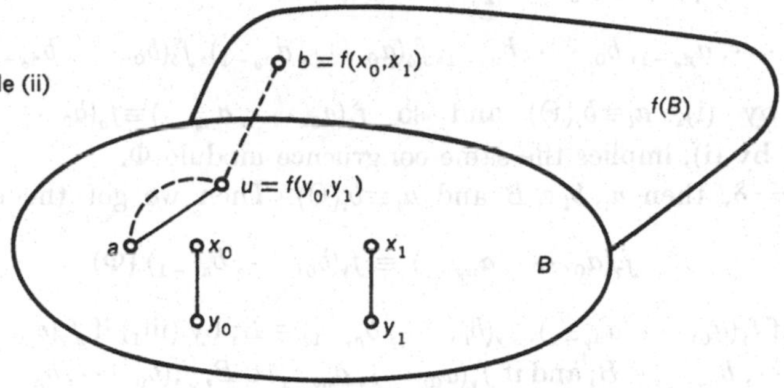

Rule (iii₁)

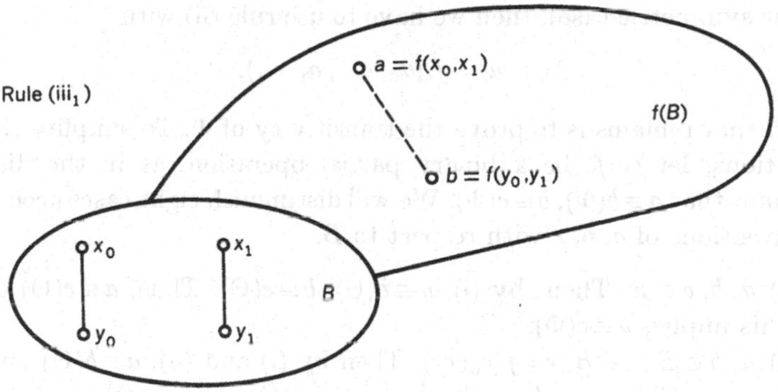

Rule (iii₂)

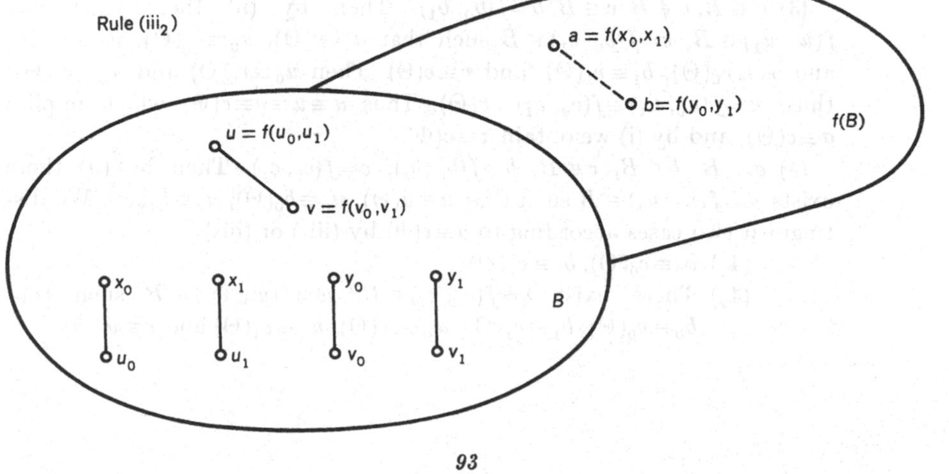

Proof of Lemma 3. Φ is reflexive since $a \equiv a(\Phi)$ follows from (i) if $a \in B$ and $a \equiv a(\Phi)$ follows from (iii_1) if $a \notin B$. Since all conditions are symmetric, Φ is symmetric. To prove the substitution property, assume that

$$a_i \equiv b_i(\Phi), \qquad 0 \leqq i < n_\delta,$$

and suppose that

$$f_\delta(a_0, \cdots, a_{n_\delta - 1}) \quad \text{and} \quad f_\delta(b_0, \cdots, b_{n_\delta - 1})$$

exist. If $\delta \neq \gamma$, then this implies

$$a_0, \cdots, a_{n_\delta - 1}, b_0, \cdots, b_{n_\delta - 1}, f_\delta(a_0, \cdots, a_{n_\delta - 1}), f_\delta(b_0, \cdots, b_{n_\delta - 1}) \in B;$$

thus, by (i), $a_i \equiv b_i(\Theta)$ and so $f_\delta(a_0, \cdots, a_{n_\delta - 1}) \equiv f_\delta(b_0, \cdots, b_{n_\delta - 1})(\Theta)$ which, by (i), implies the same congruence modulo Φ.

If $\gamma = \delta$, then $a_i, b_i \in B$ and $a_i \equiv b_i(\Theta)$. Then we get the congruence

$$f_\gamma(a_0, \cdots, a_{n_\gamma - 1}) \equiv f_\gamma(b_0, \cdots, b_{n_\gamma - 1})\ (\Phi)$$

by (i) if $f_\gamma(a_0, \cdots, a_{n_\gamma - 1}), f_\gamma(b_0, \cdots, b_{n_\gamma - 1}) \in B$; by ($\text{iii}_1$) if $f_\gamma(a_0, \cdots, a_{n_\gamma - 1})$, $f_\gamma(b_0, \cdots, b_{n_\gamma - 1}) \notin B$; and if $f_\gamma(a_0, \cdots, a_{n_\gamma - 1}) \in B$, $f_\gamma(b_0, \cdots, b_{n_\gamma - 1}) \notin B$ (and in the symmetric case), then we have to use rule (ii) with

$$u = f_\gamma(a_0, \cdots, a_{n_\gamma - 1}).$$

All that remains is to prove the transitivity of Φ. To simplify the computations, let $f = f_\gamma$ be a binary partial operation, as in the diagrams. Assume that $a \equiv b(\Phi)$, $b \equiv c(\Phi)$. We will distinguish eight cases according to the positions of a, b, c with respect to B.

(1) $a, b, c \in B$. Then, by (i), $a \equiv b(\Theta)$, $b \equiv c(\Theta)$. Thus, $a \equiv c(\Theta)$ and, by (i), this implies $a \equiv c(\Phi)$.

(2) $a, b \in B$, $c \notin B$, $c = f(c_0, c_1)$. Then by (i) and (ii), $a \equiv b(\Theta)$ and there exists $u = f(u_0, u_1) \in B$ such that $c_0 \equiv u_0(\Theta)$, $c_1 \equiv u_1(\Theta)$, and $b \equiv u(\Theta)$. Then $a \equiv u(\Theta)$ and thus (ii) implies $a \equiv c(\Phi)$, using the auxiliary element u.

(3) $a \in B$, $b \notin B$, $c \in B$, $b = f(b_0, b_1)$. Then by (ii) there exist $u = f(u_0, u_1) \in B$, $v = f(v_0, v_1) \in B$ such that $a \equiv u(\Theta)$, $u_0 \equiv b_0(\Theta)$, $u_1 \equiv b_1(\Theta)$, and $b_0 \equiv v_0(\Theta)$, $b_1 \equiv v_1(\Theta)$, and $v \equiv c(\Theta)$. Then $u_0 \equiv v_0(\Theta)$ and $u_1 \equiv v_1(\Theta)$; thus, $u = f(u_0, u_1) \equiv f(v_0, v_1) = v(\Theta)$. Thus $a \equiv u \equiv v \equiv c(\Theta)$ which implies $a \equiv c(\Theta)$, and by (i) we obtain $a \equiv c(\Phi)$.

(4) $a \in B$, $b \notin B$, $c \notin B$, $b = f(b_0, b_1)$, $c = f(c_0, c_1)$. Then by (ii) there exists $u = f(u_0, u_1) \in B$ such that $a \equiv u(\Theta)$, $u_0 \equiv b_0(\Theta)$, $u_1 \equiv b_1(\Theta)$. We distinguish two cases according to $b \equiv c(\Phi)$ by (iii_1) or (iii_2):

(4_1) $b_0 \equiv c_0(\Theta)$, $b_1 \equiv c_1(\Theta)$.

(4_2) There exist $v = f(v_0, v_1) \in B$, $w = f(w_0, w_1) \in B$ such that $b_0 \equiv v_0(\Theta)$, $b_1 \equiv v_1(\Theta)$, $w_0 \equiv c_0(\Theta)$, $w_1 \equiv c_1(\Theta)$ and $v \equiv w(\Theta)$.

In the first case, (4_1), $u_0 \equiv c_0(\Theta)$ and $u_1 \equiv c_1(\Theta)$ and thus by (ii) we get $a \equiv c(\Phi)$, using the auxiliary element u.

In the second case, (4_2), $u_0 \equiv v_0(\Theta)$, $u_1 \equiv v_1(\Theta)$ and thus $u = f(u_0, u_1) \equiv f(v_0, v_1) = v(\Theta)$. Therefore $a \equiv u \equiv v \equiv w(\Theta)$ and so $a \equiv w(\Theta)$. Thus by (ii) we get $a \equiv c(\Phi)$, using the auxiliary element w.

(5) $a \notin B, b \in B, c \in B$. The proof is similar to that of (2).

(6) $a \notin B, b \in B, c \notin B, a = f(a_0, a_1), \quad c = f(c_0, c_1)$. Then, applying (ii) twice, we get the existence of $u = f(u_0, u_1) \in B$ and of $v = f(v_0, v_1) \in B$ such that $b \equiv u(\Theta)$, $u_0 \equiv a_0(\Theta)$, $u_1 \equiv a_1(\Theta)$ and $b \equiv v(\Theta)$, $v_0 \equiv c_0(\Theta)$, $v_1 \equiv c_1(\Theta)$. Then $u \equiv v(\Theta)$ and thus $a \equiv c(\Phi)$ by (iii_2), using the auxiliary elements u and v.

(7) $a \notin B, b \notin B, c \in B$. The proof is similar to that of (4).

(8) $a, b, c \notin B, a = f(a_0, a_1), b = f(b_0, b_1), c = f(c_0, c_1)$.

We have four subcases to distinguish, according to which of (iii_1) and (iii_2) give us $a \equiv b(\Phi)$ and $b \equiv c(\Phi)$.

(8_1) We apply (iii_1) twice. Then $a_0 \equiv b_0(\Theta)$, $a_1 \equiv b_1(\Theta)$, $b_0 \equiv c_0(\Theta)$, $b_1 \equiv c_1(\Theta)$; thus we get $a \equiv c(\Phi)$ by (iii_1).

(8_2) We first apply (iii_1) and then (iii_2). Then $a_0 \equiv b_0(\Theta)$, $a_1 \equiv b_1(\Theta)$, and there exist $u = f(u_0, u_1) \in B$, $v = f(v_0, v_1) \in B$ such that $b_0 \equiv u_0(\Theta)$, $b_1 \equiv u_1(\Theta)$, $v_0 \equiv c_0(\Theta)$, $v_1 \equiv c_1(\Theta)$ and $u \equiv v(\Theta)$. Then $a_0 \equiv u_0(\Theta)$, $a_1 \equiv u_1(\Theta)$; thus, by (iii_2) $a \equiv c(\Phi)$, using the auxiliary elements u and v.

(8_3) We first apply (iii_2) and then (iii_1). The proof is similar to (8_2).

(8_4) We apply (iii_2) twice. Then there exist $u = f(u_0, u_1) \in B$, $v = f(v_0, v_1) \in B$, $w = f(w_0, w_1) \in B$, $z = f(z_0, z_1) \in B$ such that $a_0 \equiv u_0(\Theta)$, $a_1 \equiv u_1(\Theta)$, $u \equiv v(\Theta)$, $v_0 \equiv b_0(\Theta)$, $v_1 \equiv b_1(\Theta)$, $b_0 \equiv w_0(\Theta)$, $b_1 \equiv w_1(\Theta)$, $w \equiv z(\Theta)$, $z_0 \equiv c_0(\Theta)$, $z_1 \equiv c_1(\Theta)$. Then $v_0 \equiv w_0(\Theta)$ and $v_1 \equiv w_1(\Theta)$, and so $v = f(v_0, v_1) \equiv f(w_0, w_1) = w(\Theta)$. Consequently, $u \equiv v \equiv w \equiv z(\Theta)$; that is, $u \equiv z(\Theta)$ and thus we get $a \equiv c(\Phi)$, using (iii_2) and the auxiliary elements u and z.

This completes the proof of Lemma 3.

To conclude this section, we give another version of Theorem 1.

Theorem 2. *Let $\mathfrak{A}$ be a partial algebra, Θ a congruence relation on $\mathfrak{A}$, and let $\bar{a} = \langle a_0, \cdots, a_\gamma, \cdots \rangle_{\gamma < \alpha}$ be a sequence of type α of elements of A, containing each element of A exactly once. Then there exists a congruence relation Φ of $\mathfrak{P}^{(\alpha)}(\tau)$ such that $\Phi \geqq \Theta_{\bar{a}}$ and $\mathbf{x}_\gamma \equiv \mathbf{x}_\delta(\Phi)$ if and only if $a_\gamma \equiv a_\delta(\Theta)$.*

Theorem 2 is simply Theorem 1 combined with the second isomorphism theorem (Theorem 11.4).

§16. SUBALGEBRAS AND HOMOMORPHISMS OF PARTIAL ALGEBRAS

In this section we will review some of the results of Chapter 1 within the framework of partial algebras.

Since the proofs in most cases remain the same we will just rephrase the results. Some further results will be reviewed in the Exercises.

Let $\mathfrak{A}$ be a partial algebra and let $\mathscr{S}(\mathfrak{A})$ denote the family of all subsets B such that $\langle B; F\rangle$ is a subalgebra of $\mathfrak{A}$ with the void set added if there are no nullary partial operations (defined in $\mathfrak{A}$). Then Theorem 9.1 remains true; in Lemma 9.3 we have to add the condition that $p(h_0, \cdots, h_{n-1})$ is defined and equals a. The only result which fails to hold for partial algebras is Lemma 9.1.

However, congruence relations of partial algebras behave differently from congruence relations of algebras.

Lemma 10.1 remains valid and we can add that it is valid not only for congruence relations, but also for strong congruence relations. Lemma 10.2 is in general false for partial algebras, but Corollary 3 of Lemma 10.2 and Lemma 10.3 are valid. Of course, we must change the proofs, since they cannot be referred to Lemma 10.2. Since we needed only Lemmas 10.1 and 10.3 to prove Theorems 10.1 and 10.2, they remain valid.

We now proceed to prove for partial algebras the converse of Theorem 10.2.

Theorem 1 (*G. Grätzer and E. T. Schmidt* [2]). *Let $\mathfrak{A}$ be a partial algebra and let $C(\mathfrak{A})$ denote the system of all congruence relations of $\mathfrak{A}$. Then $\mathfrak{C}(\mathfrak{A}) = \langle C(\mathfrak{A}); \leqq\rangle$ is an algebraic lattice. Conversely, if $\mathfrak{L}$ is an algebraic lattice, then it is isomorphic to some $\mathfrak{C}(\mathfrak{A})$.*

Proof. The first part of Theorem 1 is just a restatement of Theorem 10.2 for partial algebras. To prove the second statement, let $\mathfrak{L}$ be an algebraic lattice. Represent this algebraic lattice $\mathfrak{L}$ as $\mathfrak{J}(\mathfrak{S})$, the lattice of all ideals of a semilattice $\mathfrak{S} = \langle S; \vee\rangle$ with 0 (Theorem 6.3).

We construct the partial algebra as follows. Let $A = S$. For $a, b \in S$, define a binary partial operation f_{ab} so that $D(f_{ab}) = \{\langle a, b\rangle, \langle 0, 0\rangle\}$, $f_{ab}(a, b) = a \vee b$, $f_{ab}(0, 0) = 0$. Further, for every $a, b \in S$ such that $b \leqq a$ we define a unary partial operation g_{ab} so that $D(g_{ab}) = \{a, 0\}$, $g_{ab}(a) = b$, $g_{ab}(0) = 0$.

For every $a, b \in S$ such that $a \neq b$, define a unary partial operation h_{ab} such that $D(h_{ab}) = \{a, b\}$ and $h_{ab}(a) = a$, $h_{ab}(b) = 0$.

Consider the partial algebra $\mathfrak{A} = \langle A; F\rangle$, where F denotes the collection of all these partial operations.

Consider an ideal I of the semilattice $\mathfrak{S}$ and define a binary relation Θ_I on A as follows:

$$x \equiv y(\Theta_I) \quad \text{if and only if} \quad x = y \quad \text{or} \quad x, y \in I.$$

We shall now verify that Θ_I is a congruence relation of $\mathfrak{A}$. It is clear that Θ_I is reflexive, symmetric, and transitive.

To prove the substitution property for f_{ab}, assume that $x_0 \equiv y_0(\Theta_I)$ and $x_1 \equiv y_1(\Theta_I)$, and that $f_{ab}(x_0, x_1)$ and $f_{ab}(y_0, y_1)$ exist and $\langle x_0, y_0\rangle \neq \langle x_1, y_1\rangle$. Then $\langle x_0, x_1\rangle = \langle a, b\rangle$ and $\langle y_0, y_1\rangle = \langle 0, 0\rangle$ (or $\langle y_0, y_1\rangle = \langle a, b\rangle$ and $\langle x_0, x_1\rangle = \langle 0, 0\rangle$). Then the conditions mean that $a, b \in I$. By applying f_{ab}, we get $a \vee b \equiv 0(\Theta_I)$, which is true since $0, a \vee b \in I$.

Similarly, the substitution property for g_{ab} is satisfied since $a \in I$, $b \leqq a$ imply $b \in I$; the substitution property for h_{ab} is satisfied since $a \neq b$, $a \equiv b(\Theta_I)$ imply $a, 0 \in I$.

Thus we have proved that:

(i) Θ_I is a congruence relation.

The following statement is trivial:

(ii) $\Theta_I \leqq \Theta_J$ if and only if $I \subseteq J$.

(iii) Let Θ be any congruence relation on $\mathfrak{A}$ and define

$$I = \{x \mid x \equiv 0(\Theta)\}.$$

Then I is an ideal.

To prove (iii), let $a, b \in I$. This means that $a \equiv 0(\Theta)$, $b \equiv 0(\Theta)$. Therefore, $a \vee b = f_{ab}(a, b) \equiv f_{ab}(0, 0) = 0(\Theta)$ and so $a \vee b \in I$.

Let $a \in I$, $b \leqq a$; then $a \equiv 0(\Theta)$ and thus $b = g_{ab}(a) \equiv g_{ab}(0) = 0(\Theta)$ and so $b \in I$, which completes the proof of (iii).

(iv) Let Θ be a congruence relation, $I = \{x \mid x \equiv 0(\Theta)\}$. Then $\Theta = \Theta_I$.

$\Theta_I \leqq \Theta$ is trivial. To prove that $\Theta_I \geqq \Theta$, let $x \equiv y(\Theta)$, $x \neq y$. Then

$$x = h_{xy}(x) \equiv h_{xy}(y) = 0(\Theta),$$

that is, $x \in I$. Similarly, $y \in I$. Thus, $x \equiv y(\Theta_I)$.

Statements (i), (ii), (iii), (iv) prove that the correspondence $I \to \Theta_I$ is an isomorphism between $\mathfrak{I}(\mathfrak{S})$ and $\mathfrak{C}(\mathfrak{A})$, completing the proof of Theorem 1.

Now we consider the problem of defining the concept of a homomorphic image of a partial algebra. Let $\mathfrak{A}$ and $\mathfrak{B}$ be partial algebras, and let φ be a homomorphism of $\mathfrak{A}$ into $\mathfrak{B}$.

Then the relative subalgebra $\langle A\varphi; F\rangle$ of $\mathfrak{B}$ is not necessarily isomorphic to the quotient algebra $\langle A/\varepsilon_\varphi; F\rangle$, not even if φ is 1–1 and onto. Consider the following trivial example. Let $A=\{x\}$, $B=\{y\}$, $F=\{f\}$, $\tau=\langle 1\rangle$, $D(f, \mathfrak{A})=\varnothing$, $D(f, \mathfrak{B})=\{y\}$, and $f(y)=y$, $\varphi: x \to y$. Then φ is a 1–1 homomorphism of $\langle A; F\rangle$ onto $\langle B; F\rangle$ but $\langle A; F\rangle \not\cong \langle A\varphi; F\rangle$ since f is not defined in $\langle A; F\rangle$, whereas it is defined in $\langle A\varphi; F\rangle$. The reason for this is that only

$$D(f_\gamma, \mathfrak{A})\varphi \subseteq D(f_\gamma, \mathfrak{A}\varphi)$$

holds in general, and we do not always have equality. Therefore, we define $\mathfrak{B}$ to be a *homomorphic image* of $\mathfrak{A}$ if there exists a homomorphism $\varphi: A \to B$ which is onto and full.

Note that an isomorphism is always a full homomorphism.

Adopting this definition, we encounter no difficulty in proving the homomorphism theorem for full homomorphisms. Also, the isomorphism theorems carry over, without any difficulty, the first isomorphism theorem (Theorem 11.2) for strong congruences, and the second isomorphism theorem (Theorem 11.4) for all congruences.

We can then define *endomorphisms*, *full endomorphisms*, and *strong endomorphisms* and consider the sets

$$E(\mathfrak{A}),\quad E_F(\mathfrak{A}),\quad \text{and}\quad E_S(\mathfrak{A})$$

of all endomorphisms, full endomorphisms, and strong endomorphisms of the partial algebra $\mathfrak{A}$, respectively.

Then $E(\mathfrak{A}) \supseteq E_F(\mathfrak{A}) \supseteq E_S(\mathfrak{A})$.

Lemma 1. *$\langle E(\mathfrak{A}); \cdot\rangle$, $\langle E_F(\mathfrak{A}); \cdot\rangle$, and $\langle E_S(\mathfrak{A}); \cdot\rangle$ are semigroups with unit element and the first contains the second and third and the second contains the third as subsemigroups.*

Finally, we will prove an embedding theorem for partial algebras which is similar to Theorem 13.3 and which characterizes the strong congruence relations.

Theorem 2. *Let $\mathfrak{A}$ be a partial algebra and let Θ be a congruence relation of $\mathfrak{A}$. The congruence relation Θ is strong if and only if $\mathfrak{A}$ can be embedded in an algebra $\mathfrak{B}$ and Θ can be extended to a congruence relation $\overline{\Theta}$ of $\mathfrak{B}$ such that*

$$[a]\Theta = [a]\overline{\Theta} \quad \textit{for all} \quad a \in A.$$

The algebra $\mathfrak{B}$ can always be chosen as $\mathfrak{P}^{(\alpha)}(\tau)/\Theta_{\mathfrak{a}}$ (see Theorem 14.2).

Remark. This condition means that $\overline{\Theta}_A = \Theta$ and any equivalence class of Θ in A is also an equivalence class of $\overline{\Theta}$ in B. Theorem 2 was announced by G. Grätzer in the Notices Amer. Math. Soc. 13 (1966), p. 146. A direct proof of Theorem 2 without the last statement can be given using the construction of Theorem 13.1.

Proof. We first prove that if such an embedding exists, then Θ is strong. Recall that a congruence relation Θ is strong if whenever $f_\gamma(a_0, \cdots, a_{n_\gamma - 1}) \in A$ and $a_i \equiv b_i(\Theta)$, then $f_\gamma(b_0, \cdots, b_{n_\gamma - 1})$ is defined in $\mathfrak{A}$.

Since $f_\gamma(b_0, \cdots, b_{n_\gamma - 1})$ is always defined in $\mathfrak{B}$, all we have to prove is that it is in A. Set $a = f_\gamma(a_0, \cdots, a_{n_\gamma - 1})$; then by assumption $[a]\Theta = [a]\overline{\Theta}$.

Since $\overline{\Theta}$ is an extension of Θ, we have that $a_i \equiv b_i(\overline{\Theta})$ and thus

$$f_\gamma(a_0, \cdots, a_{n_\gamma - 1}) \equiv f_\gamma(b_0, \cdots, b_{n_\gamma - 1})(\overline{\Theta}),$$

that is,

$$f_\gamma(b_0, \cdots, b_{n_\gamma - 1}) \in [a]\overline{\Theta} = [a]\Theta \subseteq A.$$

Thus,

$$f_\gamma(b_0, \cdots, b_{n_\gamma - 1}) \in A,$$

which was to be proved.

Now assume that Θ is a strong congruence relation and put

$$\mathfrak{B} = \mathfrak{P}^{(\alpha)}(\tau)/\Theta_{\bar{\mathfrak{a}}}.$$

We extend Θ to $\mathfrak{B}$ using Lemmas 15.1 and 15.3.

We prove that if we assume that Θ is a strong congruence relation, then $[a]\Theta = [a]\overline{\Theta}$ holds for $a \in A$.

Suppose that in Lemma 15.1, $\langle A_0; F\rangle$ is the partial algebra we start with and that we know that for each $\gamma < \alpha$,

$$[a]\Theta^0 = [a]\Theta^\gamma.$$

Then

$$\begin{aligned} [a]\overline{\Theta} &= \bigcup ([a]\Theta^\gamma \mid \gamma < \alpha) \\ &= \bigcup ([a]\Theta^0 \mid \gamma < \alpha) \\ &= [a]\Theta^0, \end{aligned}$$

so that this property is preserved under the construction of Lemma 15.1.

Now consider the construction in Lemma 15.3. Let† $a \in B$ and assume that $[a]\Theta \neq [a]\Phi$. Then there exists a $b \notin B$ such that $a \equiv b(\Phi)$. By Rule (ii) this means that $b = f(x_0, x_1)$ and that there exists a $u = f(y_0, y_1) \in B$ such that $a \equiv u(\Theta)$, $y_0 \equiv x_0(\Theta)$ and $y_1 \equiv x_1(\Theta)$. The last two congruences together with the existence of $f(y_0, y_1)$ imply (since Θ is strong) that $f(x_0, x_1)$ exists in B, that is, $b \in B$, which is a contradiction. This completes the proof of Theorem 2.

† We use the notation of Lemma 15.3.

§17. THE CHARACTERIZATION THEOREM OF CONGRUENCE LATTICES: PRELIMINARY CONSIDERATIONS

Let $\mathfrak{A} = \langle A; F \rangle$ be a unary partial algebra and let $\mathfrak{B} = \langle B; F \rangle$ denote the algebra $\mathfrak{P}^{(\alpha)}(\tau)/\Theta_{\mathfrak{a}}$ of Theorem 14.2. $\mathfrak{B}$ contains $\mathfrak{A}$ as a relative subalgebra and A generates $\mathfrak{B}$. If g and h are unary operations, we will write $gh(x)$ for $g(h(x))$ and similarly for n unary operations. If $b \in B$, then we can always represent b in the form

$$(*) \qquad b = g_1 \cdots g_n(a), \qquad a \in A \quad \text{and} \quad g_i \in F^*$$

where $F^* = F \cup \{e\}$ and e is the identity function on A, that is, $e(a) = a$ for all $a \in A$.

A representation (*) of b is *reduced* provided $b \in A$ and the representation is $b = e(b)$, or $b \notin A$ and $a \notin D(g_n, \mathfrak{A})$.

It is obvious from Theorem 14.3 that every element of B has a reduced representation.

Lemma 1. *The reduced representation is unique, that is, if* $g_1 \cdots g_r(a)$ *and* $h_1 \cdots h_s(a')$ *are both reduced representations of* $b \in B$, *then* $a = a'$, $r = s$, *and* $g_1 = h_1, \cdots, g_r = h_r$.

Proof. This follows easily from Theorem 14.4. A more direct proof is the following.

Let $b \in B$; then $b \in A_{\langle n, \gamma \rangle}$ for some $n < \omega$, $\gamma < o(\tau)$ (Lemma 14.2). We will prove the statement by transfinite induction on $\langle n, \gamma \rangle$. The statement is known for $A = A'_{\langle 0,0 \rangle}$. Assume that it has been proved for all elements of $A_{\langle m, \delta \rangle}$ with $\langle m, \delta \rangle < \langle n, \gamma \rangle$ and let $b \in A_{\langle n, \gamma \rangle}$.

We can assume by the induction hypothesis that $b \notin A'_{\langle n, \gamma \rangle}$. Thus, if $b = g_1 \cdots g_r(a)$ is any reduced representation of b, then $g_1 = f_\gamma$. Let $b = g_1' \cdots g_s'(a')$ be another reduced representation of b. Then, again, $g_1' = f_\gamma$. Thus, by Definition 14.2 and Lemma 14.3, $f_\gamma(g_2 \cdots g_r(a)) = f_\gamma(g_2' \cdots g_s'(a'))$ if and only if $g_2 \cdots g_r(a) = g_2' \cdots g_s'(a')$. Now we can apply the induction hypothesis to this element. This completes the proof.

Summarizing, we have that every element of B has a reduced representation and equality of these representations is formal equality.

Let us assume that there are in F three unary partial operations g_1, g_2, and g_3 such that $D(g_1, \mathfrak{A}) = \{a\}$, $D(g_3, \mathfrak{A}) = \{b\}$, $D(g_2, \mathfrak{A}) = \varnothing$, $g_1(a) = c$, $g_3(b) = d$, $a, b, c, d \in A$, and $a \neq b$. Form

$$A'' = A[g_1] \cup A[g_2] \cup A[g_3] \subseteq B.$$

We define in $\mathfrak{A}'' = \langle A''; F \rangle$ a relation Φ: $x \equiv y(\Phi)$ if and only if $x = y$ or $x = g_1(b)$, $y = g_2(a)$ or $x = g_2(a)$, $y = g_1(b)$, or $x = g_2(b)$, $y = g_3(a)$ or $x = g_3(a)$,

$y = g_2(b)$. Obviously, Φ is a congruence relation. Set $\mathfrak{A}' = \mathfrak{A}''/\Phi$. By identifying $[x]\Phi$ with x, we get the diagram for $\mathfrak{A}'$. Note that $D(f_\gamma, \mathfrak{A}') = D(f_\gamma, \mathfrak{A})$ if $f_\gamma \neq g_i$ and $D(g_i, \mathfrak{A}') = A$, $i = 1, 2, 3$.

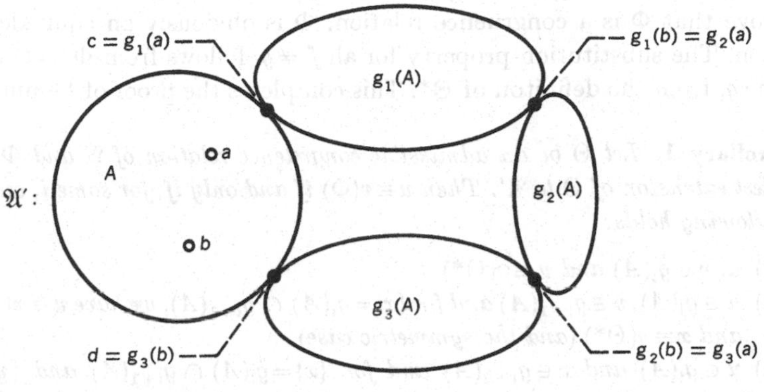

Let Θ be a congruence relation of $\mathfrak{A}$. Θ is *admissible* provided either $a \not\equiv b(\Theta)$ or $a \equiv b(\Theta)$ and $c \equiv d(\Theta)$.

Lemma 2. *Let Θ be a congruence relation of $\mathfrak{A}$. Then Θ can be extended to $\mathfrak{A}'$ if and only if Θ is admissible.*

Proof. Assume that Θ can be extended to $\mathfrak{A}'$, that is, there exists a congruence relation Φ of $\mathfrak{A}'$ such that $\Phi_A = \Theta$. If $a \equiv b(\Theta)$, then $a \equiv b(\Phi)$, and so $c = g_1(a) \equiv g_1(b) = g_2(a) \equiv g_2(b) = g_3(a) \equiv g_3(b) = d(\Phi)$, that is, $c \equiv d(\Theta)$, which was to be proved.

Assume that Θ is admissible. Define a binary relation Θ^* on A' as follows: $x \equiv y(\Theta^*)$ if $x, y \in A$ and $x \equiv y(\Theta)$, or $x, y \in g_i(A)$ for some i, $x = g_i(x')$, $y = g_i(y')$, $x', y' \in A$, and $x' \equiv y'(\Theta)$.

We claim that the transitive extension Φ of Θ^* is a congruence relation and $\Phi_A = \Theta$.

Let us agree that $g_0(x) = x$, for $x \in A$ and that the elements $g_i(a)$ and $g_i(b)$, $i = 1, 2, 3$ are called the *extreme elements* of A'. If $n \equiv i \pmod 4$, $0 \leqq i \leqq 3$, then $g_n(A)$ stands for $g_i(A)$. Then it is obvious that $g_i(A) \cap g_{i+1}(A)$ consists of one element which is an extreme element.

Since Θ is transitive on A, Θ^* is transitive on each $g_i(A)$. This implies that if $u, v \in A'$ and $u \equiv v(\Phi)$, then a nonredundant sequence $u = x_0, \cdots, x_n = v$ such that $x_{i-1} \equiv x_i(\Theta^*)$ consists of u and v and of extreme elements. Since an extreme element cannot occur twice in a nonredundant sequence, we deduce that $n \leqq 5$.

Suppose that $u, v \in g_i(A)$ and $u \equiv v(\Phi)$. Let $u = x_0, \cdots, x_n = v$ be a nonredundant sequence, as before. If $n \neq 1$, we may have $n = 3$, 4, or 5 (if u or v is an extreme element the cases $n = 3$ or $n = 4$ may occur). By possibly adding a slight redundancy, and by symmetry, we may assume that $n = 5$,

$x_1, x_4 \in g_i(A)$. Then $g_j(a) \equiv g_j(b)$ (Θ^*) for $j=1, 2$, or 3, so $a \equiv b(\Theta)$ which implies that $g_i(a) \equiv g_i(b)$ for *all* $i=1, 2, 3$ and $c \equiv d$ (Θ). Thus, we have $u \equiv x_1 \equiv x_4 \equiv v(\Theta^*)$, that is, $u \equiv v(\Theta^*)$.

This proves that $\Phi_{g_i(A)} = \Theta^*_{g_i(A)}$ and, in particular, $\Phi_A = \Theta$. It remains to prove that Φ is a congruence relation. Φ is obviously an equivalence relation. The substitution property for all $f_\gamma \neq g_i$ follows from $\Phi_A = \Theta$ and for the g_i from the definiton of Θ^*. This completes the proof of Lemma 2.

Corollary 1. *Let Θ be an admissible congruence relation of $\mathfrak{A}$ and Φ the smallest extension of Θ to $\mathfrak{A}'$. Then $u \equiv v(\Phi)$ if and only if, for some i, one of the following holds:*

(i) *$u, v \in g_i(A)$ and $u \equiv v(\Theta^*)$.*

(ii) *$u \in g_i(A)$, $v \in g_{i+1}(A)$ and for $\{x\} = g_i(A) \cap g_{i+1}(A)$, we have $u \equiv x(\Theta^*)$ and $x \equiv v(\Theta^*)$ (and the symmetric case).*

(iii) *$u \in g_i(A)$ and $v \in g_{i+2}(A)$ and for $\{x\} = g_i(A) \cap g_{i+1}(A)$ and $\{y\} = g_{i+1}(A) \cap g_{i+2}(A)$ we have $u \equiv x \equiv y \equiv v(\Theta^*)$, or the same condition for $\{x'\} = g_i(A) \cap g_{i-1}(A)$ and $\{y'\} = g_{i-1}(A) \cap g_{i-2}(A)$.*

Proof. We already know the cases (i) and (iii). To prove case (ii), it is enough to observe that in this case there are only two nonredundant sequences, namely, the one given in (ii) and $u, g_i(A) \cap g_{i-1}(A), g_{i-1}(A) \cap g_{i-2}(A), g_{i-2}(A) \cap g_{i-3}(a), v$. In the latter case, we will have $g_j(a) \equiv g_j(b)(\Theta^*)$ for $j=1, 2$ or 3. Thus, $a \equiv b(\Theta)$ and all the extreme elements are congruent to one another and to u and v. In particular, $u \equiv x \equiv v(\Theta^*)$ for the x given in (ii).

If $u = x_0, \cdots, x_n = v$ and $x_{i-1} \equiv x_i(\Theta^*)$, then let us call this a *Θ-sequence connecting u and v.*

Corollary 2. *In cases* (i) *and* (ii), *the shortest Θ-sequence connecting u and v is unique; in case* (iii), *there are one or two shortest Θ-sequences.*

Lemma 3. *Let Θ be a congruence relation of $\mathfrak{A}$. Then there exists a smallest admissible congruence relation $\Theta^0 \geqq \Theta$.*

Proof. If $a \not\equiv b(\Theta)$, then $\Theta = \Theta^0$ and if $a \equiv b(\Theta)$, then $\Theta^0 = \Theta \vee \Theta(c, d)$.

If Θ is an admissible congruence relation of $\mathfrak{A}$, then $\overline{\Theta}$ will denote the smallest extension of Θ to $\mathfrak{A}$. Note that $\overline{\Theta}$ is described by Corollary 1 to Lemma 2.

Lemma 4. *Let $u, v \in A'$. Then there exists a smallest admissible congruence relation $\Phi(u, v)$ of $\mathfrak{A}$ such that $u \equiv v(\overline{\Phi}(u, v))$.*

Proof. We distinguish three cases as in Corollary 1 to Lemma 2. Let Θ be an admissible congruence relation such that $u \equiv v(\overline{\Theta})$.

(i) $u, v \in g_i(A)$, that is, $u = g_i(u')$ and $v = g_i(v')$, $u', v' \in A$. Then $u \equiv v(\overline{\Theta})$ if and only if $u \equiv v(\Theta^*)$, which is equivalent to $u' \equiv v'(\Theta)$, that is, $\Theta(u', v') \leqq \Theta$. This implies that in this case

$$\Phi(u, v) = (\Theta(u', v'))^0. \tag{1}$$

(ii) $u \in g_i(A), v \in g_{i+1}(A), u = g_i(u'), v = g_{i+1}(v')$. Let $\{x\} = g_i(A) \cap g_{i+1}(A)$ and $x = g_i(x') = g_{i+1}(x'')$. Then $u \equiv v(\overline{\Theta})$ if and only if $u \equiv x(\Theta^*)$ and $x \equiv v(\Theta^*)$, which implies that

$$\Phi(u, v) = (\Theta(u', x') \vee \Theta(x'', v'))^0. \tag{2}$$

(iii) $u \in g_i(A)$, $v \in g_{i+2}(A)$, $u = g_i(u')$, $v = g_{i+2}(v')$. We distinguish two subcases.

First, let $i = 0$ (the case $i = 2$ is similar). Then $u \equiv v(\overline{\Theta})$ if and only if $u \equiv c = g_1(a) \equiv g_1(b) = g_2(a) \equiv v(\Theta^*)$, or, $u \equiv d = g_3(b) \equiv g_3(a) = g_2(b) \equiv v(\Theta^*)$. Let

$$\Theta_1 = \Theta(u, c) \vee \Theta(a, b) \vee \Theta(a, v')$$

and

$$\Theta_2 = \Theta(u, d) \vee \Theta(a, b) \vee \Theta(b, v').$$

Then either $\Theta_1 \leqq \Theta$ or $\Theta_2 \leqq \Theta$. Thus, if we prove that $\Theta_1{}^0 = \Theta_2{}^0$, then $\Phi(u, v) = \Theta_1{}^0$ will be established. Observe that $a \equiv b(\Theta_1{}^0)$; thus, $c \equiv d(\Theta_1{}^0)$. Therefore, $d \equiv c \equiv u(\Theta_1{}^0)$; that is, $\Theta(u, d) \leqq \Theta_1{}^0$.

Since $\Theta(b, v') \leqq \Theta_1{}^0$, we have $\Theta_2 \leqq \Theta_1{}^0$. Similarly, $\Theta_1 \leqq \Theta_2{}^0$; thus, $\Theta_1{}^0 = \Theta_2{}^0$. Therefore, in this case,

$$\Phi(u, v) = (\Theta(u, c) \vee \Theta(a, b) \vee \Theta(a, v'))^0. \tag{3}$$

Second, let $i = 1$ (the case $i = 3$ is similar). Just as in the first subcase, we form the congruence relations $\Theta_1 = \Theta(u', b) \vee \Theta(a, b) \vee \Theta(a, v')$ and $\Theta_2 = \Theta(u', a) \vee \Theta(c, d) \vee \Theta(b, v')$ and again we have that $u \equiv v(\overline{\Theta})$ implies $\Theta_1 \leqq \Theta$ or $\Theta_2 \leqq \Theta$. We will establish $\Theta_2{}^0 \leqq \Theta_1{}^0$, which will prove $\Phi(u, v) = \Theta_2{}^0$.

Indeed, $u' \equiv b \equiv a(\Theta_1)$; thus, $\Theta(u', a) \leqq \Theta_1{}^0$. Since $\Theta_1{}^0$ is admissible and $a \equiv b(\Theta_1{}^0)$, we have $\Theta(c, d) \leqq \Theta_1{}^0$. Finally, $b \equiv a \equiv v'(\Theta_1{}^0)$; thus, $\Theta(b, v') \leqq \Theta_1{}^0$. Thus, $\Theta_2 \leqq \Theta_1{}^0$, which implies that $\Theta_2{}^0 \leqq \Theta_1{}^0$. Thus, in this case,

$$\Phi(u, v) = (\Theta(u', a) \vee \Theta(c, d) \vee \Theta(b, v'))^0. \tag{4}$$

This completes the proof of Lemma 4.

We will now generalize the results of Lemmas 2 through 4.

Consider a partial algebra $\mathfrak{S} = \langle S; F \rangle$, where

$$F = \{g_i{}^\lambda \mid \lambda \in \Lambda, i = 1, 2, 3\} \cup \{f_\sigma \mid \sigma \in \Omega\},$$

and $D(g_1{}^\lambda)=\{a^\lambda\}$, $D(g_3{}^\lambda)=\{b^\lambda\}$, $D(g_2{}^\lambda)=\varnothing$, $g_1{}^\lambda(a^\lambda)=c^\lambda$, $g_3{}^\lambda(b^\lambda)=d^\lambda$ and $D(f_\sigma)=S$. In other words, every partial operation is either a member of a pathological triplet, g_1, g_2, g_3 discussed above, or it is a unary operation. We call the congruence relation Θ of $\mathfrak{S}$ *admissible* if for any $\lambda\in\Lambda$, either $a^\lambda\not\equiv b^\lambda(\Theta)$ or $a^\lambda\equiv b^\lambda(\Theta)$ and $c^\lambda\equiv d^\lambda(\Theta)$. We assume that $a^\lambda\neq b^\lambda$ for $\lambda\in\Lambda$.

Lemma 3′. *Let Θ be a congruence relation of $\langle S;F\rangle$. Then there exists a smallest admissible congruence relation $\Theta^0\geqq\Theta$.*

Proof. Define $\Theta_0=\Theta$, $\Theta_{i+1}=\Theta_i\vee\bigvee(\Theta(c^\lambda,d^\lambda)\mid\lambda\in\Lambda$ and $a^\lambda\equiv b^\lambda(\Theta_i))$. It is routine to check that $\Theta^0=\bigvee(\Theta_i\mid i<\omega)$.

Let $\mathfrak{S}^\lambda$ be the partial algebra which is constructed from $\mathfrak{S}$ using $g_1{}^\lambda$, $g_2{}^\lambda$, and $g_3{}^\lambda$ the same way as $\mathfrak{A}'$ was constructed from $\mathfrak{A}$ using g_1, g_2, and g_3. Assume that all the $\mathfrak{S}^\lambda$ are constructed in such a way that $S^\lambda\cap S^\nu=S$ if $\lambda,\nu\in\Lambda$, $\lambda\neq\nu$.

Define $S'=\bigcup(S^\lambda\mid\lambda\in\Lambda)$. Defining the operations on S' in the natural way, we get the partial algebra $\mathfrak{S}'$.

Let Θ be a congruence relation of $\mathfrak{S}$. It is obvious that if Θ can be extended to $\mathfrak{S}'$, then Θ is admissible. If Θ is admissible, then it has a smallest extension Φ_λ to $\langle S^\lambda;F\rangle$ by Lemma 2. (Note that we used the obvious fact that if Θ is admissible in the new sense, then it is admissible for any fixed $\lambda\in\Lambda$ in the old sense.)

We define a relation Φ on S' as follows: let $u\equiv v(\Phi)$ mean $u\equiv v(\Phi_\lambda)$ if $u,v\in S^\lambda$; if $u\in S^\lambda$ and $v\in S^\nu$, $\lambda,\nu\in\Lambda$, $\lambda\neq\nu$, then let $u\equiv v(\Phi)$ mean that there exists an $x\in S$ such that $u\equiv x(\Phi_\lambda)$ and $x\equiv u(\Phi_\nu)$. Φ is well defined because if $u,v\in S^\lambda$ and $u,v\in S^{\lambda'}$ with $\lambda\neq\lambda'$, $\lambda,\lambda'\in\Lambda$, then $u,v\in S^\lambda\cap S^{\lambda'}=S$. Since $(\Phi_\lambda)_S=(\Phi_{\lambda'})_S=\Theta$, we get that $u\equiv v(\Phi)$ means $u\equiv v(\Theta)$, which does not depend on λ. Φ is obviously reflexive and symmetric, and the substitution property follows from the simple observation that for any $u,v\in S'$ and operation f, if $f(u)$ and $f(v)$ are defined, then there exists a λ such that $u,v,f(u),f(v)\in S^\lambda$. Φ is also transitive. Indeed, let $u\in S^{\lambda_1}$, $v\in S^{\lambda_2}$, $w\in S^{\lambda_3}$, and $u\equiv v(\Phi)$, $v\equiv w(\Phi)$.

First, let $\lambda_1\neq\lambda_2$. Then there exists an $x\in S$ such that $u\equiv x(\Phi_{\lambda_1})$ and $x\equiv v(\Phi_{\lambda_2})$. If $\lambda_2=\lambda_3$, then $u\equiv x(\Phi_{\lambda_1})$ and $x\equiv w(\Phi_{\lambda_2})$, establishing $u\equiv w(\Phi)$. If $\lambda_2\neq\lambda_3$, then there exists a $y\in S$ such that $v\equiv y(\Phi_{\lambda_2})$ and $y\equiv w(\Phi_{\lambda_3})$. This implies that $x\equiv v\equiv y(\Phi_{\lambda_2})$ and since $x,y\in S$, we have $x\equiv y(\Theta)$. Consequently, $x\equiv y(\Phi_{\lambda_3})$. Thus, $x\equiv y\equiv w(\Phi_{\lambda_3})$. We proved that $u\equiv x(\Phi_{\lambda_1})$ and $x\equiv w(\Phi_{\lambda_3})$; thus, $u\equiv w(\Phi)$. The case $\lambda_1=\lambda_2$ can be discussed as was the case $\lambda_2=\lambda_3$.

By definition, Φ is an extension of Θ. It is also obvious that again Φ is nothing more than the transitive extension of Θ^*. (Θ^* is defined for $\mathfrak{S}'$ the same way as it was for $\mathfrak{A}'$.)

Theorem 1. *A congruence relation Θ of $\mathfrak{S}$ can be extended to $\mathfrak{S}'$ if and only if Θ is admissible. If Θ is admissible, the smallest extension of Θ to $\mathfrak{S}'$ is the transitive extension of Θ^*. Let $u, v \in S'$. Then there exists a smallest admissible congruence relation $\Phi(u, v)$ such that $u \equiv v(\overline{\Phi}(u, v))$, where $\overline{\Phi}(u, v)$ denotes the minimal extension of $\Phi(u, v)$ to S'.*

Proof. We have proved all but the last statement of Theorem 1. It has also been established for $u, v \in S^\lambda$ for some $\lambda \in \Lambda$.

To establish the last statement in the general case, it is useful to introduce the following terminology.

Let $u, v \in S'$ and let $\sigma: u = x_0, \cdots, x_n = v$ be a sequence of elements having the property that, for each i, x_{i-1} and $x_i \in g_j{}^\lambda(S)$ for some $\lambda \in \Lambda$ and $j = 1, 2, 3$. Then $x_{i-1} = g_j{}^\lambda(x'_{i-1})$ and $x_i = g_j{}^\lambda(x_i{}^*)$, where x'_{i-1} and $x_i{}^*$ are uniquely determined elements of S. We form the congruence relation

$$(\bigvee (\Theta(x'_{i-1}, x_i{}^*) \mid i = 1, \cdots, n))^0$$

and we call this congruence relation Θ^σ, the *congruence relation associated with the sequence* σ. We will again call σ a Θ*-sequence* if $x_i \equiv x_{i+1}(\Theta^*)$ and σ is nonredundant. a is an *extreme element* of S' if it is an extreme element of some S^λ. It is obvious that all members of a Θ-sequence, except the first and last one, must be extreme elements; any two consecutive members are in some $g_i{}^\lambda(S)$; and excepting the first and last elements there are at most two consecutive extreme elements of S^λ in it; if any sequence σ has these properties, we will call it a *path*.

If Θ is an admissible congruence relation of $\mathfrak{S}$ and $\sigma: u = x_0, \cdots, x_n = v$ is a Θ-sequence connecting u and v, then $\Theta^\sigma \leqq \Theta$. Hence, to prove the existence of the smallest admissible Θ such that $u \equiv v(\overline{\Theta})$, we have to find all paths $\sigma_1, \cdots$ between u and v and we have to prove that there is a smallest congruence relation of the form Θ^{σ_i}.

Let T^λ denote the set $g_1{}^\lambda(S) \cup g_2{}^\lambda(S) \cup g_3{}^\lambda(S)$.

Now let $u \in S^\lambda$, $v \in S^\nu$, $u, v \notin S$, $\lambda \neq \nu$, and take a path σ connecting u and v. The sequence σ breaks up into three parts, σ_1 in T^λ, σ_2 in S, and σ_3 in T^ν; let $\sigma_1: u = x_0, \cdots, u_0$; $\sigma_2: u_0, v_0$; $\sigma_3: v_0, \cdots, x_n = v$. Then u_0 is c^λ or d^λ and v_0 is c^ν or d^ν. If c^λ or d^λ is not u_0, then denote it by u_1 and similarly for v_1. Further, let $\sigma_1{}'$ denote the path between u and u_1 which does not contain u_0.

(‡) If σ_1 contains two extreme elements, then for the sequence σ' which consists of $\sigma_1{}'$; u_1, v_0; and σ_3 we have $\Theta^{\sigma'} \leqq \Theta^\sigma$.

Indeed, by assumption, $g_i{}^\lambda(a)$ and $g_i{}^\lambda(b)$ are in σ_1; thus, $a^\lambda \equiv b^\lambda(\Theta^\sigma)$. Hence $c^\lambda \equiv d^\lambda(\Theta^\sigma)$, that is, $u_1 \equiv u_0(\Theta^\sigma)$ and $\Theta^{\sigma'_1} \leqq \Theta^\sigma$. This, of course, implies that $\Theta^{\sigma'} \leqq \Theta^\sigma$.

Therefore, we can find a σ connecting u and v such that $\Theta^\sigma \leqq \Theta^{\sigma'}$ for any path σ' connecting u and v, in the following way: if $u \in g_1{}^\lambda(S)$, then choose $u_0 = c^\lambda$; if $u \in g_3{}^\lambda(S)$, choose $u_0 = d^\lambda$; otherwise, let $u_0 = c^\lambda$ or d^λ. We choose v_0 similarly. Then let σ_1(resp. σ_3) be the path connecting u and u_0 (resp. v and v_0) and let σ equal σ_1; u_0, v_0; σ_3. This completes the proof of Theorem 1.

In the next step we want to extend the result of Theorem 1 to the algebra $\mathfrak{B}$ which we get from $\mathfrak{S}'$ by Theorem 14.2.

Lemma 5. *Every element $b \in B$, $b \notin S'$, has a representation of the form*

$$(**) \qquad b = h_1 \cdots h_n g_i{}^\lambda(a),$$

where $n \geqq 1$, $h_1, \cdots, h_n \in F$, and $a \in S$. If $a \neq a^\lambda$, $a \neq b^\lambda$, and $a \neq c^\nu$, $a \neq d^\nu$, for all $\nu \in \Lambda$, then the representation (**) is unique. In general, if $b = h_1' \cdots h_m' g_j{}^\nu(a')$ is another representation of b, then for some p with $0 \leqq p \leqq n$, $0 \leqq p \leqq m$ we have $h_t = h_t'$ for $t \leqq p$ and $h_{p+1} \cdots h_n g_i{}^\lambda(a) = h'_{p+1} \cdots h_m' g_j{}^\nu(a') \in S'$.

Proof. Trivial from Lemma 1 and the construction of S'.

Let $T_i{}^\lambda(h_1, \cdots, h_n)$ denote the set of all elements of the form

$$h_1 \cdots h_n g_i{}^\lambda(a)$$

for $a \in S$ and

$$T^\lambda(h_1, \cdots, h_n) = T_1{}^\lambda(h_1, \cdots, h_n) \cup T_2{}^\lambda(h_1, \cdots, h_n) \cup T_3{}^\lambda(h_1, \cdots, h_n).$$

In case $n = 0$, $T_i{}^\lambda$ will stand for $g_i{}^\lambda(S)$.

Corollary 1. *$T_i{}^\lambda(h_1, \cdots, h_n)$ and $T_{i+1}^\lambda(h_1, \cdots, h_n)$, $i = 1, 2$, have exactly one element in common, namely for $i = 1$, $h_1 \cdots h_n g_1{}^\lambda(b^\lambda) = h_1 \cdots h_n g_2{}^\lambda(a^\lambda)$, for $i = 2$, $h_1 \cdots h_n g_2{}^\lambda(b^\lambda) = h_1 \cdots h_n g_3{}^\lambda(a^\lambda)$.*

Corollary 2. *Let $b \in T_i{}^\lambda(h_1, \cdots, h_n)$; then b has one and only one representation of the form*

$$b = h_1 \cdots h_n g_i{}^\lambda(a), \qquad a \in S.$$

In other words, if we already know that $b \in T_i{}^\lambda(h_1, \cdots, h_n)$, then with fixed $h_1, \cdots, h_n$, λ and i in (**), a is uniquely determined.

Let us introduce the following notation:

$$S_0 = S, \cdots, S_n = \{h(x) \mid x \in S_{n-1}, h \in F^*\}.$$

Then $S_0 \subseteq S_1 \subseteq \cdots \subseteq S_n \subseteq \cdots$ and

$$\bigcup (S_i \mid i = 1, 2, \cdots) = B.$$

Corollary 3. *$T^\lambda(h_1, \cdots, h_n)$ and S_n have one or two elements in common, namely, $h_1 \cdots h_n(c^\lambda) = h_1 \cdots h_n g_1{}^\lambda(a^\lambda)$ and $h_1 \cdots h_n(d^\lambda) = h_1 \cdots h_n g_3{}^\lambda(b^\lambda)$.*

Lemma 6. *The following equality holds:*

$$S_n = S_{n-1} \cup \bigcup (T^\lambda(h_1, \cdots, h_{n-1}) \mid \lambda \in \Lambda, h_1, \cdots, h_{n-1} \in F).$$

Proof. Observe that $S_1 = S \cup \bigcup (T^\lambda \mid \lambda \in \Lambda)$. Hence,

$$\begin{aligned} S_2 &= S_1 \cup \{h(x) \mid x \in S_1, h \in F\} \\ &= S_1 \cup \bigcup (\{h(x) \mid x \in T^\lambda\} \mid \lambda \in \Lambda, h \in F) \\ &= S_1 \cup \bigcup (h(T^\lambda) \mid \lambda \in \Lambda, h \in F) \\ &= S_1 \cup \bigcup (T^\lambda(h) \mid \lambda \in \Lambda, h \in F). \end{aligned}$$

This proves the statement for $n=2$. The proof of the general case is similar.

Next we define the relation Θ^* on B. Let Θ be an admissible congruence relation of $\mathfrak{S}$; let $u \equiv v(\Theta^*)$ if $u, v \in S$ and $u \equiv v(\Theta)$, or $u, v \in T_i^\lambda(h_1, \cdots, h_n)$ and $u' \equiv v'(\Theta)$, where u', v' are given by $u = h_1 \cdots h_n g_i^\lambda(u')$ and $v = h_1 \cdots h_n g_i^\lambda(v')$.

Then Θ^* is well defined; indeed, u and v uniquely determine u' and v' if $h_1, \cdots, h_n$, λ and i are fixed (Corollary 2 to Lemma 5). Furthermore, if $u, v \in T_i^\lambda(h_1, \cdots, h_n)$ and also $u, v \in T_j^\nu(g_1, \cdots, g_m)$, with $\lambda \neq \nu$ or $i \neq j$, then $u = v$, since if $u \neq v$, then one of the representations $u = h_1 \cdots h_n g_i^\lambda(u')$ or $v = h_1 \cdots h_n g_i^\lambda(v')$ is reduced.

Lemma 7. *Θ^* is reflexive and symmetric. It is transitive on S and on each $T_i^\lambda(h_1, \cdots, h_n)$. Finally, if $u \equiv v(\Theta^*)$, then $h(u) \equiv h(v)(\Theta^*)$ for any $h \in F$.*

Proof. All the statements are trivial since if $u \neq v$, $u, v \in T_i^\lambda(h_1, \cdots, h_n)$, then u and v uniquely determine n, $h_1, \cdots, h_n$, λ and i, and keeping these fixed u' and v' are unique.

Let Φ_n denote the transitive extension of Θ^* in S_n.

Lemma 8. *Φ_n is a congruence relation of $\mathfrak{S}_n = \langle S_n; F \rangle$. Furthermore, if $\overline{\Phi}_{n-1}$ denotes the minimal extension of Φ_{n-1} to $\mathfrak{S}_n$, then $\overline{\Phi}_{n-1} = \Phi_n$.*

Proof. The first statement of this lemma follows from the second statement since we know that Φ_0 is a congruence relation of $\mathfrak{S} = \mathfrak{S}_0$; thus, by the second statement, $\Phi_1 = \overline{\Phi}_0$ is a congruence relation of $\mathfrak{S}_1$, and so on.

We prove the second statement by induction on n.

$\overline{\Phi}_0 = \Phi_1$ was proved in Theorem 1.

Assume that $\overline{\Phi}_{k-1} = \Phi_k$ has already been proved for $k < n$. This implies that Φ_{n-1} is a congruence relation of $\mathfrak{S}_{n-1}$. It follows from Lemma 7 that

$\overline{\Phi}_{n-1} \geqq \Phi_n$. Finally, we prove that $\overline{\Phi}_{n-1} \leqq \Phi_n$. Let $u, v \in S_{n-1}[h] = S_{n-1} \cup \{h(x) \mid x \in S_{n-1}\}$. This notation is justified, since $S_{n-1} \cup \{h(x) \mid x \in S_{n-1}\}$ satisfies the requirements of Definition 14.2 by Lemma 5. Let Ψ_h denote the minimal extension of Φ_{n-1} to $S_{n-1}[h]$. We will prove that $u \equiv v(\Psi_h)$ implies $u \equiv v(\Phi_n)$. Lemma 8 follows from this since $\overline{\Phi}_{n-1}$ can be described in terms of Ψ_h in just the same way as Φ was described in terms of Φ_λ on page 104, and this description implies $\overline{\Phi}_{n-1} \leqq \Phi_n$.

So, let $u \equiv v(\Psi_h)$. Then by Lemma 15.3, we have to distinguish three cases:

(1) $u, v \in S_{n-1}$. Then $u \equiv v(\Phi_{n-1})$; thus, $u \equiv v(\Phi_n)$.

(2) $u \in S_{n-1}$, $v \notin S_{n-1}$. Then $v = h(v_1)$, and there exists a $w = h(w_1) \in S_{n-1}$ such that $u \equiv w(\Phi_{n-1})$ and $w_1 \equiv v_1(\Phi_{n-1})$. Thus, there exist sequences $u = x_0, \cdots, x_n = w$ and $w_1 = y_0, \cdots, y_m = v_1$ such that $x_{i-1} \equiv x_i(\Theta^*)$ and $y_{j-1} \equiv y_j(\Theta^*)$. By Lemma 7, $h(y_{j-1}) \equiv h(y_j)(\Theta^*)$; thus, the sequence $u = x_0, \cdots, x_n = w = h(w_1)$, $h(y_1), \cdots, h(y_m) = h(v_1) = v$ will establish that $u \equiv v(\Phi_n)$.

(3) $u, v \notin S_{n-1}$. Using the condition in Lemma 15.3 and Lemma 7, we get $u \equiv v(\Phi_n)$ in a manner similar to case (2). This completes the proof of Lemma 8.

Theorem 2. *Let $u, v \in B$. Then there exists a smallest admissible congruence relation Θ of $\mathfrak{S}$ such that $u \equiv v(\overline{\Theta})$, where $\overline{\Theta}$ denotes the smallest extension of Θ to B.*

Proof. We will use the following notation. If Θ is an admissible congruence relation of $\mathfrak{S}$, then Θ^n will denote the transitive extension of Θ^* in S_n. By Lemma 8, if $u, v \in S_n$, then $u \equiv v(\overline{\Theta})$ if and only if $u \equiv v(\Theta^n)$. Since for any $u, v \in B$ we have $u, v \in S_n$ for some n, Theorem 2 is equivalent to the following statement.

If $u, v \in S_n$, then there exists a smallest admissible congruence relation Θ such that $u \equiv v(\Theta^n)$.

We will prove this statement by induction on n. If $n = 1$, then this is simply Theorem 1. Assume that the statement has been proved for $n-1$.

If $u \equiv v(\Theta^n)$, then there exists a sequence $\sigma : u = x_0, \cdots, x_m = v$ such that $x_{i-1} \equiv x_i(\Theta^*)$. By Corollary 2 to Lemma 5 and the definition of Θ^*, we can find elements x'_{i-1} and x_i^* of S such that $x'_{i-1} \equiv x_i^*(\Theta)$ if and only if $x_{i-1} \equiv x_i(\Theta^*)$.

Thus, we can associate again with σ an admissible congruence relation Θ^σ and then necessarily $\Theta^\sigma \leqq \Theta$. Hence, again, we have only to find all paths $\sigma_1, \cdots$ connecting u and v and we have to prove that there exists a smallest congruence relation of the form Θ^{σ_i}.

Let $u \in T^\lambda(h_1, \cdots, h_{n-1})$; if $v \in T^\lambda(h_1, \cdots, h_{n-1})$, then we find Θ as in

Lemma 4. If $v \notin T^{\lambda}(h_1, \cdots, h_{n-1})$, then any path $u = x_0, \cdots, x_m = v$ breaks up into two parts $\sigma_1 : u = x_0, \cdots, u_0$ and $\sigma_2 : u_0, \cdots, x_m = v$, where

$$u_0 \in T^{\lambda}(h_1, \cdots, h_{n-1}) \cap S_{n-1},$$

that is, $u_0 = h_1 \cdots h_{n-1}(c^{\lambda})$ or $h_1 \cdots h_{n-1}(d^{\lambda})$. Hence, the principle (‡) of Theorem 1 applies in this case as well, that is, if the sequence σ_1 contains two extreme elements, then we take σ_1', the other nonredundant sequence between u and u_0, and the sequence σ', consisting of σ_1' and σ_2, will have the property that $\Theta^{\sigma'} \leqq \Theta^{\sigma}$. Thus, we can find the σ_i for which Θ^{σ_i} is minimal in the following manner. Let u_0 be that one of $h_1 \cdots h_{n-1}(c^{\lambda})$ and $h_1 \cdots h_{n-1}(d^{\lambda})$ for which $\sigma_1 : u, u_0$ is a sequence connecting u and u_0; if neither of them has this property, then u_0 is either of them. In this case, let σ_1 be the shortest path connecting u and u_0. If $v \in S_{n-1}$, we choose $v = v_0$. If $v \in T^{\nu}(k_1, \cdots, k_{n-1})$, $\nu \neq \lambda$, $v \notin S_{n-1}$, then we choose v_0 in the same manner as we have chosen u_0, and we define σ_3 the same way we defined σ_1. Since u_0 and v_0 are in S_{n-1}, there exists a smallest congruence relation Θ_1 such that $u_0 \equiv v_0(\Theta_1^{n-1})$. Let σ_2 be a nonredundant Θ_1-sequence which connects u_0 and v_0. Then the sequence σ which consists of σ_1, σ_2, and σ_3 will be the required sequence.

§18. THE CHARACTERIZATION THEOREM OF CONGRUENCE LATTICES

Theorem 1. *Let $\mathfrak{L}$ be an algebraic lattice. Then there exists a partial algebra $\mathfrak{B} = \langle B; F \rangle$ with the following properties:*

(i) *The congruence lattice of $\mathfrak{B}$ is isomorphic to $\mathfrak{L}$.*

(ii) *Every $f \in F$ is unary and f is either an operation or $D(f)$ consists of two elements.*

(iii) *B consists of all finite subsets of K containing 0, where K is the set of all compact elements of $\mathfrak{L}$.*

(iv) *Θ is a compact congruence relation of $\mathfrak{B}$ if and only if $\Theta = \Theta(a^*, \{0\})$, where $a^* = \{a, 0\}$, $a \in K$; the representation of Θ in this form is unique.*

Note that this result is a sharpening of Theorem 16.1. The proof is also quite similar.

Proof. Let K be given as in (iii). For $a \in K$, let us put $a^* = \{a, 0\}$; in particular, $0^* = \{0\}$.

We define B as the set of all finite subsets of K containing 0. Then $\langle B; \cup, \cap \rangle$ is a distributive lattice with 0^* as the zero element. It is also *relatively complemented*, which means that if $x \geqq y \geqq z$, then there exists a y_1 such that $y \cup y_1 = x$ and $y \cap y_1 = z$. This implies that there is a 1–1 correspondence between congruence relations and ideals; we obtain this correspondence by letting the congruence relation Θ correspond to the ideal $I_\Theta = \{x \mid x \equiv 0^*(\Theta)\}$. If I is an ideal, then $\Theta(I)$ will denote the congruence relation which corresponds to I. Let us define F to consist of the following operations and partial operations: for every $x \in B$, we define k_x and l_x by $k_x(y) = x \cup y$ and $l_x(y) = x \cap y$; for $a, b \in K$, $a \neq b$, $a \neq 0$, $b \neq 0$, we define g_{ab} by $D(g_{ab}) = \{\{a,b,0\}, 0^*\}$ and $g_{ab}(\{a,b,0\}) = (a \vee b)^*$, $g_{ab}(0^*) = 0^*$. Finally, for $a, b \in K$, $0 \neq b \leqq a$, we define h_{ab} by $D(h_{ab}) = \{a^*, 0^*\}$ and $h_{ab}(a^*) = b^*$, $h_{ab}(0^*) = 0^*$.

Let F denote the collection of all partial operations defined so far; let F_0 denote the collection of all operations k_x and l_x, and set $\mathfrak{B} = \langle B; F \rangle$.

A binary relation Θ is a congruence relation of $\langle B; \cup, \cap \rangle$ if and only if it is a congruence relation of $\langle B; F_0 \rangle$ (cf. Exercise 1.50). Thus, every congruence relation of $\mathfrak{B}$ is also a congruence relation of $\langle B; \cup, \cap \rangle$.

Let I be an ideal of $\langle K; \vee \rangle$ and let $\hat{I}$ denote the family of all finite subsets of I containing 0. Then $\hat{I}$ is an ideal of $\langle B; \cup, \cap \rangle$. Thus, $\hat{I}$ determines a congruence relation $\Theta(\hat{I})$. We claim that the mapping $I \to \Theta(\hat{I})$ is an isomorphism between the lattice of all ideals of $\langle K; \vee \rangle$ and the congruence lattice of $\langle B; F \rangle$. The details of the proof of this step are the same as those of Theorem 16.1, and so they can be omitted.

Now all the statements of Theorem 1 are clear; (iv) means that the compact elements correspond to the principal ideals.

In this section, let us call a partial algebra *regular* if it is of the type described on pages 103 and 104.

Lemma 1. *Let* $\langle B; F' \rangle$ *be a partial algebra satisfying* (ii) *of Theorem* 1. *Then there exists a regular partial algebra* $\langle B; F_1 \rangle$ *such that* Θ *is a congruence relation of* $\langle B; F' \rangle$ *if and only if* Θ *is an admissible congruence relation of* $\langle B; F_1 \rangle$.

Proof. Trivial. All we have to do is to replace every $f \in F'$ for which $D(f)$ consists of two elements a, b by three partial operations f_1, f_2, f_3 in the obvious manner.

Theorem 2. *Let* $\mathfrak{A} = \langle A; F \rangle$ *be a regular partial algebra having the property that if* Θ *is a compact congruence relation of* $\mathfrak{A}$, *then* Θ^0 *(the smallest admissible congruence relation containing* Θ*) is of the form* $(\Theta(a, b))^0$ *for some* $a, b \in A$. *Then there exists another regular partial algebra* $\mathfrak{A}_1 = \langle A_1; F_1 \rangle$ *such that the following conditions hold:*

(i) $A \subseteq A_1$, $F \subseteq F_1$ *and* $\langle A; F \rangle$ *is a relative subalgebra of* $\langle A_1; F \rangle$.

(ii) *Every* $f \in F$ *is fully defined on* A_1.

(iii) *Every admissible congruence relation* Θ *of* $\langle A; F \rangle$ *has one and only one extension* $\overline{\Theta}$ *to an admissible congruence relation of* $\langle A_1; F_1 \rangle$.

(iv) *Every admissible congruence relation* Φ *of* $\langle A_1; F_1\rangle$ *can be written in the form* $\Phi = \overline{\Theta}$ *for some admissible congruence relation* Θ *of* $\langle A; F\rangle$.

(v) *If* Θ *is a compact congruence relation of* $\langle A_1; F_1\rangle$, *then* Θ^0 *is of the form* $(\Theta(a, b))^0$ *for some* $a, b \in A_1$.

Proof. Let us construct the partial algebra $\langle A'; F\rangle$ as on page 104 and then let us consider the algebra $\langle A_1; F\rangle$ which we get from $\langle A'; F\rangle$ by Theorem 14.2. By Theorem 17.2, for $u, v \in A_1$, there exists a smallest admissible congruence relation Θ of $\langle A; F\rangle$ such that $u \equiv v(\overline{\Theta})$. This Θ was constructed as the least admissible congruence relation containing a compact congruence relation. Hence, by assumption

$$\Theta = (\Theta(a(u, v), b(u, v)))^0.$$

Of course, $a(u, v)$ and $b(u, v)$ are not necessarily unique but by the Axiom of Choice we can fix them.

For every $u, v \in A_1$, we define k_{uv} by $D(k_{uv}) = \{u, v\}$ and $k_{uv}(u) = a(u, v)$ $k_{uv}(v) = b(u, v)$. Let $F' = F \cup \{k_{uv} \mid u, v \in A_1\}$.

Then $\langle A_1; F'\rangle$ has the following properties:

(i′) $A \subseteq A_1$, $F \subseteq F'$, and $\langle A; F\rangle$ is a relative subalgebra of $\langle A_1; F\rangle$.

(ii′) Every $f \in F$ is fully defined on A_1.

(iii′) Every admissible congruence relation Θ of $\langle A; F\rangle$ has one and only one extension $\overline{\Theta}$ to a congruence relation of $\langle A_1; F'\rangle$.

(iv′) Every congruence relation Φ of $\langle A_1; F'\rangle$ can be written in the form $\Phi = \overline{\Theta}$ for some admissible congruence relation Θ of $\langle A; F\rangle$.

Of these, (i′) and (ii′) are trivial. To prove (iii′), first we note that by Theorem 17.1 and Theorem 15.1, every admissible congruence relation Θ of $\langle A; F\rangle$ can be extended to a congruence relation $\overline{\Theta}$ of $\langle A_1; F\rangle$. We claim that $\overline{\Theta}$ is a congruence relation of $\langle A_1; F'\rangle$, that is, the substitution property can be proved for the k_{uv}. In other words, $u \equiv v(\overline{\Theta})$ implies $a(u, v) \equiv b(u, v)(\overline{\Theta})$. Indeed, $u \equiv v(\overline{\Theta})$ implies that $\Theta \geqq \Phi(u, v) = (\Theta(a(u, v), b(u, v)))^0$, where $\Phi(u, v)$ denotes the smallest admissible congruence relation of $\langle A; F\rangle$ such that $u \equiv v(\overline{\Phi}(u, v))$. Hence, $a(u, v) \equiv b(u, v)(\Theta)$ and so $a(u, v) \equiv b(u, v)(\overline{\Theta})$.

To prove the uniqueness statement of (iii′), assume that Φ_1 and Φ_2 are both congruence relations of $\langle A_1; F'\rangle$ and that both are extensions of the admissible congruence relation Θ of $\langle A; F\rangle$. If $\Phi_1 \neq \Phi_2$, then there exist $u, v \in A_1$ such that $u \equiv v(\Phi_1)$ and $u \not\equiv v(\Phi_2)$ (or, symmetrically, $u \not\equiv v(\Phi_1)$ and $u \equiv v(\Phi_2)$). Since $u \equiv v(\Phi_1)$, we get $k_{uv}(u) \equiv k_{uv}(v)(\Phi_1)$; that is, $a(u, v) \equiv b(u, v)(\Phi_1)$. Thus, $a(u, v) \equiv b(u, v)(\Theta)$, that is, $\Theta \geqq \Phi(u, v)$. But we have that $\overline{\Theta} \leqq \Phi_2$, thus $u = v(\Phi_2)$, which is a contradiction.

(iv′) is trivial.

If we combine what we have proved so far with Lemma 1, we get the proof of (i)–(iv) of Theorem 2.

To prove (v), let Θ be a compact congruence relation of $\langle A_1; F' \rangle$, $\Theta = \bigvee (\Theta(u_i, v_i) \mid 0 \leqq i < n)$. Let Φ be a congruence relation of $\langle A; F \rangle$ defined by $\Phi = \bigvee (\Theta(a(u_i, v_i), b(u_i, v_i)) \mid 0 \leqq i < n)$. Then by assumption, $\Phi^0 = (\Theta(a, b))^0$, for some $a, b \in A$. Now it is easy to check that $\Theta = (\Theta(a, b))^0$ in $\langle A_1; F' \rangle$ implying (v).

Now we are ready to state and prove the characterization theorem for congruence lattices.

Theorem 3 (*G. Grätzer and E. T. Schmidt* [2]). *Let $\mathfrak{L}$ be an algebraic lattice. Then there exists an algebra $\mathfrak{A}$ whose congruence lattice is isomorphic to $\mathfrak{L}$.*

Proof. Consider the partial algebra $\langle B; F \rangle$ constructed in Theorem 1 and let $\langle B; F' \rangle = \langle A_0; F_0 \rangle$ denote the regular partial algebra that we get from $\langle B; F \rangle$ by applying Lemma 1. By (iv) of Theorem 1, $\langle A_0; F_0 \rangle$ satisfies the conditions of Theorem 2; hence, we can apply the construction of Theorem 2 and we get a regular partial algebra $\langle A_1; F_1 \rangle$. By (v) of Theorem 2, $\langle A_1 ; F_1 \rangle$ again satisfies the conditions of Theorem 2; hence, it can be applied again and we get the regular partial algebra $\langle A_2; F_2 \rangle$. Proceeding thus, we construct $\langle A_n; F_n \rangle$ for every nonnegative integer n. Set $A = \bigcup (A_n \mid n < \omega)$ and $\bar{F} = \bigcup (F_n \mid n < \omega)$. We claim that $\langle A; \bar{F} \rangle$ is an algebra and its congruence lattice is isomorphic to $\mathfrak{L}$.

First we note that

$$B = A_0 \subseteq A_1 \subseteq A_2 \subseteq \cdots \subseteq A_n \subseteq \cdots$$

and

$$F' = F_0 \subseteq F_1 \subseteq F_2 \subseteq \cdots \subseteq F_n \subseteq \cdots.$$

Let $f \in \bar{F}$ and let $a \in A$. Then $a \in A_n$ for some n and by (ii) of Theorem 2 we have that f is fully defined on A_n. Thus, $\langle A; \bar{F} \rangle$ is an algebra.

Finally, we observe that every admissible congruence relation of $\langle B; F' \rangle$ can be extended to a congruence relation of $\langle A; \bar{F} \rangle$ in one and only one way. Indeed, if Θ is an admissible congruence relation of $\langle B; F' \rangle$, then by Theorem 2 it has one and only one extension Θ_1 to $\langle A_1; F_1 \rangle$, and so on. Let us define the congruence relation Θ_n of $\langle A_n; F_n \rangle$ as the only extension of Θ_{n-1} to $\langle A_n; F_n \rangle$.

Set $\Theta_\omega = \bigcup (\Theta_n \mid n < \omega)$. It is obvious that Θ_ω is a congruence relation of $\langle A; \bar{F} \rangle$. The uniqueness is also obvious since if Θ has two extensions Φ_1, Φ_2 to $\langle A; \bar{F} \rangle$, then the restriction of Φ_1 and Φ_2 to some A_n would also be different, contradicting (iii) of Theorem 2.

Thus, the congruence lattice of $\langle A; \overline{F}\rangle$ is isomorphic to the lattice of admissible congruence relations of $\langle B; F'\rangle$, which in turn by Lemma 1 is isomorphic to the lattice of congruence relations of $\mathfrak{B}$, which by Theorem 1 is isomorphic to $\mathfrak{L}$, and this is what we were required to prove.

The method of the last section can be summarized as follows: we want to construct an algebra $\mathfrak{A}$ having property P; it is easier to construct a partial algebra $\mathfrak{B}$ having P; $\mathfrak{B}$ generates an algebra $\overline{\mathfrak{A}}$, however $\overline{\mathfrak{A}}$ does not have P; introducing additional partial operations on $\overline{\mathfrak{A}}$ we make it into a partial algebra which has P; and so on $\cdots$; finally a "direct limit" is formed.

This method has been successfully used by others. For instance, A. A. Iskander [1] used this method to prove that for any algebraic lattice $\mathfrak{L}$ there exists an algebra $\mathfrak{A}$ such that $\mathfrak{L} \cong \langle \mathscr{S}(\mathfrak{A}^2); \subseteq\rangle$. See also G. Grätzer and W. A. Lampe [1].

EXERCISES

1. Characterize all partial algebras in which every relative subalgebra is a subalgebra.
2. Characterize all partial algebras in which every weak subalgebra is a relative subalgebra.
3. Let $\mathfrak{A}$ and $\mathfrak{B}$ be partial algebras and φ a full homomorphism of $\mathfrak{A}$ into $\mathfrak{B}$. Prove that $\langle A\varphi; F\rangle$ is a subalgebra of $\mathfrak{B}$. Is the converse true?
4. Is it possible to distinguish within $\mathfrak{A}$ between congruence relations induced by homormorphisms and congruence relations induced by full homomorphisms?
5. Simplify Theorem 14.1 (that is, simplify the description of $\Theta_{\bar{a}}$) in case all partial operations are unary.
6. Let $\bar{a}$ be as in Theorem 14.1, and consider different representations of a polynomial symbol $\mathbf{p}$ in the form

   $$(*)\qquad \mathbf{p} = r(\mathbf{p}_0, \cdots, \mathbf{p}_{k-1}),$$

   where $\bar{a}$ can be substituted into $\mathbf{p}_i$. Is there a largest such representation (*) in the sense that if

   $$\mathbf{p} = r_1(\mathbf{p}_0', \cdots, \mathbf{p}_{n-1}')$$

   is another such representation, then the $\mathbf{p}_i$ are polynomials of $\mathbf{p}_0', \cdots, \mathbf{p}_{n-1}'$?
7. Prove that if $\langle n, \delta\rangle < \langle m, \lambda\rangle$, then in general

   $$A_{\langle n,\delta\rangle} \neq A_{\langle m,\lambda\rangle}.$$

8. Prove that for $\mathbf{p}_i \in \mathbf{P}^{(\alpha)}(\tau)$, $\{\mathbf{p}_0, \cdots, \mathbf{p}_{n-1}\}$ is $\mathscr{S}$-independent if and only if for $\mathbf{r}, \mathbf{s} \in \mathbf{P}^{(n)}(\tau)$, $r(\mathbf{p}_0, \cdots, \mathbf{p}_{n-1}) = s(\mathbf{p}_0, \cdots, \mathbf{p}_{n-1})$ implies $\mathbf{r} = \mathbf{s}$.

9. Prove that $\mathbf{p}_0, \cdots, \mathbf{p}_{n-1}$ is $\mathscr{S}$-independent if and only if $\langle[\mathbf{p}_0, \cdots, \mathbf{p}_{n-1}]; F\rangle$ is isomorphic to $\mathfrak{P}^{(n)}(\tau)$ and there is an isomorphism φ such that $\mathbf{p}_i\varphi = \mathbf{x}_i$.
10. Let $\mathfrak{B}$ be a subalgebra of $\mathfrak{P}^{(\alpha)}(\tau)$. Let us say that $\mathbf{p} \in B$ is irreducible in $\mathfrak{B}$ if $\mathbf{p} = r(\mathbf{p}_0, \cdots, \mathbf{p}_{n-1})$, $\mathbf{p}_0, \cdots, \mathbf{p}_{n-1} \in B$ implies $\mathbf{r} = \mathbf{x}_i$ and $\mathbf{p} = \mathbf{p}_i$. Prove that any sequence of irreducible polynomials is $\mathscr{S}$-independent.
11. Prove that every subalgebra of $\mathfrak{P}^{(\alpha)}(\tau)$ is isomorphic to some $\mathfrak{P}^{(\beta)}(\tau)$.
12. Let $\mathbf{p}$ be an n-ary polynomial symbol and let $\mathbf{q}_0, \cdots, \mathbf{q}_{n-1}$ be α-ary polynomial symbols. Let $\mathbf{p}(\mathbf{q}_0, \cdots, \mathbf{q}_{n-1})$ denote the α-ary polynomial symbol that we get from $\mathbf{p}$ by replacing every occurrence of $\mathbf{x}_i$ by $\mathbf{q}_i$. Prove that

$$\mathbf{p}(\mathbf{q}_0, \cdots, \mathbf{q}_{n-1}) = p(\mathbf{q}_0, \cdots, \mathbf{q}_{n-1}).$$

13. Prove Theorem 13.3 using only Lemmas 15.1 and 15.3.
14. Generalize Lemmas 7.3 and 7.4 for partial algebras.
15. Why does Lemma 8.4 fail for partial algebras?
16. Let Θ and Φ be congruence relations of the partial algebra $\mathfrak{A}$. Then $\Theta \vee \Phi$ is not necessarily a congruence relation of $\mathfrak{A}$ ($\vee$ is formed in $\mathfrak{E}(A)$).
17. Let $C_s(\mathfrak{A})$ denote the set of strong congruence relations of $\mathfrak{A}$. Show that $\mathfrak{C}_s(\mathfrak{A}) = \langle C_s(\mathfrak{A}); \leqq\rangle$ is a sublattice of $\mathfrak{E}(A)$.
18. Is $\mathfrak{C}(\mathfrak{A})$ a sublattice of $\mathfrak{E}(A)$? Is $\mathfrak{C}_s(\mathfrak{A})$ a sublattice of $\mathfrak{C}(\mathfrak{A})$?
19. Let $\mathfrak{A}$ and $\mathfrak{B}$ be partial algebras and φ a homomorphism of $\mathfrak{A}$ into $\mathfrak{B}$. When is it possible to find algebras $\mathfrak{A}_1$ and $\mathfrak{B}_1$ such that $\mathfrak{A}$ is a relative subalgebra of $\mathfrak{A}_1$, $\mathfrak{B}$ is a relative subalgebra of $\mathfrak{B}_1$, and there exists a homomorphism ψ of $\mathfrak{A}_1$ into $\mathfrak{B}_1$ with $\psi_A = \varphi$?
20. Can you generalize Ex. 1.50 to partial algebras?
21. Prove that the description of $\Theta(a, b)$ (Theorem 10.3) does not hold for partial algebras.
22. Does Lemma 10.4 hold for partial algebras? Does it hold for strong congruence relations?
23. Prove the homomorphism theorem for full homomorphisms.
24. Under what conditions can we prove the isomorphism theorem for partial algebras? Prove the necessity of the conditions.
25. Define the concept of derived partial algebra and prove Theorem 12.1 for partial algebras.
26. Characterize those subsets B of $P(A \times A)$ for which there exists a partial algebra $\mathfrak{A} = \langle A; F\rangle$ with $B = C(\mathfrak{A})$.
27. In Lemma 15.3, is it true that for given $u, v \in A$, there exists a smallest congruence relation Θ of $\mathfrak{B}$ such that $u \equiv v(\overline{\Theta})$?
28. Let $\mathfrak{L}$ be an algebraic lattice. Show that there exists a set A such that $\mathfrak{L}$ is isomorphic to some complete sublattice of $\mathfrak{E}(A)$.
29. (P. M. Whitman) Show that every lattice can be embedded into some $\mathfrak{E}(A)$.
30. For every algebra $\langle A; F\rangle$ there exists an algebra $\langle A_1; F_1\rangle$ such that $A \subseteq A_1$, $F \subseteq F_1$, $\langle A; F\rangle$ is a subalgebra of $\langle A_1; F\rangle$ and

    (i) every congruence Θ of $\langle A; F\rangle$ can be extended to a congruence relation $\overline{\Theta}$ of $\langle A_1; F_1\rangle$;

    (ii) $\Theta \rightarrow \overline{\Theta}$ is an isomorphism between $\mathfrak{C}(\langle A; F\rangle)$ and $\mathfrak{C}(\langle A_1; F_1\rangle)$;

(iii) every compact congruence relation of $\langle A_1; F_1\rangle$ is principal. (G. Grätzer and E. T. Schmidt [1] and [2].)

31. Show that the results of §17 cannot be extended to nonunary algebras (Theorem 17.2 fails to hold, in fact the extension Φ_2 of Φ_1 from $\mathfrak{S}_1$ to $\mathfrak{S}_2$ does not necessarily have the property stated in Theorem 17.2).†
32. Let $\mathfrak{G}$ be a group. Find a simple algebra $\mathfrak{A}$ such that the automorphism group of $\mathfrak{A}$ is isomorphic to $\mathfrak{G}$.
33. Let $\mathfrak{L}$ be an algebraic lattice. Find an algebra $\mathfrak{A}$ such that the congruence lattice of $\mathfrak{A}$ is isomorphic to $\mathfrak{L}$ and $\mathfrak{A}$ has *no* nontrivial automorphism (i.e., $\mathfrak{G}(\mathfrak{A}) = 1$).
34. $|C(\mathfrak{A})| = 1$ implies $|G(\mathfrak{A})| = 1$.
35. (W. A. Lampe) Let $\mathfrak{L}$ be an algebraic lattice in which there exists an element $a \neq 0$ such that $a \leqq \bigvee (x_i \mid i \in I)$ implies $a \leqq x_i$ for some $i \in I$. Then for any group $\mathfrak{G}$ there exists an algebra $\mathfrak{A}$ such that $\mathfrak{C}(\mathfrak{A}) \cong \mathfrak{L}$ and $\mathfrak{G}(\mathfrak{A}) \cong \mathfrak{G}$.
36. Let $\mathfrak{A}$ be an algebra of type τ generated by $H = \{h_\gamma \mid \gamma < \alpha\}$. There is an isomorphism φ between $\mathfrak{A}$ and $\mathfrak{P}^{(\alpha)}(\tau)$ such that $h_\gamma\varphi = \mathbf{x}_\gamma$ for $\gamma < \alpha$ if and only if one of the following conditions holds:

   (i) for $\mathbf{p}, \mathbf{q} \in \mathbf{P}^{(n)}(\tau)$, $n < \alpha$, $p(h_{\gamma_0}, \cdots, h_{\gamma_{n-1}}) = q(h_{\gamma_0}, \cdots, h_{\gamma_{n-1}})$ and $\gamma_i \neq \gamma_j$ for $i \neq j$ imply $\mathbf{p} = \mathbf{q}$;

   (ii) if $\mathfrak{B}$ is an algebra of type τ, $b_\gamma \in B$ for $\gamma < \alpha$, then there is a homomorphism ψ of $\mathfrak{A}$ into $\mathfrak{B}$ with $h_\gamma\psi = b_\gamma$, for $\gamma < \alpha$;

   (iii) there exists a homomorphism ψ from $\mathfrak{A}$ into $\mathfrak{P}^{(\alpha)}(\tau)$ with $h_\gamma\psi = \mathbf{x}_\gamma$ for $\gamma < \alpha$.
37. Let $\mathfrak{A}$ and $\mathfrak{B}$ be algebras of type τ. Prove that $\mathfrak{A}$ and $\mathfrak{B}$ have up to isomorphism a common extension if and only if either there are no nullary operations or there are nullary operations and $\langle[\varnothing]_{\mathfrak{A}}; F\rangle \cong \langle[\varnothing]_{\mathfrak{B}}; F\rangle$.
38. (K. H. Diener‡) The following condition can be added to Ex. 36:

   (iv) (ii) holds for every extension $\mathfrak{B}$ of $\mathfrak{A}$ and if there are nullary operations, $\langle[\varnothing]_{\mathfrak{A}}; F\rangle \cong \mathfrak{P}^{(0)}(\tau)$.
39. Generalize Ex. 37 for partial algebras.
40. Generalize Ex. 37 to any set of algebras (partial algebras).
41. Let $H, K \subseteq \{\mathbf{x}_\gamma \mid \gamma < \alpha\} \subseteq \mathbf{P}^{(\alpha)}(\tau)$. Prove that $H \cap K = \varnothing$ implies that $[H] \cap [K] = \varnothing$ if there are no nullary operations and $[H] \cap [K] = \mathbf{P}^{(0)}(\tau)$ otherwise.
42. Let $\mathfrak{A}$ be an infinitary partial algebra. Then there exists an infinitary algebra $\mathfrak{B}$ which contains $\mathfrak{A}$ as a relative subalgebra and has the property that every congruence relation of $\mathfrak{A}$ can be extended to $\mathfrak{B}$. (Generalize Lemmas 15.1–15.3.)
43. (W. A. Lampe) Let $\mathfrak{A}$ be an algebra, $\varphi \in E(\mathfrak{A})$, ρ_φ the right multiplication by φ on $E(\mathfrak{A})$, ε_φ and $\varepsilon_{\rho_\varphi}$ the equivalence relations induced by φ and ρ_φ on A and $E(\mathfrak{A})$ respectively. Then $\varepsilon_\varphi \to \varepsilon_{\rho_\varphi}$ is an order preserving map.

† This shows that the proofs of the Theorem of E. T. Schmidt [2], and Theorem 6 of G. Grätzer [8] are incorrect.

‡ See K. H. Diener and G. Grätzer [1].

44. (W. A. Lampe) If φ is a right-zero in $\mathfrak{E}(\mathfrak{A})$, then $\varepsilon_\varphi \geqq \varepsilon_\psi$ for all $\psi \in E(\mathfrak{A})$.
45. Let $\mathfrak{m}$ and $\mathfrak{n}$ be regular cardinals and $\mathfrak{m} < \mathfrak{n}$. Prove that every $\mathfrak{m}$-algebraic lattice is also $\mathfrak{n}$-algebraic and find an $\mathfrak{n}$-algebraic lattice which is not $\mathfrak{m}$-algebraic. (See Ex. 0.82.)
46. Describe those partial algebras $\mathfrak{A}$ in which all congruence relations are strong. ($D(f_\gamma, \mathfrak{A}) = \varnothing$ or A^{n_γ}.)
47. State and prove Theorem 11.4 for infinitary algebras.
48. An *endomorphism* of a relational system $\langle A; R\rangle$ is a mapping φ of A into A such that $r(a_0, a_1, \cdots)$ implies $r(a_0\varphi, a_1\varphi, \cdots)$ for all $r \in R$. For a unary algebra $\mathfrak{A}$ find a relational system $\langle A; R\rangle$ such that each $r \in R$ is binary and $\varphi: A \to A$ is an endomorphism of $\mathfrak{A}$ if and only if φ is an endomorphism of $\langle A; R\rangle$.
49. For every set A there exists a binary relation r such that the identity map is the only endomorphism of $\langle A; r\rangle$. (P. Vopěnka, A. Pultr and Z. Hedrlin, Comment. Math. Univ. Carolinae 6 (1965), 149–155). (Hint: for $|A| \geqq \aleph_0$ assume $A = \{\gamma \mid \gamma \leqq \delta + 1\}$ where δ is an initial ordinal. Define r by the following rules: (i) $0r2$; (ii) $\alpha r(\alpha+1)$, $\alpha \leqq \delta$; (iii) if β is a limit ordinal not cofinal with ω, then $\alpha r\beta$ if and only if α is a limit ordinal and $\alpha < \beta$; (iv) if α is a limit ordinal cofinal with ω, then $\alpha = \lim \alpha_n$, $\alpha_1 < \alpha_2 < \cdots$, and $\alpha_n = \bar{\alpha}_n + n$, where $\bar{\alpha}_n$ is a limit ordinal; set $\gamma r\alpha$ if and only if $\gamma = \alpha_n$ for some $n \geqq 2$; (v) $\alpha r(\delta+1)$ if and only if $\alpha = \delta$ or α is a nonlimit ordinal $\neq \delta + 1$.)
50. Let $\langle A; R\rangle$ be a relational system with all $r \in R$ binary. Find a binary relational system $\langle B; r\rangle$ whose endomorphism semigroup is isomorphic to the endomorphism semigroup of $\langle A; R\rangle$. (A. Pultr, Comment. Math. Univ. Carolinae 5 (1964), 227–239.) (Hint: Let $R = \{r_i \mid i \in I\}$. Set $B = A \cup \bigcup (r_i \times \{i\} \mid i \in I) \cup I \cup \{v_1, v_2, v_3, u_1, u_2\}$. Define r as follows: (i) r on I as in Ex. 49; (ii) $x_0 r\langle x_0, x_1, i\rangle r x_1$; (iii) $\langle x_0, x_1, i\rangle ri$ for $i \in I$; (iv) $v_1 r v_2 r v_3 r v_1$; (v) for $i \in I$, iru_2; (vi) $u_1 r u_2$ and $u_j r v_1$, $j = 1, 2$; (vii) for $x \in A$, xru_1.)
51. (Z. Hedrlin and A. Pultr [1]) In Theorem 12.3, $\mathfrak{A}$ can be chosen of type $\langle 1, 1\rangle$. (Hint: combine Theorem 12.3 with Ex. 48–50. In constructing $\mathfrak{A}$ from $\langle B; r\rangle$, the two unary operations should act as projection maps for r.)

PROBLEMS

11. Let $B \subseteq P(A \times A)$. When is it possible to find a partial algebra $\langle A; F\rangle$ with $B = C_s(\langle A; F\rangle)$? (See Ex. 17 and 18.)
12. Let $\mathfrak{L}_1$ and $\mathfrak{L}_2$ be lattices. Under what conditions does there exist a partial algebra $\mathfrak{A}$ with $\mathfrak{C}(\mathfrak{A}) \cong \mathfrak{L}_1$ and $\mathfrak{C}_s(\mathfrak{A}) \cong \mathfrak{L}_2$? (See Ex. 17 and 18.)
13. Relate the following four classes of lattices:
    L_0: the class of finite lattices; L_1: the class of lattices isomorphic to sublattices of finite partition lattices (i.e., lattices which are isomorphic to a sublattice of some $\langle \text{Part}(A); \leqq\rangle$ for some finite set A); L_2: the class of lattices isomorphic to strong congruence lattices of finite partial algebras; L_3: the class of lattices isomorphic to congruence lattices of finite algebras.

14. Does Theorem 14.1 hold for infinitary partial algebras?
15. Let $B \subseteq P(A \times A)$. When is it possible to find an infinitary algebra $\mathfrak{A} = \langle A; F \rangle$ with $C(\mathfrak{A}) = B$? Characterize $\mathfrak{C}(\mathfrak{A})$.
16. Characterize $\langle \mathfrak{E}(\mathfrak{A}), \mathfrak{E}_F(\mathfrak{A}), \mathfrak{E}_S(\mathfrak{A}) \rangle$ as a triplet of semigroups. (See Lemma 16.1.)
17. Characterize the congruence lattices of algebras of finite type.
18. For an integer $n > 2$ characterize the algebraic lattices $\mathfrak{L}$ which can be represented as $\langle \mathscr{S}(\mathfrak{A}^n); \subseteq \rangle$ for some algebra $\mathfrak{A}$. (See the result mentioned on p. 113).
19. For a nonvoid set A, and integer $n > 1$, characterize those subsets $B \subseteq A^n$ for which $B = \mathscr{S}(\mathfrak{A}^n)$, for some algebra $\mathfrak{A} = \langle A; F \rangle$. (For $n = 1$ this was done in Theorems 9.1 and 9.2. In contrast with Problem 18, this is open also for $n = 2$.)
20. Develop properties of algebras whose automorphism groups are transitive doubly transitive, and so on. (See, e.g., G. Grätzer [2]).

CHAPTER 3
CONSTRUCTIONS OF ALGEBRAS

It is very important to find methods of constructing new algebras from given ones. Two such methods have been discussed in Chapter 1: namely, the construction of subalgebras and homomorphic images of a given algebra. Some further methods will be discussed in this chapter while we postpone the discussion of others because we do not have the necessary background to introduce them now; e.g., free products will be discussed in Chapter 4 and the properties of prime products in Chapter 6.

§19. DIRECT PRODUCTS

Let $\mathfrak{A}_i = \langle A_i; F\rangle$, $i \in I$, be given algebras of type τ. Form the Cartesian product $\prod (A_i \mid i \in I)$ and define the operations f_γ on it as follows: if $p_0, \cdots, p_{n_\gamma - 1} \in \prod (A_i \mid i \in I)$ and $\gamma < o(\tau)$, then

$$f_\gamma(p_0, \cdots, p_{n_\gamma - 1})(i) = f_\gamma(p_0(i), \cdots, p_{n_\gamma - 1}(i)).$$

The algebra $\langle \prod (A_i \mid i \in I); F\rangle = \prod (\mathfrak{A}_i \mid i \in I)$ is called the *direct product* of the algebras $\mathfrak{A}_i$, $i \in I$.

Since $A^\varnothing$ was defined as $\{\varnothing\}$, if $I = \varnothing$, the direct product is the one element algebra $1^{(\tau)}$ of type τ.

Let $\mathfrak{A}_\gamma$, $\gamma < \alpha$ (α is an ordinal), be algebras. Then the elements of the direct product according to the convention of §4 will be α-sequences and the operations are defined componentwise. For instance, if $\alpha = 2$, i.e., we take the direct product of two algebras, then the elements are

$$\langle a_0, b_0\rangle, \cdots, \langle a_{n_\gamma - 1}, b_{n_\gamma - 1}\rangle$$

and the operation f_γ is defined by

$$f_\gamma(\langle a_0, b_0\rangle, \cdots, \langle a_{n_\gamma - 1}, b_{n_\gamma - 1}\rangle) = \langle f_\gamma(a_0, \cdots, a_{n_\gamma - 1}), f_\gamma(b_0, \cdots, b_{n_\gamma - 1})\rangle.$$

Theorem 1. *Let $\langle A_i; F\rangle$, $i \in I$, be a family of algebras. Suppose that $\langle I; \leqq\rangle$ is a well-ordered set of order type α. Let φ be an isomorphism between $\langle I; \leqq\rangle$ and the well-ordered set of all ordinals less than α. Then*

$$f \rightarrow \langle f(0\varphi^{-1}), f(1\varphi^{-1}), \cdots, f(\gamma\varphi^{-1}), \cdots, \rangle_{\gamma < \alpha}$$

is an isomorphism between $\prod(\mathfrak{A}_i \mid i \in I)$ *and* $\prod(\mathfrak{A}_\gamma \mid \gamma < \alpha)$, *where* $\mathfrak{A}_\gamma = \mathfrak{A}_{\gamma\varphi^{-1}}$.

In other words, from an algebraic point of view it is enough to consider the direct product defined in terms of α-sequences.

The proof of Theorem 1 is trivial.

If $\mathfrak{A} \cong \mathfrak{A}_i$ for all $i \in I$, then $\prod(\mathfrak{A}_i \mid i \in I)$ will be called a *direct power* of $\mathfrak{A}$ and will be denoted by $\mathfrak{A}^I$. In case $I = \{\gamma \mid \gamma < \alpha\}$ for some ordinal α, we write $\mathfrak{A}^\alpha$ for $\mathfrak{A}^I$.

Now consider $\prod(\mathfrak{A}_i \mid i \in I)$ and let $e_i^I : f \to f(i)$ $(f \in \prod(A_i \mid i \in I))$ be the i-th projection.

It is easy to verify that e_i^I is a homomorphism of $\prod(\mathfrak{A}_i \mid i \in I)$ onto $\mathfrak{A}_i$.

The congruence relation Θ_i induced by the homomorphism e_i^I can be described by

$$p \equiv q(\Theta_i)\ (p, q \in \prod(A_i \mid i \in I)) \quad \text{if and only if } p(i) = q(i),$$

and $\prod(\mathfrak{A}_i \mid i \in I)/\Theta_i \cong \mathfrak{A}_i$.

Let us consider the special case $\langle A_0 \times A_1; F \rangle$. Then we have two congruence relations, Θ_0, Θ_1. We claim that

$$\Theta_0 \vee \Theta_1 = \iota \quad \text{and} \quad \Theta_0 \wedge \Theta_1 = \omega.$$

Indeed,

$$\langle a_0, b_0 \rangle \equiv \langle a_1, b_1 \rangle (\Theta_0 \wedge \Theta_1) \quad \text{if and only if} \quad a_0 = a_1 \quad \text{and} \quad b_0 = b_1,$$

that is, $\Theta_0 \wedge \Theta_1 = \omega$.

For any $\langle a_0, b_0 \rangle$ and $\langle a_1, b_1 \rangle \in A_0 \times A_1$ we have

$$\langle a_0, b_0 \rangle \equiv \langle a_1, b_1 \rangle (\Theta_0 \vee \Theta_1)$$

because

$$\langle a_0, b_0 \rangle \equiv \langle a_0, b_1 \rangle (\Theta_0)$$

and

$$\langle a_0, b_1 \rangle \equiv \langle a_1, b_1 \rangle (\Theta_1)$$

and hence

$$\langle a_0, b_0 \rangle \equiv \langle a_0, b_1 \rangle \equiv \langle a_1, b_1 \rangle (\Theta_0 \Theta_1).$$

It is now evident that $\Theta_1 \Theta_0 = \Theta_0 \Theta_1 = \Theta_0 \vee \Theta_1$.

If $\Theta_1 \Theta_0 = \Theta_0 \Theta_1$, then Θ_0 and Θ_1 are said to be *permutable*.

Summarizing, we have the following theorem.

Theorem 2. *Let* $\mathfrak{A}$ *be isomorphic to* $\mathfrak{A}_0 \times \mathfrak{A}_1$. *Then there exist congruence relations* Θ_0, Θ_1 *of* $\mathfrak{A}$ *such that*

(i) $\mathfrak{A}/\Theta_i \cong \mathfrak{A}_i$, $i = 0, 1$;
(ii) $\Theta_0 \vee \Theta_1 = \iota$;
(iii) $\Theta_0 \wedge \Theta_1 = \omega$;
(iv) Θ_0, Θ_1 *are permutable.*

The converse also holds.

Theorem 3. *Assume we have an algebra* $\mathfrak{A}$ *and congruence relations* Θ_0, Θ_1 *on* $\mathfrak{A}$ *such that properties* (ii), (iii), (iv) *are satisfied. Then* $\mathfrak{A}/\Theta_0 \times \mathfrak{A}/\Theta_1$ *is isomorphic to* $\mathfrak{A}$.

Proof. We set up the required isomorphism in the natural way:

$$\varphi : a \to ([a]\Theta_0, [a]\Theta_1),$$

where $a \in A$ and of course $[a]\Theta_0 \in A/\Theta_0$ and $[a]\Theta_1 \in A/\Theta_1$.

φ is 1–1. Indeed, if $a \neq b$, then $a \not\equiv b(\Theta_0 \wedge \Theta_1)$. So $a \not\equiv b(\Theta_0)$ or $a \not\equiv b(\Theta_1)$. Hence

$$[a]\Theta_0 \neq [b]\Theta_0 \quad \text{or} \quad [a]\Theta_1 \neq [b]\Theta_1.$$

φ is onto, since if

$$\langle [a]\Theta_0, [b]\Theta_1 \rangle \in A/\Theta_0 \times A/\Theta_1,$$

then because $\Theta_0\Theta_1 = \iota$, there exists an element $c \in A$ such that

$$a \equiv c(\Theta_0) \quad \text{and} \quad c \equiv b(\Theta_1).$$

Then $\langle [a]\Theta_0, [b]\Theta_1 \rangle = \langle [c]\Theta_0, [c]\Theta_1 \rangle$ and thus

$$c\varphi = \langle [a]\Theta_0, [b]\Theta_1 \rangle.$$

To show that φ is a homomorphism, compute: $f_\gamma(a_0, \cdots, a_{n_\gamma - 1})\varphi =$ (by the definition of the mapping φ) $\langle [f_\gamma(a_0, \cdots, a_{n_\gamma - 1})]\Theta_0, [f_\gamma(a_0, \cdots, a_{n_\gamma - 1})]\Theta_1 \rangle =$ (by the definition of f_γ in the quotient algebra) $\langle f_\gamma([a_0]\Theta_0, \cdots, [a_{n_\gamma - 1}]\Theta_0)$, $f_\gamma([a_0]\Theta_1, \cdots, [a_{n_\gamma - 1}]\Theta_1) \rangle =$ (by the definition of f_γ in the direct product) $f_\gamma(\langle [a_0]\Theta_0, [a_0]\Theta_1 \rangle, \cdots, \langle [a_{n_\gamma - 1}]\Theta_0, [a_{n_\gamma - 1}]\Theta_1 \rangle) = f_\gamma(a_0\varphi, \cdots, a_{n_\gamma - 1}\varphi)$. This completes the proof of the theorem.

Theorems 2 and 3 are due to G. Birkhoff [6].

The last part of the proof is a special case of the following lemma.

Lemma 1. *Let* $\mathfrak{A}$ *be an algebra and* ψ_i *a homomorphism of* $\mathfrak{A}$ *into* $\mathfrak{A}_i$, *for* $i \in I$. *Let us define a mapping* ψ *of* A *into* $\prod (A_i \mid i \in I)$ by

$$(a\psi)(i) = a\psi_i.$$

Then ψ *is a homomorphism of* $\mathfrak{A}$ *into* $\prod (\mathfrak{A}_i \mid i \in I)$ *and* $\psi e_i^I = \psi_i$ *for* $i \in I$, *where* e_i^I *is the i-th projection.*

Proof. Trivial, as in the last part of the proof of Theorem 3.

Corollary. *Let* $\mathfrak{A}$ *be an algebra and* I *a nonvoid set. For* $a \in A$, *let* $p_a \in A^I$ *such that* $p_a(i) = a$ *for all* $i \in I$. *Set* $B = \{p_a \mid a \in A\}$. *Then* $\langle B; F \rangle$ *is a* *subalgebra of $\mathfrak{A}^I$ (called the* diagonal *of $\mathfrak{A}^I$) and $a \to p_a$ is an isomorphism between $\mathfrak{A}$ and $\mathfrak{B}$.*

Let $\mathfrak{A}_i$, $i \in I$, be algebras and let $I' \subseteq I$. Then there is a natural mapping $\varphi_{I'}$ from $\prod (A_i \mid i \in I)$ onto $\prod (A_i \mid i \in I')$ which is defined by letting an $f \in \prod (A_i \mid i \in I)$ correspond to its restriction $f_{I'}$ to I'. $\varphi_{I'}$ is a homomorphism of $\prod (\mathfrak{A}_i \mid i \in I)$ onto $\prod (\mathfrak{A}_i \mid i \in I')$. An important property of this homomorphism is described by the following result of C. C. Chang [1].

Theorem 4. *Let $\mathfrak{A}_i$, $i \in I$, be algebras of the same finite type (that is, $o(\tau) < \omega$), and let $B \subseteq \prod (A_i \mid i \in I)$. If B is finite, then there exists a finite $I' \subseteq I$ such that $\varphi_{I'}$ induces an isomorphism between the partial algebra $\mathfrak{B}$ and the partial algebra $\langle B\varphi_{I'}; F\rangle$.*

Remark. This means that if we are interested only in the "local" properties of a direct product, then it is enough to form *finite* direct products.

Proof. Suppose $B = \{p_0, \cdots, p_{n-1}\}$. Take all pairs p_k, p_l such that $p_k \neq p_l$. This implies that there exists an $i \in I$ such that

$$p_k(i) \neq p_l(i).$$

Pick one such i for each $p_k \neq p_l$ and take all these i as the set I_1'.

Suppose now that

$$f_\gamma(q_0, \cdots, q_{n_\gamma - 1}) = q$$

in $\mathfrak{B}$. Then also

$$f_\gamma((q_0)_{I_1'}, \cdots, (q_{n_\gamma - 1})_{I_1'}) = q_{I_1'}.$$

Finally, consider all equalities

$$f_\gamma((q_0)_{I_1'}, \cdots, (q_{n_\gamma - 1})_{I_1'}) = q_{I_1'}$$

which do *not* hold in $\mathfrak{B}$.

Then there exists an $i \in I$ such that

$$f_\gamma(q_0(i), \cdots, q_{n_\gamma - 1}(i)) \neq q(i).$$

For each such equality pick an i; these will form I_2'. It is easy to see that I_2' is also finite since B is finite and we only have a finite number of operations. Define $I' = I_1' \cup I_2'$. Obviously I' satisfies the requirements of the theorem.

Decomposability into infinite direct products will be discussed in §22. If the algebras considered are of a special type such as groups and rings (there is a "zero", 0, and a "+", such that $f_\gamma(0, \cdots, 0) = 0$ for all operations f_γ and $0+x=x+0=x$), then it is possible to find an intrinsic definition of direct products. Within this framework it is then possible to attack the problem of "common refinements" of two direct products and also the problem of isomorphism of two direct product representations of such an algebra. An elegant treatment can be found in the book of B. Jónsson and A. Tarski [1], in the finite case. For more recent results, see P. Crawley and B. Jónsson [1], C. C. Chang, B. Jónsson and A. Tarski [1] and also B. Jónsson [7].

In a very special case, the refinement problem can be translated to the problem of irreducible representations of the unit element in a modular lattice, which is a purely lattice theoretic question. For this, see Birkhoff [6].

Let $\mathfrak{A} = \mathfrak{A}_0 \times \cdots \times \mathfrak{A}_{n-1}$, let $\mathbf{p}$ and $\mathbf{q}$ be m-ary polynomial symbols, and let $a^i = \langle a_0^i, \cdots, a_{n-1}^i \rangle$, $b^i = \langle b_0^i, \cdots, b_{n-1}^i \rangle$, $a^i, b^i \in A$, for $0 \leqq i < m$. Let p, q (resp. p^i, q^i) denote the polynomials induced by $\mathbf{p}$ and $\mathbf{q}$ in $\mathfrak{A}$ (resp. $\mathfrak{A}_i$). Since the operations are defined componentwise, we have the following result:

Lemma 2. *$p(a^0, \cdots, a^{m-1}) = q(b^0, \cdots, b^{m-1})$ if and only if $p^i(a_i^0, \cdots, a_i^{m-1}) = q^i(b_i^0, \cdots, b_i^{m-1})$ for all i; equivalently, $p(a^0, \cdots, a^{m-1}) \neq q(b^0, \cdots, b^{m-1})$ if and only if for some i, $p^i(a_i^0, \cdots, a_i^{m-1}) \neq q^i(b_i^0, \cdots, b_i^{m-1})$.*

Corollary. *The mapping $\varphi: p \to \langle p^0, \cdots, p^{m-1} \rangle$ is a 1–1 homomorphism of $\mathfrak{P}^{(m)}(\mathfrak{A})$ into $\mathfrak{P}^{(m)}(\mathfrak{A}_0) \times \cdots \times \mathfrak{P}^{(m)}(\mathfrak{A}_{n-1})$.*

§20. SUBDIRECT PRODUCTS OF ALGEBRAS

If we decompose an algebra into the direct product of algebras of simpler structure, then we can prove theorems about the algebra by proving them for the components. A best possible decomposition is one such that we cannot decompose any of the components further. However, such a decomposition does not exist in general (as a trivial example, take a nonatomic Boolean algebra). To get decompositions of this kind, we have to weaken the concept of direct product.

Definition 1. *Let $(\mathfrak{A}_i \mid i \in I)$ be a family of algebras of the same type. Let*

$$A \subseteq \prod (A_i \mid i \in I)$$

be such that $\mathfrak{A}$ is a subalgebra of the direct product. $\mathfrak{A}$ is called a subdirect product *of the algebras $\mathfrak{A}_i$, $i \in I$, if $Ae_i^I = A_i$ for all $i \in I$, where e_i^I is the i-th projection.*

If $\mathfrak{A}_i = \mathfrak{B}$ for all $i \in I$, then $\mathfrak{A}$ is a subdirect power *of $\mathfrak{B}$.*

This condition is equivalent to the following: for any $a \in A_i$ there exists an $f \in A$ such that $f(i) = a$.

Let Φ_i be the congruence relation induced by e_i^I on $\prod (A_i \mid i \in I)$ and let $\Theta_i = (\Phi_i)_A$.

Theorem 1 (*G. Birkhoff* [3]).

(i) $\mathfrak{A}/\Theta_i \cong \mathfrak{A}_i$;

(ii) $\bigwedge (\Theta_i \mid i \in I) = \omega$.

Proof. Trivial.

Theorem 2 (*G. Birkhoff* [3]). *Let $\mathfrak{A}$ be an algebra; let $(\Theta_i \mid i \in I)$ be a family of congruence relations such that*

$$\bigwedge (\Theta_i \mid i \in I) = \omega.$$

Then $\mathfrak{A}$ is isomorphic to a subdirect product of the algebras $\mathfrak{A}/\Theta_i$, $i \in I$.

For each $a \in A$, we define an $f_a \in \prod (A/\Theta_i \mid i \in I)$ in the following manner

$$f_a(i) = [a]\Theta_i \ (\in A/\Theta_i).$$

Let

$$A' = \{f_a \mid a \in A\} \subseteq \prod (A/\Theta_i \mid i \in I).$$

Then the mapping

$$\varphi \colon a \to f_a$$

is an isomorphism between $\mathfrak{A}$ and $\mathfrak{A}'$; furthermore, $\mathfrak{A}'$ is a subdirect product of the algebras $\mathfrak{A}/\Theta_i$, $i \in I$.

Proof. Let us compute:

$$f_\gamma(f_{a_0}, \cdots, f_{a_{n_\gamma - 1}}) = f_{f_\gamma(a_0, \cdots, a_{n_\gamma - 1})}$$

because

$$\begin{aligned} f_\gamma(f_{a_0}, \cdots, f_{a_{n_\gamma - 1}})(i) &= f_\gamma(f_{a_0}(i), \cdots, f_{a_{n_\gamma - 1}}(i)) \\ &= f_\gamma([a_0]\Theta_i, \cdots, [a_{n_\gamma - 1}]\Theta_i) \\ &= [f_\gamma(a_0, \cdots, a_{n_\gamma - 1})]\Theta_i \\ &= f_{f_\gamma(a_0, \cdots, a_{n_\gamma - 1})}(i), \end{aligned}$$

which proves that A' is closed under the operations and that φ is a homomorphism. It is trivial that φ is onto and that $\mathfrak{A}'$ is a subdirect product.

If $f_a = f_b$, then $f_a(i) = f_b(i)$ for every $i \in I$; thus, $[a]\Theta_i = [b]\Theta_i$ and so $a \equiv b(\Theta_i)$ for every $i \in I$; therefore, $a \equiv b(\bigwedge (\Theta_i \mid i \in I))$; thus $a \equiv b(\omega)$ which means that $a = b$, proving that φ is also 1–1. This completes the proof of the theorem.

Every algebra $\mathfrak{A}$ has trivial subdirect factorizations. For instance, consider $A' \subseteq A \times A$ consisting of all pairs $\langle a, a\rangle$. Then $\mathfrak{A}$ is isomorphic to $\mathfrak{A}'$, which is a subdirect product of two copies of $\mathfrak{A}$.

Another example is given by the isomorphism

$$\mathfrak{A} \cong \mathfrak{A} \times \mathfrak{B},$$

which always holds if B has one element only.

Theorems 1 and 2 prove that to have a subdirect factorization is equivalent to the existence of congruence relations Θ_i such that $\bigwedge \Theta_i = \omega$. A *trivial* factorization is one where at least one Θ_i equals ω.

This leads us to the concept of subdirectly irreducible algebras.

Definition 2. *The algebra $\mathfrak{A}$ is called* subdirectly irreducible *if the relation*

$$\bigwedge (\Theta_i \mid i \in I) = \omega \qquad (\Theta_i \in C(\mathfrak{A}))$$

implies the existence of an $i \in I$ such that $\Theta_i = \omega$.

Corollary. *An algebra $\mathfrak{A}$ is subdirectly irreducible if and only if A has only one element or $\mathfrak{C}(\mathfrak{A})$ has one and only one atom, which is contained in every congruence relation other than ω.*

Proof. Assume that A has more than one element. Suppose we have a single atom δ which is contained in every congruence relation other than ω, and, contrary to hypothesis, that $\bigwedge \Theta_i = \omega$ and $\Theta_i > \omega$ for each i. Then $\Theta_i \geqq \delta$. Therefore, $\bigwedge \Theta_i \geqq \delta > \omega$, which is a contradiction.

Conversely, assume that $\mathfrak{A}$ is subdirectly irreducible. Let $\bigwedge (\Theta \mid \Theta \neq \omega) = \delta$. Then $\delta > \omega$ since $\mathfrak{A}$ is subdirectly irreducible. If $\Theta > \omega$, then $\Theta \geqq \delta$; hence δ is an atom and it is contained in every congruence relation other than ω.

Theorem 3 (*G. Birkhoff* [3]). *Every algebra is isomorphic to a subdirect product of subdirectly irreducible algebras.*

Proof. We have to construct a family of congruence relations $(\Theta_i \mid i \in I)$ such that: (i) $\bigwedge (\Theta_i \mid i \in I) = \omega$. By Theorem 11.3

$$\mathfrak{C}(\mathfrak{A}/\Theta_i) \cong \langle [\Theta_i); \leqq \rangle.$$

Thus the condition that the algebra $\mathfrak{A}/\Theta_i$ is subdirectly irreducible is by the corollary to Definition 2 equivalent to: (ii) there exists a congruence relation covering Θ_i, which is contained in every congruence relation properly containing Θ_i. (We assume $|A| > 1$).

We claim that a family of congruence relations satisfying (i) and (ii) is given by

$$(\Psi(a, b) \mid a \neq b,\ a, b \in A),$$

where the $\Psi(a, b)$ were constructed in Theorem 10.6.

To prove (i), assume that $x \equiv y(\bigwedge \Psi(a, b))$ and $x \neq y$. This would imply that

$$x \equiv y(\Psi(x, y)),$$

which is a contradiction.

If $\Phi > \Psi(a, b)$, then $\Phi \geqq \Theta(a, b)$, which means that $\Theta(a, b) \vee \Psi(a, b)$ covers $\Psi(a, b)$ and it is contained in every congruence relation properly containing $\Psi(a, b)$. This completes the proof of the theorem.

Given two algebras, there are many algebras which are subdirect products of the given ones. In the sequel, we will describe a method (see L. Fuchs [1]) which constructs some subdirect products.

Lemma 1. *Let $\mathfrak{A}_0$, $\mathfrak{A}_1$, and $\mathfrak{B}$ be algebras and let φ_i be a homomorphism of $\mathfrak{A}_i$ onto $\mathfrak{B}$, $i = 0, 1$. Let*

$$C = \{\langle a_0, a_1 \rangle \mid a_0\varphi_0 = a_1\varphi_1\}.$$

Then $\mathfrak{C}$ is a subdirect product of $\mathfrak{A}_0$ and $\mathfrak{A}_1$.

Proof. If $a_i\varphi_0 = b_i\varphi_1$, then

$$f_\gamma(a_0, \cdots, a_{n_\gamma - 1})\varphi_0 = f_\gamma(b_0, \cdots, b_{n_\gamma - 1})\varphi_1,$$

which implies that C is closed under the operations. Take an $a \in A_0$. Then there exists a $b \in A_1$ such that $a\varphi_0 = b\varphi_1$ since φ_1 is onto. Thus, a is the first component of the pair $\langle a, b \rangle \in C$. A similar consideration for the second component completes the proof of the lemma.

L. Fuchs [1] proved that in a number of cases this construction gives all subdirect products. His results were generalized by I. Fleischer [1].

Theorem 4 (*I. Fleischer* [1]). *Let $\mathfrak{A}$ be an algebra and assume that any two congruence relations of $\mathfrak{A}$ are permutable. Then any subdirect representation of $\mathfrak{A}$ with two factors only can be constructed by the method of Lemma* 1.

A sharper result is the following.

Theorem 5 (*I. Fleischer* [1]). *Let $\mathfrak{A}$ be isomorphic to a subdirect product of the quotient algebras $\mathfrak{A}/\Theta_0$ and $\mathfrak{A}/\Theta_1$ by the natural isomorphism*

$\varphi: a \to \langle [a]\Theta_0, [a]\Theta_1 \rangle$. *This subdirect representation of* $\mathfrak{A}$ *can be constructed by the method of Lemma* 1 *if and only if* Θ_0 *and* Θ_1 *are permutable.*

Proof. Assume that Θ_0 and Θ_1 are permutable. Let ψ_i be the natural homomorphism of $\mathfrak{A}$ onto $\mathfrak{A}/\Theta_i$ (that is, $a\psi_i = [a]\Theta_i$) and let φ_i be the natural homomorphism of $\mathfrak{A}/\Theta_i$ onto $\mathfrak{B} = \mathfrak{A}/\Theta_0 \vee \Theta_1$ (that is,

$$([a]\Theta_i)\varphi_i = [a](\Theta_0 \vee \Theta_1)).$$

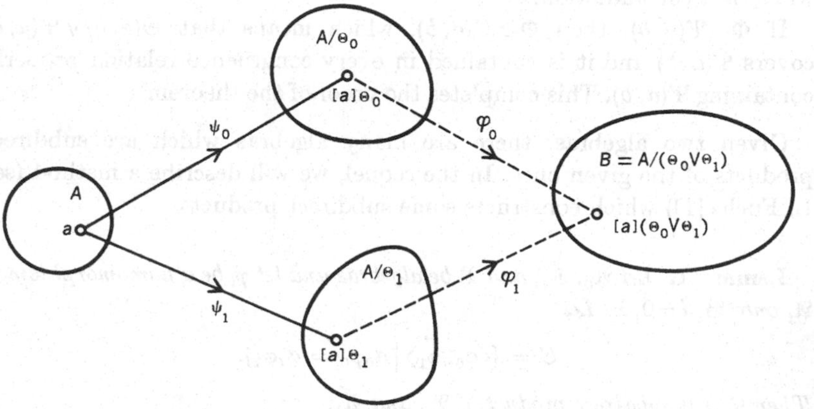

The elements of $\mathfrak{A}$ are represented in the subdirect product as pairs of the form

$$\langle [a]\Theta_0, [a]\Theta_1 \rangle \tag{1}$$

and we have to prove that these are the same as the pairs of the form

$$\langle [a]\Theta_0, [b]\Theta_1 \rangle \quad \text{with} \quad ([a]\Theta_0)\varphi_0 = ([b]\Theta_1)\varphi_1. \tag{2}$$

If we have the pair (1), then it is also of the form (2) since

$$([a]\Theta_0)\varphi_0 = [a](\Theta_0 \vee \Theta_1) = ([a]\Theta_1)\varphi_1;$$

Conversely, if we are given the pair (2), then $a \equiv b(\Theta_0 \vee \Theta_1)$ and since $\Theta_0 \vee \Theta_1 = \Theta_0\Theta_1$ by permutability there exists a $c \in A$ such that $a \equiv c(\Theta_0)$ and $c \equiv b(\Theta_1)$; that is, $[a]\Theta_0 = [c]\Theta_0$ and $[b]\Theta_1 = [c]\Theta_1$. Thus,

$$\langle [a]\Theta_0, [b]\Theta_1 \rangle = \langle [c]\Theta_0, [c]\Theta_1 \rangle,$$

which was to be proved.

To prove the converse, assume that $\mathfrak{A}$ is isomorphic to the subdirect product of $\mathfrak{A}/\Theta_0$ and $\mathfrak{A}/\Theta_1$ and that it can be constructed by the method of Lemma 1 using the homomorphisms φ_i of $\mathfrak{A}/\Theta_i$ onto $\mathfrak{B}$. Let ψ_i be the natural homomorphism of $\mathfrak{A}$ onto $\mathfrak{A}/\Theta_i$. Let $\bar{\Theta}_i$ denote the congruence relations of $\mathfrak{A}$ induced by the homomorphisms $\psi_i\varphi_i$.

Then $a \equiv b(\overline{\Theta}_i)$ if and only if $([a]\Theta_i)\varphi_i = ([b]\Theta_i)\varphi_i$. Since $\langle [a]\Theta_0, [a]\Theta_1 \rangle \in A$, it is also of the form (2), that is, $([a]\Theta_0)\varphi_0 = ([a]\Theta_1)\varphi_1$ and similarly $([b]\Theta_0)\varphi_0 = ([b]\Theta_1)\varphi_1$. Thus, $a \equiv b(\overline{\Theta}_0)$ if and only if $([a]\Theta_0)\varphi_0 = ([b]\Theta_0)\varphi_0$ in which case $([a]\Theta_1)\varphi_1 = ([b]\Theta_1)\varphi_1$, which is equivalent to $a \equiv b(\overline{\Theta}_1)$.

Hence $\overline{\Theta}_0 = \overline{\Theta}_1$. Set $\Psi = \overline{\Theta}_0 = \overline{\Theta}_1$.

Summarizing, if we have a pair of the form (2), which means that $a \equiv b(\Psi)$, then there exists a $c \in A$ such that $[a]\Theta_0 = [c]\Theta_0$ and $[b]\Theta_1 = [c]\Theta_1$. This yields $\Psi = \Theta_0\Theta_1$, proving that Θ_0 and Θ_1 are permutable.

This completes the proof of Theorem 5.

This method gives very little if more than two factors are considered; see G. H. Wenzel [1].

Let Θ_i be a congruence relation of $\mathfrak{A}_i$, $i \in I$. Define a relation $\prod(\Theta_i \mid i \in I)$ on $\prod(A_i \mid i \in I)$ as follows: if $p, q \in \prod(A_i \mid i \in I)$, then $p \equiv q(\prod(\Theta_i \mid i \in I))$ if and only if $p(i) \equiv q(i)(\Theta_i)$ for every $i \in I$.

Lemma 2. *$\prod(\Theta_i \mid i \in I)$ is a congruence relation of the direct product $\prod(\mathfrak{A}_i \mid i \in I)$.*

Proof. Trivial.

Corollary. *The following isomorphism holds:*

$$\prod(\mathfrak{A}_i \mid i \in I)/\prod(\Theta_i \mid i \in I) \cong \prod(\mathfrak{A}_i/\Theta_i \mid i \in I).$$

An isomorphism can be set up by letting $[p](\prod(\Theta_i \mid i \in I)) \to \overline{p}$, where $\overline{p}$ is defined by

$$\overline{p}(i) = [p(i)]\Theta_i.$$

A similar statement holds for homomorphisms.

Lemma 3. *Let $\mathfrak{A}_i$ and $\mathfrak{B}_i$ be algebras and let φ_i be a homomorphism of $\mathfrak{A}_i$ into $\mathfrak{B}_i$, for all $i \in I$. We define a mapping*

$$\varphi = \prod(\varphi_i \mid i \in I)$$

of $\prod(A_i \mid i \in I)$ into $\prod(B_i \mid i \in I)$ as follows: if $p \in \prod(A_i \mid i \in I)$, then $(p\varphi)(i) = p(i)\varphi_i$. Then φ is a homomorphism of $\prod(\mathfrak{A}_i \mid i \in I)$ into $\prod(\mathfrak{B}_i \mid i \in I)$.

Proof. Trivial.

Now we will prove that a subdirect product of homomorphic images is the homomorphic image of a subdirect product.

Lemma 4. *Let $\mathfrak{B}'$ be a subdirect product of the algebras $\mathfrak{A}_i'$, $i \in I$. Let φ_i be a homomorphism of $\mathfrak{A}_i$ onto $\mathfrak{A}_i'$; for $i \in I$. Then there exists a subdirect product $\mathfrak{B}$ of the algebras $\mathfrak{A}_i$ and a homomorphism φ of $\mathfrak{B}$ onto $\mathfrak{B}'$.*

Proof. Define $\mathfrak{B}$ as the complete inverse image of $\mathfrak{B}'$ under the homomorphism $\varphi = \prod (\varphi_i \mid i \in I)$. Then $\mathfrak{B}$ is a subalgebra of $\prod (\mathfrak{A}_i \mid i \in I)$ by Lemma 12.1. Take $a_i \in A_i$. Then $a_i\varphi_i \in A_i'$. Since $\mathfrak{B}'$ is a subdirect product, there exists a $p' \in B'$ such that $p'(i) = a_i\varphi_i$. Now define p by $p(i) = a_i$ and $p(j)$ = any inverse image of $p'(j)$ for $j \neq i$. Then $p \in B$ and $p(i) = a_i$. Thus $\mathfrak{B}$ is a subdirect product of the $\mathfrak{A}_i$. That $B\varphi = B'$ follows from the definition.

§21. DIRECT AND INVERSE LIMITS OF ALGEBRAS

There are two well known methods to build up algebras from families of algebras, the so-called direct and inverse limits. No systematic account of the properties of these constructions can be found in the literature, although almost all the results given below belong to the "folklore". Besides, most results are set theoretic in nature, and they can be derived from the case of groups; see for example S. Eilenberg and W. Steenrod, Foundations of algebraic topology, Princeton University Press, Princeton, N. J., 1952.†

Definition 1. *A* direct family of sets $\mathscr{A}$ *is defined to be a triplet of the following objects:*

(i) *A directed partially ordered set $\langle I; \leqq \rangle$ called the* carrier *of $\mathscr{A}$;*
(ii) *sets A_i for each $i \in I$;*
(iii) *mappings φ_{ij} for all $i \leqq j$, where φ_{ij} maps A_i into A_j such that*

$$\varphi_{ij}\varphi_{jk} = \varphi_{ik} \quad \text{if} \quad i \leqq j \leqq k$$

and φ_{ii} is the identity mapping for all $i \in I$.

For a direct family $\mathscr{A}$ consider the set‡ $\bigcup (A_i \mid i \in I)$ and define on it a binary relation $\equiv$ by $x \equiv y$ if and only if $x \in A_i$, $y \in A_j$ for some $i, j \in I$ and there exists a $z \in A_k$ such that $i \leqq k, j \leqq k$, $x\varphi_{ik} = z$, $y\varphi_{jk} = z$. It is

† The best framework for §21 would be category theory. By duality we would have to prove only one of the statements of Lemma 7 and only one of Theorems 2 and 3. Note that the categorical definitions of various algebraic constructions are given in the Exercises.

‡ We assume that the A_i are pairwise disjoint; if this is not the case, we have to form the disjoint union, say $\bigcup (A_i \times \{i\} \mid i \in I)$, and introduce $\equiv$ on this set.

obvious that this relation is reflexive and symmetric. To prove transitivity, let $y \equiv w$ also hold; then $y \in A_j$, $w \in A_l$ and there exists a $v \in A_n$, $j \leqq n$, $l \leqq n$, such that $y\varphi_{jn} = v$, $w\varphi_{ln} = v$. Let m be an upper bound for k and n. Then

$$x\varphi_{im} = x\varphi_{ik}\varphi_{km} = z\varphi_{km} = y\varphi_{jk}\varphi_{km} = y\varphi_{jm}$$

and

$$w\varphi_{lm} = w\varphi_{ln}\varphi_{nm} = v\varphi_{nm} = y\varphi_{jn}\varphi_{nm} = y\varphi_{jm},$$

which proves that $x \equiv w$.

Thus, $\equiv$ is an equivalence relation; let $\hat{x}$ denote the equivalence class containing x and let A_∞ denote the set of equivalence classes.

Definition 2. *A_∞ is called the* direct limit *of the direct family of sets $\mathscr{A}$, in symbols $A_\infty = \underset{\rightarrow}{\lim}\, \mathscr{A}$.*

Definition 3. *A* direct family of algebras *$\mathscr{A}$ is defined to be a triplet of the following objects:*

(i) *A directed partially ordered set $\langle I; \leqq \rangle$;*
(ii) *algebras $\mathfrak{A}_i = \langle A_i; F \rangle$, $i \in I$, of some fixed type;*
(iii) *homomorphisms φ_{ij} of $\mathfrak{A}_i$ into $\mathfrak{A}_j$ for all $i \leqq j$ such that*

$$\varphi_{ij}\varphi_{jk} = \varphi_{ik} \quad \textit{if} \quad i \leqq j \leqq k$$

and φ_{ii} is the identity mapping for all $i \in I$.

It is obvious that if we have a direct family of algebras, then the base sets of the algebras form a direct family of sets. Thus, we can form the direct limit A_∞.

We can define the operations f_γ on A_∞ as follows: Let $x_j \in A_{i_j}$, $0 \leqq j < n_\gamma$, and let m be an upper bound of the i_j. Then $x_j \equiv x_j' \in A_m$, where $x_j' = x_j\varphi_{i_j m}$. Define

$$f_\gamma(\hat{x}_0, \cdots, \hat{x}_{n_\gamma - 1}) = \hat{f}_\gamma(x_0', \cdots, x'_{n_\gamma - 1}).$$

We will show that this definition of f_γ does not depend on m. For, take any other upper bound m'; let $x_j'' = x_j\varphi_{i_j m'}$. Then

$$f_\gamma(x_0'', \cdots, x''_{n_\gamma - 1}) \equiv f_\gamma(x_0', \cdots, x'_{n_\gamma - 1}).$$

Indeed, let $m \leqq o$, $m' \leqq o$, and $x_j''' = x_j\varphi_{i_j o}$. Then

$$f_\gamma(x_0'', \cdots, x''_{n_\gamma - 1})\varphi_{m'o} = f_\gamma(x_0''', \cdots, x'''_{n_\gamma - 1}) = f_\gamma(x_0', \cdots, x'_{n_\gamma - 1})\varphi_{mo}$$

since φ_{mo} and $\varphi_{m'o}$ are homomorphisms.

Definition 4. *The algebra* $\mathfrak{A}_\infty = \langle A_\infty; F\rangle$ *is called the* direct limit *of the direct family of algebras and will be denoted by* $\varinjlim \mathscr{A}$.

Let us define the mapping

$$\varphi_{i\infty}: x \to \hat{x}, \quad \text{for} \quad x \in A_i.$$

Then $\varphi_{i\infty}$ is a homomorphism of $\mathfrak{A}_i$ into $\mathfrak{A}_\infty$. Furthermore, we have for all $i \leqq j$ that $\varphi_{ij}\varphi_{j\infty} = \varphi_{i\infty}$. Thus, we have:

Lemma 1. *Extend the direct family of algebras* $\mathscr{A}$ *by adding to* I *the symbol* ∞ *and partially order* $I \cup \{\infty\}$ *by* $i < \infty$ *for all* $i \in I$ *while keeping the partial ordering in* I; *add* $\mathfrak{A}_\infty$ *to the family of algebras and the mappings* $\varphi_{i\infty}$ *for* $i \in I \cup \{\infty\}$ *to the family of mappings* ($\varphi_{\infty\infty}$ *is the identity map on* A_∞). *Then the resultant system* $\mathscr{A}_\infty$ *is again a direct system of algebras.*

If all the φ_{ij} are 1–1, then so are the $\varphi_{i\infty}$. Indeed, take $x, y \in A_i$ and suppose that $x\varphi_{i\infty} = y\varphi_{i\infty}$, that is, $\hat{x} = \hat{y}$. Then there exists a $j \geqq i$ such that $x\varphi_{ij} = y\varphi_{ij}$ and since all the φ_{ij} are 1–1, we infer that $x = y$. Thus, we have proved the following lemma.

Lemma 2. *If in a direct family of algebras, all the* φ_{ij} $(i \leqq j)$ *are* 1–1, *then all the* $\varphi_{i\infty}$ *are* 1–1.

Let us give an example of a direct family of algebras.

Let $\mathfrak{A}$ be an algebra. Define I by: $B \in I$ if and only if $\langle B; F\rangle$ is a finitely generated subalgebra of $\mathfrak{A}$.

Let us define a direct family of algebras as follows: (i) the directed partially ordered set is taken to be $\langle I; \subseteq\rangle$; (ii) $\mathfrak{A}_i$ is defined as $\langle i; F\rangle$; (iii) all the φ_{ij} $(i \subseteq j)$ are defined as identity mappings $(x\varphi_{ij} = x)$.

Then the direct limit can be easily shown to be isomorphic to the algebra $\mathfrak{A}$.

Lemma 3. *Every algebra is isomorphic to a direct limit of finitely generated algebras.*

Now we define our second construction.

Definition 5. *An* inverse family of sets $\mathscr{A}$ *is defined to be a triplet of the following objects:*

(i) *A directed partially ordered set* $\langle I; \leqq\rangle$, *called the* carrier *of* $\mathscr{A}$;

(ii) *sets* A_i *for each* $i \in I$;

(iii) *mappings* φ_j^i *for all* $j \leqq i$ *such that* φ_j^i *maps* A_i *into* A_j, $\varphi_j^i\varphi_k^j = \varphi_k^i$ *if* $i \geqq j \geqq k$, *and* φ_i^i *is the identity mapping for all* $i \in I$.

For an inverse family $\mathscr{A}$, take the direct product $\prod (A_i \mid i \in I)$ and let A^∞ consist of those $p \in \prod (A_i \mid i \in I)$ for which

$$p(i)\varphi_j^i = p(j) \quad \text{if} \quad i \geqq j.$$

Definition 6. *A^∞ is called the* inverse limit *of the inverse family of sets; in notation,* $A^\infty = \varprojlim \mathscr{A}$.

Definition 7. *An* inverse family of algebras *$\mathscr{A}$ is defined to be a triplet of the following objects:*

(i) *A directed partially ordered set $\langle I; \leqq \rangle$;*
(ii) *algebras $\mathfrak{A}_i = \langle A_i; F \rangle$ for each $i \in I$;*
(iii) *homomorphisms φ_j^i of $\mathfrak{A}_i$ into $\mathfrak{A}_j$ for all $i \geqq j$ such that $\varphi_j^i \varphi_k^j = \varphi_k^i$ if $i \geqq j \geqq k$ and φ_i^i is the identity mapping for all $i \in I$.*

Clearly, the base sets form an inverse family of sets.

We can prove that $\mathfrak{A}^\infty = \langle A^\infty; F \rangle$ is a subalgebra of $\prod (\mathfrak{A}_i \mid i \in I)$, if $A^\infty \neq \varnothing$. Indeed, if

$$p = f_\gamma(p_0, \cdots, p_{n_\gamma - 1}), \qquad p_0, \cdots, p_{n_\gamma - 1} \in A^\infty \quad \text{and} \quad i \geqq j,$$

then

$$\begin{aligned} p(i)\varphi_j^i &= f_\gamma(p_0(i), \cdots, p_{n_\gamma - 1}(i))\varphi_j^i \\ &= f_\gamma(p_0(i)\varphi_j^i, \cdots, p_{n_\gamma - 1}(i)\varphi_j^i) \\ &= f_\gamma(p_0(j), \cdots, p_{n_\gamma - 1}(j)) \\ &= p(j) \end{aligned}$$

and so $p \in A^\infty$.

Definition 8. *If $A^\infty \neq \varnothing$, then $\mathfrak{A}^\infty$ is called the* inverse limit *of the inverse family of algebras; in notation,* $\mathfrak{A}^\infty = \varprojlim \mathscr{A}$.

Let us define the mapping

$$\varphi_i^\infty : p \to p(i) \quad \text{for} \quad p \in A^\infty.$$

Then φ_i^∞ is a homomorphism of $\mathfrak{A}^\infty$ into $\mathfrak{A}_i$ and we have that

$$\varphi_i^\infty \varphi_j^i = \varphi_j^\infty \quad \text{if} \quad i \geqq j.$$

Lemma 4. *If we extend the inverse family of algebras as in Lemma 1 by adding ∞ to I, the algebra $\langle A^\infty; F \rangle$ to the family of algebras and the mappings $\{\varphi_i^\infty \mid i \in I\} \cup \{\varphi_\infty^\infty\}$ (where φ_∞^∞ is the identity mapping on A^∞) to the family of mappings, then the resultant family $\mathscr{A}^\infty$ is again an inverse family of algebras.*

The assumption $A^{\infty} \neq \varnothing$ is really necessary in Definition 8. Consider the following inverse family of sets. $\langle I; \leqq \rangle$ is the partially ordered set of positive integers with their natural ordering, $A_i = I$ for each $i \in I$, and $x\varphi_j^i = 2^{i-j}x$ if $i \geqq j$. If p were in A^{∞}, then

$$p(1) = 2^i p(i+1)$$

for every i which is clearly impossible. Thus, we get that:

The inverse limit of nonvoid sets may be void.

However, the following result shows that this cannot happen if all the sets are finite and nonvoid.

Theorem 1. *The inverse limit of finite nonvoid sets is always nonvoid.*

Remark. A special case of Theorem 1 is called König's Lemma. Theorem 1 is usually derived from Tihonov's Theorem, that the product of compact spaces is also compact, see, for instance, S. Eilenberg and W. Steenrod, *loc. cit.*, p. 217. The proof, presented here, can be easily formulated so that it uses only the prime ideal theorem of Boolean algebras (Theorem 6.7).

Proof. Let the inverse family $\mathscr{A}$ be given as in Definition 5 and let each A_i be finite, nonvoid. For each finite $J \subseteq I$ set

$$B_J = \{p \mid p \in \prod (A_i \mid i \in I) \quad \text{and for} \quad i, j \in J,\ i \geqq j, \quad p(i)\varphi_j^i = p(j)\}.$$

Since $\langle I; \leqq \rangle$ is directed, each B_J is nonvoid. Also,

$$B_{J_0} \cap B_{J_1} \supseteq B_{J_0 \cup J_1};$$

therefore by the corollary to Theorem 6.7, there exists a prime dual ideal $\mathscr{D}$ of all subsets of $\prod (A_i \mid i \in I)$, such that $B_J \in \mathscr{D}$ for all finite $J \subseteq I$.

Let $\mathscr{D}_i$ denote the i-th projection of $\mathscr{D}$, that is, for $C \in \mathscr{D}$, form $C_i = Ce_i^I = \{a \mid a \in A_i$ and $a = p(i)$ for some $p \in C\}$ and $\mathscr{D}_i = \{C_i \mid C \in \mathscr{D}\}$. $\mathscr{D}_i$ is a dual ideal of $\mathfrak{P}(A_i)$; the claim that it is prime is equivalent to the statement that if $B \subseteq A_i$; $B \cap X \neq \varnothing$ for all $X \in \mathscr{D}_i$, then $B \in \mathscr{D}_i$ (Exercise 0.84). So let us assume that $B \cap X \neq \varnothing$ for all $X \in \mathscr{D}_i$ and set $\hat{B} = B(e_i^I)^{-1}$. For $D \in \mathscr{D}$, $\hat{B} \cap D \neq \varnothing$, since $\hat{B}e_i^I \cap De_i^I = B \cap D_i \neq \varnothing$, so we can pick a $p \in D$ with $p(i) \in B$ and then $p \in \hat{B} = B(e_i^I)^{-1}$. Thus $\hat{B} \in \mathscr{D}$ and $\hat{B}e_i^I = B \in \mathscr{D}_i$, which was to be proved.

Since $\mathscr{D}_i$ is a prime dual ideal of the finite Boolean algebra $\mathfrak{P}(A_i)$, there exists an $a_i \in A_i$, such that $X \in \mathscr{D}_i$ if and only if $a_i \in X$.

Define $p \in \prod (A_i \mid i \in I)$ by $p(i) = a_i$. We claim that $p \in \overleftarrow{\lim} \mathscr{A}$.

We proved that for every $i \in I$, and $D \in \mathscr{D}_i$, $D(e_i^I)^{-1} \in \mathscr{D}$. Thus if $j \leqq i$, then $\{a_i\}(e_i^I)^{-1}$ and $\{a_j\}(e_j^I)^{-1} \in \mathscr{D}$, and

$$\{a_i\}(e_i^I)^{-1} \cap \{a_j\}(e_j^I)^{-1} \cap B_{\{i,j\}} \in \mathscr{D},$$

so there exists a $q \in B_{\{i,j\}}$ with $q(i)=a_i$ and $q(j)=a_j$. Thus $a_i\varphi_j^i = a_j$, which completes the proof of $p \in \varprojlim \mathscr{A}$.

The following is an example of an inverse family of algebras.

Let $\mathfrak{B}_j$, $j \in J$, be a family of algebras and define I to be the set of all nonvoid finite subsets of J. Then define:

(i) The partially ordered set to be $\langle I; \subseteq \rangle$;
(ii) $\mathfrak{A}_i = \prod (\mathfrak{B}_j \mid j \in i)$;
(iii) for $i_0 \supseteq i_1$, $\varphi_{i_1}{}^{i_0}$ is the natural homomorphism of $\prod (\mathfrak{B}_j \mid j \in i_0)$ onto $\prod (\mathfrak{B}_j \mid j \in i_1)$. (See the definition preceding Theorem 19.4.)

Lemma 5. *The family defined above is an inverse family of algebras and the inverse limit is isomorphic to* $\prod (\mathfrak{B}_j \mid j \in J)$.

The analogue of Lemma 2 holds for inverse limits.

Lemma 6. *If in an inverse family of algebras all the* $\varphi_j{}^i$ *are* 1–1, *then all the* $\varphi_i{}^\infty$ *are also* 1–1.

Proof. Let $p, q \in A^\infty$ and $p\varphi_i{}^\infty = q\varphi_i{}^\infty$, that is, $p(i)=q(i)$. Let $j \in I$; we want to prove that $p(j)=q(j)$. Indeed, if k is any upper bound of i and j, then $p(k)\varphi_i{}^k = q(k)\varphi_i{}^k$, so $p(k)=q(k)$. Therefore $p(j)=p(k)\varphi_j{}^k = q(k)\varphi_j{}^k = q(j)$, which was to be proved.

Direct limits and especially inverse limits are very hard to visualize in general. A happy exception is when the carrier is well ordered. We will prove that in many cases we can restrict ourselves to this special case (Theorems 4 and 5). In preparation for these, we will show that certain "double" direct (inverse) limits can be represented as "simple" direct (inverse) limits (Theorems 2 and 3).

As a first step, we prove the following lemma.

Let $\mathscr{A}$ be a direct family (inverse family) of algebras as given in Definition 3 (Definition 7) and let $J \subseteq I$ such that $\langle J; \leqq \rangle$ is also a directed partially ordered set. We will denote by $\mathscr{A}_J$ the direct family (inverse family) of algebras whose carrier is $\langle J; \leqq \rangle$ and whose algebras in $\mathscr{A}_J$ are $\mathfrak{A}_i$, $i \in J$, with the associated homomorphisms φ_{ij} ($\varphi_j{}^i$) with $i, j \in J$.

Lemma 7. *Let* $\mathscr{A}$ *be a direct family (inverse family) of algebras with carrier* $\langle I; \leqq \rangle$ *and let* $J \subseteq I$ *be such that* $\langle J; \leqq \rangle$ *is a directed partially ordered set* cofinal *with* $\langle I; \leqq \rangle$ *(i.e., for every* $i \in I$ *there exists a* $j \in J$ *such that* $i \leqq j$*). Then* $\varinjlim \mathscr{A} \cong \varinjlim \mathscr{A}_J$ ($\varprojlim \mathscr{A} \cong \varprojlim \mathscr{A}_J$).

Proof. First we prove the isomorphism statement for direct limits. Set $\varinjlim \mathscr{A} = \langle A_\infty; F\rangle$, $\varinjlim \mathscr{A}_J = \langle A_\infty'; F\rangle$. An element of A_∞ is of the form $\hat{x}_I$, where the index I indicates that we take its closure in the family $\mathscr{A}$. Consider the mapping

$$\psi: \hat{x}_I \to \hat{x}_J,$$

where $x \in A_j$, $j \in J$ and the subscript J denotes closure in $\mathscr{A}_J$. The domain of ψ is A_∞ since if $\hat{x}_I \in A_\infty$, $x \in A_i$, $i \in I$, then there exists a $j \in J$ such that $i \leqq j$; since $x \equiv x\varphi_{ij}$, we get $\hat{x}_I = (\widehat{x\varphi_{ij}})_I$ and $x\varphi_{ij} \in A_j$, $j \in J$. It is obvious that ψ maps A_∞ onto A_∞' and ψ is a homomorphism. To show that ψ is 1–1, assume that $\hat{x}_J = \hat{y}_J$ and $x \in A_i$, $y \in A_j$, $i, j \in J$. Then there exists a $k \in J$ such that $i \leqq k$, $j \leqq k$ and $x\varphi_{ik} = y\varphi_{jk}$. This implies that $\hat{x}_I = \hat{y}_I$, so ψ is 1–1. This completes the proof of the isomorphism.

Now let $\mathscr{A}$ be an inverse family; set $\varprojlim \mathscr{A} = \langle A^\infty; F\rangle$ and $\varprojlim \mathscr{A}_J = \langle A_1^\infty; F\rangle$. Consider the mapping

$$\psi: p \to p_J,$$

which maps A^∞ into A_1^∞ (p is an element of $\prod (A_i \mid i \in I)$ and p_J denotes the restriction of p to J). Then ψ is a homomorphism. Now let $q \in A_1^\infty$. If $q = p\psi$ and $i \in I$, then for any $j \in J$ with $i \leqq j$ we have $p(i) = q(j)\varphi_i^j$, that is, q determines p and so ψ is 1–1. To prove that ψ is onto, for any given $q \in A_1^\infty$, define a $p \in \prod (A_i \mid i \in I)$ as follows: For a given $i \in I$ choose a $j \in J$ with $i \leqq j$ and define $p(i) = q(j)\varphi_i^j$. Note that p is well defined because if $i \leqq j' \in J$, then we can choose an upper bound n of j and j' in $\langle J; \leqq\rangle$. Then

$$q(j)\varphi_i^j = q(n)\varphi_j^n\varphi_i^j = q(n)\varphi_i^n = q(n)\varphi_{j'}^n\varphi_i^{j'} = q(j')\varphi_i^{j'},$$

and thus $p(i)$ does not depend on the choice of j.

To prove that $p \in A^\infty$, choose $i \leqq k$, $i, k \in I$. Then there exists an $l \in J$ with $k \leqq l$ since $\langle J; \leqq\rangle$ is cofinal with $\langle I; \leqq\rangle$. Let us compute:

$$p(k)\varphi_i^k = q(l)\varphi_k^l\varphi_i^k = q(l)\varphi_i^l = p(i),$$

which was to be proved.

Since $p\psi = q$ is trivial, we conclude that ψ is an isomorphism, completing the proof of the lemma.

Let $\mathscr{A}$ be a direct family of algebras with carrier $\langle I; \leqq\rangle$ and assume that $I = \bigcup (I_j \mid j \in P)$, such that $\langle I_j; \leqq\rangle$ is a directed partially ordered set of $\langle I; \leqq\rangle$ and $\langle P; \leqq\rangle$ is also a directed partially ordered set and $I_j \subseteq I_i$ if $j \leqq i$.

Let us introduce the following notations. Let $\underrightarrow{\lim}\, \mathscr{A} = \langle A_\infty; F\rangle$ and $\underrightarrow{\lim}\, \mathscr{A}_{I_j} = \langle A_\infty{}^j; F\rangle$. If $j \leqq i$, then define the mapping

$$\psi_{ji}: \hat{x}_{I_j} \to \hat{x}_{I_i}, \quad \text{where} \quad x \in A_k \quad \text{for some} \quad k \in I_j.$$

The family consisting of $\langle P; \leqq\rangle$, the algebras $\langle A_\infty{}^j; F\rangle$, and the mappings ψ_{ji} will be denoted by $\mathscr{A}/P$.

Theorem 2. *$\mathscr{A}/P$ is a direct family of algebras and we have the following isomorphism:*

$$\underrightarrow{\lim}\, \mathscr{A} \cong \underrightarrow{\lim}\, \mathscr{A}/P.$$

Proof. ψ_{ji} maps $A_\infty{}^j$ into $A_\infty{}^i$ if $j \leqq i$. It is obviously a homomorphism and all ψ_{jj} are identity mappings. Further, if $j \leqq i \leqq k$ $(j, i, k \in P)$, then

$$\hat{x}_{I_j}\psi_{ji}\psi_{ik} = \hat{x}_{I_i}\psi_{ik} = \hat{x}_{I_k} = \hat{x}_{I_j}\psi_{jk}.$$

Thus, $\psi_{ji}\psi_{ik} = \psi_{jk}$. This proves that $\mathscr{A}/P$ is a direct family of algebras. Set $\underrightarrow{\lim}\, \mathscr{A}/P = \langle A_\infty'; F\rangle$.

Let $x \in A_i$, $i \in I$ and choose a $p \in P$ such that $i \in I_p$; set $y = \hat{x}_{I_p}$ $(\in A_\infty{}^p)$ and define the mapping ψ of A_∞ into A_∞' by

$$\psi: \hat{x}_I \to \hat{y}_P.$$

We claim that ψ is the required isomorphism.

To prove that ψ is well defined, let $x \in A_i$, $z \in A_j$, $\hat{x}_I = \hat{z}_I$, $i \in I_p$, $j \in I_q$, $y = \hat{x}_{I_p}$, $w = \hat{z}_{I_q}$; we have to prove that $\hat{y}_P = \hat{w}_P$. Since $\hat{x}_I = \hat{z}_I$, for some $k \in I$ we have $i \leqq k$, $j \leqq k$ and $x\varphi_{ik} = z\varphi_{jk}$; therefore $\hat{x}_{I_r} = \hat{z}_{I_r}$ for any $r \in P$ with $p \leqq r$, $q \leqq r$, and $k \in I_r$. Thus $y\psi_{pr} = \hat{x}_{I_p}\psi_{pr} = \hat{x}_{I_r} = \hat{z}_{I_r} = \hat{z}_{I_q}\psi_{qr} = w\psi_{qr}$, which means $\hat{y}_P = \hat{w}_P$, which was to be proved.

ψ is onto since if we take $\hat{y}_P$, where $y \in A_\infty{}^p$, and $y = \hat{x}_{I_p}$, $x \in A_i$, $i \in I_p$, then

$$\psi: \hat{x}_I \to \hat{y}_P.$$

We will now prove that ψ is 1–1. Take $x \in A_i$ and $v \in A_j$ and let $i \in I_p$ and $j \in I_q$ and set

$$\hat{x}_{I_p} = y_1, \quad \hat{v}_{I_q} = y_2,$$

and assume that $(\hat{y}_1)_P = (\hat{y}_2)_P$. Then there exists an $r \in P$ with $p \leqq r$, $q \leqq r$ such that

$$\hat{x}_{I_p}\psi_{pr} = \hat{v}_{I_q}\psi_{qr},$$

that is,

$$\hat{x}_{I_r} = \hat{v}_{I_r}.$$

This implies that i and j have an upper bound m in I_r such that $x\varphi_{im} = v\varphi_{jm}$. Consequently, $\hat{x}_I = \hat{v}_I$, which was to be proved.

Since ψ obviously preserves the operations, ψ is an isomorphism. This completes the proof of Theorem 2.

Let $\mathscr{A}$ be an inverse family of algebras with carrier $\langle I; \leqq \rangle$ and assume that $I = \bigcup (I_p \mid p \in P)$ such that $\langle I_p; \leqq \rangle$ is a directed partially ordered set and $I_p \subseteq I_q$ if $p \leqq q$.

We introduce the following notation:

$$\varprojlim \mathscr{A} = \langle A^\infty; F \rangle,$$
$$\varprojlim \mathscr{A}_{I_p} = \langle A_p{}^\infty; F \rangle.$$

If $p \leqq q$, we define the mapping

$$\psi_p{}^q \colon g \to g_{I_p},$$

where $g \in A_q{}^\infty$ and g_{I_p} is the restriction of g to I_p. Let us denote the system just defined by $\mathscr{A}/P$.

Theorem 3. *$\mathscr{A}/P$ is an inverse family and we have the following isomorphism:*

$$\varprojlim \mathscr{A} \cong \varprojlim \mathscr{A}/P.$$

Proof. It is obvious that $\mathscr{A}/P$ is an inverse family. Set

$$\varprojlim \mathscr{A}/P = \langle A_1{}^\infty; F \rangle.$$

We set up the mapping

$$\psi \colon g \to f_g,$$

$g \in A^\infty$, where $f_g \in \prod (A_p{}^\infty \mid p \in P)$ is defined by

$$f_g(p) = g_{I_p} \in A_p{}^\infty.$$

To prove that $f_g \in A_1{}^\infty$, we have to show that $f_g(q)\psi_p{}^q = f_g(p)$ if $p \leqq q$, which is trivial since

$$f_g(q)\psi_p{}^q = g_{I_q}\psi_p{}^q = g_{I_p} = f_g(p).$$

If we are given f_g, then we can reconstruct g since if $i \in I$, then $i \in I_p$ for some $p \in P$ and $g(i) = (f_g(p))(i)$; thus, ψ is 1–1.

ψ is onto because if $h \in A_1{}^\infty$, then define g by

$$g(i) = (h(p))(i), \qquad i \in I_p.$$

g is well defined. For, assume that also $i \in I_q$ and let $p, q \leqq r$. Then

$$h(r)\psi_p{}^r = (h(r))_{I_p} = h(p).$$

Thus,

$$(h(r))(i) = (h(p))(i)$$

and, similarly,

$$(h(r))(i) = (h(q))(i).$$

Therefore, $(h(p))(i) = (h(q))(i)$, which was to be proved.

Now it is obvious that $g \in A^\infty$ and that $g\psi = h$.

Since ψ is obviously a homomorphism, it is an isomorphism; this completes the proof of Theorem 3.

Let $\mathscr{A}$ be a direct family (inverse family) of algebras such that the carrier $\langle I; \leqq \rangle$ is well ordered. Then we call $\mathscr{A}$ a *well-ordered direct family (inverse family)*.

A class of algebras K is called *algebraic* if it is closed under isomorphism, that is, if an algebra is isomorphic to an algebra in K, then it is contained in K.

Theorem 4. *Let K be an algebraic class. If K is closed under well-ordered direct limits, then K is closed under arbitrary direct limits.*

Proof. Take a direct family $\mathscr{A}$ with the carrier $\langle I; \leqq \rangle$. If the theorem were not true, then we could choose $\mathscr{A}$ such that $\varinjlim \mathscr{A} \notin K$ and $|I| = \mathfrak{m}$ is the smallest possible. $\mathfrak{m} < \aleph_0$ is impossible, since then $\langle I; \leqq \rangle$ is a finite directed partially ordered set, hence it has a largest element i; $\langle \{i\}; \leqq \rangle$ is cofinal with $\langle I; \leqq \rangle$, so by Lemma 7, $\varinjlim \mathscr{A} \cong \mathfrak{A}_i \in K$.

If $\mathfrak{m} \geqq \aleph_0$, then by Exercise 1.44, $I = \bigcup (I_\gamma \mid \gamma < \alpha)$, where α is an ordinal, $\langle I_\gamma; \leqq \rangle$ is directed, $I_\gamma \subseteq I_\delta$ if $\gamma \leqq \delta < \alpha$, and $|I_\gamma| < |I| = \mathfrak{m}$. Thus $\varinjlim \mathscr{A}_{I_\gamma} \in K$ (since $\mathfrak{m}$ was minimal) and by Theorem 2,

$$\varinjlim \mathscr{A} \cong \varinjlim \mathscr{A}/P,$$

where $P = \{\gamma \mid \gamma < \alpha\}$. Thus by assumption, $\varinjlim \mathscr{A}/P \in K$, and so $\varinjlim \mathscr{A} \in K$, a contradiction. The proof is thus complete.

Theorem 5. *Let K be an algebraic class. If K is closed under well-ordered inverse limits, then K is closed under arbitrary inverse limits.*

Proof. Same as that of Theorem 4, using Theorem 3 rather than Theorem 2.

Now we want to investigate equality of polynomials on direct and inverse limits.

Let $\mathscr{A}$ be a direct family of algebras $\mathfrak{A}_i$, with carrier $\langle I; \leqq \rangle$, $\lim\limits_{\rightarrow} \mathscr{A} = \mathfrak{A}_\infty$, $\mathbf{p}$ and $\mathbf{q}$ m-ary polynomial symbols, $\hat{x}_I^0, \cdots, \hat{x}_I^{m-1}, \hat{y}_I^0, \cdots, \hat{y}_I^{m-1} \in A_\infty$, $x^0 \in A_{i_0}, \cdots, x^{m-1} \in A_{i_{m-1}}, y^0 \in A_{j_0}, \cdots, y^{m-1} \in A_{j_{m-1}}$.

Lemma 8. *$p(\hat{x}_I^0, \cdots, \hat{x}_I^{m-1}) = q(\hat{y}_I^0, \cdots, \hat{y}_I^{m-1})$ in $\mathfrak{A}_\infty$ if and only if there exists an upper bound i of $i_0, \cdots, i_{m-1}, j_0, \cdots, j_{m-1}$ such that we have*

$$p(x^0\varphi_{i_0 i}, \cdots, x^{m-1}\varphi_{i_{m-1} i}) = q(y^0\varphi_{j_0 i}, \cdots, y^{m-1}\varphi_{j_{m-1} i}).$$

Or, equivalently, $p(\hat{x}_I^0, \cdots, \hat{x}_I^{m-1}) \neq q(\hat{y}_I^0, \cdots, \hat{y}_I^{m-1})$ if and only if for any upper bound j of $i_0, \cdots, i_{m-1}, j_0, \cdots, j_{m-1}$ we have

$$p(x^0\varphi_{i_0 j}, \cdots, x^{m-1}\varphi_{i_{m-1} j}) \neq q(y^0\varphi_{j_0 j}, \cdots, y^{m-1}\varphi_{j_{m-1} j}).$$

Proof. If $x = p(x^0\varphi_{i_0 i}, \cdots, x^{m-1}\varphi_{i_{m-1} i}) = q(y^0\varphi_{j_0 i}, \cdots, y^{m-1}\varphi_{j_{m-1} i}) = y$, then $x\varphi_{i\infty} = y\varphi_{i\infty}$ and so $p(\hat{x}_I^0, \cdots, \hat{x}_I^{m-1}) = q(\hat{y}_I^0, \cdots, \hat{y}_I^{m-1})$. Conversely, if $p(\hat{x}_I^0, \cdots, \hat{x}_I^{m-1}) = q(\hat{y}_I^0, \cdots, \hat{y}_I^{m-1})$, then for any upper bound k of $i_0, \cdots, i_{m-1}, j_0, \cdots, j_{m-1}$,

$$x = p(x^0\varphi_{i_0 k}, \cdots, x^{m-1}\varphi_{i_{m-1} k}) \equiv q(y^0\varphi_{j_0 k}, \cdots, y^{m-1}\varphi_{j_{m-1} k}) = y;$$

thus there exists an $i \geqq k$ with $x\varphi_{ki} = y\varphi_{ki}$.

Now let $\mathscr{A}$ be an inverse limit family of algebras, $\lim\limits_{\leftarrow} \mathscr{A} = \mathfrak{A}^\infty$, $\mathbf{p}$ and $\mathbf{q}$ m-ary polynomial symbols, $f^0, \cdots, f^{m-1}, g^0, \cdots, g^{m-1} \in A^\infty$.

Lemma 9. *$p(f^0, \cdots, f^{m-1}) = q(g^0, \cdots, g^{m-1})$ in $\mathfrak{A}^\infty$ if and only if $p(f^0(i), \cdots, f^{m-1}(i)) = q(g^0(i), \cdots, g^{m-1}(i))$ in $\mathfrak{A}_i$ for all $i \in I$. Or, equivalently, $p(f^0, \cdots, f^{m-1}) \neq q(g^0, \cdots, g^{m-1})$ in $\mathfrak{A}^\infty$ if and only if there exists an $i \in I$ such that for all $j \geqq i$, $p(f^0(j), \cdots, f^{m-1}(j)) \neq q(g^0(j), \cdots, g^{m-1}(j))$ in $\mathfrak{A}_j$.*

Proof. Trivial by Lemma 19.2.

Lemmas 8 and 9 indicate that to 1–1 direct limits correspond the onto inverse limits of polynomial algebras and to onto inverse limits correspond the 1–1 direct limits of polynomial algebras. A similar situation can be found in the next result. We prove a statement on direct limits by using inverse limits.

Theorem 6 (*G. Grätzer* [5]). *Let $\mathfrak{A}$ be a finite algebra. An algebra $\mathfrak{B}$ has a homomorphism into $\mathfrak{A}$ if and only if every finite relative subalgebra $\mathfrak{C}$ of $\mathfrak{B}$ has a homomorphism into $\mathfrak{A}$.*

Proof. The "only if" part is trivial. To prove the "if" part, for a finite relative subalgebra $\mathfrak{C}$ of $\mathfrak{B}$ let $T(C)$ denote the set of all homomorphisms of $\mathfrak{C}$ into $\mathfrak{A}$. By assumption, $T(C)$ is not void. Let I denote the set of all finite nonvoid subsets of B; then $\langle I; \subseteq \rangle$ is a directed partially ordered set. Let the inverse family $\mathfrak{A}$ consist of all $T(C)$, $C \in I$, let $\langle I; \subseteq \rangle$ be the carrier of $\mathscr{A}$, and for $C_1 \subseteq C_2$, let $\varphi_{C_1}^{C_2}$ be defined by

$$\chi\varphi_{C_1}^{C_2} = \chi_{C_1} \quad \text{for} \quad \chi \in T(C_2).$$

Then $\mathscr{A}$ is an inverse family of finite nonvoid sets, so by Theorem 1, $\varprojlim \mathscr{A}$ is nonvoid. Let $\chi \in \varprojlim \mathscr{A}$.

For $b \in B$, let $b\tilde{\chi}$ be defined as follows:

$$b\tilde{\chi} = b(\chi\varphi_{\{b\}}^{\infty}).$$

Obviously, $b\tilde{\chi} = b(\chi\varphi_C{}^{\infty})$ for every C containing b. Therefore, it is trivial to check that $\tilde{\chi}$ is a homomorphism of $\mathfrak{B}$ into $\mathfrak{A}$. This completes the proof of Theorem 5.

The following result can be proved in exactly the same way as Theorem 6.

Theorem 7 (*G. Grätzer* [5]). *Let $\mathfrak{A}$ be a finite algebra. The algebra $\mathfrak{B}$ is isomorphic to a subdirect power of $\mathfrak{A}$ if and only if for every $u, v \in B$, $u \neq v$ there exists a finite subset C_{uv} of B such that* (i) *$u, v \in C_{uv}$;* (ii) *for every finite subset C containing C_{uv} there is a mapping φ from C onto A such that $u\varphi \neq v\varphi$ and φ is a homomorphism of $\mathfrak{C}$ onto $\mathfrak{A}$.*

§22. PRODUCTS ASSOCIATED WITH THE DIRECT PRODUCT

We can construct new algebras from given ones by taking subalgebras or homomorphic images of the direct product of given algebras.

Definition 1 (*Weak direct product*). *Let $\mathfrak{A}_i = \langle A_i; F \rangle$, $i \in I$, be algebras and form the direct product $\prod (\mathfrak{A}_i \mid i \in I)$. Let $B \subseteq \prod (A_i \mid i \in I)$.*

We call $\mathfrak{B} = \langle B; F \rangle$ a weak direct product *of the algebras $\mathfrak{A}_i$ provided:*

(i) *$\mathfrak{B}$ is a subalgebra of $\prod (\mathfrak{A}_i \mid i \in I)$;*
(ii) *$f, g \in B$ imply that $\{i \mid f(i) \neq g(i)\}$ is a finite subset of I;*
(iii) *if $f \in B$, $g \in \prod (A_i \mid i \in I)$ and $\{i \mid f(i) \neq g(i)\}$ is finite, then $g \in B$.*

Note that the weak direct product does not necessarily exist (see Exercises 45, 46).

A common generalization of direct product and weak direct product, due to J. Hashimoto [1], is the following.

Definition 2. *Let $\mathfrak{A}_i = \langle A_i; F \rangle$, $i \in I$, be given algebras and let L be an ideal of the Boolean algebra*

$$\mathfrak{P}(I) = \langle P(I); \cup, \cap, ', \varnothing, I \rangle.$$

Let $B \subseteq \prod (A_i \mid i \in I)$. $\mathfrak{B} = \langle B; F \rangle$ is called an L-**restricted direct product** *of the $\mathfrak{A}_i$, $i \in I$, if*

(i) *$\mathfrak{B}$ is a subalgebra of $\prod (\mathfrak{A}_i \mid i \in I)$;*
(ii) *$f, g \in B$ imply that*

$$\{i \mid f(i) \neq g(i)\} \in L;$$

(iii) *$f \in B$ and $g \in \prod (A_i \mid i \in I)$ and $\{i \mid f(i) \neq g(i)\} \in L$ imply that $g \in B$.*

A weak direct product is an L-restricted direct product with L equal to the ideal of all finite subsets of I. If $L = P(I)$, then the L-restricted direct product is isomorphic to $\prod (\mathfrak{A}_i \mid i \in I)$. Finally, if $L = \{\varnothing\}$, then an L-restricted direct product is a one-element algebra (if the product exists).

Let $\mathfrak{B}$ be an L-restricted direct product of $\mathfrak{A}_i$, $i \in I$. If one of the $\mathfrak{A}_i$ is a one-element algebra, then it can be omitted from the direct product, so from now on we assume that $|A_i| \neq 1$ for all $i \in I$. If $i \in I$ and $\{i\} \notin L$, then $f(i) = g(i)$ for every $f, g \in B$. Hence, we can also assume that L contains all finite subsets of I. Let Θ_i be the congruence relation of $\mathfrak{B}$ under which $f \equiv g(\Theta_i)$ if and only if $f(i) = g(i)$. For any ideal $J \subseteq L$, we define a congruence relation of $\mathfrak{B}$ by $f \equiv g(\Theta_J)$ if and only if $\{i \mid f(i) \neq g(i)\} \in J$. If we introduce the notation $D(f, g) = \{i \mid f(i) \neq g(i)\}$, then $f \equiv g(\Theta_J)$ if and only if $D(f, g) \in J$. Obviously, $\Theta_i = \Theta_{P_i}$, where $P_i = \{M \mid M \in L \text{ and } i \notin M\}$.

Lemma 1. *The complete sublattice $\langle \Sigma; \leq \rangle$ of $\mathfrak{C}(\mathfrak{B})$ generated by the Θ_i consists of the Θ_J with $J \subseteq L$. Further, the correspondence $J \to \Theta_J$ is an isomorphism between $\mathfrak{I}(\mathfrak{L})$ and $\langle \Sigma; \leq \rangle$, where $\mathfrak{L} = \langle L; \subseteq \rangle$.*

Proof. If $M \in L$, then $\Theta_{(M]} = \bigcap (\Theta_i \mid i \notin M)$. Further,

$$\Theta_J = \bigvee (\Theta_{(M]} \mid M \in J).$$

Thus, if we prove that the Θ_J form a complete sublattice and $J \to \Theta_J$ is an isomorphism, then we are through.

The following statement is trivial.

(i) If $J_0 \subseteq J_1$, then $\Theta_{J_0} \leq \Theta_{J_1}$.

Next, we prove the following:

(ii) $\bigvee (\Theta_{J_\lambda} \mid \lambda \in \Lambda) = \Theta_{\bigvee (J_\lambda \mid \lambda \in \Lambda)}$.

By (i), $\leqq$ is trivial. Set $J = \bigvee (J_\lambda \mid \lambda \in \Lambda)$. Let us assume then that $f \equiv g(\Theta_J)$. Then $D(f, g) \in J$; hence, we can find $\lambda_1, \cdots, \lambda_n \in \Lambda$ and $Z_i \in J_{\lambda_i}$ such that $D(f, g) = Z_1 \cup \cdots \cup Z_n$ and $Z_i \cap Z_j = \varnothing$ if $i \neq j$.

For $k = 0, 1, \cdots, n$, define $h_k \in \prod (A_i \mid i \in I)$ by $h_0 = f$, $h_n = g$ and $h_k(i) = h_{k-1}(i)$ if $i \notin Z_k$ and $h_k(i) = g(i)$ if $i \in Z_k$. Then $D(h_{k-1}, h_k) \subseteq Z_k$. Thus, $h_{k-1} \equiv h_k(\Theta_{J_{\lambda_k}})$; therefore, $f \equiv g(\bigvee (\Theta_{J_\lambda} \mid \lambda \in \Lambda))$, which was to be proved.

(iii) $\bigwedge (\Theta_{J_\lambda} \mid \lambda \in \Lambda) = \Theta_{\bigwedge (J_\lambda \mid \lambda \in \Lambda)}$.

Indeed, $f \equiv g(\bigwedge (\Theta_{J_\lambda} \mid \lambda \in \Lambda))$ if and only if $D(f, g) \in J_\lambda$ for all $\lambda \in \Lambda$, which is equivalent to $D(f, g) \in \bigwedge (J_\lambda \mid \lambda \in \Lambda)$, which, in turn, is equivalent to $f \equiv g(\Theta_{\bigwedge (J_\lambda \mid \lambda \in \Lambda)})$.

(iv) If $J_\lambda \subset J_\nu$, then $\Theta_{J_\lambda} < \Theta_{J_\nu}$.

Let $M \in J_\nu$, $M \notin J_\lambda$, and $f \in B$. Define $g \in \prod (A_i \mid i \in I)$ by $g(i) = f(i)$ if $i \notin M$ and $g(i) \neq f(i)$ if $i \in M$ (this can be done since $|A_i| > 1$). Then $D(f, g) = M \in L$. Thus, $g \in B$. By construction, $f \equiv g(\Theta_{J_\nu})$ and $f \not\equiv g(\Theta_{J_\lambda})$.

Statements (i)–(iv) complete the proof of Lemma 1.

Definition 3. *Let Σ be a set of congruence relations of $\mathfrak{A}$. Σ is called* completely permutable *if whenever we are given $(\Theta_\lambda \mid \lambda \in \Lambda)$, $\Theta_\lambda \in \Sigma$, and set*

$$\varphi_\lambda = \bigwedge (\Theta_\nu \mid \nu \neq \lambda, \nu \in \Lambda),$$

and we are given $(x_\lambda \mid \lambda \in \Lambda)$, $x_\lambda \in A$ with $x_\lambda \equiv x_\nu(\varphi_\lambda \vee \varphi_\nu)$ for all $\lambda, \nu \in \Lambda$, then we get that there exists an $x \in A$ such that $x \equiv x_\lambda(\Theta_\lambda)$ for all $\lambda \in \Lambda$.

Corollary. *If Σ is completely permutable, then any two congruence relations of Σ are permutable.*

Proof. Indeed, let $\Lambda = \{\lambda, \nu\}$. Then $\varphi_\lambda = \Theta_\nu$ and $\varphi_\nu = \Theta_\lambda$. By complete permutability, $x_\lambda \equiv x_\nu(\Theta_\lambda \vee \Theta_\nu)$ implies that for some x, $x_\lambda \equiv x(\Theta_\lambda)$ and $x_\nu \equiv x(\Theta_\nu)$; thus, $x_\lambda \equiv x_\nu(\Theta_\lambda \Theta_\nu)$. This means that $\Theta_\lambda \Theta_\nu = \Theta_\lambda \vee \Theta_\nu$, i.e., Θ_λ and Θ_ν are permutable.

For further results on the relationship between permutability and complete permutability, see the Exercises.

Let Σ be as in Lemma 1. Then, by Lemma 1, $\langle \Sigma; \leqq \rangle$ is an algebraic lattice since it is isomorphic to $\mathfrak{I}(\mathfrak{L})$. Therefore, the compact elements of $\langle \Sigma; \leqq \rangle$ are the ones of the form Θ_J, where J is a principal ideal, $J = (M]$, $M \in L$. Let $K(\Sigma)$ denote the compact elements of $\langle \Sigma; \leqq \rangle$.

Lemma 2. *$K(\Sigma)$ is completely permutable.*

Proof. Let $\Theta_\lambda \in \Sigma$, $\lambda \in \Lambda$ and $\Theta_\lambda = \Theta_{J_\lambda}$ where $J_\lambda = (M_\lambda]$. Let $f_\lambda \in B$ for $\lambda \in \Lambda$ and $f_\lambda \equiv f_\nu(\varphi_\lambda \vee \varphi_\nu)$. Note that $f \equiv g(\Theta_\lambda)$ if and only if $D(f, g) \subseteq M_\lambda$; hence,

$$D(f_\lambda, f_\nu) \subseteq \bigcap (M_\mu \mid \mu \neq \lambda) \cup \bigcap (M_\mu \mid \mu \neq \nu).$$

Or, equivalently ($'$ is complementation in $\mathfrak{P}(I)$),

$$D(f_\lambda, f_\nu)' \supseteq \bigcup (M_\mu' \mid \mu \neq \lambda) \cap \bigcup (M_\mu' \mid \mu \neq \nu) \supseteq M_\nu' \cap M_\lambda'.$$

Summarizing:

$$\text{if } i \in M_\nu' \cap M_\lambda', \qquad \text{then } f_\lambda(i) = f_\nu(i).$$

Now we define $f \in \prod (A_i \mid i \in I)$ as follows: $f(i)$ is arbitrary if

$$i \in \bigcap (M_\lambda \mid \lambda \in \Lambda);$$

$f(i) = f_\lambda(i)$ if $i \in M_\lambda'$ for some $\lambda \in \Lambda$.

What we have proved above shows that no contradiction is obtained if i is an element of M_λ' and of M_ν' at the same time, since then $f_\lambda(i) = f_\nu(i)$.

By definition,

$$D(f, f_\nu) \subseteq M_\nu \in L,$$

which proves that $f \in B$ and that $f \equiv f_\nu(\Theta_\nu)$, completing the proof of Lemma 2.

Now we state the main result.

Theorem 1 (*J. Hashimoto* [1]). *Let $\mathfrak{A}$ be an algebra with more than one element, let I be a nonvoid set, and let L be an ideal of $\mathfrak{P}(I)$, containing all finite subsets of I. Then $\mathfrak{A}$ is isomorphic to an L-restricted direct product of algebras with more than one element if and only if $\mathfrak{C}(\mathfrak{A})$ has a complete sublattice $\langle \Sigma; \leqq \rangle$ such that the following hold:*

(i) $\omega, \iota \in \Sigma$;

(ii) $\langle K(\Sigma); \leqq \rangle \cong \langle L; \subseteq \rangle$;

(iii) $K(\Sigma)$ *is completely permutable.*

Proof. To prove the necessity of conditions (i)–(iii), observe that $\omega = \Theta_{(\varnothing]}$ and $\iota = \Theta_L$; thus, $\omega, \iota \in \Sigma$. The isomorphism in (ii) is set up by

$$\Theta_{(M]} \to M$$

by Lemma 1.

(iii) was proved in Lemma 2.

To prove the sufficiency, let us assume that $\mathfrak{C}(\mathfrak{A})$ has a complete sublattice $\langle \Sigma; \leqq \rangle$ satisfying (i), (ii), and (iii). Condition (ii) implies that $\langle I(\mathfrak{L}); \subseteq \rangle$ is isomorphic to $\langle \Sigma; \leqq \rangle$. Let this isomorphism be $J \to \Theta_J$. For

$i \in I$, consider the ideal $P_i = \{M \mid M \in L, i \notin M\}$. Then $P_i \neq L$ since $\{i\} \notin P_i$. Set $\Theta_i = \Theta_{P_i}$. Since $P_i \neq L$, $\Theta_i \neq \iota$; thus $\mathfrak{A}_i = \mathfrak{A}/\Theta_i$ has more than one element. For $x \in A$, define $f_x \in \prod (A_i \mid i \in I)$ by $f_x(i) = [x]\Theta_i$ and set $A^* = \{f_x \mid x \in A\}$.

(a) $x \to f_x$ is an isomorphism between $\mathfrak{A}$ and the subalgebra $\mathfrak{A}^*$ of $\prod (\mathfrak{A}_i \mid i \in I)$.

Indeed, the mapping is obviously an onto homomorphism and it is 1–1 since $\bigwedge (\Theta_i \mid i \in I) = \bigwedge (\Theta_{P_i} \mid i \in I) = \Theta_{\bigwedge(P_i \mid i \in I)} = \Theta_{\{\varnothing\}} = \omega$.

(b) $x \equiv y(\Theta_{(M]})$ if and only if $D(f_x, f_y) \subseteq M$.

Since $(M] = \bigwedge (P_i \mid i \notin M)$, if $x \equiv y(\Theta_{(M]})$, then $x \equiv y(\Theta_{P_i})$ for all $i \notin M$, that is, $x \equiv y(\Theta_i)$ for all $i \notin M$, which in turn implies that $D(f_x, f_y) \subseteq M$, and conversely.

(c) If $f, g \in A^*$, then $D(f, g) \in L$.

Let $f = f_x$ and $g = f_y$. Then $x \equiv y(\iota)$, i.e., $x \equiv y(\Theta_L)$. Since

$$L = \bigvee ((M] \mid M \in L),$$

we get that $x \equiv y(\Theta_{\bigvee((M] \mid M \in L)})$, that is,

$$x \equiv y(\bigvee (\Theta_{(M]} \mid M \in L)).$$

Thus, there exist a sequence $x = x_0, x_1, \cdots, x_n = y$ and $M_0, \cdots, M_{n-1} \in L$ such that $x_i \equiv x_{i+1}(\Theta_{(M_i]})$. Set $M = M_0 \cup \cdots \cup M_{n-1}$. Then $x \equiv y(\Theta_{(M]})$. Thus, by (b), $D(f_x, f_y) \subseteq M \in L$, which was to be proved.

(d) If $f \in A^*$, $g \in \prod (A_i \mid i \in I)$ and $D(f, g) \in L$, then $g \in A^*$.

If Θ is a congruence relation of $\mathfrak{A}$, let Θ^* denote the corresponding congruence relation of $\mathfrak{A}^*$, i.e., $f_x \equiv f_y(\Theta^*)$ if and only if $x \equiv y(\Theta)$. Set

$$\Sigma^* = \{\Theta^* \mid \Theta \in \Sigma\}.$$

Then $K(\Sigma^*)$ is again completely permutable.

For $i \in I$, denote $\Theta_{(\{i\}]}$ by φ_i. Then, by (b), $f_x \equiv f_y(\varphi_i^*)$ if and only if $D(f_x, f_y) \subseteq \{i\}$. By the corollary to Definition 3, Θ_i and φ_i are permutable and it is obvious that $\Theta_i \vee \varphi_i = \iota$.

Since $\bigwedge (\Theta_i \mid i \in I) = \omega$, $\mathfrak{A}^*$ is a subdirect product of the $\mathfrak{A}_i$; thus, we can find $g_i \in A^*$ with $g_i(i) = g(i)$. Then $g_i \equiv f(i)$; thus, $g_i \equiv f(\Theta_i^* \vee \varphi_i^*)$. By the permutability of Θ_i^* and φ_i^* there exists an element $h_i \in A^*$ such that $g_i \equiv h_i(\Theta_i^*)$ and $h_i \equiv f(\varphi_i^*)$.

Set $\psi_i = \Theta_{(D(f,g)-\{i\}]}$. Then $\varphi_i = \bigwedge (\psi_j \mid j \neq i)$ if $i \in D(f, g)$. We will apply the complete permutability to the congruence relations $\{\psi_i^* \mid i \in I\}$ and the elements $\{h_i \mid i \in I\}$. In order to do that, we must prove that

$$h_i \equiv h_j(\varphi_i^* \vee \varphi_j^*) \text{ for } i, j \in D(f, g).$$

Indeed, $f \equiv h_i(\varphi_i^*)$ and $f \equiv h_j(\varphi_j^*)$. Thus, $h_i \equiv h_j(\varphi_i^* \vee \varphi_j^*)$. By complete permutability, there exists an element $\bar{g} \in A^*$ such that $\bar{g} \equiv h_i(\psi_i^*)$ for all $i \in I$.

Since $\bar{g} \equiv h_i(\psi_i^*)$, $D(\bar{g}, h_i) \subseteq D(f, g) - \{i\}$. Hence, $\bar{g}(i) = h_i(i)$ if $i \in D(f, g)$. However, $h_i \equiv g_i(\Theta_i^*)$; thus, $h_i(i) = g_i(i) = g(i)$. This proves that $g(i) = \bar{g}(i)$ for $i \in D(f, g)$.

If $i \notin D(f, g)$, then $\bar{g}(i) = h_j(i)$ for all $j \in D(f, g)$. Since $f \equiv h_j(\varphi_j^*)$, $D(f, h_j) \subseteq D(f, g)$; thus, $f(i) = g(i) = h_j(i)$ for any $i \in I$. Therefore, $g(i) = \bar{g}(i)$ for $i \notin D(f, g)$.

Thus, we have proved that $g = \bar{g} \in A^*$, which completes the proof of (d).

(c) and (d) prove the sufficiency of the conditions of Theorem 1.

A construction which in a sense is the dual of an L-restricted direct product is the following.

Let L be given as in Definition 2. We define a relation on $\prod (A_i \mid i \in I)$ as follows:

$$f \equiv g(\Theta_L) \quad \text{if and only if} \quad \{i \mid f(i) \neq g(i)\} \in L.$$

Lemma 3. *Θ_L is a congruence relation of $\prod (\mathfrak{A}_i \mid i \in I)$.*

Proof. Θ_L is reflexive since $\varnothing \in L$; Θ_L is symmetric since the definition is symmetric; Θ_L is transitive since

$$\{i \mid f(i) \neq h(i)\} \subseteq \{i \mid f(i) \neq g(i)\} \cup \{i \mid g(i) \neq h(i)\}.$$

Θ_L has the substitution property because if $g_k \equiv h_k(\Theta_L)$, then

$$\{i \mid f_\gamma(h_0, \cdots, h_{n_\gamma - 1})(i) \neq f_\gamma(g_0, \cdots, g_{n_\gamma - 1})(i)\} \subseteq$$
$$\bigcup (\{i \mid h_k(i) \neq g_k(i)\} \mid k < n_\gamma) \in L.$$

Definition 4. *The quotient algebra of $\prod (\mathfrak{A}_i \mid i \in I)$ modulo Θ_L is called the* L-reduced direct product *(also called* reduced product*) of the algebras $\mathfrak{A}_i$, $i \in I$, and is denoted by*

$$\textstyle\prod_L (\mathfrak{A}_i \mid i \in I).$$

If $\mathfrak{A} = \mathfrak{A}_i$ for all $i \in I$, we will use the notation $\mathfrak{A}_L{}^I$, and we will call $\mathfrak{A}_L{}^I$ a reduced direct power.

It is hard to trace the origin of reduced direct products. Special cases have been known for a long time. For some historical notes, see T. Frayne, A. C. Morel, and D. S. Scott [1].

In the special case $L = \{\varnothing\}$, $\prod_L (\mathfrak{A}_i \mid i \in I) \cong \prod (\mathfrak{A}_i \mid i \in I)$, while if $L = P(I)$, then $\prod_L (\mathfrak{A}_i \mid i \in I) = \mathbf{1}^{(\tau)}$, the one element algebra.

Suppose Θ_i is a congruence relation of $\mathfrak{A}_i$. We define a relation which will be denoted by $\prod_L (\Theta_i \mid i \in I)$ on $\prod_L (A_i \mid i \in I)$ as follows:

$$f \equiv g(\textstyle\prod_L (\Theta_i \mid i \in I)) \quad \text{if and only if} \quad \{i \mid f(i) \not\equiv g(i)(\Theta_i)\} \in L.$$

Lemma 4. *$\prod_L (\Theta_i \mid i \in I)$ is a congruence relation of $\prod_L (\mathfrak{A}_i \mid i \in I)$.*

Proof. Same as that of Lemma 3.

Definition 5 (*J. Łoś* [2]). *A* prime product (*also called* ultra product) *is an L-reduced product where L is a prime ideal.*

The significance of prime products will be made clear in Chapter 6. These notions do not yield a new construction if L is a principal ideal.

Lemma 5. *Let $L = \{X \mid X \subseteq A\}$, where $A \subseteq I$. Then*

$$\textstyle\prod_L (\mathfrak{A}_i \mid i \in I) \cong \prod (\mathfrak{A}_i \mid i \in I - A).$$

Proof. Let $f \in \prod (A_i \mid i \in I)$ and consider the mapping

$$\psi: [f]\Theta_L \to f_{I-A}.$$

It is trivial that ψ is the required isomorphism.

Corollary. *A prime product of algebras with respect to a principal prime ideal is always isomorphic to one of the given algebras.*

Proof. Trivial since a principal prime ideal is always of the form

$$P = \{X \mid a \notin X\},$$

where $a \in I$; in this case, $A = I - \{a\}$.

Let L be an ideal of $\mathfrak{P}(I)$. Then $\mathscr{D} = \{X \mid I - X \in L\}$ is a dual ideal of $\mathfrak{P}(I)$. Furthermore,

$$D(f, g) \in L \quad \text{if and only if} \quad \{i \mid f(i) = g(i)\} \in \mathscr{D}.$$

Thus $f \equiv g(\Theta_L)$ if and only if $\{i \mid f(i) = g(i)\} \in \mathscr{D}$. Therefore, if $f \equiv g(\Theta_{\mathscr{D}})$ is defined by $\{i \mid f(i) = g(i)\} \in \mathscr{D}$, then $\Theta_L = \Theta_{\mathscr{D}}$.

Definition 6. *Let $\mathscr{D}$ be a dual ideal of $\mathfrak{P}(I)$. Then*

$$\textstyle\prod_{\mathscr{D}} (\mathfrak{A}_i \mid i \in I) = \prod (\mathfrak{A}_i \mid i \in I)/\Theta_{\mathscr{D}}$$

is called a $\mathscr{D}$-reduced product *of the $\mathfrak{A}_i$, $i \in I$. Again $\mathfrak{A}_{\mathscr{D}}{}^I$ will denote* reduced direct powers.

Corollary. *Definitions 4 and 6 give exactly the same algebras.*

If $\mathcal{D}$ is prime, then a $\mathcal{D}$-reduced product is again called a prime product; in that special case, $L = P(I) - \mathcal{D}$.

It is sometimes more convenient to use dual ideals rather than ideals.

A general associative law for reduced products is given in the following lemma (see T. Frayne, A. C. Morel, and D. S. Scott [1]).

Lemma 6. *Let π be a partition of I; for $B \in \pi$ let $\mathcal{D}_B$ be a dual ideal of $\mathfrak{P}(B)$ and let $\mathcal{D}'$ be a dual ideal of $\mathfrak{P}(\pi)$. Set*

$$\mathcal{D} = \{X \mid X \subseteq I \text{ and } \{B \mid B \in \pi \text{ and } X \cap B \in \mathcal{D}_B\} \in \mathcal{D}'\}.$$

Then $\mathcal{D}$ is a dual ideal of $\mathfrak{P}(I)$ and

$$\prod_{\mathcal{D}} (\mathfrak{A}_i \mid i \in I) \cong \prod_{\mathcal{D}'} (\prod_{\mathcal{D}_B} (\mathfrak{A}_i \mid i \in B) \mid B \in \pi).$$

Proof. An isomorphism ψ is given by

$$\psi : [f]\Theta_{\mathcal{D}} \to [g]\Theta_{\mathcal{D}'},$$

where $f \in \prod (A_i \mid i \in I)$, $g \in \prod (\prod_{\mathcal{D}_B} (A_i \mid i \in B) \mid B \in \pi)$ and g is defined by

$$g(B) = [f_B]\Theta_{\mathcal{D}_B}.$$

The proof is left as an exercise.

* * *

The third, and final, construction is a generalization of direct powers due to A. L. Foster [2].

Let $\mathfrak{A}$ be an algebra, I a set, and $\mathfrak{B} = \mathfrak{A}^I$. Any $\alpha \in B$ is a function from I into A; thus it induces a partition π_α of I: $x, y \in X \in \pi_\alpha$ is equivalent to $\alpha(x) = \alpha(y) \in A$. Thus we can associate with every $a \in \alpha(I) \subseteq A$ a subset I_a as follows: $i \in I_a$ if and only if $\alpha(i) = a$. Set $I_a = \varnothing$ if $a \notin \alpha(I)$. Then $\alpha^* : a \to I_a$ is a mapping of A onto π^*, where π is a partition of I and $\pi^* = \pi \cup \{\varnothing\}$; α^* has the properties that (i) if $a \neq b$, then $a\alpha^* \cap b\alpha^* = \varnothing$; (ii) $\bigcup (a\alpha^* \mid a \in A) = I$. We can, of course, consider α^* as a mapping of A into $P(I)$. Conversely, if α is a mapping of A into $P(I)$ with properties (i) and (ii), then we can define a mapping $\tilde{\alpha}$ of I into A by $\tilde{\alpha}: i \to a$ if $\{i\} \subseteq a\alpha$. (i) shows that $\tilde{\alpha}$ is well defined, and (ii) guarantees that the domain of $\tilde{\alpha}$ is I.

Let $\alpha_0, \cdots, \alpha_{n_\gamma - 1} \in A^I$ and $\alpha = f_\gamma(\alpha_0, \cdots, \alpha_{n_\gamma - 1})$. How can we find α^* in terms of $\alpha_0^*, \cdots, \alpha_{n_\gamma - 1}^*$? Since $\alpha(i) = f_\gamma(\alpha_0(i), \cdots, \alpha_{n_\gamma - 1}(i))$, $\alpha(i) = a$ if and only if there exist $a_0 = \alpha_0(i), \cdots, a_{n_\gamma - 1} = \alpha_{n_\gamma - 1}(i)$, $a_0, \cdots, a_{n_\gamma - 1} \in A$ with $a = f_\gamma(a_0, \cdots, a_{n_\gamma - 1})$. In other words,

$$i \in a\alpha^* \quad \text{if and only if there exist} \quad a_0, \cdots, a_{n_\gamma - 1} \in A \quad \text{with}$$

$$i \in a_0\alpha_0^*, \cdots, i \in a_{n_\gamma - 1}\alpha_{n_\gamma - 1}^* \quad \text{and} \quad a = f_\gamma(a_0, \cdots, a_{n_\gamma - 1}).$$

In formula:

$$a(f_\gamma(\alpha_0, \cdots, \alpha_{n_\gamma - 1}))^* = \bigcup (a_0\alpha_0{}^* \cap \cdots \cap a_{n_\gamma - 1}\alpha^*_{n_\gamma - 1} \mid f_\gamma(a_0, \cdots, a_{n_\gamma - 1}) = a), \quad (1)$$

and we set $f_\gamma(\alpha_0{}^*, \cdots, \alpha^*_{n_\gamma - 1}) = f_\gamma(\alpha_0, \cdots, \alpha_{n_\gamma - 1})^*$. Note that if there are no such $a_0, \cdots, a_{n_\gamma - 1}$, then we get the void union, which is $\varnothing$.

Thus we have proved the following result.

Lemma 7. *Let $\mathfrak{A}$ be an algebra and I a set. Define the set $A[\mathfrak{P}(I)]$ as the set of all mappings α of A into $P(I)$ satisfying*

(i) *if $a \neq b$, then $a\alpha \cap b\alpha = \varnothing$;*
(ii) $\bigcup (a\alpha \mid a \in A) = I$.

Define the operations f_γ on $A[\mathfrak{P}(I)]$ by (1), and let $\mathfrak{A}[\mathfrak{P}(I)]$ denote the resulting algebra. Then $\mathfrak{A}[\mathfrak{P}(I)]$ is isomorphic to $\mathfrak{A}^I$; an isomorphism is given by $\alpha \to \tilde{\alpha}$, where $\alpha \in A[\mathfrak{P}(I)]$ and $\tilde{\alpha} : I \to A$ is defined by $\tilde{\alpha}(i) = a$ if and only if $i \in a\alpha$. The inverse of $\alpha \to \tilde{\alpha}$ is $\beta \to \beta^*$ where $\beta \in A^I$, $\beta^* \in A[\mathfrak{P}(I)]$ and β^* is defined by $i \in a\beta^*$ if and only if $\beta(i) = a$.

Lemma 7 gives the motivation for the following definition.

Definition 7 (*A. L. Foster* [2]). *Let $\mathfrak{A}$ be an algebra and $\mathfrak{B}$ a Boolean algebra; we assume that if $\mathfrak{A}$ is infinite, then $\mathfrak{B}$ is complete. We define the set $A[\mathfrak{B}]$ to be the set of all mappings α of A into B satisfying the following two conditions:*

(i) *if $a \neq b$, $a, b \in A$, then $a\alpha \wedge b\alpha = 0$;*
(ii) $\bigvee (a\alpha \mid a \in A) = 1$.

We define the n-ary operation f on $A[\mathfrak{B}]$ by $f(\alpha_0, \cdots, \alpha_{n-1}) = \beta$, where β is given by

(iii) $a\beta = \bigvee (a_0\alpha_0 \wedge \cdots \wedge a_{n-1}\alpha_{n-1} \mid f(a_0, \cdots, a_{n-1}) = a)$.

The resulting algebra $\mathfrak{A}[\mathfrak{B}]$ is called the extension *of $\mathfrak{A}$ by $\mathfrak{B}$, or a* Boolean extension *of $\mathfrak{A}$.*

Corollary. *If $\mathfrak{B} \cong \mathfrak{P}(I)$, then $\mathfrak{A}[\mathfrak{B}]$ is isomorphic to $\mathfrak{A}^I$.*

It should be noted that every finite Boolean algebra is isomorphic to some $\mathfrak{P}(I)$, hence this construction gives something new only for infinite Boolean algebras. Also, every Boolean algebra is a subalgebra of some $\mathfrak{P}(I)$; thus $\mathfrak{A}[\mathfrak{B}]$ is always a subalgebra of a direct power.

Some important properties of this construction are given in Theorem 2.

Theorem 2. (i) *if f is any n-ary function on A, then Definition* 7 (iii) *defines an n-ary function $\hat{f}$ on $A[\mathfrak{B}]$;*

(ii) *if a function f on A is a composition of functions $f = g(h_0, \cdots, h_{m-1})$, then $\hat{f} = \hat{g}(\hat{h}_0, \cdots, \hat{h}_{m-1})$;*

(iii) *for $a \in A$, define $\zeta_a \in A[\mathfrak{B}]$ by $a\zeta_a = 1$, $b\zeta_a = 0$ if $b \neq a$; set $\bar{A} = \{\zeta_a \mid a \in A\}$. Then $\bar{\mathfrak{A}}$ is a subalgebra of $\mathfrak{A}[\mathfrak{B}]$ and $\varphi: a \to \zeta_a$ is an isomorphism between $\mathfrak{A}$ and $\bar{\mathfrak{A}}$; furthermore, for any n-ary function f on A, if $f(a_0, \cdots, a_{n-1}) = a$, then $\hat{f}(a_0\varphi, \cdots, a_{n-1}\varphi) = a\varphi$;*

(iv) *if $\mathbf{p}$ is an n-ary polynomial symbol, $p = (\mathbf{p})_{\mathfrak{A}}$, then $\hat{p} = (\mathbf{p})_{\mathfrak{A}[\mathfrak{B}]}$;*

(v) *if f is an n-ary algebraic function on $\mathfrak{A}$ which we get from the polynomial $(\mathbf{p})_{\mathfrak{A}}$ by substituting some x_i by a_i, then $\hat{f}$ is an algebraic function which we get from the polynomial $(\mathbf{p})_{\mathfrak{A}[\mathfrak{B}]}$ by substituting the same x_i by ζ_{a_i};*

(vi) *if f and g are n-ary functions on A, then $f = g$ if and only if $\hat{f} = \hat{g}$.*

Remark. Theorem 2 (i) makes Definition 7 legitimate, since it proves that for the β defined by Definition 7 (iii), we have $\beta \in A[\mathfrak{B}]$.

Proof. The proofs of (i)–(vi) are simple computations. To simplify the notations, we will sometimes assume that the functions considered are binary.

(i) Let f be $+$, $\alpha_0 + \alpha_1 = \beta$; to show $\beta \in A[\mathfrak{B}]$ we have to verify Definition 7 (i) and (ii). If $a \neq b$, then

$$\begin{aligned} a\beta \wedge b\beta &= \bigvee (a_0\alpha_0 \wedge a_1\alpha_1 \mid a_0 + a_1 = a) \wedge \bigvee (b_0\alpha_0 \wedge b_1\alpha_1 \mid b_0 + b_1 = b) \\ &= \bigvee (a_0\alpha_0 \wedge a_1\alpha_1 \wedge b_0\alpha_0 \wedge b_1\alpha_1 \mid a_0 + a_1 = a \text{ and } b_0 + b_1 = b) \\ &= 0, \end{aligned}$$

since either $a_0 \neq b_0$ and so $a_0\alpha_0 \wedge b_0\alpha_0 = 0$ or $a_1 \neq b_1$ and then $a_1\alpha_1 \wedge b_1\alpha_1 = 0$. Now we compute:

$$\begin{aligned} \bigvee (a\beta \mid a \in A) &= \bigvee (a_0\alpha_0 \wedge a_1\alpha_1 \mid a_0 + a_1 = a \text{ and } a \in A) \\ &= \bigvee (a_0\alpha_0 \wedge a_1\alpha_1 \mid a_0, a_1 \in A) \\ &= \bigvee (a_0\alpha_0 \mid a_0 \in A) \wedge \bigvee (a_1\alpha_1 \mid a_1 \in A) \\ &= 1 \wedge 1 = 1. \end{aligned}$$

(ii) Let $f(x_0, x_1) = h_0(x_0, x_1) + h_1(x_0, x_1)$ and $\alpha_0, \alpha_1 \in A[\mathfrak{B}]$. Then if $a \in A$,

$$\begin{aligned} a\hat{f}(\alpha_0, \alpha_1) &= \bigvee (a_0\alpha_0 \wedge a_1\alpha_1 \mid f(a_0, a_1) = a) \\ &= \bigvee (a_0\alpha_0 \wedge a_1\alpha_1 \mid h_0(a_0, a_1) + h_1(a_0, a_1) = a). \quad (2) \end{aligned}$$

On the other hand,

$$\begin{aligned}
a(\hat{h}_0(\alpha_0, \alpha_1) \hat{+} \hat{h}_1(\alpha_0, \alpha_1)) \\
&= \bigvee (a_0\hat{h}_0(\alpha_0, \alpha_1) \wedge a_1\hat{h}_1(\alpha_0, \alpha_1) \mid a_0 + a_1 = a) \\
&= \bigvee (\bigvee (b_0\alpha_0 \wedge b_1\alpha_1 \mid h_0(b_0, b_1) = a_0) \wedge \\
&\qquad \bigvee (c_0\alpha_0 \wedge c_1\alpha_1 \mid h_1(c_0, c_1) = a_1 \mid a_0 + a_1 = a) \\
&= \bigvee (b_0\alpha_0 \wedge b_1\alpha_1 \wedge c_0\alpha_0 \wedge c_1\alpha_1 \mid h_0(b_0, b_1) = a_0, h_1(c_0, c_1) = a_1 \\
&\qquad \text{and } a_0 + a_1 = a) \\
&= (\text{if } b_0 \neq c_0, \text{ then } b_0\alpha_0 \wedge c_0\alpha_0 = 0; \text{ if } b_1 \neq c_1, \text{ then } b_1\alpha_1 \wedge c_1\alpha_1 = 0, \\
&\qquad \text{so we can assume that } b_0 = c_0 \text{ and } b_1 = c_1) \\
&= \bigvee (b_0\alpha_0 \wedge b_1\alpha_1 \mid h_0(b_0, b_1) + h_1(b_0, b_1) = a). \qquad (3)
\end{aligned}$$

(2) and (3) prove (ii).

(iii) and (vi). $\zeta_a \in A[\mathfrak{B}]$ is trivial. It is easy to see that for any n-ary function f on A, we have

$$\zeta_{f(a_0, \cdots, a_{n-1})} = \hat{f}(\zeta_{a_0}, \cdots, \zeta_{a_{n-1}}),$$

proving (iii). This also proves (vi), since if $f \neq g$, then, for example, $f(a_0, \cdots, a_{n-1}) \neq g(a_0, \cdots, a_{n-1})$ and so

$$\hat{f}(\zeta_{a_0}, \cdots, \zeta_{a_{n-1}}) \neq \hat{g}(\zeta_{a_0}, \cdots, \zeta_{a_{n-1}}).$$

(iv) Let $\alpha_0, \cdots, \alpha_{n-1} \in A[\mathfrak{B}]$, $a \in A$. Then

$$\begin{aligned}
a\hat{e}_i^n(\alpha_0, \cdots, \alpha_{n-1}) &= \bigvee (a_0\alpha_0 \wedge \cdots \wedge a_{n-1}\alpha_{n-1} \mid e_i^n(a_0, \cdots, a_{n-1}) = a) \\
&= (e_i^n(a_0, \cdots, a_{n-1}) = a \quad \text{if and only if } a_i = a) \\
&= \bigvee (a_0\alpha_0 \wedge \cdots \wedge a_{n-1}\alpha_{n-1} \mid a_i = a) \\
&= \bigvee (a_0\alpha_0 \mid a_0 \in A) \wedge \cdots \wedge \bigvee (a_{i-1}\alpha_{i-1} \mid a_{i-1} \in A) \\
&\qquad \wedge a\alpha_i \wedge \cdots \wedge \bigvee (a_{n-1}\alpha_{n-1} \mid a_{n-1} \in A) \\
&= 1 \wedge \cdots \wedge 1 \wedge a\alpha_i \wedge \cdots \wedge 1 = a\alpha_i,
\end{aligned}$$

so $\hat{e}_i^n = (\mathbf{x}_i)_{\mathfrak{A}[\mathfrak{B}]}$. Now (iv) follows from (ii).

(v) Trivial from (i)–(iv).

This completes the proof of Theorem 2.

Corollary 1. $\mathfrak{P}^{(n)}(\mathfrak{A}) \cong \mathfrak{P}^{(n)}(\mathfrak{A}[\mathfrak{B}])$.

Corollary 2. *If we identify $\mathfrak{A}$ with $\hat{\mathfrak{A}}$, then $\mathfrak{A}[\mathfrak{B}]$ is an extension of $\mathfrak{A}$ with the property that if f and g are algebraic functions on $\mathfrak{A}$, then $f = g$ on $\mathfrak{A}$ if and only if $\hat{f} = \hat{g}$ on $\mathfrak{A}[\mathfrak{B}]$.*

We want now to relate the construction $\mathfrak{A}[\mathfrak{B}]$ to constructions we already know. Since we can do it only for finite algebras, from now on we assume that $\mathfrak{A}$ is a finite algebra, $|A| > 1$ (the case $|A| = 1$ is trivial).

We introduce an operation (transformation) in the direct power $\mathfrak{A}^I$.

Definition 8. *Let* $\alpha, \beta, \gamma, \delta \in A^I$; $T(\alpha, \beta, \gamma, \delta) = \varepsilon$ *is defined as follows*

$$\varepsilon(i) = \begin{cases} \gamma(i) & \textit{if} \quad \alpha(i) = \beta(i) \\ \delta(i) & \textit{if} \quad \alpha(i) \neq \beta(i). \end{cases}$$

Definition 9. *A subalgebra* $\mathfrak{C}$ *of* $\mathfrak{A}^I$ *is called a* normal subdirect power *if the following conditions are satisfied:*

(i) $\mathfrak{C}$ *contains the diagonal;*
(ii) *if* $\alpha, \beta, \gamma, \delta \in C$, *then* $T(\alpha, \beta, \gamma, \delta) \in C$.

Remark. It is obvious from (i) that a normal subdirect power is indeed a subdirect power.

Now we prove that the extension of a finite algebra by a Boolean algebra is always isomorphic to a normal subdirect power.

Theorem 3. *Let* $\mathfrak{A} = \langle A; F \rangle$ *be a finite algebra,* $\mathfrak{B}$ *a Boolean algebra, and* $\mathfrak{B}'$ *an atomic Boolean algebra which contains* $\mathfrak{B}$ *as a subalgebra. Let I denote the set of atoms of* $\mathfrak{B}'$. *For* $\alpha \in A[\mathfrak{B}]$ *define* $\tilde{\alpha} : I \to A$ *by* $\tilde{\alpha}(i) = a$ *if and only if* $i \leqq a\alpha$. *Set* $A(I) = \{\tilde{\alpha} \mid \alpha \in A[\mathfrak{B}]\} \subseteq A^I$. *Then* $\mathfrak{A}(I) = \langle A(I); F \rangle$ *is a normal subdirect power of* $\mathfrak{A}$ *and* $\alpha \to \tilde{\alpha}$ *is an isomorphism between* $\mathfrak{A}[\mathfrak{B}]$ *and* $\mathfrak{A}(I)$.

Proof. $\mathfrak{B}$ is isomorphic to a subalgebra of $\mathfrak{P}(I)$ and therefore (since $\mathfrak{A}$ is finite) $\mathfrak{A}[\mathfrak{B}]$ is isomorphic to a subalgebra of $\mathfrak{A}[\mathfrak{P}(I)]$; $\varphi : \alpha \to \tilde{\alpha}$ is by Lemma 7 an isomorphism between $\mathfrak{A}[\mathfrak{P}(I)]$ and $\mathfrak{A}^I$; thus φ also sets up an isomorphism between $\mathfrak{A}[\mathfrak{B}]$ and $\mathfrak{A}(I)$. Thus it remains to prove that (i) and (ii) of Definition 9 hold for $\mathfrak{A}(I)$.

For $a \in A$, let $\varepsilon_a \in A^I$ be defined by $\varepsilon_a(i) = a$ for all $i \in I$. ε_a is an element of the diagonal.

(i) is obvious since $\zeta_a\varphi = \varepsilon_a$ (ζ_a was defined in Theorem 2 (iii).)

To prove (ii) let $\tilde{\alpha}, \tilde{\beta}, \tilde{\gamma}, \tilde{\delta} \in A(I)$. Set $x = \bigvee (a\alpha \wedge a\beta \mid a \in A)$ and define $\varepsilon : A \to B$ by

$$a\varepsilon = (a\gamma \wedge x) \vee (a\delta \wedge x') \text{ for } a \in A.$$

$\varepsilon \in A[\mathfrak{B}]$, since if $a, b \in A$, $a \neq b$, then

$$a\varepsilon \wedge b\varepsilon = [(a\gamma \wedge x) \vee (a\delta \wedge x')] \wedge [(b\gamma \wedge x) \vee (b\delta \wedge x')] = 0,$$

since $a\gamma \wedge b\gamma = a\delta \wedge b\delta = x \wedge x' = 0$. Also,

$$\begin{aligned} \bigvee (a\varepsilon \mid a \in A) &= \bigvee (a\gamma \wedge x \mid a \in A) \vee \bigvee (a\delta \wedge x' \mid a \in A) \\ &= x \vee x' = 1. \end{aligned}$$

We claim that $T(\tilde{\alpha}, \tilde{\beta}, \tilde{\gamma}, \tilde{\delta}) = \tilde{\varepsilon}$. Let $i \in I$; if $\tilde{\alpha}(i) = \tilde{\beta}(i)$, then $i \leqq a\alpha$ and $i \leqq a\beta$ for some $a \in A$ and so $i \leqq x$. Thus $i \leqq a\varepsilon$ if and only if $i \leqq a\gamma$, and so $\tilde{\varepsilon}(i) = \tilde{\gamma}(i)$. Similarly, if $\tilde{\alpha}(i) \neq \tilde{\beta}(i)$, then $\tilde{\varepsilon}(i) = \tilde{\delta}(i)$. This completes the proof of Theorem 3.

The converse of Theorem 3 also holds.

Theorem 4. *Let $\mathfrak{A}(I)$ be a normal subdirect power of $\mathfrak{A}$. Then $\mathfrak{P}(I)$ has a subalgebra $\mathfrak{B}$ such that $\mathfrak{A}(I)$ is isomorphic to $\mathfrak{A}[\mathfrak{B}]$.*

Proof. We set

$$B = \{X \mid X \subseteq I \text{ and there exist } a \in A,\ \alpha \in A(I) \text{ with } X = \{i \mid \alpha(i) = a\}\}.$$

Then $I = \{i \mid \varepsilon_a(i) = a\}$ for any $a \in A$, so $I \in B$. If $X = \{i \mid \alpha(i) = a\} \in B$ and $b \neq a$, set $\beta = T(\alpha, \varepsilon_a, \varepsilon_a, \varepsilon_b)$. Then $I - X = \{i \mid \beta(i) = b\} \in B$. Also if $Y = \{i \mid \gamma(i) = c\}$ ($\gamma \in A(I)$, $c \in A$), then first we define $\delta = T(\gamma, \varepsilon_c, \varepsilon_a, \varepsilon_b)$ and observe that $Y = \{i \mid \delta(i) = a\}$. Now put $\chi = T(\alpha, \varepsilon_a, \alpha, \delta)$. Then $X \cup Y = \{i \mid \chi(i) = a\}$. Thus $\mathfrak{B}$ is a subalgebra of $\mathfrak{P}(I)$.

For every $\alpha \in A(I)$, we define a mapping α^* of A into B:

$$a\alpha^* = \{i \mid \alpha(i) = a\}.$$

We claim that $\alpha^* \in A[\mathfrak{B}]$. Indeed, if $a \neq b$, then

$$a\alpha^* \cap b\alpha^* = \{i \mid \alpha(i) = a, \text{ and } \alpha(i) = b\} = \varnothing.$$

Also,

$$\bigcup (a\alpha^* \mid a \in A) = I,$$

since $i \in \alpha(i)\alpha^*$, for all $i \in I$.

Then $\varphi \colon \alpha \to \alpha^*$ maps $A(I)$ into $A[\mathfrak{B}]$. We claim that φ is an isomorphism. φ is a 1–1 homomorphism by Lemma 7.

It remains to prove that φ is onto. Let $\chi \in A[\mathfrak{B}]$, $A = \{a_0, \cdots, a_{n-1}\}$, $a_i\chi = X_i$. Then $X_i = \{j \mid \alpha^i(j) = a^i\}$ for some $a^i \in A$, $\alpha^i \in A(I)$. Set $\alpha_0 = T(\alpha^0, \varepsilon_{a^0}, \varepsilon_{a_0}, \varepsilon_{a_1})$ and for $0 < k < n$, $\alpha_k = T(\alpha^k, \varepsilon_{a^k}, \varepsilon_{a_k}, \varepsilon_{a_0})$.

Now we define β_k, for $k < n$, by recursion. $\beta_0 = \alpha_0$ and

$$\beta_k = T(\alpha_k, \varepsilon_{a_k}, \varepsilon_{a_k}, \beta_{k-1}) \quad \text{for} \quad 0 < k < n.$$

It is simple to check by induction on k that if $i \leqq k$, and $j \in X_i$ then $\beta_k(j) = a_i$. Thus $\beta^*_{n-1} = \chi$ and so φ is onto. This completes the proof of Theorem 4.

In the special case of f-algebras, Theorems 3 and 4 were proved by A. L. Foster [2]. For the general case, see M. I. Gould and G. Grätzer [1].

§23. OPERATORS ON CLASSES OF ALGEBRAS

$\mathbf{X}$ is an *operator* if for every class K of algebras, $\mathbf{X}(K)$ is also a class of algebras. If $\mathbf{X}$ and $\mathbf{Y}$ are operators, so is $\mathbf{XY}$ defined by

$$\mathbf{XY}(K) = \mathbf{X}(\mathbf{Y}(K)).$$

$\mathbf{X}^2$ will stand for $\mathbf{XX}$.

The most frequently used operators are $\mathbf{I}$, $\mathbf{S}$, $\mathbf{H}$, $\mathbf{P}$, $\mathbf{P}^*$, $\mathbf{P}_S$, $\mathbf{P}_S^*$, $\underset{\rightarrow}{\mathbf{L}}$, $\underset{\leftarrow}{\mathbf{L}}$, defined as follows:

$\mathbf{I}(K)$: isomorphic copies of algebras of K;
$\mathbf{S}(K)$: subalgebras of algebras of K;
$\mathbf{H}(K)$: homomorphic images of algebras of K;
$\mathbf{P}(K)$: direct products of nonvoid families of algebras of K;
$\mathbf{P}^*(K)$: direct products of algebras of K;
$\mathbf{P}_S(K)$: subdirect products of nonvoid families of algebras of K;
$\mathbf{P}_S^*(K)$: subdirect products of algebras of K;
$\underset{\rightarrow}{\mathbf{L}}(K)$: direct limits of algebras of K;
$\underset{\leftarrow}{\mathbf{L}}(K)$: inverse limits of algebras of K.

Thus K is an algebraic class if $\mathbf{I}(K) = K$.

Lemma 1. *If* $\mathbf{X}$ *is any of the operators introduced above, then* $\mathbf{XI} = \mathbf{IX}$. *If* $\mathbf{X} = \mathbf{I}, \mathbf{H}, \mathbf{S}, \mathbf{P}, \mathbf{P}^*, \mathbf{P}_S, \mathbf{P}_S^*$, *then* $\mathbf{X}^2 = \mathbf{X}$.

Proof. All these statements are trivial.

Theorem 1. *Let K be a class of algebras. Then:*

(i) $\mathbf{SH}(K) \subseteq \mathbf{HS}(K)$;
(ii) $\mathbf{PH}(K) \subseteq \mathbf{HP}(K)$;
(iii) $\mathbf{PS}(K) \subseteq \mathbf{SP}(K)$;
(iv) $\mathbf{P}_S\mathbf{H}(K) \subseteq \mathbf{HP}_S(K)$.

Proof. All these inclusions were proved before.

Call K an *equational class* (also called *variety*, and *primitive class*) if K is nonvoid and $\mathbf{H}(K) \subseteq K$, $\mathbf{S}(K) \subseteq K$, $\mathbf{P}(K) \subseteq K$.

The motivation for calling such a class an equational class will be given in the next chapter.

Theorem 2. *Let K be a class of algebras. Then* $\mathbf{HSP}(K)$ *is an equational class containing K and it is the smallest equational class containing K.*

Proof. It is obvious that if $K \subseteq K_1$ and K_1 is an equational class, then $\mathbf{HSP}(K) \subseteq K_1$. To prove that $\mathbf{HSP}(K)$ is equational we make the following computations, using Lemma 1 and Theorem 1:

$$\begin{aligned} &\mathbf{HHSP}(K) = \mathbf{HSP}(K); \\ &\mathbf{SHSP}(K) \subseteq \mathbf{HSSP}(K) = \mathbf{HSP}(K); \\ &\mathbf{PHSP}(K) \subseteq \mathbf{HPSP}(K) \subseteq \mathbf{HSPP}(K) = \mathbf{HSP}(K), \end{aligned}$$

completing the proof of Theorem 2.

Equational classes of groups have been thoroughly investigated. (See H. Neumann's book, Varieties of Groups, Ergebnisse der Mathematik, Springer-Verlag, Berlin-West, 1967.)

Theorem 3 (*S. R. Kogalovskiĭ* [10]). *Let K be a class of algebras. Then* $\mathbf{HP}_S(K)$ *is the smallest equational class containing K.*

Proof. It is enough to prove that

$$\mathbf{S}(K) \subseteq \mathbf{HP}_S(K)$$

and then the statement follows as in the proof of Theorem 2.

Let $\mathfrak{A} \in \mathbf{S}(K)$, that is, $\mathfrak{A}$ is a subalgebra of some $\mathfrak{B} \in K$. Take a set I with $|I| = \aleph_0$ and define a subset C of B^I by the rule:

For $f \in B^I$, $f \in C$ if and only if for some $b \in A$, $\{i \mid f(i) \neq b\}$ is finite.

Then $\mathfrak{C} \in \mathbf{P}_S(\mathfrak{B})$. Now we define a congruence relation Θ of $\mathfrak{C}$:

$$f \equiv g(\Theta) \text{ if and only if } \{i \mid f(i) \neq g(i)\} \text{ is finite.}$$

Then for every $f \in C$ there is exactly one $b \in A$ such that $f \equiv f_b(\Theta)$, where $f_b(i) = b$ for all $i \in I$ and $f_b \not\equiv f_c(\Theta)$ if $b, c \in A$, $b \neq c$. Thus $\mathfrak{A} \in \mathbf{H}(\mathfrak{C})$ and so

$$\mathfrak{A} \in \mathbf{HP}_S(K),$$

which was to be proved.

See also B. M. Schein [2].

EXERCISES

1. Let $\mathfrak{A}_i$ be algebras, $i \in I$ and $I = \bigcup (I_j \mid j \in J)$, $I_j \cap I_{j'} = \varnothing$ if $j \neq j'$. Then
$$\prod (\mathfrak{A}_i \mid i \in I) \cong \prod (\prod (\mathfrak{A}_i \mid i \in I_j) \mid j \in J).$$
2. Prove the following generalization of Theorems 19.2 and 19.3.

The direct decompositions of $\mathfrak{A}$ into n algebras are determined by n congruence relations $\Theta_0, \cdots, \Theta_{n-1}$ with the following properties:

(i) $\Theta_0 \wedge \cdots \wedge \Theta_{n-1} = \omega$;
(ii) $(\Theta_0 \wedge \cdots \wedge \Theta_{i-1}) \vee \Theta_i = \iota$, $i = 1, \cdots, n-1$;
(iii) $\Theta_0 \wedge \cdots \wedge \Theta_{i-1}$ and Θ_i are permutable, $i = 1, \cdots, n-1$.

3. Prove that (iii) in Ex. 2 cannot be replaced by

(iii′) Θ_i and Θ_j are permutable, $0 \leqq i, j \leqq n-1$.

4. Prove that the sublattice of the congruence lattice generated by the Θ_i of Ex. 2 is isomorphic to the Boolean lattice of 2^n elements. Prove that any two congruence relations in this sublattice are permutable.
5. Let $\boldsymbol{0}$ be a nullary operation and assume that in $\mathfrak{A}, f_\gamma(0, \cdots, 0) = 0$ for every $\gamma < o(\tau)$. Prove that $\mathfrak{B}$ is isomorphic to a subalgebra of $\mathfrak{A} \times \mathfrak{B}$.
6. Let K be an equational class of algebras, having $\boldsymbol{0}$ and $+$ as operations, such that $0 + a = a + 0 = a$ in every algebra of K. Prove that if $\mathfrak{C} = \mathfrak{A} \times \mathfrak{B}$, then $\mathfrak{C}$ has two subalgebras $\mathfrak{A}'$ and $\mathfrak{B}'$ such that (i) $a \to \langle a, 0\rangle$ is an isomorphism between $\mathfrak{A}$ and $\mathfrak{A}'$; (ii) $b \to \langle 0, b\rangle$ is an isomorphism between $\mathfrak{B}$ and $\mathfrak{B}'$; (iii) $A' \cap B' = \{0\}$; (iv) every $c \in C$ has a unique representation $c = a + b$, $a \in A'$, $b \in B'$.
7. Is the converse of Ex. 6 true?
8. Let $\mathfrak{C}$ be a subalgebra of $\prod (\mathfrak{A}_i \mid i \in I)$, Θ a congruence relation on $\mathfrak{C}$, and $\mathfrak{B}$ a finite relative subalgebra of $\mathfrak{C}/\Theta$. Is it true that there exists a finite $I' \subseteq I$ such that $\mathfrak{C}\varphi_{I'}$ has a quotient algebra $\mathfrak{C}\varphi_{I'}/\Theta$, containing $\mathfrak{B}$ as a relative subalgebra? (Theorem 19.4 is the special case $\Theta = \omega$.)
9. Show that $\mathfrak{P}^{(m)}(\mathfrak{A}_0 \times \mathfrak{A}_1) \cong \mathfrak{P}^{(m)}(\mathfrak{A}_0) \times \mathfrak{P}^{(m)}(\mathfrak{A}_1)$ may not hold.
10. Let $\mathfrak{B}$ be a homomorphic image of $\mathfrak{A}$. Then $\mathfrak{P}^{(m)}(\mathfrak{B})$ is a homomorphic image of $\mathfrak{P}^{(m)}(\mathfrak{A})$.
11. Let $\mathfrak{A}$ be isomorphic to a subdirect product of the $\mathfrak{B}_i$, $i \in I$ and let $\mathfrak{B}_i$ be isomorphic to a subdirect product of the $\mathfrak{C}_j$, $j \in I_i$. Then $\mathfrak{A}$ is isomorphic to a subdirect product of the $\mathfrak{C}_j$, $j \in \bigcup (I_i \mid i \in I)$.
12. $\mathfrak{A}$ is isomorphic to a subdirect product of the $\mathfrak{B}_i$, $i \in I$ if and only if for $a, b \in A$, $a \neq b$, there exists an $i \in I$ and a homomorphism φ_i of $\mathfrak{A}$ onto $\mathfrak{B}_i$ with $a\varphi_i \neq b\varphi_i$, and every i is chosen for some a, b.
13. Is it true that an algebra $\mathfrak{A}$ is subdirectly irreducible if and only if $|A| = 1$ or $\mathfrak{C}(\mathfrak{A})$ has exactly one atom?
14. Let $\mathfrak{F}$ be a semilattice with 0 (as in §6), I an ideal of $\mathfrak{F}$ and $a \in F$, $a \notin I$. Then there exists an ideal J such that $a \notin$ J, J $\supseteq I$ and J is maximal with respect to these two properties.
15. Let $S(I, a)$ denote the set of all ideals J defined in Ex. 14. Prove that J $\in S(I, a)$ for some I and a if and only if one of the following two conditions hold:

(i) J is *completely meet irreducible*, that is, if $J = \bigcap (J_i \mid i \in K)$ and $J_i \in I(\mathfrak{F})$, then $J = J_i$ for some $i \in K$;

(ii) there is an ideal L such that $L \supset J$ and if for the ideal N, $N \supset J$ then $N \supseteq L$.

16. Prove that $I = \bigcap (J \mid J \in S(I, a), a \in \mathfrak{F}, a \notin I)$.
17. Use Ex. 16 to prove Theorem 20.3.
18. (G. Birkhoff and O. Frink [1]) Every congruence relation is the complete meet of completely meet irreducible congruence relations.
19. Apply Ex. 14–16 to subalgebra lattices.
20. $\mathfrak{A}$ is a *diagonal subdirect power* of $\mathfrak{B}$ if $\mathfrak{A}$ is a subalgebra of some direct power of $\mathfrak{B}$, containing the diagonal. Prove that $\mathfrak{A}$ is isomorphic to a diagonal subdirect power of $\mathfrak{B}$ if and only if $\mathfrak{A}$ has a subalgebra $\mathfrak{B}'$ isomorphic to $\mathfrak{B}$ and for $a, b \in A$, $a \neq b$, there exists an endomorphism φ of $\mathfrak{A}$ such that $A\varphi = B'$, $a\varphi \neq b\varphi$ and $c\varphi = c$ for all $c \in B'$.
21. (R. P. Dilworth) Let $\mathfrak{L}$ be a lattice. It is a direct product of a finite number of simple lattices if and only if any two congruence relations permute and the congruence lattice is a finite Boolean lattice.
22. An algebra is a subdirect product of simple algebras if and only if ω is the intersection of dual atoms of the congruence lattice (Θ is a *dual atom* if $\Theta < \iota$ and $\Theta \leqq \Phi < \iota$ implies that $\Theta = \Phi$).
23. (T. Tanaka [1]) Let $\mathfrak{A}$ be an algebra with a Boolean congruence lattice. Prove that $\mathfrak{A}$ is a subdirect product of simple algebras. Does the converse hold provided that $\mathfrak{C}(\mathfrak{A})$ is distributive? (Combine Ex. 22 with Ex. 0.79.)
24. Let $\mathfrak{L}$ be a distributive lattice, $a \in L$ and a is neither 0 nor 1. Define Θ, Φ: $x \equiv y(\Theta)$ if and only if $x \wedge a = y \wedge a$; $x \equiv y(\Phi)$ if and only if $x \vee a = y \vee a$. Prove that Θ and Φ are congruence relations and that $\Theta \wedge \Phi = \omega$.
25. (G. Birkhoff) Every distributive lattice of more than one element is a subdirect product of two-element lattices. (Combine Ex. 24 with Theorem 20.3.)
26. (M. H. Stone) Every Boolean algebra is a subdirect product of two-element Boolean algebras.
27. Prove that every semilattice $\langle F; \vee \rangle$ with more than one element is a subdirect product of two-element semilattices.
28. Which semigroups with more than one element are subdirect powers of a given two-element semigroup?
29. Let $\mathfrak{A}$ be the direct product of the algebras $\mathfrak{A}_i$, $i \in I$. Then the following statements hold.

    (i) for every $i \in I$ there is a homomorphism φ_i of $\mathfrak{A}$ onto $\mathfrak{A}_i$;

    (ii) let $\mathfrak{B}$ be an algebra and let ψ_i, $i \in I$, be a family of homomorphisms, $\psi_i \colon B \to A_i$; then there exists a unique homomorphism $\psi \colon B \to A$ such that $\psi\varphi_i = \psi_i$ for all $i \in I$.

30. If an algebra has properties (i) and (ii) of Ex. 29, then it is isomorphic to the direct product of the $\mathfrak{A}_i$, $i \in I$.
31. If the congruence relations on $\mathfrak{A}$ permute, then the same holds for every homomorphic image of $\mathfrak{A}$.
32. Let $\mathscr{A}$ be a direct family of algebras $\mathfrak{A}_i$ with the carrier $\langle I; \leqq \rangle$. Let $B \subseteq \prod (A_i \mid i \in I)$ be defined by $f \in B$ if and only if there exists a $k \in I$ such that for all $k \leqq i \leqq j$, $f(i)\varphi_{ij} = f(j)$. Prove that $\mathfrak{B}$ is a subalgebra.

33. Define a relation Θ on $\mathfrak{B}$ of Ex. 32 as follows: $f \equiv g(\Theta)$ if there exists an $i \in I$ such that $f(j) = g(j)$ for all $j \geqq i$. Prove that
    (i) Θ is a congruence relation of $\mathfrak{B}$;
    (ii) $\mathfrak{B}/\Theta$ is isomorphic to $\varinjlim \mathscr{A}$.
34. Prove that an equational class is closed under direct limits.
35. Let $\mathscr{A}$ be a direct family of algebras. If all $\varphi_{i\infty}$ are 1–1, then all $\varphi_{ij} (i \leqq j)$ are 1–1.
36. Let $\mathscr{A}$ be a direct family of algebras. If all the φ_{ij} $(i \leqq j)$ are onto, then all the $\varphi_{i\infty}$ are onto.
37. Prove that the converse of Ex. 36 is false.
38. Prove that the converse of Lemma 21.6 is false.
39. Let $\mathscr{A}$ be an inverse family of algebras with carrier $\langle I; \leqq \rangle$. If all the φ_i^{∞} are onto, then all φ_j^{i} are onto.
40. Construct an inverse family $\mathscr{A}$ of nonvoid sets such that all maps are onto, and the inverse limit is void (G. Higman and A. H. Stone [1]). (Hint: Use the following sets and maps. Let ω_1 be the first uncountable ordinal. For $\alpha < \omega_1$ set $A_\alpha = \{\gamma \mid \gamma < \alpha\}$, $B_\alpha = \{f \mid f \in [0, 1)^{A_\alpha}$ and f is monotone$\}$, where $[0, 1)$ is the real interval $[0, 1)$ and f is monotone if $\gamma < \delta$ implies $f(\gamma) < f(\delta)$. For $\alpha < \beta < \omega_1$ let φ_α^{β}: $B_\beta \to B_\alpha$ be defined by $f\varphi_\alpha^{\beta} = f_{A_\alpha}$.)
41. Prove that the converse of Ex. 39 is false.
42. Prove that the converse of Ex. 39 holds if $|I| \leqq \aleph_0$.
43. Let $\mathscr{A}$ be an inverse family of algebras. Is it true that all the φ_j^{i} are 1–1 and onto if and only if all the φ_i^{∞} are 1–1 and onto?
44. Give an example of algebras $\mathfrak{A}_i$, $i \in I$, which have no weak direct product.
45. A weak direct product of the algebras $\mathfrak{A}_i$, $i \in I$, exists if and only if there exists a $p \in \prod (A_i \mid i \in I)$ such that for each $\gamma < o(\tau)$,
$$\{i \mid p(i) \neq f_\gamma(p(i), \cdots, p(i))\}$$
is finite.
46. Find conditions for the existence of L-restricted direct products.
47. $\mathfrak{A}$ is a *weak subdirect product* of the $\mathfrak{A}_i$, $i \in I$, if $\mathfrak{A}$ is a subalgebra of $\mathfrak{B}$, which is a weak direct product of the $\mathfrak{A}_i$ and $Ae_i^{I} = A_i$ for $i \in I$. Give necessary and sufficient conditions for an algebra $\mathfrak{A}$ to be isomorphic to a weak subdirect product.
48. (J. Hashimoto [1]) Define *L-restricted subdirect products*. Give necessary and sufficient conditions for $\mathfrak{A}$ to be isomorphic to an L-restricted subdirect product.
49. A subdirect power $\mathfrak{A}$ of $\mathfrak{B}$ is *bounded* if for every $\varphi \in A$, $I\varphi$ is finite (where I is the exponent of $\mathfrak{B}$). Find conditions for $\mathfrak{A}$ to be a bounded diagonal subdirect power of $\mathfrak{B}$.
50. Generalize Ex. 49 to L-bounded, where L is an ideal of $\mathfrak{P}(I)$.
51. Why is it necessary to assume in Definition 22.2 that L is an ideal?
52. (J. Hashimoto [1]) Let $\Theta_0, \cdots, \Theta_{n-1}$ be congruence relations of $\mathfrak{A}$ and let $\varphi_i = \bigwedge (\Theta_j \mid j \neq i)$. Prove that if $\varphi_0, \cdots, \varphi_{n-1}$ are permutable, then $\Theta_0, \cdots, \Theta_{n-1}$ are completely permutable.

53. Does the result of the previous exercise extend to the case when we have infinitely many Θ_i?
54. (J. Hashimoto [1]) Let $\mathfrak{A}$ be an algebra and $\langle \Sigma; \leqq \rangle$ a complete sublattice of $\mathfrak{C}(\mathfrak{A})$. Let Ω denote the set of compact elements of $\langle \Sigma; \leqq \rangle$. Prove that Σ is completely permutable if and only if Ω is completely permutable.
55. Give necessary and sufficient conditions on $\mathfrak{A}$ for $\mathfrak{A}$ to be isomorphic to the weak direct product of the algebras $\mathfrak{A}_i$, $i \in I$.
56. A dual ideal $\mathscr{D}$ of $\mathfrak{P}(I)$ is called *$\mathfrak{m}$-complete,* where $\mathfrak{m}$ is an infinite cardinal, if $X_j \in \mathscr{D}$, $j \in J$, $|J| < \mathfrak{m}$ imply $\bigcap (X_j \mid j \in J) \in \mathscr{D}$. Let $\mathscr{D}$ be an $\mathfrak{m}$-complete dual ideal of $\mathfrak{P}(I)$, $I = \bigcup (I_k \mid k \in K)$, where $I_k \cap I_{k'} = \varnothing$ if $k \neq k'$, and $|K| < \mathfrak{m}$. Set $\mathscr{D}_k = \mathscr{D} \cap P(I_k)$ for $k \in K$. Then

$$\prod_{\mathscr{D}}(\mathfrak{A}_i \mid i \in I) \cong \prod (\prod_{\mathscr{D}_k}(\mathfrak{A}_i \mid i \in I_k) \mid k \in K).$$

(see e.g., T. Frayne, A. C. Morel, and D. S. Scott [1]).
57. Let $\mathscr{D}$ be a dual ideal of $\mathfrak{P}(I)$, $J \in \mathscr{D}$, and $\mathscr{D}_1 = \mathscr{D} \cap P(J)$. Then

$$\prod_{\mathscr{D}}(\mathfrak{A}_i \mid i \in I) \cong \prod_{\mathscr{D}_1}(\mathfrak{A}_i \mid i \in J).$$

58. If $\mathscr{D}$ and $\mathscr{D}'$ are dual ideals of $\mathfrak{P}(I)$ and $\mathscr{D} \subseteq \mathscr{D}'$, then $\prod_{\mathscr{D}'}(\mathfrak{A}_i \mid i \in I)$ is a homomorphic image of $\prod_{\mathscr{D}} (\mathfrak{A}_i \mid i \in I)$.
59. Prove that if A is infinite, Theorem 22.3 fails to hold, since $\mathfrak{A}[\mathfrak{B}]$ is not a subalgebra of $\mathfrak{A}[\mathfrak{P}(I)]$.
60. Prove that if A is infinite, Theorem 22.4 fails to hold.
61. Let us call $\mathfrak{A}$ an *f_*-algebra,* if there are two elements $0, 1 \in A$ and two binary algebraic functions $+, \cdot$ such that $0 \cdot a = a \cdot 0 = 0$, $1 \cdot a = a \cdot 1 = a$, $a + 0 = 0 + a = a$ for every $a \in A$. Prove that if $\mathfrak{A}$ is an f_*-algebra, then so is $\mathfrak{A}[\mathfrak{B}]$ with ζ_0 and ζ_1, and also every diagonal subdirect power of $\mathfrak{A}$ is an f_*-algebra with e_0 and e_1.
62. Let $\mathfrak{A}$ be an f_*-algebra. For $a \in A$ define in $\mathfrak{A}^I$ a unary operation P_a by

$$P_a(\alpha)(i) = \begin{cases} 1 & \text{if} \quad \alpha(i) = a \\ 0 & \text{if} \quad \alpha(i) \neq a. \end{cases}$$

Let $\mathfrak{B}$ be a diagonal subdirect power of $\mathfrak{A}$. Prove that $\mathfrak{B}$ is a normal subdirect power if and only if $\alpha \in B$ implies $P_a(\alpha) \in B$ for all $a \in A$. (This is essentially a result of A. L. Foster [2]; see M. I. Gould and G. Grätzer [1].)
63. A *bounded normal subdirect power* is a normal subdirect power which is bounded (see Ex. 49). A *bounded extension* of the algebra $\mathfrak{A}$ by an arbitrary Boolean algebra $\mathfrak{B}$ is defined as in Definition 22.7, with the restriction that we consider only mappings $\alpha: A \to B$ for which $\{a \mid a\alpha \neq 0\}$ is finite. Prove Theorems 22.3 and 22.4 for infinite algebras, bounded extensions, and bounded normal subdirect powers. (This generalizes a result of A. L. Foster [2]; see M. I. Gould and G. Grätzer [1].)
64. Let $\mathfrak{A}$ be a finite algebra and $\mathfrak{B}$ a Boolean algebra. By Theorem 22.3 we can represent $\mathfrak{A}[\mathfrak{B}]$ as a normal subdirect power $\mathfrak{A}(I)$. In Theorem 22.4 we construct a subalgebra $\mathfrak{B}'$ of $\mathfrak{P}(I)$ such that $\mathfrak{A}(I)$ is isomorphic to $\mathfrak{A}[\mathfrak{B}']$. Prove that $\mathfrak{B} \cong \mathfrak{B}'$.

65. Let $\mathfrak{A}$ be a finite algebra, let $\mathfrak{B}, \mathfrak{B}'$ be Boolean algebras where $\mathfrak{B}'$ is a homomorphic image of $\mathfrak{B}$. Prove that $\mathfrak{A}[\mathfrak{B}']$ is a homomorphic image of $\mathfrak{A}[\mathfrak{B}]$.
66. Extend Ex. 64 and 65 to infinite algebras using bounded extensions.
67. Which of the operators of §23 satisfy the law

$$\mathbf{X}(K \cup L) = \mathbf{X}(K) \cup \mathbf{X}(L)$$

for any two classes K, L of algebras?
68. Show by examples that if we reverse the inclusions in (i)–(iv) of Theorem 23.1, then we get false statements.
69. Prove that statement (iii) of Theorem 23.1 is equivalent to the Axiom of Choice.

* * *

The following notion (due to B. H. Neumann [2]) will be used in Ex. 70–76: Let K be a class of algebras. Let $\mathfrak{A} \in \mathbf{C}(K)$ ($\mathbf{C}$ for covering) if there exist algebras $\mathfrak{A}_i \in K$, $i \in I$ and 1–1 homomorphisms φ_i of $\mathfrak{A}_i$ into $\mathfrak{A}$ such that $A = \bigcup (A_i\varphi_i \mid i \in I)$. The results of Ex. 70–76 are from G. Grätzer [9].

70. Prove that $\mathbf{SC}(K) = \mathbf{CS}(K)$.
71. Prove that $\mathbf{HC}(K) = \mathbf{CH}(K)$ if there are no nullary operations.
72. Prove that $\mathbf{PC}(K) \subseteq \mathbf{CP}(K)$ and that the reverse inclusion is false.
73. Prove that $\mathbf{P}_S\mathbf{C}(K) \nsubseteq \mathbf{CP}_S(K)$.
74. Prove that $\mathbf{CP}_S(K) \nsubseteq \mathbf{P}_S\mathbf{C}(K)$.
75. Prove the previous three exercises for $\mathbf{P}^*$ and $\mathbf{P}_S{}^*$.
76. Let K be an equational class. Prove that $\mathbf{C}(K)$ is also an equational class.
77. Characterize $\mathbf{P}^*(K) - \mathbf{P}(K)$ and $\mathbf{P}_S{}^*(K) - \mathbf{P}_S(K)$.
78. Let $R \subseteq A \times A$ be an equivalence relation on A. Prove that R is a congruence relation of $\mathfrak{A} = \langle A; F\rangle$ if and only if $\langle R; F\rangle$ is a subalgebra of $\mathfrak{A}^2$.
79. Let $\varphi \subseteq A \times B$ be a mapping of A into B. Prove that φ is a homomorphism of $\mathfrak{A} = \langle A; F\rangle$ into $\mathfrak{B} = \langle B; F\rangle$ if and only if $\langle \varphi; F\rangle$ is a subalgebra of $\mathfrak{A} \times \mathfrak{B}$.
80. Let $\mathbf{X}$ and $\mathbf{Y}$ be operators. $\mathbf{X} = \mathbf{Y}$ if $\mathbf{X}(K) = \mathbf{Y}(K)$ for every class of algebras K. If H is a set of operators, the semigroup $\mathfrak{E}(H)$ generated by H consists of all finite products $\mathbf{X}_0 \cdots \mathbf{X}_{n-1}$ where $\mathbf{X}_i \in H$ with equality as defined above. Prove that for $H = \{\mathbf{I}, \mathbf{H}, \mathbf{S}\}$, $\mathfrak{E}(H)$ has 6 elements.
81. Find an algebraic class K such that $\mathbf{I}\overleftarrow{\mathbf{L}}\mathbf{I}\overleftarrow{\mathbf{L}}(K) \neq \mathbf{I}\overleftarrow{\mathbf{L}}(K)$.
82. Find an algebraic class K such that $\mathbf{I}\overrightarrow{\mathbf{L}}^{1-1}\mathbf{I}\overrightarrow{\mathbf{L}}^{1-1}(K) \neq \mathbf{I}\overrightarrow{\mathbf{L}}^{1-1}(K)$; where $\overrightarrow{\mathbf{L}}^{1-1}$ is the operator for $1-1$ direct limits. (A. H. Kruse [2].)
83. Let $\mathfrak{A}_0$ and $\mathfrak{A}_1$ be partial algebras. Let $\langle\langle a_0, b_0\rangle, \cdots, \langle a_{n_\gamma-1}, b_{n_\gamma-1}\rangle\rangle$ be in $D(f_\gamma, \mathfrak{A}_0 \times \mathfrak{A}_1)$ if and only if $\langle a_0, \cdots, a_{n_\gamma-1}\rangle \in D(f_\gamma, \mathfrak{A}_0)$ and $\langle b_0, \cdots, b_{n_\gamma-1}\rangle \in D(f_\gamma, \mathfrak{A}_1)$. The resulting partial algebra is $\mathfrak{A}_0 \times \mathfrak{A}_1 = \mathfrak{A}$. Prove that the mappings $\varphi_0\colon \langle a, b\rangle \to a$ and $\varphi_1\colon \langle a, b\rangle \to b$ are homomorphisms of $\mathfrak{A}$ onto $\mathfrak{A}_0$ and $\mathfrak{A}_1$, respectively, but the φ_i need not be full or strong.
84. Prove Theorem 19.2 for partial algebras.

85. Prove that under the conditions of Theorem 19.3 for a partial algebra $\mathfrak{A}$ we can only conclude that $\mathfrak{A}$ is isomorphic to a weak subalgebra of $\mathfrak{A}/\Theta_0 \times \mathfrak{A}/\Theta_1$.
86. Does Theorem 20.3 hold for partial algebras?
87. Prove the results of §19 and §20 for infinitary algebras.
88. Is it always possible to define the direct limit of infinitary algebras as an infinitary algebra?
89. Prove Theorem 23.1 and 23.2 for infinitary algebras.
90. Which inclusions of Theorem 23.1 fail to hold for partial algebras?
91. The *spectrum* $S = Sp(K)$ of an equational class K is the set of all integers n such that there is an n-element algebra in K. Prove that $1 \in S$ and $S \cdot S \subseteq S$ (i.e., S is closed under multiplication).
92. Let S be a set of positive integers and let $1 \in S$ and $S \cdot S \subseteq S$. Then there exists an equational class K such that $S = Sp(K)$. (G. Grätzer [12].) (Hint: Use Ex. 6.53 or 6.54.)
93. Let $\mathscr{A}$ be a direct family of algebras $\mathfrak{A}_i$ such that all φ_{ij} are 1–1. Prove that $\mathfrak{P}^{(\alpha)}(\lim\limits_{\rightarrow} \mathscr{A})$ is isomorphic to a subalgebra of $\lim\limits_{\leftarrow} \mathscr{A}_1$, where $\mathscr{A}_1$ has the same carrier as $\mathscr{A}$, the algebras in $\mathscr{A}_1$ are $\mathfrak{P}^{(\alpha)}(\mathfrak{A}_i)$, and if $i \geqq j$, then $\varphi_j{}^i$ is defined by $(\mathbf{p})\mathfrak{A}_i\varphi_j{}^i = (\mathbf{p})\mathfrak{A}_j$.
94. Get a similar isomorphism for inverse limit families of algebras for which all $\varphi_j{}^i$ are onto.
95. Express Ex. 93 and 94 in terms of operators.
96. Let $\mathfrak{A}_i$, $i \in I$ be partial algebras, $\mathfrak{A} = \prod (\mathfrak{A}_i \mid i \in I)$ (see Ex. 84), and $e_i{}^I$ the i-th projection. If $\mathfrak{B}$ is a relative subalgebra of $\mathfrak{A}$ and $Be_i{}^I = A_i$ for $i \in I$, then $\mathfrak{B}$ is said to be a *subdirect product* of the $\mathfrak{A}_i$. Prove that every partial algebra is isomorphic to a subdirect product of subdirectly irreducible partial algebras. (See H. E. Pickett [1]; this result is not explicitly stated, but it follows easily from Theorem 7 and Theorem 8, F, E_1, and D_1 part c, or directly from Theorem 5, using the fact that Lemma 10.3 holds for partial algebras.)
97. For a given infinite cardinal $\mathfrak{m}$, find an equational class K such that the one element algebra is the only finite algebra in K and every infinite algebra in K is of cardinality $\geqq \mathfrak{m}$ and there is an algebra of cardinality $\mathfrak{m}$ in K.
98. Let $\mathscr{A}$ be a direct family of algebras as in Definition 21.3. Consider the following two properties for an algebra $\mathfrak{B}$ and for a family $(\varphi_i \mid i \in I)$ of homomorphisms, where φ_i is a homomorphism of $\mathfrak{A}_i$ into $\mathfrak{B}$:

    (i) for $i, j \in I$, $i \leqq j$, we have that $\varphi_i = \varphi_{ij}\varphi_j$;

    (ii) if $\mathfrak{C}$ is an algebra, and $(\psi_i \mid i \in I)$ is a family of homomorphisms where ψ_i is a homomorphism of $\mathfrak{A}_i$ into $\mathfrak{C}$ such that for $i, j \in I$, $i \leqq j$, we have that $\psi_i = \varphi_{ij}\psi_j$, then there exists a *unique* homomorphism φ of $\mathfrak{B}$ into $\mathfrak{C}$ with $\varphi_i\varphi = \psi_i$ for all $i \in I$.

    Prove that conditions (i) and (ii) characterize the direct limit of $\mathscr{A}$ along with the homomorphisms $(\varphi_{i\infty} \mid i \in I)$.
99. State and prove a characterization of inverse limits along with $(\varphi_i{}^\infty \mid i \in I)$.

100. Let $\mathscr{D}$ be a dual ideal of $\mathfrak{P}(I)$ and let $(\mathfrak{A}_i \mid i \in I)$ be a family of algebras. For $H \in \mathscr{D}$ set

$$\mathfrak{B}_H = \prod (\mathfrak{A}_i \mid i \in H)$$

and for $H, K \in \mathscr{D}, H \supseteq K$, let φ_{HK} be the natural homomorphism of $\mathfrak{B}_H$ onto $\mathfrak{B}_K$. Prove that the algebras $\mathfrak{B}_H$, with the homomorphisms φ_{HK}, form a direct family $\mathscr{A}$ over the carrier $\langle \mathscr{D}, \supseteq \rangle$. Prove that

$$\lim_{\rightarrow} \mathscr{A} \cong \prod_{\mathscr{D}} (\mathfrak{A}_i \mid i \in I).$$

101. Prove that the statement: "$S(I, a) \neq \varnothing$ for any $a \notin I$" (notation of Ex. 15) is equivalent to the Axiom of Choice.

102. Prove that the statement of Theorem 20.3 is equivalent to the Axiom of Choice. (G. Grätzer, Notices Amer. Math. Soc. 14(1967), 133; this solves a problem proposed in H. Rubin and J. E. Rubin, Equivalents of the Axiom of Choice. North-Holland, Amsterdam, 1963, p. 15). (Hint: Prove that Theorem 18.3 can be proved without the Axiom of Choice. Use Theorem 11.3 to verify that Theorem 20.3 is equivalent to Ex. 16, which, in turn, is equivalent to "$S(I, a) \neq \varnothing$ for any $a \in I$", so a reference to Ex. 101 completes the proof.)

103. Let AC denote the Axiom of Choice, $BR(K)$ that G. Birkhoff's Representation Theorem (Theorem 20.3) holds in the class K, let PI denote the Prime Ideal Theorem (Theorem 6.7) and D, L, G, R be the class of distributive lattices, groups and rings, respectively. Prove the implications in the following diagram:

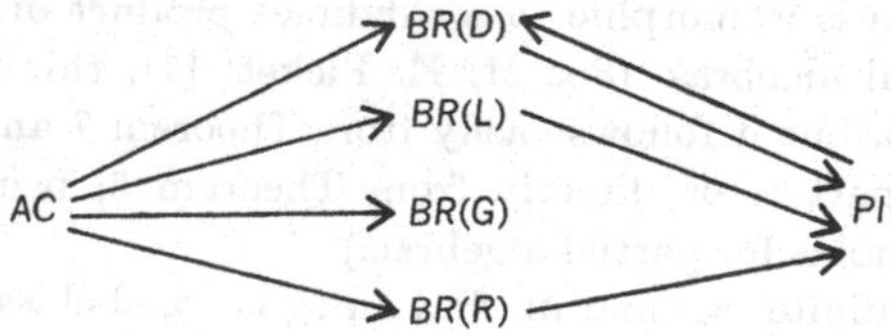

PROBLEMS

21. It follows from Theorem 23.2, from Corollary 1 to Theorem 22.2, and from Theorem 26.3, that if $\mathfrak{A}$ is an algebra and $\mathfrak{B}$ is a Boolean algebra, then $\mathfrak{A}[\mathfrak{B}] \in \mathbf{HSP}(\{\mathfrak{A}\})$. Find an explicit expression of $\mathfrak{A}$ in terms of $\mathbf{H}$, $\mathbf{S}$ and $\mathbf{P}$. (If $\mathfrak{A}$ is finite, this was done in Theorems 22.3 and 22.4.)

22. Given the algebras $\mathfrak{A}$ and $\mathfrak{C}$, find necessary and sufficient conditions for the existence of a Boolean algebra $\mathfrak{B}$ with $\mathfrak{A}[\mathfrak{B}] \cong \mathfrak{C}$. When is this $\mathfrak{B}$ unique (up to isomorphism)?

23. For what classes K of algebras is it true that $K_1 = \{\mathfrak{A}[\mathfrak{B}] \mid \mathfrak{A} \in K, \mathfrak{B}$ an arbitrary Boolean algebra$\}$ is an equational class? (This is the case if $K = \{\mathfrak{A}\}$, where $\mathfrak{A}$ is a primal algebra; see §27.)

24. Find all subsets H of $\{\mathbf{I}, \mathbf{H}, \mathbf{S}, \mathbf{P}, \mathbf{P}^*, \mathbf{P}_S, \mathbf{P}_S^*, \underrightarrow{\mathbf{L}}, \underleftarrow{\mathbf{L}}, \mathbf{C}\}$ which generate finite semigroups (see Ex. 80), and describe these semigroups. (Some recent results: The semigroup generated by **H**, **S**, and **P** has been described by D. Pigozzi, Notices Amer. Math. Soc. 13 (1966), 829. The finiteness of the semigroups with the generating sets $\{\mathbf{H}, \mathbf{S}, \mathbf{P}, \mathbf{P}_S\}$, $\{\mathbf{H}, \mathbf{S}, \mathbf{P}, \mathbf{C}\}$ has been proved by E. M. Nelson, Master Thesis, McMaster University, 1966.)
25. For those H of Problem 24 which generate infinite semigroups, give a "normal form" theorem, i.e., call certain products "normal", prove that all normal products represent distinct operators, and that every product of the operators in H equals a normal product.
26. For an algebraic class K, define K^α for every ordinal α as follows: $K^0 = K$, $K^{\alpha+1} = \mathbf{I}\underrightarrow{\mathbf{L}}(K^\alpha)$, $K^{\lim \alpha_\nu} = \bigcup (K^{\alpha_\nu} \mid \nu)$. Is it possible to find for every ordinal α a class such that $K^{\alpha+1} \neq K^\alpha$? Is it true that for every class K there exists an α with $K^\alpha = K^{\alpha+1}$? If this is so, which ordinals can occur as a smallest such α? (See Ex. 82.)
27. The same as Problem 26 but for inverse limits. (See Ex. 81.)
28. Describe the finite semigroups of Problem 24 without the Axiom of Choice.
29. Let K be an equational class, $\mathfrak{A}$ a finite algebra, and

$$K_{\mathfrak{A}} = \{\mathfrak{B} \mid \mathfrak{B} \in K \text{ and } \mathfrak{A} \notin \mathbf{IS}(\mathfrak{B})\}.$$

Under what conditions is $K_{\mathfrak{A}}$ an equational class?
30. Which implications can be reversed in Ex. 103? Is $BR(G) \Rightarrow PI$ or $PI \Rightarrow BR(G)$ true?
31. Is the statement that $\mathbf{HSP}(K)$ is an equational class equivalent to the Axiom of Choice?

CHAPTER 4
FREE ALGEBRAS

One of the most useful concepts in algebra is that of the free algebra. We devote three chapters to the study of free algebras, Chapters 4, 5, and 8. In this first chapter on this topic, we first examine in detail the basic problems of existence and construction of free algebras, and the connection of free algebras with identities. Then we apply free algebras to problems of equational completeness and also to the word problem. We also discuss free algebras generated by partial algebras.

§24. DEFINITION AND BASIC PROPERTIES

Given a class K of algebras† and a set X, it is sometimes very useful to know the most general algebra in K generated by X. For instance, if K is the class of semigroups and $X=\{x\}$, then there are many semigroups generated by X. The simplest one is $\langle X;\cdot\rangle$, in which $x\cdot x=x$; and the most general is $\langle Y;\cdot\rangle$, where $Y=\{x, x^2, x^3, \cdots, x^n, \cdots\}$ and $x^n\cdot x^m=x^{n+m}$. The second semigroup is the "most general" because every semigroup which can be generated by one element is a homomorphic image of $\langle Y;\cdot\rangle$. Such algebras will be called free and a formal definition follows.

Definition 1 (*G. Birkhoff* [2]). *Let K be a class of algebras, let $\mathfrak{A}\in K$ and let $X=(x_i \mid i\in I)$ be a family of elements of $\mathfrak{A}$ such that $\mathfrak{A}$ is generated by $\{x_i \mid i\in I\}$. $\mathfrak{A}$ is said to be a* free algebra *over K, with the* free generating family *X if for any $\mathfrak{B}\in K$, and for any mapping $\psi\colon I\to B$, there is a homomorphism φ of $\mathfrak{A}$ into $\mathfrak{B}$ such that $i\psi=x_i\varphi$ for all $i\in I$.*

The diagram on the next page illustrates this definition.

The case $I=\varnothing$ is allowed if and only if there are nullary operations.

$\{x_i \mid i\in I\}$ is a *free generating set* of $\mathfrak{A}$ over K.

Thus a free algebra over K must belong to K. However, in a class K, there need not be any free algebra. Also, if $\mathfrak{A}\in K$ and $\mathfrak{A}$ is free over K, then $\mathfrak{A}$ may have more than one free generating set.

† A "class of algebras" will always be assumed to be "a class of algebras of the same type".

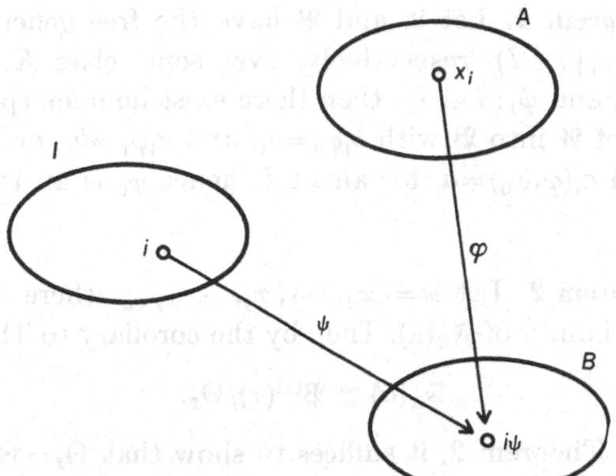

We will use the notation $\mathfrak{F}_K(\mathfrak{m})$ for an algebra, free over K, which has a free generating family $(x_i \mid i \in I)$ with $|I| = \mathfrak{m}$. If $I = \{\gamma \mid \gamma < \alpha\}$ for some ordinal α, we will write $\mathfrak{F}_K(\alpha)$. Thus if $\mathfrak{F}_K(0)$ exists, then there are nullary operations.

The following two observations are very important.

Corollary 1. *The homomorphism φ in Definition 1 is unique.*

This is obvious from Theorem 12.2 since $[\{x_i \mid i \in I\}] = A$.

Corollary 2. *If K contains an algebra with more than one element and $\mathfrak{A}$ has a free generating family $(x_i \mid i \in I)$ over K, and $i, j \in I$, $i \neq j$, then $x_i \neq x_j$.*

Indeed, take a $\mathfrak{B} \in K$ with $|B| \neq 1$. Then there is a mapping ψ of I into B with $i\psi \neq j\psi$ and so $x_i\varphi \neq x_j\varphi$, which implies $x_i \neq x_j$.

If for $i, j \in I$, $i \neq j$, we have $x_i \neq x_j$, then Definition 1 can be somewhat more simply stated: any mapping $p\colon x_i \to b_i$ $(b_i \in B)$ can be extended to a homomorphism φ of $\mathfrak{A}$ into $\mathfrak{B}$.

Two basic properties of free algebras are given in the following theorems.

Theorem 1. *If $\mathfrak{F}_K(\mathfrak{m})$ exists, then it is unique up to isomorphism.*

Theorem 2. *If $\mathfrak{F}_K(\alpha)$ exists, then*

$$\mathfrak{F}_K(\alpha) \cong \mathfrak{P}^{(\alpha)}(\tau)/\Theta_K,$$

where Θ_K is the congruence relation defined in §8.

Theorem 2 implies Theorem 1. However, we give a proof of Theorem 1 independent of Theorem 2.

Proof of Theorem 1. Let $\mathfrak{A}$ and $\mathfrak{B}$ have the free generating families $(a_i \mid i \in I)$ and $(b_i \mid i \in I)$, respectively, over some class K. Consider the maps $\psi_0: i \to a_i$ and $\psi_1: i \to b_i$; then there exist homomorphisms φ_0 of $\mathfrak{B}$ into $\mathfrak{A}$ and φ_1 of $\mathfrak{A}$ into $\mathfrak{B}$ with $b_i\varphi_0 = a_i$, and $a_i\varphi_1 = b_i$ for all $i \in I$. Thus $b_i(\varphi_0\varphi_1) = b_i$ and $a_i(\varphi_1\varphi_0) = a_i$ for all $i \in I$, hence φ_0 is an isomorphism of $\mathfrak{B}$ with $\mathfrak{A}$.

Proof of Theorem 2. Let $\bar{x} = \langle x_0, \cdots, x_\gamma, \cdots \rangle_{\gamma<\alpha}$ where $(x_\gamma \mid \gamma < \alpha)$ is a free generating family of $\mathfrak{F}_K(\alpha)$. Then by the corollary to Theorem 8.2

$$\mathfrak{F}_K(\alpha) \cong \mathfrak{P}^{(\alpha)}(\tau)/\Theta_{\bar{x}}.$$

Thus, to verify Theorem 2, it suffices to show that $\Theta_{\bar{x}} = \Theta_K$. Obviously, $\Theta_{\bar{x}} \geqq \Theta_K$. To show the reverse inequality, we have to verify that $\Theta_{\bar{x}} \leqq \Theta_{\bar{a}}$, where $\bar{a} = \langle a_0, \cdots, a_\gamma, \cdots \rangle_{\gamma<\alpha}$, $a_\gamma \in A$, $\mathfrak{A} \in K$, which is trivial from the existence of a homomorphism φ with $x_\gamma\varphi = a_\gamma$ for all $\gamma < \alpha$.

Corollary 1. *$\Theta_{\bar{x}} = \Theta_K$ is a necessary and sufficient condition for the algebra generated by $\bar{x}$ to be free.*

Corollary 2. *$\mathfrak{F}_K(\alpha) \cong \mathfrak{F}_K(\beta)$ if $\bar{\alpha} = \bar{\beta}$.*

Corollary 3. *$\mathfrak{P}^{(\alpha)}(\tau)$ is isomorphic to $\mathfrak{F}_{K(\tau)}(\alpha)$.*

Lemma 1. *Let $\beta < \alpha$ and let K be a class of algebras for which $\mathbf{S}(K) \subseteq K$. If $\mathfrak{F}_K(\alpha)$ exists, then so does $\mathfrak{F}_K(\beta)$.*

Proof. Let $x_0, \cdots, x_\gamma, \cdots, \gamma < \alpha$, be a free generating family of $\mathfrak{F}_K(\alpha)$ and consider the subalgebra $\mathfrak{B}$ generated by $x_0, \cdots, x_\gamma, \cdots, \gamma < \beta$. Then $\mathfrak{B} \in K$. Let

$$\psi: \gamma \to a_\gamma, \qquad \gamma < \beta,$$

be a mapping of $\{\gamma \mid \gamma < \beta\}$ into $\mathfrak{A} \in K$. Extend this mapping arbitrarily to a mapping $\bar{\psi}$ of all γ with $\gamma < \alpha$; then there is a homomorphism φ with $\gamma\bar{\psi} = x_\gamma\varphi$ for $\gamma < \alpha$. Then φ_B is the required homomorphism. This shows that

$$\mathfrak{B} \cong \mathfrak{F}_K(\beta).$$

More generally, we have the following statement.

Corollary 1. *If $\mathbf{S}(K) \subseteq K$ and $\mathfrak{F}_K(\mathfrak{m})$ exists, then any non-void subset of the free generating set generates a free algebra.*

Corollary 2. *Let α be a limit ordinal. Then $\mathfrak{F}_K(\alpha)$ is isomorphic to a 1–1 direct limit of the $\mathfrak{F}_K(\beta)$, $\beta < \alpha$.*

Theorem 3. *Consider an algebra* $\mathfrak{Q}$ *and let* $[X]=Q$. *Define a class* $K(\mathfrak{Q}, X)$ *of algebras in the following manner:* $\mathfrak{A} \in K(\mathfrak{Q}, X)$ *if and only if every mapping* $p: X \to A$ *can be extended to a homomorphism of* $\mathfrak{Q}$ *into* $\mathfrak{A}$. *Then* $K(\mathfrak{Q}, X)$ *is an equational class.*

Proof. $\mathbf{S}[K(\mathfrak{Q}, X)] \subseteq K(\mathfrak{Q}, X)$ is trivial.

Let $\mathfrak{A} \in K(\mathfrak{Q}, X)$, $\mathfrak{B}$ be a homomorphic image of $\mathfrak{A}$ under the homomorphism ψ, and p be a mapping of X into B; take any mapping $\bar{p}: X \to A$ with $x\bar{p}\psi = xp$. Let φ be an extension of $\bar{p}$ to a homomorphism. Then $\varphi\psi$ extends p to a homomorphism, that is, $\mathfrak{B} \in K(\mathfrak{Q}, X)$. $K(\mathfrak{Q}, X)$ is closed under direct products by Lemma 19.1.

The following results are immediate consequences of Theorem 3.

Theorem 4. *The generating set* X *of the algebra* $\mathfrak{Q}$ *is free with respect to some class* K *if and only if*

$$\mathfrak{Q} \in K(\mathfrak{Q}, X).$$

Proof. By definition, every free algebra must belong to the class over which it is free. Conversely, if $\mathfrak{Q} \in K(\mathfrak{Q}, X)$, then $\mathfrak{Q}$ is free over $K(\mathfrak{Q}, X)$.

Theorem 5. *Let* $\mathfrak{Q}$ *be an algebra. Then the following three conditions are equivalent:*

(i) $\mathfrak{Q}$ *is a free algebra over some class* K;
(ii) $\mathfrak{Q}$ *is a free algebra over an equational class* K;
(iii) $\mathfrak{Q}$ *is free over the class consisting of* $\mathfrak{Q}$ *only.*

Corollary 1. X *is a free generating set of* $\mathfrak{Q}$ *with respect to some class* K *if and only if any mapping of* X *into* Q *can be extended to an endomorphism.*

Remark. By (iii) of Theorem 5, we can talk about free algebras and free generating sets without specifying the class K.

Corollary 2. *Let* $\mathfrak{Q}$ *be a free algebra with the free generating set* X *over some class* K. *Then*

$$K \subseteq K(\mathfrak{Q}, X).$$

Theorem 2 gives a representation of the free algebra as a quotient algebra of the polynomial algebra. The following result, which is due to B. H. Neumann [2] characterizes the congruence relations which occur in such representations.

Theorem 6. *Let $\mathfrak{F}(\alpha)$ be a free algebra with free generating family $(x_\gamma \mid \gamma < \alpha)$, and let Θ be a congruence relation of $\mathfrak{F}(\alpha)$. Then $\mathfrak{F}(\alpha)/\Theta$ is a free algebra with the free generating family $([x_\gamma]\Theta \mid \gamma < \alpha)$ if and only if Θ is fully invariant.*

Proof. Assume that Θ is fully invariant. Let p be a mapping of the $[x_\gamma]\Theta$ into $F(\alpha)/\Theta$ and

$$p\colon [x_\gamma]\Theta \to [a_\gamma]\Theta, \qquad \gamma < \alpha.$$

Let

$$\tilde{p}\colon x_\gamma \to a_\gamma, \qquad \gamma < \alpha$$

and let φ be an extension of $\tilde{p}$ to an endomorphism of $\mathfrak{F}(\alpha)$. By Lemma 12.6, $\bar{\varphi}\colon [x]\Theta \to [x\varphi]\Theta$ is an extension of p to an endomorphism.

By Corollary 1 to Theorem 5, this means that $\mathfrak{F}(\alpha)/\Theta$ is free.

Now assume that $\mathfrak{F}(\alpha)/\Theta$ is freely generated by the $[x_\gamma]\Theta$ and let φ be an arbitrary endomorphism of $\mathfrak{F}(\alpha)$.

In order to show that Θ is fully invariant by Lemma 12.6 it suffices to show that

$$\bar{\varphi}\colon [a]\Theta \to [a\varphi]\Theta$$

is an endomorphism of $\mathfrak{F}(\alpha)/\Theta$.

Consider the mapping

$$\bar{p}\colon [x_\gamma]\Theta \to [x_\gamma\varphi]\Theta.$$

Since $\mathfrak{F}(\alpha)/\Theta$ is free, $\bar{p}$ can be extended to an endomorphism φ'.

Then $([x_\gamma]\Theta)\bar{\varphi} = ([x_\gamma]\Theta)\varphi'$. Now, if $a = p(x_{\gamma_0}, \cdots, x_{\gamma_n-1})$, then

$$\begin{aligned}([a]\Theta)\varphi' &= p([x_{\gamma_0}]\Theta, \cdots, [x_{\gamma_n-1}]\Theta)\varphi' = p(([x_{\gamma_0}]\Theta)\varphi', \cdots, ([x_{\gamma_n-1}]\Theta)\varphi') \\ &= p([x_{\gamma_0}\varphi]\Theta, \cdots, [x_{\gamma_n-1}\varphi]\Theta) = [p(x_{\gamma_0}\varphi, \cdots, x_{\gamma_n-1}\varphi)]\Theta \\ &= [p(x_{\gamma_0}, \cdots, x_{\gamma_n-1})\varphi]\Theta = [a\varphi]\Theta = ([a]\Theta)\bar{\varphi}.\end{aligned}$$

Hence, $\varphi' = \bar{\varphi}$, i.e., $\bar{\varphi}$ is an endomorphism, which was to be proved.

Corollary. *There is a 1–1 correspondence between free algebras with α generators and the fully invariant congruence relations Θ of $\mathfrak{P}^{(\alpha)}(\tau)$.*

§25. CONSTRUCTION OF FREE ALGEBRAS

The following method of constructing free algebras is due to G. Birkhoff [2].

Let K be a class of algebras. Let $\{\Theta_i \mid i \in I\}$ be the set of all congruence relations of $\mathfrak{P}^{(\alpha)}(\tau)$ such that $\mathfrak{P}^{(\alpha)}(\tau)/\Theta_i$ is isomorphic to an algebra in

$\mathbf{S}(K)$. Form the direct product of all the $\mathfrak{P}^{(\alpha)}(\tau)/\Theta_i$ and consider the subalgebra $\mathfrak{A}$ generated by the $x_\gamma, \gamma < \alpha$, defined by

$$x_\gamma(i) = [\mathbf{x}_\gamma]\Theta_i.$$

Let $\bar{x}$ denote the sequence of the x_γ. If $\mathbf{p}, \mathbf{q} \in \mathbf{P}^{(\alpha)}(\tau)$, then $p(\bar{x}) = q(\bar{x})$ if and only if $p(\overline{x(i)}) = q(\overline{x(i)})$ for all $i \in I$, which is equivalent to $p(\bar{a}) = q(\bar{a})$ for all $\bar{a} = \langle a_0, \cdots, a_\gamma, \cdots \rangle_{\gamma < \alpha}$, $a_\gamma \in B$, $\mathfrak{B} \in K$. This means that $\Theta_{\bar{x}} = \Theta_K$; thus, $\mathfrak{A} \cong \mathfrak{P}^{(\alpha)}(\tau)/\Theta_K$.

Theorem 1 (*G. Birkhoff* [2]). *If the free algebra with* α *generators exists, then it is isomorphic to the algebra* $\mathfrak{A}$ *as constructed above.*

Corollary 1. *Let* K *be a class such that* $\mathbf{P}(K) \subseteq K$, $\mathbf{S}(K) \subseteq K$. *Then all the free algebras over* K *exist.*

Corollary 2. *If* K *is an equational class, then all free algebras over* K *exist.*

The following method of constructing free algebras (which follows from a more general idea due to G. Grätzer [10]—see Chapter 8) leads to the existence of algebras which resemble free algebras and are called maximally free algebras.

Definition 1. *Let* K *be a class of algebras and let* α *be an ordinal. The algebra* $\mathfrak{A}$ *with the generating family* x_γ, $\gamma < \alpha$, *is called a* maximally free algebra (*with respect to* K) *if*

(i) $\mathfrak{A} \in K$;

(ii) *if* $\mathfrak{B} \in K$ *and* $\mathfrak{B}$ *is generated by* y_γ, $\gamma < \alpha$, *and if* φ *is a homomorphism of* $\mathfrak{B}$ *into* $\mathfrak{A}$ *with*

$$y_\gamma\varphi = x_\gamma, \qquad \gamma < \alpha,$$

then φ *is an isomorphism.*

Corollary. *Every free algebra is maximally free.*

Proof. Trivial.

Definition 2. *Let* K *be a class of algebras and let* $\mathfrak{A}_i$, $i \in I$, *be a family of maximally free algebras with the generating families* $x_\gamma{}^i$, $\gamma < \alpha$. *We say that* K *is* covered *by this family of maximally free algebras if whenever* $b_\gamma \in B$, $\gamma < \alpha$, $\mathfrak{B} \in K$, *then* at least one *of the mappings*

$$p_i \colon x_\gamma{}^i \to b_\gamma, \qquad \gamma < \alpha,$$

can be extended to a homomorphism of $\mathfrak{A}_i$ *into* $\mathfrak{B}$.

Corollary. *Assume that the class K is covered by a family $\mathfrak{A}_i$, $i \in I$, of maximally free algebras with the generating families x_γ^i, $\gamma < \alpha$. The free algebra $\mathfrak{F}_K(\alpha)$ exists if and only if for any $i, j \in I$, there is an isomorphism φ between $\mathfrak{A}_i$ and $\mathfrak{A}_j$ such that*

$$x_\gamma^i \varphi = x_\gamma^j, \qquad \gamma < \alpha.$$

This shows that to require the existence of a covering family of maximally free algebras is a natural generalization of the requirement of the existence of free algebras.

Theorem 2. *Let K be a class of algebras, closed under taking inverse limits and under the formation of subalgebras. Then for any ordinal α, for which there is an algebra in K of at least $\bar{\alpha}$ elements, K has a covering family of maximally free algebras with α generators.*

Theorem 2 follows trivially from the following statement.

Lemma 1. *Let the conditions of Theorem 2 hold. Let $\mathfrak{A} \in K$ be generated by $(a_\gamma \mid \gamma < \alpha)$. Then there exists a maximally free algebra $\mathfrak{B}$ with a generating family $(b_\gamma \mid \gamma < \alpha)$, and there exists a homomorphism φ of $\mathfrak{B}$ onto $\mathfrak{A}$ such that $b_\gamma \varphi = a_\gamma$ for all $\gamma < \alpha$.*

Proof. Let $\mathfrak{A} \in K$ be generated by α elements: a_γ, $\gamma < \alpha$. Set $\bar{a} = \langle a_0, \cdots, a_\gamma, \cdots \rangle_{\gamma < \alpha}$. Let P be the family of all congruence relations Θ of $\mathfrak{P}^{(\alpha)}(\tau)$ such that $\mathfrak{P}^{(\alpha)}(\tau)/\Theta$ is isomorphic to an algebra in K, and $\Theta \leqq \Theta_{\bar{a}}$.

We will prove that we can apply Zorn's Lemma to the dual of the partially ordered set $\langle P; \leqq \rangle$. Indeed, let

$$\{\Theta_i \mid i \in I\}$$

be a chain in $\langle P; \leqq \rangle$ and define a partial ordering on I such that $j \leqq i$ if and only if $\Theta_j \geqq \Theta_i$. There is a natural homomorphism φ_j^i of $\mathfrak{P}^{(\alpha)}(\tau)/\Theta_i$ onto $\mathfrak{P}^{(\alpha)}(\tau)/\Theta_j$ if $i \geqq j$ defined by

$$[\mathbf{p}]\Theta_i \to [\mathbf{p}]\Theta_j,$$

where $\mathbf{p} \in \mathbf{P}^{(\alpha)}(\tau)$ (see Theorem 11.4).

Now consider the inverse system consisting of the algebras $\mathfrak{P}^{(\alpha)}(\tau)/\Theta_i$ and the homomorphisms φ_j^i, and let $\mathfrak{B}$ denote the inverse limit, which obviously exists. Consider the subalgebra $\mathfrak{C}$ of $\mathfrak{B}$ generated by the x_γ with $x_\gamma(i) = [\mathbf{x}_\gamma]\Theta_i$ and set

$$\bar{x} = \langle x_0, \cdots, x_\gamma, \cdots \rangle_{\gamma < \alpha}.$$

Then $\Theta_{\bar{x}} \in P$ since φ_i^∞ is a homomorphism of $\mathfrak{B}$ onto $\mathfrak{P}^{(\alpha)}(\tau)/\Theta_i$ and so $\Theta_{\bar{x}} \leqq \Theta_i$. This shows that Zorn's Lemma can be applied to the dual of

$\langle P; \leqq\rangle$, and thus we get the existence of minimal elements in $\langle P; \leqq\rangle$. Any such minimal element Θ will give rise to a maximally free algebra $\mathfrak{P}^{(\alpha)}(\tau)/\Theta$ which can be homomorphically mapped onto $\mathfrak{A}$.

§26. IDENTITIES AND FREE ALGEBRAS

Free algebras can be very neatly characterized in terms of identities. The identities (as, for instance, commutativity and associativity) are the simplest and most frequently used axioms in algebra. A formal definition is the following:

Definition 1. *Let* $\mathbf{p}, \mathbf{q} \in \mathbf{P}^{(n)}(\tau)$. *The n-ary* identity $\mathbf{p}=\mathbf{q}$ *is said to be* satisfied *in a class K of algebras of type* τ *if*

$$\mathbf{p} \equiv \mathbf{q}(\Theta_K).$$

Remark. In other words, $\mathbf{p}=\mathbf{q}$ is an *identity* in K if $\mathbf{p}$ and $\mathbf{q}$ induce the same polynomials in each algebra in K, or, equivalently, $p(a_0, \cdots, a_{n-1}) = q(a_0, \cdots, a_{n-1})$ for all $a_0, \cdots, a_{n-1} \in A$, $\mathfrak{A} \in K$.

Lemma 1. *Let K be a class of algebras and let* $\mathbf{p}, \mathbf{q} \in \mathbf{P}^{(n)}(\tau)$; *assume that* $\mathfrak{F}_K(n)$ *exists. Then* $\mathbf{p}=\mathbf{q}$ *is an identity of K if and only if* $p(x_0, \cdots, x_{n-1}) = q(x_0, \cdots, x_{n-1})$, *where the* x_i *are the free generators of* $\mathfrak{F}_K(n)$.

Proof. This is immediate by Corollary 1 to Theorem 24.2, but the following is a direct proof.

Since $\mathfrak{F}_K(n) \in K$, the condition is obviously necessary. It is also sufficient because if $a_0, \cdots, a_{n-1} \in A$, $\mathfrak{A} \in K$, then there exists a homomorphism φ of $\mathfrak{F}_K(n)$ into $\mathfrak{A}$ with $x_i\varphi = a_i$, and thus

$$\begin{aligned} p(a_0, \cdots, a_{n-1}) &= p(x_0\varphi, \cdots, x_{n-1}\varphi) \\ &= p(x_0, \cdots, x_{n-1})\varphi \\ &= q(x_0, \cdots, x_{n-1})\varphi \\ &= q(x_0\varphi, \cdots, x_{n-1}\varphi) \\ &= q(a_0, \cdots, a_{n-1}). \end{aligned}$$

Since ω-ary polynomials provide a common notational system for all n-ary polynomials, it is convenient to consider identities of the type $\mathbf{p}=\mathbf{q}$, where $\mathbf{p}, \mathbf{q} \in \mathbf{P}^{(\omega)}(\tau)$. The definition of satisfiability in a class is the same as in Definition 1. Then we have the following theorem.

Theorem 1. *The identities which are satisfied in K are the same as those satisfied in* $\mathfrak{F}_K(\omega)$, *provided the latter exists, which in turn are the same which* *satisfy* $p(x_0, \cdots, x_n, \cdots) = q(x_0, \cdots, x_n, \cdots)$, *where* $x_0, \cdots, x_n, \cdots$ *are the free generators of* $\mathfrak{F}_K(\omega)$. *The n-ary identities which are satisfied in* K *are the same as those satisfied in* $\mathfrak{F}_K(n)$ *provided the latter exists.*

Let K be a class and let $Id(K)$ denote the set of all identities satisfied in K. Then Theorem 1 implies that $Id(K)$ determines the structure of $\mathfrak{F}_K(\omega)$, and conversely.

Corollary 1. *Let* K, K' *be classes of algebras and assume that* $\mathfrak{F}_K(\omega)$ *and* $\mathfrak{F}_{K'}(\omega)$ *exist. Then*

$$Id(K) = Id(K')$$

if and only if

$$\mathfrak{F}_K(\omega) \cong \mathfrak{F}_{K'}(\omega).$$

Corollary 2. *Let* K, K' *be classes of algebras and suppose that* $\mathfrak{F}_K(\alpha)$, $\mathfrak{F}_{K'}(\alpha)$ *exist for some* $\alpha \geqq \omega$. *Then*

$$Id(K) = Id(K')$$

if and only if

$$\mathfrak{F}_K(\alpha) \cong \mathfrak{F}_{K'}(\alpha).$$

Corollary 2 follows from Corollary 1 using the observation that we can use any ω free generators of $\mathfrak{F}_K(\alpha)$ in place of the free generators of $\mathfrak{F}_K(\omega)$.

If we start with a class of algebras K, we get a set of identities: $Id(K)$. Conversely, if we start with a set† of identities Σ, we get a class of algebras Σ^* satisfying these identities. We will now characterize those sets of identities which can be represented as $Id(K)$ and those classes of algebras which can be represented as Σ^*.

Definition 2. *A set of identities* Σ *is called* closed *provided:*

(i) $\mathbf{x}_i = \mathbf{x}_i$ *is in* Σ *for* $i < \omega$;
(ii) *if* $\mathbf{p} = \mathbf{q}$ *is in* Σ, *then so is* $\mathbf{q} = \mathbf{p}$;
(iii) *if* $\mathbf{p} = \mathbf{q}$ *and* $\mathbf{q} = \mathbf{r}$ *are in* Σ, *then so is* $\mathbf{p} = \mathbf{r}$;
(iv) *if* $\mathbf{p}_i = \mathbf{q}_i$ *is in* Σ *for* $i = 0, \cdots, n_\gamma - 1$, *then so is*

$$\mathbf{f}_\gamma(\mathbf{p}_0, \cdots, \mathbf{p}_{n_\gamma - 1}) = \mathbf{f}_\gamma(\mathbf{q}_0, \cdots, \mathbf{q}_{n_\gamma - 1});$$

(v) *if* $\mathbf{p} = \mathbf{q}$ *is in* Σ *and we get* $\mathbf{p}'$ *and* $\mathbf{q}'$ *from* $\mathbf{p}$ *and* $\mathbf{q}$ *by replacing all occurrences of* $\mathbf{x}_i$ *by an arbitrary polynomial symbol* $\mathbf{r}$, *then* $\mathbf{p}' = \mathbf{q}'$ *is also in* Σ.

Theorem 2 (*G. Birkhoff* [2]). *A set* Σ *of identities can be represented in the form* $\Sigma = Id(K)$ *if and only if* Σ *is closed.*

† A set of identities will always be assumed to be of the same type.

Proof. It is obvious that if $\Sigma = Id(K)$, then Σ is closed. Conversely, suppose that Σ is closed and define the relation Θ on $\mathbf{P}^{(\omega)}(\tau)$ by $\mathbf{p} \equiv \mathbf{q}(\Theta)$ if and only if $\mathbf{p} = \mathbf{q}$ is in Σ. The reflexivity of Θ follows from (i) and (iv), while (ii) and (iii) guarantee that Θ is symmetric and transitive, and (iv) gives the substitution property. Since any endomorphism of $\mathfrak{P}^{(\omega)}(\tau)$ is uniquely determined if we are given the image $\mathbf{r}_i$ of $\mathbf{x}_i$ and the image of $\mathbf{p}$ can be constructed by replacing the $\mathbf{x}_i$ by $\mathbf{r}_i$, we infer by rule (v) that Θ is fully invariant. By Theorem 24.6, $\mathfrak{P}^{(\omega)}(\tau)/\Theta$ is a free algebra; thus, by Theorem 1, the identities $\mathbf{p} = \mathbf{q}$ satisfied in it are the same as those for which

$$p(x_0, \cdots, x_n, \cdots) = q(x_0, \cdots, x_n, \cdots),$$

where the x_i are the free generators, which by construction are the same as the identities for which $\mathbf{p} \equiv \mathbf{q}(\Theta)$, i.e., which are included in Σ. We conclude that $\Sigma = Id(K)$, where

$$K = \{\mathfrak{P}^{(\omega)}(\tau)/\Theta\}.$$

Remark. Theorem 2 is the "completeness theorem" of rules (i)–(v) of Definition 2. For a set of identities Σ, let us say that Σ implies the identity $\mathbf{p} = \mathbf{q}$ if whenever Σ is satisfied in an algebra, then so is $\mathbf{p} = \mathbf{q}$. It is obvious that if $\mathbf{p} = \mathbf{q}$ is provable from Σ, then Σ implies $\mathbf{p} = \mathbf{q}$ in the above sense. Now, Theorem 2 asserts that if Σ implies $\mathbf{p} = \mathbf{q}$, then $\mathbf{p} = \mathbf{q}$ is provable from Σ, using the rules (i)–(v) of Definition 2. Thus (i)–(v) form "a complete set of rules of inference", because whatever follows from Σ can be proved by (i)–(v).

The following result justifies the terminology "equational class".

Theorem 3 (*G. Birkhoff* [2]). *A class K can be represented as $K = \Sigma^*$ for some set of identities Σ if and only if K is an equational class.*

Proof. Since identities are obviously preserved under the formation of subalgebras, homomorphic images, and direct products, Σ^* is always an equational class.

Conversely, let K be an equational class and set $\Sigma = Id(K)$. By Corollary 2 to Theorem 25.1, $\mathfrak{F}_K(\alpha)$ exists for any α and for $\alpha \geqq \omega$ the identities satisfied by the free generators are the same as the identities in Σ. This implies that if $\mathfrak{A}$ is any algebra in Σ^* with $|A| = \bar{\alpha}$, then any identity satisfied by the generators of $\mathfrak{F}_K(\alpha)$ is satisfied by the elements of $\mathfrak{A}$; hence, by Theorem 12.2, $\mathfrak{A}$ is a homomorphic image of $\mathfrak{F}_K(\alpha)$. Consequently, $\Sigma^* \subseteq K$ and by definition $K \subseteq \Sigma^*$; thus, $K = \Sigma^*$, which was to be proved.

Corollary 1. *Let K, K' be equational classes. Then $K = K'$ if and only if*

$$Id(K) = Id(K'),$$

or, equivalently, if and only if

$$\mathfrak{F}_K(\omega) \cong \mathfrak{F}_{K'}(\omega).$$

Corollary 2. *Let K, K' be equational classes. Then $K \subseteq K'$ if and only if*

$$Id(K) \supseteq Id(K'),$$

or, equivalently, if and only if $\mathfrak{F}_K(\omega)$ is a homomorphic image of $\mathfrak{F}_{K'}(\omega)$.

Definition 3. *The class K is said to be* generated by the algebra $\mathfrak{A}$ *if* $K = \mathbf{HSP}(\{\mathfrak{A}\})$.

Corollary 3 (*A. Tarski* [1]). *A class K is equational if and only if it is generated by a suitable algebra $\mathfrak{A}$.*

Proof. If K is an equational class, we can always take

$$\mathfrak{A} = \mathfrak{F}_K(\omega).$$

Then set

$$K' = \mathbf{HSP}(\{\mathfrak{F}_K(\omega)\}).$$

Since $Id(K') = Id(K)$, we get, by Corollary 1, that $K = K'$.

All equational classes of algebras of a given type τ form a lattice $\mathfrak{L}(\tau)$ under inclusion, called the *lattice of equational classes.* (It is, of course, not legitimate to form a set whose elements are classes. In this instance, however, we can get around this difficulty by defining $\mathfrak{L}(\tau)$ to be the dual of the lattice of fully invariant congruence relations of $\mathfrak{P}^{(\omega)}(\tau)$.) $\mathfrak{L}(\tau)$ is a complete lattice, whose zero is the class of all one element algebras (determined by $\mathbf{x}_0 = \mathbf{x}_1$) and whose unit element is $K(\tau)$ (defined by $\mathbf{x}_0 = \mathbf{x}_0$). In the next section we will study the atoms of $\mathfrak{L}(\tau)$.

The properties of free algebras can be utilized to find identities which characterize equational classes with certain properties, which are preserved under homomorphisms. The following theorem illustrates this method.

Theorem 4 (*A. I. Mal'cev* [3]). *For an equational class K the following two properties are equivalent:*

(i) *For every $\mathfrak{A} \in K$, the congruences of $\mathfrak{A}$ permute;*

(ii) *there exists a ternary polynomial symbol* $\mathbf{p}$ *such that the following two identities hold in K:*

$$\mathbf{x}_0 = \mathbf{p}(\mathbf{x}_0, \mathbf{x}_1, \mathbf{x}_1) \quad \text{and} \quad \mathbf{x}_1 = \mathbf{p}(\mathbf{x}_0, \mathbf{x}_0, \mathbf{x}_1).$$

Proof. Let us assume (i). Then in $\mathfrak{F}_K(3) = \mathfrak{P}^{(3)}(\tau)/\Theta_K$,

$$x_0 \equiv x_2(\Theta(x_1, x_2) \cdot \Theta(x_0, x_1)),$$

so

$$x_0 \equiv [\mathbf{p}]\Theta_K(\Theta(x_1, x_2)) \quad \text{and} \quad [\mathbf{p}]\Theta_K \equiv x_2(\Theta(x_0, x_1)),$$

for some $\mathbf{p} \in \mathbf{P}^{(3)}(\tau)$. Since $\mathfrak{F}_K(3)/\Theta(x_0, x_1) \cong \mathfrak{F}_K(3)/\Theta(x_1, x_2) \cong \mathfrak{F}_K(2)$, we get that

$$x_0 = p(x_0, x_1, x_1) \quad \text{and} \quad x_1 = p(x_0, x_0, x_1)$$

in $\mathfrak{F}_K(2)$; thus Lemma 1 implies (ii).

Conversely, if (ii) holds, $\mathfrak{A} \in K$, $\Theta, \Phi \in C(\mathfrak{A})$, $a, b, c \in A$ and $a \equiv b(\Theta)$, $b \equiv c(\Phi)$, then

$$\begin{aligned} a = p(a, b, b) &\equiv p(a, b, c)(\Phi), \\ p(a, b, c) &\equiv p(b, b, c) = c(\Theta), \end{aligned}$$

proving that $\Theta\Phi = \Phi\Theta$.

This method is applied, e.g., in B. Csákány [1]–[3], B. Jónsson [8], and A. F. Pixley [1] (see Exercises 5.69 and 5.70).

§27. EQUATIONAL COMPLETENESS AND IDENTITIES OF FINITE ALGEBRAS

Every set of identities is satisfied by some algebra, namely by the one-element algebra.

Definition 1. *A set of identities* Σ *is called* strictly consistent *if there exists an algebra* $\mathfrak{A} \in \Sigma^*$ *such that* $|A| > 1$.

Lemma 1. *Let* $\overline{\Sigma}$ *denote the smallest closed set of identities that contains* Σ. *Then* Σ *is strictly consistent if and only if* $\mathbf{x}_0 = \mathbf{x}_1 \notin \overline{\Sigma}$.

Proof. This follows from the construction used in Theorem 26.2. Indeed, if $\mathbf{x}_0 = \mathbf{x}_1 \notin \overline{\Sigma}$, then the generators of $\mathfrak{P}^{(\omega)}(\tau)$ will not be congruent under Θ and then

$$\mathfrak{P}^{(\omega)}(\tau)/\Theta \in \Sigma^*$$

and has more than one element. The converse is trivial.

If we start with a closed strictly consistent set of identities Σ_1, then in most cases we can add some further identities and we get Σ_2 which is also closed and strictly consistent. Does this process ever terminate? To study this problem, let us make the following definition.

Definition 2. *Let Σ be a strictly consistent set of identities. Σ is called* equationally complete *if whenever $\Sigma \subseteq \Sigma'$, where Σ' is strictly consistent, then $\Sigma = \Sigma'$.*

Let us note that an equationally complete set of identities be always closed.

Definition 3. *An equational class K of algebras is* equationally complete *provided $Id(K)$ is equationally complete.*

Let us note that K is equationally complete if K is an equational class such that if K_1 is an equational class, $K \supseteq K_1$ and K_1 contains at least one algebra with more than one element, then $K = K_1$. Thus K is equationally complete if it is an atom in $\mathfrak{L}(\tau)$.

Definition 4. *An algebra $\mathfrak{A}$ is* equationally complete *if the equational class generated by $\mathfrak{A}$ is equationally complete.*

Let us note that an equivalent definition is to require that $Id(\mathfrak{A})$ be an equationally complete set of identities.

Example. Consider the class K of distributive lattices. Set $\Sigma = Id(K)$. We will verify that K is equationally complete. Suppose $\mathbf{p} = \mathbf{q} \notin \Sigma$. Assume that contrary to our assumption,

$$\Sigma' = \Sigma \cup \{\mathbf{p} = \mathbf{q}\}$$

is strictly consistent. Then there exists a lattice $\mathfrak{L}$ such that $|L| \neq 1$ and Σ' is satisfied in it. Since every lattice with more than one element has a two-element sublattice, the two-element lattice therefore also satisfies Σ'. Since every distributive lattice with more than one element is a subdirect product of copies of the two-element lattice, we get that every distributive lattice with more than one element satisfies Σ'. Consequently, Σ' is contained in Σ, contrary to our assumption. This example shows that: (i) Σ is an equationally complete set of identities; (ii) the class of distributive lattices is an equationally complete class; and (iii) any distributive lattice with more than one element is an equationally complete algebra.

Theorem 1. *Let Σ be a strictly consistent set of identities. Then Σ is contained in an equationally complete set of identities.*

Proof. Let K, K' be equational classes. In the proof of Theorem 26.2, we associated fully invariant congruence relations Θ, Θ' on $\mathfrak{P}^{(\omega)}(\tau)$ with

$Id(K)$, $Id(K')$, respectively, and by Corollary 2 to Theorem 26.3 we know that $K \supseteq K'$ if and only if $\Theta \leqq \Theta'$.

Thus, to prove Theorem 1, we have to exhibit a fully invariant congruence relation Θ' such that $\Theta \leqq \Theta'$, $\Theta' \neq \iota$, and Θ' is maximal with respect to these properties. This follows from Theorem 12.4, since $\Theta' \neq \iota$ is equivalent to $\mathbf{x}_0 \not\equiv \mathbf{x}_1(\Theta')$.

Corollary 1. *If K is an equational class which contains an algebra with more than one element, then K contains an equationally complete class.*

Corollary 2. *Equationally complete classes correspond in a 1–1 manner to maximal fully invariant congruence relations of $\mathfrak{P}^{(\omega)}(\tau)$ which separate the generators.*

Let us assume that $\overline{o(\tau)} \leqq \aleph_0$. Then there are $\aleph_0$ ω-ary polynomials and $\aleph_0$ identities. Therefore, there are $\mathfrak{c} = 2^{\aleph_0}$ sets of identities. How many of these can be equationally complete? This question is answered in the following theorem.

Theorem 2 (*J. Kalicki* [2]). *Let $\tau = \langle 2 \rangle$. Then there are $\mathfrak{c}$ equationally complete sets of identities.*

Proof. We will consider algebras of type $\langle 2 \rangle$ and the operation will be denoted by $+$. For simplicity's sake, we will write $2x$ for $x + x$ and $2^n x$ for $2(2^{n-1}x)$. Let I be the set of positive integers. Fix two subsets $N, M \subseteq I$. We define a set of identities $X(M, N)$ as follows:

(i) $2\mathbf{x}_0 + \mathbf{x}_0 = 2\mathbf{x}_1 + \mathbf{x}_1$;
(ii) $2^m\mathbf{x}_0 + \mathbf{x}_0 = 2\mathbf{x}_0 + \mathbf{x}_0$ if $m \in M$;
(iii) $2^n\mathbf{x}_0 + \mathbf{x}_0 = \mathbf{x}_0$ if $n \in N$.

Let us say that M and N are *complementary* if $M \cup N = I - \{1\}$ and $M \cap N = \varnothing$.

Lemma 2. *If M and N are complementary, then $X(M, N)$ is strictly consistent.*

Proof. We exhibit an algebra with more than one element which satisfies $X(M, N)$. Let $A = \{a_1, a_2, \cdots\}$. Define $+$ on A as follows:

$a_{i+1} + a_i = a_1$;
$a_{i+m} + a_i = a_1$ if $m \in M$;
$a_{i+n} + a_i = a_i$ if $n \in N$;
$a_i + a_i = a_{i+1}$;
$a_i + a_j = a_j + a_i$.

Observe that $a_i + a_i = a_{i+1}$ implies that $2^m a_n = a_{m+n}$. (Proof is by induction on m.)

Compute: $2a_i + a_i = a_{i+1} + a_i = a_1$. This verifies axiom (i).

Further, for $m \in M$, $2^m a_i + a_i = a_{i+m} + a_i = a_1 = a_{i+1} + a_i = 2a_i + a_i$. This verifies axiom (ii).

Finally, for $n \in N$, $2^n a_i + a_i = a_{i+n} + a_i = a_i$, which verifies axiom (iii). This completes the proof of Lemma 2.

Lemma 3. *Let M, N and M', N' be complementary and suppose $M \neq M'$. Then $X(M, N) \cup X(M', N')$ is not strictly consistent.*

Proof. By assumption, there exists an m_0 in M such that $m_0 \notin M'$ (or the other way around); then $m_0 \in M$ and $m_0 \in N'$. Therefore $2^{m_0}\mathbf{x}_0 + \mathbf{x}_0 = 2\mathbf{x}_0 + \mathbf{x}_0$ and $2^{m_0}\mathbf{x}_0 + \mathbf{x}_0 = \mathbf{x}_0$; hence $2\mathbf{x}_0 + \mathbf{x}_0 = \mathbf{x}_0$.

Similarly, $2\mathbf{x}_1 + \mathbf{x}_1 = \mathbf{x}_1$. Then, by axiom (i), $\mathbf{x}_0 = \mathbf{x}_1$, which means that the set is not strictly consistent.

Proof of Theorem 2. There are $\mathfrak{c}$ complementary sets; hence, there are $\mathfrak{c}$ strictly consistent $X(M, N)$. By Theorem 1, each can be extended to an equationally complete set, and, by Lemma 3, no two such extensions can coincide. This concludes the proof of the theorem.

Lemma 4 (*A. Tarski* [1]). *If K is an equationally complete class and $\mathfrak{A} \in K$, $|A| > 1$, then $\mathfrak{A}$ is an equationally complete algebra.*

Proof. Consider the equational class generated by $\mathfrak{A}$. It is contained in K; therefore, it equals K.

Corollary. *Let K be an arbitrary equational class which contains an algebra with more than one element; then there exists an equationally complete algebra $\mathfrak{A}$ in K.*

Proof. By Theorem 1 and Lemma 4.

In contrast with Theorem 2, the identities of a finite algebra have only finitely many equationally complete extensions.

Theorem 3 (*D. Scott* [1]). *Let $\mathfrak{A}$ be a finite algebra with more than one element, that is, $1 < |A| < \aleph_0$. Then $Id(\mathfrak{A})$ has finitely many equationally complete extensions.*

Proof. Set $K = \mathbf{HSP}(\{\mathfrak{A}\})$. Then $Id(K) = Id(\{\mathfrak{A}\})$. Therefore, $\mathfrak{F}_K(n) \cong \mathfrak{P}^{(n)}(\mathfrak{A})$ for any nonnegative integer n. Thus, $\mathfrak{F}_K(n)$ is finite. Take $K' \subseteq K$, where K' is equationally complete. By Lemma 4, K' is generated by $\mathfrak{F}_{K'}(n)$, for any n, which, by Corollary 2 to Theorem 26.3, is a homomorphic image of $\mathfrak{F}_K(n)$. Hence, there are no more equationally complete classes contained in K than there are homomorphic images of $\mathfrak{F}_K(n)$. Since the latter is finite, this completes the proof of Theorem 3.

The fact that $\mathfrak{F}_K(n)$ is finite if A is finite implies that if $\mathfrak{A}$ is of finite type, then we can list a finite number of n-ary identities such that every n-ary identity follows from these. In other words, the n-ary identities have a *finite basis*. The next theorem shows that this is not true of all identities.

Theorem 4 (*R. C. Lyndon* [2]). *There exists a finite algebra* $\mathfrak{A}$ *of finite type such that* $Id(\mathfrak{A})$ *has no finite basis.*

The proof of Theorem 4 is sketched in the exercises.

R. C. Lyndon's example has seven elements and is of type $\langle 0, 2\rangle$. This was improved by V. V. Višin [1] and V. L. Murskiĭ [1] who found four-, and three-element algebras, respectively, of type $\langle 2\rangle$ having the same property. P. Perkins (Ph.D. thesis, Stanford University, 1966) has shown that $\mathfrak{A}$ can be chosen to be a six-element semigroup.

An interesting class of finite algebras, called primal algebras, is considered next, to illustrate the above results.

Definition 5. *An algebra* $\mathfrak{A}$ *is called* primal *if* A *is finite, containing more than one element, and every function on* A *is a polynomial.*

The two-element Boolean algebra is primal. Many other examples will be given in the exercises. A Boolean algebra with more than two elements is never primal, although it can be made primal if we add its elements as nullary polynomials. (Primal algebras are also called *functionally complete* and *functionally strictly complete* algebras.)

Lemma 5. *A primal algebra* $\mathfrak{A}$ *is simple and has no proper subalgebras.*

Proof. If Θ is a congruence relation of $\mathfrak{A}$, $a \equiv b(\Theta)$, $a \neq b$, and $c, d \in A$, then let p be a unary polynomial with $p(a) = c$ and $p(b) = d$. Then $a \equiv b(\Theta)$ implies $c = p(a) \equiv p(b) = d(\Theta)$, so $\Theta = \iota$. The second statement is obvious.

Lemma 6. *Every primal algebra* $\mathfrak{A}$ *is equationally complete.*

Proof. Let K be the equational class generated by $\mathfrak{A}$; then†

$$\mathfrak{F}_K(0) \cong \mathfrak{P}^{(0)}(\mathfrak{A}) \cong \mathfrak{A}.$$

Thus if $K_1 \subseteq K$ and K_1 is equational, then $\mathfrak{F}_{K_1}(0)$ is a homomorphic image of $\mathfrak{A}$ by Corollary 2 to Theorem 26.3. Therefore the result follows from Lemma 5.

Primal algebras behave very much like the two element Boolean algebra.

Theorem 5. *Let $\mathfrak{A}$ be a primal algebra and let K be the equational class generated by $\mathfrak{A}$. Then the following conditions are equivalent on an algebra $\mathfrak{B}$ with $|B| > 1$:*

(i) $\mathfrak{B} \in K$;
(ii) $Id(\mathfrak{A}) = Id(\mathfrak{B})$;
(iii) $Id(\mathfrak{A}) \subseteq Id(\mathfrak{B})$;
(iv) *$\mathfrak{B}$ is isomorphic to a normal subdirect power of $\mathfrak{A}$;*
(v) *$\mathfrak{B}$ is isomorphic to an extension of $\mathfrak{A}$ by a Boolean algebra.*

Remark. The equivalence of (i) and (iv) is due to L. I. Wade [1]. The equivalence of (i)–(iii) follows from P. C. Rosenbloom [1] who also obtained (v) for finite algebras. The equivalence of all five conditions is stated in A. L. Foster [3].

Proof. (i), (ii) and (iii) are equivalent by Lemmas 4 and 6. (iv) and (v) are equivalent by Theorems 22.3 and 22.4. Obviously, (iv) implies (i); thus it suffices to prove that (ii) implies (iv).

We are given $\mathfrak{B}$ with $Id(\mathfrak{A}) = Id(\mathfrak{B})$ and $|B| \neq 1$. Since $\mathfrak{A}$ and $\mathfrak{B}$ generate the same equationally complete class K, we get that

$$\mathfrak{A} \cong \mathfrak{P}^{(0)}(\mathfrak{A}) \cong \mathfrak{F}_K(0) \cong \mathfrak{P}^{(0)}(\mathfrak{B}) \in \mathbf{IS}(\mathfrak{B}),$$

and so we can assume that $\mathfrak{A}$ is a subalgebra of $\mathfrak{B}$. Thus, $\mathfrak{B}$ is a normal subdirect power of $\mathfrak{A}$ if and only if the following conditions are satisfied:

(α) There exists a family $(\varphi_i \mid i \in I)$ of endomorphisms of $\mathfrak{B}$ such that $B\varphi_i = A$ and $a\varphi_i = a$ for $a \in A$, $i \in I$;

(β) for $u, v \in B$, $u \neq v$, there exists an $i \in I$ with $u\varphi_i \neq v\varphi_i$;

† If there are no nullary operations, interpret $\mathfrak{F}_K(0)$ and $\mathfrak{P}^{(0)}(\mathfrak{A})$ as the smallest subalgebras of $\mathfrak{F}_K(1)$ and $\mathfrak{P}^{(1)}(\mathfrak{A})$, respectively, which exist since there are operations constant in K; indeed, for $a \in A$, there is an operation $f\colon A^n \to \{a\}$, and then the identity

$$f(\mathbf{x}_0, \cdots, \mathbf{x}_{n-1}) = f(\mathbf{x}_n, \cdots, \mathbf{x}_{2n-1})$$

holds in $\mathfrak{A}$ and thus in K. Therefore f is constant in K. In other words, $\mathfrak{P}^{(0)}(\mathfrak{A})$ will stand for $\langle P^{(1,0)}(\mathfrak{A}); F\rangle$.

(γ) there exists a 4-ary function T on B such that for $a, b, c, d \in B$, $i \in I$, $T(a, b, c, d)\varphi_i = T(a\varphi_i, b\varphi_i, c\varphi_i, d\varphi_i)$ and

$$T(a\varphi_i, b\varphi_i, c\varphi_i, d\varphi_i) = \begin{cases} c\varphi_i & \text{if} \quad a\varphi_i = b\varphi_i \\ d\varphi_i & \text{if} \quad a\varphi_i \neq b\varphi_i. \end{cases}$$

By Theorems 12.2 and 21.7, (α) and (β) follow from the following statement:

(δ) If $b_0, \cdots, b_{n-1} \in B$, $b_0 \neq b_1$, then there exist $a_0, \cdots, a_{n-1} \in A$, $a_0 \neq a_1$ such that for $\mathbf{p}, \mathbf{q} \in \mathbf{P}^{(n)}(\tau)$, $p(b_0, \cdots, b_{n-1}) = q(b_0, \cdots, b_{n-1})$ implies $p(a_0, \cdots, a_{n-1}) = q(a_0, \cdots, a_{n-1})$.

In order to prove (δ), choose the nullary polynomial symbols 0, 1, the unary polynomial symbol $\mathbf{f}$, the binary polynomial symbols $+$, $-$, $\vee$, $\cdot$ such that $\langle A; +, -, 0\rangle$ is an abelian group and $\langle A; \vee, 0\rangle$ is a semilattice with zero and $\langle A; f, \cdot, 0, 1\rangle$ satisfies $f(0) = 0$ and $f(x) = 1$ if $x \neq 0$, and $x \cdot 1 = x = 1 \cdot x$, $x \cdot 0 = 0 = 0 \cdot x$, $0 \neq 1$. Since abelian groups and semilattices with 0 are defined by identities, we get that $\langle B; +, -, 0\rangle$ is also an abelian group and $\langle B; \vee, 0\rangle$ is a semilattice with 0 and $x \cdot 1 = x$, $x \cdot 0 = 0$ for all $x \in B$.

Suppose that (δ) fails to hold. Then for every sequence $a_0, \cdots, a_{n-1}$ of elements of A with $a_0 \neq a_1$ there exist polynomial symbols $\mathbf{p}$ and $\mathbf{q}$ such that $p(b_0, \cdots, b_{n-1}) = q(b_0, \cdots, b_{n-1})$ and $p(a_0, \cdots, a_{n-1}) \neq q(a_0, \cdots, a_{n-1})$. Let $\mathbf{p}_i$, $\mathbf{q}_i$, $0 \leqq i < k$ be the polynomial symbols we pick, one for every such finite sequence of elements of A, and set

$$\mathbf{r} = (\mathbf{p}_0 - \mathbf{q}_0) \vee \cdots \vee (\mathbf{p}_{k-1} - \mathbf{q}_{k-1}).$$

Then $r(b_0, \cdots, b_{n-1}) = 0$ and $r(a_0, \cdots, a_{n-1}) \neq 0$ whenever $a_0 \neq a_1$, $a_0, \cdots, a_{n-1} \in A$. Then consider the identity

$$(*) \qquad \mathbf{x}_0 = \mathbf{x}_1 + (\mathbf{x}_0 - \mathbf{x}_1)f(r(\mathbf{x}_0, \cdots, \mathbf{x}_{n-1})).$$

(*) holds in $\mathfrak{A}$, that is

$$a_0 = a_1 + (a_0 - a_1)f(r(a_0, \cdots, a_{n-1})).$$

Indeed, if $a_0 = a_1$ we get $a_0 = a_1 + 0 \cdot f(r(a_0, \cdots, a_{n-1})) = a_1$. If $a_0 \neq a_1$, then $r(a_0, a_1, \cdots, a_{n-1}) \neq 0$, so $f(r(a_0, \cdots, a_{n-1})) = 1$ and we get

$$a_0 = a_1 + (a_0 - a_1) \cdot 1 = a_1 + (a_0 - a_1) = a_0.$$

Since $Id(\mathfrak{A}) = Id(\mathfrak{B})$, (*) holds in $\mathfrak{B}$. Let us substitute $b_0, \cdots, b_{n-1}$ in (*). Since $r(b_0, \cdots, b_{n-1}) = 0$, $f(r_0(b_0, \cdots, b_{n-1})) = 0$, so

$$b_0 = b_1 + (b_0 - b_1) \cdot 0 = b_1,$$

which contradicts $b_0 \neq b_1$. This contradiction proves (δ), and therefore (α) and (β).

(γ) is almost trivial. Choose a polynomial symbol $\mathbf{T}$ which acts on $\mathfrak{A}$ as follows: $T(a, b, c, d) = c$ if $a = b$ and $T(a, b, c, d) = d$ if $a \neq b$. Then $T = (\mathbf{T})_{\mathfrak{B}}$ is a polynomial on $\mathfrak{B}$ so $T(a, b, c, d)\varphi_i = T(a\varphi_i, b\varphi_i, c\varphi_i, d\varphi_i)$ for $a, b, c, d \in B$, which implies (γ). This completes the proof of Theorem 5.

It is implicit in P. C. Rosenbloom [1] that the identities of a primal algebra of finite type have a finite basis, see also A. Yaqub [1].

§28. FREE ALGEBRAS GENERATED BY PARTIAL ALGEBRAS†

It is useful to consider algebras which are as freely generated by a set as possible within certain limitations. One way of prescribing this limitation is to assume that the generating set is a partial algebra. This leads us to the following definition.

Definition 1. *Let K be a class of algebras and let $\mathfrak{A}$ be a partial algebra. The algebra $\mathfrak{F}_K(\mathfrak{A})$ is called the* algebra freely generated by the partial algebra $\mathfrak{A}$ over K *if the following conditions are satisfied:*

(i) $\mathfrak{F}_K(\mathfrak{A}) \in K$;

(ii) *$\mathfrak{F}_K(\mathfrak{A})$ is generated by A' and there exists an isomorphism χ: $A' \to A$ between $\mathfrak{A}'$ and $\mathfrak{A}$, where $\mathfrak{A}'$ is a relative subalgebra of $\mathfrak{F}_K(\mathfrak{A})$;*

(iii) *If φ is a homomorphism of $\mathfrak{A}$ into $\mathfrak{C} \in K$, then there exists a homomorphism ψ of $\mathfrak{F}_K(\mathfrak{A})$ into $\mathfrak{C}$ such that ψ is an extension of $\chi\varphi$.*

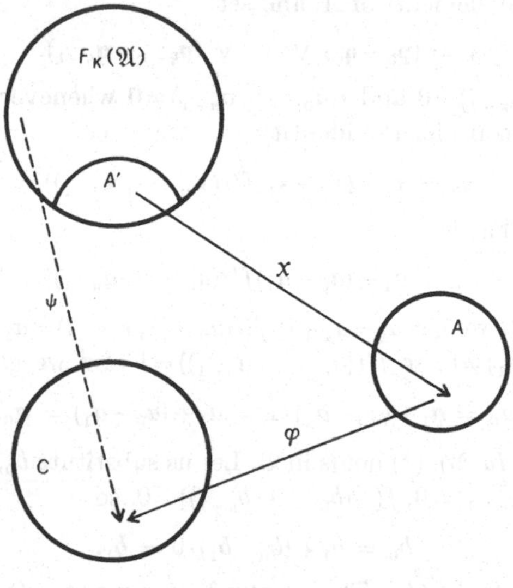

† Most of the results of sections 28 and 29 are special cases of well-known results in category theory; see Exercises of Chapter 3 in P. Freyd, Abelian Categories, Harper & Row, 1964 (see also A. A. Iskander [2] and J. Słomiński [9]).

Note that ψ is necessarily unique.

Corollary 1. *If* $\mathfrak{A}$ *is an algebra and* $\mathfrak{A} \in K$, *then* $\mathfrak{F}_K(\mathfrak{A}) \cong \mathfrak{A}$.

Corollary 2. $\mathfrak{F}_K(\mathfrak{A})$ *is unique up to isomorphism.*

Corollary 3. *If* $D(f_\gamma, \mathfrak{A}) = \varnothing$ *for all* $\gamma < o(\tau)$ *and K contains an algebra with more than one element, then*

$$\mathfrak{F}_K(\mathfrak{A}) \cong \mathfrak{F}_K(\mathfrak{m}),$$

where $\mathfrak{m} = |A|$.

In this section, we will give two sufficient conditions for the existence of $\mathfrak{F}_K(\mathfrak{A})$, the first of which constructs it from $\mathfrak{F}_K(\mathfrak{m})$ while the second is based on G. Birkhoff's idea of the construction of the free algebra (see §25).

Theorem 1. *Let K be a class of algebras and let* $\mathfrak{A}$ *be a partial algebra. Let us make the following assumptions:*

(i) $\mathfrak{A}$ *is (isomorphic to) a relative subalgebra of an algebra* $\mathfrak{B}$ *in* K;
(ii) $\mathfrak{F}_K(\mathfrak{m})$ *exists for some* $\mathfrak{m} \geqq |A|$;
(iii) $\mathbf{H}(K) \subseteq K$.

Then $\mathfrak{F}_K(\mathfrak{A})$ *exists.*

Proof. (ii) and (iii) imply that $\mathfrak{F}_K(\mathfrak{m})$ exists for $\mathfrak{m} = |A|$. Let α be an ordinal with $\bar{\alpha} = \mathfrak{m}$. Let $A = \{a_\gamma \mid \gamma < \alpha\}$. Let x_γ, $\gamma < \alpha$, denote a free generating set of $\mathfrak{F}_K(\alpha)$. Let T be a set of pairs of elements of $\mathfrak{F}_K(\alpha)$ defined as follows: $\langle x, y \rangle \in T$ if $x = p(x_{i_0}, \cdots, x_{i_{n-1}})$, $y = q(x_{j_0}, \cdots, x_{j_{m-1}})$, and $p(a_{i_0}, \cdots, a_{i_{n-1}}) = q(a_{j_0}, \cdots, a_{j_{m-1}})$, where $\mathbf{p}$ and $\mathbf{q}$ are polynomial symbols. Set $\Theta = \bigvee (\Theta(x, y) \mid \langle x, y \rangle \in T)$.

We claim that $\mathfrak{F}_K(\alpha)/\Theta$ is the free algebra generated by $\mathfrak{A}$. It follows from (ii) and (iii) that it is in K.

Let $a_\gamma' = [x_\gamma]\Theta$ and $A' = \{a_\gamma' \mid \gamma < \alpha\}$. Let φ be a homomorphism of $\mathfrak{A}$ into $\mathfrak{C} \in K$; set $a_\gamma \varphi = c_\gamma$, $\gamma < \alpha$. Define T' in terms of the c_γ as T was defined in terms of a_γ, and let Φ be the corresponding congruence relation of $\mathfrak{F}_K(\alpha)$. The assumption that φ is a homomorphism implies that $\Theta \leqq \Phi$. Thus by the second isomorphism theorem there exists a homomorphism ψ of $\mathfrak{F}_K(\alpha)/\Theta$ into $\mathfrak{C}$ with $a_\gamma' \psi = c_\gamma$. Applying this observation to $\mathfrak{B}$ and $c_\gamma = a_\gamma \varphi$, we get a homomorphism χ of $\mathfrak{A}'$ onto $\mathfrak{A}$, while it is trivial by the construction of Θ that χ^{-1} is a homomorphism of $\mathfrak{A}$ onto $\mathfrak{A}'$. Thus, χ is an isomorphism and then $\chi\psi = \varphi$ is trivial, completing the proof of Theorem 1.

Corollary 1. *Let K be an equational class. Then* (i) *is necessary and sufficient for the existence of* $\mathfrak{F}_K(\mathfrak{A})$.

Corollary 2. *Let* $K = K(\tau)$, $A = \{a_\gamma \mid \gamma < \alpha\}$ *and* $\bar{a} = \langle a_0, \cdots, a_\gamma, \cdots \rangle_{\gamma < \alpha}$; *then* $\mathfrak{F}_K(\mathfrak{A})$ *always exists and is isomorphic to* $\mathfrak{P}^{(\alpha)}(\tau)/\Theta_{\bar{a}}$.

Proof. Over $K(\tau)$ the free algebra is $\mathfrak{P}^{(\alpha)}(\tau)$ and the congruence relation Θ constructed in the proof of Theorem 1 is the same as $\Theta_{\bar{a}}$ of Theorem 14.1. For $\mathfrak{P}^{(\alpha)}(\tau)/\Theta_{\bar{a}}$ (i) of Definition 1 is obvious and (ii) follows from Theorem 14.2, with $A' = A^*$, $\chi: [\mathbf{x}_\gamma]\Theta_{\bar{a}} \to a_\gamma$. To verify (iii) set $c_\gamma = a_\gamma\varphi$, $\bar{c} = \langle c_0, \cdots, c_\gamma, \cdots \rangle$. Then $\Theta_{\bar{a}} \leqq \Theta_{\bar{c}}$ by the definition of homomorphisms of partial algebras and so by the Second Isomorphism Theorem (Theorem 11.4), $\psi': [x]\Theta_{\bar{a}} \to [x]\Theta_{\bar{c}}$ is a homomorphism of $\mathfrak{P}^{(\alpha)}(\tau)/\Theta_{\bar{a}}$ onto $\mathfrak{P}^{(\alpha)}(\tau)/\Theta_{\bar{c}}$, which in turn is isomorphic to a subalgebra of $\mathfrak{C}$; such an isomorphism ε can be given so as to satisfy $\varepsilon: [\mathbf{x}_\gamma]\Theta_{\bar{c}} \to c_\gamma$. Thus $\psi'\varepsilon = \psi$ satisfies (iii).

Thus $\mathfrak{P}^{(\alpha)}(\tau)/\Theta_{\bar{a}}$ can be called the algebra *absolutely freely* generated by $\mathfrak{A}$.

Theorem 2. *Let* K *be a class of algebras and let* $\mathfrak{A}$ *be a partial algebra. Assume that the following conditions hold:*

(i) $\mathfrak{A}$ *is isomorphic to a relative subalgebra of an algebra in* K;
(ii) $\mathbf{S}(K) \subseteq K$ *and* $\mathbf{P}(K) \subseteq K$.

Then $\mathfrak{F}_K(\mathfrak{A})$ *exists.*

Proof. We proceed the same way as in the construction of free algebras in §25—namely, let $\bar{\alpha} = |A|$, $A = \{a_\gamma \mid \gamma < \alpha\}$, and take all congruence relations Θ_i, $i \in I$, of $\mathfrak{P}^{(\alpha)}(\tau)$ for which $a_\gamma \to [\mathbf{x}_\gamma]\Theta_i$ is a homomorphism of $\mathfrak{A}$ into $\mathfrak{P}^{(\alpha)}(\tau)/\Theta_i$. Again, we take the direct product of all these algebras and then we take the subalgebra generated by the "diagonal" elements g_γ defined by $g_\gamma(i) = [\mathbf{x}_\gamma]\Theta_i$. The details of the proof are similar to the proof of Theorem 25.1 and are therefore omitted.

We conclude this section with the following definition and theorem.

Definition 2 (*A. I. Mal'cev* [5]). *Let* K *be a class of algebras and let* $\mathbf{p}_i$, $\mathbf{q}_i$, $i \in I$, *be* α*-ary polynomial symbols. The algebra* $\mathfrak{A}$ *is said to be* freely generated in the class K by the elements a_γ, $\gamma < \alpha$, with respect to the set Ω of equations

$$\mathbf{p}_i = \mathbf{q}_i, \qquad i \in I,$$

if the following conditions hold:

(i) $\mathfrak{A} \in K$;
(ii) $\mathfrak{A}$ *is generated by the elements* a_γ, $\gamma < \alpha$, *and* $p_i(\bar{a}) = q_i(\bar{a})$, $i \in I$, *where*

$$\bar{a} = \langle a_0, \cdots, a_\gamma, \cdots \rangle_{\gamma < \alpha};$$

(iii) *If* $\mathfrak{B} \in K$, $b_\gamma \in B$, $\gamma < \alpha$, $\bar{b} = \langle b_0, \cdots, b_\gamma, \cdots \rangle_{\gamma < \alpha}$, *and* $p_i(\bar{b}) = q_i(\bar{b})$,

$i \in I$, then the mapping p: $a_\gamma \to b_\gamma$, $\gamma < \alpha$, can be extended to a homomorphism φ of $\mathfrak{A}$ into $\mathfrak{B}$.

The algebra $\mathfrak{A}$ will be denoted by $\mathfrak{F}_K(\alpha, \Omega)$.

Corollary 1. *$\mathfrak{F}_K(\alpha, \Omega)$ is unique up to isomorphism.*

Corollary 2. *$\mathfrak{F}_K(\alpha, \varnothing)$ is isomorphic to $\mathfrak{F}_K(\alpha)$.*

Note that we did not require in Definition 2 that the elements a_γ be distinct. This implies that all the results of this section carry over without such assumptions as (i) in Theorem 1 or Theorem 2. We will rephrase only one theorem; the others will be treated as exercises.

Theorem 3. *Let K be an equational class. Then $\mathfrak{F}_K(\alpha, \Omega)$ always exists. The elements a_μ, a_δ (μ, $\delta < \alpha$) are distinct if and only if there exists a $\mathfrak{B} \in K$, $b_\gamma \in B$, $\gamma < \alpha$, such that $b_\mu \neq b_\delta$ and $p_i(\bar{b}) = q_i(\bar{b})$ for all $i \in I$, where $\bar{b} = \langle b_0, \cdots, b_\gamma, \cdots \rangle_{\gamma < \alpha}$.*

§29. FREE PRODUCTS OF ALGEBRAS

Let $\mathfrak{Q}$ be a partial algebra, let K be a class of algebras satisfying conditions (ii) and (iii) of Theorem 28.1 or condition (ii) of Theorem 28.2, and let us apply the constructions given in these theorems to $\mathfrak{Q}$. It is easy to see that we will get an algebra $\mathfrak{Q}'$ which has properties (i) and (iii) of Definition 28.1, with the exception that $\mathfrak{Q}'$ is only a homomorphic image of $\mathfrak{Q}$. This $\mathfrak{Q}'$ is what we can call a maximal homomorphic image of $\mathfrak{Q}$ in K.

Definition 1. *Let K be a class of algebras and let $\mathfrak{Q}$ be a partial algebra. The partial algebra $\mathfrak{Q}'$ is called the* maximal homomorphic image of $\mathfrak{Q}$ *in K if $\mathfrak{Q}'$ is a relative subalgebra of an algebra in K and there exists a homomorphism χ of $\mathfrak{Q}$ onto $\mathfrak{Q}'$ such that if φ is any homomorphism of $\mathfrak{Q}$ into an algebra $\mathfrak{B} \in K$, then $\varphi = \chi\varphi'$ for some homomorphism φ' of $\mathfrak{Q}'$ into $\mathfrak{B}$.*

It is easily proved that $\mathfrak{Q}'$ is unique up to isomorphism.

Thus, rephrasing the results of §28, we have the following statement:

Theorem 1. *Let K be a class of algebras and let $\mathfrak{Q}$ be a partial algebra. Assume that either $\mathfrak{F}_K(\mathfrak{m})$ exists for some $\mathfrak{m} \geqq |Q|$ and $\mathbf{H}(K) \subseteq K$ or that $\mathbf{S}(K) \subseteq K$ and $\mathbf{P}(K) \subseteq K$. Then the maximal homomorphic image $\mathfrak{Q}'$ of $\mathfrak{Q}$ in K exists and $\mathfrak{F}_K(\mathfrak{Q}')$ exists.*

We will apply this result to study the problem of the existence of free products.

Definition 2. *Let K be a class of algebras and let $\mathfrak{A}_i$, $i \in I$, be algebras of K. Then $\mathfrak{A}$ is the* free product *in K of the algebras $\mathfrak{A}_i$ if:*

(i) $\mathfrak{A} \in K$;

(ii) *there exist* 1–1 *homomorphisms ψ_i of $\mathfrak{A}_i$ into $\mathfrak{A}$, for $i \in I$;*

(iii) $A = [\bigcup (A_i\psi_i \mid i \in I)]$;

(iv) *if $\mathfrak{B}$ is an algebra in K and φ_i is a homomorphism of $\mathfrak{A}_i$ into $\mathfrak{B}$ for $i \in I$, then there exists a homomorphism $\varphi: A \to B$ such that $\varphi_i = \psi_i\varphi$.*

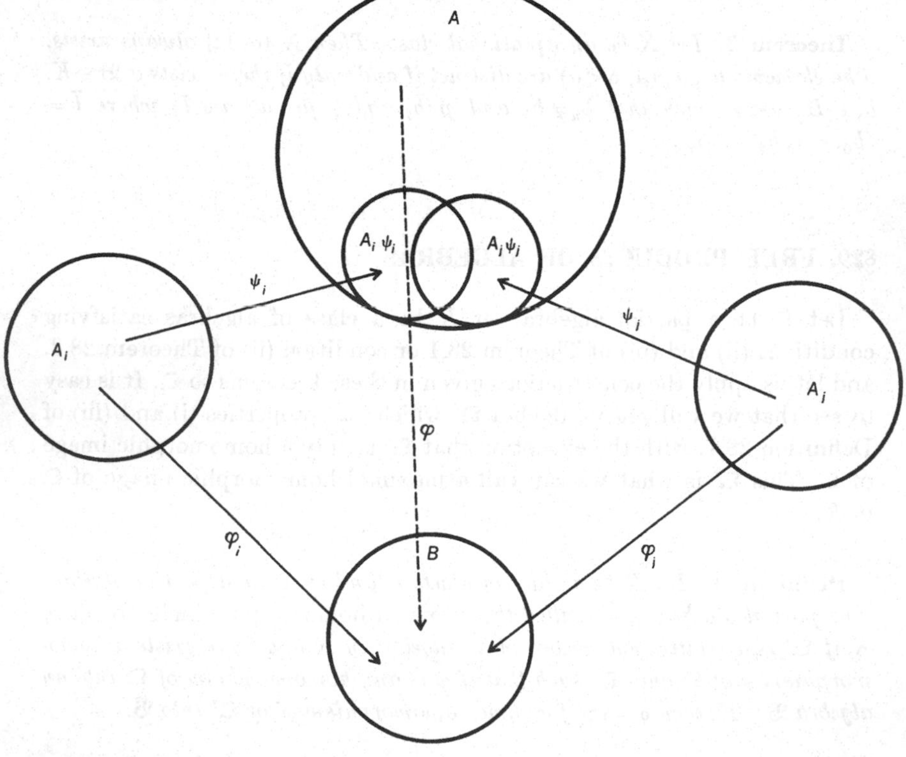

Note that φ is necessarily unique.

A free product consists of $\mathfrak{A}$ together with the homomorphisms ψ_i, $i \in I$. If it exists, then it is unique up to isomorphism.

Suppose $\mathfrak{A}$, ψ_i, $i \in I$, and $\mathfrak{A}'$, ψ_i', $i \in I$, are both free products in K of the algebras $\mathfrak{A}_i$. Then there exists an isomorphism χ of $\mathfrak{A}$ with $\mathfrak{A}'$ such that $\psi_i\chi = \psi_i'$ for $i \in I$.

The concept of free product of algebras is due to R. Sikorski [1].

It is advantageous to define the free product of partial algebras as well, which is done in the same way as in Definition 2 with the only exception that the $\mathfrak{A}_i$ are partial algebras and therefore we drop the condition that they are in K.

Let $\mathfrak{A}_i$, $i \in I$, be partial algebras and choose isomorphic copies $\mathfrak{A}_i'$ which are pairwise disjoint in case no operation is nullary; if there are nullary operations, let the $\mathfrak{A}_i'$ be chosen in such a way that if $a \in A_i' \cap A_j'$ $(i \neq j)$ then there exists a nullary polynomial symbol $\mathbf{p}$ for which $(\mathbf{p})_{\mathfrak{A}_i}$ and $(\mathbf{p})_{\mathfrak{A}_j}$ exist and $a = (\mathbf{p})_{\mathfrak{A}_i'} = (\mathbf{p})_{\mathfrak{A}_j'}$; and conversely, if for a nullary polynomial symbol $\mathbf{p}$ $(\mathbf{p})_{\mathfrak{A}_i'}$ and $(\mathbf{p})_{\mathfrak{A}_j'}$ exist, then $(\mathbf{p})_{\mathfrak{A}_i'} = (\mathbf{p})_{\mathfrak{A}_j'} \in A_i' \cap A_j'$; this can be done if and only if the following condition holds:

(*) If $\mathbf{p}$ and $\mathbf{q}$ are nullary polynomials, $i \in I$, and $(\mathbf{p})_{\mathfrak{A}_i}$, $(\mathbf{q})_{\mathfrak{A}_i}$ exist and $(\mathbf{p})_{\mathfrak{A}_i} = (\mathbf{q})_{\mathfrak{A}_i}$, then for every $j \in I$, $(\mathbf{p})_{\mathfrak{A}_j}$ and $(\mathbf{q})_{\mathfrak{A}_j}$ exist, and $(\mathbf{p})_{\mathfrak{A}_j} = (\mathbf{q})_{\mathfrak{A}_j}$.

Let χ_i be the isomorphisms between the $\mathfrak{A}_i$ and the $\mathfrak{A}_i'$. Form the partial algebra $\mathfrak{Q}$, where $Q = \bigcup (A_i' \mid i \in I)$ and the operations are defined only in A_i'.

Theorem 2. *The free product of the $\mathfrak{A}_i$, $i \in I$, in the class K exists if and only if (*) and the following conditions are satisfied:*

(i) *$\mathfrak{Q}$ has a maximal homomorphic image $\bar{\mathfrak{Q}}$ in K; let χ denote the homomorphism of $\mathfrak{Q}$ onto $\bar{\mathfrak{Q}}$;*

(ii) *if $a, b \in A_i$, $a \neq b$, then $a\chi_i\chi \neq b\chi_i\chi$;*

(iii) *$\mathfrak{F}_K(\bar{\mathfrak{Q}})$ exists.*

Proof. Let us assume that the conditions are satisfied and let $\mathfrak{A}$ denote the algebra $\mathfrak{F}_K(\bar{\mathfrak{Q}})$. Then $\psi_i = \chi_i\chi$ is a 1–1 homomorphism of $\mathfrak{A}_i$ into $\mathfrak{A}$. Since $\bar{Q} = Q\chi = (\bigcup (A_i\chi_i \mid i \in I))\chi = \bigcup (A_i\psi_i \mid i \in I)$, it follows that $A = [\bigcup (A_i\psi_i \mid i \in I)]$. Hence, to prove that $\mathfrak{A}$ is the free product, it remains only to establish (iv) of Definition 2.

Assume that the hypothesis of (iv) of Definition 2 is satisfied.

Define a mapping $\psi: Q \to B$ by $a\psi = a\chi_i^{-1}\varphi_i$ if $a \in A_i\chi_i$. Then ψ is a homomorphism of $\mathfrak{Q}$ into $\mathfrak{B}$. Thus it factors through χ, that is, there exists a homomorphism ψ' of $\bar{Q}$ into B such that $\psi = \chi\psi'$. Since $\mathfrak{A}$ is freely generated by the partial algebra $\bar{\mathfrak{Q}}$, ψ' can be extended to a homomorphism φ of $\mathfrak{A}$ into $\mathfrak{B}$. We only have to check that $\varphi_i = \psi_i\varphi$. Indeed, if $a \in A_i$, then $a\psi_i\varphi = a\chi_i\chi\varphi = a\chi_i\chi\psi' = a\chi_i\psi = a\varphi_i$. This proves the "if" part of the theorem.

To prove the "only if" part, assume that the free product $\mathfrak{A}$ exists. Then (*) is trivial. Define

$$\bar{Q} = \bigcup (A_i\psi_i \mid i \in I),$$

let $\bar{\mathfrak{Q}}$ be the relative subalgebra of $\mathfrak{A}$ defined by $\bar{Q}$ and define the mapping $\chi: Q \to \bar{Q}$ by $(a\chi_i)\chi = a\psi_i$. Then $\bar{\mathfrak{Q}}$ is a homomorphic image of $\mathfrak{Q}$ and $\mathfrak{A}$ is generated by $\bar{Q}$. It remains to prove that $\mathfrak{A}$ is freely generated by $\bar{\mathfrak{Q}}$ and that $\bar{\mathfrak{Q}}$ is a maximal homomorphic image of $\mathfrak{Q}$ in K. A homomorphism ψ of $\mathfrak{Q}$ into an algebra $\mathfrak{B}$ in K is equivalent to a family of homomorphisms φ_i of the partial algebras $\mathfrak{A}_i$ into $\mathfrak{B}$, the two being connected by the relation $\varphi_i = \chi_i\psi$. Since $\mathfrak{A}$ is a free product of the $\mathfrak{A}_i$, there exists a homomorphism φ of $\mathfrak{A}$ into $\mathfrak{B}$ such that $\varphi_i = \psi_i\varphi$. To prove that $\bar{\mathfrak{Q}}$ is the maximal homomorphic image, we must verify that $\psi = \chi\varphi$. Indeed, if $a \in A_i$, then $a\chi_i\psi = a\varphi_i = a\psi_i\varphi = a\chi_i\chi\varphi$, from which we infer that $\psi = \chi\varphi$.

To prove that $\mathfrak{A}$ is freely generated by $\bar{\mathfrak{Q}}$, take an arbitrary homomorphism λ of $\bar{\mathfrak{Q}}$ into $\mathfrak{B}$ and set $\varphi_i = \psi_i\lambda$. Since $\mathfrak{A}$ is the free product of the $\mathfrak{A}_i$, there is a homomorphism φ of $\mathfrak{A}$ into $\mathfrak{B}$ such that $\varphi_i = \psi_i\varphi$. The above relations imply that φ and λ coincide on $\bar{Q}$; hence φ is the required extension of λ. This completes the proof of the theorem.

Corollary 1 (*R. Sikorski* [1]). *Let K be a class of algebras and let $\mathfrak{A}_i \in K$, $i \in I$. The free product of the $\mathfrak{A}_i$ exists if the following conditions are satisfied:*

(i) *$\mathfrak{F}_K(\mathfrak{m})$ exists for some $\mathfrak{m} \geqq \Sigma\ (|A_i| \mid i \in I)$;*

(ii) $\mathbf{H}(K) \subseteq K$;

(iii) *there exists an algebra $\mathfrak{B} \in K$ and a family of 1–1 homomorphisms ψ_i of $\mathfrak{A}_i$ into $\mathfrak{B}$ for all $i \in I$.*

Corollary 2 (*D. J. Christensen and R. S. Pierce* [1]). *In Corollary* 1, *we can replace* (i) *and* (ii) *by* $\mathbf{P}(K) \subseteq K$ *and* $\mathbf{S}(K) \subseteq K$.

Both corollaries are evident from Theorems 1 and 2.

It should be noted that the proofs of Corollaries 1 and 2 are unnecessarily difficult within this framework. Obviously, to find convenient sufficient conditions for the existence of free products we do not have to go through the painful process of finding necessary and sufficient conditions.

Examples and further results concerning free products will be given in the exercises.

§30. WORD PROBLEM

In this section, all algebras considered are of type τ, where τ is a fixed finite type, that is, $o(\tau) < \omega$.

Let Σ be a finite set of identities and let R be another finite set of identities

$$\mathbf{p}_i = \mathbf{q}_i,$$

$i = 0, \cdots, k-1$, where $\mathbf{p}_i$ and $\mathbf{q}_i$ are n-ary polynomial symbols for some n.

If $\mathfrak{A}$ is an algebra, $a_0, \cdots, a_{n-1} \in A$, and

$$p_i(a_0, \cdots, a_{n-1}) = q_i(a_0, \cdots, a_{n-1}),$$

$i=0, \cdots, k-1$, then we write $R(a_0, \cdots, a_{n-1})$.

The *word problem* is the following. Given Σ and R and two n-ary polynomial symbols $\mathbf{p}$ and $\mathbf{q}$, find an *effective process*† to decide whether $p(a_0, \cdots, a_{n-1}) = q(a_0, \cdots, a_{n-1})$ whenever $\mathfrak{A}$ is an algebra satisfying Σ and $a_0, \cdots, a_{n-1} \in A$ satisfies $R(a_0, \cdots, a_{n-1})$. In short, the word problem is solvable if there is an effective process which decides whether Σ and $R(a_0, \cdots, a_{n-1})$ imply $p(a_0, \cdots, a_{n-1}) = q(a_0, \cdots, a_{n-1})$. (In case of semigroups, polynomial symbols are usually called words. This explains why this problem is called "word problem".)

We say that the word problem is *solvable for* Σ if it is solvable for any R, $\mathbf{p}$ and $\mathbf{q}$.

The *embedding problem* for Σ is the following. Given Σ, find an *effective process* to decide whether a finite partial algebra $\mathfrak{A}$ can be weakly embedded in an algebra satisfying Σ, where *weak embedding* means the existence of a 1–1 homomorphism.

Theorem 1 (*T. Evans* [1], [3]). *The word problem for Σ is solvable if and only if the embedding problem for Σ is solvable.*

The proof is based on the following lemma.

Lemma 1. *Let R be a finite set of n-ary identities. Then a finite partial algebra $\mathfrak{A}$ and elements $z_0, \cdots, z_{n-1} \in A$ can be effectively constructed such that $R(z_0, \cdots, z_{n-1})$ and the following condition holds. If $\mathfrak{B}$ is an algebra and $b_0, \cdots, b_{n-1} \in B$, then $R(b_0, \cdots, b_{n-1})$ if and only if the mapping $z_i \to b_i$, $i=0, \cdots, n-1$, can be extended to a homomorphism of $\mathfrak{A}$ into $\mathfrak{B}$.*

Proof. First we define inductively the *components* of a polynomial symbol:

(i) the component of $\mathbf{x}_i$ is $\mathbf{x}_i$;

(ii) the components of $\mathbf{p} = \mathbf{f}_\gamma(\mathbf{p}_0, \cdots, \mathbf{p}_{n_\gamma - 1})$ are $\mathbf{p}$ and the components of $\mathbf{p}_0, \cdots, \mathbf{p}_{n_\gamma - 1}$.

† An "effective process" or "algorithm" is a finite set of rules which, if followed, will give us the answer in a finite number of steps. The Euclidean algorithm is, for instance, an effective process to determine the greatest common divisor of two integers. Since everybody knows an effective process when he sees one, a precise definition has to be given only when we want to prove the nonexistence of an effective process. For a discussion of various definitions of an effective process, see E. Mendelson, Introduction to Mathematical Logic, D. Van Nostrand Co., Princeton, N.J., 1964, Chapter 5.

Note that the components of a polynomial symbol are themselves polynomial symbols.

Let A_1 be the set of all components of $\mathbf{p}_i$ and $\mathbf{q}_i$, where $\mathbf{p}_i = \mathbf{q}_i$ is in R, $i = 0, \cdots, k-1$, and all the $\mathbf{x}_i$, $i < n$. For every $\mathbf{q} \in A_1$, we introduce the symbol $z_{\mathbf{q}}$; we will write z_i for $z_{\mathbf{x}_i}$. We define the partial operations f_γ on these symbols as follows: if $\mathbf{r} = \mathbf{f}_\gamma(\mathbf{r}_0, \cdots, \mathbf{r}_{n_\gamma - 1})$, $\mathbf{r}$, $\mathbf{r}_i \in A_1$, then

$$z_{\mathbf{r}} = f_\gamma(z_{\mathbf{r}_0}, \cdots, z_{\mathbf{r}_{n_\gamma - 1}}).$$

For every i, we identify the symbols which correspond to $p_i(z_0, \cdots, z_{n-1})$ and $q_i(z_0, \cdots, z_{n-1})$. A is the set we get from A_1 by performing these identifications, with the induced partial operations.

$\mathfrak{A}$ is finite, the construction is obviously effective, and, by construction, $R(z_0, \cdots, z_{n-1})$. The condition of Lemma 1 is obviously satisfied.

Corollary. *Under the conditions of Lemma* 1, $R(b_0, \cdots, b_{n-1})$ *if and only if a homomorphic image of* $\mathfrak{A}$ *can be weakly embedded in* $\mathfrak{B}$ *such that* $\bar{z}_i \to b_i$ *for* $0 \leq i < n$ *under this embedding*, where $\bar{z}_i$ is the image of z_i.

Proof of Theorem 1. Let us assume that the embedding problem for Σ is solved and let R, $\mathbf{p}$, $\mathbf{q}$ be given. Augment R by $\mathbf{p} = \mathbf{p}$ and $\mathbf{q} = \mathbf{q}$ and construct (effectively) the partial algebra $\mathfrak{A}$ of Lemma 1 for this augmented set. If $p(z_0, \cdots, z_{n-1}) = q(z_0, \cdots, z_{n-1})$ in $\mathfrak{A}$, then Σ and $R(z_0, \cdots, z_{n-1})$ imply $p(z_0, \cdots, z_{n-1}) = q(z_0, \cdots, z_{n-1})$. Otherwise let $\mathfrak{A}_0, \cdots, \mathfrak{A}_{k-1}$ be all homomorphic images of $\mathfrak{A}$ such that

$$p(\bar{z}_0, \cdots, \bar{z}_{n-1}) \neq q(\bar{z}_0, \cdots, \bar{z}_{n-1}),$$

where $\bar{z}_i$ denotes the homomorphic image of z_i. Decide for each $\mathfrak{A}_i$ whether or not it is weakly embeddable in an algebra satisfying Σ. If for some i the answer is yes, then Σ and R do not imply $\mathbf{p} = \mathbf{q}$; if for all i the answer is no, then Σ and R do imply $\mathbf{p} = \mathbf{q}$.

To prove the converse, assume the word problem for Σ is solved. Let $\mathfrak{A}$ be a finite partial algebra; let $a_0, \cdots, a_{n-1}$ be the elements of A. Let R be the finite set of n-ary identities consisting of all identities of the form

$$\mathbf{x}_i = \mathbf{f}_\gamma(\mathbf{x}_{i_0}, \cdots, \mathbf{x}_{i_{n_\gamma - 1}}),$$

where

$$a_i = f_\gamma(a_{i_0}, \cdots, a_{i_{n_\gamma - 1}})$$

holds in $\mathfrak{A}$.

Let Θ be the minimal congruence relation of $\mathfrak{P}^{(n)}(\tau)$ such that

$$\varphi: a_i \to [\mathbf{x}_i]\Theta$$

is a homomorphism of $\mathfrak{A}$ into $\mathfrak{P}^{(n)}(\tau)/\Theta$ and the latter satisfies Σ. (Θ is the join of all $\Theta(p(\mathbf{r}_0, \cdots, \mathbf{r}_{m-1}), q(\mathbf{r}_0, \cdots, \mathbf{r}_{m-1}))$, where $\mathbf{p} = \mathbf{q}$ is in Σ and of all

$\Theta(\mathbf{p}, \mathbf{q})$, where $\mathbf{p}=\mathbf{q}$ is in R.) $\mathfrak{A}$ can be weakly embedded in an algebra satisfying Σ if and only if φ is 1–1. Therefore, we can decide the weak embeddability by checking whether Σ and R imply $\mathbf{x}_i=\mathbf{x}_j$ for $i \neq j$, which completes the proof of Theorem 1.

Let $\mathfrak{B}$ be an algebra satisfying Σ and let $\mathfrak{A}$ be a relative subalgebra of $\mathfrak{B}$. Then $\mathfrak{A}$ satisfies the following two conditions:

(i) if $a_0, \cdots, a_{n-1} \in A$, $p(a_0, \cdots, a_{n-1})$ and $q(a_0, \cdots, a_{n-1})$ are defined in $\mathfrak{A}$, where $\mathbf{p}, \mathbf{q} \in \mathbf{P}^{(n)}(\tau)$ and $\mathbf{p}=\mathbf{q} \in \Sigma$, then

$$p(a_0, \cdots, a_{n-1}) = q(a_0, \cdots, a_{n-1});$$

(ii) if $a_0, \cdots, a_{n-1} \in A$, $\mathbf{p}, \mathbf{q} \in \mathbf{P}^{(n)}(\tau)$, $\mathbf{q}=f_\gamma(\mathbf{q}_0, \cdots, \mathbf{q}_{n_\gamma-1})$, $\mathbf{q}_0, \cdots, \mathbf{q}_{n_\gamma-1} \in \mathbf{P}^{(n)}(\tau)$, $p(a_0, \cdots, a_{n-1})$, $q_i(a_0, \cdots, a_{n-1})$, $i=0, \cdots, n_\gamma-1$, are all defined in $\mathfrak{A}$ and $\mathbf{p}=\mathbf{q} \in \Sigma$, then $q(a_0, \cdots, a_{n-1})$ is also defined.

A partial algebra satisfying (i) and (ii) will be called a *partial Σ-algebra.* Let $\mathfrak{A}$ be a partial algebra. We will say that $\mathfrak{A}$ *can be strengthened to a partial Σ-algebra* if $\mathfrak{A}$ is a weak subalgebra of a partial Σ-algebra $\langle A; F' \rangle$, that is, if we can make $\mathfrak{A}$ a partial Σ-algebra by extending the domain of the partial operations.

Lemma 2. *There exists an effective process which decides whether or not a finite partial algebra $\mathfrak{A}$ can be strengthened to a partial Σ-algebra.*

Proof. If $\mathfrak{A}$ violates (ii), then we define $q(a_0, \cdots, a_{n-1})=p(a_0, \cdots, a_{n-1})$. Since we have a finite number of elements and a finite number of operations, in a finite number of steps we get a partial algebra $\langle A; F' \rangle$ which does not violate (ii). Thus, $\mathfrak{A}$ can be strengthened into a partial Σ-algebra if and only if $\langle A; F' \rangle$ does not violate (i).

Theorem 2 (*T. Evans* [1]). *Let us assume that every finite partial Σ-algebra can be weakly embedded in an algebra satisfying Σ. Then the word problem for Σ is solvable.*

Indeed, by the assumption and Lemma 2, the embedding problem for Σ is solvable and thus by Theorem 1 the word problem is also solvable.

EXERCISES

1. Give an upper bound for $|F_K(\alpha)|$ in terms of τ and α.
2. Determine the free lattices with 1 and 2 generators and the free Boolean algebras with 0 and n (>0) generators.

3. Determine the free semigroup and the free group on n generators.
4. Let K be the equational class generated by the cyclic group of order p (prime). Determine $\mathfrak{F}_K(\alpha)$.
5. Let K be the class of Abelian groups. Prove that $\mathfrak{F}_K(\alpha)$ is isomorphic to the weak direct product of α copies of the group of integers.
6. Let K be an algebraic class. Set $E(K) = \{\alpha \mid \mathfrak{F}_K(\alpha) \text{ exists}\}$. Let us assume that K is closed under the formation of direct limits and subalgebras. Prove that $E(K)$ is either the class of all ordinals or $E(K) = \{k \mid k \leqq n\}$, where n is a finite ordinal.
7. Let K be a class of algebras. Let $\mathfrak{A}, \mathfrak{B} \in K$, φ a homomorphism of $\mathfrak{A}$ onto $\mathfrak{B}$, and ψ a homomorphism of $\mathfrak{F}_K(\mathfrak{m})$ into $\mathfrak{B}$. Then† there exists a homomorphism χ of $\mathfrak{F}_K(\mathfrak{m})$ into $\mathfrak{A}$ such that $\chi\varphi = \psi$.
8. Prove that (i)–(v) of Definition 26.2 are independent, that is, if $\text{i} \leqq a \leqq \text{v}$, then (i)–(v) without ($a$) do not imply ($a$).
9. Prove that the dual of $\mathfrak{L}(\tau)$ (the lattice of equational classes) is an algebraic lattice.
10. Prove that the class of Boolean algebras is equationally complete.
11. (J. Kalicki and D. Scott [1]) Prove that the following algebras $\mathfrak{A}_i$, $i = 1, \cdots, 5$ of type $\langle 2 \rangle$ are equationally complete and determine the class K_i generated by $\mathfrak{A}_i$. $A_i = \{0, 1\}$ for $1 \leqq i \leqq 4$ and $+_i$ is defined as follows:

$+_1$	0	1
0	0	0
1	0	1

$+_2$	0	1
0	0	0
1	1	1

$+_3$	0	1
0	0	1
1	0	1

$+_4$	0	1
0	0	0
1	0	0

$\mathfrak{A}_5$ has p elements for some prime p, $A_5 = \{0, \cdots, p-1\}$ and $+_5$ is addition modulo p.

12. (J. Kalicki and D. Scott [1]) Let K be an equationally complete class of algebras of type $\langle 2 \rangle$ in which the associative law holds. Prove that $K = K_i$ for some K_i in Ex. 11.
13. (D. Scott [1]) Prove that the algebra $\mathfrak{A}$ is equationally complete if and only if the following conditions are satisfied for some $n > 1$:

(i) $\mathfrak{A}$ has more than one element;

(ii) if $\mathfrak{B}$ is a homomorphic image, with more than one element, of $\mathfrak{P}^{(n)}(\mathfrak{A})$, then $\mathfrak{P}^{(n)}(\mathfrak{A})$ is isomorphic to $\mathfrak{P}^{(n)}(\mathfrak{B})$;

(iii) $\mathfrak{A}$ is in the class generated by $\mathfrak{P}^{(n)}(\mathfrak{A})$.

† In contrast with the categorical characterizations of algebraic constructions, this exercise is *not* a categorical characterization of free algebras. To give an example, let $\mathfrak{R}$ be a ring with identity, $\mathfrak{R}_n$ the ring of $n \times n$ matrices over $\mathfrak{R}$, and K the class of unitary left $\mathfrak{R}_n$ modules $(n > 1)$. Then $\mathfrak{R}^n$ as an $\mathfrak{R}_n$ module is in K and satisfies the conditions of Ex. 7 but it is not isomorphic to $\mathfrak{F}_K(\mathfrak{m})$ for any $\mathfrak{m}$, though all $\mathfrak{F}_K(\mathfrak{m})$ exist. Algebras satisfying the conditions of Ex. 7 are called *projective*.

14. (D. Scott [1]) Let $\mathfrak{A}$ have k elements and $m=k^2$. Prove that $Id(\mathfrak{A})$ has at most 2^{m^m} equationally complete extensions.
15. For a class K of algebras, let $Id_n(K)$ denote all identities $\mathbf{p}=\mathbf{q}$, with $\mathbf{p}, \mathbf{q} \in \mathbf{P}^{(n)}(\tau)$, satisfied by the algebras in K. Prove that K is an equational class if and only if $K=\bigcap (Id_n(K)^* \mid n<\omega)$.
16. For an equational class K, set $v(K)=\min\{n \mid (Id_n(K))^*=K\}$. Prove that $Id(K)$ has a finite basis if and only if $v(K)<\omega$.
17. Prove that an algebra $\mathfrak{A}$ belongs to $(Id_n(K))^*$ if and only if every subalgebra of $\mathfrak{A}$ generated by n elements belongs to $(Id_n(K))^*$.
18. For an equational class K, $\mathbf{C}(K)=K$ if and only if $v(K)\leqq 1$. (For the definition of $\mathbf{C}$, see the text preceding Ex. 3.70. The result is from G. Grätzer [9].)
19. Find an equational class K for which $v(K)=\omega$.
20. Let K be the class of lattices. Prove that $v(K)=3$.
21. (T. Evans [2]) Let K be an equational class of algebras of type τ. Let $\overline{o(\tau)}\leqq \aleph_0$ and let us assume that every countable algebra in K can be embedded in an algebra of K generated by m $(<\omega)$ elements. Prove that for some $n<\omega$, $\mathfrak{F}_K(\omega)$ is isomorphic to a subalgebra of $\mathfrak{F}_K(n)$.
22. Let K be an equational class. Prove that $K=\mathbf{HSP}(\{\mathfrak{F}_K(n) \mid n<\omega\})$.
23. Let K be equationally complete. Prove that $K=\mathbf{HSP}(\{\mathfrak{F}_K(2)\})$.
24. For an equational class K, set $f(K)=\min\{n \mid K=\mathbf{HSP}(\{\mathfrak{F}_K(n)\})\}$. Let K_1 be the class of distributive lattices and K_2 the class of Boolean algebras. Prove that $f(K_1)=2$ and $f(K_2)=0$.
25. Let K be the class of lattices. Prove that $f(K)=3$.
26. Find an equational class K for which $v(K)\neq f(K)$.
27. Prove that Ex. 3.92 does not hold for equational classes defined by a finite set of identities. (See the reference given in Ex. 3.92.)
28. Let K be an equational class defined by a finite set of identities. Then there exists an equational class K_1 defined by four identities such that $\mathrm{Sp}(K)=\mathrm{Sp}(K_1)$. (Hint: if K is defined by the identities $\mathbf{p}_i=\mathbf{q}_i$, $0\leqq i<n$, then take two new binary operation symbols $\vee$ and $-$; define K_1 by the identities

$$\mathbf{x}_0-(\mathbf{x}_1-(\mathbf{x}_2-(\mathbf{x}_0-\mathbf{x}_1)))=\mathbf{x}_2,$$
$$\mathbf{x}_0\vee(\mathbf{x}_1\vee\mathbf{x}_2)=(\mathbf{x}_1\vee\mathbf{x}_0)\vee(\mathbf{x}_2\vee\mathbf{x}_2),$$
$$\mathbf{x}_0\vee(\mathbf{x}_0-\mathbf{x}_0)=\mathbf{x}_0,$$
$$(\cdots((\mathbf{p}_0-\mathbf{q}_0)\vee(\mathbf{p}_1-\mathbf{q}_1))\vee\cdots)\vee(\mathbf{p}_{n-1}-\mathbf{q}_{n-1})=\mathbf{x}_0-\mathbf{x}_0.$$

It is proved in G. Grätzer [12], that these four identities can be replaced by two, which was improved to "one", by B. H. Neumann [3].)

* * *

In Ex. 29–33 $\mathfrak{A}$ denotes an algebra of type $\langle 0, 2\rangle$, f_0 will be denoted by $\mathbf{0}$, f_1 by $\cdot$. $A=\{0, e, b_1, b_2, c, d_1, d_2\}$; $\cdot$ is defined by $ce=c$, $cb_j=d_j$, $d_je=d_j$, $d_jb_k=d_j$ $(j, k=1, 2)$, $xy=0$ in all other cases and $(\mathbf{0})_{\mathfrak{A}}=0$. See R. C. Lyndon [2].

29. Prove that the following identities Σ hold in $\mathfrak{A}$:

(A_1) $\mathbf{0}\mathbf{x}_0=\mathbf{0}$
(A_2) $\mathbf{x}_0\mathbf{0}=\mathbf{0}$;

(A_3) $\mathbf{x}_0(\mathbf{x}_1\mathbf{x}_2)=\mathbf{0}$;

(B_n) $((\cdots((\mathbf{x}_0\mathbf{x}_1)\mathbf{x}_2)\cdots)\mathbf{x}_{n-1})\mathbf{x}_0=\mathbf{0}$ $\quad(n=1,2,\cdots)$;

(C_n) $((\cdots((\mathbf{x}_0\mathbf{x}_1)\mathbf{x}_2\cdots)\mathbf{x}_{n-1})\mathbf{x}_1=(\cdots((\mathbf{x}_0\mathbf{x}_1)\mathbf{x}_2)\cdots)\mathbf{x}_{n-1}$ $\quad(n=2,3,\cdots)$.

30. If $\mathbf{p}=\mathbf{q}$ holds in $\mathfrak{A}$ and $\mathbf{p},\mathbf{q}\in\mathbf{P}^{(n)}(\tau)$, then $\mathbf{p}=\mathbf{q}$ follows from (A_1), (A_2), (A_3), (B_m), (C_m) with $m\leqq n$.
31. Every polynomial symbol $\mathbf{p}$ equals $\mathbf{0}$ or a left-associated product, $(\cdots(\mathbf{x}_{i_0}\mathbf{x}_{i_1})\mathbf{x}_{i_2}\cdots)\mathbf{x}_{i_{m-1}}$ as a consequence of Σ.
32. A left-associated product $\mathbf{p}=(\cdots(\mathbf{x}_{i_0}\cdot\mathbf{x}_{i_1})\mathbf{x}_{i_2}\cdots)\mathbf{x}_{i_{m-1}}$ has property P_n if $\mathbf{p}\in\mathbf{P}^{(n)}$ and each $\mathbf{x}_j$ $(0<j<n)$ which occurs in $\mathbf{p}$, occurs at least once to the left of the second occurrence of $\mathbf{x}_0$. Prove that if $\mathbf{p}=\mathbf{q}$ follows from Σ_n and $\mathbf{p}$ has property P_n, then so has $\mathbf{q}$, where Σ_n is (A_i), (B_m), (C_m), $i=1,2,3$, $m<n$.
33. Prove Theorem 27.4. (Hint: the left-hand side of B_n has property P_n and $\mathbf{0}$ does not have P_n.)

* * *

34. (W. Sierpinski [1]) Prove that every n-ary function is a composition of binary functions.
35. Let $\mathfrak{A}$ be a finite algebra. $\mathfrak{A}$ is primal if and only if $\mathfrak{P}^{(2)}(\mathfrak{A})$ is isomorphic to $\mathfrak{A}^{|A|^2}$. (Sioson [3].)
36. (Foster [2]) Let $\langle G;\cdot\rangle$ be a group, $G=\{g_0,\cdots,g_{n-1}\}$ and $A=G\cup\{0\}$. Then $\langle A;\cdot,'\rangle$ is a primal algebra if the operations are defined as follows: $0\cdot g_i=g_i\cdot 0=0$, $0'=g_0$, $g_k{}'=g_{k+1}$ if $k<n-1$ and $g_{n-1}'=0$.
37. Every finite field of prime order $\langle F;+,\cdot,1\rangle$ is primal.
38. (E. L. Post [1]) Let $\mathfrak{B}=\langle B;\vee,\wedge,',0,1\rangle$ be a two element Boolean algebra. Let f be an n-ary function on B. f preserves the 0 if $f(0,\cdots,0)=0$; f preserves 1 if $f(1,\cdots,1)=1$; f is linear if $f(ax_0,\cdots,ax_{n-1})=af(x_0,\cdots,x_{n-1})$ and $f(x_0,\cdots,x_i+y,\cdots,x_n)=f(x_0,\cdots,x_i,\cdots,x_n)+f(x_0,\cdots,y,\cdots,x_n)$ (where $a+b=(a'\wedge b)\vee(a\wedge b')$); f is self-dual if $f(x_0',\cdots,x_{n-1}')'=f(x_0,\cdots,x_{n-1})$; f is monotone if $x_i\leqq y_i$ $0\leqq i<n$ imply $f(x_0,\cdots,x_{n-1})\leqq f(y_0,\cdots,y_{n-1})$. Prove that $\langle B;f_0,\cdots,f_k\rangle$ is primal if and only if some f_i does not preserve 0, some f_j does not preserve 1, some f_k is not linear, some f_l is not self-dual, and some f_m is not monotone $(0\leqq i,j,k,l,m<n)$.
39. Find an equationally complete algebra which is not primal.
40. Define a *weakly primal algebra* $\mathfrak{A}$ as a finite algebra with more than one element with the property that every function on A is an algebraic function. Prove that every weakly primal algebra is simple. (See A. L. Foster [2], where the term "functionally complete" algebra is used.)
41. Prove that every finite Boolean algebra with more than one element is weakly primal.
42. If p and q are algebraic functions, $p=q$ is called an *algebraic identity*. Prove the following variant of Theorem 27.5: Let $\mathfrak{A}$ be a weakly primal algebra and $\mathfrak{B}$ an extension of $\mathfrak{A}$. If every algebraic identity which holds in $\mathfrak{A}$ holds in $\mathfrak{B}$ as well, then $\mathfrak{B}$ is a normal subdirect power of $\mathfrak{A}$.
43. Let $\mathfrak{A}$ be an algebra. The representations of $\mathfrak{A}$ as a normal subdirect power

of some algebra are in 1–1 correspondence with sets $\{\varepsilon_i \mid i \in I\}$ of endomorphisms of $\mathfrak{A}$ satisfying the following conditions:

(i) $\varepsilon_i\varepsilon_j = \varepsilon_i$ for all $i, j \in I$;

(ii) if $x, y \in A$, $x \neq y$, then $x\varepsilon_i \neq y\varepsilon_i$ for some $i \in I$;

(iii) there exists a 4-ary function T on A such that if $T(a, b, c, d) = e$, then $e\varepsilon_i = d\varepsilon_i$ for $a\varepsilon_i \neq b\varepsilon_i$ and $e\varepsilon_i = c\varepsilon_i$ for $a\varepsilon_i = b\varepsilon_i$. (M. I. Gould and G. Grätzer [1].)

44. (Mal'cev [5]) Let K be an algebraic class. Prove that all $\mathfrak{F}_K(\alpha, \Omega)$ exist if and only if $\mathbf{S}(K) \subseteq K$ and $\mathbf{P}(K) \subseteq K$.
45. Prove that the maximal homomorphic image $\mathfrak{Q}'$ of the partial algebra $\mathfrak{Q}$ in the class K is unique up to isomorphism.
46. Prove Theorem 29.1.
47. Assume that $\mathfrak{F}_K(1)$ and $\mathfrak{F}_K(\mathfrak{m})$ exist. Prove that $\mathfrak{F}_K(\mathfrak{m})$ is the free product of $\mathfrak{m}$ copies of $\mathfrak{F}_K(1)$ over K. (R. Sikorski [1].)
48. Prove directly Corollaries 1 and 2 of Theorem 29.2.
49. Let K_1 and K be algebraic classes, $K_1 \subseteq K$. Let $\mathfrak{A}_i \in K$ for $i \in I$ and let $\mathfrak{A}$ be the free product of the $\mathfrak{A}_i$ over K. Assume that $\tilde{\mathfrak{A}}$ and $\tilde{\mathfrak{A}}_i$ are maximal homomorphic images of $\mathfrak{A}$ and $\mathfrak{A}_i$ in K_1, respectively. Prove that the free product of the $\tilde{\mathfrak{A}}_i$ in K_1 exists, and it is isomorphic to $\tilde{\mathfrak{A}}$. (A. I. Mal'cev [10]).
50. Show that the embedding problem is trivial for lattices and apply Theorem 30.2 to show that the word problem is solvable.
51. Show that the word problem is solvable for distributive lattices and apply Theorem 30.1 to show that the embedding problem is solvable.
52. Show that all the results of Chapter 4, with the exception of §30, hold for infinitary algebras.
53. Can (i) of Theorem 28.1 be replaced by the following condition: (i′) For $a, b \in A$, $a \neq b$, there exists a homomorphism φ of $\mathfrak{A}$ into some algebra $\mathfrak{B}$ in K, such that $a\varphi \neq b\varphi$?
54. Let K be an equational class and $\mathfrak{A}_i \in K$ for $i \in I$. The free product of the $\mathfrak{A}_i$, $i \in I$, in K exists if and only if all $\mathfrak{A}_i$ can be embedded in some $\mathfrak{B} \in K$.
55. Assume that $\mathbf{P}(K) \subseteq K$ and each $\mathfrak{A}_i \in K$, $i \in I$ has a one element subalgebra. Prove that the condition of Ex. 54 is satisfied.
56. Show that the homomorphism φ of Definition 29.2 is unique.
57. Prove that the free product is "associative". That is, if

$$I = \bigcup (I_j \mid j \in J),\ I_j \neq \varnothing,\ I_i \cap I_j = \varnothing\ \ (i \neq j)$$

and if $\mathfrak{B}_j$ is the free product of the algebras $\mathfrak{A}_i$, $i \in I_j$, then the free product $\mathfrak{B}$ of the algebras $\mathfrak{B}_j$, $j \in J$, is isomorphic to the free product of the algebras $\mathfrak{A}_i$, $i \in I$ (whenever all these products exist).

58. Let $\mathfrak{B}_i$, $i \in I$ be Boolean algebras of more than one element and $\mathfrak{B}$ the free product of the $\mathfrak{B}_i$, $i \in I$. Let $T(\mathfrak{B}_i)$ and $T(\mathfrak{B})$ denote the Boolean space (Stone space) associated with $\mathfrak{B}_i$ and $\mathfrak{B}$, respectively (see, e.g., G. Birkhoff [6]). Show that $T(\mathfrak{B})$ is the topological product of the $T(\mathfrak{B}_i)$, $i \in I$.
59. Show that the homomorphism φ of Definition 29.2 is onto if and only if $B = [\bigcup (A_i\varphi_i \mid i \in I)]$.

60. Let K be a class of algebras. The *K-product* $\mathfrak{A}$ of a family of algebras $(\mathfrak{A}_i \mid i \in I)$ of K is defined as follows:

(i) $\mathfrak{A} \in K$;
(ii) there exist homomorphisms ψ_i of $\mathfrak{A}$ onto $\mathfrak{A}_i$;
(iii) if $\mathfrak{B} \in K$ and φ_i is a homomorphism of $\mathfrak{B}$ into $\mathfrak{A}_i$ for $i \in I$, then there exists a unique homomorphism φ of $\mathfrak{B}$ into $\mathfrak{A}$ such that $\varphi_i = \varphi\psi_i$.

Show that $\mathbf{P}(K) \subseteq K$ implies the existence of the K-product of any family of algebras in K.

61. Give an example of a class in which K-products exist, but direct products do not.

62. Consider the following definition of K-products. The algebra $\mathfrak{A}$ is the K-product of the algebras $(\mathfrak{A}_i \mid i \in I)$ of the class K if:

(i) $\mathfrak{A} \in K$;
(ii) there exists a family $(\Theta_i \mid i \in I)$ of congruences defined on $\mathfrak{A}$ such that $\bigcap (\Theta_i \mid i \in I) = \omega$ and $\mathfrak{A}/\Theta_i \cong \mathfrak{A}_i$ for $i \in I$;
(iii) if $\mathfrak{B} \in K$ and ψ_i is a homomorphism of $\mathfrak{B}$ into $\mathfrak{A}/\Theta_i$ for $i \in I$, then there exists a homomorphism φ of $\mathfrak{B}$ into $\mathfrak{A}$ such that $\psi_i = \varphi\varphi_i$, where φ_i is the natural mapping of A onto A/Θ_i.

We say that the homomorphisms φ_i of $\mathfrak{C}$ into $\mathfrak{A}_i$, $i \in I$ *separate points* if for $x, y \in C$, $x\varphi_i = y\varphi_i$ for all $i \in I$ implies $x = y$. Show that the definition of K-product in Ex. 60, and the fact that the φ_i in that definition separate points, is equivalent to the above definition of the K-product.

63. (G. C. Hewitt [1]) Let K be a class of algebras. Assume that $\mathbf{H}(K) \subseteq K$ and that free products in K exist. Let $\mathfrak{A}_i \in K$ for $i \in I$ and $\mathfrak{C} \in K$ and φ_i be a homomorphism of $\mathfrak{C}$ onto $\mathfrak{A}_i$ for $i \in I$. Prove that the K-product of $(\mathfrak{A}_i \mid i \in I)$ exists.

64. (G. C. Hewitt [1]) Let us assume $\mathbf{H}(K) \subseteq K$, $\mathbf{S}(K) \subseteq K$ and that every algebra in K has a one element subalgebra. Then a given subset of K has a free product in K if and only if it has a K-product.

PROBLEMS

32. Characterize the lattices $\mathfrak{L}(\tau)$, i.e., give necessary and sufficient condition on a lattice $\mathfrak{L}$ to be isomorphic to some $\mathfrak{L}(\tau)$. Characterize the lattice of equational classes of lattices, groups and rings.

33. Determine the number of atoms of $\mathfrak{L}(\tau)$.

34.† Let K be an equational class which can be defined by a finite set of identities. Let $D(K)$ be the set of those integers n such that K can be

† In January 1968, the author recieved a prepublication copy of "Equational Logic and Equational Theories of Algebras," by A. Tarski, which is to appear in the Proceedings of the 1966 Hannover Logic Colloqium. In this paper solutions to Problems 34 and 35 are announced.

defined by an independent set Σ of n identities. (The *independence* of Σ means that $\Sigma_1{}^* \neq K$ for $\Sigma_1 \subset \Sigma$.) Characterize $D(K)$.

35.† Characterize $D(K)$ for the class of all rings, lattices.

36. Find an algebraic characterization of the smallest element of $D(K)$.

37. Which pairs can be represented as $\langle v(K), f(K)\rangle$ for some equational class? (See Ex. 15–26.)

38. Characterize the spectra of equational classes defined by finite sets of identities. (See Ex. 3.92, 3.93, 4.27, and 4.28.)

39. Find classes K of "small" finite algebras such that $\{\mathfrak{A} \mid \mathfrak{A} \in K,\ Id(\mathfrak{A})$ has a finite basis$\} = \{\mathfrak{A} \mid \mathfrak{A} \in K,\ \mathbf{HSP}(\mathfrak{A}) = \mathbf{IP}_S\mathbf{HS}(\mathfrak{A})\}$.

40. Characterize equational classes K in which $\langle \Theta, \Phi\rangle \to \Theta \times \Phi$ is an isomorphism between $\mathfrak{C}(\mathfrak{A}) \times \mathfrak{C}(\mathfrak{B})$ and $\mathfrak{C}(\mathfrak{A} \times \mathfrak{B})$ for any $\mathfrak{A}, \mathfrak{B} \in K$.

41. Let K be an equational class, $\mathfrak{A}_i \in K$, $i \in I$ and P the set of all finite nonvoid subsets of I. For $s \in P$ let $\mathfrak{A}_s$ be the free product of the $\mathfrak{A}_i$, $i \in s$. Let us assume that (a) all $\mathfrak{A}_s$ exist; (b) every algebra K has a one-element subalgebra determined by the polynomial symbol $\mathbf{0}$. Then there is a natural homomorphism from $\mathfrak{A}_s$ into $\mathfrak{A}_t$ if $t \subseteq s$ so we can form the inverse limit $\mathfrak{A}$ (for groups this construction is due to G. Higman)
   (α) Describe the structure of $\mathfrak{A}$;
   (β) Let $\mathfrak{A}_i \cong \mathfrak{F}_K(1)$ for all $i \in I$; prove that $\mathfrak{A}$ is not free over K.

42. Let $\langle c_1, c_2, \cdots, c_n, \cdots\rangle$ be a sequence of nonnegative integers. Under what conditions can one find an equational class K such that $c_n = |F_K(n)|$?

43. Characterize the congruence lattices of free algebras.

44. Let K be an equational class and $L = \{\mathfrak{C}(\mathfrak{A}) \mid \mathfrak{A} \in K\}$. Characterize L as a class of lattices.

45. Find a finite (modular) lattice whose identities have no finite basis.

46. Is there a finite algebra $\mathfrak{A}$ such that all elements of $\mathfrak{A}$ are algebraic constants and the identities of $\mathfrak{A}$ have no finite basis?

47. If the identities of $\mathfrak{A}$ and $\mathfrak{B}$ have (no) finite basis, under what conditions is the same true of $\mathfrak{A} \times \mathfrak{B}$?

48. Let K be an equational class of infinitary algebras. Define $C(K) = \{\mathfrak{m} \mid$ there exists $\mathfrak{A} \in K,\ |A| = \mathfrak{m}\}$. Characterize $C(K)$ as a class of cardinals. (See the author's abstract, Notices Amer. Math. Soc. 14 (1967), 73.)

† See footnote on p. 194.

CHAPTER 5

INDEPENDENCE

In this chapter we investigate the bases of free algebras. A set of elements H of an algebra $\mathfrak{A}$ is independent if the subalgebra generated by H is free over the equational class generated by $\mathfrak{A}$. The basic results are given in §31. Then we investigate classes of algebras in which independence has special properties. Some invariants of finite algebras are then considered after which we discuss the system of independent sets of an algebra. Two generalizations of independence are also considered.

§31. INDEPENDENCE AND BASES

Let K be a class of algebras and let us consider the free algebra $\mathfrak{F}_K(\alpha)$ with $\bar{\alpha}$ generators over K having the free generating family $(x_\gamma \mid \gamma < \alpha)$.

Lemma 1. *$\{x_\gamma \mid \gamma < \alpha\}$ is $\mathscr{S}^+$-independent in $\mathfrak{F}_K(\alpha)$, provided $|F_K(\alpha)| > 1$.*

Proof. Let $\{x_i \mid i \in I\} \subset \{x_\gamma \mid \gamma < \alpha\}$ and suppose $x_\delta \notin \{x_i \mid i \in I\}$. If $\{x_i \mid i \in I\}$ is a generating set, then we have $x_\delta = p(x_{i_0}, \cdots, x_{i_{n-1}})$, where $i_j \in I$ and $\mathbf{p}$ is an n-ary polynomial symbol. Consider an arbitrary mapping $x_\gamma \to a_\gamma (\in F_K(\alpha))$, $\gamma < \alpha$, $\gamma \neq \delta$, and map x_δ onto any element different from $p(a_{i_0}, \cdots, a_{i_{n-1}})$. This mapping cannot be extended to an endomorphism of $\mathfrak{F}_K(\alpha)$ because, if φ were such an extension, then we would have

$$x_\delta\varphi = p(x_{i_0}\varphi, \cdots, x_{i_{n-1}}\varphi) = p(a_{i_0}, \cdots, a_{i_{n-1}}),$$

contradicting $p(a_{i_0}, \cdots, a_{i_{n-1}}) \neq x_\delta\varphi$.

It remains to show that if $\alpha = 1$, then $\{x_0\}$ is $\mathscr{S}^+$-independent. Indeed, if $\{x_0\}$ is $\mathscr{S}^+$-dependent, then x_0 is constant in $\mathfrak{F}_K(1)$, implying that there exists exactly one endomorphism of $\mathfrak{F}_K(1)$ contradicting $|F_K(1)| > 1$.

Lemma 2. *Let $\mathfrak{A}$ be an algebra, $|A| > 1$, and let $\{x_i \mid i \in I\}$ and $\{y_j \mid j \in J\}$ be $\mathscr{S}^+$-independent generating sets of $\mathfrak{A}$. If $|I| \geqq \aleph_0$, then $|J| \geqq \aleph_0$ and $|I| = |J|$.*

Proof. For any $y_j, j \in J$, there exists a finite $Y_j \subseteq \{x_i \mid i \in I\}$ such that $y_j \in [Y_j]$.

Let $Y = \bigcup (Y_j \mid j \in J) \subseteq \{x_i \mid i \in I\}$. Obviously, Y is a generating set; therefore, $Y = \{x_i \mid i \in I\}$. If J is finite, so is Y, which is a contradiction. If $|J| \geqq \aleph_0$, then $|Y| \leqq |J|$; thus, $|I| \leqq |J|$. Similarly, $|J| \leqq |I|$, which proves that $|I| = |J|$.

Combining Lemmas 1 and 2, we have the following result.

Theorem 1. *Let $\mathfrak{F}_K$ be a free algebra over the class K. If $\{x_i \mid i \in I\}$ and $\{y_j \mid j \in J\}$ are free generating sets of $\mathfrak{F}_K$ and $|I| \geqq \aleph_0$, then $|I| = |J|$.*

In a special case, we can extend the results of Theorem 1 to finite free generating sets as well.

Theorem 2. *Let $\mathfrak{F}_K$ be a free algebra with more than one element over the class K which has a finite algebra of more than one element. If $\{x_i \mid i \in I\}$ and $\{y_j \mid j \in J\}$ are free generating sets of $\mathfrak{F}_K$, then $|I| = |J|$.*

Proof. If $|I| \geqq \aleph_0$, then $|I| = |J|$ by Theorem 1. Assume that $|I| < \aleph_0$. Let $\mathfrak{A}$ be a finite algebra in K of m elements, $m \neq 1$.

The number of homomorphisms of $\mathfrak{F}_K$ into $\mathfrak{A}$ is the same as the number of mappings of $\{x_i \mid i \in I\}$ into $\mathfrak{A}$, that is, $m^{|I|}$. Similarly, we get that the number of homomorphisms is $m^{|J|}$, which implies that $|I| = |J|$.

Definition 1. *Let $\mathfrak{A}$ be an algebra. The set $\{a_i \mid i \in I\} \subseteq A$ is called* independent *if $I = \varnothing$ or it is a free generating set of the algebra*

$$\langle [\{a_i \mid i \in I\}]; F \rangle$$

over the equational class generated by $\mathfrak{A}$. $\{a_i \mid i \in I\}$ is dependent *if it is not independent. An element $a \in A$ is* self-dependent *if $\{a\}$ is dependent.*

The following is a restatement of Lemma 1.

Corollary. *Independence implies $\mathscr{S}^+$-independence in any algebra of more than one element.*

Definition 1 (in an equivalent form) is due to E. Marczewski [2], who observed that many different concepts of independence used in different branches of mathematics are special cases of this definition; for examples, see the exercises at the end of this chapter.

Theorem 3. *Let $\mathfrak{A}$ be an algebra with more than one element and let a_i, $i \in I$, be distinct elements of A. Then the following conditions are equivalent:*

(i) *$\{a_i \mid i \in I\}$ is an independent set;*
(ii) *every finite subset of $\{a_i \mid i \in I\}$ is independent;*
(iii) *every mapping $a_i \to b_i \in A$, $i \in I$, can be extended to a homomorphism of $\langle[\{a_i \mid i \in I\}]; F\rangle$ into $\mathfrak{A}$;*
(iv) *if $\mathbf{p}, \mathbf{q} \in \mathbf{P}^{(n)}(\tau)$ and $p(a_{i_0}, \cdots, a_{i_{n-1}}) = q(a_{i_0}, \cdots, a_{i_{n-1}})$, $i_j \in I$, then $p(b_0, \cdots, b_{n-1}) = q(b_0, \cdots, b_{n-1})$ for all $b_i \in A$;*
(v) *if p and q are n-ary polynomials over $\mathfrak{A}$ and $p(a_{i_0}, \cdots, a_{i_{n-1}}) = q(a_{i_0}, \cdots, a_{i_{n-1}})$, then $p = q$.*

Proof. (iv) and (v) are equivalent by the definition of a polynomial. The equivalence of (i) and (iii) follows from Theorem 24.5. (iii) and (iv) are equivalent by Theorem 12.2. Finally, (iii) implies (ii) and (ii) implies (iv) are trivial. Thus, all the conditions are equivalent.

Corollary. *$\{a_0, \cdots, a_{n-1}\}$ is independent if and only if $p(a_0, \cdots, a_{n-1}) = q(a_0, \cdots, a_{n-1})$ implies $p(b_0, \cdots, b_{n-1}) = q(b_0, \cdots, b_{n-1})$ for all $b_i \in A$.*

Definition 2. *A generating set $\{a_i \mid i \in I\}$, $I \neq \varnothing$ is a* basis *of the algebra $\mathfrak{A}$ if it is independent. $\varnothing$ is a basis of $\mathfrak{A}$ if all elements of $\mathfrak{A}$ are constants.*

In other words, a basis is a free generating set of the algebra over the equational class generated by it.

Corollary. *If an algebra has an infinite basis, then all bases are infinite and have the same cardinal number.*

Proof. By Theorem 1 and Definition 1.

Theorem 4 (*E. Marczewski* [5]). *Let $\mathfrak{A}$ be an algebra with more than one element. If $\mathfrak{A}$ has bases with different cardinal numbers, then all bases of $\mathfrak{A}$ are finite and the numbers n for which there is a basis of n elements form an arithmetic progression.*

Proof. It suffices to prove that if $\mathfrak{A}$ has bases $\{a_0, \cdots, a_{p-1}\}$, $\{b_0, \cdots, b_{q-1}\}$ and $\{c_0, \cdots, c_{q+r-1}\}$, then there exists a basis consisting of $p+r$ elements. Set $C = [c_0, \cdots, c_{q-1}]$. Then there exists an isomorphism φ of $\mathfrak{A}$ with $\mathfrak{C}$ such that $b_i\varphi = c_i$. We will prove that

$$\{a_0\varphi, \cdots, a_{p-1}\varphi, c_q, \cdots, c_{q+r-1}\}$$

is a basis of $\mathfrak{A}$. Obviously, this is a generating set. To show that these elements are independent, consider the mapping ψ:

$$a_0\varphi \to d_0,$$
$$\cdots$$
$$a_{p-1}\varphi \to d_{p-1},$$
$$c_q \to d_p,$$
$$\cdots$$
$$c_{q+r-1} \to d_{p+r-1},$$

where $d_0, \cdots, d_{p-1}, d_p, \cdots, d_{p+r-1} \in A$. Since $\{a_0, \cdots, a_{p-1}\}$ is independent, there exists an endomorphism χ of $\mathfrak{A}$ such that $a_i\chi = d_i$, $i = 0, \cdots, p-1$. Set $\rho = \varphi^{-1}\chi$. Then

$$(a_i\varphi)\rho = (a_i\varphi)\varphi^{-1}\chi = a_i\chi = d_i.$$

Set $d_i' = c_i\rho$, $i = 0, \cdots, q-1$, and consider the mapping δ:

$$c_0 \to d_0'$$
$$\cdots$$
$$c_{q-1} \to d_{q-1}',$$
$$c_q \to d_p,$$
$$\cdots$$
$$c_{q+r-1} \to d_{p+r-1}.$$

Since $\{c_0, \cdots, c_{q+r-1}\}$ is independent, δ can be extended to an endomorphism δ'.

Note that $c_i\delta' = c_i\rho$, $0 \leqq i < q$; thus, δ' and ρ coincide on C. Therefore,

$$(a_i\varphi)\delta' = d_i,$$

that is, δ' extends ψ to an endomorphism, which concludes the proof of Theorem 4.

To illustrate Theorem 4, let K be the equational class of all algebras $\mathfrak{A} = \langle A; F\rangle$, where $F = \langle h, g_0, g_1\rangle$ is of type $\langle 2, 1, 1\rangle$, satisfying the identities

$$\mathbf{h}(\mathbf{g}_0(\mathbf{x}_0), \mathbf{g}_1(\mathbf{x}_0)) = \mathbf{x}_0,$$
$$\mathbf{g}_0(\mathbf{h}(\mathbf{x}_0, \mathbf{x}_1)) = \mathbf{x}_0,$$
$$\mathbf{g}_1(\mathbf{h}(\mathbf{x}_0, \mathbf{x}_1)) = \mathbf{x}_1.$$

First we show that K contains an algebra having more than one element. Let A be an infinite set; then there exists a 1–1 mapping φ of A onto A^2. Let

$$x\varphi = \langle g_0(x), g_1(x)\rangle$$

and

$$\langle x, y\rangle\varphi^{-1} = h(x, y).$$

This defines h, g_0, and g_1 on A and $\mathfrak{A}$ obviously satisfies the identities. Let $\mathfrak{F}_K(1)$ be the free algebra with one generator x over K. We claim that $g_0(x)$ and $g_1(x)$ form a basis for $\mathfrak{F}_K(1)$. They indeed generate it since $h(g_0(x), g_1(x))=x$. To prove the independence, consider the mapping $g_0(x)\rightarrow a$, $g_1(x)\rightarrow b$. Set $c=h(a, b)$. Then there exists an endomorphism ψ with $x\psi=c$. Compute: $g_0(x)\psi=g_0(x\psi)=g_0(c)=g_0(h(a, b))=a$ and similarly $g_1(x)\psi=b$.

Thus, $\mathfrak{F}_K(1)$ has bases consisting of one and two elements, respectively; therefore by Theorem 4, it has a basis consisting of n elements for each positive integer n.

The proof of Theorem 4 yields the following result, which is well known for vector spaces:

Exchange Theorem. *Let $\mathfrak{A}$ be an algebra. Let X, Y, and Z be subsets of A, $X \cap Y=\varnothing$, such that $X \cup Y$ and Z are independent and $[Y]=[Z]$. Then $X \cup Z$ is independent.*

Theorem 5. (*A. Goetz and C. Ryll-Nardzewski* [1]). *Let $\mathfrak{A}$ be an algebra which has a basis consisting of $n>0$ elements. Then $\mathfrak{A}$ has a basis consisting of $m>0$ elements if and only if there exist n-ary polynomial symbols $\mathbf{g}_0, \cdots, \mathbf{g}_{m-1}$ and m-ary polynomial symbols $\mathbf{h}_0, \cdots, \mathbf{h}_{n-1}$ such that the following identities hold in $\mathfrak{A}$:*

$$\mathbf{g}_i(\mathbf{h}_0(\mathbf{x}_0, \cdots, \mathbf{x}_{m-1}), \cdots, \mathbf{h}_{n-1}(\mathbf{x}_0, \cdots, \mathbf{x}_{m-1})) = \mathbf{x}_i,$$
$$\mathbf{h}_i(\mathbf{g}_0(\mathbf{x}_0, \cdots, \mathbf{x}_{n-1}), \cdots, \mathbf{g}_{m-1}(\mathbf{x}_0, \cdots, \mathbf{x}_{n-1})) = \mathbf{x}_i.$$

Proof. Let $\{a_0, \cdots, a_{n-1}\}$ be the n-element basis. If $\{b_0, \cdots, b_{m-1}\}$ is also a basis, then $b_i=g_i(a_0, \cdots, a_{n-1})$ and $a_i=h_i(b_0, \cdots, b_{m-1})$. The identities for $\mathbf{g}_i$ and $\mathbf{h}_j$ are obviously satisfied. On the other hand, if we have the $\mathbf{g}_i$ and $\mathbf{h}_j$ satisfying the identities, then we define the m-element basis by $b_i=g_i(a_0, \cdots, a_{n-1})$. One can prove that $\{b_0, \cdots, b_{m-1}\}$ forms a basis in the same way as in the example.

The converse of Theorem 4 was proved by S. Świerczkowski [3]. Namely, he proved that for any arithmetic progression $n+kd$, $k=0, 1, 2, \cdots, n>0$, $d>0$, there exists an algebra $\mathfrak{A}$ such that $\{m \mid \mathfrak{A}$ has an m element basis$\}=\{n+kd \mid k=0, 1, \cdots\}$. The special case $n<1+d$ was settled first by A. Goetz and C. Ryll-Nardzewski [1].

§32. INDEPENDENCE IN SPECIAL CLASSES OF ALGEBRAS

In this section we will study various classes of algebras in which independence has several properties of independence in vector spaces. The investigation of these classes of algebras was proposed by E. Marczewski [3] and was carried out mostly by Marczewski himself, K. Urbanik, and W. Narkiewicz.

Definition 1. *Let* $\langle A; F\rangle = \mathfrak{A}$ *be an algebra and let* $p, q \in P^{(n)}(\mathfrak{A})$. *We say that p and q can be* distinguished by x_i *if there exist* $a_0, \cdots, a_{n-1}, a_i' \in A$ *such that*

$$p(a_0, \cdots, a_i, \cdots, a_{n-1}) = q(a_0, \cdots, a_i, \cdots, a_{n-1})$$

and

$$p(a_0, \cdots, a_i', \cdots, a_{n-1}) \neq q(a_0, \cdots, a_i', \cdots, a_{n-1}).$$

Corollary. *If the n-ary polynomials p and q over* $\mathfrak{A}$ *cannot be distinguished by any* x_i, *then either* $p=q$ *or* $p(a_0, \cdots, a_{n-1}) \neq q(a_0, \cdots, a_{n-1})$ *for all* $a_0, \cdots, a_{n-1} \in A$.

Example: Let $\mathfrak{S} = \langle S; +, \cdot, 0, 1\rangle$ be a field. A vector space $\mathfrak{V} = \langle V; +, \{f_s \mid s \in S\}\rangle$ over $\mathfrak{S}$ is defined as usual; f_s is a unary operation, namely, left multiplication by s. Every n-ary polynomial of $\mathfrak{V}$ can be written in the form

$$\sum_{i=0}^{n-1} s_i x_i, \qquad s_i \in S.$$

It is obvious that $\sum_{i=0}^{n-1} s_i x_i$ can be distinguished from $\sum_{i=0}^{n-1} t_i x_i$ by x_j if and only if $s_j \neq t_j$. In this case, two polynomials are equal if and only if they cannot be distinguished by any x_i.

Definition 2. *An algebra* $\mathfrak{A} = \langle A; F\rangle$ *is called a* v-algebra *if for every* $n > 0$ *and* $p, q \in P^{(n)}(\mathfrak{A})$ *that can be distinguished by* x_{n-1}, *there exists an* $r \in P^{(n-1)}(\mathfrak{A})$ *such that* $p(a_0, \cdots, a_{n-1}) = q(a_0, \cdots, a_{n-1})$ *if and only if* $a_{n-1} = r(a_0, \cdots, a_{n-2})$. *If there are no nullary operations,* $P^{(0)}(\mathfrak{A})$ *stands for* $P^{(1,0)}(\mathfrak{A})$.

Example: In the case of a vector space $\mathfrak{V}$, the condition means that $s_{n-1} \neq t_{n-1}$ and $r = 1/(s_{n-1} - t_{n-1}) \sum_{i=0}^{n-2} (t_i - s_i)x_i$. Thus $\mathfrak{V}$ is a v-algebra.

Definition 3 (*W. Narkiewicz* [1]). *The algebra* $\mathfrak{A}$ *is called a* v^*-algebra *if it has the following two properties:*

(i) *if a is not constant in* $\mathfrak{A}$, *then* $\{a\}$ *is independent;*

(ii) *if* $n > 1$, $\{a_0, \cdots, a_{n-1}\}$ *is independent, and* $\{a_0, \cdots, a_{n-1}, a_n\}$ *is dependent, then*†

$$a_n \in [a_0, \cdots, a_{n-1}].$$

Note that if the algebra has more than one element and $\{a\}$ is independent, then a is not constant in $\mathfrak{A}$. Thus, in the case of v^*-algebras with more than one element, a is not constant in $\mathfrak{A}$ if and only if $\{a\}$ is independent.

Definition 4 (*W. Narkiewicz* [3]). *An algebra* $\mathfrak{A}$ *is called a* v^{**}-algebra *if* $\mathscr{S}^+$*-independence implies independence.*

Let us recall that if $|A| > 1$, then independence always implies $\mathscr{S}^+$-independence. Thus, an algebra with more than one element is a v^{**}-algebra if $\mathscr{S}^+$-independence is equivalent to independence.

Examples of v-, v^*-, and v^{**}-algebras will be given in the Exercises.

Theorem 1. *A* v*-algebra is also a* v^**-algebra and a* v^**-algebra is also a* v^{**}*-algebra.*

Proof. Obviously a one-element algebra is a v-algebra, a v^*-algebra, and a v^{**}-algebra, all conditions being vacuously satisfied. Therefore, in the proof, we can assume that the algebra $\mathfrak{A}$ has more than one element.

Let $\mathfrak{A}$ be a v-algebra. If $a \in A$ and $\{a\}$ is dependent, then there exist $p, q \in P^{(1)}(\mathfrak{A})$ such that $p \neq q$ and $p(a) = q(a)$. This implies that p and q can be distinguished by x_0. Thus, there exists an $r \in P^{(1,0)}(\mathfrak{A})$ such that $p(x_0) = q(x_0)$ if and only if $x_0 = r$. Therefore, $r = a$; that is, a is constant in $\mathfrak{A}$, proving (i) of Definition 3.

If $n > 1$, $\{a_0, \cdots, a_{n-1}\}$ is independent, and $\{a_0, \cdots, a_{n-1}, a_n\}$ is dependent, then there exist $p, q \in P^{(n+1)}(\mathfrak{A})$ such that $p \neq q$ and

$$p(a_0, \cdots, a_n) = q(a_0, \cdots, a_n).$$

We claim that p and q can be distinguished by x_n. If this were not true, then we would have $p(a_0, \cdots, a_{n-1}, x) = q(a_0, \cdots, a_{n-1}, x)$ for every $x \in A$ and in particular $p(a_0, \cdots, a_{n-1}, a_{n-1}) = q(a_0, \cdots, a_{n-1}, a_{n-1})$. This implies by the independence of $\{a_0, \cdots, a_{n-1}\}$ that $p(b_0, \cdots, b_{n-1}, b_{n-1}) = q(b_0, \cdots, b_{n-1}, b_{n-1})$ for all $b_0, \cdots, b_{n-1} \in A$. Since for some $b_0, \cdots, b_{n-1}, b_n$, we have $p(b_0, \cdots, b_{n-1}, b_n) \neq q(b_0, \cdots, b_{n-1}, b_n)$, we get that p and q can be distinguished by x_n.

Thus, for some $r \in P^{(n)}(\mathfrak{A})$, we have $a_n = r(a_0, \cdots, a_{n-1})$, which proves (ii) of Definition 3.

Now let $\mathfrak{A}$ be a v^*-algebra and let $\{a_0, \cdots, a_{n-1}\}$ be $\mathscr{S}^+$-independent. Let $n = 1$; $\{a_0\}$ is $\mathscr{S}^+$-independent, so, by definition, a_0 is not constant in

† To simplify the notation we shall write $[a_0, \cdots, a_{n-1}]$ for $[\{a_0, \cdots, a_{n-1}\}]$

$\mathfrak{A}$, which implies by (i) of Definition 3 that $\{a_0\}$ is independent. Let $n>1$. Now $a_0 \notin [a_1, \cdots, a_{n-1}]$, therefore a_0 is not constant in $\mathfrak{A}$. Thus, by (*i*) of Definition 3, $\{a_0\}$ is independent. Let $0 \leq i < n-1$ and assume that $\{a_0, \cdots, a_i\}$ is independent. If $\{a_0, \cdots, a_i, a_{i+1}\}$ is dependent, then, by (ii) of Definition 3, $a_{i+1} \in [a_0, \cdots, a_i]$, contradicting $\mathscr{S}^+$-independence. Thus, $\{a_0, \cdots, a_{n-1}\}$ is independent, which was to be proved.

Examples show that there are v^*-algebras which are not v-algebras and v^{**}-algebras which are not v^*-algebras (see Exercises).

Theorem 2. *The property of being a v-, v*-, or v**-algebra is preserved under the formation of subalgebras.*

Proof. Let $\mathfrak{B}$ be a subalgebra of $\mathfrak{A}$. As in the proof of Theorem 1, we can assume that $|B| \neq 1$. If $\mathbf{p}$ and $\mathbf{q}$ are n-ary polynomial symbols and the induced polynomials of $\mathfrak{B}$ can be distinguished by x_{n-1}, then the same holds in $\mathfrak{A}$ and the statement holds for v-algebras trivially.

Let $\mathfrak{A}$ be a v^*-algebra. We observe that if $\{a_0, \cdots, a_{n-1}\} \subseteq B$, then $\{a_0, \cdots, a_{n-1}\}$ is independent in $\mathfrak{B}$ if and only if it is independent in $\mathfrak{A}$. (Proof is by induction on n as in the proof of Theorem 1.) This implies Theorem 2 for v^*-algebras, while for v^{**}-algebras it is trivial.

Corollary. *Suppose there are $n>0$ independent elements in $\mathfrak{A}$. If $\mathfrak{A}$ is a v-, v*-, v**-*algebra, *then so is* $\mathfrak{P}^{(n)}(\mathfrak{A})$.

The same statement holds for arbitrary n in case of v-algebras.

Theorem 3 (*E. Marczewski* [3]). *If $\mathfrak{A}$ is a v-algebra, then $\mathfrak{P}^{(n)}(\mathfrak{A})$ is a v-algebra for all $n>0$.*

Proof. Let $\mathbf{p}$ and $\mathbf{q}$ be n-ary polynomial symbols. Let p and q denote the polynomials of $\mathfrak{A}$ induced by $\mathbf{p}$ and $\mathbf{q}$ and let p^o and q^o denote the polynomials of $\mathfrak{P}^{(n)}(\mathfrak{A})$ induced by $\mathbf{p}$ and $\mathbf{q}$. We claim that if p^o and q^o can be distinguished by x_{n-1}, then the same holds for p and q. Indeed, by assumption, there exist n-ary polynomials $r_0, \cdots, r_{n-1}, r'_{n-1}$ over $\mathfrak{A}$ such that

$$p^o(r_0, \cdots, r_{n-1}) = q^o(r_0, \cdots, r_{n-1}),$$

and

$$p^o(r_0, \cdots, r'_{n-1}) \neq q^o(r_0, \cdots, r'_{n-1}).$$

Thus, there exist $a_0, \cdots, a_{n-1} \in A$ such that if we set $b_i = r_i(a_0, \cdots, a_{n-1})$ and $b'_{n-1} = r'_{n-1}(a_0, \cdots, a_{n-1})$, then $p(b_0, \cdots, b_{n-1}) = q(b_0, \cdots, b_{n-1})$ and $p(b_0, \cdots, b'_{n-1}) \neq q(b_0, \cdots, b'_{n-1})$, which was to be proved.

Thus, there exists an $(n-1)$-ary polynomial symbol $\mathbf{r}$ such that $p(c_0, \cdots, c_{n-1}) = q(c_0, \cdots, c_{n-1})$ if and only if $c_{n-1} = r(c_0, \cdots, c_{n-2})$. (For

$n=1$, $r \in P^{(1,0)}(\mathfrak{A})$.) It is obvious that the polynomial of $\mathfrak{P}^{(n)}(\mathfrak{A})$ induced by $\mathbf{r}$ establishes that $\mathfrak{P}^{(n)}(\mathfrak{A})$ is a v-algebra.

For v^*-algebras, one can prove the existence of bases in the same way as it is done for vector spaces.

Theorem 4. *Let $\mathfrak{A}$ be a v^*-algebra. Any maximal independent subset of $\mathfrak{A}$ is a basis.*

Corollary. *Every v^*-algebra has a basis.*

Of course, the existence of maximal independent subsets follows from the Teichmüller-Tukey Lemma; therefore, the corollary follows trivially from Theorem 4.

Proof of Theorem 4. Let $B \neq \varnothing$ be a maximal independent subset of A and assume $[B] \neq A$ and $|A| \neq 1$. Take an $a \in A$, $a \notin [B]$. Then $B \cup \{a\}$ is dependent. Therefore, we can find $b_0, \cdots, b_{n-1} \in B$ such that

$$\{b_0, \cdots, b_{n-1}, a\}$$

is dependent. If $n=0$, then a is self-dependent; thus, a is constant in $\mathfrak{A}$, which contradicts $a \notin [B]$. If $n \neq 0$, then, by (ii) of Definition 3,

$$a \in [b_0, \cdots, b_{n-1}] \subseteq [B],$$

a contradiction; since the case $B = \varnothing$ is obvious, the proof of Theorem 4 is complete.

The following example shows that Theorem 4 does not hold for v^{**}-algebras. Let A be the set of integers and $F = \{f_a \mid a > 0\}$, where $f_a(x) = x + a$. Then $\mathfrak{A} = \langle A; F \rangle$ is a v^{**}-algebra in which any two distinct elements are dependent and $[a] = \{x \mid x \geqq a\} \neq A$; thus, $\mathfrak{A}$ has no basis.

For v- and v^*-algebras, one can prove that any two bases have the same number of elements by using the same technique as for vector spaces, namely, by use of the† EIS property. It is rather surprising that, although the EIS property is not valid for v^{**}-algebras, the invariance of the number of elements in a basis can still be proved.

Theorem 5 (*W. Narkiewicz* [3]). *Let $\mathfrak{A}$ be a v^{**}-algebra, $|A| > 1$. If $\mathfrak{A}$ has a basis, then all bases have the same cardinal number.*

Proof (by G. H. Wenzel). If the theorem were false there would be a smallest integer m such that there exists a v^{**}-algebra $\mathfrak{A}$, bases $B_0 = \{a_0, \cdots, a_{n-1}\}$ and $B_1 = \{b_0, \cdots, b_{m-1}\}$ of $\mathfrak{A}$ such that $n > m$. Obviously, $m > 0$. Then we have the polynomial symbols $\mathbf{g}_i$ and $\mathbf{h}_j$ of Theorem 31.5.

† See Exercise 12.

Set

$$c_i = g_i(a_0, \cdots, a_{n-2}, a_{n-2}), 0 \leq i < m.$$

Then using the identities of Theorem 31.5 we get that

$$h_k(c_0, \cdots, c_{m-1}) = \begin{cases} a_k & \text{if } 0 \leq k < n-1, \\ a_{n-2} & \text{if } k = n-1. \end{cases}$$

Now consider the algebra

$$\mathfrak{A}' = \langle [a_0, \cdots, a_{n-2}]; F \rangle.$$

By Theorem 2, $\mathfrak{A}'$ is a v^{**}-algebra and $C_0 = \{a_0, \cdots, a_{n-2}\}$ is a basis of $\mathfrak{A}'$, $C_1 = \{c_0, \cdots, c_{m-1}\}$ is a generating set of $\mathfrak{A}'$. Since m was chosen to be minimal, C_1 is also a basis of $\mathfrak{A}'$, which implies that C_1 is independent in $\mathfrak{A}$. But this is impossible since

$$h_{n-1}(c_0, \cdots, c_{m-1}) = h_{n-2}(c_0, \cdots, c_{m-1}),$$

while

$$a_{n-1} = h_{n-1}(b_0, \cdots, b_{m-1}) \neq h_{n-2}(b_0, \cdots, b_{m-1}) = a_{n-2},$$

completing the proof of Theorem 5.

In the name v-algebra, v stands for vector spaces. This is justified by the first example. But more is true. It is proved in K. Urbanik [1] that if $\mathfrak{A}$ is a v-algebra and $P^{(3)}(\mathfrak{A}) \neq P^{(3,1)}(\mathfrak{A})$, then there exists a field $\mathfrak{K}$ such that A is a linear space over $\mathfrak{K}$, and there exists a linear subspace A_0 of A, such that all polynomials over $\mathfrak{A}$ are of the form $\Sigma\lambda_k x_k + a$, where $\lambda_i \in K$, $a \in A_0$ and if $P^{(0)}(\mathfrak{A}) = \varnothing$, then $\Sigma\lambda_k = 1$.

Similar representation theorems were given by K. Urbanik [2], [5], [6], [7], W. Narkiewicz [3], and G. Grätzer [2], completely describing all v-, and v^*-algebras. There are also some results on v^{**}-algebras; but the problem is far from settled.

§33. SOME INVARIANTS OF FINITE ALGEBRAS

In this section, $\mathfrak{A} = \langle A; F \rangle$ will always be a *finite algebra with more than one element*.† Following E. Marczewski [6], we now introduce a list of notations (our notations are different but the concepts are the same):

$g^*(\mathfrak{A})$ = smallest integer such that every subset of A consisting of $g^*(\mathfrak{A})$ elements generates $\mathfrak{A}$, unless all elements of A are constants in $\mathfrak{A}$, when we put $g^*(\mathfrak{A}) = 0$;

† For one element algebras we put $g^* = g = i = i_* = 1, p = \infty$. Most of the results of this section are then valid also for one element algebras.

$g(\mathfrak{A})$ = smallest number of generators of $\mathfrak{A}$, unless all elements of A are constants in $\mathfrak{A}$, when we put $g(\mathfrak{A})=0$;

$i(\mathfrak{A})$ = maximum number of independent elements in $\mathfrak{A}$;

$i_*(\mathfrak{A})$ = maximum number $\leqq |A|$ such all subsets of A of $i_*(\mathfrak{A})$ elements are independent;

$p(\mathfrak{A})$ = ∞ or the maximum number p such that all p-ary polynomials are trivial, that is, equal to one of the e_i^p; we set $p(\mathfrak{A})=0$ if there exist nontrivial unary polynomials and there are no constants in $\mathfrak{A}$, and we set $p(\mathfrak{A})=-1$ if there are constants in $\mathfrak{A}$.

The numbers introduced are invariant under isomorphism. Even more is true: the above numbers are invariant under equivalence of algebras.

If there is no danger of confusion, we will write g^* for $g^*(\mathfrak{A})$, and so on.

Lemma 1 (*S. Świerczkowski* [2]). $p \geqq i_* - 1$.

Proof. The statement is obvious for $i_*=0$ (since $p \geqq -1$ always holds) and for $i_*=1$ (since no element can be constant if $i_*=1$). So we can assume that $i_*>1$. Let $p \in P^{(k)}(\mathfrak{A})$, $k<i_*$.

Let $a_0, \cdots, a_{k-1}$ be distinct elements of A. If $p(a_0, \cdots, a_{k-1}) \neq a_i$, $i=0, \cdots, k-1$, then $\{a_0, \cdots, a_{k-1}, p(a_0, \cdots, a_{k-1})\}$ has not more than i_* elements and is dependent, which is a contradiction.

Thus, $p(a_0, \cdots, a_{k-1})=a_i$ for some $0 \leqq i < k$. Since $\{a_0, \cdots, a_{k-1}\}$ is independent, this implies that $p=e_i^k$, which was to be proved.

Theorem 1 (*S. Świerczkowski* [4]). *Let $i_*=|A| \neq 2$. Then $\mathfrak{A}$ is trivial. In other words, if A has more than two elements and A is independent, then $\mathfrak{A}$ is trivial.*

Obviously, if $\mathfrak{A}$ is trivial, then $i_*=|A|$.

If $|A|=2$, then the statement is false. Let $A=\{a, b\}$, $f(u, u, v)= f(u, v, u)=f(v, u, u)=u$, where $u=a$ or b, $v=a$ or b. Then $\langle A; f\rangle$ is nontrivial and $i_*=2$. All two-element algebras with $i_*=2$ have been described up to equivalence by E. Marczewski and K. Urbanik [1].

The proof of Theorem 1 is based on the following lemma.

Lemma 2. *Let S be a finite set and let n be a positive integer $\geqq 3$. Let P denote the set of all partitions of S into not more than n blocks. Assume that for every $\pi \in P$ a (nonvoid) block $\hat{\pi}$ of π is chosen such that $\pi_1 \leqq \pi_2$ implies $\hat{\pi}_1 \subseteq \hat{\pi}_2$. Then $\bigcap(\hat{\pi} \mid \pi \in P) \neq \varnothing$.*

Proof. It suffices to verify that for $\pi_1, \pi_2 \in P$ there exists a $\pi_3 \in P$ such that $\hat{\pi}_1 \cap \hat{\pi}_2 = \hat{\pi}_3$.

Let $\pi_1, \pi_2 \in P$. Let the partition δ_i consist of the two blocks $\hat{\pi}_i$ and $S-\hat{\pi}_i$, $i=1, 2$. Then $\pi_i \leqq \delta_i$; thus, $\hat{\pi}_i \subseteq \hat{\delta}_i$, that is, $\hat{\pi}_i = \hat{\delta}_i$. If $\hat{\pi}_1 \cap \hat{\pi}_2 = \varnothing$, then we can define the partition δ which consists of the blocks $\hat{\pi}_1$, $\hat{\pi}_2$, and $S-(\hat{\pi}_1 \cup \hat{\pi}_2)$. Then $\delta \leqq \delta_i$; thus, $\hat{\delta} \subseteq \hat{\delta}_i = \hat{\pi}_i$, which implies $\hat{\delta} = \varnothing$, a contradiction. Therefore, $\hat{\pi}_1 \cap \hat{\pi}_2 \neq \varnothing$. Let π_3 consist of the blocks $\hat{\pi}_1 \cap \hat{\pi}_2$, $\hat{\pi}_2 - \hat{\pi}_1$ and $S-\hat{\pi}_2$. Then $\pi_3 \leqq \delta_2$; thus, $\hat{\pi}_3 \subseteq \hat{\delta}_2 = \hat{\pi}_2$, which implies that $\hat{\pi}_3 = \hat{\pi}_1 \cap \hat{\pi}_2$ or $\hat{\pi}_2 - \hat{\pi}_1$. The second possibility leads to the contradiction $\hat{\pi}_3 \cap \hat{\pi}_1 = \varnothing$; thus, $\hat{\pi}_3 = \hat{\pi}_1 \cap \hat{\pi}_2$, which was to be proved.

Proof of Theorem 1. If the theorem is false, then there exists a $p \in P^{(k)}(\mathfrak{A})$ which is nontrivial such that all l-ary polynomials are trivial, for $l<k$.

By Lemma 1, $k \geqq |A|$. If $k=|A|$ and $A=\{a_0, \cdots, a_{k-1}\}$, then $p(a_0, \cdots, a_{k-1}) = a_i$ for some i; thus, $p=e_i^k$, which is a contradiction. Therefore, $k>|A| \geqq 3$.

We will show that $k \geqq 4$ leads again to a contradiction (even if $|A|=2$). Set $S=\{0, \cdots, k-1\}$ and $n=k-1 \geqq 3$. If π is any partition of S into at most n blocks, then define p^π as the function which we get from $p(x_0, \cdots, x_{k-1})$ by identifying x_i and x_j if i and j are in the same block of π. By Corollary 1 to Lemma 8.2, p^π is an n-ary polynomial; since $n<k$, p^π is trivial, that is, $p^\pi = e_i^k$ for some i. All the i for which $p^\pi = e_i^k$ form a block, say, $\hat{\pi}$ of π. Obviously, $\pi_1 \leqq \pi_2$ implies $\hat{\pi}_1 \subseteq \hat{\pi}_2$. By Lemma 2, there exists an $i \in S$ such that $i \in \hat{\pi}$ for all π. This implies $p=e_i^k$ since whenever we make a substitution there are at least two equal elements inasmuch as $k>|A|$.

Corollary. *If $i_*(\mathfrak{A}) = |A| = 2$ and $p(\mathfrak{A}) > 2$, then $\mathfrak{A}$ is trivial.*

Theorem 2 (*S. Świerczkowski* [4]). *If $n>3$, $|A| \geqq n$, and any n elements of A form a basis, then $\mathfrak{A}$ is trivial.*

Proof. Let $\mathfrak{A}$ satisfy the assumptions of Theorem 2. If we can prove that $|A|=n$, then by Theorem 1 we are through. It suffices to prove that $P^{(n)}(\mathfrak{A})$ is trivial. Assume that $p \in P^{(n)}(\mathfrak{A})$ is nontrivial.

By Lemma 1, $p(x_0, x_1, x_1, x_3, \cdots, x_{n-1})$ is a trivial polynomial; thus, for any substitution, $p(a_0, a_1, a_1, a_3, \cdots, a_{n-1}) = a_0$ or a_1 or $\cdots$ or a_{n-1}. We can assume that $p(a_0, a_1, a_1, a_3, \cdots, a_{n-1}) \neq a_0$; otherwise, we permute x_0 and x_3. Thus, we have

$$p(a_0, a_1, a_1, a_3, \cdots, a_{n-1}) = a_1 \quad \text{or} \quad a_3 \quad \text{or} \quad \cdots \quad \text{or} \quad a_{n-1}.$$

Take n distinct elements $a_0, \cdots, a_{n-1}$ and define $a_n = p(a_0, \cdots, a_{n-1})$. Since p is nontrivial and $\{a_0, \cdots, a_{n-1}\}$ is independent, $a_n \neq a_0, \cdots, a_{n-1}$.

The elements $a_1, \cdots, a_n$ generate the algebra; hence $a_0 = q(a_1, \cdots, a_n)$. Thus, $a_0 = q(a_1, \cdots, a_{n-1}, p(a_0, \cdots, a_{n-1}))$.

Since $\{a_0, \cdots, a_{n-1}\}$ is independent, the above equality holds for any substitution, in particular, if $a_1 = a_2$. Then we get

$$a_0 = q(a_1, a_1, a_3, \cdots, a_{n-1}, p(a_0, a_1, a_1, a_3, \cdots, a_{n-1})).$$

Since

$$p(a_0, a_1, a_1, a_3, \cdots, a_{n-1}) = a_1 \quad \text{or} \quad a_3 \quad \text{or} \quad \cdots \quad \text{or} \quad a_{n-1},$$

we get that a_0 is in the subalgebra generated by $a_1, a_2, \cdots, a_{n-1}$, contradicting the independence.

Theorem 2 is false for $n = 2$ or 3 (see the Exercises).

Theorem 3 (*E. Marczewski and S. Świerczkowski*). *$g \geqq i$, that is, if $\mathfrak{A}$ can be generated by n elements and there are m independent elements, then $n \geqq m$.*

Proof. Let $\{a_0, \cdots, a_{n-1}\}$ be a generating set, let $\{b_0, \cdots, b_{m-1}\}$ be independent, and suppose $n < m$. Set $B = [b_0, \cdots, b_{n-1}]$. Then $B \neq A$ since $b_{m-1} \notin B$. On the other hand, the mapping $b_i \rightarrow a_i$, $i = 0, \cdots, n-1$, can be extended to a homomorphism of $\mathfrak{B}$ into $\mathfrak{A}$. Such a homomorphism is onto since $a_0, \cdots, a_{n-1}$ generate $\mathfrak{A}$. Thus, $|B| \geqq |A|$, which is a contradiction.

Corollary. $|A| \geqq g^* \geqq g \geqq i \geqq i_*$.

Theorem 4 (*E. Marczewski* [6]). *If $\mathfrak{A}$ is nontrivial, then $p = i_*$ or $p = i_* - 1$.*

Corollary. *If $\mathfrak{A}$ is nontrivial, then $|A| \geqq g^* \geqq g \geqq i \geqq i_* \geqq p$.*

Proof. First we prove that $i_* \geqq p$. This is trivial if $p \leqq 0$. Let $1 \leqq p \leqq |A|$. Since all p-ary polynomials are trivial, we get that all p-element subsets are independent, that is, $i_* \geqq p$.

If $|A| < p$, then all subsets of A are independent; thus, if $|A| \neq 2$, then, by Theorem 1, $\mathfrak{A}$ is trivial, contrary to our assumption. Therefore, $2 = |A| < p$ and by the corollary to Theorem 1, $\mathfrak{A}$ is trivial, which is a contradiction.

Since $p \geqq i_* - 1$ was verified in Lemma 1, this completes the proof of Theorem 4.

Theorem 5 (*S. Świerczkowski* [6]). *If $\mathfrak{A}$ is nontrivial and $i_* > 3$, then $i_* = p$.*

Proof. Set $l = i_*$. Then every l-element subset of A is independent by definition. Take $\{a_0, \cdots, a_{l-1}\} \subseteq A$ and set $[a_0, \cdots, a_{l-1}] = A_0$. We claim that any l elements of $\mathfrak{A}_0$ form a basis of $\mathfrak{A}_0$. Indeed, if $\{b_0, \cdots, b_{l-1}\} \subseteq A_0$, then $b_0, \cdots, b_{l-1}$ are independent in $\mathfrak{A}$; thus, $\langle [b_0, \cdots, b_{l-1}]; F\rangle \cong \mathfrak{A}_0$. This implies that $|[b_0, \cdots, b_{l-1}]| = |A_0|$; thus, $A_0 = [b_0, \cdots, b_{l-1}]$. By Theorem 2, $\mathfrak{A}_0$ is trivial. Let $q \in P^{(l)}(\mathfrak{A})$. Since $\mathfrak{A}_0$ is trivial, q restricted to A_0 is trivial; thus, $q(a_0, \cdots, a_{l-1}) = a_i$ for some $0 \leqq i \leqq l-1$. The independence of $\{a_0, \cdots, a_{l-1}\}$ implies $q = e_i^l$. Thus, $P^{(l)}(\mathfrak{A})$ is trivial, that is, $p \geqq i_*$. By Theorem 4, $p \leqq i_*$; thus, $p = i_*$, as stated.

Theorem 6 (*E. Marczewski* [6]). *If three of the numbers g^*, g, i, i_* are equal, then all of them are equal.*

Proof. Since $g^* \geqq g \geqq i \geqq i_*$, if three of them take the value k, then $i = g = k$, that is, there is a generating set $a_0, \cdots, a_{k-1}$ and there are k independent elements $b_0, \cdots, b_{k-1}$. If $g^* = g$, then $\{b_0, \cdots, b_{k-1}\}$ is a basis; thus, the number of endomorphisms of $\mathfrak{A}$ is $|A|^k$. If there is a generating set of k elements which is not a basis, then the number of endomorphisms is less than $|A|^k$, which is a contradiction. Thus, $i_* = i$.

If we assume that $i_* = i$, then $g^* = g$ can be deduced in a similar fashion.

§34. THE SYSTEM OF INDEPENDENT SETS OF AN ALGEBRA

Let $\mathfrak{A}$ be an algebra and let $\mathscr{S}$ denote the system of independent sets in A. The problem of characterizing $\mathscr{S}$ as a system of sets is as yet unsolved. In this section we will discuss some results which may contribute to the solution of this problem.

First of all it is advantageous to consider the system $\mathscr{J}$ of all *finite* independent sets rather than $\mathscr{S}$. Since a set is independent if and only if every finite subset of it is independent, the characterization problems of $\mathscr{S}$ and $\mathscr{J}$ are equivalent.

It is obvious that $\mathscr{J}$ is a *hereditary system* of finite sets, that is, $X \in \mathscr{J}$ and $Y \subseteq X$ imply $Y \in \mathscr{J}$.

Let $A = \{a, b, c\}$ and $\mathscr{J} = \{\varnothing, \{a\}, \{b\}, \{c\}, \{a, c\}, \{b, c\}\}$. Let us assume that $\mathscr{J}$ is the system of independent sets of some algebra $\langle A; F\rangle$. Since $\{a, c\}$ is independent $a \to a$, $c \to b$ can be extended to a homomorphism φ of $\langle [a, c]; F\rangle$ onto $\langle [a, b]; F\rangle$. But $\{a, b\}$ is not independent, hence φ is not an isomorphism, thus $[a, c] = [b, c] = A$ and φ is not 1–1, so $3 \geqq |[a, c]| \geqq |[a, b]|$, $[a, b] = \{a, b\}$. Thus for some $p \in P^{(2)}(\mathfrak{A})$, $p(a, c) = b$. Obviously, $p(c, a) = b$, $p(b, c) = a$, and $p(a, b) = a$ or b, say $p(a, b) = a$. Then

$$p(a, p(a, c)) = a,$$

and by the independence of $\{a, c\}$, $p(b, p(b, c)) = b$, that is, $p(b, a) = b$. Thus we have shown that $p(a, c) = p(c, a)$ and $p(a, b) \neq p(b, a)$, contradicting the independence of $\{a, c\}$. Therefore, $\mathcal{J}$ is an example of a hereditary system of finite subsets of a set A which is not the system of finite independent sets of any algebra defined on A.

Theorem 1 (*S. Świerczkowski* [5]). *Let A be a set and let $\mathcal{J} \neq \varnothing$ be a hereditary system of finite subsets of A; then there exists a set $B \supseteq A$ and an algebra $\mathfrak{B}$ such that for all finite $X \subseteq B$, X is independent if and only if $X \in \mathcal{J}$.*

Theorem 1′ (*S. Świerczkowski* [5]). *Let A be a set, $|A| \geqq 2$, and let $\mathcal{J} \neq \varnothing$ be a hereditary system of finite subsets of A. Set $S = \{x \mid x \in A$ and $\{x\} \notin \mathcal{J}\}$. If $|S| \geqq |\mathcal{J}|$, then there exists an algebra $\mathfrak{A}$ such that for all finite $X \subseteq A$, X is independent if and only if $X \in \mathcal{J}$.*

Obviously Theorem 1′ implies Theorem 1. It is also trivial that the condition of Theorem 1′ is not a necessary condition; for instance, if $\mathfrak{A}$ is a lattice, then we always have $|S| < |\mathcal{J}|$ since $S = \varnothing$.

Proof of Theorem 1′. The case $\mathcal{J} = \{\varnothing\}$ is trivial; all we have to do is to make all elements of A constants in $\mathfrak{A}$.

Now assume $\mathcal{J} \neq \{\varnothing\}$. Fix an $a \in S$. Since $|S| \geqq |\mathcal{J}|$, there exists a 1–1 mapping φ of $\mathcal{J} - \{\varnothing\}$ into $S - \{a\}$. Let m be the greatest integer such that there exists an m-element set in $\mathcal{J}$; set $m = \infty$ if no such greatest integer exists.

For every positive integer $n \leqq \max\{m, |A - S|\}$, we define an n-ary operation f_n as follows:

If $n \leqq m$, then

$$f_n(a_0, \cdots, a_{n-1}) = \begin{cases} \{a_0, \cdots, a_{n-1}\}\varphi \text{ if all the } a_i \text{ are distinct} \\ \quad \text{and } \{a_0, \cdots, a_{n-1}\} \in \mathcal{J}, \\ a \text{ otherwise;} \end{cases}$$

if $m < |A - S|$ and $m < n \leqq |A - S|$, then

$$f_n(a_0, \cdots, a_{n-1}) = \begin{cases} a_{n-1} \text{ if all the } a_i \text{ are distinct and } a_i \notin S, \\ a \text{ otherwise;} \end{cases}$$

We define $F = \{f_i \mid i = 1, 2, \cdots, \max\{m, |A - S|\}\} \cup \{g\}$, where g is a nullary operation, $g = a$.

Let us make the following observations on the algebra $\mathfrak{A} = \langle A; F \rangle$:

(i) If $n \leqq m$, then $f_n(a_0, \cdots, a_{n-1}) \in S$;

(ii) if $n \leqq m$ and $p \in P^{(n)}(\mathfrak{A})$, then either p is trivial or $p=a$ or p can be expressed in the form

$$p(x_0, \cdots, x_{n-1}) = f_l(x_{i_0}, \cdots, x_{i_{l-1}}), \tag{*}$$

where $\{i_0, \cdots, i_{l-1}\} \subseteq \{0, \cdots, n-1\}$.

Proof of (ii). Let Q_n denote the set of functions described in (ii). Then $Q_n \subseteq P^{(n)}(\mathfrak{A})$. Since $e_i^n \in Q_n$, to prove $Q_n = P^{(n)}(\mathfrak{A})$ it suffices to show that if $p_0, \cdots, p_{t-1} \in Q_n$, then $p=f_t(p_0, \cdots, p_{t-1}) \in Q_n$.

If all the p_i are trivial polynomials, then p is of the form (*) or $p=a$. If at least one p_i is not trivial, then, by (i), it takes values in S and therefore by the definition of f_t we have that $p=a$.

Now let $\{a_0, \cdots, a_{r-1}\} \notin \mathscr{I}$. If $r \leqq m$, then $f_r(a_0, \cdots, a_{r-1})=a$. Since $f_r=g$ does not hold identically, $\{a_0, \cdots, a_{r-1}\}$ is dependent. If $m<r \leqq |A-S|$, then $f_r(a_0, \cdots, a_{r-1})=a_{r-1}$ or a; thus, $\{a_0, \cdots, a_{r-1}\}$ is again dependent, since neither $f_r=e^r_{r-1}$ nor $f_r=g$ holds identically. Finally, if $\max\{m, |A-S|\}<r$, then $a_i \in S$ for some i. Therefore, $a=f_1(a_i)=g$; thus, $\{a_i\}$ is dependent and consequently $\{a_0, \cdots, a_{r-1}\}$ is dependent.

Now let $\{a_0, \cdots, a_{r-1}\} \in \mathscr{I}$. Then $r \leqq m$. Let $p, q \in P^{(r)}(\mathfrak{A})$ and

$$p(a_0, \cdots, a_{r-1}) = q(a_0, \cdots, a_{r-1}).$$

To prove that $\{a_0, \cdots, a_{r-1}\}$ is independent, we have to verify that $p=q$.

If $p=e_i^r$, then $p(a_0, \cdots, a_{r-1}) \notin S$. It follows from (ii) that $q=e_j^r$. Obviously, $i=j$; thus, $p=q$.

If $p=g$ and $q \neq g$, then $q(a_0, \cdots, a_{r-1})$ equals some a_i or $\{a_{i_0}, \cdots, a_{i_{s-1}}\}\varphi$, where $\{i_0, \cdots, i_{s-1}\} \subseteq \{0, \cdots, r-1\}$; in both cases, $q(a_0, \cdots, a_{r-1}) \neq a$, which is a contradiction.

The only remaining case is when p and q are both of the form (*); that is,

$$p(x_0, \cdots, x_{r-1}) = f_l(x_{i_0}, \cdots, x_{i_{l-1}})$$

and

$$q(x_0, \cdots, x_{r-1}) = f_t(x_{j_0}, \cdots, x_{j_{t-1}}).$$

Then $p(a_0, \cdots, a_{r-1}) = \{a_{i_0}, \cdots, a_{i_{l-1}}\}\varphi$ and

$$q(a_0, \cdots, a_{r-1}) = \{a_{j_0}, \cdots, a_{j_{t-1}}\}\varphi.$$

Thus, $\{a_{i_0}, \cdots, a_{i_{l-1}}\}\varphi = \{a_{j_0}, \cdots, a_{j_{t-1}}\}\varphi$. Since φ is 1–1, this implies $l=t$ and $\{i_0, \cdots, i_{l-1}\} = \{j_0, \cdots, j_{t-1}\}$, that is, $p=q$. This completes the proof of Theorem 1′.

It was observed by S. Fajtlowicz [1] that the same proof can be used to yield a more general result:

Corollary. *It is enough to assume in Theorem* 1′ *that there exists a mapping* φ *of* $\mathscr{J}$ *into* S *such that if* $X \neq Y$, $X \cup Y \in \mathscr{J}$, *then* $X\varphi \neq Y\varphi$.

Another sufficient condition is given in S. Fajtlowicz [1], and some necessary conditions are given in K. Urbanik [3] and S. Fajtlowicz [1].

§35. GENERALIZATIONS OF THE NOTION OF INDEPENDENCE

The concept of independence which was discussed in this chapter is in a certain sense too restrictive. For example, if it is applied to abelian groups (see Exercises), then it yields only the classical concept of independence, that is, a necessary condition of independence is that every element be torsion free. However, it proves to be very useful to have a concept of independence where every one-element set is independent (see, for instance, the book of L. Fuchs, Abelian Groups, Akadémiai Kiadó, Budapest, 1958). In this section, we are going to consider two generalizations of Marczewski's concept of independence which have this property in abelian groups.

Definition 1. *Let* $\mathfrak{A}$ *be an algebra and let* $I \subseteq A$. *Then* I *is called* locally independent *if* $I = \varnothing$ or I *is independent in* $\langle [I]; F \rangle$.

This definition is due to J. Schmidt [4] and it is called by him *algebraische Unabhängigkeit in sich.*

Theorem 31.3 remains valid for local independence except that in conditions (iii) and (iv) we have to require that $b_i \in [I]$ and in (v) $p = q$ means that they are equal in $[I]$.

From the new form of (iv) it follows that local independence is a property of finite character.

Corollary 1. *Every locally independent set can be extended to a maximal locally independent set.*

Corollary 2. *In an algebra with more than one element, independence implies local independence, which in turn implies* $\mathscr{S}^+$*-independence.*

We start off by proving the analogue of Theorem 34.1′.

Theorem 1 (*J. Schmidt* [4]). *Let* $\mathscr{S} \neq \varnothing$ *be a hereditary system of finite character of subsets of* A. *Assume that either* $\mathscr{S}$ *contains all one-element* *subsets or there are at least two one-element subsets which are not in $\mathscr{S}$. Then there exists an algebra $\mathfrak{A}$ such that $\mathscr{S}$ is the system of all locally independent sets of $\mathfrak{A}$.*

Proof. For every $k \geqq 1$ and $a \in A$, define the k-ary operation f_a^k by the rule

$$f_a^k(a_0, \cdots, a_{k-1}) = \begin{cases} a_0 & \text{if } \ a_0, \cdots, a_{k-1} \in I \ \text{ for some } \ I \in \mathscr{J}, \\ a & \text{otherwise.} \end{cases}$$

Set $F = \{f_a^k \mid a \in A, k = 1, 2, \cdots\}$. We claim that $\mathfrak{A} = \langle A; F \rangle$ is effective in Theorem 1.

First we note that if $B \in \mathscr{S}$, then $[B] = B$, and if $B \notin \mathscr{S}$, then $[B] = A$. Indeed, if $B \in \mathscr{S}$ and $a_0, \cdots, a_{k-1} \in B$, then $f_a^k(a_0, \cdots, a_{k-1}) = a_0 \in B$. On the other hand, if $B \notin \mathscr{S}$, then since $\mathscr{S}$ is of finite character there exists $\{a_0, \cdots, a_{k-1}\} \subseteq B$ such that $\{a_0, \cdots, a_{k-1}\} \notin \mathscr{S}$ (note that $k \geqq 1$ since $\varnothing \in \mathscr{S}$). Thus, $f_a^k(a_0, \cdots, a_{k-1}) = a$, proving $[B] = A$.

Now, if $B \in \mathscr{S}$, then to prove its local independence it is enough to verify that every mapping φ of B into itself is an endomorphism of $\mathfrak{B}$, which follows from the following equation:

$$f_a^k(a_0, \cdots, a_{k-1})\varphi = a_0\varphi = f_a^k(a_0\varphi, \cdots, a_{k-1}\varphi).$$

Finally, assume $B \notin \mathscr{S}$ is locally independent and again take

$$\{a_0, \cdots, a_{k-1}\} \subseteq B, \{a_0, \cdots, a_{k-1}\} \notin \mathscr{S}, k \geqq 1.$$

If $k \geqq 2$, then every permutation φ of $a_0, \cdots, a_{k-1}$ can be extended to an endomorphism $\bar{\varphi}$ of $\mathfrak{A}$. However,

$$a\bar{\varphi} = f_a^k(a_0, \cdots, a_{k-1})\bar{\varphi} = f_a^k(a_0\varphi, \cdots, a_{k-1}\varphi) = a$$

would imply that $a_i\varphi = a_i$, which is a contradiction.

If $k = 1$, then $\{a_0\} \notin \mathscr{S}$; hence, by the assumption on $\mathscr{S}$, there exists a $b \in A$, $a_0 \neq b$, such that $\{b\} \notin \mathscr{S}$. Let the mapping φ be defined by $a_0\varphi = b$. Let $\bar{\varphi}$ be the extension of φ to an endomorphism of $\mathfrak{A}$. Then, for all $a \in A$, we have

$$a\bar{\varphi} = f_a^1(a_0)\bar{\varphi} = f_a^1(a_0\bar{\varphi}) = f_a^1(b) = a,$$

contradicting $a_0\bar{\varphi} = b$. This concludes the proof of Theorem 1.

The exceptional case when $\mathscr{S}$ does not contain exactly one one-element subset occurs, for instance, in the algebra $\langle A; f_0, f_1 \rangle$, where $A = \{0, 1, 2\}$ and the unary operations f_0, f_1 are defined by the following table:

	0	1	2
f_0	1	2	1
f_1	1	1	2

In this case, the subalgebras are $\{1, 2\}$ and A while the locally independent sets are $\varnothing$, $\{1\}$, $\{2\}$. This exceptional case is discussed in the following theorem.

Theorem 2 (*J. Schmidt* [4]). *Let $\mathfrak{A}$ be an algebra, let $a \in A$, and assume that $\{a\}$ is locally dependent. Then there exists a $B \subseteq A$ such that $1 \leqq |B| \leqq 2$, $a \notin B$ and B is locally dependent.*

Corollary. *Let $\mathcal{S}$ be a hereditary family of finite character of subsets of A and assume that there exists one and only one one-element subset $\{a\}$ which is not in $\mathcal{S}$. If $\mathcal{S}$ is the family of locally independent sets of some algebra $\langle A; F\rangle$, then there exists a $B \subseteq A$ such that B has two elements, $a \notin B$ and $B \notin \mathcal{S}$.*

Proof of Theorem 2. Let us assume that Theorem 2 is false, that is, if $a \notin B$, and $1 \leqq |B| \leqq 2$, then B is locally independent.

Since the one-element set $\{a\}$ is locally dependent, there exist unary polynomials p, q and $b \in [a]$ such that $p(a) = q(a)$, $p(b) \neq q(b)$; $p(b) \in \{a, b\}$ for otherwise $\{b, p(b)\}$ is a two-element locally dependent set (see Corollary 2 to Definition 1) which does not contain a, contradicting our assumption. Similarly, $q(b) \in \{a, b\}$. Let, for example, $p(b) = b$, $q(b) = a$. Thus, $a \in [b]$ and so $[a] = [b]$. Since $a \neq b$, $\{b\}$ is locally independent. Hence $p(a) = a$ and there exists an endomorphism φ of $\langle [a]; F\rangle$ such that $b\varphi = a$ and $a\varphi = q(b)\varphi = q(b\varphi) = q(a) = a$. Finally, since $b \in [a]$, we have $b = r(a)$ for some unary polynomial r. Thus, $b\varphi = r(a)\varphi = r(a\varphi) = r(a) = b$, which contradicts $b\varphi = a$. This completes the proof of Theorem 2.

* * *

The other generalization of the concept of independence is based on the following notion (for the remainder of this section see G. Grätzer [6]).

Definition 2. *Let K be a class of algebras. Let $\mathfrak{A} \in K$ and $a \in A$. Then the mapping $e_0^1 \to a$ can be extended to a homomorphism of $\mathfrak{P}^{(1)}(K)$ into $\mathfrak{A}$. Let $\mathcal{O}_K(a)$ denote the induced congruence relation. Then $\mathcal{O}_K(a)$ will be called the* order of a in the class K.

Examples show (see Exercises) that by specializing this concept to groups, semigroups, and modules over rings, we get the known concepts used in these special cases.

Lemma 1. *Let $\mathfrak{A}, \mathfrak{B} \in K$, $a \in A$, $b \in B$. There exists a homomorphism φ of $\langle [a]; F\rangle$ into $\mathfrak{B}$ such that $a\varphi = b$, if and only if $\mathcal{O}_K(a) \leqq \mathcal{O}_K(b)$.*

Proof. Suppose that such a homomorphism φ exists and let φ_1 and φ_2 be the homomorphisms extending $e_0^1 \to a$ and $e_0^1 \to b$, respectively.

Then $\varphi_2 = \varphi_1\varphi$. Thus, if $u \equiv v(\Theta_K(a))$, that is, $u\varphi_1 = v\varphi_1$, then $u\varphi_2 = u\varphi_1\varphi = v\varphi_1\varphi = v\varphi_2$ and so $u \equiv v(\Theta_K(b))$. Conversely, if $\Theta_K(a) \leq \Theta_K(b)$, then the existence of such a homomorphism follows from the second isomorphism theorem (Theorem 11.4).

Definition 3. *Let K be a class of algebras, $\mathfrak{A} \in K$, $I \subseteq A$. The set I will be called a* weakly independent set of $\mathfrak{A}$ in K *if every mapping $\varphi: I \to B$, $\mathfrak{B} \in K$, satisfying the condition $\Theta_K(a) \leq \Theta_K(a\varphi)$, $a \in I$, can be extended to a homomorphism of $\langle [I]; F \rangle$ into $\mathfrak{B}$.*

In other words, if φ can be extended to a homomorphism, then, by Lemma 1, $\Theta_K(a) \leq \Theta_K(a\varphi)$; thus, weak independence means that if the extendibility of a mapping to a homomorphism is not ruled out by Lemma 1, then it can indeed be extended.

If we take the special case $K = \{\mathfrak{A}\}$ and I consists of elements a with $\Theta_K(a) = \omega$, then weak independence is equivalent to independence.

Theorem 3. *Let K be a class of algebras, $\mathfrak{A} \in K$, $I \subseteq A$. Then I is a weakly independent set of $\mathfrak{A}$ in K if and only if $a_0, \cdots, a_{k-1} \in I$, $b_0, \cdots, b_{k-1} \in B$, $\mathfrak{B} \in K$ and $\Theta_K(a_i) \leq \Theta_K(b_i)$, $i = 0, \cdots, k-1$, imply that whenever $p(a_0, \cdots, a_{k-1}) = q(a_0, \cdots, a_{k-1})$, then $p(b_0, \cdots, b_{k-1}) = q(b_0, \cdots, b_{k-1})$, where $p, q \in P^{(k)}(K)$, $k = 1, 2, \cdots$.*

Proof. Just as in Theorem 31.3.

Corollary. *The set I is weakly independent if and only if every finite subset of I is weakly independent.*

Of course, this is trivial directly from Definition 3 also. To formulate the analogue of Theorem 31.3 for weak independence, we need the following concept.

Definition 4. *Let $\Theta_0, \cdots, \Theta_{n-1}$ be congruence relations of $\mathfrak{P}^{(1)}(K)$. Set $A_i = \langle [e_i^n]; F \rangle$. Then $\mathfrak{A}_i \cong \mathfrak{P}^{(1)}(K)$. Thus, we can consider Θ_i as a congruence relation of $\mathfrak{A}_i$. We say that $\sum_{i=0}^{n-1} \Theta_i$ exists and equals Θ if there exists a congruence relation Θ of $\mathfrak{P}^{(n)}(K)$ which is the smallest with respect to the following two properties:*

(i) $\Theta_{A_i} \geq \Theta_i$;
(ii) *$\mathfrak{P}^{(n)}(K)/\Theta$ is isomorphic to a subalgebra of an algebra in K.*

Theorem 4. *Let $\mathfrak{A} \in K$, $a_0, \cdots, a_{n-1}, \in A$. Then $\{a_0, \cdots, a_{n-1}\}$ is weakly independent if and only if $\sum_{i=0}^{n-1} \Theta_K(a_i)$ exists and*

$$\mathfrak{P}^{(n)}(K)/\Sigma\Theta_K(a_i) \cong \langle [a_0, \cdots, a_{n-1}]; F \rangle.$$

Proof. Assume that $\{a_0, \cdots, a_{n-1}\}$ is weakly independent and let φ be the homomorphism of $\mathfrak{P}^{(n)}(K)$ into $\mathfrak{A}$ for which $e_i^n\varphi = a_i$, $i = 0, \cdots, n-1$. Let Θ be the congruence relation induced by φ. We claim that $\Theta = \Sigma\mathcal{O}_K(a_i)$. It is obvious that $\Theta_{A_i} \geqq \mathcal{O}_K(a_i)$; further,

$$\mathfrak{P}^{(n)}(K)/\Theta \cong \langle [a_0, \cdots, a_{n-1}]; F\rangle,$$

which is a subalgebra of $\mathfrak{A}$ in K.

Now let Φ be an arbitrary congruence relation such that $\Phi_{A_i} \geqq \mathcal{O}_K(a_i)$ and $\mathfrak{P}^{(n)}(K)/\Phi$ is isomorphic to a subalgebra of $\mathfrak{B} \in K$. Let b_i be the element which corresponds to

$$[e_i^n]\Phi$$

under this isomorphism. Then $\mathcal{O}_K(b_i) = \Phi_{A_i} \geqq \mathcal{O}_K(a_i)$. Thus, by weak independence, there exists a homomorphism χ of $\langle [a_0, \cdots, a_{n-1}]; F\rangle$ onto $\langle [b_0, \cdots, b_{n-1}]; F\rangle$ with $a_i\chi = b_i$. Then, if ψ denotes the homomorphism of $\mathfrak{P}^{(n)}(K)$ into $\mathfrak{B}$ with $e_i^n\psi = b_i$, then $\psi = \varphi\chi$. Thus, Φ, which equals the congruence relation induced by ψ, is greater than or equal to the congruence relation induced by φ, which is Θ. We conclude that $\Phi \geqq \Theta$, which was to be proved.

To prove the converse, let us assume that $\sum_{i=0}^{n-1} \mathcal{O}_K(a_i)$ exists and that $\mathfrak{P}^{(n)}(K)/\Sigma\mathcal{O}_K(a_i) \cong \langle [a_0, \cdots, a_{n-1}]; F\rangle$. Let $\mathfrak{B} \in K$, $b_0, \cdots, b_{n-1} \in B$, and $\mathcal{O}_K(a_i) \leqq \mathcal{O}_K(b_i)$, $i = 0, \cdots, n-1$. Let φ_1, φ_2 be the homomorphisms of $\mathfrak{P}^{(n)}(K)$ satisfying $e_i^n\varphi_1 = a_i$ and $e_i^n\varphi_2 = b_i$, $i = 0, \cdots, n-1$, respectively.

Let Θ_1, Θ_2 denote the induced congruence relations. Then, by Definition 4, $\Theta_1 = \Sigma\mathcal{O}_K(a_i)$. Since Θ_2 satisfies the conditions of Definition 4, we get that $\Theta_1 \leqq \Theta_2$. Thus, by the second isomorphism theorem (Theorem 11.4), the homomorphism $[x]\Theta_1 \to [x]\Theta_2$ establishes that $\{a_0, \cdots, a_{n-1}\}$ is weakly independent.

Some further results on weak independence will be given in the Exercises.

EXERCISES

1. (B. Jónsson and A. Tarski [2]) Let K be an equational class and $F(K)$ the class of finite algebras in K. If $Id(F(K)) \subseteq Id(K)$, then every n element generating set of $\mathfrak{F}_K(n)$ is a basis of $\mathfrak{F}_K(n)$.
2. Let R denote the set of real numbers. Show that independence in $\langle R; +\rangle$ is equivalent to linear independence of real numbers over the integers, and independence in $\langle R; +, \cdot\rangle$ is equivalent to the algebraic independence of real numbers.

3. Let $\langle V; +\rangle$ be a vector space over the field $\mathfrak{F}$ and let

$$\mathfrak{B} = \langle V; +, \{f_a \mid a \in F\}\rangle$$

be the algebra defined by $f_a(x) = a \cdot x$. Prove that independence in $\mathfrak{B}$ is the same as linear independence of vectors.
4. Describe independence in $\langle P(X); \cup\rangle$.
5. Describe independence in semilattices. Compare the results of Ex. 4 and 5.
6. Let $A \subseteq P(X)$. Prove that A is independent in the Boolean algebra $\mathfrak{P}(X)$ if and only if whenever $\{X_0, \cdots, X_{n-1}\} \subseteq A$, the sets $X_0^{i_0} \cap \cdots \cap X_{n-1}^{i_{n-1}}$ are nonvoid, where $i_0, \cdots, i_{n-1} = 0$ or 1, $X_k{}^0 = X_k$ and $X_k{}^1 = X_k'$.
7. When does $\mathfrak{P}(X)$ have a basis?
8. (W. Nitka [1]) If a finite algebra $\mathfrak{A}$ is generated by one self-dependent element, then every element of $\mathfrak{A}$ is self-dependent.
9. Is the statement of Ex. 8 true for infinite algebras?
10. (A. Goetz and C. Ryll-Nardzewski [1]) If an algebra $\mathfrak{A}$ has bases of one and n elements, $n > 1$, then $\mathfrak{A}$ can be generated by a self-dependent element.
11. (A. Goetz and C. Ryll-Nardzewski [1]) If the algebra $\mathfrak{A}$ has bases consisting of one and $n > 1$ elements, then for every $k \geqq 1$, there exists a minimal generating set of $\mathfrak{A}$ consisting of k self-dependent elements.
12. (J. Płonka [1]) The algebra $\mathfrak{A}$ has the EIS (exchange of independent sets) property if the conclusion of the Exchange Theorem holds with $[Y] = [Z]$ replaced by $Z \subseteq [Y]$. Prove that the algebra $\mathfrak{A} = \langle A; \cdot\rangle$ does not have the EIS property, where $A = \{0, 1, 2, 3, 4, 5, 6\}$, $0 \cdot 1 = 1 \cdot 0 = 3$, $0 \cdot 2 = 2 \cdot 0 = 4$, $1 \cdot 2 = 2 \cdot 1 = 5$ and $x \cdot y = 6$ otherwise. (Płonka also proves that this result is the best possible: every algebra $\mathfrak{A}$ with $|A| \leqq 6$ has property EIS.)
13. (J. Płonka [1]) Every unary algebra has the EIS property.
14. (J. Płonka [1]) Every semilattice has the EIS property.
15. (J. Płonka [1]) Let $A_0, \cdots, A_{n-1}$ be nonvoid sets, $A = A_0 \times \cdots \times A_{n-1}$. Define the algebra $\mathfrak{A} = \langle A; d\rangle$ of type $\langle n\rangle$ by

$$d(\langle a_0{}^0, \cdots, a_{n-1}^0\rangle, \cdots, \langle a_0^{n-1}, \cdots, a_{n-1}^{n-1}\rangle) = \langle a_0{}^0, a_1{}^1, \cdots, a_{n-1}^{n-1}\rangle.$$

Prove that $\mathfrak{A}$ has a basis if and only if $|A_0| = \cdots = |A_{n-1}|$.
16. (J. Płonka [1]) An *n-dimensional diagonal algebra* $\mathfrak{A} = \langle A, d\rangle$ is an algebra of type $\langle n\rangle$ satisfying $d(a, \cdots, a) = a$ and

$$d(d(a_0{}^0, \cdots, a_{n-1}^0), \cdots, d(a_0^{n-1}, \cdots, a_{n-1}^{n-1})) = d(a_0{}^0, a_1{}^1, \cdots, a_{n-1}^{n-1}).$$

Prove that every n-dimensional diagonal algebra is isomorphic to an algebra constructed in Ex. 15. (The case $n = 2$ is due to N. Kimura.)
17. (A. Goetz and C. Ryll-Nardzewski [1]) Let $\mathfrak{A} = \langle A; f, g_0, \cdots, g_{n-1}\rangle$ be the free algebra, with the free generator x, over the equational class defined by

$$\mathbf{f}(\mathbf{g}_0(\mathbf{x}_0), \cdots, \mathbf{g}_{n-1}(\mathbf{x}_0)) = \mathbf{x}_0$$

and

$$\mathbf{g}_i(\mathbf{f}(\mathbf{x}_0, \cdots, \mathbf{x}_{n-1})) = \mathbf{x}_i, \quad 0 \leqq i < n.$$

Prove that $\{a_0, \cdots, a_{k-1}\}$ is a basis of $\mathfrak{A}$ if and only if

$$\{g_0(a_0), \cdots, g_{n-1}(a_0), a_1, \cdots, a_{k-1}\}$$

is a basis of $\mathfrak{A}$.

18. (A. Goetz and C. Ryll-Nardzewski [1]) Let us call the bases X and Y of $\mathfrak{A}$ (in Ex. 17) *equivalent* if one can get Y from X by applying the construction of Ex. 17 a finite number of times. Show that every basis X of $\mathfrak{A}$ is equivalent to a homogeneous basis (A *homogeneous basis* is the set of *all* elements of the form $g_{i_0} \cdots g_{i_{k-1}}(x)$, with a fixed k.)
19. (A. Goetz and C. Ryll-Nardzewski [1]) Prove that for every $m = 1 + s(n-1)$, $\mathfrak{A}$ (of Ex. 17) has a basis of m elements and, conversely, if $\mathfrak{A}$ has a basis of m elements, then $m = 1 + s(n-1)$ for some s.
20. (A. Goetz and C. Ryll-Nardzewski [1]) Let $0 < k < n$ and $\mathfrak{A}$ the algebra of Ex. 17. Prove that $\mathfrak{P}^{(k)}(\mathfrak{A})$ has a basis of m elements if and only if $m = k + s(n-1)$ for some nonnegative integer s.
21. (E. Marczewski [3]) Let X and Y be independent in $\mathfrak{A}$. Then $[X] \cap [Y] = [X \cap Y]_{\mathscr{S}^+}$.
22. (E. Marczewski [5]) Let X be a basis of $\mathfrak{A}$. Prove that if $a \in A$ and a is not constant in $\mathfrak{A}$, then a has a representation $a = p(a_0, \cdots, a_{n-1})$, $a_0, \cdots, a_{n-1} \in X$, $p \in P^{(n)}(\mathfrak{A})$, $n \geqq 1$, such that (i) p depends on all its variables, (ii) $\{a_0, \cdots, a_{n-1}\}$ is unique and (iii) p is unique up to the permutation of variables.
23. (E. Marczewski [3]) Let $\langle V; +\rangle$ be a vector space over the field $\langle F; +, \cdot, 0\rangle$ and let $\langle W; +\rangle$ be a subspace of $\langle V; +\rangle$. Set

$$G = \{f_a \mid a \in F\} \cup \{g_w \mid w \in W\},$$

where $f_a(x) = ax$ and $g_w(x) = x + w$. Prove that $\langle V, G\rangle$ is a v-algebra.
24. (K. Urbanik [1]) Let $\mathfrak{G}$ be a group of mappings of the set A into itself. Let us assume that every $\alpha \in G$ has at most one fixed point if $\alpha \neq \varepsilon$ (the identity) and let B be a subset of A containing all the fixed points such that $B\alpha \subseteq B$ for all $\alpha \in G$. Let $T = \{t_\alpha \mid \alpha \in G\} \cup \{g_w \mid w \in B\}$, where $t_\alpha(x) = x\alpha$ and g_w is nullary, $g_w = w$. Prove that $\langle A; T\rangle$ is a v-algebra.
25. (W. Narkiewicz [1]) Take the algebra $\mathfrak{A}$ of Ex. 24 without assuming that every $\alpha \neq \varepsilon$ has at most one fixed point. Prove that $\mathfrak{A}$ is a v^*-algebra and if there is an $\alpha \in G$, $\alpha \neq \varepsilon$ with at least two fixed points, then $\mathfrak{A}$ is not a v-algebra.
26. (W. Narkiewicz [3]) Let $\mathfrak{R} = \langle R; +, \cdot, 0\rangle$ be a ring with identity 1 and without zero-divisors, such that for every $a, b \in R$, there exists $c \in R$ with $a = cb$ or $b = ca$. Let $\langle A; +\rangle$ be a left-module over $\mathfrak{R}$ and $\langle A_0; +\rangle$ a submodule of $\langle A; +\rangle$ with the property that every $a \in A_0$ is left-divisible by every $r \in R$, $r \neq 0$. Let T be the set of all n-ary operations ($n \geqq 0$) f of the form

$$f(x_0, \cdots, x_{n-1}) = \sum_{i=0}^{n-1} \lambda_i x_i + a, \qquad \lambda_0, \cdots, \lambda_{n-1} \in R, \quad a \in A_0.$$

Then $\mathfrak{A} = \langle A; T\rangle$ is a v^{**}-algebra.

27. The algebra $\mathfrak{A}$ of Ex. 26 is a v^*-algebra if and only if $\mathfrak{R}$ is a division-ring.
28. (W. Narkiewicz [3]) Define $T_0 \subseteq T$ (as in Ex. 26) by

$$f(x_0, \cdots, x_{n-1}) = \sum_{i=0}^{n-1} \lambda_i x_i + a \in T_0 \quad \text{if} \quad \sum_{i=0}^{n-1} \lambda_i = 1.$$

Prove that $\mathfrak{A}_0 = \langle A; T_0 \rangle$ is a v^{**}-algebra.
29. The algebra $\mathfrak{A}_0$ of Ex. 28 is a v^*-algebra if and only if $\mathfrak{R}$ is a division-ring.
30. (G. H. Wenzel) $\mathfrak{A}$ is a *v'-algebra* if $\mathfrak{A}$ is a v^{**}-algebra in which every maximally independent set is a basis. Prove that a v^{**}-algebra $\mathfrak{A}$ is a v'-algebra if and only if every independent set is contained in a basis.
31. (G. H. Wenzel) A v'-algebra $\mathfrak{A}$ is a v^*-algebra if and only if every finitely generated subalgebra of $\mathfrak{A}$ is a v'-algebra.
32. (G. H. Wenzel) A v^{**}-algebra $\mathfrak{A}$ is a v^*-algebra if and only if whenever $\mathfrak{T}$ and $\mathfrak{U}$ are finitely generated subalgebras of $\mathfrak{A}$ satisfying $T \subseteq U$ and $\mathfrak{T} \cong \mathfrak{U}$, then $T = U$.
33. (G. H. Wenzel) Let $\mathfrak{A}$ be a v'-algebra. Let $\mathfrak{T}$ and $\mathfrak{U}$ be finitely generated subalgebras of $\mathfrak{A}$, with bases T_1 and U_1, respectively, satisfying $T \subseteq U$ and $\mathfrak{T} \cong \mathfrak{U}$. Let $T_1 \cup C$ be a basis of $\mathfrak{A}$. Then $U_1 \cup C$ is also a basis of $\mathfrak{A}$.
34. (K. Urbanik [7] and G. H. Wenzel) $\mathfrak{A}$ is a v^*-algebra if and only if $\mathfrak{A}$ is a v'-algebra.
35. Let $\mathfrak{A}$ be a finite algebra. $\mathfrak{A}$ has a basis if and only if $i = g$.
36. Let U be a finite unary algebra. Then $i_* = 0$, 1, or $|A|$.
37. If $i_* \geq 1$ in the finite algebra $\mathfrak{A}$, then $\mathfrak{A}$ is *idempotent*, i.e., $f(x_0, \cdots, x_0) = x_0$ for every operation f. (In other words, $\mathfrak{P}^{(1)}(\mathfrak{A}) = \{e_0^1\}$).
38. (G. H. Wenzel†) Let $\mathfrak{A}$ be a finite v^{**}-algebra. Then $i_* \geq 1$ if and only if $\mathfrak{A}$ is idempotent.
39. Let $\mathfrak{A}$ be a finite algebra, $\mathfrak{B}$ a subalgebra of $\mathfrak{A}$, $B \neq A$. Then $g^* \geq |B| + 1$.
40. Set $A = \{0, 1, \cdots, n-1\}$, $\mathfrak{A} = \langle A; + \rangle$, where $+$ is addition modulo n. Show that $g^*(\mathfrak{A}) = d(n) + 1$, $i_* = 0$, $p = -1$. ($d(n)$ is the greatest divisor of n which is different from n.)
41. Prove that for $\mathfrak{P}(I)$, $g^* = 2^{n-1} + 1$, $g = k + 1$, $i = k$ if $|I| = n$, $2^k < n < 2^{k+1}$; and if $n = 2^k$, then $g^* = 2^{n-1} + 1$, $g = k$, $i = k$.
42. Set $P = \{a, b\}$ and define the ternary operations p_* and p^* by

$$p_*(x_0, x_1, x_1) = p_*(x_1, x_0, x_1) = p_*(x_1, x_1, x_0) = x_0,$$
$$p^*(x_0, x_1, x_1) = p^*(x_1, x_0, x_1) = p^*(x_1, x_1, x_0) = x_1.$$

The *Post-algebras* are $\mathfrak{P} = \langle P; p_*, p^* \rangle$, $\mathfrak{P}_* = \langle P; p_* \rangle$ and $\mathfrak{P}^* = \langle P; p^* \rangle$. (See Post [1]; also Marczewski and Urbanik [2].) Show that $g^* = g = i = i_* = p = 2$ for all three Post-algebras.
43. (S. Świerczkowski [4]) Set $S = \{a, b, c, d\}$, $\mathfrak{S} = \langle S; f \rangle$, where f is the ternary operation which associates with every three distinct elements of S the remaining one and $f = p_*$ of Ex. 42 otherwise. Show that $g^* = g = i = i_* = 3$ and $p = 2$; thus Theorem 33.2 does not hold for $n = 3$.

† Wenzel's results are taken from his Ph.D. Thesis, Pennsylvania State University, 1966. The results are being published in G. H. Wenzel [2] and [3].

44. (E. Marczewski [6]) Set $A=\{0,\cdots,n-1\}$, $\mathfrak{A}=\langle A;f\rangle$, where f is an $(n-1)$-ary operation which associates with $n-1$ distinct elements the remaining one and $f(x_0,\cdots,x_{n-2})=x_0$ otherwise. Compute g^*, g, i, i_* and p.
45. Let $\langle B; +,\cdot, 0, 1\rangle$ be a finite Boolean ring, $|B|=2^n$, and $\mathfrak{B}=\langle B; +\rangle$. Then $g^*=2^{n-1}+1$, $g=i=n$, $i_*=0$ and $p=-1$ for $\mathfrak{B}$.
46. (G. H. Wenzel) Set $\mathfrak{B}'=\langle B; g\rangle$, where B is as in Ex. 45 and $g(x_0,x_1,x_2)=x_0+x_1+x_2$. Show that for $\mathfrak{B}'$, $g^*=2^{n-1}+1$, $g=i=n+1$, $i_*=\min\{3, 2^n\}$, $p=2$.
47. Show that for $\mathfrak{P}'(I)=\langle P(I)-\{I\}; \cap\rangle$, $g^*=2^n-1$, $g=i=n$, $i_*=p=1$, where $n=|I|$.
48. (G. H. Wenzel) Let $\mathfrak{Z}_q$ be the cyclic group with q elements (q is a prime), $Z=(Z_q)^n$ and $\mathfrak{Z}=\langle Z; g\rangle$, where $g(x_0,\cdots,x_q)=x_0+\cdots+x_q$. Show that for $\mathfrak{Z}$, $g^*=q^{n-1}+1$, $g=i=n+1$, $i_*=2$, $p=1$.
49. Let $\mathfrak{A}$ be a finite algebra with $i_*\geqq 1$. Then $|[a]|=|[b]|$ for all $a, b\in A$.
50. Let $\mathfrak{A}$ be a finite unary algebra with $i_*\geqq 1$. Then $g^*=|A|-|[a]|+1$, for any $a\in A$.
51. (G. H. Wenzel) Prove that $\langle n, m\rangle$ can be represented as $\langle |A|, i_*(\mathfrak{A})\rangle$ for some finite algebra $\mathfrak{A}$ if and only if $n\geqq m\geqq 0$ and $\langle n, m\rangle\neq\langle 0, 0\rangle$.
52. (S. Świerczkowski [6]) Show that the pair $\langle n, m\rangle$ can be represented as $\langle i_*(\mathfrak{A}), p(\mathfrak{A})\rangle$ for some finite nontrivial algebra $\mathfrak{A}$ if and only if $n=m\geqq 0$ or $\langle n, m\rangle\in\{\langle 3, 2\rangle, \langle 2, 1\rangle, \langle 1, 0\rangle, \langle 0, -1\rangle\}$.
53. (K. Urbanik [3]) Show that Theorem 33.1 is false for infinite algebras. (Hint: Let N be the set of positive integers, $n>3$ and $f\colon A^n\to A$ satisfying $f(i_0,\cdots,i_{n-1})>i_j$, $j=0, 1,\cdots,n-1$, and take the algebra $\langle N; f\rangle$.)
54. Let $\langle G; +, 0\rangle$ be an abelian group, and let $H\subseteq G$. Given necessary and sufficient condition for the local independence of H.
55. Let $\langle R; +,\cdot, 0, 1\rangle$ be a ring with identity. Show that $\{a\}$ is locally independent for every $a\in R$.
56.† Let K be the class of groups, $\mathfrak{G}\in K$, $a\in G$. Show that $\mathcal{O}_K(a)$ is uniquely determined by $|[a]|$, which is called the order of a in group theory.
57. Lee K be the class of semigroups, $\mathfrak{S}\in K$, $a\in S$. Show that $\mathcal{O}_K(a)$ can be described by a pair $\langle n, m\rangle$ of nonnegative integers.
58. Interpret $\mathcal{O}_K(a)$, if $a\in A$, and $\mathfrak{A}$ is a left-module over a ring $\mathfrak{R}$, and K is the class of all left $\mathfrak{R}$-modules.
59. Let K be a class of algebras, $\mathfrak{A}, \mathfrak{B}, \mathfrak{C}\in K$, $\mathfrak{C}=\mathfrak{A}\times\mathfrak{B}$, $a\in A$, $b\in B$. Then $\mathcal{O}_K(\langle a, b\rangle)=\mathcal{O}_K(a)\wedge\mathcal{O}_K(b)$.
60. Let K be a class of algebras satisfying $\mathbf{H}(K)\subseteq K$ and $\mathbf{S}(K)\subseteq K$. Let $a_0,\cdots,a_{n-1}\in A$, $\mathfrak{A}\in K$. Then $\sum_{i=0}^{n-1}\mathcal{O}_K(a_i)$ exists.
61. For every integer n, there exists a class K of algebras and $\mathfrak{A}\in K$, such that $\mathfrak{A}$ has a *weak basis* (i.e., a generating set which is weakly independent) of k elements if and only if $k\leqq n$.
62. Show that weak independence does not imply $\mathscr{S}^+$-independence.
63.‡ $\mathfrak{A}$ is a *2-algebra* if $2=g^*=g=i=i_*$, i.e., every two element subset of A

† For Ex. 56–62, see G. Grätzer [6].

‡ For Ex. 63–66, see G. Grätzer [2].

is a basis of $\mathfrak{A}$. Show that $\mathfrak{A}$ is a 2-algebra if and only if $\langle A; P^{(2)}(\mathfrak{A})\rangle$ is a 2-algebra.

64. Let $\mathfrak{A}$ be an algebra which is generated by any two distinct elements and let $\mathfrak{G}$ be the automorphism group of $\mathfrak{A}$. If $\mathfrak{A}$ is a 2-algebra, then $\mathfrak{G}$ is *doubly-transitive* and *minimal* (i.e., for $a, b, c, d \in A$, $a \neq b$, $c \neq d$, there exists exactly one $\alpha \in G$ with $a\alpha = c$ and $b\alpha = d$). If $\mathfrak{G}$ is doubly-transitive and $|A| > 2$, then $\mathfrak{A}$ is a 2-algebra. (See also S. Świerczkowski [4].)

65. Let $\langle A; -, \cdot, 0, 1\rangle$ be an algebra satisfying the following axioms: (i) $\langle A; \cdot, 0\rangle$ is a semigroup with 0 as a zero; (ii) $\langle A - \{0\}; \cdot, 1\rangle$ is a group; (iii) $a - 0 = a$; (iv) $a(b - c) = ab - ac$; (v) $a - (b - c) = a$ if $a = b$, and $a - (b - c) = (a - b) - (a - b)(b - a)^{-1}c$ if $a \neq b$ (if $a \neq b$, then $b - a \neq 0$ and so by (ii), $(b - a)^{-1}$ is the inverse of $(b - a)$). Let F be the set of all operations f of the form $f(x_0, x_1) = x_0 - (x_0 - x_1)a$, $a \in A$. Then $\mathfrak{A} = \langle A; F\rangle$ is a 2-algebra. (For further results on algebras satisfying (i)–(v), see V. D. Belousov [1].)

66. Prove that the algebra $\mathfrak{A}$ of Ex. 65 is a v^*-algebra and it is a v-algebra if and only if for some $u \in A$, $\langle A; +, \cdot, 0, 1\rangle$ is a division ring, where $x_0 + x_1 = x_0 - x_1 \cdot u$.

67. (E. Narkiewicz [3]) Prove that a finite v^{**}-algebra is a v^*-algebra.

68. Let us say that the *Chinese remainder theorem* holds in the algebra $\mathfrak{A}$ if for any $\Theta_0, \cdots, \Theta_{n-1} \in C(\mathfrak{A})$, $a_0, \cdots, a_{n-1} \in A$, the condition $a_i \equiv a_j(\Theta_i \vee \Theta_j)$ for $0 \leqq i, j < n$ implies the existence of an $a \in A$ with $a \equiv a_i(\Theta_i)$, $i = 0, \cdots, n-1$. Prove that the Chinese remainder theorem holds in the algebra $\mathfrak{A}$ if and only if $\mathfrak{C}(\mathfrak{A})$ is distributive and the congruence relations permute (i.e., $\Theta \vee \Phi = \Theta\Phi$ for $\Theta, \Phi \in C(\mathfrak{A})$). (For rings this is stated in O. Zariski and P. Samuel, Commutative Algebra, vol. 1, 1958, Van Nostrand, Princeton, N.J., p. 279.)

69. (A. F. Pixley [1]) Let K be an equational class. For every $\mathfrak{A} \in K$, $\mathfrak{C}(\mathfrak{A})$ is distributive and the congruences of $\mathfrak{A}$ permute if and only if there are ternary polynomial symbols $\mathbf{p}$ and $\mathbf{q}$ such that the following identities hold in K: $\mathbf{p}(\mathbf{x}_0, \mathbf{x}_0, \mathbf{x}_2) = \mathbf{x}_2$, $\mathbf{p}(\mathbf{x}_0, \mathbf{x}_2, \mathbf{x}_2) = \mathbf{x}_0$, $\mathbf{q}(\mathbf{x}_0, \mathbf{x}_0, \mathbf{x}_2) = \mathbf{x}_0$, $\mathbf{q}(\mathbf{x}_0, \mathbf{x}_1, \mathbf{x}_0) = \mathbf{x}_0$ and $\mathbf{q}(\mathbf{x}_0, \mathbf{x}_0, \mathbf{x}_2) = \mathbf{x}_0$.

70. (B. Jónsson [8]) Let K be an equational class. $\mathfrak{C}(\mathfrak{A})$ is distributive for all $\mathfrak{A} \in K$ if and only if for some integer $n \geqq 2$, there exist ternary polynomial symbols $\mathbf{p}_0, \cdots, \mathbf{p}_n$ such that the following identities hold in K:

$$\mathbf{p}_0(\mathbf{x}_0, \mathbf{x}_1, \mathbf{x}_2) = \mathbf{x}_0, \ \mathbf{p}_n(\mathbf{x}_0, \mathbf{x}_1, \mathbf{x}_2) = \mathbf{x}_2, \qquad \mathbf{p}_i(\mathbf{x}_0, \mathbf{x}_1, \mathbf{x}_0) = \mathbf{x}_0,$$

$$\mathbf{p}_i(\mathbf{x}_0, \mathbf{x}_0, \mathbf{x}_1) = \mathbf{p}_{i+1}(\mathbf{x}_0, \mathbf{x}_0, \mathbf{x}_1) \quad \text{for } i \text{ even},$$

$$\mathbf{p}_i(\mathbf{x}_0, \mathbf{x}_1, \mathbf{x}_1) = \mathbf{p}_{i+1}(\mathbf{x}_0, \mathbf{x}_1, \mathbf{x}_1) \quad \text{for } i \text{ odd}.$$

71. Let $\mathfrak{A}$ be a finite algebra and K the equational class generated by $\mathfrak{A}$. Is there an effective process which decides whether or not $\mathfrak{C}(\mathfrak{B})$ is distributive for all $\mathfrak{B} \in K$?

72. Let K be an equational class and π a partition of positive integers such that i and j are in the same block if and only if $\mathfrak{F}_K(i) \cong \mathfrak{F}_K(j)$. Characterize the partitions of positive integers which we get this way.

PROBLEMS

49. Characterize those algebras which have one or both of the following two properties: (A) every maximally independent set is a minimal generating set; (B) every minimal generating set is a maximally independent set.
50. Characterize the six-tuples which can be represented as

$$\langle |A|, g^*(\mathfrak{A}), g(\mathfrak{A}), i(\mathfrak{A}), i_*(\mathfrak{A}), p(\mathfrak{A})\rangle$$

for some finite algebra $\mathfrak{A}$.
51. Characterize the six-tuple of Problem 50 for v-, v^*-, and v^{**}-algebras.
52. Characterize the system of independent sets in an algebra.
53. Characterize the system of locally independent sets in an algebra.
54. Characterize the system of weakly independent sets in an algebra.
55. Characterize the set of cardinal numbers of weak bases of an algebra. Can an algebra $\langle A; F\rangle$ without algebraic constants have a finite and an infinite weak basis? Does $|F| < \aleph_0$ change the situation?
56. Let K be an equational class of algebras, each of which has a one-element subalgebra determined by a nullary operation, and for $\{a_0, \cdots, a_{n-1}\} \subseteq A$, $\mathfrak{A} \in K$, $\{a_0, \cdots, a_{n-1}\}$ is weakly independent if and only if

$$\langle [a_0, \cdots, a_{n-1}]; F\rangle \cong \langle [a_0]; F\rangle \times \cdots \times \langle [a_{n-1}]; F\rangle.$$

Is K equivalent to the class of all $\mathfrak{R}$-modules? If not, what additional conditions are needed?
57. Let $\langle A; -, \cdot, 0, 1\rangle$ be an algebra as in Ex. 65. Is it always possible to find a near-field $\langle A; +, -, \cdot, 0, 1\rangle$? (See G. Grätzer [2], Theorems 3 and 4, and V. D. Belousov [1].)
58. Find necessary and sufficient conditions on the finite algebra $\mathfrak{A}$, under which $\mathbf{IP}_S\mathbf{HS}(\mathfrak{A})$ is the equational class generated by $\mathfrak{A}$.
59. Let $\mathfrak{B}$, $\mathfrak{B}^*$, and $\mathfrak{B}^{**}$ denote the class of all v-, v^*-, and v^{**}-algebras, respectively. Find all operators $\mathbf{X}$ for which $\mathbf{X}(K) \subseteq K$, where $K = \mathfrak{B}$, $\mathfrak{B}^*$, or $\mathfrak{B}^{**}$.
60. Define and discuss $|A|$, g^*, $\cdots$, p for infinite algebras.
61. Define and discuss $|A|$, g^*, $\cdots$, p for infinitary algebras and relate these to the characteristic.

CHAPTER 6

ELEMENTS OF MODEL THEORY

In this chapter we will develop a language which can describe the so-called elementary or first order properties of algebras. It will be useful to consider structures, that is, sets on which relations as well as operations are defined. We use a semantical approach, that is, our discussion will be based on the concept of satisfiability of a formula in a structure. Sections 36 and 37 contain the rudiments of first order logic along with a rather long illustration of satisfiability in Boolean set algebras, which is not only instructive but also very useful in applications (§47). Elementary extension and elementary equivalence play a very important role; they are discussed in §38. The elementary properties of prime products and applications of prime products are presented in §§39–41. Classes of algebras defined by first order sentences are discussed in §42. Many algebraic applications of the results of this chapter will be given in Chapters 7 and 8.

§36. STRUCTURES AND THE FIRST ORDER LOGIC

A *type* of structures τ is a pair

$$\langle\langle n_0, \cdots, n_\gamma, \cdots\rangle_{\gamma < o_0(\tau)}, \langle m_0, \cdots, m_\gamma, \cdots\rangle_{\gamma < o_1(\tau)}\rangle,$$

where $o_0(\tau)$ and $o_1(\tau)$ are ordinals, n_γ and m_γ are nonnegative integers. For every $\gamma < o_0(\tau)$ we have a symbol $\mathbf{f}_\gamma$ of an n_γ-ary operation and for every $\gamma < o_1(\tau)$ we have a symbol $\mathbf{r}_\gamma$ of an m_γ-ary relation.

A *first order structure*, or simply, *structure* $\mathfrak{A}$ is a triplet $\langle A; F, R\rangle$, where A is a nonvoid set, for every $\gamma < o_0(\tau)$ we realize $\mathbf{f}_\gamma$ as an n_γ-ary operation $(\mathbf{f}_\gamma)_\mathfrak{A}$ on A, for every $\gamma < o_1(\tau)$ we realize $\mathbf{r}_\gamma$ as an m_γ-ary relation $(\mathbf{r}_\gamma)_\mathfrak{A}$ on A, and

$$F = \langle(\mathbf{f}_0)_\mathfrak{A}, \cdots, (\mathbf{f}_\gamma)_\mathfrak{A}, \cdots\rangle, \qquad \gamma < o_0(\tau),$$
$$R = \langle(\mathbf{r}_0)_\mathfrak{A}, \cdots, (\mathbf{r}_\gamma)_\mathfrak{A}, \cdots\rangle, \qquad \gamma < o_1(\tau).$$

If there is no danger of confusion, we will write f_γ for $(\mathbf{f}_\gamma)_\mathfrak{A}$ and r_γ for $(\mathbf{r}_\gamma)_\mathfrak{A}$.

The *order* of τ is $o_0(\tau)+o_1(\tau)$. $K(\tau)$ will denote the class of all structures of type τ.

If $o_1(\tau)=0$, that is, there are no relations, we call $\mathfrak{A}$ an *algebra* and if $o_0(\tau)=0$, that is, there are no operations, then we call $\mathfrak{A}$ a *relational system*. (In other words, we identify $\langle A; F, \varnothing\rangle$ with $\langle A; F\rangle$ and $\langle A; \varnothing, R\rangle$ with $\langle A; R\rangle$.)

Usually, structures will be denoted by capital German letters while the corresponding capital italic letters will denote the sets on which they are defined.

If we say "let $\mathfrak{A}$ and $\mathfrak{B}$ be structures", or "let K be a class of structures", or "let $\mathfrak{A}_i$, $i \in I$, be structures" it will always be assumed that all the structures considered are of a fixed type τ, unless we explicitly state that this is not the case.

The basic concepts of universal algebras (subalgebra, homomorphism, and so on) can be generalized for structures. We now briefly review these, and we leave it to the reader to verify that many properties of these concepts which were established for universal algebras in Chapters 1–3, hold also for structures.

Let $\mathfrak{A}$ and $\mathfrak{B}$ be structures. $\mathfrak{B}$ is called a *substructure of* $\mathfrak{A}$ if $\langle B; F\rangle$ is a subalgebra of $\langle A; F\rangle$ and $(\mathbf{r}_\gamma)_{\mathfrak{B}}$ is the restriction of $(\mathbf{r}_\gamma)_{\mathfrak{A}}$ to B. If $\mathfrak{B}$ is a substructure of $\mathfrak{A}$, then $\mathfrak{A}$ is called an *extension* of $\mathfrak{B}$. If $\mathfrak{A}$ is a structure, $B \subseteq A$, $\langle B; F\rangle$ a subalgebra of $\langle A; F\rangle$, then $\mathfrak{B}$ will always denote the corresponding substructure. In particular, if $\mathfrak{A}$ is a relational system, then for every $\varnothing \neq B \subseteq A$, we have a substructure $\mathfrak{B}$.

If φ is a mapping of A into B, then φ is called a *homomorphism* of $\mathfrak{A}$ into $\mathfrak{B}$ if it is a homomorphism of $\langle A; F\rangle$ into $\langle B; F\rangle$ and $r_\gamma(a_0, \cdots, a_{m_\gamma - 1})$, $a_0, \cdots, a_{m_\gamma - 1} \in A$, imply that $r_\gamma(a_0\varphi, \cdots, a_{m_\gamma - 1}\varphi)$. If φ is onto and $r_\gamma(b_0, \cdots, b_{m_\gamma - 1})$, $b_0, \cdots, b_{m_\gamma - 1} \in B$, imply the existence of $a_0, \cdots, a_{m_\gamma - 1} \in A$ such that $r_\gamma(a_0, \cdots, a_{m_\gamma - 1})$ and $a_0\varphi = b_0, \cdots, a_{m_\gamma - 1}\varphi = b_{m_\gamma - 1}$, then $\mathfrak{B}$ is a *homomorphic image* of $\mathfrak{A}$.

The structure $\mathfrak{A} = \langle A; F, R\rangle$ is an *expansion* of $\mathfrak{A}_1 = \langle A; F, R_1\rangle$ if $(\mathbf{r}_\gamma)_{\mathfrak{A}} \supseteq (\mathbf{r}_\gamma)_{\mathfrak{A}_1}$ for all $\gamma < o_1(\tau)$. In other words, the base sets and the operations are identical, but the relations are expanded.

$\mathfrak{B}$ is an *expanded homomorphic image* of $\mathfrak{A}$, if $\mathfrak{B}$ is an expansion of a homomorphic image of $\mathfrak{A}$. This means simply that there is a homomorphism φ of $\mathfrak{A}$ *onto* $\mathfrak{B}$.

φ is an *isomorphism* of $\mathfrak{A}$ and $\mathfrak{B}$ if $\mathfrak{B}$ is a homomorphic image of $\mathfrak{A}$ with respect to φ and φ is 1–1.

φ is an *embedding* of $\mathfrak{A}$ into $\mathfrak{B}$ if φ is a homomorphism and $\mathfrak{A}$ is isomorphic under φ to the image of $\mathfrak{A}$ in $\mathfrak{B}$.

The *direct product* $\mathfrak{A}$ of the structures $\mathfrak{A}_i$, $i \in I$ (denoted by $\prod(\mathfrak{A}_i \mid i \in I)$) is defined on the set $\prod(A_i \mid i \in I)$, and operations and relations are defined componentwise; in particular, for $p_0, \cdots, p_{m_\gamma - 1} \in \prod(A_i \mid i \in I)$, $(\mathbf{r}_\gamma)_{\mathfrak{A}}(p_0, \cdots, p_{m_\gamma - 1})$ if and only if $(\mathbf{r}_\gamma)_{\mathfrak{A}_i}(p_0(i), \cdots, p_{m_\gamma - 1}(i))$ for all $i \in I$. The mappings $e_i^I\colon p \to p(i)$ $(p \in \prod(A_i \mid i \in I))$ are the *projections*; e_i^I is a homomorphism of $\mathfrak{A}$ onto $\mathfrak{A}_i$, but $\mathfrak{A}_i$ is not necessarily a homomorphic image of $\mathfrak{A}$.

Inverse limits are defined the same way as for algebras, as a suitable substructure of the direct product.

If $\mathfrak{B}$ is a substructure of $\prod(\mathfrak{A}_i \mid i \in I)$ and $Be_i{}^I = A_i$ for all $i \in I$, then $\mathfrak{A}$ is a *subdirect product* of the $\mathfrak{A}_i$, $i \in I$.

Let $\langle I; \leqq \rangle$ be a directed partially ordered set, and let $\mathfrak{A}_i$ be a structure for $i \in I$. Let us assume that if $i \leqq j$, $i, j \in I$, then $\mathfrak{A}_i$ is a substructure of $\mathfrak{A}_j$. Set $A = \bigcup (A_i \mid i \in I)$. Then we can define the operations and relations on A in the natural way. The resulting structure $\mathfrak{A}$ is the *union* of the $\mathfrak{A}_i$, $i \in I$. (This is a special case of a 1–1 *direct limit*, which is defined similarly.)

The simplest properties of algebras are the identities discussed in Chapter 4. We get more complicated properties by combining them using the connectives "and", "or", "if $\cdots$ then $\cdots$", and the prefixes "not", "for all x" and "there exists an x". To formalize this we introduce the language $L(\tau)$, which can be used for structures of type τ.

Definition 1. *The* first order language $L(\tau)$ *contains the following objects or symbols:*

(i) *(individual) variables* $\mathbf{x}_0, \mathbf{x}_1, \cdots, \mathbf{x}_n, \cdots$ *for* $n < \omega$;

(ii) *operational symbols* $\mathbf{f}_\gamma$ *for all* $\gamma < o_0(\tau)$;

(iii) *relational symbols* $\mathbf{r}_\gamma$ *for all* $\gamma < o_1(\tau)$;

(iv) *symbol for equality*, $=$;

(v) *logical connectives*, $\vee$ (or), $\neg$ (*not*), *and the quantifier* $\exists$ (*existential quantifier*);

(vi) *parentheses* ( ) *and commas.*

(v) is obviously a very scanty list, for it does not contain "and", "implies" and "for all x". The reason for this is simple: these can be expressed in terms of the other listed connectives. However, it would be inconvenient not to have these additional logical connectives. Therefore they will be introduced in Definition 4. The list in (v) can be made even more scanty by replacing $\vee$ and $\neg$ by the single symbol $\mid$ called the *Sheffer stroke*, which stands for "neither $\cdots$, nor $\cdots$". This would make some of the proofs shorter, but intuitively less clear.

We will also use $\mathbf{x}, \mathbf{y}, \mathbf{z}, \cdots$ to stand for individual variables for the sake of simplicity. These should always be interpreted as some $\mathbf{x}_i$ different from the ones which occur in a given context.

We will use polynomial symbols $\mathbf{p}$ as defined in §8, but we note that in logic they are called *terms*. If $\mathbf{p}$ is built up from $\mathbf{x}_{i_0}, \cdots, \mathbf{x}_{i_{n-1}}$, we will indicate this by writing $\mathbf{p}(\mathbf{x}_{i_0}, \cdots, \mathbf{x}_{i_{n-1}})$ and we call $\mathbf{x}_{i_0}, \cdots, \mathbf{x}_{i_{n-1}}$ the *variables* of $\mathbf{p}$.

It is very important that we have (iv). Sometimes, to emphasize that we have it, the system developed here is called "first order logic with equality". However, some authors prefer not to have (iv); see, e.g., A. Robinson [2].

Next we define the atomic formulas of our language.

Definition 2. *The atomic formulas of $L(\tau)$ are the following:*

(i) *If* $\mathbf{p}$ *and* $\mathbf{q}$ *are polynomial symbols, then* $\mathbf{p}=\mathbf{q}$ *is an atomic formula;*
(ii) $\mathbf{r}_\gamma(y_0, \cdots, y_{m_\gamma - 1})$ *is an atomic formula for all* $\gamma < o_1(\tau)$, *where* y_i *is either an* $\mathbf{x}_j$ *or a nullary operational symbol.*

Definition 3. *The formulas of $L(\tau)$ are defined by the following four rules of formation:*

(i) *Every atomic formula is a formula;*
(ii) *if* Φ *is a formula, then so is* $(\neg \Phi)$;
(iii) *if* Φ_0 *and* Φ_1 *are formulas, then so is* $(\Phi_0 \vee \Phi_1)$;
(iv) *if* Φ *is a formula, then so is* $((\exists \mathbf{x}_k)\Phi)$.

The set of all formulas of $L(\tau)$ will be denoted by $\mathbf{L}(\tau)$.

The parentheses in formulas will be omitted whenever there is no danger of confusion; in particular, the outermost pair of parentheses will almost always be omitted.

In (iv), Φ is called the *scope* of $(\exists \mathbf{x}_k)$.

Definition 4.

(i) $((\mathbf{x}_k)\Phi)$ *stands for* $(\neg(\exists \mathbf{x}_k)(\neg \Phi))$;
(ii) $(\Phi_0 \wedge \Phi_1)$ *stands for* $(\neg((\neg \Phi_0) \vee (\neg \Phi_1)))$;
(iii) $(\Phi_0 \to \Phi_1)$ *stands for* $((\neg \Phi_0) \vee \Phi_1)$;
(iv) $(\Phi_0 \leftrightarrow \Phi_1)$ *stands for* $((\Phi_0 \to \Phi_1) \wedge (\Phi_1 \to \Phi_0))$;
(v) $(\bigvee(\Phi_i \mid 0 \leqq i < n))$ *and* $(\Phi_0 \vee \cdots \vee \Phi_{n-1})$ *stand for*
$$(\cdots(\Phi_0 \vee \Phi_1) \vee \cdots) \vee \Phi_{n-1},$$
$(\bigwedge(\Phi_i \mid 0 \leqq i < n))$ *and* $(\Phi_0 \wedge \cdots \wedge \Phi_{n-1})$ *stand for*
$$(\cdots(\Phi_0 \wedge \Phi_1) \wedge \cdots) \wedge \Phi_{n-1}.$$

$(\mathbf{x}_k)\Phi$ will have the usual meaning "for all $\mathbf{x}_k$, Φ" and will be called *universal quantifier*, and sometimes will be referred to by the symbol $\forall$. $\Phi_0 \wedge \Phi_1$ is "Φ_0 and Φ_1" and $\Phi_0 \to \Phi_1$ will mean "Φ_0 implies Φ_1".

We will use $\equiv$ to mean "is defined by".

Finally, we define some concepts clarifying the relationship of a variable and a formula:

Definition 5.

(i) *An occurrence of the variable* $\mathbf{x}_k$ *in the formula* Φ *is* bound *if it occurs in* $(\exists \mathbf{x}_k)$ *or is within the scope of a quantifier* $(\exists \mathbf{x}_k)$. *If an occurrence of* $\mathbf{x}_k$ *is not bound, then it is called* free.

(ii) *A* sentence Φ *is a formula in which no variable has a free occurrence.*

(iii) *A formula* Φ *is* open *if every occurrence of every variable is free.*

We will often write $\Phi(\mathbf{x}_{i_0}, \cdots, \mathbf{x}_{i_{n-1}})$ to indicate that Φ has some of the $\mathbf{x}_{i_0}, \cdots, \mathbf{x}_{i_{n-1}}$ as free variables. Then if $\mathbf{p}_0, \cdots, \mathbf{p}_{n-1}$ are polynomial symbols $\Phi(\mathbf{p}_0, \cdots, \mathbf{p}_{n-1})$ will denote the formula

$$\Phi \wedge (\mathbf{x}_{i_0} = \mathbf{p}_0) \wedge \cdots \wedge (\mathbf{x}_{i_{n-1}} = \mathbf{p}_{n-1}).$$

§37. SATISFIABILITY AND THE CASE OF BOOLEAN SET ALGEBRAS

In §36 we introduced the language $L(\tau)$ and the set of formulas $\mathbf{L}(\tau)$. There are two approaches to the further development of the language. The first one is the *syntactical* approach in which we concentrate on the formal aspects of the language, by formally specifying which formulas are always "true" (logical axioms, e.g., $\Phi \to \Phi$), and by specifying how we can derive from given formulas new ones which are "true" if the given ones are "true" (rules of inference). The second one is the *semantical* approach which is based on the concept of satisfiability of a formula by elements of a structure. The semantical approach is set-theoretic in nature, and therefore it is not very suitable for studying the foundations of mathematics. However, our purpose in studying the first order language is to give algebraic applications. Since all the results needed for that can be developed using only the semantical approach, the syntax of $L(\tau)$ will be completely neglected. The interested reader should consult E. Mendelson, Introduction to Mathematical Logic, D. Van Nostrand Co., Princeton, N. J., 1964 (especially Chapter 2).

Notation. If A is a set, A^ω will denote the set of all sequences a of type ω of elements of A. Sequences will be denoted by small grotesque letters $\mathsf{a}, \mathsf{b}, \mathsf{c}, \cdots$. The k-th entry of a will be denoted by a_k.

If $\mathsf{a} \in A^\omega$, $k < \omega$, $b \in A$, then $\mathsf{a}(k/b)$ will denote the sequence which we get from a by replacing the k-th entry by b.

Definition 1 (*A. Tarski*). *Let* Φ *be a formula in* $L(\tau)$, *let* $\mathfrak{A}$ *be a structure of type* τ, *and let* $\mathsf{a} \in A^\omega$. *The formula* Φ *is* satisfied *by* a *in* $\mathfrak{A}$ *if:*

(i) Φ *is an atomic formula of the form*

$$\mathbf{p}(\mathbf{x}_{i_0}, \cdots, \mathbf{x}_{i_{n-1}}) = \mathbf{q}(\mathbf{x}_{j_0}, \cdots, \mathbf{x}_{j_{m-1}})$$

and

$$p(a_{i_0}, \cdots, a_{i_{n-1}}) = q(a_{j_0}, \cdots, a_{j_{m-1}})$$

or Φ is of the form $\mathbf{r}_\gamma(y_0, \cdots, y_{m_\gamma-1})$ where each y_i is either $\mathbf{x}_{j_i}$ or a nullary operational symbol $\mathbf{f}_{\gamma_i}$ and $r_\gamma(b_0, \cdots, b_{m_\gamma-1})$, where $b_i = a_{j_i}$ if y_i is $\mathbf{x}_{j_i}$ and b_i is $(\mathbf{f}_{\gamma_i})_{\mathfrak{A}}$ if y_{γ_i} is $\mathbf{f}_{\gamma_i}$.

(ii) *Φ is of the form $\neg\Phi_0$ and a does not satisfy Φ_0;*

(iii) *Φ is of the form $\Phi_0 \vee \Phi_1$ and a satisfies Φ_0 or a satisfies Φ_1;*

(iv) *Φ is of the form $(\exists\mathbf{x}_k)\Phi_0$ and Φ_0 is satisfied by $\mathsf{a}(k/b)$ for some $b \in A$.*

Remarks. (1) By Definition 36.3, Definition 1 defines satisfiability for every formula.

(2) In Definition 1 (i) the $\mathbf{x}_{i_j}$ need not be different; however, *each* occurrence of $\mathbf{x}_{i_j}$ is replaced by a_{i_j}.

(3) In checking whether a formula Φ is satisfied by $\mathsf{a} \in A^\omega$ in $\mathfrak{A}$ we have to "tear down" Φ to the "pieces" it was built up from. Since in this process (when (iv) is applied) the number of free variables may increase, it is not convenient to start with an $\mathsf{a} \in A^k$, where k is the number of free variables in Φ. However, if the free variables in Φ are at most $\mathbf{x}_{i_0}, \cdots, \mathbf{x}_{i_{n-1}}$, $\Phi = \Phi(\mathbf{x}_{i_0}, \cdots, \mathbf{x}_{i_{n-1}})$, then we can say that $\langle b_{i_0}, \cdots, b_{i_{n-1}}\rangle$ *satisfies* Φ in $\mathfrak{A}$ if there is an $\mathsf{a} \in A^\omega$ satisfying Φ such that $a_{i_j} = b_{i_j}$, $j = 0, \cdots, n-1$. If this is the case, we will say that $\Phi(b_{i_0}, \cdots, b_{i_{n-1}})$ is *true*. In particular, if $\Phi(\mathbf{x}_i)$ is free at most in $\mathbf{x}_i$, then "$\Phi(a)$ is true", or "a satisfies Φ" will mean that $a \in A$ and $\langle a\rangle$ satisfies Φ. It should be emphasized that $\Phi(b_{i_0}, \cdots, b_{i_{n-1}})$ is *not* a formula in the language $L(\tau)$.

We will illustrate the concept of satisfiability by analyzing the example of *Boolean set algebras*, that is, the Boolean algebras of the form $\mathfrak{P}(A) = \langle P(A); \cap, \cup, ', \varnothing, A\rangle$. (This part of §37 may be omitted, or only paged over at first reading. Theorems 1 and 2 will be applied only in §47.) In the corresponding language we use $\vee$, $\wedge$, $'$, $\mathbf{0}$, and $\mathbf{1}$ as operation symbols, where $(\mathbf{0})_{\mathfrak{P}(A)} = \varnothing$ and $(\mathbf{1})_{\mathfrak{P}(A)} = A$.

In this section up to Remark 2 following Theorem 2, "formula" will always mean "formula in the language associated with Boolean algebras".

First we give examples of what formulas can express.

There is a formula $P^{(1)}(\mathbf{x})$ which is satisfied by $a \in P(A)$ if and only if $|a| = 1$. Indeed,

$$P^{(1)}(\mathbf{x}) \equiv (\mathbf{y})((\mathbf{x} \subseteq \mathbf{y} \vee \mathbf{x} \wedge \mathbf{y} = \mathbf{0}) \wedge \mathbf{x} \neq \mathbf{0}).$$

In this formula we used the abbreviations $\mathbf{x} \subseteq \mathbf{y}$ for $\mathbf{x} = \mathbf{x} \wedge \mathbf{y}$ and $\mathbf{x} \neq \mathbf{0}$ for $\neg(\mathbf{x} = \mathbf{0})$.

There are formulas $P^{(n)}(\mathbf{x})$ and $P^{(\geqq n)}(\mathbf{x})$ for every $n \geqq 0$ which are satisfied by $a \in P(A)$ if and only if $|a| = n$, $|a| \geqq n$, respectively. Indeed, if $n > 0$

$$\begin{aligned} P^{(n)}(\mathbf{x}) \equiv\ & (\exists\mathbf{x}_0)\cdots(\exists\mathbf{x}_{n-1})(P^{(1)}(\mathbf{x}_0) \wedge P^{(1)}(\mathbf{x}_1) \wedge \cdots \wedge P^{(1)}(\mathbf{x}_{n-1}) \wedge \\ & \mathbf{x}_0 \wedge \mathbf{x}_1 = \mathbf{0} \wedge \mathbf{x}_0 \wedge \mathbf{x}_2 = \mathbf{0} \wedge \cdots \wedge \mathbf{x}_{n-2} \wedge \mathbf{x}_{n-1} = \mathbf{0} \wedge \\ & \mathbf{x} = \mathbf{x}_0 \vee \mathbf{x}_1 \vee \cdots \vee \mathbf{x}_{n-1}). \end{aligned}$$

We get $P^{(\geqq n)}(\mathbf{x})$ by replacing $\mathbf{x}=\mathbf{x}_0 \vee \mathbf{x}_1 \vee \cdots \vee \mathbf{x}_{n-1}$ by $\mathbf{x} \supseteq \mathbf{x}_0 \vee \mathbf{x}_1 \vee \cdots \vee \mathbf{x}_{n-1}$. Finally, $P^{(0)}(\mathbf{x})$ is $\mathbf{x}=\mathbf{0}$ and $P^{(\geqq 0)}(\mathbf{x})$ is, e.g., $\mathbf{x}=\mathbf{x}$.

An *atomic Boolean polynomial symbol* in $\mathbf{x}_0, \cdots, \mathbf{x}_{n-1}$ is one of the form $\mathbf{x}_0^{i_0} \wedge \mathbf{x}_1^{i_1} \wedge \cdots \wedge \mathbf{x}_{n-1}^{i_{n-1}}$, where $i_j=0$ or 1 and $\mathbf{x}_j{}^0$ stands for $\mathbf{x}_j$ and $\mathbf{x}_j{}^1$ stands for $\mathbf{x}_j'$. Note that there are 2^n atomic Boolean polynomial symbols in $\mathbf{x}_0, \cdots, \mathbf{x}_{n-1}$. (Compare Exercise 1.13.)

To express the meaning of combinations of $P^{(n)}(\mathbf{x})$ and $P^{(\geqq n)}(\mathbf{x})$ we introduce the concept of an n-ary table of conditions.

Definition 2.

(i) *An n-ary table of conditions T is a pair $\langle p, U\rangle$, where p is a mapping of $P(\{0, \cdots, n-1\})$ into the set of natural numbers and U is a subset of $P(\{0, \cdots, n-1\})$.*

(ii) $\langle a_0, \cdots, a_{n-1}\rangle$ satisfies *T in $\mathfrak{P}(A)$ $(a_0, \cdots, a_{n-1} \in P(A))$ if for every $r \subseteq \{0, \cdots, n-1\}$*

$$\left|\bigcap (a_i \mid i \in r) \cap \bigcap (a_i' \mid i \notin r)\right| \begin{cases} = p(r) & \text{if } r \in U, \\ \geqq p(r) & \text{if } r \notin U. \end{cases}$$

(iii) *Finally, for every n, we formally introduce an n-ary table of conditions F, which is not satisfied by any sequence.*

Remark. Intuitively, we get an n-ary table of conditions as follows: we divide the set of atomic Boolean polynomial symbols in $\mathbf{x}_0, \cdots, \mathbf{x}_{n-1}$ into two sets

$$C_0 = \{\mathbf{x}_0^{i_0} \wedge \cdots \wedge \mathbf{x}_{n-1}^{i_{n-1}} \mid \{j \mid i_j = 0\} \in U\},$$
$$C_1 = \{\mathbf{x}_0^{i_0} \wedge \cdots \wedge \mathbf{x}_{n-1}^{i_{n-1}} \mid \{j \mid i_j = 0\} \notin U\}.$$

Each atomic Boolean polynomial symbol $\mathbf{p}$ is then assigned to a formula $P^{(k)}$ if $\mathbf{p} \in C_0$ or to a formula $P^{(\geqq k)}$ if $\mathbf{p} \in C_1$.

For $n=1$, a unary table of conditions T is in effect a $P^{(k)}(\mathbf{x}_0)$ or $P^{(\geqq k)}(\mathbf{x}_0)$. For $n=2$, we give the following example:

	$\mathbf{x}_0$	$\mathbf{x}_0'$
$\mathbf{x}_1$	$\geqq 2$	$\geqq 0$
$\mathbf{x}_1'$	5	7

In this example we have $P^{(\geqq 2)}(\mathbf{x}_0 \wedge \mathbf{x}_1)$, $P^{(\geqq 0)}(\mathbf{x}_0' \wedge \mathbf{x}_1)$, $P^{(5)}(\mathbf{x}_0 \wedge \mathbf{x}_1')$, $P^{(7)}(\mathbf{x}_0' \wedge \mathbf{x}_1')$. Thus $U=\{\{0\}, \varnothing\}$ and the mapping p is given by

	$\varnothing$	$\{0\}$	$\{1\}$	$\{0, 1\}$
p	7	5	0	2

Theorem 1. *Let $T_0, \cdots, T_{M-1}$ be n-ary tables of conditions. Then there exists a formula Φ which is free at most in $\mathbf{x}_0, \cdots, \mathbf{x}_{n-1}$ such that Φ is satisfied by $\langle a_0, \cdots, a_{n-1}\rangle$ in $\mathfrak{P}(A)$ if and only if $\langle a_0, \cdots, a_{n-1}\rangle$ satisfies T_i for some i, $0 \leqq i < M$.*

The proof of Theorem 1 is obvious since any table of conditions can easily be expressed using $P^{(m)}(\mathbf{x})$ and $P^{(\geqq m)}(\mathbf{x})$, with the exception of F, which is equivalent to $\mathbf{0}=\mathbf{1} \wedge \mathbf{0} \neq \mathbf{1}$.

Theorem 2 (*T. Skolem*†). *Let Φ be a formula free at most in $\mathbf{x}_0, \cdots, \mathbf{x}_{n-1}$. Then there exist a positive integer M and n-ary tables of conditions $T_0, \cdots, T_{M-1}$ such that $\mathbf{a}=\langle a_0, \cdots, a_{n-1}\rangle$, $a_i \in P(A)$, satisfies Φ in $\mathfrak{P}(A)$ if and only if $\mathbf{a}$ satisfies some T_i with $0 \leqq i < M$.*

Proof. Let Γ denote the set of all formulas for which the statement of the theorem holds.

(i) Every atomic formula is in Γ. Indeed, if Φ is an atomic formula,

$$\mathbf{p}(\mathbf{x}_{i_0}, \cdots, \mathbf{x}_{i_{m-1}}) = \mathbf{q}(\mathbf{x}_{j_0}, \cdots, \mathbf{x}_{j_{m'-1}}),$$

free at most in $\mathbf{x}_0, \cdots, \mathbf{x}_{n-1}$, then it is equivalent to $\mathbf{w}=\mathbf{0}$, where

$$\mathbf{w} \equiv (\mathbf{p} \wedge \mathbf{q}') \vee (\mathbf{p}' \wedge \mathbf{q}).$$

Since $\mathbf{w}$ is a Boolean polynomial symbol, it is equivalent to a join of atomic Boolean polynomial symbols,

$$\bigvee(\bigwedge(\mathbf{x}_i \mid i \in r) \wedge \bigwedge(\mathbf{x}_i' \mid i \notin r) \mid r \in U).$$

Since $w(a_0, \cdots, a_{n-1}) = \varnothing$ if and only if

$$\bigcap(a_i \mid i \in r) \cap \bigcap(a_i' \mid i \notin r) = \varnothing \quad \text{for all } r \in U,$$

if we define $p(r)=0$ for all $r \in P(\{0, \cdots, n-1\})$, then with $M=1$, $T_0=\langle p, U\rangle$ satisfies the requirements of the theorem.

(ii) If $\Phi \in \Gamma$, then $\neg\Phi \in \Gamma$. Since $P^{(n)}(\mathbf{x})$ can be negated by "$P^{(0)}(\mathbf{x})$ or $P^{(1)}(\mathbf{x})$ or $\cdots$ or $P^{(n-1)}(\mathbf{x})$ or $P^{(\geqq n+1)}(\mathbf{x})$" and $P^{(\geqq n)}(\mathbf{x})$ can be negated by "$P^{(0)}(\mathbf{x})$ or $\cdots$ or $P^{(n-1)}(\mathbf{x})$", thus the negation of every table of conditions T can be expressed as a finite set of tables of conditions. Let $S(T)$ denote the set of all tables of conditions we get by the above process. If Φ is associated with $T_0, \cdots, T_{M-1}$, then $\neg\Phi$ is associated with $S(T_0) \cap S(T_1) \cap \cdots \cap S(T_{M-1})$.

(iii) If $\Phi_0, \Phi_1 \in \Gamma$, then $\Phi_0 \vee \Phi_1 \in \Gamma$. We can assume that Φ_0 and Φ_1 are both free at most in $\mathbf{x}_0, \cdots, \mathbf{x}_{n-1}$. If Φ_0 is equivalent to the tables of

† Skrifter utgit av Videnskapsselskapet, Kristiania, I. Klasse No. 3, Oslo, 1919.

conditions $T_0, \cdots, T_{M-1}$ and Φ_1 to $T_0', \cdots, T'_{M'-1}$, then $\Phi_0 \vee \Phi_1$ is equivalent to $T_0, \cdots, T_{M-1}, T_0', \cdots, T'_{M'-1}$, which concludes the proof of (iii).

(iv) If $\Phi \in \Gamma$, then $(\exists \mathbf{x}_k)\Phi \in \Gamma$.

Example: Let $\Phi(\mathbf{x}_0, \mathbf{x}_1)$ denote the formula which is equivalent to the table of conditions

	$\mathbf{x}_0$	$\mathbf{x}_0'$
$\mathbf{x}_1$	$\geqq 3$	2
$\mathbf{x}_1'$	0	$\geqq 5$

Since there exists an a_1 such that $|a_0 \cap a_1| \geqq 3$ and $|a_0 \cap a_1'| = 0$ if and only if $|a_0| \geqq 3$, and similarly $|a_0' \cap a_1| = 2$ and $|a_0' \cap a_1'| \geqq 5$ yield $|a_0'| \geqq 7$, we get that $(\exists \mathbf{x}_1)\Phi(\mathbf{x}_0, \mathbf{x}_1)$ is equivalent to the table of conditions

$\mathbf{x}_0$	$\mathbf{x}_0'$
$\geqq 3$	$\geqq 7$

The same method applies in the general case and the formal proof is left to the reader.

Remark 1. If we assume that $\mathfrak{a} = \langle a_0, \cdots, a_{n-1} \rangle$ is a partition of A, that is, $a_i \cap a_j = \varnothing$ if $i \neq j$ and $a_0 \cup a_1 \cup \cdots \cup a_{n-1} = A$, then a table of conditions for such a sequence simply imposes conditions on $|a_0|, \ldots, |a_{n-1}|$.

Remark 2. Theorem 2 is constructive, that is, given a formula Φ, we can construct in a finite number of steps M and $T_0, \cdots, T_{M-1}$ by following the procedures in the proof.

Remark 3. The proof of Theorem 2 is an illustration of a method which will be called *induction on formulas*. This method is applied when we want to describe the meaning of sentences of a certain theory. The trick is to do more: describe the meaning of all formulas and prove this statement using steps (i)–(iii) as in Theorem 2. The same inductive proof could not be applied to sentences only, because in step (iv), if $(\exists \mathbf{x}_k)\Phi$ is a sentence, Φ need not be a sentence. Other examples of this method can be found in almost every section of Chapters 6 and 7 and also in the Exercises. The applicability of this method is restricted; it can only be applied when not only the valid sentences but also the satisfiability of formulas admit a good description.

Next we introduce some basic concepts of the first order logic.

Definition 3. *Let Φ and Ψ be formulas in $L(\tau)$.*

(i) *We write $\Phi \models \Psi$ if, for any structure $\mathfrak{A}$ and $\mathsf{a} \in A^\omega$, if Φ is satisfied by a in $\mathfrak{A}$, then Ψ is also satisfied by a in $\mathfrak{A}$.*

(ii) *If $\Phi \models \Psi$ and $\Psi \models \Phi$, then we write $\Phi \Leftrightarrow \Psi$ and we say that Φ and Ψ are* equivalent.

(iii) *If Φ is satisfied by every $\mathsf{a} \in A^\omega$, then we write $_{\mathfrak{A}}\models \Phi$.*

(iv) *If $_{\mathfrak{A}}\models \Phi$ holds for all structures $\mathfrak{A}$, then we write $\models \Phi$.*

(v) *If Σ is a family of formulas, then $\Sigma \models \Psi$ means that if a satisfies Φ in $\mathfrak{A}$ for all Φ in Σ then a satisfies Ψ in $\mathfrak{A}$.*

(vi) *If Σ and Ω are families of formulas, then $\Sigma \models \Omega$ means that $\Sigma \models \Phi$ for all $\Phi \in \Omega$.*

(vii) *$\Sigma \Leftrightarrow \Omega$ means that $\Sigma \models \Omega$ and $\Omega \models \Sigma$.*

(viii) *If Σ is a set of formulas and $_{\mathfrak{A}}\models \Phi$ for all $\Phi \in \Sigma$, then $\mathfrak{A}$ is a* model *of Σ or $\mathfrak{A}$* satisfies Σ.

These notations will be most frequently used for sentences. If Φ is a sentence, then $_{\mathfrak{A}}\models \Phi$ means that Φ holds in $\mathfrak{A}$, which is equivalent to saying that Φ is satisfied by some or by all $\mathsf{a} \in A^\omega$ in $\mathfrak{A}$.

The next result establishes the rules of satisfiability for the connectives $\wedge$, $\rightarrow$, and $\leftrightarrow$, and for the universal quantifier.

Theorem 3. *Let $\mathfrak{A}$ be a structure and $\mathsf{a} \in A^\omega$.*

(i) *$(\mathbf{x}_k)\Phi$ is satisfied by a if and only if $\mathsf{a}(k/b)$ satisfies Φ for all $b \in A$.*

(ii) *$\Phi_0 \rightarrow \Phi_1$ is satisfied by a if and only if the satisfaction of Φ_0 by a implies the satisfaction of Φ_1 by a.*

(iii) *$\Phi_0 \leftrightarrow \Phi_1$ is satisfied by a if and only if both Φ_0 and Φ_1 are satisfied by a or neither Φ_0 nor Φ_1 is satisfied by a.*

(iv) *$\Phi_0 \wedge \Phi_1$ is satisfied by a if and only if Φ_0 and Φ_1 are satisfied by a.*

The proofs of these and of most of the other rules which will be given in the sequel are very easy. As an illustration, we will prove (i).

$(\mathbf{x}_k)\Phi$ is satisfied by a if and only if $\neg((\exists \mathbf{x}_k)(\neg \Phi))$ is satisfied by a. This is equivalent, by Definition 1, to the nonsatisfaction of $(\exists \mathbf{x}_k)(\neg \Phi)$ by a. Again by Definition 1, the last statement is equivalent to the condition that no $\mathsf{a}(k/b)$ satisfies $\neg \Phi$, that is, $\mathsf{a}(k/b)$ satisfies Φ for all $b \in A$.

Definition 4. *A formula Ψ is in the* prenex normal form *if it is of the form*

$$(Q_0\mathbf{x}_{i_0})(Q_1\mathbf{x}_{i_1})\cdots(Q_{n-1}\mathbf{x}_{i_{n-1}})\Phi,$$

where the Q_i are quantifiers (that is, $\exists$ or $\forall$) and Φ is a formula which does not *contain any quantifier. In particular, the case $n=0$, i.e., the case of no quantifiers, is permitted. In this case, Ψ is simply an open formula.*

$$(Q_0\mathbf{x}_{i_0})\cdots(Q_{n-1}\mathbf{x}_{i_{n-1}})$$

is called the prefix *and* Φ *is called the* matrix *of* Ψ.

Theorem 4. *Every formula is equivalent to a formula in the prenex normal form.*

Again, the proof is a trivial induction which is based on the following rules:

Let Φ be an arbitrary formula and let Ψ be a formula which does not contain the variable $\mathbf{x}$. Then

(i) $\neg(\mathbf{x})\Phi \Leftrightarrow (\exists\mathbf{x})(\neg\Phi)$.
(ii) $\neg(\exists\mathbf{x})\Phi \Leftrightarrow (\mathbf{x})(\neg\Phi)$.
(iii) $(\mathbf{x})\Phi(\mathbf{x})\wedge\Psi \Leftrightarrow (\mathbf{x})(\Phi(\mathbf{x})\wedge\Psi)$.
(iv) $(\mathbf{x})\Phi(\mathbf{x})\vee\Psi \Leftrightarrow (\mathbf{x})(\Phi(\mathbf{x})\vee\Psi)$.
(v) $(\exists\mathbf{x})\Phi(\mathbf{x})\wedge\Psi \Leftrightarrow (\exists\mathbf{x})(\Phi(\mathbf{x})\wedge\Psi)$.
(vi) $(\exists\mathbf{x})\Phi(\mathbf{x})\vee\Psi \Leftrightarrow (\exists\mathbf{x})(\Phi(\mathbf{x})\vee\Psi)$.

If Ψ is an arbitrary formula, then

(vii) $(\mathbf{x})\Phi(\mathbf{x})\wedge(\mathbf{x})\Psi(\mathbf{x}) \Leftrightarrow (\mathbf{x})(\Phi(\mathbf{x})\wedge\Psi(\mathbf{x}))$

and

(viii) $(\mathbf{x})\Phi(\mathbf{x})\vee(\mathbf{x})\Psi(\mathbf{x}) \Leftrightarrow (\mathbf{x})(\mathbf{y})(\Phi(\mathbf{x})\vee\Psi(\mathbf{y}))$,

where $\mathbf{y}$ is any variable which occurs neither in Φ nor in Ψ, and $\Psi(\mathbf{y})$ is the formula obtained from $\Psi(\mathbf{x})$ by replacing all free occurrences of $\mathbf{x}$ by $\mathbf{y}$.

If Φ is a sentence (or, in general, a formula) in the prenex normal form and all the quantifiers are universal, then Φ is called a *universal sentence* (resp. *universal formula*). Let us note that in most algebra books the universal sentences are given in the form of open formulas. For instance, the commutative law is usually given as $\mathbf{xy}=\mathbf{yx}$, and by that is meant $(\mathbf{x})(\mathbf{y})(\mathbf{xy}=\mathbf{yx})$.

Finally, let us note that a sentence defines a property of structures. That is, if Φ is a sentence and $\mathfrak{A}$ is a structure, then either Φ holds or Φ fails to hold for $\mathfrak{A}$. On the other hand, a formula $\Phi(\mathbf{x}_0,\cdots,\mathbf{x}_{n-1})$ which is free at most in $\mathbf{x}_0,\cdots,\mathbf{x}_{n-1}$ defines an n-ary relation on the elements of a structure. The relations thus defined are called *first order relations*. Not all relations are first order relations. For instance, let $\mathfrak{A}$ be a structure of type τ where $o_0(\tau)$ and $o_1(\tau)$ are finite, $|A|\geqq\aleph_0$. Then the number of n-ary relations on A is $2^{|A|}$, while the number of first order relations definable in $L(\tau)$ cannot exceed the number of all formulas in $L(\tau)$, which is $\aleph_0$.

This concludes our discussion of the language $L(\tau)$. Now we are ready to do something interesting.

§38. ELEMENTARY EQUIVALENCE AND ELEMENTARY EXTENSIONS

In this section, let τ be a fixed type; every structure will be assumed to be of type τ unless otherwise specified.

Definition 1. *The structures* $\mathfrak{A}$ *and* $\mathfrak{B}$ *are* elementarily equivalent *(in notation:* $\mathfrak{A} \equiv \mathfrak{B}$*) if and only if every sentence which holds in* $\mathfrak{A}$ *holds in* $\mathfrak{B}$ *as well.*

Corollary 1. *If* $\mathfrak{A}$ *and* $\mathfrak{B}$ *are elementarily equivalent, then a sentence holds in* $\mathfrak{A}$ *if and only if it holds in* $\mathfrak{B}$.

Indeed, if the sentence Φ holds in $\mathfrak{B}$ and Φ does not hold in $\mathfrak{A}$, then $\neg\Phi$ holds in $\mathfrak{A}$. Thus, by definition, $\neg\Phi$ holds in $\mathfrak{B}$, which yields a contradiction.

Corollary 2. *If* $\mathfrak{A}$ *and* $\mathfrak{B}$ *are isomorphic, then they are elementarily equivalent.*

This statement is so obvious that it hardly needs a proof. The reader will not find any difficulty in proving it using induction on formulas. The following more general (but equally trivial) statement, has to be verified.

Let $\mathfrak{A}$ and $\mathfrak{B}$ be isomorphic, let φ be an isomorphism, and Φ a formula. Then $\mathbf{a} \in A^\omega$ satisfies Φ in $\mathfrak{A}$ if and only if $\mathbf{b} \in B^\omega$ satisfies Φ in $\mathfrak{B}$, where $b_i = a_i\varphi$, $i = 0, 1, \cdots$.

In general, isomorphism is much stronger than elementary equivalence. However, in the finite case, the two concepts coincide.

Consider the sentence Φ:

$$(\exists \mathbf{x}_0) \cdots (\exists \mathbf{x}_{n-1})(\mathbf{x}_n)(\mathbf{x}_0 \neq \mathbf{x}_1 \wedge \mathbf{x}_0 \neq \mathbf{x}_2 \wedge \cdots \wedge \mathbf{x}_{n-2} \neq \mathbf{x}_{n-1} \wedge (\mathbf{x}_n = \mathbf{x}_0 \vee \cdots \vee \mathbf{x}_n = \mathbf{x}_{n-1})).$$

Since Φ holds in $\mathfrak{A}$ if and only if A has exactly n elements, we obtain:

Corollary 3. *If* $\mathfrak{A}$ *and* $\mathfrak{B}$ *are elementarily equivalent and* $\mathfrak{A}$ *has n elements, then* $\mathfrak{B}$ *also has n elements.*

Definition 2 (*A. Tarski*). *Let* $\mathfrak{A}$ *and* $\mathfrak{B}$ *be structures.* $\mathfrak{B}$ *is said to be an* elementary extension *of* $\mathfrak{A}$ *if* $A \subseteq B$ *and for each formula* Φ *and each*

$\mathbf{a} \in A^{\omega}$ *the formula* Φ *is satisfied by* $\mathbf{a}$ *in* $\mathfrak{B}$ *if and only if* Φ *is satisfied by* $\mathbf{a}$ *in* $\mathfrak{A}$.

If $\mathfrak{B}$ *is an elementary extension of* $\mathfrak{A}$, *then* $\mathfrak{A}$ *is said to be an* elementary substructure of $\mathfrak{B}$. *An embedding* φ *of* $\mathfrak{A}$ *into* $\mathfrak{B}$ *is called an* elementary embedding *if the image of* $\mathfrak{A}$ *under* φ *is an elementary substructure of* $\mathfrak{B}$.

Corollary 1. *If* $\mathfrak{B}$ *is an elementary extension of* $\mathfrak{A}$, *then* $\mathfrak{A}$ *is a substructure of* $\mathfrak{B}$.

The proof is trivial. We have only to apply the definition to the formulas

$$\Phi(\mathbf{x}_0, \cdots, \mathbf{x}_{m_\gamma - 1}) \equiv \mathbf{r}_\gamma(\mathbf{x}_0, \cdots, \mathbf{x}_{m_\gamma - 1})$$

and

$$\Phi(\mathbf{x}_0, \cdots, \mathbf{x}_{n_\gamma}) \equiv \mathbf{f}_\gamma(\mathbf{x}_0, \cdots, \mathbf{x}_{n_\gamma - 1}) = \mathbf{x}_{n_\gamma}.$$

As a matter of fact, $\mathfrak{A}$ is a substructure of $\mathfrak{B}$ if and only if $\mathfrak{A}$ and $\mathfrak{B}$ satisfy the criterion for elementary extension for atomic formulas.

Corollary 2. *If* $\mathfrak{B}$ *is an elementary extension of* $\mathfrak{A}$, *then* $\mathfrak{A}$ *and* $\mathfrak{B}$ *are elementarily equivalent.*

Theorem 1 (*A. Tarski*). *Let* $\mathfrak{A}_\gamma$, $\gamma < \delta$, *be structures with the property that if* $\gamma_0 < \gamma_1 < \delta$, *then* $\mathfrak{A}_{\gamma_0}$ *is an elementary substructure of* $\mathfrak{A}_{\gamma_1}$. *Define* $\mathfrak{A}$ *as the union of the structures* $\mathfrak{A}_\gamma$. *Then* $\mathfrak{A}$ *is an elementary extension of* $\mathfrak{A}_0$.

Proof. Let Γ denote the set of all formulas Φ with the property that, for any $\gamma < \delta$ and $\mathbf{a} \in A_\gamma^{\omega}$, Φ is satisfied by $\mathbf{a}$ in $\mathfrak{A}_\gamma$ if and only if Φ is satisfied by $\mathbf{a}$ in $\mathfrak{A}$.

(i) If Φ is an atomic formula, then $\Phi \subset \Gamma$. This is the remark following Corollary 1.

(ii) If $\Phi \in \Gamma$, then $\neg\Phi \in \Gamma$. Indeed, $\neg\Phi$ is satisfied by $\mathbf{a}$ in $\mathfrak{A}_\gamma$ if and only if Φ is not satisfied by $\mathbf{a}$ in $\mathfrak{A}_\gamma$. Since $\Phi \in \Gamma$, this is equivalent to the condition that Φ is not satisfied by $\mathbf{a}$ in $\mathfrak{A}$, that is, $\neg\Phi$ is satisfied by $\mathbf{a}$ in $\mathfrak{A}$. This shows that $\neg\Phi \in \Gamma$.

(iii) $\Phi_0, \Phi_1 \in \Gamma$ imply that $\Phi_0 \vee \Phi_1 \in \Gamma$. This is trivial (like (ii)).

(iv) $\Phi \in \Gamma$ implies that $(\exists \mathbf{x}_k)\Phi \in \Gamma$. Indeed, if $(\exists \mathbf{x}_k)\Phi$ is satisfied by $\mathbf{a} \in A_\gamma^{\omega}$ in $\mathfrak{A}_\gamma$, then Φ is satisfied by $\mathbf{a}(k/b)$ for some $b \in A_\gamma$. Since $\Phi \in \Gamma$, we get that Φ is satisfied by $\mathbf{a}(k/b)$ in $\mathfrak{A}$, which implies that $(\exists \mathbf{x}_k)\Phi$ is satisfied by $\mathbf{a}$ in $\mathfrak{A}$. Conversely, if $(\exists \mathbf{x}_k)\Phi$ is satisfied by $\mathbf{a} \in A_\gamma^{\omega}$ in $\mathfrak{A}$, then Φ is satisfied by $\mathbf{a}(k/b)$ for some $b \in A$. Let $b \in A_\eta$, $\gamma \leqq \eta$. Then $\mathbf{a}(k/b) \in A_\eta^{\omega}$. Since $\Phi \in \Gamma$, we get that $\mathbf{a}(k/b)$ satisfies Φ in $\mathfrak{A}$ if and only if it satisfies Φ in $\mathfrak{A}_\eta$. Thus, $\mathbf{a}(k/b)$ satisfies Φ in $\mathfrak{A}_\eta$. Therefore, $(\exists \mathbf{x}_k)\Phi$ is satisfied by $\mathbf{a}$ in $\mathfrak{A}_\eta$. Since $\mathfrak{A}_\eta$ is an elementary extension of $\mathfrak{A}_\gamma$, we get that $\mathbf{a}$ satisfies $(\exists \mathbf{x}_k)\Phi$ in $\mathfrak{A}_\gamma$. This proves that $(\exists \mathbf{x}_k)\Phi \in \Gamma$.

By the definition of a formula, we get that Γ is the set of all formulas. Thus, $\mathfrak{A}$ is an elementary extension of $\mathfrak{A}_\gamma$, $\gamma < \delta$, and, in particular, of $\mathfrak{A}_0$.

The proof is again an example of induction on formulas.

A useful criterion for recognizing elementary extensions is given in the following theorem.

Theorem 2 (*A. Tarski*). *Let $\mathfrak{A}$ and $\mathfrak{B}$ be structures. Then $\mathfrak{B}$ is an elementary extension of $\mathfrak{A}$ if and only if*

(i) *$\mathfrak{B}$ is an extension of $\mathfrak{A}$;*

(ii) *for every formula Φ, $k \geqq 0$, $\mathsf{a} \in A^\omega$, if a satisfies $(\exists \mathbf{x}_k)\Phi$ in $\mathfrak{B}$, then there exists an element $b \in A$ such that $\mathsf{a}(k/b)$ satisfies Φ in $\mathfrak{B}$.*

Proof. Let $\mathfrak{B}$ be an elementary extension of $\mathfrak{A}$. Then (i) is Corollary 1 to Definition 2. To prove (ii), assume that a satisfies $(\exists \mathbf{x}_k)\Phi$ in $\mathfrak{B}$. Then a satisfies $(\exists \mathbf{x}_k)\Phi$ in $\mathfrak{A}$. Thus, Φ is satisfied by $\mathsf{a}(k/b)$ in $\mathfrak{A}$ by some $b \in A$. Again, since $\mathfrak{B}$ is an elementary extension of $\mathfrak{A}$, Φ is satisfied by $\mathsf{a}(k/b)$ in $\mathfrak{B}$.

Now assume that (i) and (ii) hold and let Γ denote the set of all formulas Φ such that Φ is satisfied by $\mathsf{a} \in A^\omega$ in $\mathfrak{A}$ if and only if Φ is satisfied by a in $\mathfrak{B}$. By (i), Γ contains all atomic formulas, and Φ_0, $\Phi_1 \in \Gamma$ imply $\neg\Phi_0$, $\Phi_0 \vee \Phi_1 \in \Gamma$. (This is trivial, as in the proof of Theorem 1.)

It remains to prove that $\Phi \in \Gamma$ implies $(\exists \mathbf{x}_k)\Phi \in \Gamma$. Indeed, $(\exists \mathbf{x}_k)\Phi$ is satisfied by $\mathsf{a} \in A^\omega$ in $\mathfrak{B}$ implies, by (ii), the existence of $b \in A$ such that $\mathsf{a}(k/b)$ satisfies Φ in $\mathfrak{B}$. Since $\Phi \in \Gamma$, $\mathsf{a}(k/b)$ satisfies Φ in $\mathfrak{A}$. Thus, $(\exists \mathbf{x}_k)\Phi$ is satisfied by a in $\mathfrak{A}$. Conversely, if $(\exists \mathbf{x}_k)\Phi$ is satisfied by a in $\mathfrak{A}$, then for some $b \in A$, Φ is satisfied by $\mathsf{a}(k/b)$ in $\mathfrak{A}$. Thus, since $\Phi \in \Gamma$, we get that Φ is satisfied by $\mathsf{a}(k/b)$ in $\mathfrak{B}$. This completes the proof of Theorem 2.

An elementary extension of $\mathfrak{A}$ is an extension of $\mathfrak{A}$ which is elementarily equivalent to $\mathfrak{A}$. However, the converse is not true. (See Exercise 21.)

Now we apply Theorem 2 to get the celebrated result of Löwenheim and Skolem in the stronger form due to Tarski. (This is the "downward" theorem; the "upward" form will be given in §40.)

Theorem 3 (*Löwenheim-Skolem-Tarski*). *Let $\mathfrak{A}$ be a structure. Let $B \subseteq A$ satisfy $\overline{o(\tau)} \leqq |B|$, $\aleph_0 \leqq |B|$. Then there exists $C \subseteq A$ such that $B \subseteq C$, $|B| = |C|$, and $\mathfrak{C}$ is an elementary substructure of $\mathfrak{A}$.*

Proof. Let $A = \{a_\gamma \mid \gamma < \alpha\}$. Set $C_0 = B$, and for each $n \geqq 0$ define $C_{n+1} = \{b \mid b \in A$ and there exist a formula Φ, $k \geqq 0$, and $\mathsf{a} \in C_n{}^\omega$ such that b is the first element with the property that $\mathsf{a}(k/b)$ satisfies Φ in $\mathfrak{A}\}$, where "first element" refers to the given well ordering of A.

Using the formula $\mathbf{x}_0 = \mathbf{x}_1$, we get $C_0 \subseteq C_1 \subseteq C_2 \subseteq \cdots$. Set

$$C = \bigcup (C_i \mid i < \omega).$$

Obviously $\mathfrak{C}$ is a substructure which satisfies the conditions of Theorem 2. Thus, $\mathfrak{A}$ is an elementary extension of $\mathfrak{C}$. Since the number of formulas does not exceed $|B|$, it follows that $|C_n| = |C_{n+1}|$ for all n. Thus, $|B| \leq |C| \leq |B|\aleph_0 = |B|$, which proves that $|B| = |C|$.

Corollary. *Assume that* $\overline{o(\tau)} \leq \aleph_0$. *Then every infinite structure* $\mathfrak{A}$ *has a countable elementary substructure.*

It is obvious that this fails to hold in general if $\overline{o(\tau)} > \aleph_0$. Indeed, if there are $\mathfrak{p}$ nullary operational symbols and they represent distinct elements of $\mathfrak{A}$, then they represent distinct elements in any substructure $\mathfrak{B}$ of $\mathfrak{A}$, thus $|B| \geq \mathfrak{p}$. However, though $|B| \geq \mathfrak{p}$ is necessary it is not sufficient for Theorem 3 to hold, as will be shown in the next section.

The following two theorems will show interesting examples of elementary extensions.

Theorem 4 (*R. Vaught*). *Let K be a class of algebras of type τ. Assume that the free algebras $\mathfrak{F}_K(\mathfrak{m})$ and $\mathfrak{F}_K(\mathfrak{n})$ exist and that $\aleph_0 \leq \mathfrak{n} \leq \mathfrak{m}$. Let $\{x_i \mid i \in I\}$ be the set of free generators of $\mathfrak{F}_K(\mathfrak{m})$ ($|I| = \mathfrak{m}$) and let $J \subseteq I$, $|J| = \mathfrak{n}$. Let $\mathfrak{F}'$ be the subalgebra generated by $\{x_i \mid i \in J\}$. Then $\mathfrak{F}' \cong \mathfrak{F}_K(\mathfrak{n})$ and $\mathfrak{F}_K(\mathfrak{m})$ is an elementary extension of $\mathfrak{F}'$.*

Proof. We have already proved that $\mathfrak{F}' \cong \mathfrak{F}_K(\mathfrak{n})$. Let Φ be free at most in $\mathbf{x}_0, \cdots, \mathbf{x}_{n-1}$, and let $\mathsf{b} \in (F')^\omega$ satisfy $(\exists \mathbf{x}_k)\Phi$ in $\mathfrak{F}_K(\mathfrak{m})$. Then there exists an $a \in F_K(\mathfrak{m})$ such that $\mathsf{b}(k/a)$ satisfies Φ in $\mathfrak{F}_K(\mathfrak{m})$. Choose $J' \subseteq J$ and $I' \subseteq I$, I', J' finite, such that $b_0, \cdots, b_{n-1} \in [\{x_i \mid i \in J'\}]$ and $a \in [\{x_i \mid i \in I'\}]$. Since J is infinite, there exists a permutation φ of I such that

$$I'\varphi \subseteq J \quad \text{and} \quad i\varphi = i \quad \text{for } i \in J'.$$

Let $\bar{\varphi}$, be the automorphism of $\mathfrak{F}_K(\mathfrak{m})$ with $x_i\bar{\varphi} = x_{i\varphi}$ for $i \in I$. Then $b_i\bar{\varphi} = b_i$, $i = 0, \cdots, n-1$ and $a\bar{\varphi} \in F'$. Thus $\mathsf{b}(k/a\bar{\varphi})$ satisfies Φ in $\mathfrak{F}_K(\mathfrak{m})$. By Theorem 2, $\mathfrak{F}_K(\mathfrak{m})$ is an elementary extension of $\mathfrak{F}'$, which was to be proved.

Corollary. *If* $\mathfrak{m}, \mathfrak{n} \geq \aleph_0$, *then* $\mathfrak{F}_K(\mathfrak{m})$ *and* $\mathfrak{F}_K(\mathfrak{n})$ *are elementarily equivalent.*

Theorem 5. *Let I be a set. Consider the Boolean algebra $\mathfrak{P}(I)$. Let $\mathfrak{C}$ be an arbitrary subalgebra of $\mathfrak{P}(I)$ containing all finite subsets of I. Then $\mathfrak{P}(I)$ is an elementary extension of $\mathfrak{C}$.*

Proof. This follows from Theorem 2 and the corollary to Theorem 37.2.

Let K be the class of all Boolean algebras of the form $\mathfrak{P}(I)$. It is easy to see that $\mathbf{I}(K)$ is the class of all complete and atomic Boolean algebras.

Corollary. *Let τ be the type of Boolean algebras and let $\Phi \in \mathbf{L}(\tau)$ be a sentence which holds for an algebra of type τ if and only if it is an atomic Boolean algebra. If $\Psi \in \mathbf{L}(\tau)$ is a sentence that holds for all algebras in K, then $\Phi \models \Psi$.*

In other words, a first order property of all atomic complete Boolean algebras is a consequence of the fact that the Boolean algebras considered are atomic. (See G. Grätzer, Notices Amer. Math. Soc. 12 (1965), p. 126. This is implicit in A. Tarski, Bull. Amer. Math. Soc. 65 (1949), p. 64.)

Proof. Let us assume that Ψ violates the statement of the corollary. Then there exists an atomic Boolean algebra $\mathfrak{B} = \langle B; \vee, \wedge, ', 0, 1\rangle$ satisfying $\neg\Psi$. Let I be the set of all atoms of $\mathfrak{B}$ and define $r(a) = \{p \mid p \leqq a, p \in I\}$ for all $a \in B$. Then $\varphi: a \rightarrow r(a)$ is an embedding of $\mathfrak{B}$ into $\mathfrak{A} = \mathfrak{P}(I)$. Let $C = B\varphi$. Then $\mathfrak{C}$ is a subalgebra of $\mathfrak{A}$ containing all finite subsets of I. Thus, by Theorem 5, $\mathfrak{A}$ and $\mathfrak{C}$ are elementarily equivalent. Since $\mathfrak{B}$ and $\mathfrak{C}$ are isomorphic, $\neg\Psi$ holds in $\mathfrak{C}$ and thus in $\mathfrak{A}$ as well, contrary to our assumption.

* * *

Sometimes it is useful to consider several languages simultaneously. Starting with a type τ, we get larger types η, ν, etc., by adding new relational symbols and operational symbols. Then if we consider a structure $\mathfrak{A}'$ of type η, we get from that a structure $\mathfrak{A}$ of type τ by deleting all the adjoined operations and relations. This structure $\mathfrak{A}$ is called a *τ-reduct*, or an *$L(\tau)$ reduct* of $\mathfrak{A}'$, or simply a reduct of $\mathfrak{A}'$.

The following statements are then trivial.

Lemma 1. *Let $\mathfrak{A}$ and $\mathfrak{B}$ be structures of type τ and let $\mathfrak{B}$ be an extension of $\mathfrak{A}$. Then $\mathfrak{B}$ is an elementary extension of $\mathfrak{A}$ if and only if every reduct $\mathfrak{B}'$ of $\mathfrak{B}$ of finite type is an elementary extension of the corresponding reduct $\mathfrak{A}'$ of $\mathfrak{A}$.*

Lemma 2. *Let $\mathfrak{A}$ and $\mathfrak{B}$ be structures of type τ. Then $\mathfrak{A}$ is elementarily equivalent to $\mathfrak{B}$ if and only if every reduct $\mathfrak{A}'$ of $\mathfrak{A}$ of finite type is elementarily equivalent to the corresponding reduct $\mathfrak{B}'$ of $\mathfrak{B}$.*

Both Lemma 1 and Lemma 2 follow from the observation that in a formula we use only a finite number of relational and operational symbols.

A class of very useful languages can be constructed as follows. Let $\mathfrak{A}$ be a structure of type τ, β an ordinal, and $\mathsf{a} \in A^\beta$. We define a new type $\tau' = \tau_\mathsf{a}$ as follows: $o_0(\tau') = o_0(\tau) + \beta$ and $o_1(\tau') = o_1(\tau)$; for every $\gamma < \beta$, $\mathbf{f}_{o_0(\tau)+\gamma}$ is a nullary operational symbol, that is, a constant. We will write $\mathbf{k}_\gamma$ for $\mathbf{f}_{o_0(\tau)+\gamma}$. We will denote (quite irregularly) by $\langle \mathfrak{A}, \mathsf{a} \rangle$ a structure of type τ' on the set A, in which the "old" operations and relations are defined as in $\mathfrak{A}$ and $(\mathbf{k}_\gamma)_{\langle \mathfrak{A}, \mathsf{a} \rangle} = a_\gamma$ for $\gamma < \beta$. Intuitively speaking, $\langle \mathfrak{A}, \mathsf{a} \rangle$ is the structure that we get from $\mathfrak{A}$ by introducing a given sequence of elements as constants. Note that if $\mathfrak{B}$ is an extension of $\mathfrak{A}$, then $\langle \mathfrak{B}, \mathsf{a} \rangle$ is again a structure of type τ'.

If a is a well ordering of $\mathfrak{A}$, then in $\langle \mathfrak{A}, \mathsf{a} \rangle$ every element is a constant. The corresponding language will be denoted by $L_\mathfrak{A}(\tau)$ and is called the *diagram language* of $\mathfrak{A}$. This concept is due to A. Robinson [2].

Theorem 6. *Let $\mathfrak{A}$ and $\mathfrak{B}$ be structures of type τ, let $\mathfrak{B}$ be an extension of $\mathfrak{A}$, and let a be a well ordering of A. Then $\mathfrak{B}$ is an elementary extension of $\mathfrak{A}$ if and only if $\langle \mathfrak{B}, \mathsf{a} \rangle$ is elementarily equivalent to $\langle \mathfrak{A}, \mathsf{a} \rangle$.*

In other words, $L_\mathfrak{A}(\tau)$ is so much stronger than $L(\tau)$ that elementary extension in the weaker language is equivalent to elementary equivalence in the stronger one.

Proof. By the definition of an elementary extension, if $\mathfrak{B}$ is an elementary extension of $\mathfrak{A}$, then $\langle \mathfrak{B}, \mathsf{a} \rangle$ is elementarily equivalent to $\langle \mathfrak{A}, \mathsf{a} \rangle$.

Conversely, assume that $\langle \mathfrak{B}, \mathsf{a} \rangle$ and $\langle \mathfrak{A}, \mathsf{a} \rangle$ are elementarily equivalent. Let Φ be a formula in $L(\tau)$ and assume that $\mathsf{b} \in A^\omega$ satisfies Φ in $\mathfrak{A}$. Let $\mathbf{x}_{i_0}, \cdots, \mathbf{x}_{i_{n-1}}$ be all the distinct free variables in Φ. Replacing $\mathbf{x}_{i_j}$ by $\mathbf{k}_{i_j}$, $j = 0, \cdots, n-1$, we get the sentence $\Phi' \in \mathbf{L}(\tau_\mathsf{a})$.

Since b satisfies Φ, Φ' holds in $\langle \mathfrak{A}, \mathsf{a} \rangle$ and thus it holds in $\langle \mathfrak{B}, \mathsf{a} \rangle$. Thus, b satisfies Φ in $\mathfrak{A}$ if and only if it satisfies Φ in $\mathfrak{B}$. That is, $\mathfrak{B}$ is an elementary extension of $\mathfrak{A}$.

Corollary. *Let $\mathfrak{B}$ be an extension of $\mathfrak{A}$. Then $\mathfrak{B}$ is an elementary extension of $\mathfrak{A}$ if and only if $\langle \mathfrak{A}, \mathsf{a} \rangle$ is elementarily equivalent to $\langle \mathfrak{B}, \mathsf{a} \rangle$ for all finite sequences a of elements of A.*

§39. PRIME PRODUCTS

Prime products of algebras were defined in §22. We are going to define the same concept for structures.

Definition 1. *Let $(\mathfrak{A}_i \mid i \in I)$ be a family of structures and let $\mathscr{D}$ be a prime* *dual ideal in the Boolean algebra* $\mathfrak{P}(I)$. (*Briefly*, $\mathscr{D}$ *is* prime over I.) *Set* $A = \prod (A_i \mid i \in I)$. *For* $f, g \in A$, *define* $f \equiv g(\mathscr{D})$ *if and only if*

$$\{i \mid f(i) = g(i)\} \in \mathscr{D}.$$

Let $f^{\vee}$ *denote the equivalence class containing* f. *The* prime product $\prod (\mathfrak{A}_i \mid i \in I)$ *of the* $\mathfrak{A}_i$ *with respect to* $\mathscr{D}$ *is defined on the set* $\prod_{\mathscr{D}} (A_i \mid i \in I)$ *of all* $f^{\vee}$, $f \in \prod (A_i \mid i \in I)$; *the operations are defined by* $f_\gamma(f_0^{\vee}, \cdots, f_{n_\gamma-1}^{\vee}) = f^{\vee}$ *if and only if*

$$\{i \mid f_\gamma(f_0(i), \cdots, f_{n_\gamma-1}(i)) = f(i)\} \in \mathscr{D}$$

and the relations by $r_\gamma(f_0^{\vee}, \cdots, f_{m_\gamma-1}^{\vee})$ *holds if and only if*

$$\{i \mid r_\gamma(f_0(i), \cdots, f_{m_\gamma-1}(i))\} \in \mathscr{D}.$$

Of course, it has to be verified that the operations and relations are well defined; this can be done in the same way as in §22.

Intuitively, Definition 1 can be restated as follows: Let us say that a subset of I is of *measure one* (*measure zero*) if it is (is not) contained in $\mathscr{D}$. Then we define equality and the operations and relations componentwise, except on a set of measure zero.

Prime products are also called *ultra products* in the literature (since dual ideals are also called *filters* and prime dual ideals: *ultrafilters*).

Definition 1 and Theorem 1 are due to J. Łoś [2], although the concept of prime product in special cases was already considered by T. Skolem. The importance of this concept was observed by A. Tarski who pointed out that the compactness theorem (see Theorem 2) can be derived from Theorem 1. Influenced by this discovery, T. Frayne, A. C. Morel, D. S. Scott, H. J. Keisler, S. Kochen, and others developed the theory of prime products.

The prime product of the $\mathfrak{A}_i$ with respect to $\mathscr{D}$ will be denoted by $\prod_{\mathscr{D}} (\mathfrak{A}_i \mid i \in I)$.

Let $\mathfrak{f} \in (\prod (A_i \mid i \in I))^{\omega}$, that is, $\mathfrak{f} = \langle f_0, f_1, \cdots \rangle$. Then $\mathfrak{f}(i)$ will denote the sequence $\langle f_0(i), f_1(i), \cdots \rangle$, and $\mathfrak{f}^{\vee}$ will denote the sequence $\langle f_0^{\vee}, f_1^{\vee}, \cdots \rangle$. Obviously, $\mathfrak{f}(i) \in A_i^{\omega}$ and $\mathfrak{f}^{\vee} \in (\prod_{\mathscr{D}} (A_i \mid i \in I))^{\omega}$.

Corollary. *Let* Φ *be an atomic formula and* $\mathfrak{f} \in (\prod (A_i \mid i \in I))^{\omega}$. *Then* $\mathfrak{f}^{\vee}$ *satisfies the formula* Φ *in* $\prod_{\mathscr{D}} (\mathfrak{A}_i \mid i \in I)$ *if and only if*

$$\{i \mid \mathfrak{f}(i) \text{ satisfies } \Phi \text{ in } \mathfrak{A}_i\} \in \mathscr{D}.$$

Proof. Trivial by definition.

The most important property of prime products is that if Φ is a first order sentence that holds in every $\mathfrak{A}_i$, then it holds in $\prod_{\mathscr{D}} (\mathfrak{A}_i \mid i \in I)$. The proof again uses induction on formulas. Thus the basic result, Theorem 1, is a statement about arbitrary formulas.

Theorem 1. *Let* $\mathfrak{A} = \prod_{\mathscr{D}} (\mathfrak{A}_i \mid i \in I)$. *Let* Φ *be a formula and let* $\mathfrak{f} \in (\prod (A_i \mid i \in I))^\omega$. *Then* Φ *is satisfied by* $\mathfrak{f}^\mathrm{v}$ *in* $\mathfrak{A}$ *if and only if*

$$\{i \mid \mathfrak{f}(i) \textit{ satisfies } \Phi \textit{ in } \mathfrak{A}_i\} \in \mathscr{D}.$$

Intuitively, we get satisfaction if the components satisfy Φ on a set of measure 1.

Proof. For every formula Φ and $\mathfrak{f} \in (\prod (A_i \mid i \in I))^\omega$ set $S(\Phi, \mathfrak{f}) = \{i \mid i \in I,\ \Phi \text{ is satisfied by } \mathfrak{f}(i) \text{ in } \mathfrak{A}_i\}$, and $\Gamma = \{\Phi \mid \Phi \text{ is satisfied by } \mathfrak{f}^\mathrm{v} \text{ in } \mathfrak{A}$ if and only if $S(\Phi, \mathfrak{f}) \in \mathscr{D}\}$.

(i) If Φ is atomic, then $\Phi \in \Gamma$ by the corollary to Definition 1.

(ii) If $\Phi \in \Gamma$, then $\neg\Phi \in \Gamma$. First note that $S(\neg\Phi, \mathfrak{f}) = I - S(\Phi, \mathfrak{f})$. Now, $\neg\Phi$ is satisfied by $\mathfrak{f}^\mathrm{v}$ if and only if Φ is not satisfied by $\mathfrak{f}^\mathrm{v}$, which is equivalent to $S(\Phi, \mathfrak{f}) \notin \mathscr{D}$ since $\Phi \in \Gamma$. This is equivalent to $I - S(\Phi, \mathfrak{f}) = S(\neg\Phi, \mathfrak{f}) \in \mathscr{D}$ which was to be proved.

(iii) If $\Phi_0, \Phi_1 \in \Gamma$, then $\Phi_0 \vee \Phi_1 \in \Gamma$. Using the identity

$$S(\Phi_0, \mathfrak{f}) \cup S(\Phi_1, \mathfrak{f}) = S(\Phi_0 \vee \Phi_1, \mathfrak{f}),$$

we can reason as in (ii).

(iv) If $\Phi \in \Gamma$, then $(\exists \mathbf{x}_k)\Phi \in \Gamma$. $(\exists \mathbf{x}_k)\Phi$ is satisfied by $\mathfrak{f}^\mathrm{v}$ if and only if Φ is satisfied in $\mathfrak{A}$ by some $\mathfrak{f}^\mathrm{v}(k/g^\mathrm{v})$, $g \in \prod (A_i \mid i \in I)$, that is, if and only if $S(\Phi, \mathfrak{f}(k/g)) \in \mathscr{D}$. This is equivalent to $\{i \mid \mathfrak{f}(i)(k/g(i)) \text{ satisfies } \Phi \text{ in } \mathfrak{A}_i\} \in \mathscr{D}$. Since $S((\exists \mathbf{x}_k)\Phi, \mathfrak{f}) \supseteq \{i \mid \mathfrak{f}(i)(k/g(i)) \text{ satisfies } \Phi \text{ in } \mathfrak{A}_i\}$, we get that

$$S((\exists \mathbf{x}_k)\Phi, \mathfrak{f}) \in \mathscr{D}.$$

Conversely, if $S((\exists \mathbf{x}_k)\Phi, \mathfrak{f}) \in \mathscr{D}$, then define $g(i) \in A_i$ for $i \in S((\exists \mathbf{x}_k)\Phi, \mathfrak{f})$ such that $\mathfrak{f}(i)(k/g(i))$ satisfies Φ in $\mathfrak{A}_i$.

Define $g(i)$ in an arbitrary manner for $i \notin S((\exists \mathbf{x}_k)\Phi, \mathfrak{f})$. Then $S((\exists \mathbf{x}_k)\Phi, \mathfrak{f}) \subseteq S(\Phi, \mathfrak{f}(k/g))$. Thus, $S(\Phi, \mathfrak{f}(k/g)) \in \mathscr{D}$. Since $\Phi \in \Gamma$, this means that Φ is satisfied by $\mathfrak{f}^\mathrm{v}(k/g^\mathrm{v})$ in $\mathfrak{A}$, which implies that $(\exists \mathbf{x}_k)\Phi$ is satisfied by $\mathfrak{f}^\mathrm{v}$ in $\mathfrak{A}$.

By the definition of formulas, this completes the proof of Theorem 1.

Corollary 1. *Let* Φ *be a sentence and let* $\mathfrak{A}_i$, $i \in I$, *be structures in which* Φ *holds. Then* Φ *holds in* $\prod_{\mathscr{D}} (\mathfrak{A}_i \mid i \in I)$ *for any* $\mathscr{D}$ *prime over* I. *Further, the sentence* Φ *holds in* $\prod_{\mathscr{D}} (\mathfrak{A}_i \mid i \in I)$ *if and only if* $\{i \mid \Phi \textit{ holds in } \mathfrak{A}_i\} \in \mathscr{D}$, *where* $\mathscr{D}$ *is prime over* I.

A typical application is the following.

Corollary 2. *Let* $\mathscr{D}$ *be prime over* I *and let* n *be a positive integer. Then* $\prod_{\mathscr{D}}(\mathfrak{A}_i \mid i \in I)$ *has* n *elements if and only if* $\{i \mid \mathfrak{A}_i \textit{ has } n \textit{ elements}\} \in \mathscr{D}$.

Indeed, in §38 we constructed a sentence Φ that holds if and only if the structure has n elements. The second part of Corollary 1 applied to this sentence Φ yields Corollary 2.

A very important consequence of Theorem 1 is the so-called *compactness theorem:*

Theorem 2. *Let Σ be a set of sentences. Assume that for each finite subset Σ_1 of Σ there is a structure in which all sentences of Σ_1 are satisfied. Then there exists a structure in which all sentences of Σ are satisfied. In other words, if each finite subset of Σ has a model, then Σ has a model.*

Remark. For languages of finite type, Theorem 2 is a trivial corollary to Gödel's completeness theorem. It was observed by A. I. Mal'cev and later by L. A. Henkin that the result can be extended to languages of arbitrary type.

The reason for the name "compactness theorem" is that it implies the compactness of a certain topological space (see Exercises).

Proof. We can assume that Σ is closed under conjunction. Let $\Sigma = \{\Phi_i \mid i \in I\}$. Let $\mathfrak{A}_i$ be a structure in which Φ_i is satisfied. Set $I_j = \{i \mid i \in I$ and Φ_j holds in $\mathfrak{A}_i\}$. Now if $\Phi_j \wedge \Phi_k = \Phi_l$ then $I_j \cap I_k = I_l$. Thus, by the corollary to Theorem 6.7, there exists a $\mathscr{D}$ prime over I such that $I_j \in \mathscr{D}$ for all $j \in I$.

Set $\mathfrak{A} = \prod_{\mathscr{D}} (\mathfrak{A}_i \mid i \in I)$. By Corollary 1 to Theorem 1, Φ_j holds in $\mathfrak{A}$ if and only if $\{i \mid \Phi_j$ holds in $\mathfrak{A}_i\} \in \mathscr{D}$, that is, if and only if $I_j \in \mathscr{D}$, which holds by construction. Thus, every $\Phi \in \Sigma$ holds in $\mathfrak{A}$, which completes the proof of Theorem 2.

Corollary 2 settles the question: When is a prime product finite?

The next result gives some further information on the cardinality of prime products of finite structures. We can restrict ourselves to considering nonprincipal prime dual ideals since if $\mathscr{D}$ is principal, that is, $\mathscr{D} = \{X \mid X \subseteq I$ and $p \in X\}$ for some $p \in I$, then $\prod_{\mathscr{D}} (\mathfrak{A}_i \mid i \in I) \cong \mathfrak{A}_p$ (apply the dual of the corollary to Lemma 22.5); thus, in this case the cardinality of the prime product is the same as that of $\mathfrak{A}_p$.

Theorem 3†. *Let $\mathfrak{A}_i$, $i \in I$, be finite structures and let $\mathscr{D}$ be a nonprincipal prime dual ideal over I. Then $\prod_{\mathscr{D}} (\mathfrak{A}_i \mid i \in I)$ is either finite or of power $\geqq 2^{\aleph_0}$.*

Proof. First we construct a family $\mathscr{F}$ of functions of positive integers into positive integers such that: $|\mathscr{F}| = 2^{\aleph_0}$; if $f, g \in \mathscr{F}$ and $f \neq g$, then $\{i \mid f(i) = g(i)\}$ is finite; $f(n) < 2^n$ for all $n > 0$.

† For Theorems 3 and 4, see, e.g., T. Frayne, A. C. Morel, and D. S. Scott [1].

Let A be the set of all sequences of type ω of zeros and ones, starting with a 1. If $\varphi \in A$, define $f_\varphi(n) = \sum_{i<n} \varphi(i)2^i$ for $n > 0$ and set $\mathscr{F} = \{f_\varphi \mid \varphi \in A\}$. Since every positive integer n has a unique representation of the form $\sum x_i 2^i$, where $x_i = 0$ or 1, we get that $\mathscr{F}$ has all the required properties.

Let us assume that $\prod_{\mathscr{D}} (\mathfrak{A}_i \mid i \in I)$ is not finite.

By Corollary 2, we can find a sequence $0 < k_1 < k_2 < k_3 < \cdots$ such that $I_j = \{i \mid \mathfrak{A}_i$ has at least 2^{k_j} but less than $2^{k_{j+1}}$ elements$\}$ is nonvoid and $I_j \notin \mathscr{D}$ for $j = 1, 2, \cdots$.

Thus, if $i \in I_j$, then we can pick in $\mathfrak{A}_i$ at least 2^{k_j} distinct elements, $a_1^i, \cdots, a_{2^{k_j}}^i$. Define for every $f \in \mathscr{F}$ an $\bar{f} \in \prod (\mathfrak{A}_i \mid i \in I)$ by $\bar{f}(i) = a^i_{f(j)}$, where $i \in I_j$.

If $f \neq g$, then $\{i \mid \bar{f}(i) = \bar{g}(i)\} \subseteq \bigcup (I_l \mid l \leqq n) \notin \mathscr{D}$, for some $n < \omega$; thus, $\bar{f}^{\vee} \neq \bar{g}^{\vee}$. Therefore, we have found $2^{\aleph_0}$ distinct elements in the prime product.

The prime products of finite structures, and, in particular, those of finite relational systems, are very important, as is shown by the following result.

Theorem 4. *Let $\mathfrak{A}$ be a relational system. Let $\{\mathfrak{A}_i, i \in I\}$ be the set of all finite subsystems of $\mathfrak{A}$. Then there exists a $\mathscr{D}$ prime over I such that $\mathfrak{A}$ can be embedded in $\prod_{\mathscr{D}} (\mathfrak{A}_i \mid i \in I)$.*

Proof. Let $g \in \prod (A_i \mid i \in I)$. Set $L_i = \{j \mid A_j \supseteq A_i\}$. Now if $A_i \cup A_k = A_l$, then $L_i \cap L_k = L_l$. Thus, by the corollary to Theorem 6.7 there exists a $\mathscr{D}$ prime over I such that $L_i \in \mathscr{D}$ for all $i \in I$.

For $a \in A$, define $f_a \in \prod (A_i \mid i \in I)$ by

$$f_a(i) = \begin{cases} a & \text{if} \quad a \in A_i, \\ g(i) & \text{if} \quad a \notin A_i. \end{cases}$$

If $a \neq b$ and $a, b \in A_j$, then $\{i \mid f_a(i) \neq f_b(i)\} \supseteq L_j \in \mathscr{D}$; thus, $f_a^{\vee} \neq f_b^{\vee}$. This shows that the mapping $\varphi: a \to f_a^{\vee}$ is 1–1. It remains to show that $r_\gamma(a_0, \cdots, a_{m_\gamma - 1})$ if and only if $r_\gamma(f^{\vee}_{a_0}, \cdots, f^{\vee}_{a_{m_\gamma - 1}})$.

Indeed, if $r_\gamma(a_0, \cdots, a_{m_\gamma - 1})$, then $B = \{i \mid r_\gamma(f_{a_0}(i), \cdots, f_{a_{m_\gamma - 1}}(i))\} \supseteq L_j$, where $\{a_0, \cdots, a_{m_\gamma - 1}\} = A_j$. Thus, $B \in \mathscr{D}$ and, by Corollary 1 to Theorem 1, $r_\gamma(f^{\vee}_{a_0}, \cdots, f^{\vee}_{a_{m_\gamma - 1}})$.

Conversely, if $r_\gamma(f^{\vee}_{a_0}, \cdots, f^{\vee}_{a_{m_\gamma - 1}})$, then $B = \{i \mid r_\gamma(f_{a_0}(i), \cdots, f_{a_{m_\gamma - 1}}(i))\} \in \mathscr{D}$. Let $A_j = \{a_0, \cdots, a_{m_\gamma - 1}\}$. Pick an $i \in B \cap L_j$. Then we have $r_\gamma(f_{a_0}(i), \cdots, f_{a_{m_\gamma - 1}}(i))$, that is, $r_\gamma(a_0, \cdots, a_{m_\gamma - 1})$ in $\mathfrak{A}_i$. Thus, $r_\gamma(a_0, \cdots, a_{m_\gamma - 1})$ in $\mathfrak{A}$, which completes the proof.

Theorem 4 was formulated for relational systems because in this case every finite subset is a subsystem. For algebras and structures, the following two variants hold.

Corollary 1. *Every structure can be embedded in some prime product of its finitely generated substructures.*

Corollary 2. *Let $\mathfrak{A}$ be an algebra and let K be a class of algebras that is closed under the formation of prime products. Assume that every finite relative subalgebra of $\mathfrak{A}$ can be embedded in some algebra in K. Then $\mathfrak{A}$ can be embedded in some algebra in K.*

The following result is an interesting application of Theorem 4.

Theorem 5. *Let $\mathfrak{A}$ be a relational system of type τ. Let τ' denote the type obtained from τ by adding an n-ary relational symbol* **r** *(thus, $o(\tau') = o(\tau) + 1$). Let Σ denote a set of universal sentences in $L(\tau')$. Let us assume that on every finite subsystem of $\mathfrak{A}$ we can define r such that Σ is satisfied. Then r can be defined on $\mathfrak{A}$ so as to satisfy Σ.*

Proof. By defining r on every finite subsystem $\mathfrak{A}_i$, we have it defined on $\mathfrak{B} = \prod_{\mathscr{D}} (\mathfrak{A}_i \mid i \in I)$ of Theorem 4. By Theorem 1, Σ will be satisfied in $\mathfrak{B}$, and since Σ consists of universal sentences it will be satisfied in every subsystem, which completes the proof.

We conclude this section by applying prime products to a special type of equational classes.

Theorem 6 (*B. Jónsson* [8]). *Let K_0 be a class of algebras and let K be the equational class generated by K_0. Let us assume that the congruence lattice of every algebra in K is distributive. Then*

$$K = \mathbf{IP}_S\mathbf{HSP}_P(K_0),$$

where $\mathbf{P}_P$ is the operator of the formation of prime products.

Remark. By Theorem 23.2, $K = \mathbf{HSP}(K_0)$, which involves only three operators, while the expression in Theorem 6 involves five. Despite this, Theorem 6 is a very strong statement. To illustrate its power, take $K_0 = \{\mathfrak{A}\}$, where $\mathfrak{A}$ is a primal algebra. Corollary 1 to Theorem 1 and Exercise 20 yield $\mathbf{P}_P(K_0) \subseteq \mathbf{I}(K_0)$. By Lemma 27.5, $\mathbf{HSI}(\mathfrak{A}) = \mathbf{I}(K_0)$; thus by Exercise 5.70 ($n = 2$) and Theorem 6, $K = \mathbf{IP}_S\mathbf{I}(\mathfrak{A})$, which is the only nontrivial part of Theorem 27.5. Other applications of Theorem 6 will be given in the Exercises.

Proof of Theorem 6. By the proof of Theorem 20.3 every $\mathfrak{A} \in K$ is isomorphic to a subdirect product of subdirectly irreducible algebras in K. Therefore it is enough to prove that if $\mathfrak{A} \in K$ and $\mathfrak{A}$ is subdirectly irreducible, then $\mathfrak{A} \in \mathbf{HSP}_P(K_0)$.

So let $\mathfrak{A} \in K$ be subdirectly irreducible. We can assume that $|A| > 1$. Since $\mathfrak{A} \in K = \mathbf{HSP}(K_0)$, there exist $\mathfrak{A}_i \in K_0$ for $i \in I$, a subalgebra $\mathfrak{B}$ of $\prod (\mathfrak{A}_i \mid i \in I)$ and a congruence relation Φ on $\mathfrak{B}$ such that $\mathfrak{B}/\Phi \cong \mathfrak{A}$.

For $J \subseteq I$ let Θ_J be the congruence relation on $\prod (\mathfrak{A}_i \mid i \in I)$ defined by $f \equiv g(\Theta_J)$ if and only if $\{i \mid f(i) \neq g(i)\} \subseteq I - J$. Set

$$\mathscr{E} = \{J \mid J \subseteq I \text{ and } (\Theta_J)_B \leqq \Phi\}.$$

$\mathscr{E}$ has the following properties:

(*) $I \in \mathscr{E}$; also $J \in \mathscr{E}$, $J \subseteq L$ imply $L \in \mathscr{E}$.

(**) $M, N \subseteq I$, $M \cup N \in \mathscr{E}$ imply that $M \in \mathscr{E}$ or $N \in \mathscr{E}$.

(*) is obvious. To prove (**), let $M \cup N \in \mathscr{E}$, that is,

$$(\Theta_{M \cup N})_B \leqq \Phi.$$

A trivial calculation shows that $\Theta_{M \cup N} = \Theta_M \wedge \Theta_N$ and

$$(\Theta_M \wedge \Theta_N)_B = (\Theta_M)_B \wedge (\Theta_N)_B,$$

thus

$$((\Theta_M)_B \wedge (\Theta_N)_B) \vee \Phi = \Phi.$$

Since $\mathfrak{C}(\mathfrak{B})$ is distributive, we get

$$\Phi = (\Phi \vee (\Theta_M)_B) \wedge (\Phi \vee (\Theta_N)_B).$$

$\mathfrak{B}/\Phi$ is subdirectly irreducible; therefore Φ is meet-irreducible and so $\Phi = \Phi \vee (\Theta_M)_B$ or $\Phi = \Phi \vee (\Theta_N)_B$. Thus $(\Theta_M)_B \leqq \Phi$, or $(\Theta_N)_B \leqq \Phi$; that is, $M \in \mathscr{E}$ or $N \in \mathscr{E}$. Finally, $\varnothing \notin \mathscr{E}$ since $|A| > 1$.

Let $\mathscr{D}$ be a dual ideal of $\mathfrak{P}(I)$ maximal with respect to the property $\mathscr{D} \subseteq \mathscr{E}$. We prove that $\mathscr{D}$ is prime over I. Indeed, if $\mathscr{D}$ were not prime, then there would exist $J \subseteq I$ with $J \notin \mathscr{D}$, $I - J \notin \mathscr{D}$. If for every $L \in \mathscr{D}$, $J \cap L \in \mathscr{E}$, then (*) would imply that $\mathscr{D}$ and J generate a dual ideal contained in $\mathscr{E}$, and properly containing $\mathscr{D}$, thus contradicting the maximality of $\mathscr{D}$. Thus there exist $L_0 \in \mathscr{D}$ and $L_1 \in \mathscr{D}$ with $J \cap L_0 \notin \mathscr{E}$ and $(I - J) \cap L_1 \notin \mathscr{E}$. Set $L = L_0 \cap L_1$. Then $L \in \mathscr{E}$, $J \cap L \notin \mathscr{E}$, $(I - J) \cap L \notin \mathscr{E}$, and $L = (J \cap L) \cup ((I - J) \cap L)$, which contradicts (**).

Putting $\Theta_{\mathscr{D}} = \bigcup (\Theta_J \mid J \in \mathscr{D})$, we have $(\Theta_{\mathscr{D}})_B \leqq \Phi$. Thus

$$\prod\nolimits_{\mathscr{D}} (\mathfrak{A}_i \mid i \in I) \in \mathbf{P}_P(K_0),$$

$\mathfrak{B}/(\Theta_{\mathscr{D}})_B \in \mathbf{ISP}_P(K_0)$, and by the second isomorphism theorem $\mathfrak{A}$ is isomorphic to

$$(\mathfrak{B}/(\Theta_{\mathscr{D}})_B)/(\Phi/(\Theta_{\mathscr{D}})_B) \in \mathbf{HSP}_P(K_0),$$

which was to be proved.

Remark. In the proof we used only that $\mathfrak{A}$ is "finitely" subdirectly irreducible. Hence we proved that all "finitely" subdirectly irreducible algebras belong to $\mathbf{HSP}_P(K_0)$.

§40. PRIME POWERS

If $\mathfrak{A}_i = \mathfrak{A}$ for all $i \in I$ and $\mathscr{D}$ is prime over I, then $\prod_{\mathscr{D}} (\mathfrak{A}_i \mid i \in I)$ will be called a *prime power* of $\mathfrak{A}$ and will be denoted by $\mathfrak{A}_{\mathscr{D}}{}^I$.

$\mathfrak{A}$ has a *natural embedding* into $\mathfrak{A}_{\mathscr{D}}{}^I$. Indeed, for $a \in A$, define $f_a{}^{\mathrm{v}} \in A_{\mathscr{D}}{}^I$ by $f_a(i) = a$ for each $i \in I$. If $a \neq b$, then $\{i \mid f_a(i) \neq f_b(i)\} = I \in \mathscr{D}$; thus, $f_a{}^{\mathrm{v}} \neq f_b{}^{\mathrm{v}}$. The rest of the proof is similar.

Theorem 1. *The natural embedding φ of $\mathfrak{A}$ into $\mathfrak{A}_{\mathscr{D}}{}^I$, where $\mathscr{D}$ is prime over I, is an elementary embedding. That is, φ is an embedding and the image of $\mathfrak{A}$ under φ is an elementary substructure of $\mathfrak{A}_{\mathscr{D}}{}^I$.*

Proof. Let $\tilde{\mathfrak{A}}$ be the image of $\mathfrak{A}$ under φ. As we noted above $\tilde{\mathfrak{A}}$ is a substructure of $\mathfrak{A}_{\mathscr{D}}{}^I$. Let Φ be a formula and assume that $(\exists \mathbf{x}_k)\Phi$ is satisfied by $\mathfrak{f}^{\mathrm{v}} \in \tilde{A}^{\omega}$ in $\mathfrak{A}_{\mathscr{D}}{}^I$. Then there exists a $g^{\mathrm{v}} \in A_{\mathscr{D}}{}^I$ such that $\mathfrak{f}^{\mathrm{v}}(k/g^{\mathrm{v}})$ satisfies Φ in $\mathfrak{A}_{\mathscr{D}}{}^I$. Thus, by Theorem 39.1, $\{i \mid \mathfrak{f}(i)(k/g(i))$ satisfies Φ in $\mathfrak{A}_i\} \in \mathscr{D}$.

Therefore, for some $i \in I$, $\mathfrak{f}(i)(k/g(i))$ satisfies Φ in $\mathfrak{A}$. Define $(g^*)^{\mathrm{v}} \in A_{\mathscr{D}}{}^I$ by $g^*(j) = g(i)$ for all $j \in I$. Then $(g^*)^{\mathrm{v}} \in \tilde{A}$ and $\mathfrak{f}^{\mathrm{v}}(k/(g^*)^{\mathrm{v}})$ satisfies Φ in $\mathfrak{A}_{\mathscr{D}}{}^I$. Thus, by Theorem 38.2, $\tilde{\mathfrak{A}}$ is an elementary substructure of $\mathfrak{A}_{\mathscr{D}}{}^I$.

Corollary 1. *A structure is elementarily equivalent to any of its prime powers.*

Corollary 2. *Any prime power of a finite structure is isomorphic to the structure itself.*

Now we can attack the problem that is the counterpart of the Löwenheim-Skolem-Tarski Theorem (Theorem 38.3)—namely, if we are given a structure $\mathfrak{A}$ and a cardinal number $\mathfrak{m} \geqq |A|$, we want to know if $\mathfrak{A}$ has an elementary extension of cardinality $\mathfrak{m}$ (the upward theorem).

Theorem 2. *Let $\mathfrak{A}$ be an infinite structure and let $\mathfrak{m}$ be an infinite cardinal number. Then, for every set I of cardinality $\mathfrak{m}$, there exists a $\mathscr{D}$ prime over I such that $|A_{\mathscr{D}}{}^I| \geqq 2^{\mathfrak{m}}$. That is, $\mathfrak{A}_{\mathscr{D}}{}^I$ is isomorphic to an elementary extension of $\mathfrak{A}$ of cardinality $\geqq 2^{\mathfrak{m}}$.*

The proof is based on the following observation.†

† This proof is from T. Frayne, A. C. Morel, and D. S. Scott [1].

Lemma 1. *Let $(A_i \mid i \in I)$ be a family of sets and let $\{J_k \mid k \in K\}$ be the set of all finite subsets of I. Then there exists a $\mathscr{D}_1$ prime over K such that $\prod (A_i \mid i \in I)$ can be embedded into $\prod_{\mathscr{D}_1} (\prod (A_i \mid i \in J_k) \mid k \in K)$.*

Proof. Set $L_i = \{k \mid k \in K, i \in J_k\}$ for $i \in I$. Note that

$$L_{i_1} \cap L_{i_2} \cap \cdots \cap L_{i_n} = \{k \mid \{i_1, \cdots, i_n\} \subseteq J_k\}.$$

Thus, $L_{i_1} \cap \cdots \cap L_{i_n} \neq \varnothing$. Therefore, there exists a $\mathscr{D}_1$ prime over K such that $L_i \in \mathscr{D}_1$ for all $i \in I$.

Let $f \in \prod (A_i \mid i \in I)$ and define $f\varphi \in \prod (\prod (A_i \mid i \in J_k) \mid k \in K)$ by $f\varphi(k) = f_{J_k}$, the restriction of f to J_k.

If $f \neq g$, then $f(i) \neq g(i)$ for some $i \subset I$ and therefore $f_{J_k} \neq g_{J_k}$ if $i \in J_k$. Therefore,

$$\{k \mid f\varphi(k) \neq g\varphi(k)\} \supseteq L_i \in \mathscr{D}_1,$$

which proves that $(f\varphi)^{\vee} \neq (g\varphi)^{\vee}$.

Now we prove Theorem 2.

In Lemma 1, put $A_i = A$ for all $i \in I$. Since A is infinite, $|A| = |A^{J_k}|$. Thus, if $\mathfrak{m} \leqq |I|$, then

$$\begin{aligned} 2^{\mathfrak{m}} &\leqq |A^I| \leqq \text{(by Lemma 1)} \\ &\leqq |\textstyle\prod_{\mathscr{D}_1} (A^{J_k} \mid k \in K)| = (\text{since } |A| = |A^{J_k}|) \\ &= |A^K_{\mathscr{D}_1}|. \end{aligned}$$

Since $|I| = |K|$, we can find a $\mathscr{D}$ prime over I satisfying $|A^K_{\mathscr{D}_1}| = |A_{\mathscr{D}}{}^I|$, concluding the proof of Theorem 2.

Corollary 1. *Let $\mathfrak{A}$ be an infinite structure of countable type τ and let $\aleph_0 \leqq \mathfrak{m} \leqq |A| \leqq \mathfrak{n}$. Then $\mathfrak{A}$ has an elementary substructure of cardinality $\mathfrak{m}$ and a proper elementary extension of cardinality $\mathfrak{n}$.*

Corollary 2. *Let $\mathfrak{A}$ be an infinite structure of type τ and let $\mathfrak{m}$ be a cardinal number with $\overline{\overline{o(\tau)}} \leqq \mathfrak{m}$ and $|A| \leqq \mathfrak{m}$. Then $\mathfrak{A}$ has a proper elementary extension of cardinality $\mathfrak{m}$.*

We get Corollaries 1 and 2 by combining Theorem 2 and Theorem 38.3. Of course, the question arises as to whether the condition $\overline{\overline{o(\tau)}} \leqq \mathfrak{m}$ is necessary to get the conclusion of Corollary 2. The following result shows that if we drop this condition, then the conclusion of Corollary 2 fails to hold.

Theorem 3 (*M. O. Rabin* [1]). *There exists a countable structure* $\mathfrak{A}$ *such that every proper elementary extension of* $\mathfrak{A}$ *has a power* $\geqq \mathfrak{c}$, *the power of the continuum.*

Proof. Let A be the set of nonnegative integers. We consider every element of A as a constant of our language. Let $\mathscr{F}$ be the family of functions constructed in Theorem 39.3, and let $f \in \mathscr{F}_0$ if $f \in \mathscr{F}$ and f arises from a 0, 1 sequence that is not eventually 0. For each $f \in \mathscr{F}_0$ define a unary relational symbol $\mathbf{r}_f$ such that $r_f(a)$, $a \in A$, if and only if $a = f(n)$ for some $n \in A$. Let $\mathfrak{A}$ be the relational system defined on A with the constants and relations described above and with the natural $\leqq$ relation. Then $\mathfrak{A}$ has the following properties each of which can be easily described by infinitely many first order sentences:

(i) For every $a \in A$, if $a \leqq n$, then there are at most n elements $\leqq a$ each of which is a constant.

(ii) for every $a \in A$ and $f \in \mathscr{F}_0$, there is a smallest $b \in A$ such that $a \leqq b$ and $r_f(b)$.

(iii) for every $f, g \in \mathscr{F}_0$ with $f \neq g$, there are finitely many elements $a \in A$ such that $r_f(a)$ and $r_g(a)$, and each of these elements is a constant.

Now, let $\mathfrak{B}$ be a proper elementary extension of $\mathfrak{A}$ and let $a \in B$, $a \notin A$. Since property (i) holds for $\mathfrak{B}$ as well, but all the constants are in $\mathfrak{A}$, we get immediately that $n < a$ for all $n \in A$.

Thus, if, for each $f \in \mathscr{F}_0$, a_f denotes the smallest element b with $a \leqq b$ and $r_f(b)$ (which exists by (ii)), then $a_f \notin A$. If $f \neq g$, then $a_f \neq a_g$ since $r_f(\mathrm{b})$ and $r_g(b)$ imply $b \in A$, by (iii). Therefore, the subset of B defined by $\{a_f \mid f \in \mathscr{F}_0\}$ has the same power as $\mathscr{F}_0$, that is, $\mathfrak{c}$.

M. O. Rabin [1] and H. J. Keisler [4] have some further results on this problem.

§41. TWO ALGEBRAIC CHARACTERIZATIONS OF ELEMENTARY EQUIVALENCE

In the last section, we found an algebraic construction, the prime power, which produced elementary extensions.

In this section we will show that the prime power combined with direct limits yields an algebraic characterization of elementary equivalence (Theorem 1). Further, if we assume the Generalized Continuum Hypothesis, then the prime power itself yields the characterization (corollary to Theorem 2).

Both results are essentially due to H. J. Keisler [2] and S. Kochen [1]. The prime limit construction is taken from Kochen's work (Keisler's construction is somewhat different). The second construction is taken from Keisler's work. (Kochen announced in an abstract that the corollary to Theorem 2 can be derived from a more general result which was shown to be incorrect; later on, Kochen corrected the mistake in his construction).

We start with the following lemma due to T. Frayne.

Lemma 1. *Two structures $\mathfrak{A}$ and $\mathfrak{B}$ are elementarily equivalent if and only if $\mathfrak{B}$ can be elementarily embedded into some prime power $\mathfrak{A}_{\mathscr{D}}{}^{I}$ of $\mathfrak{A}$.*

Proof. If $\mathfrak{B}$ has an elementary embedding into some $\mathfrak{A}_{\mathscr{D}}{}^{I}$, then it is elementarily equivalent to $\mathfrak{A}_{\mathscr{D}}{}^{I}$ and, by Theorem 40.1, to $\mathfrak{A}$ as well.

Let $L^{(\alpha)}(\tau)$ denote the language for the type τ which differs from $L(\tau)$ only in that it contains α individual variables $\mathbf{x}_0, \cdots, \mathbf{x}_\gamma, \cdots, \gamma<\alpha$ (thus, $L(\tau)=L^{(\omega)}(\tau)$).

Let $\mathfrak{A}$ and $\mathfrak{B}$ be elementarily equivalent and let $B=\{b_\gamma \mid \gamma<\alpha\}$. Let Γ denote the set of all formulas $\Phi(\mathbf{x}_{\gamma_0}, \cdots, \mathbf{x}_{\gamma_{n-1}})$, $\gamma_0, \cdots, \gamma_{n-1}<\alpha$, free at most in $\mathbf{x}_{\gamma_0}, \cdots, \mathbf{x}_{\gamma_{n-1}}$, for which $\Phi(b_{\gamma_0}, \cdots, b_{\gamma_{n-1}})$ holds in $\mathfrak{B}$. If $\Phi(\mathbf{x}_{\gamma_0}, \cdots, \mathbf{x}_{\gamma_{n-1}}) \in \Gamma$, then the sentence $(\exists \mathbf{x}_{\gamma_0}), \cdots, (\exists \mathbf{x}_{\gamma_{n-1}})\Phi$ holds in $\mathfrak{B}$, and therefore it holds in $\mathfrak{A}$. Thus, there exists a sequence of elements $\langle a_{\Phi,\gamma} \mid \gamma<\alpha\rangle$ of $\mathfrak{A}$ such that $\Phi(a_{\Phi,\gamma_0}, \cdots, a_{\Phi,\gamma_{n-1}})$ holds in $\mathfrak{A}$.

We set $I=\Gamma$ and we consider a prime dual ideal $\mathscr{D}$ over I that contains for all $\Phi \in \Gamma$ the set

$$I_\Phi = \{\Psi \mid \Psi(\mathbf{x}_{\gamma_0}, \cdots, \mathbf{x}_{\gamma_{m-1}}) \in \Gamma \text{ and } \Phi(a_{\Psi,\gamma_0}, \cdots, a_{\Psi,\gamma_{m-1}}) \text{ holds in } \mathfrak{A}\}.$$

Such a $\mathscr{D}$ exists since the sets I_Φ have the finite intersection property. Indeed, $\Phi_0 \vee \Phi_1 \vee \cdots \vee \Phi_{k-1} \in I_{\Phi_0} \cap \cdots \cap I_{\Phi_{k-1}}$.

For $\gamma<\alpha$, define $f_\gamma \in A^I$ by $f_\gamma(\Phi)=a_{\Phi,\gamma}$. Then we claim the mapping $\varphi\colon b_\gamma \to f_\gamma{}^{\mathrm{v}}$ is an elementary embedding of $\mathfrak{B}$ into $\mathfrak{A}_{\mathscr{D}}{}^{I}$. Indeed, if $\Phi(\mathbf{x}_{\gamma_0}, \cdots, \mathbf{x}_{\gamma_{n-1}})$ is a formula and $\Phi(b_{\gamma_0}, \cdots, b_{\gamma_{n-1}})$ holds in $\mathfrak{B}$, then $\Phi \in \Gamma$. Thus, $I_\Phi \in \mathscr{D}$; that is, $\{\Psi \mid \Phi(a_{\Psi,\gamma_0}, \cdots, a_{\Psi,\gamma_{n-1}}) \text{ holds in } \mathfrak{A}\}=$ $\{\Psi \mid \Phi(f_{\gamma_0}(\Psi), \cdots, f_{\gamma_{n-1}}(\Psi)) \text{ holds in } \mathfrak{A}\} \in \mathscr{D}$. By Theorem 39.1, this means that $\Phi(f_{\gamma_0}^{\mathrm{v}}, \cdots, f_{\gamma_{n-1}}^{\mathrm{v}})$, that is, $\Phi(b_{\gamma_0}\varphi, \cdots, b_{\gamma_{n-1}}\varphi)$ holds in $\mathfrak{A}_{\mathscr{D}}{}^{I}$. If we specialize Φ to atomic formulas and negations of atomic formulas, then we get that φ is an embedding; and applying the result again with arbitrary Φ, we see that φ is an elementary embedding.

Corollary. *The mapping $\psi\colon A \to B$ is an elementary embedding of $\mathfrak{A}$ into $\mathfrak{B}$ if and only if $\mathfrak{B}$ has an elementary embedding φ into some prime power $\mathfrak{A}_{\mathscr{D}}{}^{I}$ of $\mathfrak{A}$ such that $\chi=\psi\varphi$, where χ is the natural embedding of $\mathfrak{A}$ into $\mathfrak{A}_{\mathscr{D}}{}^{I}$.*

The situation is illustrated in the following diagram:

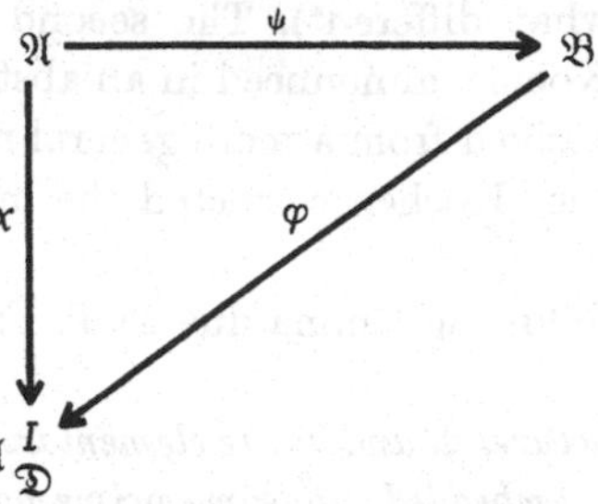

Proof. A trivial combination of Lemma 1 with Theorem 38.6.

Definition 1. *Let $\mathfrak{A}$ be a structure, $\mathfrak{A}_0=\mathfrak{A}$, $\mathfrak{A}_1=(\mathfrak{A}_0)^{I_1}_{\mathscr{D}_1}$, $\mathfrak{A}_2=(\mathfrak{A}_1)^{I_2}_{\mathscr{D}_2}, \cdots$, where $\mathscr{D}_n$ is prime over I_n, and let χ_n be the natural embedding of $\mathfrak{A}_n$ into $\mathfrak{A}_{n+1}$. The structures $\mathfrak{A}_i$, $i=0, 1, 2, \cdots$ with the mappings*

$$\chi_{nm} = \chi_n\chi_{n+1}\cdots\chi_{m-1}$$

for $n<m$, $\chi_{nn}=$the identity map on A_n, form a direct family. Let $\mathfrak{A}_\omega$ denote the direct limit of this direct family. $\mathfrak{A}_\omega$ is called a prime limit *of $\mathfrak{A}$.*

Corollary. *Any prime limit $\mathfrak{A}_\omega$ is isomorphic to an elementary extension of $\mathfrak{A}$.*

This follows by a combination of Theorem 40.1 and Theorem 38.1.

Now we state the first algebraic characterization of elementary equivalence.

Theorem 1. *Two structures are elementarily equivalent if and only if they have isomorphic prime limits.*

Proof. The "if" part is trivial. To prove the "only if" part, let us assume that $\mathfrak{A}$ and $\mathfrak{B}$ are elementarily equivalent. Let $\mathfrak{A}=\mathfrak{A}_0$, $\mathfrak{B}=\mathfrak{B}_0$. By Lemma 1, there exists an elementary embedding φ_1 of $\mathfrak{B}_0$ into some $\mathfrak{A}_1=\mathfrak{A}^{I_1}_{\mathscr{D}_1}$. Let χ_1 denote the natural embedding of $\mathfrak{A}_0$ into $\mathfrak{A}_1$. Applying the corollary to Lemma 1 to $\mathfrak{B}_0$, $\mathfrak{A}_1$, and φ_1, we obtain $\mathfrak{B}_1=(\mathfrak{B}_0)^{J_1}_{\mathscr{E}_1}$ and an elementary embedding ψ_1 of $\mathfrak{A}_1$ into $\mathfrak{B}_1$. If η_1 denotes the natural embedding of $\mathfrak{B}_0$ into $\mathfrak{B}_1$, then $\eta_1=\varphi_1\psi_1$.

By applying the corollary to Lemma 1 to $\mathfrak{A}_1$, $\mathfrak{B}_1$, and ψ_1, we get $\mathfrak{A}_2=(\mathfrak{A}_1)^{I_2}_{\mathscr{D}_2}$, an elementary embedding φ_2 of $\mathfrak{B}_1$ into $\mathfrak{A}_2$, and a natural embedding χ_2 of $\mathfrak{A}_1$ into $\mathfrak{A}_2$ such that $\chi_2=\psi_1\varphi_2$.

We construct in a similar manner $\mathfrak{A}_n$, $\mathfrak{B}_n$, φ_n, ψ_n, χ_n, η_n for each $n<\omega$ (see the diagram).

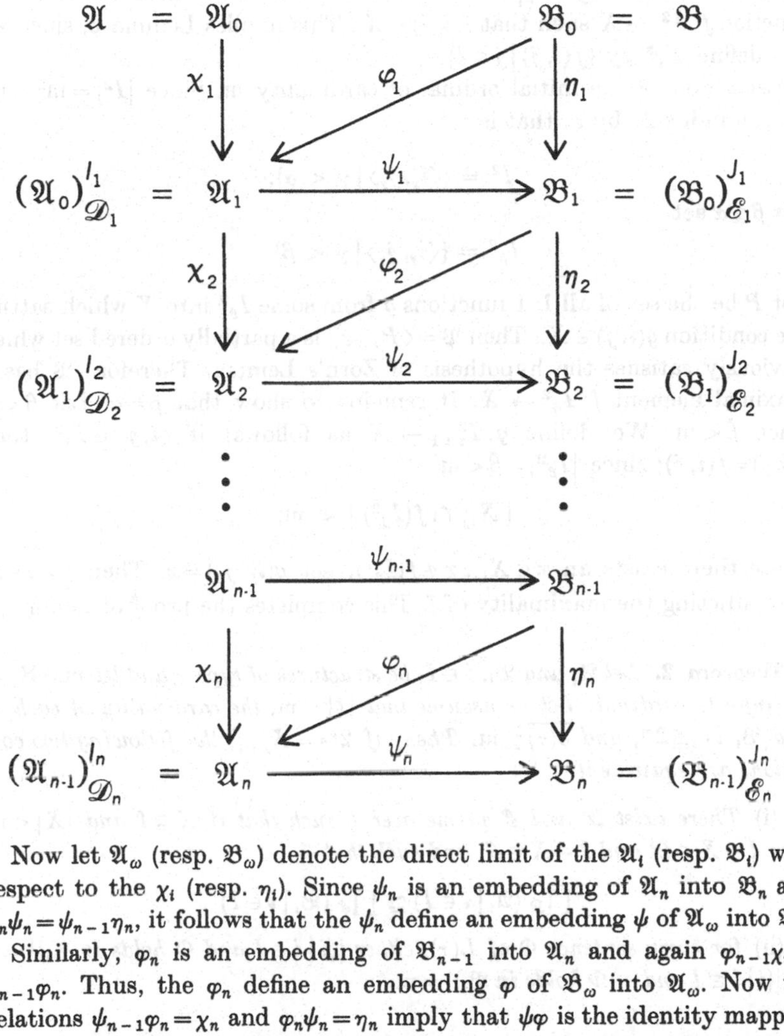

Now let $\mathfrak{A}_\omega$ (resp. $\mathfrak{B}_\omega$) denote the direct limit of the $\mathfrak{A}_i$ (resp. $\mathfrak{B}_i$) with respect to the χ_i (resp. η_i). Since ψ_n is an embedding of $\mathfrak{A}_n$ into $\mathfrak{B}_n$ and $\chi_n\psi_n=\psi_{n-1}\eta_n$, it follows that the ψ_n define an embedding ψ of $\mathfrak{A}_\omega$ into $\mathfrak{B}_\omega$. Similarly, φ_n is an embedding of $\mathfrak{B}_{n-1}$ into $\mathfrak{A}_n$ and again $\varphi_{n-1}\chi_n=\eta_{n-1}\varphi_n$. Thus, the φ_n define an embedding φ of $\mathfrak{B}_\omega$ into $\mathfrak{A}_\omega$. Now the relations $\psi_{n-1}\varphi_n=\chi_n$ and $\varphi_n\psi_n=\eta_n$ imply that $\psi\varphi$ is the identity mapping on $\mathfrak{A}_\omega$ and $\varphi\psi$ is the identity mapping on $\mathfrak{B}_\omega$. Thus, ψ is an isomorphism of $\mathfrak{A}_\omega$ with $\mathfrak{B}_\omega$.

We prepare for the proof of the second characterization theorem by proving the following lemma.

Lemma 2. *Let X_i, $i\in I$, be sets, with $|X_i|=\mathfrak{m}$ for all $i\in I$, and suppose*

$|I| = \mathfrak{m}$, where $\mathfrak{m}$ is an infinite cardinal. Then we can find $X_i^* \subseteq X_i$ such that $|X_i^*| = \mathfrak{m}$ and if $i, j \in I$, $i \neq j$, then $X_i^* \cap X_j^* = \varnothing$.

Proof. Set $X = \bigcup (X_i \mid i \in I)$. We will prove that there exists a 1–1 function $f\colon I^2 \to X$ such that $f(i, j) \in X_i$. This implies Lemma 2, since we can define X_i^* by $\{f(i, j) \mid j \in I\}$.

Let α denote the initial ordinal of cardinality $\mathfrak{m}$. Since $|I^2| = \mathfrak{m}^2 = \mathfrak{m}$, we can index I^2 by α, that is

$$I^2 = \{\langle i_\gamma, j_\gamma\rangle \mid \gamma < \alpha\}.$$

For $\beta \leqq \alpha$ set

$$I_\beta{}^2 = \{\langle i_\gamma, j_\gamma\rangle \mid \gamma < \beta\}.$$

Let P be the set of all 1–1 functions g from some $I_\beta{}^2$ into X which satisfy the condition $g(i, j) \in X_i$. Then $\mathfrak{P} = \langle P; \subseteq\rangle$ is a partially ordered set which obviously satisfies the hypothesis of Zorn's Lemma. Therefore $\mathfrak{P}$ has a maximal element $f\colon I_\beta{}^2 \to X$. It remains to show that $\beta = \alpha$. Let $\beta < \alpha$. Then $\bar{\beta} < \mathfrak{m}$. We define $g\colon I^2_{\beta+1} \to X$ as follows: if $\langle i, j\rangle \in I_\beta{}^2$, then $g(i, j) = f(i, j)$; since $|I_\beta{}^2| = \bar{\beta} < \mathfrak{m}$,

$$| X_{i_\beta} \cap f(I_\beta{}^2) | < \mathfrak{m},$$

hence there exists an $x \in X_{i_\beta}$, $x \notin f(I_\beta{}^2)$; set $g(i_\beta, j_\beta) = x$. Then $f \subset g \in P$, contradicting the maximality of f. This completes the proof of Lemma 2.

Theorem 2. *Let $\mathfrak{A}_i$ and $\mathfrak{B}_i$, $i \in I$, be structures of type τ and let $\mathfrak{m} = \aleph_\alpha$ be an infinite cardinal. Let us assume that $|I| = \mathfrak{m}$, the cardinality of each $\mathfrak{A}_i$ and $\mathfrak{B}_i$ is $\leqq 2^{\mathfrak{m}}$, and $\overline{o(\tau)} \leqq \mathfrak{m}$. Then, if $2^{\aleph_\alpha} = \aleph_{\alpha+1}$, the following two conditions are equivalent:*

(i) *There exist $\mathscr{D}$ and $\mathscr{E}$ prime over I such that if $X \subseteq I$ and $|X| < \mathfrak{m}$, then $I - X \in \mathscr{D}$ and $I - X \in \mathscr{E}$, and such that*

$$\prod_{\mathscr{D}} (\mathfrak{A}_i \mid i \in I) \cong \prod_{\mathscr{E}} (\mathfrak{B}_i \mid i \in I);$$

(ii) *for every sentence Φ of $L(\tau)$, either $|\{i \mid i \in I$ and Φ holds in $\mathfrak{A}_i\}| = \mathfrak{m}$ or $|\{i \mid i \in I$ and $\neg\Phi$ holds in $\mathfrak{B}_i\}| = \mathfrak{m}$.*

Remark. Condition (ii) means that if a sentence Φ holds for fewer than $\mathfrak{m}$ of the $\mathfrak{A}_i$, then $\neg\Phi$ holds for $\mathfrak{m}$ of the $\mathfrak{B}_i$. Another equivalent statement is that there is no sentence Φ such that $\{i \mid i \in I$ and Φ holds in $\mathfrak{A}_i\} \in \bar{P}_{\mathfrak{m}}(I)$ and $\{i \mid i \in I$ and $\neg\Phi$ holds in $\mathfrak{B}_i\} \in \bar{P}_{\mathfrak{m}}(I)$, where $P_{\mathfrak{m}}(I)$ is the set of subsets of I having cardinality less than $\mathfrak{m}$ and $\bar{P}_{\mathfrak{m}}(I)$ is the set of subsets X of I for which $I - X \in P_{\mathfrak{m}}(I)$. The first condition in (i) is $\bar{P}_{\mathfrak{m}}(I) \subseteq \mathscr{D}$ and $\bar{P}_{\mathfrak{m}}(I) \subseteq \mathscr{E}$.

Proof of the Theorem. Suppose (i) holds. Let Φ be a sentence and let $X=\{i \mid i \in I$ and Φ holds in $\mathfrak{A}_i\}$. If $|X|<\mathfrak{m}$, then $I-X \in \bar{P}_{\mathfrak{m}}(I) \subseteq \mathscr{D}$ and thus, by Theorem 39.1, $\neg\Phi$ holds in $\prod_{\mathscr{D}}(\mathfrak{A}_i \mid i \in I)$. If also we have $|Y|<\mathfrak{m}$, where $Y=\{i \mid i \in I, \neg\Phi$ holds in $\mathfrak{B}_i\}$, then similarly Φ holds in $\prod_{\mathscr{E}}(\mathfrak{B}_i \mid i \in I)$, contradicting the isomorphism. Thus, (i) implies (ii).

Now assume that (ii) holds. Let β be the initial ordinal of cardinality $\mathfrak{m}$ and let β^+ be the initial ordinal of cardinality $2^{\mathfrak{m}}$. The assumption $2^{\aleph_\alpha}=\aleph_{\alpha+1}$ implies that if $\gamma<\beta^+$, then $\bar{\gamma} \leqq \mathfrak{m}$. Putting $A=\prod(A_i \mid i \in I)$ and $B=\prod(B_i \mid i \in I)$ we have that $|A| \leqq 2^{\mathfrak{m}}$, and $|B| \leqq 2^{\mathfrak{m}}$. Thus, they can be written in the forms $A=\{a_\gamma \mid \gamma<\beta^+\}$ and $B=\{b_\gamma \mid \gamma<\beta^+\}$.

If ξ is an ordinal and $\xi \leqq \beta^+$, then $L(\tau \oplus \xi)$ will denote the language which we get from $L(\tau)$ by adding the nullary operations $\mathbf{k}_\gamma=\mathbf{f}_{o_0(\tau)+\gamma}$ for all $\gamma<\xi$.

For all $\xi<\beta^+$, we will inductively construct elements $c_\xi \in A$ and $d_\xi \in B$ such that the following three conditions are satisfied:

(1) If λ is a limit ordinal, n is a nonnegative integer, and $\lambda+2n<\xi$, then $c_{\lambda+2n}=a_{\lambda+n}$;

(2) if λ is a limit ordinal, n is a nonnegative integer, and $\lambda+2n+1<\xi$, then $d_{\lambda+2n+1}=b_{\lambda+n}$;

(3) Let Φ be a sentence of $L(\tau \oplus \xi)$ and for each $i \in I$ let $\mathfrak{A}_{i,\xi}$ (resp. $\mathfrak{B}_{i,\xi}$) be the structure $\mathfrak{A}_i$ (resp. $\mathfrak{B}_i$) augmented by the new constants $(\mathbf{k}_\lambda)_{\mathfrak{A}_{i,\xi}}=c_\lambda(i)$ for $\lambda<\xi$ (resp. $(\mathbf{k}_\lambda)_{\mathfrak{B}_{i,\xi}}=d_\lambda(i)$ for $\lambda<\xi$). (Recall that $c_\lambda \in \prod(A_i \mid i \in I)$, so $c_\lambda(i) \in A_i$.) Then either $|\{i \mid \Phi$ holds in $\mathfrak{A}_{i,\xi}\}|=\mathfrak{m}$ or $|\{i \mid \neg\Phi$ holds in $\mathfrak{B}_{i,\xi}\}|=\mathfrak{m}$.

Let $P(\xi)$ denote the statement that c_δ and d_δ have been constructed for $\delta<\xi$.

P(0) holds since (1) and (2) are vacuously satisfied, and since $L(\tau \oplus 0)=L(\tau)$, $\mathfrak{A}_{i,0}=\mathfrak{A}_i$, and $\mathfrak{B}_{i,0}=\mathfrak{B}_i$, (3) reduces to our assumption (ii).

If ξ is a limit ordinal and $P(\delta)$ holds for $\delta<\xi$, then c_δ and d_δ are defined for $\delta<\xi$; (1) and (2) are obvious and (3) follows from the fact that any formula of $L(\tau \oplus \xi)$ is also a formula of some $L(\tau \oplus \delta)$ with $\delta<\xi$.

If ξ is not a limit ordinal, let λ be the greatest limit ordinal $\leqq \xi$. Then ξ can be expressed as $\xi=\lambda+2n$ $(0<n<\omega)$ or $\xi=\lambda+2n+1$ $(0 \leqq n<\omega)$ (see Exercise 0.58).

First, we consider the case $\xi=\lambda+2n+1$. We have to define $c_{\lambda+2n}$ and $d_{\lambda+2n}$. We set $c_{\lambda+2n}=a_{\lambda+n}$.

Let Γ denote the set of all formulas $\Phi(\mathbf{x}_0)$ of $L(\tau \oplus (\lambda+2n))$ having one free variable $\mathbf{x}_0$ and such that

$$\{i \mid i \in I \text{ and } \Phi(\mathbf{k}_{\lambda+2n}) \text{ holds in } \mathfrak{A}_{i,\xi}\} \in \bar{P}_{\mathfrak{m}}(I).$$

$|\Gamma| \leqq \mathfrak{m}$ since $\overline{\overline{o(\tau \oplus \xi)}}=\overline{\overline{o(\tau)}}+\bar{\xi}$, and $\overline{\overline{o(\tau)}} \leqq \mathfrak{m}$ is assumed, and $\bar{\xi} \leqq \mathfrak{m}$ follows from $\xi<\beta^+$. Thus the set of all formulas is of cardinality $\mathfrak{m}$.

For $\Phi \in \Gamma$, let

$$X(\Phi) = \{i \mid i \in I, (\exists \mathbf{x}_0)\Phi(\mathbf{x}_0) \text{ holds in } \mathfrak{B}_{i,\lambda+2n}\}.$$

Since $\{i \mid i \in I$ and $(\exists \mathbf{x}_0)\Phi(\mathbf{x}_0)$ holds in $\mathfrak{A}_{i,\lambda+2n}\} \in \bar{P}_{\mathfrak{m}}(I)$, we get $\{i \mid i \in I$ and $\neg(\exists \mathbf{x}_0)\Phi(\mathbf{x}_0)$ holds in $\mathfrak{B}_{i,\lambda+2n}\} \notin \bar{P}_{\mathfrak{m}}(I)$ by applying (3) for $\lambda+2n$ to the sentence $(\exists \mathbf{x}_0)\Phi(\mathbf{x}_0)$ (see the remark following the statement of the theorem). Thus, $I - X(\Phi) \notin \bar{P}_{\mathfrak{m}}(I)$ and therefore $|X(\Phi)| = \mathfrak{m}$.

By Lemma 2, we can find for all $\Phi \in \Gamma$ an $X^*(\Phi) \subseteq X(\Phi)$ such that $|X^*(\Phi)| = \mathfrak{m}$ and if $\Phi \neq \Psi$, $\Phi, \Psi \in \Gamma$, then $X^*(\Phi) \cap X^*(\Psi) = \varnothing$.

Thus, we can choose an $f \in B$ such that, for $i \in X^*(\Phi)$, we have that $f(i)$ satisfies $\Phi(x_0)$ in $\mathfrak{B}_{i,\lambda+2n}$. Set $d_{\lambda+2n} = f$.

Then (1) and (2) are obviously satisfied. To show (3), let Φ be a sentence in $L(\tau \oplus \xi)$ and assume that

$$\{i \mid i \in I \text{ and } \Phi \text{ holds in } \mathfrak{A}_{i,\xi}\} \in \bar{P}_{\mathfrak{m}}(I).$$

If Φ does not contain the constant $\mathbf{k}_{\lambda+2n}$, then Φ is a formula in $L(\tau \oplus (\lambda+2n))$ and we get $\{i \mid i \in I$ and $\neg\Phi$ holds in $\mathfrak{B}_{i,\xi}\} \notin \bar{P}_{\mathfrak{m}}(I)$ by applying (3) for $\lambda+2n$.

If Φ contains the constant $\mathbf{k}_{\lambda+2n}$, then it is equivalent to a sentence

$$(\exists \mathbf{x}_0)(\Phi_0(\mathbf{x}_0) \wedge \mathbf{x}_0 = \mathbf{k}_{\lambda+2n})$$

for some formula Φ_0 in $L(\tau \oplus (\lambda+2n))$. Obviously, $\Phi_0 \in \Gamma$ and thus $|X^*(\Phi_0)| = \mathfrak{m}$. Since $X^*(\Phi_0)$ is disjoint from $\{i \mid i \in I$ and $\neg\Phi$ holds in $\mathfrak{B}_{i,\xi}\}$, the latter cannot be in $\bar{P}_{\mathfrak{m}}(I)$, which was to be proved.

Second, if $\xi = \lambda+2n$ $(0 < n < \omega)$, then we set $d_{\lambda+2n-1} = b_{\lambda+n-1}$ and proceed as in the first case.

This proves the existence of the c_ξ and d_ξ for $\xi < \beta^+$.

Now, for any sentence Φ in $L(\tau \oplus \beta^+)$, set

$$Y(\Phi) = \{i \mid i \in I \text{ and } \Phi \text{ holds in } \mathfrak{A}_{i,\beta^+}\}$$

and

$$Z(\Phi) = \{i \mid i \in I \text{ and } \Phi \text{ holds in } \mathfrak{B}_{i,\beta^+}\}.$$

Property (3) of the sequence shows that either $|Y(\Phi)| = \mathfrak{m}$ or $|Z(\neg\Phi)| = \mathfrak{m}$.

Let $\mathscr{E}_0 = \{Z(\Phi) \mid \Phi$ is a sentence in $L(\tau \oplus \beta^+)$ and $Y(\Phi) \in \bar{P}_{\mathfrak{m}}(I)\}$. Noting that for any pair of sentences Φ_0, Φ_1 of $L(\tau \oplus \beta^+)$ we have $Y(\Phi_0) \cap Y(\Phi_1) = Y(\Phi_0 \wedge \Phi_1)$ and $Z(\Phi_0) \cap Z(\Phi_1) = Z(\Phi_0 \wedge \Phi_1)$ it is easily verified that $\mathscr{E}_0$ is closed under finite intersections. Furthermore, if $Z(\Phi) \in \mathscr{E}_0$, then $Y(\Phi) \in \bar{P}_{\mathfrak{m}}(I)$. Thus, $|Y(\neg\Phi)| = |I - Y(\Phi)| < \mathfrak{m}$ and so, by (3), $|Z(\Phi)| = \mathfrak{m}$. So if $J \in \mathscr{E}_0$, $K \in \bar{P}_{\mathfrak{m}}(I)$, and $J \cap K = \varnothing$, then $J \subseteq I - K$, contradicting $|J| = \mathfrak{m}$ and $|I - K| < \mathfrak{m}$. This shows that $\mathscr{E}_0 \cup \bar{P}_{\mathfrak{m}}(I)$ has the finite intersection property. Thus, it is contained in a prime dual ideal $\mathscr{E}$.

Set $\mathscr{D}_0 = \{Y(\Phi) \mid Z(\Phi) \in \mathscr{E}\}$. If $J = Y(\Phi) \in \mathscr{D}_0$, then $Z(\Phi) \in \mathscr{E}$ and $Z(\neg\Phi) \notin \mathscr{E}$ (since $\mathscr{E}$ is proper). In particular, $Z(\neg\Phi) \notin \mathscr{E}_0$, so

$$Y(\neg\Phi) \notin \bar{P}_{\mathfrak{m}}(I).$$

By the definition of $Y(\neg\Phi)$, this proves that $|J| = \mathfrak{m}$.

Now we proceed as we did for $\mathscr{E}_0$, and we find a prime dual ideal $\mathscr{D}$ containing $\mathscr{D}_0$ and $\bar{P}_{\mathfrak{m}}(I)$.

Set $\mathfrak{A}_{\mathscr{D}} = \prod_{\mathscr{D}} (\mathfrak{A}_i \mid i \in I)$ and $\mathfrak{B}_{\mathscr{E}} = \prod_{\mathscr{E}} (\mathfrak{B}_i \mid i \in I)$. As usual, if

$$f \in \prod (A_i \mid i \in I),$$

then $f^{\vee}$ will denote the corresponding element of the prime product, and the same for $g \in \prod (B_i \mid i \in I)$. $\mathfrak{A}_{\mathscr{D},\beta^+}$ and $\mathfrak{B}_{\mathscr{E},\beta^+}$ will denote $\mathfrak{A}_{\mathscr{D}}$ and $\mathfrak{B}_{\mathscr{E}}$ augmented by the constants $c_\lambda{}^{\vee}$ and $d_\lambda{}^{\vee}$ $(\lambda < \beta^+)$, respectively.

Let Φ be a sentence in $L(\tau \oplus \beta^+)$. Then $Y(\Phi) \in \mathscr{D}$ if and only if $Y(\neg\Phi) \notin \mathscr{D}$, which is equivalent to $Z(\neg\Phi) \notin \mathscr{E}$ and hence to $Z(\Phi) \in \mathscr{E}$. Thus, Φ holds in $\mathfrak{A}_{\mathscr{D},\beta^+}$ if and only if it holds in $\mathfrak{B}_{\mathscr{E},\beta^+}$, by Corollary 1 to Theorem 39.1. Thus, $\mathfrak{A}_{\mathscr{D},\beta^+}$ and $\mathfrak{B}_{\mathscr{E},\beta^+}$ are elementarily equivalent.

Consider the mapping $\varphi\colon c_\xi{}^{\vee} \to d_\xi{}^{\vee}$, $\xi < \beta^+$. This mapping is well defined since if $\xi, \zeta < \beta^+$, then $c_\xi{}^{\vee} = c_\zeta{}^{\vee}$ implies $Y(\mathbf{k}_\xi = \mathbf{k}_\zeta) \in \mathscr{D}$. Thus, $Z(\mathbf{k}_\xi = \mathbf{k}_\zeta) \in \mathscr{E}$, which means that $d_\xi{}^{\vee} = d_\zeta{}^{\vee}$.

Since every element of A is a c_ξ and every element of B is a d_ξ, by (1) and (2), the mapping φ is defined on the whole of $A_{\mathscr{D}}$ and it is onto $B_{\mathscr{E}}$. An argument similar to the previous paragraph shows that φ is also 1–1. By applying the elementary equivalence of $\mathfrak{A}_{\mathscr{D},\beta^+}$ and $\mathfrak{B}_{\mathscr{E},\beta^+}$ to sentences which are atomic formulas, we infer that φ is an isomorphism, completing the proof of the theorem.

The analogue of Theorem 1 is the following special case of Theorem 2.

Corollary. *Assuming the Generalized Continuum Hypothesis, two structures are elementarily equivalent if and only if they have isomorphic prime powers.*

It follows from this corollary that the relation "having isomorphic prime powers" is transitive. It is interesting to note that not even this statement has yet been proved without the Generalized Continuum Hypothesis.

§42. ELEMENTARY AND AXIOMATIC CLASSES

We introduce the following notation. Let K be a class of structures of type τ. Then K^* denotes the set of all those sentences in $L(\tau)$ that hold in every structure in K. If $K = \{\mathfrak{A}\}$, we write $\mathfrak{A}^*$ for K^*. Similarly, if Σ is a set of sentences in $L(\tau)$, then Σ^* will denote the class of all structures of type τ in which every sentence of Σ holds. Thus $\mathfrak{A} \in \Sigma^*$, if and only if $\mathfrak{A}$ is a model of Σ. Again, if $\Sigma = \{\Phi\}$, we write Φ^* for Σ^*. Σ is called *consistent* if Σ^* is not void.

Definition 1. *A class K of structures is called* axiomatic *if $K = \Sigma^*$ for some set of sentences Σ. If $K = \Phi^*$, for some sentence Φ, then K is called* elementary.

Remark. The terminology in the literature varies a great deal. A. Tarski, who originally introduced these concepts, called an elementary class an *arithmetic class* and an axiomatic class an *arithmetic class in the wider sense*. Some papers use elementary class for axiomatic class and so on. In the same way, K^* and Σ^* are denoted variously—frequently used notations are $Th(K)$ and $M(\Sigma)$.

Some formal laws concerning K^* and Σ^* are listed below.

Lemma 1. *For classes K, K_1 of structures, and sets of sentences Σ, Σ_1, we have*

(i) $K \subseteq K^{**}$;
(ii) $\Sigma \subseteq \Sigma^{**}$;
(iii) $K \subseteq K_1$ *implies* $K_1^* \subseteq K^*$;
(iv) $\Sigma \subseteq \Sigma_1$ *implies* $\Sigma_1^* \subseteq \Sigma^*$.

Proof. Exercise.

Corollary. *K is an axiomatic class if and only if $K = K^{**}$.*

Indeed, if $K = K^{**}$, then $K = \Sigma^*$ with $\Sigma = K^*$. On the other hand, if $K = \Sigma^*$, then, by (ii), $K^* = \Sigma^{**} \supseteq \Sigma$ and so, by (iii), $K^{**} \subseteq \Sigma^* = K$. Therefore, $K^{**} \subseteq K$ and by (i) $K = K^{**}$.

Definition 2. *A nonvoid set Σ of sentences is called* closed *if Σ is closed under conjunction and $\Phi \in \Sigma$ and $\Phi \models \Psi$ imply $\Psi \in \Sigma$.*

Remark. Although it is by no means trivial, this definition is closely related to Definition 26.2. The difference is that for identities we listed the "logical axioms" and "rules of inference", whereas now we use the model-theoretic implication. If we modified Definition 2 by replacing $\Phi \models \Psi$ by any usual rules of inference of first order logic, we would get a definition which is clearly a generalization of the previous one.

Theorem 1. *The following three conditions on a set Σ of sentences are equivalent;*

(i) Σ *is closed*;
(ii) $\Sigma = \Sigma^{**}$;
(iii) $\Sigma = K^*$ *for some class K.*

Proof. (i) implies (ii). Let Σ be closed. By (ii) of Lemma 1, it is enough to prove that $\Sigma^{**} \subseteq \Sigma$. If $\Phi \in \Sigma^{**}$, then $\Sigma \models \Phi$. If $\Psi \models \Phi$ for some $\Psi \in \Sigma$, then $\Phi \in \Sigma$ since Σ is closed. Let us assume that $\Psi \models \Phi$ for *no* $\Psi \in \Sigma$. Then for every $\Psi \in \Sigma$ there is a structure in which Ψ and $\neg \Phi$ are satisfied. Since Σ is closed under conjunction, this implies that every finite subset of $\Sigma \cup \{\neg \Phi\}$ has a model. Thus, by Theorem 39.2, $\Sigma \cup \{\neg \Phi\}$ also has a model, contradicting $\Sigma \models \Phi$.

(ii) implies (iii). To prove this, set $K = \Sigma^*$.

(iii) implies (i). If $\Sigma = K^*$, then Σ is obviously closed under conjunction. If $\Phi \in \Sigma$ and $\Phi \models \Psi$, then Φ holds in every structure in K and, by $\Phi \models \Psi$, so does Ψ, that is, $\Psi \in \Sigma$. This completes the proof of Theorem 1.

Using our newly introduced terminology, we can put the compactness theorem in the following form.

Theorem 2. *Let K_i, $i \in I$, be axiomatic classes. If the finite intersections of the K_i are never void, then the intersection of all K_i is not void either.*

Proof. Let $K_i = \Sigma_i^*$. Then $\bigcap (K_i \mid i \in I) = (\bigcup (\Sigma_i \mid i \in I))^*$. Since every finite subset of $\bigcup (\Sigma_i \mid i \in I)$ has a model, so does $\bigcup (\Sigma_i \mid i \in I)$. Thus, $\bigcap (K_i \mid i \in I) \neq \varnothing$.

The problem of giving an algebraic characterization of axiomatic classes was first solved by J. Łoś [2] using prime products. The following characterization is from T. Frayne, A. C. Morel, and D. S. Scott [1], and from S. Kochen [1].

Theorem 3. *A class K of structures is an axiomatic class if and only if the following two conditions are satisfied:*

(i) *K is closed under the formation of prime products;*
(ii) *K is closed under elementary equivalence.*

Proof. If K is an axiomatic class, then (ii) is obvious and (i) follows from Corollary 1 to Theorem 39.1.

Now assume that (i) and (ii) hold for K. By the corollary to Lemma 1, it is enough to prove that if $\mathfrak{A} \in K^{**}$, then $\mathfrak{A} \in K$. Let $\mathfrak{A} \in K^{**}$ and $\mathfrak{A}^* = \{\Phi_i \mid i \in I\}$. For each Φ_i we can choose a $\mathfrak{B}_i \in K$ in which Φ_i holds, since otherwise $\neg \Phi_i \in K^*$, contradicting $\mathfrak{A} \in K^{**}$.

For $i \in I$, set $L_i = \{j \mid j \in I \text{ and } \Phi_i \in \mathfrak{B}_j^*\}$. Then the family $(L_i \mid i \in I)$ has the finite intersection property. Thus, it is contained in some $\mathscr{D}$ prime over I. Then $\mathfrak{B} = \prod_{\mathscr{D}} (\mathfrak{B}_i \mid i \in I)$ is in K by (i), and is elementarily equivalent to $\mathfrak{A}$ by construction. Hence $\mathfrak{A} \in K$, by (ii), which was to be proved.

Using the results of §41, we can get stronger versions of Theorem 3.

Corollary 1. *In Theorem 3, condition* (ii) *can be replaced by*
(ii′) *K is closed under isomorphism and under taking elementary substructures.*

Proof. By Lemma 41.1.

Corollary 2. *In Theorem 3, condition* (ii) *can be replaced by*
(ii″) *K is closed under isomorphism, and K and* $K(\tau)-K$ *are closed under the formation of prime limits.*

Proof. By Theorem 41.1.

Corollary 3. *Under the Generalized Continuum Hypothesis, condition* (ii) *of Theorem* 3 *can be replaced by*
(ii‴) *K is closed under isomorphism and* $K(\tau)-K$ *is closed under the formation of prime powers.*

Proof. By the corollary to Theorem 41.2.

Now we apply Theorem 3 to characterize elementary classes.

Theorem 4. *Let K be a class of structures of type* τ. *The following conditions on K are equivalent:*

(a) *K is an elementary class;*
(b) *K and* $K(\tau)-K$ *are axiomatic classes;*
(c) *The following three conditions hold:*
  (i) *K is closed under the formation of prime products;*
  (ii) *K is closed under elementary equivalence;*
  (iii) $K(\tau)-K$ *is closed under the formation of prime products.*

Proof. (a) implies (c) since, if $K=\Phi^*$ for some sentence Φ, then $K(\tau)-K=(\neg\Phi)^*$; thus, the conditions in (c) follow from Theorem 3.

(c) implies (b) since, if K is closed under elementary equivalence, then so is $K(\tau)-K$; thus, the conditions of Theorem 3 are satisfied for K as well as for $K(\tau)-K$.

(b) implies (a). Let Σ_1 and Σ_2 be sets of sentences such that $K=\Sigma_1{}^*$ and $K(\tau)-K=\Sigma_2{}^*$. Obviously, $\Sigma_1\cup\Sigma_2$ is not consistent; thus, by the compactness theorem, there is a finite subset Σ_3 of Σ_2 such that $\Sigma_1\cup\Sigma_3$ is not consistent. Let Φ be the conjunction of the sentences in Σ_3. Then Φ^* is disjoint from K and $\Phi^*\supseteq K(\tau)-K$. Thus, $\Phi^*=K(\tau)-K$ and $K=(\neg\Phi)^*$, which proves (a).

The results of §41 can again be used to sharpen Theorem 4.

Corollary 1. *In condition* (c), (ii) *can be replaced by*
(ii′) *K is closed under isomorphism, and K and $K(\tau)-K$ are closed under the formation of prime limits.*

Corollary 2. *Under the Generalized Continuum Hypothesis, condition* (ii) *of* (c) *can be replaced by*
(ii″) *K is closed under isomorphism.*

Prime products are useful in studying classes which are more general than axiomatic classes.

Definition 3. *Let the type μ be a subtype of the type τ, that is, $o_0(\mu) \leqq o_0(\tau)$, $o_1(\mu) \leqq o_1(\tau)$, and two sequences of ordinals $\alpha_0 < \alpha_1 < \cdots < \alpha_\gamma < \cdots (\gamma < o_0(\mu))$, $\beta_0 < \beta_1 < \cdots < \beta_\gamma < \cdots (\gamma < o_1(\mu))$ are fixed such that $\mathrm{f}_{\alpha_\gamma}$ in the type τ is the same as f_γ in μ and $\mathrm{r}_{\beta_\gamma}$ in τ is the same as r_γ in μ. Let K_1 be an axiomatic (resp. elementary) class of structures of type τ. For $\mathfrak{A} \in K_1$, form $\mathfrak{A}^{(\mu)}$, the μ-reduct of $\mathfrak{A}$. Let K be the class of all $\mathfrak{A}^{(\mu)}$ with $\mathfrak{A} \in K_1$. Then K is called a* pseudo-axiomatic (*resp.* pseudo-elementary) class.

For instance, if K_1 is the class of all rings and we drop the $+$, then we get the class K of all semigroups upon which a ring can be built. K is an example of a pseudo-elementary class which is not even an axiomatic class. (S. R. Kogalovskiĭ, Dokl. Akad. Nauk. SSSR 140 (1961), 1005-1007). Further examples will be given in the Exercises.

Theorem 5. *A pseudo-axiomatic class K is closed under the formation of prime products.*

Proof. If $K_1 = \Sigma^*$, K is the class of all μ-reducts of members of K_1, $\mathfrak{A}_i \in K_1$, for all $i \in I$, and $\mathscr{D}$ is prime over I, then obviously the μ-reduct of $\prod_{\mathscr{D}} (\mathfrak{A}_i \mid i \in I)$ is $\prod_{\mathscr{D}} (\mathfrak{A}_i^{(\mu)} \mid i \in I)$, from which the statement follows.

Unfortunately, the condition of Theorem 5 does not characterize pseudo-axiomatic classes.

Definition 4. *Let Σ be a set of formulas of $L^{(\alpha)}(\tau)$ and let K be the class of all structures $\mathfrak{A}$ of type τ for which there is a sequence $\mathfrak{f} \in A^\alpha$ simultaneously satisfying all the formulas in Σ. Such a class K is called an* α-quasi-axiomatic class. *If there exists an $\alpha \geqq \omega$ such that K is α-quasi-axiomatic, then K is called a* quasi-axiomatic class.

Prove that every quasi-axiomatic class is pseudo-axiomatic.

Note that in a sentence we can use only a finite number of variables and it does not matter what they are. However, in a system of formulas, when we discuss simultaneous satisfaction the names of the free variables are very important. A typical example of a quasi-axiomatic class is the class of all structures which contain a substructure which is isomorphic to a given structure. The following characterization of quasi-axiomatic classes is due to T. Frayne, A. C. Morel, and D. S. Scott [1].

Theorem 6. *A class K of structures is a quasi-axiomatic class if and only if the following conditions hold:*

(i) *K is closed under isomorphism;*
(ii) *K is closed under the formation of prime products;*
(iii) *K is closed under elementary extensions;*
(iv) *there is a cardinal number $\mathfrak{m}$ such that every structure in K has an elementary substructure in K of cardinality less than or equal to $\mathfrak{m}$.*

Proof. If K is an α-quasi-axiomatic class, then (i) is trivial and (ii) follows from the fact that Theorem 39.1 holds for formulas in $L^{(\alpha)}(\tau)$ as well. Theorem 38.6 shows that (iii) holds, while Theorem 38.3 shows that (iv) is satisfied. ($\mathfrak{m}$ can be taken to be the cardinality of $\mathbf{L}^{(\alpha)}(\tau)$.)

Now assume that K satisfies conditions (i)–(iv). Let $(\mathfrak{A}_i \mid i \in I)$ be a family of structures in K, each of cardinality at most $\mathfrak{m}$, such that every structure in K of cardinality at most $\mathfrak{m}$ is isomorphic to some $\mathfrak{A}_i$. Let $\prod (A_i \mid i \in I) = \{a_\gamma \mid \gamma < \alpha\}$ and let a be the corresponding α-termed sequence. Let Σ be the set of all formulas Φ in $L^{(\alpha)}(\tau)$ which are satisfied by $\mathsf{a}(i) = \langle a_0(i), a_1(i), \cdots \rangle$ in $\mathfrak{A}_i$ for each $i \in I$. We claim that Σ defines the α-quasi-axiomatic class K.

Indeed, if $\mathfrak{A} \in K$, then, by (iv), $\mathfrak{A}$ has an elementary part $\mathfrak{B}$ which is isomorphic to some $\mathfrak{A}_i$. Thus, there is a sequence f which simultaneously satisfies all the formulas in Σ.

Conversely, suppose that some $\mathsf{b} \in A^\alpha$ simultaneously satisfies in $\mathfrak{A}$ all the formulas of Σ. Let Σ' be the set of all formulas in $L^{(\alpha)}(\tau)$ which are satisfied by b in $\mathfrak{A}$. By definition, $\Sigma \subseteq \Sigma'$. For Φ in Σ', set $J_\Phi = \{i \mid i \in I$ and $\mathsf{a}(i)$ satisfies Φ in $\mathfrak{A}_i\}$. Then $J_\Phi \neq \varnothing$ since otherwise $\neg \Phi \in \Sigma \subseteq \Sigma'$. Since $J_{\Phi \wedge \Psi} = J_\Phi \cap J_\Psi$ for $\Phi, \Psi \in \Sigma'$, there is a $\mathscr{D}$ prime over I containing all the J_Φ. If $a_\xi{}^\vee = a_\zeta{}^\vee$ $(\xi, \zeta < \alpha)$ in $\prod_{\mathscr{D}} (\mathfrak{A}_i \mid i \in I)$, then $\mathbf{x}_\xi = \mathbf{x}_\zeta$ is in Σ'; thus, $b_\xi = b_\zeta$. Therefore the mapping $\varphi: a_\xi{}^\vee \to b_\xi$ is well defined. As in Lemma 41.1, we can prove that φ is an elementary embedding of $\prod_{\mathscr{D}} (\mathfrak{A}_i \mid i \in I) \in K$ into $\mathfrak{A}$. Thus, $\mathfrak{A}$ is an elementary extension of a structure in K; by (iii), $\mathfrak{A} \in K$, which completes the proof of Theorem 6.

All classes considered in this section are closed under the formation of prime products. This condition alone is enough to establish an interesting property.

Theorem 7. *Let K be a class which is closed under the formation of prime products. Let $\mathfrak{A}$ be a structure every finite "partial substructure" of which can be embedded in a structure in K. Then $\mathfrak{A}$ can be embedded in a structure in K.*

Proof. Since every structure can be embedded in some prime product of "partial substructures" (see Theorem 39.4), the result follows immediately.

For quasi-axiomatic classes, this result was first established by L. A. Henkin [2].

The next result shows how close the classes which are closed under the formation of prime products are to axiomatic classes.

Theorem 8. *Let the class K be closed under isomorphism, and under the formation of prime products. Form the class $\hat{K}$ consisting of the elementary substructures of all structures $\mathfrak{A} \in K$. Then $\hat{K}$ is an axiomatic class. In fact, $\hat{K}$ is the smallest axiomatic class containing K.*

Proof. $\hat{K}$ obviously satisfies (ii′) of Corollary 1 to Theorem 3. To prove that $\hat{K}$ satisfies (i) of Theorem 3, we first prove the following result.

Lemma 2. *If $\mathfrak{A}_i$ is an elementary substructure of $\mathfrak{B}_i$ for all $i \in I$, and $\mathscr{D}$ is prime over I, then*

$$\mathfrak{A} = \prod_{\mathscr{D}} (\mathfrak{A}_i \mid i \in I)$$

is an elementary substructure of

$$\mathfrak{B} = \prod_{\mathscr{D}} (\mathfrak{B}_i \mid i \in I).$$

Proof. Let $f_0^{\vee}, \cdots, f_{n-1}^{\vee} \in A$. Assume that $\Phi(f_0^{\vee}, \cdots, f_{n-1}^{\vee})$ holds in $\mathfrak{A}$ for some formula Φ. By Theorem 39.1 this is equivalent to

$$\{i \mid \Phi(f_0(i), \cdots, f_{n-1}(i)) \text{ holds in } \mathfrak{A}_i\} \in \mathscr{D}$$

which, in turn, is equivalent to $\Phi(f_0^{\vee}, \cdots, f_{n-1}^{\vee})$ in $\mathfrak{B}$. This completes the proof of the lemma.

Now let $\mathfrak{A}_i \in \hat{K}$ for $i \in I$. Then by the definition of $\hat{K}$ there exist $\mathfrak{B}_i \in K$, $i \in I$, such that $\mathfrak{A}_i$ is an elementary substructure of $\mathfrak{B}_i$. Thus,

$$\prod_{\mathscr{D}} (\mathfrak{B}_i \mid i \in I) \in K$$

and so, by Lemma 2, $\prod_{\mathscr{D}} (\mathfrak{A}_i \mid i \in I) \in \hat{K}$, which was to be proved.

Some other interesting classes of algebras will be discussed in the next chapter.

There are many problems concerning elementary and axiomatic classes which are purely logical in nature and therefore cannot be studied within our purely semantical framework. For an excellent review article on decidability and undecidability of elementary theories, see Ju. L. Eršov, I. A. Lavrov, A. D. Taĭmanov and M. A. Taĭclin, Elementary theories, Usp. Mat. Nauk 20 (1965), 37–108 (English translation was published in Russian Mathematical Surveys).

EXERCISES

1. Let $\mathfrak{A}$ be a relational system and Θ an equivalence relation on A. Show that there exists a relational system $\mathfrak{B}$ and a homomorphism φ of $\mathfrak{A}$ onto $\mathfrak{B}$, such that φ induces Θ.
2. Show that the projection e_i^I of a direct product $\prod(\mathfrak{A}_i \mid i \in I)$ onto $\mathfrak{A}_i$ is an onto homomorphism but $\mathfrak{A}_i$ is not always a homomorphic image of $\prod(\mathfrak{A}_i \mid i \in I)$.
3. Let $\mathfrak{A}$ and $\mathfrak{A}'$ be structures on the same set A. Show that it is possible that $x \to x$ is a homomorphism of $\mathfrak{A}$ onto $\mathfrak{A}'$, but $\mathfrak{A}$ and $\mathfrak{A}'$ are not isomorphic.
4. Change Definition 36.2(ii) to "$\mathbf{r}_\gamma(\mathbf{p}_0, \cdots, \mathbf{p}_{m_\gamma - 1})$ is an atomic formula, if $\mathbf{p}_0, \cdots, \mathbf{p}_{m_\gamma - 1}$ are polynomial symbols". Is the resulting language stronger than $L(\tau)$?
5. Can an atomic formula be "identically false"?
6. Describe the open formulas which are sentences.
7. Build up $L(\tau)$ using the Sheffer stroke rather than $\vee$ and $\neg$ and modify Definition 37.1 accordingly.
8. Let 0 denote the void-type ($o(\tau) = 0$). Let K be a finite set or the complement of a finite set of positive integers. Find a sentence Φ_K of $L(0)$ such that a set A is a model of Σ if and only if $|A| \in K$.
9. Let K be an arbitrary set of positive integers. Find a set of sentences Σ of $L(0)$, such that a finite set A is a model of Σ if and only if $|A| \in K$.
10. Prove that every sentence Ψ of $L(0)$ is equivalent to a Ψ_K, described in Ex. 8 (Find the meaning of any formula of $L(0)$ and use induction on formulas.)
11. A chain $\langle A; \leqq \rangle$ is *dense* if $|A| > 1$ and $a, b \in A$, $a < b$ imply $a < c < b$, for some $c \in A$. Describe dense chains by a first order sentence in $L(\tau)$, $\tau = \langle 0, \langle 2 \rangle \rangle$.
12. Find all properties of dense chains which can be described by a first order sentence. (Use induction on formulas.)
13. Describe atomless Boolean algebras by a first order sentence. Find all properties of atomless Boolean algebras that can be expressed by a first order sentence. (There is no such property.)
14. (S. Feferman and R. Vaught [1]) Describe the meaning of sentences in $\langle P(I); \cup, \cap, ', \varnothing, I, Fin \rangle$, where Fin is a unary relation defined as

follows: for $A \subseteq I$, $Fin(A)$ if and only if A is finite. (Use induction on formulas. Extend the definition of n-ary tables of conditions by "$\geqq k$ and finite" and "infinite".)

15. Let $\mathfrak{A}$ and $\mathfrak{B}$ be structures of type τ ($o_0(\tau) \leqq \omega$) with the following properties: (i) $\mathfrak{A}$ has a substructure $\mathfrak{A}_0$ and $\mathfrak{B}$ has a substructure $\mathfrak{B}_0$, such that $\mathfrak{A}_0$ and $\mathfrak{B}_0$ are isomorphic; (ii) if $\mathfrak{A}_1$ is a substructure of $\mathfrak{A}$, $|A_1| < |A|$, and $\mathfrak{A}_1 \cong \mathfrak{B}_1$, $\mathfrak{B}_1$ a substructure of $\mathfrak{B}$, then for every $a \in A$, there exists a $b \in B$ (and for every $b \in B$ there exists an $a \in A$), such that the substructure generated by $\mathfrak{A}_1$ and a is isomorphic to the substructure generated by $\mathfrak{B}_1$ and b; (iii) is the same as (ii) with $\mathfrak{A}$ and $\mathfrak{B}$ interchanged. What additional conditions are needed to prove that $\mathfrak{A}$ and $\mathfrak{B}$ are isomorphic?
16. Apply the method of Ex. 15 to prove that up to isomorphism there are only three countable dense chains and only one countable atomless Boolean algebra.
17. Prove the *disjunctive normal form* theorem: every open formula Φ is equivalent to a formula $\Phi_0 \vee \cdots \vee \Phi_{n-1}$, where every Φ_i is a conjunction of atomic formulas and negations of atomic formulas.
18. Formulate and prove the conjunctive normal form theorem.
19. Prove that $(\Phi \wedge (\Phi \rightarrow \Psi)) \models \Psi$, $\Phi \models (\exists \mathbf{x}_k)\Phi$ and $(\mathbf{x}_k)\Phi \models \Phi$. Furthermore, if $\mathbf{k}$ does not occur in Φ or $\Psi(\mathbf{x})$ and $\Phi \models \Psi(\mathbf{k})$, then $\Phi \models (\mathbf{x})\Psi(\mathbf{x})$.
20. Prove the converse of Corollary 2 to Definition 38.1. (Hint: Use Lemma 38.2.)
21. Find structures $\mathfrak{A}$ and $\mathfrak{B}$ such that $\mathfrak{A}$ is a substructure of $\mathfrak{B}$, $\mathfrak{A} \equiv \mathfrak{B}$, and $\mathfrak{A}$ is not an elementary substructure of $\mathfrak{B}$.
22. Find structures $\mathfrak{A}, \mathfrak{A}_0, \mathfrak{A}_1, \cdots, \mathfrak{A}_n, \cdots$ such that $\mathfrak{A}$ is the union of the $\mathfrak{A}_i$, $\mathfrak{A}_i$ is a substructure of $\mathfrak{A}_{i+1}$, $i = 0, 1, \cdots$, $\mathfrak{A}_i \equiv \mathfrak{A}_j$, $i, j = 0, 1, \cdots$, and $\mathfrak{A} \equiv \mathfrak{A}_0$ does not hold.
23. Let $\mathfrak{C}_0$ and $\mathfrak{C}_1$ be dense chains, $C_0 \subseteq C_1$. Then $\mathfrak{C}_0 \equiv \mathfrak{C}_1$ and $\mathfrak{C}_0$ is an elementary substructure of $\mathfrak{C}_1$ if and only if the following conditions hold:

    (i) $\mathfrak{C}_0$ has a least element if and only if $\mathfrak{C}_1$ has a least element, and if the least elements exist, they are equal;

    (ii) $\mathfrak{C}_0$ has a greatest element if and only if $\mathfrak{C}_1$ has a greatest element, and if the greatest elements exist, they are equal.
24. If $\mathfrak{A}$ and $\mathfrak{A}'$ have a common elementary extension and $\mathfrak{A}'$ is an extension of $\mathfrak{A}$, then $\mathfrak{A}'$ is an elementary extension of $\mathfrak{A}$.
25. $\mathfrak{B}$ is isomorphic to an elementary extension of $\mathfrak{A}$ if and only if $\langle \mathfrak{A}, \mathbf{a} \rangle^* \supseteq \langle \mathfrak{B}, \mathbf{b} \rangle^*$, for some $\mathbf{b} \in B^\alpha$, where $\mathbf{a} \in A^\alpha$ is a well ordering of A.
26. (R. L. Vaught) Let K be a class of algebras, $\mathfrak{A}_i \in K$ for $i \in I$. Assume that for each $i \in I$ $\mathfrak{A}_i \cong \mathfrak{A}_j$ for infinitely many $j \in J$, where $J \subseteq I$, and J is infinite. Let $\mathfrak{A}$ be the free product over K of the $\mathfrak{A}_i$, $i \in I$, and $\mathfrak{B}$ the free product over K of the $\mathfrak{A}_i$, $i \in J$. Then $\mathfrak{A}$ is isomorphic to an elementary extension of $\mathfrak{B}$.
27. Can the conditions of Ex. 26 be relaxed so as to allow finitely many isomorphic algebras?
28. Prove the formulas (i)–(viii) in the proof of Theorem 37.4 and carry out the proof of Theorem 37.4.
29. Prove Theorem 39.1 from its Corollary 1. (This is another example of

induction on formulas. We wanted to get the statement of Corollary 1 to Theorem 39.1, but we had to prove the more general Theorem 39.1. This exercise shows that Theorem 39.1 is only apparently stronger than its Corollary 1.)

30. Prove that the following statement is equivalent to the compactness theorem: let Σ be a set of sentences and $\Sigma \models \Phi$; then $\Sigma_0 \models \Phi$ for some finite subset Σ_0 of Σ.
31. Let Σ be a set of sentences. If $\Sigma \Leftrightarrow \Phi$, then $\Sigma_0 \Leftrightarrow \Phi$ for some finite subset Σ_0 of Σ.
32. Prove that the following statement is equivalent to the compactness theorem: let Σ be an infinite set of sentences such that for every $\Sigma_0 \subseteq \Sigma$ with $|\Sigma_0| < |\Sigma|$, there exists a structure in which every sentence of Σ_0 is satisfied; then there exists a structure in which every sentence of Σ is satisfied.
33. Let K be a class of fields closed under the formation of prime products. Let us assume that for every natural number n, there exists a prime $p > n$ such that K contains a field of characteristic p. Then K contains a field of characteristic 0.
34. Is the class of finite lattices closed under the formation of prime products?
35. Prove that the compactness theorem implies the prime ideal theorem for distributive lattices (Theorem 6.7).
36. Use the compactness theorem to prove that every torsion-free abelian group can be embedded in a *divisible abelian group* (i.e., an abelian group in which $nx = a$ has a solution for any positive integer n and group element a).
37. Let $\mathfrak{A} = \langle A; R\rangle$ be a relational system and suppose $\leqq$ is in R, and $\langle A; \leqq\rangle$ is a partially ordered set. Then there is a relational system $\mathfrak{B} = \langle B; R\rangle$ and a 1–1 homomorphism φ of $\mathfrak{A}$ into $\mathfrak{B}$ such that $\langle B; \leqq\rangle$ is a chain.
38. Prove that the compactness theorem implies the Axiom of Choice for families of finite sets.
39. (B. Jónsson [8]) Let K be a finite set of finite algebras. If the congruence lattice of every $\mathfrak{A} \in \mathbf{HSP}(K)$ is distributive, then
$$\mathbf{HSP}(K) = \mathbf{IP}_S\mathbf{HS}(K).$$
40. (B. Jónsson [8]) Let K be an equational class of algebras such that the congruence lattice of every $\mathfrak{A} \in K$ is distributive. Let $\mathfrak{A}, \mathfrak{B} \in K$, $\mathfrak{A}$ and $\mathfrak{B}$ finite and subdirectly irreducible. Then $\mathfrak{A} \cong \mathfrak{B}$ if and only if $Id(\mathfrak{A}) = Id(\mathfrak{B})$.
41. (A. L. Foster [4]) Let the class of algebras $\{\mathfrak{A}_i \mid i \in I\}$ be *independent* in the sense that for every choice of polynomials $p_i \in P^{(\omega)}(\mathfrak{A}_i)$, $i \in I$, there is a polynomial symbol $\mathbf{p}$, such that $\mathbf{p}_{\mathfrak{A}_i} = p_i$ for all $i \in I$. Prove that I is finite.
42. (F. M. Sioson [3]) $\{\mathfrak{A}_0, \cdots, \mathfrak{A}_{n-1}\}$ is independent if and only if
$$(\mathbf{p})_{\mathfrak{A}_0 \times \cdots \times \mathfrak{A}_{n-1}} \to \langle \mathbf{p}_{\mathfrak{A}_0}, \cdots, \mathbf{p}_{\mathfrak{A}_{n-1}}\rangle$$
is an isomorphism between $\mathfrak{P}^{(\omega)}(\mathfrak{A}_0 \times \cdots \times \mathfrak{A}_{n-1})$ and $\mathfrak{P}^{(\omega)}(\mathfrak{A}_0) \times \cdots \times \mathfrak{P}^{(\omega)}(\mathfrak{A}_{n-1})$. (Use Lemma 19.2)
43. (A. L. Foster [4]) Prove that the algebras $\mathfrak{A}_0, \cdots, \mathfrak{A}_{n-1}$ are independent

if and only if there is a polynomial symbol $\mathbf{p}$ such that $(\mathbf{p})_{\mathfrak{A}_i} = e_i^n$, $i = 0, \cdots, n-1$.

44. Prove the *Chinese remainder theorem:* if $m_0, \cdots, m_{n-1}$ are relatively prime in pairs and $a_0, \cdots, a_{n-1}$ are any natural numbers, then there is a natural number b with $b \equiv a_i (\text{mod } m_i)$, $i = 0, \cdots, n-1$.
45. Does the Chinese remainder theorem imply that the rings $\mathfrak{Z}_i$ of integers modulo m_i, $0 \leqq i < n$ are independent if $m_0, \cdots, m_{n-1}$ are relatively prime in pairs?
46. Show that Theorem 27.5 follows from Ex. 39.
47. Show that the following generalization of Theorem 27.5 (A. L. Foster [3]) also follows from Ex. 39: The algebra $\mathfrak{A}$ is subdirectly representable in the primal cluster K, if and only if $Id(K_1) \subseteq Id(\mathfrak{A})$ for some finite $K_1 \subseteq K$. (K is a *primal cluster* if K is a class of primal algebras and every finite subclass of K is independent. $\mathfrak{A}$ is *subdirectly representable* in K if $\mathfrak{A}$ is a subdirect product of algebras in K_1 for some finite $K_1 \subseteq K$.)
48. Prove the statement of Ex. 47 using the method of the proof of Theorem 27.5 (M. I. Gould and G. Grätzer [1]).
49. Let $\mathfrak{A}_0, \cdots, \mathfrak{A}_{n-1}$ be independent primal algebras, $\mathfrak{A}' = \mathfrak{A}_0 \times \cdots \times \mathfrak{A}_{n-1}$. Show that if Θ is a congruence relation of $\mathfrak{A}'$, $\langle a_0, \cdots, a_i, \cdots, a_{n-1} \rangle \equiv \langle b_0, \cdots, b_i, \cdots, b_{n-1} \rangle (\Theta)$ and $a_i \neq b_i$, then $\langle a_0, \cdots, a_{i-1}, a, a_{i+1}, \cdots, a_{n-1} \rangle \equiv \langle b_0, \cdots, b_{i-1}, b, b_{i+1}, \cdots, b_{n-1} \rangle (\Theta)$, for any $a, b \in A_i$.
50. Under the conditions of Ex. 49, show that $\mathfrak{A}'/\Theta \cong \mathfrak{A}_{i_0} \times \cdots \times \mathfrak{A}_{i_{k-1}}$ if $\Theta \neq \iota$, where $k \geqq 1$ and $\{i_0, \cdots, i_{k-1}\} \subseteq \{0, \cdots, n-1\}$.
51. Under the conditions of Ex. 49, show that if $\mathfrak{A}'/\Theta$ is simple, then $\mathfrak{A}'/\Theta \cong \mathfrak{A}_i$ for some $0 \leqq i < n$.
52. (A. L. Foster [3]) If K is a primal cluster, $K_1 \subseteq K$, K_1 is finite, and $\mathfrak{A}$ is subdirectly representable in K_1 but in no $K_2 \subset K_1$, then K_1 is the *support* of $\mathfrak{A}$ in K. Show that if $|A| > 1$, then it has a unique support.
53. (A. L. Foster [4]) For each integer $n > 1$, define the *n-field* $\mathfrak{F}_n = \langle F_n; \times, ^\frown \rangle$, where $F_n = \{0, 1, a, \cdots, a^{n-2}\}$, $\langle F_n; \times \rangle$ is a cyclic group of order $n-1$, generated by a, $0 \times b = b \times 0 = 0$ for all $b \in F_n$ and $0^\frown = 1$, $1^\frown = a, \cdots, (a^{n-2})^\frown = 0$. Show that $K = \{\mathfrak{F}_n \mid n \geqq 2\}$ is a primal cluster.
54. (A. L. Foster [5]) A *basic Post algebra* of order n, $\mathfrak{P}_n = \langle P_n; \times, ^\frown \rangle$ is defined as follows: $P_n = \{0, 1, a_2, \cdots, a_{n-2}\}$, $0 \times b = b \times 0 = 0$, $1 \times b = b \times 1 = b$, for $b \in P_n$, $a_i \times a_j = a_{\max(i,j)}$, $0^\frown = 1$, $1^\frown = a_2$, $a_2^\frown = a_3, \cdots, a_{n-1}^\frown = 0$. Prove that $K_1 = \{\mathfrak{P}_n \mid n \geqq 2\}$ is a primal cluster.
55. (A. L. Foster [5]) Show that $K \cup (K_1 - \{\mathfrak{P}_2\})$ is again a primal cluster, where K is given in Ex. 53 and K_1 is given in Ex. 54.
56. (A. L. Foster [7]) Let $\mathfrak{A}$ and $\mathfrak{B}$ be nonisomorphic primal algebras. Then for $b_0, b_1 \in B$ there exist $a \in A$ and polynomial symbols $\mathbf{p}_0, \mathbf{p}_1$ such that $(\mathbf{p}_0)_{\mathfrak{B}} = b_0$, $(\mathbf{p}_1)_{\mathfrak{B}} = b_1$, $(\mathbf{p}_0)_{\mathfrak{A}} = (\mathbf{p}_1)_{\mathfrak{A}} = a$.
57. (E. S. O'Keefe [1]) Let $\mathfrak{A}$ and $\mathfrak{B}$ be nonisomorphic primal algebras. Then there exist unary polynomial symbols $\mathbf{p}_0, \cdots, \mathbf{p}_{n-1}$ such that

$$\mathbf{p}_0^{\mathfrak{A}} = \mathbf{p}_1^{\mathfrak{A}} = \cdots = \mathbf{p}_{n-1}^{\mathfrak{A}}$$

and

$$P^{(1)}(\mathfrak{B}) = \{\mathbf{p}_0^{\mathfrak{B}}, \cdots, \mathbf{p}_{n-1}^{\mathfrak{B}}\}.$$

58. (A. L. Foster [4]) Let $\mathfrak{A}$ be a finite algebra, and K a primal cluster. Then $\mathfrak{A}$ is subdirectly representable in K if and only if $\mathfrak{A}$ is directly representable in K, that is $\mathfrak{A} \cong \mathfrak{A}_0{}^{m_0} \times \cdots \times \mathfrak{A}_{n-1}^{m_{n-1}}$, where $\mathfrak{A}_0, \cdots, \mathfrak{A}_{n-1} \in K$.
59. Let $\mathfrak{A}$ and $\mathfrak{A}'$ be elementarily equivalent structures. Then there exists a structure $\mathfrak{A}''$ such that $\mathfrak{A}$ and $\mathfrak{A}'$ are respectively isomorphic to the elementary substructures $\bar{\mathfrak{A}}$ and $\bar{\mathfrak{A}}_1$ of $\mathfrak{A}''$.
60. Let $\mathfrak{A} \equiv \mathfrak{A}'$, a_i elements of A for $i \in I$, and R_j relations on A', $j \in J$. Then there exists an $\mathfrak{A}'' \equiv \mathfrak{A}$, and elements b_i, $i \in I$ of A'', relations S_j on A'' for $j \in J$, such that $\mathfrak{A}$ with the a_i as constants is elementarily equivalent to $\mathfrak{A}''$ with b_i as constants, and $\mathfrak{A}'$ with the R_j as added relations is elementarily equivalent with $\mathfrak{A}''$ with the S_j added.
61. (M. Morley and R. Vaught [1]) Let $\mathfrak{A}$ be a structure of type τ. Let $\mathfrak{A}^+$ denote the structure $\mathfrak{A}$ with all first order relations of $L(\tau)$ added as relations. Show that $\mathfrak{A}$ is an elementary substructure of $\mathfrak{B}$ if and only if $\mathfrak{A}^+$ is a substructure of $\mathfrak{B}^+$, and this,,in turn, is equivalent to $\mathfrak{A}^+$ being an elementary substructure of $\mathfrak{B}^+$.
62. (S. Kochen [1]) Characterize the property that $\mathfrak{A}$ is an elementary substructure of $\mathfrak{B}$ in terms of isomorphic prime limits.
63. Give an upper bound for the cardinalities of the prime limits in Theorem 41.1.
64. (S. Kochen [1]) Let Φ be a formula in $L(\tau)$ and let $\Phi(\mathfrak{A})$ denote the first order relation of A defined by Φ. For any structure $\mathfrak{A}$, let $U_\Phi(\mathfrak{A})$ be the structure $\langle \mathfrak{A}, \Phi(\mathfrak{A}) \rangle$ obtained by augmenting $\mathfrak{A}$ with the new relation $\Phi(\mathfrak{A})$. Then show that:
    (i) the operation U_Φ commutes with the prime product operation;
    (ii) the operation U_Φ commutes with the prime limit operation (for any fixed sequence of prime dual ideals).
    (iii) U_Φ preserves isomorphisms.
65. (S. Kochen [1]) Let G be a function which associates with every structure $\mathfrak{A}$ of type τ an n-ary relation on A. Define $U_G(\mathfrak{A})$ to be $\langle \mathfrak{A}, G(\mathfrak{A}) \rangle$ similar to that in Ex. 64. Then if U_G satisfies (i)–(iii) of Ex. 64, show that for some formula Φ in $L(\tau)$, $G(\mathfrak{A}) = \Phi(\mathfrak{A})$ for all $\mathfrak{A}$ of type τ.
66. Apply Ex. 65 to prove the following result of Beth [1]: Let Σ_0 be a set of sentences, $\Sigma_1(\mathbf{r})$ a set of sentences containing the n-ary relational symbol $\mathbf{r}$ which does not occur in Σ_0, let $\mathbf{r}'$ be an n-ary relational symbol which occurs neither in Σ_0 nor in $\Sigma_1(\mathbf{r})$, and let $\Sigma_1(\mathbf{r}')$ denote the set of sentences which we get from $\Sigma_1(\mathbf{r})$ by replacing every occurrence of $\mathbf{r}$ by $\mathbf{r}'$. If
$$\Sigma_0 \cup \Sigma_1(\mathbf{r}) \cup \Sigma_1(\mathbf{r}') \models (\mathbf{x}_0) \cdots (\mathbf{x}_{n-1})(\mathbf{r}(\mathbf{x}_0, \cdots, \mathbf{x}_{n-1}) \leftrightarrow \mathbf{r}'(\mathbf{x}_0, \cdots, \mathbf{x}_{n-1})),$$
then there is a formula $\Phi(\mathbf{x}_0, \cdots, \mathbf{x}_{n-1})$ in which neither $\mathbf{r}$ nor $\mathbf{r}'$ occurs, such that
$$\Sigma_0 \cup \Sigma_1(\mathbf{r}) \models (\mathbf{x}_0) \cdots (\mathbf{x}_{n-1})(\Phi(\mathbf{x}_0, \cdots, \mathbf{x}_{n-1}) \leftrightarrow \mathbf{r}(\mathbf{x}_0, \cdots, \mathbf{x}_{n-1})).$$
(If $\mathbf{r}$ can be implicitly defined it can also be explicitly defined.)
67. (A. Robinson [13]) Let $\mathfrak{A}$ be a substructure of $\mathfrak{A}'$. Then the following are equivalent:
    (i) There is no extension of $\mathfrak{A}'$ which is an elementary extension of $\mathfrak{A}$.

(ii) There is a universal sentence Φ in $L_{\mathfrak{A}}(\tau)$ which holds in $\langle \mathfrak{A}, \mathfrak{a} \rangle$ but not in $\langle \mathfrak{A}', \mathfrak{a} \rangle$, where $\mathfrak{a}$ is a well-ordering of A.

68. (J. Łoś [1]) A set of sentences Σ is $\mathfrak{m}$-*categorical*, where $\mathfrak{m}$ is an infinite cardinal, if any two models of Σ of cardinality $\mathfrak{m}$ are isomorphic and there are models of Σ of cardinality $\mathfrak{m}$. Find a Σ which is $\aleph_0$-categorical, but Σ is not $\mathfrak{m}$-categorical for any $\mathfrak{m} > \aleph_0$. (Use Ex. 16.)
69. (J. Łoś [1]) Find a set of sentences which is $\mathfrak{m}$-categorical for all $\mathfrak{m} > \aleph_0$, but not $\aleph_0$-categorical. (Hint: look at divisible torsion free abelian groups.)
70. (J. Łoś [1]) Find sets of sentences categorical in every (in no) power.†
71. (R. L. Vaught [3]) A set Σ of sentences of $L(\tau)$ is called *complete* if for every sentence Φ of $L(\tau)$, either $\Sigma \models \Phi$ or $\Sigma \models \neg\Phi$. Let Σ be a set of sentences such that every model of Σ is infinite. Show that if Σ is categorical for some infinite $\mathfrak{m}$ such that $\mathfrak{m} \geqq \overline{\overline{o(\tau)}}$, then Σ is complete. (Use Corollary 2 to Theorem 40.2.)
72. Find applications of Ex. 71 to known algebraic theories (e.g., torsion-free divisible abelian groups).
73. (A. Robinson [9]) 1. Let $\mathfrak{A}$ be a structure of type τ. Let $L_{\mathfrak{A}}(\tau)$ be the diagram language (defined in terms of a well ordering $\mathfrak{a}$ of A), and $\mathfrak{A}' = \langle \mathfrak{A}, \mathfrak{a} \rangle$. The *diagram of* $\mathfrak{A}$ is the set $D(\mathfrak{A})$ of all sentences Φ in $L_{\mathfrak{A}}(\tau)$ such that Φ holds in $\mathfrak{A}'$ and Φ is either an atomic formula or the negation of an atomic formula.

    Show that for any structure $\mathfrak{B}$ of type τ, $\mathfrak{B}$ is the τ-reduct of some model of $D(\mathfrak{A})$ if and only if $\mathfrak{B}$ contains a substructure isomorphic to $\mathfrak{A}$. (Thus, $D(\mathfrak{A})$ characterizes the extensions of $\mathfrak{A}$).

    2. The *complete diagram* $CD(\mathfrak{A})$ *of* $\mathfrak{A}$ is the set $(\mathfrak{A}')^*$. Show that $CD(\mathfrak{A})$ characterizes the elementary extensions of $\mathfrak{A}$ as τ-reducts of models of $CD(\mathfrak{A})$.
74. (A. Robinson [9]) A consistent set Σ of sentences is called *model-complete* if (i) or (ii) below holds. Prove that (i) and (ii) are equivalent.

    (i) if $\mathfrak{A}$ is a substructure of $\mathfrak{B}$, and $\mathfrak{A}$ and $\mathfrak{B}$ are models of Σ, then $\mathfrak{A}$ is an elementary substructure of $\mathfrak{B}$;

    (ii) if $\mathfrak{A}$ is a model of Σ, $D(\mathfrak{A})$ as in Ex. 73, then $D(\mathfrak{A}) \cup \Sigma$ is complete.
75. (A. Robinson [9]) 1. $\mathfrak{A}_0$ is a *prime-model* of Σ if every model of Σ contains a substructure isomorphic to $\mathfrak{A}_0$. Show that there may be many non-isomorphic prime-models. 2. Let $\mathfrak{A}$ be a prime-model of Σ, and $D(\mathfrak{A})$ the diagram of $\mathfrak{A}$. For any sentence Φ of $L(\tau)$, if $\Sigma \cup D(\mathfrak{A}) \models \Phi$, then $\Sigma \models \Phi$.
76. (A. Robinson [9]) If Σ is model-complete and has a prime-model, then it is complete.
77. (H. J. Keisler [2]) Let K_0 and K_1 be classes of structures. If GCH (Generalized Continuum Hypothesis) holds, then $K_0{}^* \cup K_1{}^*$ is consistent if and only if $\mathbf{IP}_P(K_0) \cap \mathbf{P}_P(K_1) \neq \varnothing$.
78. (H. J. Keisler [2]) Let K_0 and K_1 be classes of structures. If GCH holds,

† M. Morley [1] proved the amazing result that if a set of sentences Σ is categorical in some $\mathfrak{m} > \aleph_0$, then Σ is categorical in *every* $\mathfrak{m} > \aleph_0$. The proof of this very deep result uses the concept of special structures, developed in M. Morley and R. Vaught [1].

and $\mathbf{IP}_P(K_0) \subseteq K_0$, $\mathbf{IP}_P(K_1) \subseteq K_1$, $K_0 \cap K_1 = \varnothing$, then there is a class K closed under elementary equivalence with $K_0 \subseteq K$, $K \cap K_1 = \varnothing$.

79. (J. R. Büchi and W. Craig [1]) Prove that the following two statements are equivalent.

(i) If K_0 and K_1 are pseudo-elementary classes and $K_0 \cap K_1 = \varnothing$, then there exists an elementary class K with $K_0 \subseteq K$ and $K \cap K_1 = \varnothing$.

(ii)* If Φ and Ψ are sentences and $\Phi \models \Psi$, then there is a sentence Θ such that $\Phi \models \Theta$, $\Theta \models \Psi$, and every operational or relational symbol which occurs in Θ also occurs in Φ and Ψ.

80. Prove Craig's Lemma, i.e., (ii) of Ex. 79. (Combine Ex. 78 and 79 with Theorem 42.5.)
81. Let Σ be a closed set of sentences in $L(\tau)$. Let $C(\Sigma)$ denote the system of closed subsets of Σ. Show that $C(\Sigma)$ is an algebraic closure system.
82. Let K be an axiomatic class. Under what conditions does there exist an $\mathfrak{A} \in K$ with $\mathfrak{A}^* = K^*$?
83. Prove that Theorem 42.2 implies the compactness theorem.
84. Apply the compactness theorem to prove the inverse limit theorem for finite sets (Theorem 21.1).
85. Let $F(\tau)$ be the set of all sentences in $L(\tau)$. On $F(\tau)$, "$\Leftrightarrow$" is an equivalence relation. Let $T(\tau)$ denote the set of equivalence classes under "$\Leftrightarrow$", and for $H \subseteq F(\tau)$ let $\hat{H}$ denote the union of all equivalence classes which intersect H; in particular, we set $\hat{\Phi}$ for $\widehat{\{\Phi\}}$. For $H \subseteq T(\tau)$ the following closure operation is introduced:

$$\bar{H} = \bigcap(\{\hat{\Psi} \mid \Phi \models \Psi\} \mid \Phi \in F(\tau) \text{ and } H \subseteq \{\hat{\Psi} \mid \Phi \models \Psi\}).$$

Prove that this makes $T(\tau)$ a topological space, which is a T_1*-space* (i.e., for $p \in T(\tau)$, $\overline{\{p\}} = \{p\}$) and which is *totally disconnected* (i.e., for $p, q \in T(\tau)$, if $p \neq q$, there exists an $A \subseteq T(\tau)$ with $p \in A$, $q \notin A$ and A both open and closed).

86. Show that the compactness theorem is equivalent to the statement that $T(\tau)$ is *compact*. (That is, if A_i, $i \in I$ are open sets, and $\bigcup(A_i \mid i \in I) = T(\tau)$, then $\bigcup(A_i \mid i \in I') = T(\tau)$, for some finite $I' \subseteq I$.)
87. Prove that the compactness theorem is equivalent to the statement that the system of closed sets of $T(\tau)$ is an algebraic closure system.
88. (L. Fuchs [2]) An algebra $\mathfrak{A} = \langle A; F\rangle$ of type τ is said to have a *natural partial ordering* $\leq$ with respect to

$$(*) \qquad \langle\langle\Phi_0, \gamma_0{}^0, \cdots, \gamma^0_{n_0-1}\rangle, \cdots, \langle\Phi_\delta, \gamma_0{}^\delta, \cdots, \gamma^\delta_{n_\delta-1}\rangle, \cdots\rangle_{\delta < o(\tau)}$$

if the following conditions are satisfied:

(i) Φ_δ is a formula in $L(\tau)$, free at most in $\mathbf{x}_0$;

(ii) $\gamma_i{}^\delta$ is one of the following symbols: $\uparrow$, $\downarrow$, or $\updownarrow$;

(iii) if for $a_0, \cdots, a_{n_\gamma-1}$, $a \in A$, $a = f_\delta(a_0, \cdots, a_{n_\delta-1})$ and $\Phi_\delta(a_i)$ for $i < n_\delta$, then $\Phi(a)$;

* This is the statement of Craig's Lemma; see W. Craig [1].

(iv) if $\gamma_i^\delta = \uparrow$, $a_0, \cdots, a_{n_\delta - 1}, a_i' \in A$ and $\Phi_\delta(a_i)$, $\Phi_\delta(a_i')$, then

$$(**) \qquad f_\delta(a_0, \cdots, a_i, \cdots, a_{n_\delta - 1}) \leqq f_\delta(a_0, \cdots, a_i', \cdots, a_{n_\delta - 1});$$

if $\gamma_i^\delta = \downarrow$ or $\updownarrow$, then we have $\geqq$ or $=$ in (**), respectively.

$\mathfrak{A}$ is an O-*algebra* for (*) if $\leqq$ can be defined on A so as to satisfy (i)–(iv). Prove that if all the Φ_γ are universal formulas, then $\mathfrak{A}$ is an O-algebra for (*) if and only if every finitely generated subalgebra of $\mathfrak{A}$ is an O-algebra for (*).

89. Let α and β be infinite ordinals, $\bar{\alpha} < \bar{\beta}$. Find a β-quasi-axiomatic class which is not α-quasi-axiomatic.
90. Find a quasi-axiomatic class which is not an axiomatic class.

PROBLEMS

62. Let Φ be a sentence in $L(\tau)$ and $Sp(\Phi)$ (the *spectrum* of Φ) the set of positive integers n, such that there is an n-element algebra in Φ^*. Characterize $Sp(\Phi)$ as a set of positive integers.†
63.‡ Let S be a set of positive integers, $1 \in S$, $S \cdot S \subseteq S$. Set $Ir(S) = \{a \mid a \in S,$ $a = x \cdot y$, $x, y \in S$ imply $x = 1$ or $y = 1\}$. Is it true that S is the spectrum of a single identity if and only if $Ir(S)$ is the spectrum of a first order sentence?
64. Let K be an axiomatic class of algebras and $F(K)$ the set of all cardinals $\mathfrak{m}$ for which $\mathfrak{F}_K(\mathfrak{m})$ exists. Characterize $F(K)$. (Note that if $\mathfrak{m}$ is infinite and $\mathfrak{m} \in F(K)$, then $F(K)$ contains all infinite cardinals.)
65. Same as Problem 64 for elementary classes. (One can conjecture that the solutions for Problems 62 and 65 are identical.)
66. Establish a connection (sort of a distributive law) between free products and prime products.
67. Let K be an axiomatic class in which every algebra satisfies (i) and (iii) of Ex. 88 for (*). If $\mathfrak{F}_K(\mathfrak{m})$ exists, is it always an O-algebra for (*)?
68. Let K be the class of all O-algebras for (*) in Ex. 88. Is K an axiomatic class?
69. Find all types τ for which any class of pairwise nonisomorphic algebras is a primal cluster. (For type $\langle n \rangle$ this was proved by E. S. O'Keefe [1], for types $\langle n, 1 \rangle$ by F. M. Sioson [2].)
70. Find all types τ for which any class of primal algebras, in which every two element subset is independent, is a primal cluster.

† Proposed by H. Scholz, J. Symbolic Logic 17 (1952), 160. References: G. Asser, Z. Math. Logik Grundlagen Math. 1 (1955), 252–263; A. Mostowski, Z. Math. Logik Grundlagen Math. 2 (1956), 210–214.

‡ The "if" part follows from a result announced by J. H. Bennett, J. Symbolic Logic 30 (1965), 264, combined with B. H. Neumann [3]. The "only if" part follows from a result of G. Grätzer and R. McKenzie, Notices Amer. Math. Soc. 14 (1967), 697, provided one can prove that the complement of the spectrum of a first order sentence is also the spectrum of a first order sentence.

CHAPTER 7
ELEMENTARY PROPERTIES OF ALGEBRAIC CONSTRUCTIONS

It is well known that a homomorphic image of a commutative ring (group, semigroup, and so on) is again commutative. Similarly, a subring of a ring without proper divisors of zero is again a ring without proper divisors of zero. The proofs of results of this kind are usually trivial because the validity of such a result usually depends only on the form of the sentence describing the given property. For instance, if a universal sentence holds in a structure $\mathfrak{A}$, then it holds in every substructure $\mathfrak{B}$ of $\mathfrak{A}$.

In this chapter we want to relate the formal properties of sentences, that is, those that can be determined by mere inspection (for instance, being a universal sentence is such a property), to the property of being preserved by a given algebraic construction (or constructions).

Intuitively, it seems clear that a first order sentence is preserved under the formation of subalgebras (substructures) if and only if it is a universal sentence. However, this is not quite true, for if Φ is a universal sentence, then $\Psi \equiv \Phi \wedge (\exists \mathbf{x})(\mathbf{x}=\mathbf{x})$ is not universal and yet Ψ is preserved under the formation of subalgebras. But Ψ is equivalent to Φ. This suggests that the results we want to get are of the following form: Φ is preserved under a given algebraic construction (constructions) if and only if Φ is equivalent to a sentence having certain formal properties. One such result, due to G. Birkhoff, was proved in Chapter 4, namely that a sentence is preserved under the formation of subalgebras, homomorphic images, and direct products, if and only if it is equivalent to a conjunction of identities. In this chapter we will investigate the following constructions: subalgebras, extensions, homomorphisms, chain unions, direct products, and subdirect products.

§43. EXTENSIONS AND SUBSTRUCTURES

A sentence (or, in general, a formula) Ψ which is in the prenex normal form is called an *existential sentence* (resp. *existential formula*) if the prefix contains only existential quantifiers, that is, $\Psi = (\exists \mathbf{x}_{\iota_0}) \cdots (\exists \mathbf{x}_{\iota_{n-1}})\Phi$, where Φ contains no quantifiers.

It is obvious that if Ψ is an existential sentence that holds in the structure $\mathfrak{A}$ and if $\mathfrak{B}$ is an extension of $\mathfrak{A}$ (that is, $\mathfrak{A}$ is a substructure of $\mathfrak{B}$), then Ψ holds in $\mathfrak{B}$. To prove the converse, we establish the following stronger result.

Theorem 1. *Let $\mathfrak{A}$ and $\mathfrak{B}$ be structures. $\mathfrak{A}$ has an elementary extension that contains a substructure isomorphic to $\mathfrak{B}$ if and only if every existential sentence that holds in $\mathfrak{B}$ also holds in $\mathfrak{A}$.*

Remark. This result was first stated by L. Henkin [5]. However, it is an obvious consequence of Theorem 3, below, which was proved by A. Tarski [3] and J. Łoś.

Proof. Let $\mathfrak{A}$ have an elementary extension $\mathfrak{A}_1$ which contains a substructure $\mathfrak{B}_1$ which is isomorphic to $\mathfrak{B}$. If Ψ is an existential sentence which holds in $\mathfrak{B}$, then it holds in $\mathfrak{B}_1$ and thus in $\mathfrak{A}_1$; since $\mathfrak{A}_1 \equiv \mathfrak{A}$, Ψ holds in $\mathfrak{A}$, as required.

To prove the converse, assume that every existential sentence which holds in $\mathfrak{B}$ also holds in $\mathfrak{A}$.

Let a be a well ordering of A of type α. We will use frequently the language $L(\tau_{\mathsf{a}}) = L_{\mathfrak{A}}(\tau)$ introduced in §38. The constant corresponding to $a \in A$ is denoted by $\mathbf{k}_a$. For every extension $\mathfrak{C}$ of $\mathfrak{A}$, $\langle \mathfrak{C}, \mathsf{a} \rangle$ will be denoted by $\mathfrak{C}'$ and will be called the *natural extension* of $\mathfrak{C}$ in $L_{\mathfrak{A}}(\tau)$.

Now, for the given structures $\mathfrak{A}$, $\mathfrak{B}$ we consider the diagram languages $L_{\mathfrak{A}}(\tau)$, $L_{\mathfrak{B}}(\tau)$ and it is assumed that the new constants of $L_{\mathfrak{A}}(\tau)$ and $L_{\mathfrak{B}}(\tau)$ are chosen to be distinct. Finally, $L_{\mathfrak{A},\mathfrak{B}}(\tau)$ is the language that contains all the new constants of $L_{\mathfrak{A}}(\tau)$ and $L_{\mathfrak{B}}(\tau)$.

Let $\mathfrak{A}'$ be the natural extension of $\mathfrak{A}$ in $L_{\mathfrak{A}}(\tau)$ and let $\mathfrak{B}'$ be the natural extension of $\mathfrak{B}$ in $L_{\mathfrak{B}}(\tau)$. Set $\Sigma = (\mathfrak{A}')^*$ in $L_{\mathfrak{A}}(\tau)$, and let Γ denote all sentences in $L_{\mathfrak{B}}(\tau)$ that hold in $\mathfrak{B}'$ and are either atomic formulas or negations of atomic formulas. Then $\Sigma \cup \Gamma$ is a set of sentences of $L_{\mathfrak{A},\mathfrak{B}}(\tau)$.

We claim that $(\Sigma \cup \Gamma)^*$ is not void. By the compactness theorem (Theorem 39.2) it suffices to prove that if $\Gamma_0 \subseteq \Gamma$ and Γ_0 is finite, then $(\Sigma \cup \Gamma_0)^*$ is not void.

Let Φ be the conjunction of the sentences in Γ_0. We can obtain Φ from an open formula $\Psi(\mathbf{x}_0, \cdots, \mathbf{x}_{n-1})$ by replacing the variable $\mathbf{x}_i$ by the constant $\mathbf{k}_{b_i}$, $b_i \in B$, $i = 0, \cdots, n-1$. Thus, $\Psi(\mathbf{k}_{b_0}, \cdots, \mathbf{k}_{b_{n-1}})$ holds in $\mathfrak{B}'$, which implies that $(\exists \mathbf{x}_0) \cdots (\exists \mathbf{x}_{n-1})\Psi$ holds in $\mathfrak{B}$, which by our assumption on $\mathfrak{A}$ implies that it holds in $\mathfrak{A}$. Therefore, there exist $a_0, \cdots, a_{n-1} \in A$ such that $\Psi(a_0, \cdots, a_{n-1})$.

Interpret $\mathbf{k}_{b_i}$ on $\mathfrak{A}'$ as a_i and $\mathbf{k}_b$ for $b \neq b_i$, $i = 0, \cdots, n-1$, arbitrarily. The resulting structure $\mathfrak{A}''$ obviously satisfies $\Sigma \cup \Gamma_0$.

Thus, $(\Sigma \cup \Gamma)^*$ is nonvoid. Let $\mathfrak{C} \in (\Sigma \cup \Gamma)^*$. Let $\mathfrak{C}_1, \mathfrak{C}_2, \mathfrak{C}_3$ be the $L(\tau)$, $L_{\mathfrak{A}}(\tau)$, $L_{\mathfrak{B}}(\tau)$ reducts of $\mathfrak{C}$, respectively.

Since $\mathfrak{C}_2 \in \Sigma^*$, by Theorem 38.6, we can assume that $\mathfrak{A}$ is an elementary substructure of $\mathfrak{C}_1$. The mapping $\varphi: b \to (\mathbf{k}_b)_{\mathfrak{C}_3}$, is an isomorphism of $\mathfrak{B}'$ into $\mathfrak{C}_3$ since $\mathfrak{C}_3$ satisfies Γ. Thus, $\mathfrak{C}_1$ is an elementary extension of $\mathfrak{A}$ which contains $\mathfrak{B}\varphi$, an isomorphic copy of $\mathfrak{B}$ as a substructure, completing the proof of Theorem 1.

Corollary 1. *Let K be an axiomatic class. Form K_1, the class of extensions of structures in K and K_2, the class of elementary substructures of structures in K_1. Then K_2 is an axiomatic class; moreover $K_2 = \Sigma^*$, where Σ is the set of all existential sentences which hold in K.*

Remark. It follows immediately from Theorem 42.8 that K_2 is an axiomatic class.

Proof. $K_2 \subseteq \Sigma^*$ is obvious. To prove the reverse inclusion, let $\mathfrak{A} \in \Sigma^*$ and let $\Gamma = \{\Phi \mid \Phi \in \mathfrak{A}^*$ and $\neg\Phi$ is equivalent to an existential sentence$\}$.

Then Γ is closed under conjunction. If $\Phi \in \Gamma$, then $\Phi \in \mathfrak{B}^*$ for some $\mathfrak{B} \in K$; otherwise, $\neg\Phi \in K^*$ and $\neg\Phi$ being existential would imply $\neg\Phi \in \mathfrak{A}^*$, which is a contradiction.

By the compactness theorem, there exists a $\mathfrak{C} \in (K^* \cup \Gamma)^*$. Note that $\mathfrak{C} \in K^{**} = K$. Further, if Φ is an existential sentence which holds in $\mathfrak{C}$, then Φ holds in $\mathfrak{A}$; otherwise, $\neg\Phi \in \Gamma$ would contradict $\mathfrak{C} \in \Gamma^*$. Thus, by Theorem 1, $\mathfrak{A}$ has an elementary extension $\mathfrak{A}_1$ which contains an isomorphic copy of $\mathfrak{C}$ as a substructure. Therefore, $\mathfrak{C} \in K$ implies $\mathfrak{A}_1 \in K_1$ and $\mathfrak{A} \in K_2$, which was to be proved.

Now we are ready to prove the result that was formulated at the beginning of this section.

Corollary 2. *A sentence Φ is equivalent to an existential sentence if and only if, whenever Φ holds in a structure $\mathfrak{A}$, then Φ holds in every extension of $\mathfrak{A}$.*

Proof. Assume that Φ is preserved under extension. Set $K = \Phi^*$. Then K is an elementary class and, using the notation of Corollary 1, $K_2 = \Sigma^*$. Thus, $\Phi \Leftrightarrow \Sigma$. Since (up to equivalence) Σ is closed under conjunction, by the compactness theorem there exists a $\Psi \in \Sigma$ such that $\Phi \Leftrightarrow \Psi$, which was to be proved.

Corollary 1 combined with Corollary 2 takes the following form.

Corollary 3. *Let K be an axiomatic class. K is closed under extensions if and only if $K = \Sigma^*$ for some set Σ of existential sentences.*

However, in most cases we are not interested in classes that are closed under arbitrary extensions. For instance, if K is a class of groups, then we would like to know when K is closed under *group extensions*.

Corollary 4 (*A. Robinson* [10]). *Let K and M be axiomatic classes and $K \subseteq M$. Let Σ denote the set of all existential sentences that hold in K. Then the following two conditions are equivalent:*

(i) *If $\mathfrak{A} \in K$ and $\mathfrak{A}$ is a substructure of $\mathfrak{B}$ that is in M, then $\mathfrak{B} \in K$.*
(ii) $K = \Sigma^* \cap M$.

Proof. (ii) implies (i) since if $K = \Sigma^* \cap M$, $\mathfrak{A} \in K$, $\mathfrak{B}$ is an extension of $\mathfrak{A}$ and $\mathfrak{B} \in M$, then, by Corollary 1, $\mathfrak{B} \in \Sigma^*$. Thus, $\mathfrak{B} \in \Sigma^* \cap M = K$.

(i) implies (ii). Let us assume (i). Since $K \subseteq \Sigma^* \cap M$, it suffices to prove the reverse inclusion. Let $\mathfrak{C} \in \Sigma^* \cap M$. Since $\mathfrak{C} \in \Sigma^*$, by Corollary 1, $\mathfrak{C}$ has an elementary extension $\mathfrak{C}_1$ which is an extension of some $\mathfrak{A} \in K$. Obviously, $\mathfrak{C}_1 \in M$; thus, by (i), $\mathfrak{C}_1 \in K$. Since K is an axiomatic class, this implies $\mathfrak{C} \in K$, which was to be proved.

A *positive existential sentence* Ψ is an existential sentence

$$\Psi \equiv (\exists x_{i_0}) \cdots (\exists x_{i_{n-1}}) \Phi$$

such that the only logical connectives in the matrix Φ are $\vee$ and $\wedge$.

By analyzing the proof of Theorem 1, we get the following result.

Theorem 2 (*H. J. Keisler* [1]). *Let $\mathfrak{A}$ and $\mathfrak{B}$ be structures. $\mathfrak{A}$ has an elementary extension that contains an expanded homomorphic image of $\mathfrak{B}$ as a substructure if and only if every positive existential sentence that holds in $\mathfrak{B}$ also holds in $\mathfrak{A}$.*

Proof. In the proof of Theorem 1, replace Γ by the set of all atomic sentences. Then everything goes through as in Theorem 1, except that the mapping φ will only be a homomorphism.

The analogues of the four corollaries will be stated as exercises.

Next, we want to determine those sentences that are preserved under the formation of substructures. The following statement is trivial.

The sentence Φ is preserved under the formation of substructures if and only if $\neg \Phi$ is preserved under extensions.

Indeed, if Φ is preserved under the formation of substructures, $\neg \Phi$ holds in $\mathfrak{A}$, $\mathfrak{B}$ is an extension of $\mathfrak{A}$, and $\neg \Phi$ does not hold in $\mathfrak{B}$, then Φ holds in $\mathfrak{B}$; thus, Φ holds in $\mathfrak{A}$, which is a contradiction.

Since the negation of an existential sentence can be written in the form of a universal sentence, we get by Corollary 2 to Theorem 1 the following result which is due to J. Łoś [3] and A. Tarski [3].

A sentence Φ is preserved under the formation of substructures if and only if Φ is equivalent to a universal sentence.

Because of the importance of the result, we want to establish it independently of Theorem 1.

Theorem 3. *Let $\mathfrak{A}$ and $\mathfrak{B}$ be structures. $\mathfrak{A}$ is isomorphic to a substructure of an elementary extension of $\mathfrak{B}$ if and only if every universal sentence that holds in $\mathfrak{B}$ also holds in $\mathfrak{A}$.*

Proof. The "only if" part is again trivial. To prove the "if" part, let $\mathfrak{A}$ and $\mathfrak{B}$ be given as in the theorem. We form $L_{\mathfrak{A}}(\tau)$, $L_{\mathfrak{B}}(\tau)$, $\mathfrak{A}'$, $\mathfrak{B}'$, and $L_{\mathfrak{A},\mathfrak{B}}(\tau)$. Let $\Sigma = (\mathfrak{B}')^*$ and let Γ be the set of all sentences in $L_{\mathfrak{A}}(\tau)$ that hold in $\mathfrak{A}'$ and are atomic formulas or negations of atomic formulas.

We will prove that $(\Sigma \cup \Gamma)^* \neq \varnothing$. Let $\Gamma_0 \subseteq \Gamma$, Γ_0 finite. Then the conjunction of all sentences in Γ_0 can be written in the form

$$\neg \Psi(\mathbf{k}_{b_0}, \cdots, \mathbf{k}_{b_{n-1}}),$$

where $\Psi(\mathbf{x}_0, \cdots, \mathbf{x}_{n-1})$ is an open formula. This implies that $(\exists \mathbf{x}_0) \cdots (\exists \mathbf{x}_{n-1}) \neg \Psi$ holds in $\mathfrak{A}'$ and also in $\mathfrak{A}$, that is, $\neg (\mathbf{x}_0) \cdots (\mathbf{x}_{n-1}) \Psi$ holds in $\mathfrak{A}$ and thus in $\mathfrak{B}$ as well. Therefore, we can find elements $b_0, \cdots, b_{n-1} \in B$ such that $\neg \Psi(b_0, \cdots, b_{n-1})$, and then we can proceed as in Theorem 1.

Now we take a $\mathfrak{C} \in (\Sigma \cup \Gamma)^*$ and the mapping $\varphi: (\mathbf{k}_a)_{\mathfrak{C}} \to a$. $\mathfrak{C}_1$, the $L(\tau)$ reduct of $\mathfrak{C}$, will be an elementary extension of $\mathfrak{B}$ and the elements $(\mathbf{k}_a)_{\mathfrak{C}}$ will form a substructure whose isomorphism with $\mathfrak{A}$ is realized by φ.

Definition 1. *A class K of structures is called a* universal class *if $K = \Sigma^*$, where Σ is a set of universal sentences. If $\Sigma = \{\Phi\}$, K is called an* elementary universal class.

Remark. These concepts were introduced by A. Tarski [3]; he used the terminology "universal class in the wider sense" (UC_Δ) for "universal class" and "universal class" (UC) for "elementary universal class".

Corollary. *Let K be an axiomatic class and let $K_1 = \mathbf{S}(K)$, the class of all substructures of structures in K; then K_1 is a universal class and $K_1 = \Sigma^*$, where Σ is the set of all universal sentences that hold in K.*

If, in the proof of Theorem 3, Γ consists only of the negations of atomic sentences, then φ will not be 1–1 and the domain of φ will not be a substructure, but φ will have all the properties of a homomorphism and somewhat more. This leads us to the following definition.

Definition 2 (*H. J. Keisler* [1]). *Let* $\mathfrak{A}$ *and* $\mathfrak{B}$ *be structures. Suppose* $B_1 \subseteq B$ *such that* B_1 *contains all the constants, and let* φ *be a mapping of* B_1 *onto* A *such that for* $b_0, \cdots, b_{n-1} \in B_1$, $\mathbf{p}, \mathbf{q} \in \mathbf{P}^{(n)}(\tau_0)$, *if* $p(b_0, \cdots, b_{n-1}) = q(b_0, \cdots, b_{n-1})$ *in* $\mathfrak{B}$, *then* $p(b_0\varphi, \cdots, b_{n-1}\varphi) = q(b_0\varphi, \cdots, b_{n-1}\varphi)$ *in* $\mathfrak{A}$ *and for* $b_0, \cdots, b_{m_\gamma-1} \in B_1$, $r_\gamma(b_0, \cdots, b_{m_\gamma-1})$ *implies* $r_\gamma(b_0\varphi, \cdots, b_{m_\gamma-1}\varphi)$. *Then* φ *is called an* abridgment of $\mathfrak{B}$ onto $\mathfrak{A}$ *and* $\mathfrak{A}$ *is called an* abridgment of $\mathfrak{B}$.

Replacing Γ in the proof of Theorem 3 as indicated, we get the following result.

Theorem 4. *Let* $\mathfrak{A}$ *and* $\mathfrak{B}$ *be structures.* $\mathfrak{A}$ *is an abridgment of an elementary extension of* $\mathfrak{B}$ *if and only if every positive universal sentence which holds in* $\mathfrak{B}$ *also holds in* $\mathfrak{A}$.

Corollary 1. *Let* K *be an axiomatic class and let* K_1 *be the class of abridgments of structures in* K. *Then* K_1 *is a universal class and* $K_1 = \Sigma^*$, *where* Σ *is the set of all positive universal sentences which hold in* K.

If $\mathfrak{A}$ is a derived structure of $\mathfrak{B}$, that is, $\mathfrak{A} \in \mathbf{HS}(\mathfrak{B})$, then $\mathfrak{A}$ is an abridgment of $\mathfrak{B}$. The converse is not true. However, if $\mathfrak{A}$ and $\mathfrak{B}$ are algebras and $\mathfrak{A}$ is an abridgment of $\mathfrak{B}$, using $\varphi: B_1 \to A$, then φ satisfies the requirements of Theorem 12.2, whence φ can be extended to a homomorphism of $\langle [B_1]; F \rangle$ onto $\mathfrak{A}$, and $\mathfrak{A} \in \mathbf{HS}(\mathfrak{B})$.

Corollary 2. *Let* K *be an axiomatic class of algebras. Then* $\mathbf{HS}(K)$ *is a universal class and* $\mathbf{HS}(K) = \Sigma^*$, *where* Σ *is the set of all positive universal sentences which hold in* K.

§44. GENERALIZED ATOMIC SETS OF FORMULAS

All the concepts and proofs in §43 are based directly or indirectly on the concept of an atomic formula. H. J. Keisler [1] observed, on the one hand, that the validity of the results depends not so much on the way atomic formulas were defined as on some properties of the set of formulas which are equivalent to atomic formulas and, on the other hand, that these new results can be used in situations where the old results do not apply (see §45). This led him to the definition of a generalized atomic set of formulas.

Definition 1. *A set* $\mathscr{F}$ *of formulas of* $L(\tau)$ *is called* generalized atomic (*G.A.*) *if the following conditions hold:*

(i) *if* $\varphi(\mathbf{x}_0, \cdots, \mathbf{x}_{n-1}) \in \mathscr{F}$ *and* $\mathbf{x}_0 \neq \mathbf{x}_i$, $i = 1, \cdots, n-1$, *then*
   (a) *for any variable* $\mathbf{y}$, $\Phi(\mathbf{y}, \mathbf{x}_1, \cdots, \mathbf{x}_{n-1}) \in \mathscr{F}$;
   (b) *for any constant* $\mathbf{k}$ *of* $L(\tau)$, $\Phi(\mathbf{k}, \mathbf{x}_1, \cdots, \mathbf{x}_{n-1}) \in \mathscr{F}$;

(ii) *if* $\Phi \in \mathcal{F}$ *and* $\Phi \Leftrightarrow \Psi$, *then* $\Psi \in \mathcal{F}$;
(iii) $\mathbf{x}_0 = \mathbf{x}_1 \in \mathcal{F}$;
(iv) *the identically false formula is an element of* $\mathcal{F}$.

In other words, $\mathcal{F}$ is closed under substitution and equivalence and it contains the two formulas given in (iii) and (iv).

A simple example of a G.A. set is the set of all formulas which are equivalent to atomic formulas, or which are identically false; this will be denoted by $(L(\tau))$. Obviously, $\mathbf{L}(\tau)$, the set of all formulas, is another example.

Since the intersection of any family of G.A. sets is also a G.A. set, we get that any set of formulas $\mathcal{F}$ is contained in a smallest G.A. set $[\mathcal{F}]$, which we call the G.A. set *generated by* $\mathcal{F}$.

Definition 2. *Let* $\mathbf{X}$ *be a subset of* $\{\vee, \wedge, \neg, \exists, \forall\}$ *and let* $\mathcal{F}$ *be a G.A. set. Let* $\mathcal{F}_1$ *denote the set of all formulas which can be formed from the formulas in* $\mathcal{F}$ *using only the connectives and quantifiers of* $\mathbf{X}$. *The G.A. set* $[\mathcal{F}_1]$ *will be denoted by* $\mathbf{X}\mathcal{F}$.

Corollary. $\mathbf{X}$ *is a closure operator on the set of all G.A. sets, and* $\Phi \in \mathbf{X}\mathcal{F}$ *if and only if* Φ *is equivalent to a formula in* $\mathcal{F}_1$.

We will use the notation $\mathbf{B} = \{\vee, \wedge, \neg\}$. If $\mathbf{X}$ and $\mathbf{Y}$ are closure operators as described in Definition 2, then we write $\mathbf{XY}\mathcal{F} = \mathbf{X}(\mathbf{Y}\mathcal{F})$. Note that these operators do not commute in general. If $\mathbf{X}$ is a one-element set, for instance, $\mathbf{X} = \{\vee\}$, then we will write $\vee\mathcal{F}$ for $\mathbf{X}\mathcal{F}$.

Normal Form Theorem. *Let* $\mathcal{F}$ *be a G.A. set. Then:*

(i) *disjunctive normal form theorem:* $\mathbf{B}\mathcal{F} = \vee\wedge\neg\mathcal{F}$;
(ii) *conjunctive normal form theorem:* $\mathbf{B}\mathcal{F} = \wedge\vee\neg\mathcal{F}$;
(iii) *prenex normal form theorem:* $\{\wedge, \vee, \neg, \forall, \exists\}\mathcal{F} = \{\forall, \exists\}\mathbf{B}\mathcal{F}$.

The proof of these normal form theorems is the same as in the special case of atomic formulas.

It is interesting to observe that, up to equivalence, the types of formulas we defined so far can be very easily described.

For instance, atomic formulas: $(L(\tau))$; open formulas, $\mathbf{B}(L(\tau))$; universal sentences: sentences that belong to $\{\forall, \mathbf{B}\}(L(\tau))$, and so on.

In order to generalize the results of §43, we have to generalize the concepts of homomorphism, subalgebra, and so on, so as to apply to arbitrary G.A. sets.

Definition 3. *Let $\mathscr{F}$ be a G.A. set and let $\mathfrak{A}$, $\mathfrak{B}$ be structures, $A \subseteq B$. Then $\mathfrak{A}$ is called an* $\mathscr{F}$-subsystem *of $\mathfrak{B}$ (and $\mathfrak{B}$ is called an* $\mathscr{F}$-extension *of $\mathfrak{A}$) if whenever $\Phi(\mathbf{x}_0, \cdots, \mathbf{x}_{n-1}) \in \mathscr{F}$, where Φ is free at most in $\mathbf{x}_0, \cdots, \mathbf{x}_{n-1}$, and $a_0, \cdots, a_{n-1} \in A$, then $\Phi(a_0, \cdots, a_{n-1})$ holds in $\mathfrak{A}$ if and only if it holds in $\mathfrak{B}$.*

Definition 4. *Let $\mathscr{F}$ be a G.A. set, let $\mathfrak{A}$, $\mathfrak{B}$ be structures, and let φ be a mapping of A onto B. Then φ is called an* $\mathscr{F}$-homomorphism *if whenever $\Phi(\mathbf{x}_0, \cdots, \mathbf{x}_{n-1}) \in \mathscr{F}$, where Φ is free at most in $\mathbf{x}_0, \cdots, \mathbf{x}_{n-1}$, and $a_0, \cdots, a_{n-1} \in A$, then $\Phi(a_0, \cdots, a_{n-1})$ holds in $\mathfrak{A}$ implies $\Phi(a_0\varphi, \cdots, a_{n-1}\varphi)$ holds in $\mathfrak{B}$.*

Corollary. *An $(L(\tau))$-homomorphism is an onto homomorphism, and conversely.*

Definition 5. *Let $\mathscr{F}$ be a G.A. set, let $\mathfrak{A}$, $\mathfrak{B}$ be structures and let φ be a mapping of a subset A_0 of A onto B. Then φ is called an* $\mathscr{F}$-abridgment of $\mathfrak{A}$ onto $\mathfrak{B}$ *(and $\mathfrak{B}$ is called an* $\mathscr{F}$-abridgment of $\mathfrak{A}$*) if whenever $\varphi(\mathbf{x}_0, \cdots, \mathbf{x}_{n-1}) \in \mathscr{F}$, where φ is free at most in $\mathbf{x}_0, \cdots, \mathbf{x}_{n-1}$, and $a_0, \cdots, a_{n-1} \in A_0$, then $\Phi(a_0, \cdots, a_{n-1})$ holds in $\mathfrak{A}$ implies that $\Phi(a_0\varphi, \cdots, a_{n-1}\varphi)$ holds in $\mathfrak{B}$ and A_0 contains all the constants in $\mathfrak{A}$.*

Now we are ready to state the generalizations of the results of §43. We will formulate only those which will be needed later on.

Theorem 1. *Let $\mathscr{F}$ be a G.A. set and let $\mathfrak{A}$, $\mathfrak{B}$ be structures. $\mathfrak{A}$ has an elementary extension which contains an $\mathscr{F}$-subsystem that is isomorphic to $\mathfrak{B}$ if and only if $\mathfrak{A} \in (\mathfrak{B}^* \cap \exists\mathbf{B}\mathscr{F})^*$.*

Corollary. *Let K be an axiomatic class and let $\mathscr{F}$ be a G.A. set. Form K_1, the class of $\mathscr{F}$-extensions of structures in K, and K_2, the class of elementary substructures of structures of K_1. Then*

$$K_2 = (K^* \cap \exists\mathbf{B}\mathscr{F})^*.$$

Theorem 2. *Let $\mathfrak{A}$, $\mathfrak{B}$ be structures and let $\mathscr{F}$ be a G.A. set. $\mathfrak{B}$ has an $\mathscr{F}$-homomorphism onto a substructure of some elementary extension of $\mathfrak{A}$ if and only if $\mathfrak{A} \in (\mathfrak{B}^* \cap \exists\wedge\vee\mathscr{F})^*$.*

Corollary. *Let K be an axiomatic class and let $\mathscr{F}$ be a G.A. set. Form the class K_1 of all structures $\mathfrak{A}$ such that there exists a $\mathfrak{B} \in K$ that has an $\mathscr{F}$-homomorphism onto a substructure of $\mathfrak{A}$. Let K_2 be the class of all elementary substructures of structures in K_1. Then*

$$K_2 = (K^* \cap \exists\wedge\vee\mathscr{F})^*.$$

Theorem 3. *Let $\mathfrak{A}$, $\mathfrak{B}$ be structures and let $\mathscr{F}$ be a G.A. set. $\mathfrak{A}$ is isomorphic to an $\mathscr{F}$-subsystem of an elementary extension of $\mathfrak{B}$ if and only if $\mathfrak{A} \in (\mathfrak{B}^* \cap \forall\mathbf{B}\mathscr{F})^*$.*

Corollary. *Let K be an axiomatic class and $\mathscr{F}$ a G.A. set. Let K_1 be the class of all $\mathscr{F}$-subsystems of structures in K. Then*

$$K_1 = (K^* \cap \forall\mathbf{B}\mathscr{F})^*.$$

Theorem 4. *Let $\mathfrak{A}$, $\mathfrak{B}$ be structures and $\mathscr{F}$ a G.A. set. $\mathfrak{A}$ is an $\mathscr{F}$-abridgment of an elementary extension of $\mathfrak{B}$ if and only if $\mathfrak{A} \in (\mathfrak{B}^* \cap \forall\wedge\vee\mathscr{F})^*$.*

Corollary. *Let K be an axiomatic class and let $\mathscr{F}$ be a G.A. set. Let K_1 be the class of $\mathscr{F}$-abridgments of structures in K. Then*

$$K_1 = (K^* \cap \forall\wedge\vee\mathscr{F})^*.$$

The proofs of these results are the same as the proofs of the analogous results in §43, with slight modifications. For instance, in the proof of Theorem 2, the step in which we formed the set Γ of all atomic sentences in $L_{\mathfrak{B}}(\tau)$ that held in $\mathfrak{B}'$ should be replaced by "form the G.A. set $\mathscr{F}'$ generated by $\mathscr{F}$ in $L_{\mathfrak{B}}(\tau)$ and set $\Gamma = \mathscr{F}' \cap (\mathfrak{B}')^*$."

§45. CHAIN UNIONS AND HOMOMORPHISMS

In this section we are going to apply the results of §44 to get "classical results", that is, results that do not involve the concept of G.A. set. This will substantiate the statement of §44 to the effect that the results of §44 can be applied in situations in which the results of §43 are not powerful enough.

Let the $\mathfrak{A}_i$ be structures, $i = 0, 1, 2, \cdots$, where $\mathfrak{A}_i$ is a substructure of $\mathfrak{A}_{i+1}$, and let $\mathfrak{A}$ be the union of the ω-chain $\{\mathfrak{A}_i \mid i = 0, 1, \cdots\}$.

Let Φ be a sentence that holds for all the $\mathfrak{A}_i$. If this implies that Φ holds for $\mathfrak{A}$ as well, we will say that Φ is preserved under the formation of chain unions. In this section we will prove a theorem due to C. C. Chang [3], J. Łoś and R. Suszko [3], which characterize such sentences. The proof given below is due to H. J. Keisler [1] and is based on some results of §44.

In preparation for the proof, we introduce the following concepts.

We will write $\langle \mathfrak{A}_0, \mathfrak{A}_1, \mathfrak{A}_2 \rangle^0$ to denote that $\mathfrak{A}_0$ is a substructure of $\mathfrak{A}_1$, $\mathfrak{A}_1$ is a substructure of $\mathfrak{A}_2$, and $\mathfrak{A}_0$ is an elementary substructure of $\mathfrak{A}_2$. If K is a class of structures, then $\hat{K} = \{\mathfrak{A}_0 \mid \text{there exist } \mathfrak{A}_1 \in K \text{ and } \mathfrak{A}_2 \text{ arbitrary such that } \langle \mathfrak{A}_0, \mathfrak{A}_1, \mathfrak{A}_2 \rangle^0\}$.

Let $\mathscr{E}$ be the G.A. set of all formulas that are equivalent to existential formulas.

Theorem 1 (*H. J. Keisler* [1]). *Let K be a class of structures. Then $\hat{K}$ is the same as the class of all $\mathscr{E}$-subsystems of structures in K.*

Proof. Let $\mathfrak{B} \in K$ and let $\mathfrak{A}$ be an $\mathscr{E}$-subsystem of $\mathfrak{B}$. We want to prove that $\mathfrak{A} \in \hat{K}$. Let $L_{\mathfrak{A}}(\tau)$ be the diagram language of $\mathfrak{A}$ and let $\mathfrak{B}'$ be the natural extension of $\mathfrak{B}$ in $L_{\mathfrak{A}}(\tau)$. Then an existential sentence Ψ of $L_{\mathfrak{A}}(\tau)$ holds in $\mathfrak{B}'$ if and only if it holds in $\mathfrak{A}'$. This implies that every universal sentence of $L_{\mathfrak{A}}(\tau)$ which holds in $\mathfrak{A}'$ also holds in $\mathfrak{B}'$. Thus, by Theorem 43.3, $\mathfrak{A}'$ has an elementary extension $\mathfrak{A}_1{}'$ which is isomorphic to a structure $\mathfrak{A}_2{}'$, containing $\mathfrak{B}'$ as a substructure. Let $\varphi\colon A_1{}' \to A_2{}'$ be the isomorphism between $\mathfrak{A}_1{}'$ and $\mathfrak{A}_2{}'$. Since

$$A = \{(\mathbf{k}_a)_{\mathfrak{A}_1'} \mid a \in A\}$$

and $A\varphi = \{(\mathbf{k}_a)_{\mathfrak{A}_2'} \mid a \in A\}$, we get that $\mathfrak{A}'\varphi$ is an elementary substructure of $\mathfrak{A}_2{}'$; thus, $\langle \mathfrak{A}\varphi, \mathfrak{B}, \mathfrak{A}_2\rangle^0$, that is, $\mathfrak{A} \in \hat{K}$, which was to be proved.

To prove the converse, let $\langle \mathfrak{A}, \mathfrak{B}, \mathfrak{A}_1\rangle^0$ and $\mathfrak{B} \in K$. We will prove that $\mathfrak{A}$ is an $\mathscr{E}$-subsystem of $\mathfrak{B}$, which will conclude the proof of Theorem 1.

Indeed, let $\Phi(\mathbf{x}_0, \cdots, \mathbf{x}_{n-1})$ be an existential formula free at most in $\mathbf{x}_0, \cdots, \mathbf{x}_{n-1}$; then

$$\Phi(\mathbf{x}_0, \cdots, \mathbf{x}_{n-1}) \equiv (\exists \mathbf{y}_0) \cdots (\exists \mathbf{y}_{m-1})\Psi(\mathbf{x}_0, \cdots, \mathbf{x}_{n-1}, \mathbf{y}_0, \cdots, \mathbf{y}_{m-1}),$$

where Ψ is an open formula. Let $a_0, \cdots, a_{n-1} \in A$. Then $\Phi(a_0, \cdots, a_{n-1})$ implies that $\Psi(a_0, \cdots, a_{n-1}, b_0, \cdots, b_{m-1})$ for some $b_0, \cdots, b_{m-1} \in A$. Thus, $\Phi(a_0, \cdots, a_{n-1})$ holds in $\mathfrak{B}$. In a similar manner, if $\Phi(a_0, \cdots, a_{n-1})$ holds in $\mathfrak{B}$, this implies that $\Phi(a_0, \cdots, a_{n-1})$ holds in $\mathfrak{A}_1$, which, in turn, implies that $\Phi(a_0, \cdots, a_{n-1})$ holds in $\mathfrak{A}$, since $\mathfrak{A}$ is an elementary substructure of $\mathfrak{A}_1$. Thus $\Phi(a_0, \cdots, a_{n-1})$ holds in $\mathfrak{A}$ if and only if it holds in $\mathfrak{B}$, completing the proof.

A *universal-existential* sentence is a sentence of the form

$$(\mathbf{x}_0) \cdots (\mathbf{x}_{n-1})(\exists \mathbf{y}_0) \cdots (\exists \mathbf{y}_{m-1})\Psi,$$

where Ψ does not contain quantifiers. Up to equivalence, the universal-existential sentences are the sentences in $\forall\exists\mathbf{B}\,(L(\tau))$.

Corollary. *Let K be an axiomatic class and let Σ be the set of all universal-existential sentences that hold in K. Then $\hat{K} = \Sigma^*$.*

Proof. This is obvious from Theorem 1 and from the Corollary to Theorem 44.3.

Theorem 2. *Let K be an axiomatic class. Let K_1 be the class of unions of ω-chains of structures in K, and let K_2 be the class of elementary substructures* *of structures in K_1. Let Σ be the set of all universal-existential sentences that hold in K. Then K_2 is an axiomatic class and $K_2 = \Sigma^*$.*

Proof. Let $\mathfrak{A} \in K_1$. Then $\mathfrak{A}$ is the union of structures $\mathfrak{A}_i \in K$, where $\mathfrak{A}_i$ is a substructure of $\mathfrak{A}_{i+1}$, $i = 0, 1, \cdots$. Let $\Phi \in \Sigma$,

$$\Phi \equiv (\mathbf{x}_0) \cdots (\mathbf{x}_{n-1})(\exists \mathbf{y}_0) \cdots (\exists \mathbf{y}_{m-1})\Psi(\mathbf{x}_0, \cdots, \mathbf{x}_{n-1}, \mathbf{y}_0, \cdots, \mathbf{y}_{m-1}),$$

where Ψ is an open formula. Let $a_0, \cdots, a_{n-1} \in A$. Then $a_0, \cdots, a_{n-1} \in A_i$ for some i. Since Φ holds in $\mathfrak{A}_i$, there are elements $b_0, \cdots, b_{m-1} \in A_i$ such that $\Psi(a_0, \cdots, a_{n-1}, b_0, \cdots, b_{m-1})$, which implies that Φ holds in $\mathfrak{A}$.

Thus, $K_1 \subseteq \Sigma^*$, which obviously implies that $K_2 \subseteq \Sigma^*$.

Now let $\mathfrak{A} \in \Sigma^*$. By the corollary to Theorem 1, $\mathfrak{A} \in \hat{K}$, that is, there exist $\mathfrak{A}_1 \in K$ and $\mathfrak{A}_2$ arbitrary such that $\langle \mathfrak{A}, \mathfrak{A}_1, \mathfrak{A}_2 \rangle^0$. Since $\mathfrak{A}$ is an elementary substructure of $\mathfrak{A}_2$, we have $\mathfrak{A}_2 \in \Sigma^*$; thus, $\mathfrak{A}_2 \in \hat{K}$ and therefore there exist $\mathfrak{A}_3 \in K$ and $\mathfrak{A}_4$ arbitrary such that $\langle \mathfrak{A}_2, \mathfrak{A}_3, \mathfrak{A}_4 \rangle^0$. Proceeding thus, we construct an ω-chain of structures $\mathfrak{A} = \mathfrak{A}_0, \mathfrak{A}_1, \mathfrak{A}_2, \cdots$ such that $\mathfrak{A}_{2i+1} \in K$, $i = 0, 1, \cdots$, and $\mathfrak{A}_{2i}$ is an elementary substructure of $\mathfrak{A}_{2i+2}$, $i = 0, 1, \cdots$. Let $\mathfrak{B}$ be the union of the $\mathfrak{A}_i$. Then $\mathfrak{B}$ is the union of the $\mathfrak{A}_{2i+1}$; thus, $\mathfrak{B} \in K_1$. On the other hand, $\mathfrak{B}$ is the union of the $\mathfrak{A}_{2i}$; thus, by Theorem 38.1, $\mathfrak{A}$ is an elementary substructure of $\mathfrak{B}$, that is, $\mathfrak{A} \in K_2$, which completes the proof of Theorem 2.

Corollary. *The following four conditions on a sentence Φ are equivalent:*

(i) *Φ is equivalent to a universal-existential sentence;*
(ii) *Φ is preserved under the formation of ω-chain unions;*
(iii) *Φ is preserved under the formation of chain unions;*
(iv) *Φ is preserved under the formation of direct unions.*

Proof. (i) implies (iv) can be proved by using the argument of the first part of the proof of Theorem 2. (iv) implies (iii), (iii) implies (ii) are trivial. (ii) implies (i) can be established by taking the class $K = \Phi^*$ as in the proof of Corollary 2 to Theorem 43.1.

It was observed by R. C. Lyndon† and also by E. Marczewski [1] that *positive sentences* (that is, sentences given in the prenex normal form, whose matrix does not contain the negation sign)‡ are preserved under homomorphisms. We are going to prove the converse of this statement: if a sentence is preserved under homomorphisms, then it is equivalent to a positive sentence. This was conjectured by E. Marczewski [1] and the result was announced without proof by J. Łoś [2] and A. I. Mal'cev [6].

† Review of A. Horn [1], J. Symbolic Logic 16 (1951), 216.

‡ In general, a formula having this property is called a *positive formula*.

The first proof was published by R. C. Lyndon [4]. Using the results of §44, we will give the very elegant proof due to H. J. Keisler [1].

Theorem 3. *Let $\mathfrak{A}$ and $\mathfrak{B}$ be structures. $\mathfrak{B}$ has an elementary extension which has a homomorphism onto an elementary extension of $\mathfrak{A}$ if and only if every positive sentence which holds in $\mathfrak{B}$ also holds in $\mathfrak{A}$.*

Proof. The "only if" part is trivial (for a more general statement, see the Exercises). To prove the "if" part, let $\mathbf{P}$ denote the set of sentences that are equivalent to positive sentences, that is, $\mathbf{P}=\{\wedge, \vee, \forall, \exists\}(L(\tau))$. By assumption, $\mathfrak{A} \in (\mathfrak{B}^* \cap \mathbf{P})^*$. Since $\mathbf{P}=\exists\wedge\vee\mathbf{P}$, we have that

$$\mathfrak{A} \in (\mathfrak{B}^* \cap \exists\wedge\vee\mathbf{P})^*;$$

thus, by Theorem 44.2, $\mathfrak{A}$ has an elementary extension $\mathfrak{A}_1$ such that there exists a $\mathbf{P}$-homomorphism φ_1 of $\mathfrak{B}$ into $\mathfrak{A}_1$. We now form the diagram language $L_{\mathfrak{B}}(\tau)$ of $\mathfrak{B}$ and the natural extension $\mathfrak{B}'$ of $\mathfrak{B}$. Let $\mathbf{P}'$ be the G.A. set generated by $\mathbf{P}$ in $L_{\mathfrak{B}}(\tau)$ (obviously $\mathbf{P}'$ is the set of all formulas which are equivalent to positive sentences in $L_{\mathfrak{B}}(\tau)$).

We interpret the new constants $\mathbf{k}_b$, $b \in B$, in $\mathfrak{A}_1$ by $b\varphi_1$; $\mathfrak{A}_1'$ denotes the resultant structure.

If Φ is a sentence in $\mathbf{P}'$, then it follows from the way we defined the constants in $\mathfrak{A}_1'$ that if Φ holds in $\mathfrak{B}'$ then it also holds in $\mathfrak{A}_1'$. Moreover, $\forall\wedge\vee\mathbf{P}'=\mathbf{P}'$; thus,

$$\mathfrak{A}_1' \in (\mathfrak{B}'^* \cap \forall\wedge\vee\, \mathbf{P}')^*.$$

Thus, we can apply Theorem 44.4, and we get the existence of an elementary extension $\mathfrak{B}_1'$ of $\mathfrak{B}'$, a subset B_{10} of B_1', and a mapping $\theta_1\colon B_{10} \to A_1$ such that θ_1 is a $\mathbf{P}'$-abridgment of $\mathfrak{B}_1'$ onto $\mathfrak{A}_1'$. By the definition of a $\mathbf{P}'$-abridgment, B_{10} contains all the constants. Thus, $B_{10} \supseteq B$. Furthermore, a $\mathbf{P}'$-abridgment carries constants into constants; thus,

$$\theta_1\colon (\mathbf{k}_b)_{\mathfrak{B}_1'} \to (\mathbf{k}_b)_{\mathfrak{A}_1'},$$

that is, $\theta_1\colon b \to (\mathbf{k}_b)_{\mathfrak{A}_1'}$. Therefore, θ_1 is an extension of φ_1.

Then we consider the language $L_{\mathfrak{B}_{10}}(\tau)$, the structures $\mathfrak{B}_1''$, $\mathfrak{A}_1''$ (defined in terms of θ_1), and $\mathbf{P}''$. Then again we will find that $\exists\wedge\vee\mathbf{P}''=\mathbf{P}''$ and $\mathfrak{A}_1'' \in (\mathfrak{B}_1''^* \cap \exists\wedge\vee\mathbf{P}'')^*$, and thus we can again apply Theorem 44.2.

Repeating these arguments, we construct the structures

$$\mathfrak{A} = \mathfrak{A}_0, \mathfrak{A}_1, \mathfrak{A}_2, \cdots \quad\text{and}\quad \mathfrak{B} = \mathfrak{B}_0, \mathfrak{B}_1, \mathfrak{B}_2, \cdots,$$

each of which is of type τ, and an ascending chain of mappings

$$\varphi_1 \subseteq \theta_1 \subseteq \varphi_2 \subseteq \theta_2 \subseteq \cdots$$

such that

(i) $\mathfrak{A}_{i+1}$ is an elementary extension of $\mathfrak{A}_i$, $i=0, 1, \cdots$;

(ii) $\mathfrak{B}_{i+1}$ is an elementary extension of $\mathfrak{B}_i$, $i=0, 1, \cdots$;
(iii) φ_i is a **P**-homomorphism of $\mathfrak{B}_{i-1}$ into $\mathfrak{A}_i$, $i=1, 2, \cdots$;
(iv) θ_i is a **P**-abridgment of $\mathfrak{B}_i$ onto $\mathfrak{A}_i$, $i=1, 2, \cdots$.

Let $\mathfrak{A}_\omega$ and $\mathfrak{B}_\omega$ be the unions of the chains of the $\mathfrak{A}_i$ and of the $\mathfrak{B}_i$ respectively. By Theorem 38.1, each $\mathfrak{A}_i$ is an elementary substructure of $\mathfrak{A}_\omega$ and each $\mathfrak{B}_i$ is an elementary substructure of $\mathfrak{B}_\omega$.

Let $\theta = \bigcup (\theta_i \mid 1 \leq i < \omega) = \bigcup (\varphi_i \mid 1 \leq i < \omega)$. Then θ is defined on the whole of B_ω and $B_\omega\theta = A_\omega$. We claim that θ is a **P**-homomorphism. Indeed, let $\Phi(\mathbf{x}_0, \cdots, \mathbf{x}_{n-1}) \in \mathbf{P}$, where Φ is free at most in $\mathbf{x}_0, \cdots, \mathbf{x}_{n-1}$, let $b_0, \cdots, b_{n-1} \in B_\omega$, and let $\Phi(b_0, \cdots, b_{n-1})$ hold in $\mathfrak{B}_\omega$. For some i, $b_0, \cdots, b_{n-1} \in B_i$ and $\Phi(b_0, \cdots, b_{n-1})$ holds in $\mathfrak{B}_i$, since $\mathfrak{B}_i$ is an elementary substructure of $\mathfrak{B}_\omega$. Therefore

$$\Phi(b_0\varphi_{i+1}, b_1\varphi_{i+1}, \cdots, b_{n-1}\varphi_{i+1})$$

holds in $\mathfrak{A}_{i+1}$ since φ_{i+1} is a **P**-homomorphism of $\mathfrak{B}_i$ into $\mathfrak{A}_{i+1}$; since $\varphi_{i+1} \subseteq \theta$ and $\mathfrak{A}_{i+1}$ is an elementary substructure of $\mathfrak{A}_\omega$, we get that $\Phi(b_0\theta, \cdots, b_{n-1}\theta)$ holds in $\mathfrak{A}_\omega$, which was to be proved.

Thus, θ is a **P**-homomorphism; in particular, θ is a homomorphism of $\mathfrak{B}_\omega$ onto $\mathfrak{A}_\omega$, which completes the proof of Theorem 1.

Corollary 1. *Let K be an axiomatic class. Form K_1, the class of all expanded homomorphic images of structures in K, and K_2, the class of all elementary substructures of structures in K_1. Then K_2 is an axiomatic class and $K_2 = (K^* \cap \{\vee, \wedge, \forall, \exists\}(L(\tau)))^*$, that is, $K_2 = \Sigma^*$, where Σ is the set of all positive sentences which hold in K.*

Corollary 2. *A sentence Φ is preserved under the formation of expanded homomorphic images if and only if Φ is equivalent to a positive sentence.*

For algebras, expanded homomorphic image is the same as homomorphic image. Thus for algebras, Corollary 2 states that Φ is preserved under the formation of homomorphic images if and only if Φ is equivalent to a positive sentence.

This appears to be the proper place to point out that, in his paper, H. J. Keisler [1] gave several other applications of his new method. Among other results, he characterized the sentences which are preserved under the formation of direct limits and inverse limits. However, his main result is a characterization of axiomatic classes K which can be represented as $K = \Sigma^*$, where each sentence Φ in Σ is in prenex normal form and the number of alternations of $\forall$ and $\exists$ in the prefix of Φ is at most n, where n is a fixed integer.

§46. DIRECT PRODUCTS

Our first goal is to determine those universal sentences which are preserved under direct products.

Let Σ be a set of universal sentences. We assume that each $\Phi \in \Sigma$ is in prenex normal form and since all sentences considered are universal, we will write down only the matrix of a sentence, that is, we think of an open formula as a universal sentence.

Definition 1. *A set Σ of universal sentences, written as a set of open formulas, is in* normal form *if every $\Phi \in \Sigma$ is of the form $\Phi = \theta_0 \vee \cdots \vee \theta_{n-1}$, where each θ_i is an atomic formula or the negation of an atomic formula.*

Corollary. *Every set Σ of universal sentences is equivalent to a set Σ' of universal sentences in normal form.*

Proof. For all $\Phi \in \Sigma$, take the conjunctive normal form of Φ: $\Phi = \Phi_0 \wedge \cdots \wedge \Phi_{m-1}$ and replace Φ by $\{\Phi_0, \cdots, \Phi_{m-1}\}$.

Definition 2. *Let Σ be a set of universal sentences in normal form. Let $\Phi \in \Sigma$, $\Phi = \theta_0 \vee \cdots \vee \theta_{n-1}$. We say that Φ* can be reduced (with respect to Σ) *if $n > 1$ and for $\Phi' = \theta_0 \vee \cdots \vee \theta_{i-1} \vee \theta_{i+1} \vee \cdots \vee \theta_{n-1}$ for some $0 \leqq i < n$, Σ is equivalent to $(\Sigma - \{\Phi\}) \cup \{\Phi'\}$. Φ is* reduced *if it cannot be reduced. Σ is* reduced *if each $\Phi \in \Sigma$ is reduced.*

Remark. Obviously, Φ' is always stronger than Φ; therefore,

$$(\Sigma - \{\Phi\}) \cup \{\Phi'\}$$

always implies Σ. Thus, to see that Φ can be reduced, it suffices to show that $\Sigma \models \Phi'$.

Lemma 1. *Let Σ be a set of universal sentences in normal form. Then there exists a set of universal sentences Σ' in normal form which is reduced and which is equivalent to Σ.*

Proof. For each $\Phi \in \Sigma$, choose a reduced form Φ' and set $\Sigma' = \{\Phi' \mid \Phi \in \Sigma\}$. By definition, $\Sigma \models \Phi'$ for all $\Phi \in \Sigma$; thus, $\Sigma \models \Sigma'$ and obviously $\Sigma' \models \Sigma$. Therefore, $\Sigma \Leftrightarrow \Sigma'$. Furthermore, Σ' is reduced since if $\Phi' \in \Sigma'$ can be further reduced to Φ'', then $\Sigma \Leftrightarrow \Sigma' \models \Phi''$. Thus $\Sigma \models \Phi''$, contrary to the choice of Φ'.

This very simple proof is due to M. Makkai.

The next result will prove that our definition of reduced normal form is equivalent to the one used by W. Peremans [2].

Lemma 2. *Let Σ be in normal form, $\Phi \in \Sigma$, $\Phi \equiv \theta_0 \vee \cdots \vee \theta_{n-1}$. Assume that the free variables in Φ and in each θ_i are at most $\mathbf{x}_0, \cdots, \mathbf{x}_{m-1}$. Φ is reduced if and only if $n=1$, or for all i, $0 \leq i < n$, there exists a structure $\mathfrak{A} \in \Sigma^*$ and $a_0, \cdots, a_{m-1} \in A$ such that $\theta_i(a_0, \cdots, a_{m-1})$ in $\mathfrak{A}$ and $\neg\theta_j(a_0, \cdots, a_{m-1})$ in $\mathfrak{A}$ for $j \neq i$. In general, if for $\Phi \in \Sigma$ and for fixed i, Φ satisfies the above condition, then θ_i occurs in all the reduced forms of Φ.*

Proof. Let Φ be reduced and $n > 1$. Put $\Phi' \equiv \theta_0 \vee \cdots \vee \theta_{i-1} \vee \theta_{i+1} \vee \cdots \vee \theta_{n-1}$. Then Σ does not imply Φ'; thus, there exists a structure $\mathfrak{A} \in \Sigma^*$ which does not satisfy Φ'. This means that we can find elements $a_0, \cdots, a_{m-1} \in A$ such that $\neg\theta_j(a_0, \cdots, a_{m-1})$ for $j \neq i$. However, $\Phi(a_0, \cdots, a_{m-1})$; thus, $\neg\theta_j(a_0, \cdots, a_{m-1})$ for $j \neq i$ implies $\theta_i(a_0, \cdots, a_{m-1})$, verifying the condition of the theorem.

Since the converse is included in the second statement of Lemma 2, it suffices to prove that statement.

Thus, let us assume that $\Phi \in \Sigma$ and for fixed i the condition is satisfied, and set $\Phi' = \theta_{i_0} \vee \cdots \vee \theta_{i_{k-1}}$ with $i \notin \{i_0, \cdots, i_{k-1}\}$. Then $i_j \neq i$ for all j. Thus, $\neg\theta_{i_j}(a_0, \cdots, a_{m-1})$, which implies $\neg\Phi'(a_0, \cdots, a_{m-1})$. Therefore, $\Sigma \models \Phi'$ does not hold and so Φ' cannot be a reduced form of Φ.

Theorem 1. *Let Σ be a reduced set of universal sentences in normal form. Then Σ is preserved under the formation of direct products if and only if, for all $\Phi \in \Sigma$, $\Phi \equiv \theta_0 \vee \cdots \vee \theta_{n-1}$, where at most one of the θ_i is an atomic formula.*

Proof. Assume that Σ is preserved under the formation of direct products and for $\Phi \in \Sigma$, $\Phi \equiv \theta_0 \vee \cdots \vee \theta_{n-1}$, where θ_0 and θ_1 are both atomic. Then, by Lemma 2, there exist $\mathfrak{A} \in \Sigma^*$ and $a_0, \cdots, a_{m-1} \in A$ such that $\theta_0(a_0, \cdots, a_{m-1})$ and $\neg\theta_i(a_0, \cdots, a_{m-1})$ for $i > 0$. Also, there exist $\mathfrak{B} \in \Sigma^*$, and $b_0, \cdots, b_{m-1} \in B$, such that $\theta_1(b_0, \cdots, b_{m-1})$ and $\neg\theta_j(b_0, \cdots, b_{m-1})$ for $j \neq 1$.

We will prove that $\mathfrak{A} \times \mathfrak{B} \notin \Sigma^*$, which will be a contradiction. In fact, we will prove that $\neg\Phi(\langle a_0, b_0\rangle, \cdots, \langle a_{m-1}, b_{m-1}\rangle)$.

It suffices to prove that $\neg\theta_i(\langle a_0, b_0\rangle, \cdots, \langle a_{m-1}, b_{m-1}\rangle)$ for all i.

Let $i=0$ and $\theta_0 \equiv (\mathbf{p}_0 = \mathbf{q}_0)$, where $\mathbf{p}_0$, $\mathbf{q}_0$ are polynomial symbols, or $\theta_0 \equiv \mathbf{r}_\gamma$. By the definition of operations and relations in direct products, $\theta_0(\langle a_0, b_0\rangle, \cdots, \langle a_{m-1}, b_{m-1}\rangle)$ if and only if $\theta_0(a_0, \cdots, a_{m-1})$ and $\theta_0(b_0, \cdots, b_{m-1})$, and the latter fails to hold.

The same proof applies for $i=1$ and for all i for which θ_i is atomic.

Similarly, if θ_i is the negation of an atomic formula, then

$$\theta_i(\langle a_0, b_0\rangle, \cdots, \langle a_{m-1}, b_{m-1}\rangle)$$

if and only if either $\theta_i(a_0, \cdots, a_{m-1})$ or $\theta_i(b_0, \cdots, b_{m-1})$ and, by assumption both fail to hold.

To prove the converse, let $\Phi \equiv \theta_0 \vee \cdots \vee \theta_{n-1}$, with at most one θ_i atomic, and let $\mathfrak{A}_j, j \in J$, be structures satisfying Φ. We will prove that $\mathfrak{A} = \prod (\mathfrak{A}_j \mid j \in J)$ also satisfies Φ.

Let $t_0, \cdots, t_{m-1} \in A$. We will prove $\Phi(t_0, \cdots, t_{m-1})$. If $n = 1$, the statement is trivial. Let $n > 1$ and assume that all the θ_i are negations of atomic formulas, that is, $\theta_i \equiv (\mathbf{p}_i \neq \mathbf{q}_i)$ or $\theta_i \equiv \neg \mathbf{r}_\gamma$. In the first case, for $j \in J$ $\Phi(t_0(j), \cdots, t_{m-1}(j))$ holds in $\mathfrak{A}_j$; thus

$$p_i(t_0(j), \cdots, t_{m-1}(j)) \neq q_i(t_0(j), \cdots, t_{m-1}(j))$$

in $\mathfrak{A}_j$ for some i. Therefore, $p_i(t_0, \cdots, t_{m-1}) \neq q_i(t_0, \cdots, t_{m-1})$, which implies $\Phi(t_0, \cdots, t_{m-1})$. The same reasoning applies in the second case.

Finally, let $\Phi \equiv \theta_0 \vee \cdots \vee \theta_{n-1}$ and let θ_0 be atomic. Since for all $j \in J$, $\Phi(t_0(j), \cdots, t_{m-1}(j))$; it follows that for some i_j, $\theta_{i_j}(t_0(j), \cdots, t_{m-1}(j))$. If, for some j, $i_j > 0$, then we can argue as above. If, for all $j \in J$, $i_j = 0$, then $\theta_0(t_0, \cdots, t_{m-1})$, which implies $\Phi(t_0, \cdots, t_{m-1})$. This completes the proof of the theorem.

Definition 3. *A* basic Horn formula Φ *is a formula of the form* $\theta_0 \vee \cdots \vee \theta_{n-1}$, *where* θ_i *is an atomic formula or the negation of an atomic formula, and at most one of the* θ_i *is atomic. A* Horn formula *is a formula in prenex normal form whose matrix is a conjunction of basic Horn formulas.*

The following result was proved by J. C. C. McKinsey [1].

Corollary 1. *Let* Φ *be a universal sentence.* Φ *is preserved under direct products if and only if* Φ *is equivalent to a universal Horn sentence.*

Proof. By Theorem 1, Φ is preserved under direct products if and only if $\Phi \Leftrightarrow \Sigma$, where Σ is a set of basic Horn formulas. By the compactness theorem, there exist $\Psi_0, \cdots, \Psi_{k-1} \in \Sigma$ such that $\Phi \Leftrightarrow \Psi_0 \wedge \cdots \wedge \Psi_{k-1}$, which was to be proved.

Let us note that in the first part of the proof of Theorem 1, we used only the fact that Σ is preserved under the formation of direct products of two factors. This gives the following result of A. Tarski [4].

Corollary 2. *Let* K *be a universal class. If* K *is closed under the formation of direct products of two structures, then* K *is closed under direct products of arbitrary nonvoid families of structures in* K.

This will be generalized to arbitrary axiomatic classes in the next section.

Corollary 1 was extended by A. Horn [1] (see also A. Tarski [4]): every Horn sentence is preserved under direct products. This gave rise to the conjecture that a sentence Φ is preserved under the formation of direct products if and only if it is equivalent to a Horn sentence. This was disproved by C. C. Chang and A. C. Morel [1]. They proved the following result.

Theorem 2. *Let $\mathfrak{A}_i$, $i \in I$, be structures and let $\mathscr{D}$ be a proper dual ideal of $\mathfrak{P}(I)$. Let Φ be a Horn formula and let $\mathfrak{f}^{\mathrm{v}} \in (\prod_{\mathscr{D}} (A_i \mid i \in I))^{\omega}$. Then $S(\Phi, \mathfrak{f}) \in \mathscr{D}$ implies that $\mathfrak{f}^{\mathrm{v}}$ satisfies Φ in $\prod_{\mathscr{D}} (\mathfrak{A}_i \mid i \in I)$.*

Recall that $S(\Phi, \mathfrak{f}) = \{i \mid i \in I$ and $\mathfrak{f}(i)$ satisfies Φ in $\mathfrak{A}_i\}$. This result is analogous to Theorem 39.1, in which we assumed that $\mathscr{D}$ was prime.

Proof. Let Γ be the set of all Horn formulas for which the statement of Theorem 2 holds.

(i) If $\Phi \equiv \theta_0 \wedge \cdots \wedge \theta_{n-1}$, where all the θ_i are atomic, then $S(\Phi, \mathfrak{f}) \in \mathscr{D}$ if and only if $\mathfrak{f}^{\mathrm{v}}$ satisfies Φ in $\prod_{\mathscr{D}} (\mathfrak{A}_i \mid i \in I)$. Thus $\Phi \in \Gamma$.

(ii) If $\Phi_0, \Phi_1 \in \Gamma$, then (up to equivalence) $\Phi_0 \wedge \Phi_1$, $(\exists \mathbf{x}_k)\Phi_0$, $(\mathbf{x}_k)\Phi_0 \in \Gamma$.

The proofs of (i)–(ii) are analogous to the proof of Theorem 39.1.

(iii) If Φ is a basic Horn formula, then $\Phi \in \Gamma$.

We will distinguish three cases. (1) If Φ is atomic, this was proved in (i). (2) Assume that $\Phi \equiv (\neg \theta_0) \vee \cdots \vee (\neg \theta_{n-1})$, where the θ_i are atomic, $S(\Phi, \mathfrak{f}) \in \mathscr{D}$, and still $\mathfrak{f}^{\mathrm{v}}$ does not satisfy Φ in $\prod_{\mathscr{D}} (\mathfrak{A}_i \mid i \in I)$. Then, $\mathfrak{f}^{\mathrm{v}}$ satisfies $\neg \Phi$ (which is equivalent to $\theta_0 \wedge \cdots \wedge \theta_{n-1}$) in $\prod_{\mathscr{D}} (\mathfrak{A}_i \mid i \in I)$; thus, by (ii), $S(\neg \Phi, \mathfrak{f}) \in \mathscr{D}$. However, $S(\Phi, \mathfrak{f}) \cap S(\neg \Phi, \mathfrak{f}) = \varnothing$. Thus, $\varnothing \in \mathscr{D}$, contradicting the fact that $\mathscr{D}$ is proper. (3) Assume that $\Phi \equiv (\neg \theta_0) \vee \cdots \vee (\neg \theta_{n-1}) \vee \theta_n$, where all the θ_i are atomic, $S(\Phi, \mathfrak{f}) \in \mathscr{D}$, and $\mathfrak{f}^{\mathrm{v}}$ does not satisfy Φ in $\prod_{\mathscr{D}} (\mathfrak{A}_i \mid i \in I)$. Set $\Psi \equiv \theta_0 \wedge \cdots \wedge \theta_{n-1}$. Since $\mathfrak{f}^{\mathrm{v}}$ satisfies $\neg \Phi$ (which is equivalent to $\Psi \wedge \neg \theta_n$), $\mathfrak{f}^{\mathrm{v}}$ satisfies Ψ; thus, by (i), $S(\Psi, \mathfrak{f}) \in \mathscr{D}$; However, $S(\theta_n, \mathfrak{f}) \supseteq S(\Phi, \mathfrak{f}) \cap S(\Psi, \mathfrak{f}) \in \mathscr{D}$; therefore, $S(\theta_n, \mathfrak{f}) \in \mathscr{D}$ and, by (i), $\mathfrak{f}^{\mathrm{v}}$ satisfies θ_n and thus Φ in $\prod_{\mathscr{D}} (\mathfrak{A}_i \mid i \in I)$, contrary to our assumption. This completes the proof of Theorem 2.

Corollary. *Every Horn sentence is preserved under the formation of reduced direct products.*

This follows from Theorem 2 and the fact that $I \in \mathscr{D}$.

The converse of the corollary to Theorem 2 was proved in H. J. Keisler [9]. The proof is a beautiful combination of the method of the proof of

Theorem 41.2 with the "special models" of M. Morley and R. Vaught [1]. Keisler's proof uses the Generalized Continuum Hypothesis. F. Galvin recently eliminated the Generalized Continuum Hypothesis from the proof (Ann. Math. Logic 1 (1970), 389–422).

Theorem 3. *There exists a sentence Φ that is preserved under the formation of direct products and which is not equivalent to a Horn sentence.*

Proof. Let $\tau = \langle 2, 2, 1, 0, 0\rangle$ be the type of Boolean algebras. Let Φ be a sentence in $L(\tau)$ that is satisfied by a structure $\mathfrak{A}$ of type τ if and only if $\mathfrak{A}$ is a Boolean algebra having at least one atom. Obviously, Φ is preserved under the formation of direct products. However, Φ is not preserved under the formation of reduced direct products. Indeed, if $\mathfrak{A}_i$, $i \in I$, are two-element Boolean algebras, I is infinite, and $\mathscr{D}$ is the dual ideal consisting of the complements of finite subsets of I, then $\prod_{\mathscr{D}} (\mathfrak{A}_i \mid i \in I)$ has no atom (see Exercises) while each $\mathfrak{A}_i$ had one.

Now, Theorem 3 is obvious since, if Φ were equivalent to a Horn sentence, then by the corollary to Theorem 2, it would be preserved under the formation of reduced direct products.

Although the conjecture, stated before Theorem 2, is false in general, it is true for certain types of sentences. Such types are the following: universal sentences (Corollary 1 to Theorem 1); existential sentences (R. C. Lyndon [7]); universal-existential sentences (J. Weinstein, Doctoral dissertation, University of Wisconsin, 1965). Since the counterexample of Theorem 3 is an existential-universal sentence, Weinstein's result is the best possible.

§47. DIRECT PRODUCTS (*Continued*)

For every type of algebraic construction discussed so far, we found a type of formula such that a sentence is preserved under the algebraic construction if and only if the sentence is equivalent to a sentence of the given type. The only exception was the direct product. Recently J. Weinstein (Doctoral dissertation, University of Wisconsin, 1965) has found a characterization of sentences preserved under direct product. In this section a related theorem of S. Feferman and R. L. Vaught [1] will be discussed which in a certain sense serves as a substitute.

To grasp the idea involved, take the algebras $\mathfrak{A}_i$, $i \in I$, $\mathfrak{A} = \prod (\mathfrak{A}_i \mid i \in I)$, and $t_0, \cdots, t_{n-1} \in A$. If $\mathbf{p}$ and $\mathbf{q}$ are n-ary polynomial symbols, then $p(t_0, \cdots, t_{n-1}) = q(t_0, \cdots, t_{n-1})$ holds in $\mathfrak{A}$ if and only if

$$\{i \mid p(t_0(i), \cdots, t_{n-1}(i)) = q(t_0(i), \cdots, t_{n-1}(i))\} = I;$$

we can write the latter condition as $S(\mathbf{p}=\mathbf{q}, \mathrm{t})=I$, where $\mathrm{t}=\langle t_0, \cdots, t_{n-1}\rangle \in A^n$. Similarly, $p(t_0, \cdots, t_{n-1}) \neq q(t_0, \cdots, t_{n-1})$ if and only if,

$$|S(\mathbf{p}\neq\mathbf{q}, \mathrm{t})| \geqq 1.$$

In other words, in these and in many other cases, we find that the satisfiability of a formula Γ by a sequence of elements of $\mathfrak{A}$ is equivalent to the satisfiability of a formula Φ in the language of Boolean set algebras by certain sets which we construct from Γ using constructions like $S(\Gamma, \mathrm{t})$. The theorem of S. Feferman and R. L. Vaught states in effect that this is always the case.

To formulate the theorem, we will make use of two languages: $L(\tau)$ and $L(\sigma)$, where τ is the type of the structures considered and $\sigma=\langle 2, 2, 1, 0, 0\rangle$ is the type of Boolean algebras $\langle B; \vee, \wedge, ', 0, 1\rangle$; the variables in $L(\tau)$ are $\mathbf{x}_0, \mathbf{x}_1, \cdots$ and the variables in $L(\sigma)$ are $\mathbf{X}_0, \mathbf{X}_1, \cdots$.

Definition 1. *An* acceptable sequence *is a sequence* $\zeta=\langle\Phi, \theta_0, \cdots, \theta_m\rangle$, *where* $\Phi \in \mathbf{L}(\sigma)$, $\theta_0, \cdots, \theta_m \in \mathbf{L}(\tau)$, *and* Φ *is free at most in* $\mathbf{X}_0, \cdots, \mathbf{X}_m$. *We will say that* $\mathbf{x}_k$ *is* free *in* ζ *if* $\mathbf{x}_k$ *is free in some* θ_i.

Theorem 1 (*S. Feferman and R. L. Vaught* [1]). *With every formula* $\Gamma \in \mathbf{L}(\tau)$ *we can associate an acceptable sequence* $\zeta=\langle\Phi, \theta_0, \cdots, \theta_m\rangle$ *such that* $\mathbf{x}_k$ *is free in* ζ *if and only if it is free in* Γ, *and the following holds:*

If $\mathfrak{A}_i$, $i \in I$, *are structures of type* τ, $\mathfrak{A}=\prod(\mathfrak{A}_i \mid i \in I)$, *and* $\mathrm{t} \in A^\omega$, *then* Γ *is satisfied by* t *in* $\mathfrak{A}$ *if and only if* $\Phi(S(\theta_0, \mathrm{t}), \cdots, S(\theta_m, \mathrm{t}))$ *holds in* $\mathfrak{P}(I)$.

Proof. Let Ω denote those formulas in $L(\tau)$ for which the theorem holds.

(1) If Γ is atomic, then $\Gamma \in \Omega$.

Set $\Phi \equiv (\mathbf{X}_0=\mathbf{1})$, $\theta_0 \equiv \Gamma$, and $\zeta=\langle\Phi, \theta_0\rangle$. Then ζ is an acceptable sequence since Φ is free only in $\mathbf{X}_0$. Furthermore, $\mathbf{x}_k$ is free in ζ if and only if it is free in Γ since $\theta_0 \equiv \Gamma$. Finally, Γ is satisfied by t if and only if $S(\Gamma, \mathrm{t})=I$, which is equivalent to $\Phi(S(\theta_0, \mathrm{t}))$ in $\mathfrak{P}(I)$, completing the proof of (1).

(2) If $\Gamma_1, \Gamma_2 \in \Omega$, then $\Gamma_1 \vee \Gamma_2 \in \Omega$.

Let Γ_1 and Γ_2 be associated with $\zeta'=\langle\Phi', \theta_0', \cdots, \theta_{m'}'\rangle$ and $\zeta''=\langle\Phi'', \theta_0'', \cdots, \theta_{m''}''\rangle$, respectively. Set

$$\Phi \equiv \Phi' \vee \Phi''(\mathbf{X}_{m'+1}, \cdots, \mathbf{X}_{m'+m''+1})$$

and

$$\zeta = \langle\Phi, \theta_0', \cdots, \theta_{m'}', \theta_0'', \cdots, \theta_{m''}''\rangle.$$

It is obvious that ζ satisfies the requirements of Definition 1 and Theorem 1.

(3) If $\Gamma \in \Omega$, then $\neg\Gamma \in \Omega$.

Let Γ be associated with $\zeta = \langle \Phi, \theta_0, \cdots, \theta_m \rangle$. It is obvious that $\neg\Gamma$ is associated with $\langle \neg\Phi, \theta_0, \cdots, \theta_m \rangle$.

The following two steps ((4) and (5)) prepare the final step (6). Steps (4) and (5) give two important properties of the concept introduced in the following definition.

Definition 2. *An acceptable sequence* $\zeta = \langle \Phi, \theta_0, \cdots, \theta_m \rangle$ *is called a* partitioning sequence *if* $\theta_0 \vee \theta_1 \vee \cdots \vee \theta_m$ *and, for all* $i \neq j$, $\neg(\theta_i \wedge \theta_j)$ *are provable formulas, that is, formulas which always hold (so-called* tautologies*).*

(4) If $\zeta = \langle \Phi, \theta_0, \cdots, \theta_m \rangle$ is a partitioning sequence and $\mathbf{t} \in A^\omega$ (where $\mathfrak{A}$ is given in the theorem), then $S(\theta_0, \mathbf{t}), \cdots, S(\theta_m, \mathbf{t})$ is a partition of I.

Indeed, $S(\theta_0, \mathbf{t}) \cup S(\theta_1, \mathbf{t}) \cup \cdots \cup S(\theta_m, \mathbf{t}) =$ (by the definition of satisfaction) $= S(\theta_0 \vee \cdots \vee \theta_m, \mathbf{t}) = I$, since every sequence satisfies a provable formula. Similarly, if $i \neq j$, then $S(\theta_i, \mathbf{t}) \cap S(\theta_j, \mathbf{t}) = S(\theta_i \wedge \theta_j, \mathbf{t}) = I - S(\neg(\theta_i \wedge \theta_j), \mathbf{t}) = I - I = \varnothing$.

(5) Let $\zeta' = \langle \Phi', \theta_0', \cdots, \theta'_{m'} \rangle$ be an acceptable sequence. Then there exists a partitioning sequence $\zeta = \langle \Phi, \theta_0, \cdots, \theta_m \rangle$ such that ζ and ζ' are free in the same variables and for any $\mathbf{t} \in A^\omega$ (where $\mathfrak{A}$ is given in the theorem) $\Phi'(S(\theta_0', \mathbf{t}), \cdots, S(\theta'_{m'}, \mathbf{t}))$ if and only if $\Phi(S(\theta_0, \mathbf{t}), \cdots, S(\theta_m, \mathbf{t}))$.

Let $m = 2^{m'+1} - 1$ and let $q_0, \cdots, q_m$ be a list of all subsets of $\{0, \cdots, m'\}$. Set

$$\theta_k = \bigwedge(\theta_j' \mid j \in q_k) \wedge \bigwedge(\neg\theta_j' \mid j \notin q_k)$$

for all $k \leqq m$. Further, set

$$s_l = \{k \mid k \leqq m \text{ and } l \in q_k\}$$

for $l \leqq m'$. Finally, define

$$\Phi = \Phi'(\bigvee(\mathbf{X}_k \mid k \in s_0), \cdots, \bigvee(\mathbf{X}_k \mid k \in s_{m'}))$$

and $\zeta = \langle \Phi, \theta_0, \cdots, \theta_m \rangle$. Obviously, ζ and ζ' have the same free variables.

Then, using the rules of computation with sets (see Exercises 0.19–0.21), we get

$$\begin{aligned} \bigcup(S(\theta_k, \mathbf{t}) \mid k \in s_l) &= \bigcup(S(\bigwedge(\theta_j' \mid j \in q_k) \wedge \bigwedge(\neg\theta_j' \mid j \notin q_k), \mathbf{t}) \mid k \in s_l) \\ &= \bigcup(\bigcap(S(\theta_j', \mathbf{t}) \mid j \in q_k) \cap \bigcap(I - S(\theta_j', \mathbf{t}) \mid j \notin q_k) \mid k \in s_l) \\ &= S(\theta_l', \mathbf{t}). \end{aligned}$$

Similar computations show that ζ is a partitioning sequence.

(6) If $\Gamma' \in \Omega$, then $\Gamma \equiv (\exists \mathbf{x}_k)\Gamma' \in \Omega$.

By assumption, Γ' is associated with an acceptable sequence $\zeta' = \langle \Phi', \theta_0', \cdots, \theta_m' \rangle$. By (5), we can assume that ζ' is a partitioning sequence. We define $\zeta = \langle \Phi, \theta_0, \cdots, \theta_m \rangle$ by setting $\theta_j = (\exists \mathbf{x}_k)\theta_j'$ and

$$\Phi = (\exists \mathbf{Y}_0) \cdots (\exists \mathbf{Y}_m)(\mathrm{Part}_m(\mathbf{Y}_0, \cdots, \mathbf{Y}_m) \wedge \bigwedge (\mathbf{Y}_j \subseteq \mathbf{X}_j \mid j \leq m) \wedge \Phi'(\mathbf{Y}_0, \cdots, \mathbf{Y}_m)),$$

where $\mathbf{Y}_0, \cdots, \mathbf{Y}_m$ are variables not used in Φ' and $\mathrm{Part}_m \in \mathbf{L}(\sigma)$, expresses that $\mathbf{Y}_0, \cdots, \mathbf{Y}_m$ form a partition.

It is obvious that ζ is an acceptable sequence and that the free variables of ζ are the same as those of ζ', excepting $\mathbf{x}_k$. Thus, $\mathbf{x}_l$ is free in Γ if and only if it is free in ζ. It remains to prove that $\Gamma(\mathrm{t})$ in $\mathfrak{A}$ if and only if $\Phi(S(\theta_0, \mathrm{t}), \cdots, S(\theta_m, \mathrm{t}))$ in $\mathfrak{P}(I)$.

$\Gamma(\mathrm{t})$ is equivalent to $\Gamma'(\mathrm{t}(k/g))$ for some $g \in A$, which in turn is equivalent to the following condition:

(*) $\Phi'(S(\theta_0', \mathrm{t}(k/g)), \cdots, S(\theta_m', \mathrm{t}(k/g)))$ for some $g \in A$.

On the other hand, $\Phi(S(\theta_0, \mathrm{t}), \cdots, S(\theta_m, \mathrm{t}))$ is equivalent to the following condition:

(**) There exist subsets $Y_0, \cdots, Y_m$ of I such that

(a) $Y_0, \cdots, Y_m$ form a partition of I;
(b) $Y_j \subseteq S((\exists \mathbf{x}_k)\theta_j', \mathrm{t})$ for $j \leq m$;
(c) $\Phi'(Y_0, \cdots, Y_m)$.

To complete the proof, it suffices to prove that (*) is equivalent to (**).

(*) implies (**). Set $Y_j = S(\theta_j', \mathrm{t}(k/g))$. Then (a) is trivial by (4). Furthermore, $Y_j = S(\theta_j', \mathrm{t}(k/g)) \subseteq \bigcup (S(\theta_j', \mathrm{t}(k/h)) \mid h \in A) = S((\exists \mathbf{x}_k)\theta_j', \mathrm{t})$, by the definition of satisfaction, proving (b). Finally, (c) is nothing but (*).

(**) implies (*). Define g as follows. Take $i \in I$; by (a), $i \in Y_j$ for exactly one $j \leq m$; by (b), $Y_j \subseteq S((\exists \mathbf{x}_k)\theta_j', \mathrm{t})$. Thus, $(\exists \mathbf{x}_k)\theta_j'$ is satisfied by $\mathrm{t}(i)$ in $\mathfrak{A}_i$. This means that there exists an $a_i \in A_i$ such that θ_j' is satisfied by $\mathrm{t}(i)(k/a_i)$. Set $g(i) = a_i$.

Using the identity $\mathrm{t}(i)(k/g(i)) = \mathrm{t}(k/g)(i)$, we get that if $i \in Y_j$, then θ_j' is satisfied by $\mathrm{t}(k/g)(i)$. That is,

$$Y_j \subseteq S(\theta_j', \mathrm{t}(k/g))$$

for $j \leq m$.

This, however, implies that $Y_j = S(\theta_j', \mathrm{t}(k/g))$ since, by (a), the Y_j form a partition and, by (4), the $S(\theta_j', \mathrm{t}(k/g))$ also form a partition. This equality and (c) yield (*). This completes the proof of Theorem 1.

Corollary 1. *In the statement of Theorem* 1, ζ *can always be chosen as a partitioning sequence.*

This is obvious from (5).

If θ is a sentence, then we have nothing to substitute. Thus, $S(\theta, \mathbf{t})$ does not depend on $\mathbf{t}$. In this case, set $S(\theta)=S(\theta, \mathbf{t})$ for any $\mathbf{t}$.

Corollary 2. *Let* Γ *be a sentence which is associated with* $\zeta=\langle\Phi, \theta_0, \cdots, \theta_m\rangle$ *as in Theorem* 1. *Then all the* θ_i *are sentences, and* Γ *holds in* $\mathfrak{A}$ *if and only if* $\Phi(S(\theta_0), \cdots, S(\theta_m))$ *holds in* $\mathfrak{P}(I)$.

Since a variable is free in Γ if and only if it is free in some θ_i, we get that all the θ_i are necessarily sentences. The rest of Corollary 2 is only a restatement of Theorem 1 in this special case.

Now we are ready to give the first application of Theorem 1.

Theorem 2. *Let* $\mathfrak{A}_i$ *be an elementary substructure of* $\mathfrak{B}_i$ *for all* $i \in I$. *Then* $\prod (\mathfrak{A}_i \mid i \in I)$ *is an elementary substructure of* $\prod (\mathfrak{B}_i \mid i \in I)$.

Proof. Form $\mathfrak{A}$ and $\mathfrak{B}$ and take $\mathbf{t} \in A^{\omega}$. Then $S(\theta_i, \mathbf{t})$ for the $\mathfrak{A}_i$ is the same as $S(\theta_i, \mathbf{t})$ for the $\mathfrak{B}_i$, by the definition of elementary substructure. The result now follows from Theorem 1.

The following special case of Theorem 2 is due to A. Mostowski [1]: The formation of direct powers preserves elementary equivalence. Theorem 2 in its present form is a special case of results of S. Feferman and R. L. Vaught [1].

If we combine Theorem 1 with Skolem's result (Theorems 37.1 and 37.2) we get the following statement.

Theorem 3. *For any sentence* $\Gamma \in \mathbf{L}(\tau)$, *there exists a natural number* N *such that whenever* Γ *holds in* $\mathfrak{A}=\prod (\mathfrak{A}_i \mid i \in I)$, *then there exists an* $I' \subseteq I$, $|I'| \leq N$, *with the property that for all* I'' *with* $I' \subseteq I'' \subseteq I$ *we have that* Γ *holds in* $\prod (\mathfrak{A}_i \mid i \in I'')$.

Remark. For direct powers, this statement was first proved by A. Mostowski [1]. A straightforward generalization of Mostowski's work was carried out by R. L. Vaught who proved Theorem 3 with the corollaries given below. The idea of the present proof is due to S. Feferman (see S. Feferman and R. L. Vaught [1]).

Proof of Theorem 3. Let Γ be associated with the partitioning sequence $\zeta=\langle\Phi, \theta_0, \cdots, \theta_m\rangle$ (see Theorem 1 and Corollary 1). Step (4) combined with Remark 1 to Theorem 37.2, yields the following statement.

There exist a natural number M, subsets $q_0, \cdots, q_m$ of $\{0, \cdots, m\}$, and sequences of nonnegative integers $r_0^k, \cdots, r_m^k$, $k=0, \cdots, M$, such that Γ holds in $\mathfrak{A} = \prod (\mathfrak{A}_i \mid i \in I)$ if and only if for some $0 \leqq k \leqq M$, the set $S(\theta_i)$ has exactly r_i^k elements if $i \in q_k$ and $S(\theta_i)$ has at least r_i^k elements if $i \notin q_k$.

Let $n_k = r_0^k + \cdots + r_m^k$ and $N = \max\{n_0, \cdots, n_M\}$.

By the statement given above, we can find pairwise disjoint subsets $X_0, \cdots, X_m$ of I such that $|X_i| = r_i^k$ and $X_i \subseteq S(\theta_i)$ for $i = 0, \cdots, m$. Set $I' = X_0 \cup \cdots \cup X_m$. Obviously, $|I'| = n_k \leqq N$. Furthermore, if $I' \subseteq I'' \subseteq I$ and $\mathfrak{B} = \prod (\mathfrak{A}_i \mid i \in I'')$, then $S(\theta_i)$ for $\mathfrak{B}$ has exactly r_i^k elements if $i \in q_k$, since all the $j \in I$ for which θ_i holds in $\mathfrak{A}_j$ are already contained in I'. Similarly, if $i \notin q_k$, then $S(\theta_i)$ for $\mathfrak{B}$ has at least r_i^k elements since we already have at least r_i^k elements in I'. Thus, Γ holds in $\mathfrak{B}$, completing the proof of Theorem 3.

Corollary 1. *Let $\mathfrak{A}_0, \mathfrak{A}_1, \cdots, \mathfrak{A}_n, \cdots$ be structures of type τ and let Γ be a sentence in $L(\tau)$. If Γ holds in $\mathfrak{A}_0$, $\mathfrak{A}_0 \times \mathfrak{A}_1$, $\mathfrak{A}_0 \times \mathfrak{A}_1 \times \mathfrak{A}_2, \cdots$, then Γ also holds in the infinite direct product $\mathfrak{A}_0 \times \mathfrak{A}_1 \times \cdots \times \mathfrak{A}_n \times \cdots$.*

This statement was conjectured by J. Łoś who proved it for universal sentences. The following result is a generalization of Corollary 2 to Theorem 46.1.

Corollary 2. *Let K be an axiomatic class which is closed under the formation of direct products of two structures. Then K is closed under arbitrary direct products of nonvoid families of structures in K.*

Corollary 3. *Let $\mathfrak{A}_i$, $i \in I$, be structures of type τ. Then there exists a subset I' of I with $|I'| \leqq \max\{\aleph_0, \overline{\overline{o(\tau)}}\}$ such that, for all $I' \subseteq I'' \subseteq I$, $\prod (\mathfrak{A}_i \mid i \in I'')$ is elementarily equivalent to $\prod (\mathfrak{A}_i \mid i \in I)$.*

Proof. Let $\Sigma = \{\Gamma \mid \Gamma$ is a sentence in $L(\tau)$ and Γ holds in $\prod (\mathfrak{A}_i \mid i \in I)\}$. By Theorem 3, we can associate with each Γ in Σ a finite subset I_Γ. Obviously, $I' = \bigcup (I_\Gamma \mid \Gamma \in \Sigma)$ satisfies the assumptions of Corollary 3.

The ideas of the proof of Theorem 1 apply not only to direct products but also to a very wide range of product formations of algebraic systems, including weak direct products, restricted direct products, reduced direct products, and many other product formations not discussed in this book as, for instance, ordinal products, Hahn products, and cardinal sums. To get these generalizations, S. Feferman and R. L. Vaught take a system $\mathfrak{P} = \langle P(I); \subseteq, r_0, r_1, \cdots \rangle$ to replace the Boolean set algebra where $r_0, r_1, \cdots$ are arbitrary relations. Then, roughly speaking, they associate with every acceptable sequence $\zeta = \langle \Phi, \theta_0, \cdots, \theta_m \rangle$ a relation on $A = \prod (A_i \mid i \in I)$ by the rule:

$$\mathbf{t} = \langle t_0, \cdots, t_{p-1} \rangle$$

are in relation if

$$\Phi(S(\theta_0, \mathbf{t}), \cdots, S(\theta_m, \mathbf{t}))$$

holds in $\mathfrak{P}$.

The set A equipped with some of these relations they call a *generalized product*. In fact, step (1) of the proof of Theorem 1 shows that the direct product is a generalized product. The proof of Theorem 1 for generalized products is almost identical with the one given above except that step (1) is then trivial by definition.

§48. SUBDIRECT PRODUCTS

Sentences which are preserved under the formation of subdirect products were characterized by R. C. Lyndon [5]. The result is that every such sentence is equivalent to a special type of Horn sentence which one can build up by conjunction and universal quantification from formulas of the form $\Phi \to \Theta$, where Φ is a positive formula and Θ is an atomic formula.

Another form of Lyndon's result is that if we are given an axiomatic class K and we form K_1: subdirect products of structures in K, and K_2: elementary substructures of structures in K_1, then $K_2 = \Sigma^*$, where Σ is the set of all Horn sentences of the special type described above which hold in K. Lacking a simple proof of this result, we will deal in this section with a very special case of this situation, namely when K consists of a single finite structure of finite type. We will see that even this simple special case raises many unsolved problems.

In this section, $\mathfrak{A}$ will always stand for a fixed finite structure ($|A| < \aleph_0$) of finite type ($o(\tau) < \omega$). If $\mathfrak{A}$ is a two-element Boolean algebra, a two-element distributive lattice, or a group of order p (where p is a prime), then $\mathbf{IP}_S(\mathfrak{A})$ is the class of Boolean algebras, distributive lattices, elementary abelian p-groups (that is, groups in which every element is of order p), respectively. These and other examples suggest that $\mathbf{IP}_S(\mathfrak{A})$ is always an elementary class. However, this is not true in general (see Exercises).

The purpose of this section is to give a sufficient condition on $\mathfrak{A}$ for $\mathbf{IP}_S(\mathfrak{A})$ to be an elementary class; this is due to G. Grätzer [5].

A very convenient condition for $\mathfrak{B} \in \mathbf{IP}_S(\mathfrak{A})$ is given in Theorem 21.7. However, the condition of Theorem 21.7 cannot in general be expressed by a first order formula. We need an extra condition in order to do that.

Definition 1. *A* simple existential sentence $\Psi[\mathbf{x}_0, \cdots, \mathbf{x}_{n-1}]$ *over* $\mathfrak{A}$ *is an existential sentence in the diagram language* $L_{\mathfrak{A}}(\tau)$ *of* $\mathfrak{A}$ *whose matrix is a conjunction of formulas of the form* $\mathbf{f}_\gamma(b_0, \cdots, b_{n_\gamma - 1}) = b$ *or of the form* $\mathbf{r}_\gamma(b_0, \cdots, b_{m_\gamma - 1})$, *where each* b_i, *as well as* b, *is either an* $\mathbf{x}_i$, $0 \leqq i < n$, *or a constant. (The notation* $\Psi[\mathbf{x}_0, \cdots, \mathbf{x}_{n-1}]$ *indicates that* $\mathbf{x}_0, \cdots, \mathbf{x}_{n-1}$ *are all the variables which occur in* Ψ; *of course, all are bound variables.) We will say that* $\Psi[\mathbf{x}_0, \cdots, \mathbf{x}_{n-1}]$ holds *in* $\mathfrak{A}$ *if it holds in the natural extension* $\mathfrak{A}'$ *of* $\mathfrak{A}$ *in* $L_{\mathfrak{A}}(\tau)$.

If $\Psi[\mathbf{x}_0, \cdots, \mathbf{x}_{n-1}]$ *and* $\Psi'[\mathbf{x}_{i_0}, \cdots, \mathbf{x}_{i_{k-1}}]$ *are simple existential sentences over* $\mathfrak{A}$, *we say that* Ψ implies Ψ' *if* $\{\mathbf{x}_{i_0}, \cdots, \mathbf{x}_{i_{k-1}}\} \subseteq \{\mathbf{x}_0, \cdots, \mathbf{x}_{n-1}\}$ *and whenever a sequence* $\mathbf{a} \in A^\omega$ *satisfies the matrix of* Ψ *in* $\mathfrak{A}$ *it also satisfies the matrix of* Ψ' *in* $\mathfrak{A}$.

In case $\mathfrak{A}$ is an algebra, we can think of Ψ as a "set of equations" in the usual sense which holds in $\mathfrak{A}$ if "it can be solved" for the "unknowns" $\mathbf{x}_0, \cdots, \mathbf{x}_{n-1}$.

Definition 2. *Let* N *be a nonnegative integer. The structure* $\mathfrak{A}$ satisfies the condition (U_N) *if, whenever* $\Psi[\mathbf{x}_0, \cdots, \mathbf{x}_{n-1}]$ *is a simple existential sentence which does not hold in* $\mathfrak{A}$, *then there exists a simple existential sentence* $\Psi'[\mathbf{x}_{i_0}, \cdots, \mathbf{x}_{i_{k-1}}]$ *implied by* Ψ *which does not hold in* $\mathfrak{A}$ *either and* $k \leqq N$.

For instance, if a set of equations over a field does not have a solution, then one can always derive from it the statement $0 = 1$, proving that every finite field satisfies the condition (U_0).

Lemma 1. *Let* $\mathfrak{A}$ *be a finite structure satisfying condition* (U_N) *for some nonnegative integer* N. *Let* $\mathfrak{B}_1$ *and* $\mathfrak{B}_2$ *be finite partial algebras and let* $\mathfrak{B}_1$ *be a relative subalgebra of* $\mathfrak{B}_2$. *Finally, let* φ *be a homomorphism of* $\mathfrak{B}_1$ *onto* $\mathfrak{A}$. *Then* φ *can be extended to* $\mathfrak{B}_2$ *if and only if it can be extended to all* $\mathfrak{B}_3$ *with* $B_1 \subseteq B_3 \subseteq B_2$ *and* $|B_3 - B_1| \leqq N$.

Proof. The "only if" part is trivial. To prove the "if" part, let $B_2 - B_1 = \{d_0, \cdots, d_{n-1}\}$. Take all equalities $f_\gamma(c_0, \cdots, c_{n_\gamma - 1}) = c$ and relations $r_\gamma(c_0, \cdots, c_{m_\gamma - 1})$ with $c_i, c \in B_2$ which hold in $\mathfrak{B}_2$. Substitute every d_i which occurs by $\mathbf{x}_i$ and every $b \in B_1$ by $b\varphi \in A$. Take the simple existential sentence $\Psi[\mathbf{x}_0, \cdots, \mathbf{x}_{n-1}]$ whose matrix is the conjunction of all these formulas. Obviously, Ψ holds in $\mathfrak{A}$ if and only if φ can be extended to $\mathfrak{B}_2$. If there were no such extension, then Ψ would not hold in $\mathfrak{A}$ and thus, by condition (U_N), Ψ would imply a $\Psi_1[\mathbf{x}_{i_0}, \cdots, \mathbf{x}_{i_{k-1}}]$ with $k \leqq N$. Then take $B_3 = B_1 \cup \{d_{i_0}, \cdots, d_{i_{k-1}}\}$ and we get that φ has no extension to $\mathfrak{B}_3$, contrary to our assumption. This completes the proof of Lemma 1.

Theorem 1. *If the structure K satisfies condition* (U_N), *then, in Theorem 21.7, it suffices to consider those* $\mathfrak{B}_2$ *that satisfy* $|B_2 - B_1| \leqq N$.

Lemma 2. *Let* $\mathfrak{A}$ *be a finite structure. Then there exists a first order formula* $\Phi_l(\mathbf{x}_0, \cdots, \mathbf{x}_{l-1})$ *which is free in* $\mathbf{x}_0, \cdots, \mathbf{x}_{l-1}$ *such that if* $\mathfrak{B}$ *is a partial algebra,* $B = \{b_0, \cdots, b_{l-1}\}$, *then* $\Phi_l(b_0, \cdots, b_{l-1})$ *holds in* $\mathfrak{B}$ *if and only if there exists a homomorphism* Φ *of* $\mathfrak{B}$ *onto* $\mathfrak{A}$ *with* $b_0\varphi = b_1\varphi$.

Proof. Let $A = \{a_0, \cdots, a_{m-1}\}$. Φ_l can be described as follows: There exist elements $y_0{}^0, \cdots, y_{l-1}^0, \cdots, y_0^{m-1}, \cdots, y_{l-1}^{m-1}$ such that

$$\{y_0{}^0, \cdots, y_{l-1}^0\}, \cdots, \{y_0^{m-1}, \cdots, y_{l-1}^{m-1}\}$$

form a partition; x_0 and x_1 belong to distinct classes under this partition; the partition is a congruence relation; and, finally, the mapping $y_i{}^j \rightarrow a_j$ is a homomorphism.

Since we have a finite number of operational and relational symbols and A is also finite, these conditions can easily be expressed by first order formulas.

Using similar ideas, but a slightly more difficult argument, one can construct a first order formula which will hold for the elements of a finite structure $\mathfrak{B}_2$ with a given substructure $\mathfrak{B}_1$ if and only if $\mathfrak{B}_1$ has a homomorphism onto $\mathfrak{A}$ which can be extended to a homomorphism of $\mathfrak{B}_2$ onto $\mathfrak{A}$.

These remarks, combined with Lemma 2, give our final result.

Theorem 2. *Let* $\mathfrak{A}$ *be a finite structure which satisfies condition* (U_N) *for some nonnegative integer* N. *Then* $\mathbf{IP}_S(\mathfrak{A})$ *is an elementary class.*

EXERCISES

1. (L. A. Henkin [5]) A class K of structures has the *finite persistence* property if for every elementary class M satisfying (i) $M \cap K \neq \varnothing$, (ii) if $\mathfrak{A} \in M \cap K$, and $\mathfrak{B} \in K$ is an extension of $\mathfrak{A}$, then $\mathfrak{B} \in M$, we have that $M \cap K$ contains a finite structure. K has the *finite embedding property* if for every $\mathfrak{A} \in K$, every finite relative substructure of $\mathfrak{A}$ can be embedded in a finite structure in K. Prove that the finite persistence property implies the finite embedding property.
2. (L. A. Henkin [5]) If K is an axiomatic class which has the finite embedding property, then K has the finite persistence property.
3. Prove that the class of Boolean algebras has the finite persistence property.

4. Prove that the class of abelian groups has the finite embedding property (L. A. Henkin [2]) and therefore the finite persistence property (A. Robinson [1]).
5. The class K of well-ordered sets has the finite embedding property but not the finite persistence property. Therefore, K is not axiomatic.
6. Let Φ_0 and Φ_1 be existential sentences. Then there exists an existential sentence Φ with $\Phi_0 \vee \Phi_1 \Leftrightarrow \Phi$.
7. For a class K of structures of type τ, the condition that K is closed under isomorphisms, prime products, and extensions and $K(\tau) - K$ is closed under prime powers, is equivalent to $K = \Sigma^*$, where Σ is a set of existential sentences.
8. Prove that K is an elementary universal class if and only if K is both an elementary and a universal class.
9. Let $\Phi(\mathbf{x}_0, \cdots, \mathbf{x}_{n-1})$ be a positive formula free in $\mathbf{x}_0, \cdots, \mathbf{x}_{n-1}$, $\mathfrak{A}$, $\mathfrak{B}$ structures, $\varphi: A \to B$ a homomorphism. If $\Phi(a_0, \cdots, a_{n-1})$ in $\mathfrak{A}$ for $a_0, \cdots, a_{n-1} \in A$, then $\Phi(a_0\varphi, \cdots, a_{n-1}\varphi)$ in $\mathfrak{B}$.
10. Let K be an axiomatic class and form the class K_1 of homomorphic images of structures in K, the class K_2 of extensions of structures in K_1, and the class K_3 of elementary substructures of structures in K_2. Then $K_3 = \Sigma^*$, where Σ is the set of all positive existential sentences which hold in K.
11. A sentence Φ is preserved under extensions and homomorphisms if and only if it is equivalent to a positive existential sentence.
12. Formulate and prove the analogue of Corollary 4 to Theorem 43.1 for Theorem 43.2.
13. Find an axiomatic class K such that the class K_1 of all extensions of structures in K is not an axiomatic class.
14. Find structures $\mathfrak{A}$ and $\mathfrak{B}$ such that $\mathfrak{A}$ is an abridgment of $\mathfrak{B}$ but $\mathfrak{A}$ is not a derived structure of $\mathfrak{B}$.
15. Prove the corollary to Definition 44.2.
16. Prove the Normal Form Theorems for G.A. sets.
17. Do we get the Normal Form Theorems if we drop one of the conditions of Definition 44.1?
18. (H. J. Keisler [1]) Let K be a class of structures, $\mathscr{F}$ and $\mathscr{G}$ be G.A. sets. If $\mathscr{F} \subseteq \forall\mathbf{B}\mathscr{G}$, then an $\mathscr{F}$-homomorphic image of a $\mathscr{G}$-substructure of a structure in K is an $\mathscr{F}$-abridgment of a structure in K.
19. (H. J. Keisler [1]) If the G.A. set $\mathscr{F}$ contains existential formulas only, then every $\mathscr{F}$-abridgment is an $\mathscr{F}$-homomorphic image of a substructure.
20. Prove Theorems 44.1–44.4.
21. Specialize Theorem 44.3 to $\mathscr{F} = \mathbf{L}(\tau)$. Show that this special case yields Ex. 6.59.
22. (H. J. Keisler [1]) Let K be an axiomatic class, $\mathscr{F}$ a G.A. set, and K_1 the class of all $\exists\mathbf{B}\mathscr{F}$-substructures of structures in K. Prove that $\mathfrak{A}_0 \in K$ if and only if there exist structures $\mathfrak{A}_1$ and $\mathfrak{A}_2$ such that $\mathfrak{A}_1 \in K$, $\mathfrak{A}_0$ is an $\mathscr{F}$-substructure of $\mathfrak{A}_1$, $\mathfrak{A}_1$ is an $\mathscr{F}$-substructure of $\mathfrak{A}_2$, and $\mathfrak{A}_0$ is an elementary substructure of $\mathfrak{A}_2$.

23. (H. J. Keisler [1]) Generalize Theorem 45.2 for closure under $\mathcal{F}$-*union*, which means that if $\mathfrak{A}_i \in K$, $i < \omega$, $\mathfrak{A}_i$ is an $\mathcal{F}$-substructure of $\mathfrak{A}_{i+1}$, $\mathfrak{A}$ is the union of the $\mathfrak{A}_i$ and all $\mathfrak{A}_i$ are $\mathcal{F}$-substructures of $\mathfrak{A}$, then $\mathfrak{A} \in K$.
24. Show that the condition "and all $\mathfrak{A}_i$ are $\mathcal{F}$-substructures of $\mathfrak{A}$" in Ex. 23 is not implied by the other conditions.
25. (H. J. Keisler [1]) Generalize Theorem 45.3 and its corollaries to $\mathcal{F}$-homomorphisms.
26. K is a universal class if and only if $\mathbf{P}_P(K) \subseteq K$ and $\mathbf{IS}(K) \subseteq K$.
27. $\mathbf{ISP}_P(K)$ is a universal class.
28. (R. Fraïssé [5]) Let K be an elementary universal class of relational systems of type τ. Then there are finite relational systems $\mathfrak{A}_0, \cdots, \mathfrak{A}_{n-1}$ such that $\mathfrak{A} \in K$ if and only if no $\mathfrak{A}_i$ is isomorphic to a subsystem of $\mathfrak{A}$.
29. (R. Fraïssé [5]) Show that if $\mathfrak{A}_0, \cdots, \mathfrak{A}_{n-1}$ are finite relational systems, then the class K of all relational systems into which no $\mathfrak{A}_i$ can be embedded is an elementary universal class. (For $n = 1$, this is due to A. Tarski [3].)
30. Generalize Ex. 28 and 29 to structures.
31. Characterize elementary universal classes by $\mathbf{P}_P$, and $\mathbf{S}$, and $\mathbf{I}$.
32. (A. Horn [1]) Find all universal sentences Φ such that if Φ holds for $\mathfrak{A}_0 \times \cdots \times \mathfrak{A}_{n-1}$, then Φ holds for all $\mathfrak{A}_i$
33. Replace "for all" by "for some" in Ex. 32.
34. (A. Horn [1]) Find all universal sentences Φ such that Φ holds for $\mathfrak{A}_0 \times \cdots \times \mathfrak{A}_{n-1}$ if and only if it holds for all $\mathfrak{A}_i$.
35. Replace "for all" by "for some" in Ex. 34.
36. Do we have to use some form of the Axiom of Choice in the proof of Lemma 46.1 ? (No!)
37. Prove Corollary 2 to Theorem 46.1 using Theorem 19.4.
38. Prove that in the proof of Theorem 46.3 we could take $\tau = \langle 2 \rangle$ and Φ to be a universal-existential sentence.
39. Use Ex. 6.14 to extend the result of §47 to weak direct products.
40. (S. Feferman and R. L. Vaught [1]) Let $\langle I; \leqq \rangle$ be a chain and suppose that $\langle A_i; \leqq \rangle$ are also chains for $i \in I$. Define $\leqq$ on $\prod (A_i \mid i \in I)$ by: $f < g$ if $f(i) < g(i)$ for some $i \in I$ and $f(i') = g(i')$ for all $i' < i$. Then

$$\langle \prod (A_i \mid i \in I); \leqq \rangle$$

is called the *ordinal product* of the $\mathfrak{A}_i$. Prove the result of §47 for ordinal products, using the algebra $\mathfrak{P}^*(I) = \langle P(I); \cup, \cap, ', \leqq \rangle$.
41. Let the $\mathfrak{A}_i$ be graphs for $i \in I$. Following G. Sabidussi, define the product by taking $\prod (A_i \mid i \in I)$ and requiring that $f, g \in \prod (A_i \mid i \in I)$ are connected by an edge if for some $i \in I$, $f(i)$ and $g(i)$ are connected by an edge and $f(i') = g(i')$ for all $i' \neq i$. Prove the results of §47 for this product.
42. Let $\mathfrak{A}$ be the two element Boolean algebra. Does $\mathfrak{A}$ satisfy (U_2) ?
43. Let $\mathfrak{A}$ be the n element chain as a distributive lattice. Does $\mathfrak{A}$ have (U_n) ? (See F. W. Anderson and R. L. Blair, Math. Annalen 143 (1961), 187–211.)
44. Let $\mathfrak{A}$ be a group of order p (p is a prime). Does $\mathfrak{A}$ have (U_0) ?
45. Let $\mathfrak{A}$ be a finite field. Does $\mathfrak{A}$ have (U_0) ?

46. Let $\mathfrak{A} = \langle A; f \rangle$ be of type $\langle 1 \rangle$, where $A = \{0, 1\}$, $f(0) = 1$, $f(1) = 0$. Does $\mathfrak{A}$ have (U_n) for some N?
47. Write down the formulas needed in the proof of Lemma 48.2.
48. Prove Theorem 21.6 using the compactness theorem (R. C. Lyndon).
49. Let $K = \{\mathfrak{A}_0, \cdots, \mathfrak{A}_{n-1}\}$ and suppose that each $\mathfrak{A}_i$ has (U_N). Then $\mathbf{IP}_S(K)$ is an elementary class.
50. Show that $\mathbf{IP}_S(\mathfrak{A})$ is an elementary class, if $\mathfrak{A}$ is the algebra given in Ex. 46.
51. Let $\mathfrak{A}$ be an algebra with $|A| > \aleph_0$. Then $\mathbf{IP}_S(\mathfrak{A})$ is not an axiomatic class.
52. Let $\mathfrak{A}$ be $\omega + \omega^* + \omega$ as a lattice, where ω^* is the dual of ω. Show that $\mathbf{IP}_S(\mathfrak{A})$ is not an axiomatic class.
53. Show that Ex. 49 cannot be extended to countably many algebras.
54. Let $\mathfrak{A}$ be a field with $|A| \geqq \aleph_0$. Show that there is no axiomatic class K and cardinal $\mathfrak{m}$ such that $\mathfrak{B} \in K$ and $\mathfrak{B} \in \mathbf{IP}_S(\mathfrak{A})$ are equivalent if $|B| \geqq \mathfrak{m}$. (This settles Problem 5 of G. Grätzer [5], raised by R. C. Lyndon.)
55.† Let $\mathfrak{A} = \langle A; + \rangle$ be defined by $A = \{0, 1, 2\}$, $x + x = 0$ for all $x \in A$, and $x + y = x$, for all $x, y \in A$, $x \neq y$. Let $B \subseteq A^\omega$ be defined by:
$f \in B$ if and only if $\{n \mid f(n) \neq 0\}$ is finite, $f(0) \neq 2$, and $f(0) = 1$ implies that $f(n) = 1$ for at most one $n \neq 0$.
Then $\mathfrak{B}$ is subalgebra of $\mathfrak{A}^\omega$.
56. Let $\mathfrak{A}$ and $\mathfrak{B}$ be given as in Ex. 55. Prove that $\mathfrak{B} \in \mathbf{IP}_S(\mathfrak{A})$. (Hint: let $f = \langle 1, 0, 0, \cdots \rangle$, $g = \langle 0, 0, \cdots \rangle$. Then there is no homomorphism φ of $\mathfrak{B}$ onto $\mathfrak{A}$ with $f\varphi \neq g\varphi$.)
57. Let $\mathfrak{A}$ be given as in Ex. 55. Prove that $\mathfrak{C} \in (\mathbf{P}_S(\mathfrak{A}))^{**}$ if and only if for any positive integer n, the following condition holds:
For all $x, y \in C$ with $x \neq y$, there exists a $z \in C$, such that for all $x_0, \cdots, x_{n-1} \in C$ the relative subalgebra $\langle \{x, y, z, x_0, \cdots, x_{n-1}\}; + \rangle$ has a homomorphism φ onto $\mathfrak{A}$ such that $x\varphi \neq y\varphi$ and $\{x, y, z\}\varphi = A$.
(A similar statement holds for any finite algebra $\mathfrak{A}$.) (Hint: Form a suitable prime limit of $\mathfrak{C}$ and apply Theorem 21.7.)
58. Let $\mathfrak{A}$ be given as in Ex. 55. Then $\mathbf{IP}_S(\mathfrak{A})$ is not an axiomatic class. (Hint: Combine Ex. 55–57 to prove that $\mathfrak{B} \in \mathbf{P}_S(\mathfrak{A})^{**} - \mathbf{IP}_S(\mathfrak{A})$, where $\mathfrak{B}$ was given in Ex. 55.)
59. (A. Robinson [2]) An axiomatic class K is *convex* if $\mathfrak{A}$, $\mathfrak{B}$ are substructures of $\mathfrak{C}$, $\mathfrak{A}, \mathfrak{B}, \mathfrak{C}, \in K$, $A \cap B = D \neq \varnothing$, imply that $\mathfrak{D} \in K$. Prove that if K is convex, then K is closed under chain union. (A complete description of convex axiomatic classes is given in M. O. Rabin [3].)
60. (M. O. Rabin [3]) Prove that a convex axiomatic class is closed under complete nonvoid intersections.
61. Let $\Phi(\mathbf{x})$ be a formula free in $\mathbf{x}$ and K an axiomatic class. If $\Phi(\mathbf{x})$ is preserved under the formation of subalgebras, homomorphic images and direct products, then there is a universal formula $\Psi(\mathbf{x})$ free in $\mathbf{x}$, such that

$$\Sigma \models (\mathbf{x})(\Phi(\mathbf{x}) \leftrightarrow \Psi(\mathbf{x})).$$

† Ex. 55–58 are due to R. McKenzie (unpublished).

PROBLEMS

71. Let $\mathscr{F}$ and $\mathscr{G}$ be G.A. sets. Under what conditions are the corresponding homomorphism and subalgebra concepts equivalent?
72. Let $\mathscr{F}$ be a G.A. set. Under what conditions is every $\mathscr{F}$-abridgment a derived structure?
73. Which first order properties are preserved under free products?
74. Which sentences Φ have the property that the substructures satisfying Φ of a structure $\mathfrak{A}$ form a sublattice of the lattice of all substructures of $\mathfrak{A}$?
75. Let τ be the type of lattices. Which sentences in $L(\tau)$ are preserved under the formation of the lattices of all ideals of a lattice?
76. Let $\mathfrak{K}$ be a finite lattice. Find conditions on $\mathfrak{K}$ under which if $\mathfrak{K}$ is isomorphic to a sublattice of $\mathfrak{I}(\mathfrak{L})$, then $\mathfrak{K}$ is isomorphic to a sublattice of $\mathfrak{L}$. (See G. Grätzer [14].)
77. Let $\mathfrak{A}$ be a finite structure with $o(\tau) < \omega$. Find necessary and sufficient conditions for $\mathbf{IP}_S(\mathfrak{A})$ to be an elementary (axiomatic) class.
78. What can be said about $\mathbf{IP}_S(\mathfrak{A})$ if either $o(\tau) \geq \omega$ or $|A| \geq \aleph_0$?
79. Let $K = \{\mathfrak{A}_0, \mathfrak{A}_1, \cdots, \mathfrak{A}_n, \cdots\}$, where each $\mathfrak{A}_i$ satisfies (U_N). Under what conditions is $\mathbf{IP}_S(K)$ axiomatic?

CHAPTER 8
FREE Σ-STRUCTURES

Given a set Σ of first order sentences, concepts of Σ-substructure and Σ-homomorphism will be defined which reduce to the classical concepts if Σ is universal. Σ-substructures and Σ-homomorphic images always satisfy Σ. In terms of these concepts we define free Σ-structures. §51 contains the uniqueness theorem; this was trivial in the classical case, but for free Σ-structures it is a consequence of the compactness theorem. In §52 we examine the connections among the assertions of the existence of free Σ-structures on different numbers of generators. Construction of free Σ-structures is discussed in §53, and in §54 we discuss the problem of when it is possible to define free Σ-structures in terms of the classical concepts, possibly over a richer language. In this chapter Σ (and the type) will be kept fixed unless otherwise specified; we always assume that every Φ ∈ Σ is given in a prenex normal form. A *Σ-structure* is a structure satisfying Σ. Most of the results of this chapter were published in G. Grätzer [3], [10], and [11].

§49. Σ-INVERSES AND Σ-SUBSTRUCTURES

If Σ is universal, then every substructure of a Σ-structure satisfies Σ, so there is no problem: a Σ-substructure can be defined as a substructure. As an example of a set of sentences which is not universal, let us take Σ_1 which defines lattices as partially ordered sets:

$$\Phi_1\colon (\mathbf{x})(\mathbf{y})(\mathbf{z})(((\mathbf{x} \leqq \mathbf{y} \wedge \mathbf{y} \leqq \mathbf{x}) \leftrightarrow \mathbf{x} = \mathbf{y}) \wedge ((\mathbf{x} \leqq \mathbf{y} \wedge \mathbf{y} \leqq \mathbf{z}) \rightarrow \mathbf{x} \leqq \mathbf{z}))$$
$$\Phi_2\colon (\mathbf{x})(\mathbf{y})(\exists\mathbf{z})(\mathbf{u})(\mathbf{x} \leqq \mathbf{z} \wedge \mathbf{y} \leqq \mathbf{z} \wedge ((\mathbf{x} \leqq \mathbf{u} \wedge \mathbf{y} \leqq \mathbf{u}) \rightarrow \mathbf{z} \leqq \mathbf{u}))$$
$$\Phi_3\colon (\mathbf{x})(\mathbf{y})(\exists\mathbf{z})(\mathbf{u})(\mathbf{z} \leqq \mathbf{x} \wedge \mathbf{z} \leqq \mathbf{y} \wedge ((\mathbf{u} \leqq \mathbf{x} \wedge \mathbf{u} \leqq \mathbf{y}) \rightarrow \mathbf{u} \leqq \mathbf{z})).$$

Then a sublattice of $\langle L; \leqq\rangle$ is not any sub-partially ordered set $\langle H; \leqq\rangle$ satisfying Σ_1, but it has the additional property that if $x, y \in H$, then the z_1 required by Φ_2 and the z_2 required by Φ_3 are also in H.

In this example x and y uniquely determine z in Φ_2 and Φ_3. What should we do if this is not the case? Let us consider Σ_2 which defines complemented lattices with 0 and 1:

$$\Phi_1\colon (\mathbf{x})(\mathbf{y})(\mathbf{z})(\mathbf{x}\vee\mathbf{x}=\mathbf{x}\ \wedge\ \mathbf{x}\wedge\mathbf{x}=\mathbf{x}\ \wedge\ \mathbf{x}\vee\mathbf{y}=\mathbf{y}\vee\mathbf{x}\ \wedge$$
$$\mathbf{x}\wedge\mathbf{y}=\mathbf{y}\wedge\mathbf{x}\ \wedge\ \mathbf{x}\vee(\mathbf{y}\vee\mathbf{z})=(\mathbf{x}\vee\mathbf{y})\vee\mathbf{z}\ \wedge\ \mathbf{x}\wedge(\mathbf{y}\wedge\mathbf{z})$$
$$=(\mathbf{x}\wedge\mathbf{y})\wedge\mathbf{z}\ \wedge\ \mathbf{x}\vee(\mathbf{x}\wedge\mathbf{y})=\mathbf{x}\ \wedge\ \mathbf{x}\wedge(\mathbf{x}\vee\mathbf{y})=\mathbf{x}),$$
$$\Phi_2\colon (\mathbf{x})(\exists\mathbf{y})(\mathbf{x}\vee\mathbf{y}=1\ \wedge\ \mathbf{x}\wedge\mathbf{y}=0).$$

Set $L=\{0,1,p_1,p_2,p_3\}$, $p_i\vee p_j=1$ and $p_i\wedge p_j=0$ if $i\neq j$. If $\langle H;\vee,\wedge,0,1\rangle$ is a substructure of $\mathfrak{L}$ satisfying Σ_2 and $p_1\in H$ then, by Φ_2, p_2, or $p_3\in H$. Thus $\langle\{p_1,p_2,0,1\};\vee,\wedge,0,1\rangle$, $\langle\{p_1,p_3,0,1\};\vee,\wedge,0,1\rangle$ and $\mathfrak{L}$ are the substructures of $\mathfrak{L}$ satisfying Σ_2 which contain p_1.

Now any good substructure concept should have the property that every subset B is contained in a smallest substructure of that kind. By symmetry, in the above example, the substructure generated by $\{p_1\}$ has to be $\mathfrak{L}$. That is, along with x, the substructure has to contain all y required by Φ_2. This leads us to the concept of Σ-substructure: $\mathfrak{B}$ is a Σ-substructure of $\mathfrak{A}$, if $\mathfrak{B}$ is a substructure, and if whenever $x_0,\cdots,x_{n-1}\in B$ and there is a $\Phi\in\Sigma$ whose prefix begins with $(\mathbf{x}_0)\cdots(\mathbf{x}_{n-1})(\exists\mathbf{y})\cdots$, then all $y\in A$ satisfying Φ are in B. To illustrate this take the following two axioms:

$$(\mathbf{x})(\exists\mathbf{y})\Psi_1(\mathbf{x},\mathbf{y}) \tag{1}$$

$$(\mathbf{x})(\exists\mathbf{y})(\mathbf{u})(\exists\mathbf{v})\Psi_2(\mathbf{x},\mathbf{y},\mathbf{u},\mathbf{v}) \tag{2}$$

where Ψ_1 and Ψ_2 do not contain quantifiers.

For (1) this definition means that if $a\in B$, $b\in A$ and $\Psi_1(a,b)$, then $b\in B$. For (2) this means that if $a\in B$, $b\in A$ and

$$(\mathbf{u})(\exists\mathbf{v})\Psi_2(a,b,\mathbf{u},\mathbf{v})\ \text{holds in}\ \mathfrak{A}, \tag{3}$$

then $b\in B$ and furthermore if for $a,c\in B$ and $d\in A$ there exists a $b\in A$ such that (3) holds and $\Psi_2(a,b,c,d)$, then $d\in B$.

A Φ sequence for (1) means a, b with $\Psi_1(a,b)$; a Φ sequence for (2) will be a,b,c,d satisfying (3) and $\Psi_2(a,b,c,d)$.

To give a rigorous definition of Σ-substructure first we have to define the concept of inverse.

Definition 1. *Let $\Phi\in\Sigma$ be of the following form:*

$$(\mathbf{x}_0)\cdots(\mathbf{x}_{n_0-1})(\exists\mathbf{y}_0)(\mathbf{x}_{n_0})\cdots(\mathbf{x}_{n_1-1})(\exists\mathbf{y}_1)\cdots(\exists\mathbf{y}_k)(\mathbf{x}_{n_k})\cdots(\mathbf{x}_{n-1})$$
$$\Psi(\mathbf{x}_0,\cdots,\mathbf{x}_{n_0-1},\mathbf{y}_0,\mathbf{x}_{n_0},\cdots,\mathbf{x}_{n_1-1},\mathbf{y}_1,\cdots,\mathbf{y}_k,\mathbf{x}_{n_k},\cdots,\mathbf{x}_{n-1}), \tag{4}$$

where $0\leqq n_0\leqq n_1\leqq\cdots\leqq n_k\leqq n$; $0=n_0$ means that no universal quantifier precedes $\exists\mathbf{y}_0$, $n_0=n_1$ means that there is no universal quantifier between $\exists\mathbf{y}_0$ and $\exists\mathbf{y}_1$, and so on; $n_k=n$ means that no universal quantifier follows $\exists\mathbf{y}_k$; Ψ contains no quantifiers. Set $e(\Phi)=k+1$. The concepts of Φ–l inverse *and* Φ–l sequence *are defined for all* $0 \leq l < e(\Phi)$ *by induction on* l. *Let* $\mathfrak{A}$ *be a* Σ*-structure,* $a_0, a_1, \cdots$ *and* $b_0, b_1, \cdots \in A$.

(i) b_0 *is a* Φ–0 *inverse of* $a_0, \cdots, a_{n_0-1}$ *in* $\mathfrak{A}$ *if*

$$(\mathbf{x}_{n_0}) \cdots (\mathbf{x}_{n_1-1})(\exists \mathbf{y}_1) \cdots (\exists \mathbf{y}_k)(\mathbf{x}_{n_k}) \cdots (\mathbf{x}_{n-1})$$
$$\Psi(a_0, \cdots, a_{n_0-1}, b_0, \mathbf{x}_{n_0}, \cdots, \mathbf{x}_{n_1-1}, \mathbf{y}_1, \cdots, \mathbf{y}_k, \mathbf{x}_{n_k}, \cdots, \mathbf{x}_{n-1})$$

holds in $\mathfrak{A}$; *in this case,* $a_0, \cdots, a_{n_0-1}, b_0$ *is a* Φ–0 *sequence;*

(ii) b_l *is a* Φ–l *inverse of* $a_0, \cdots, a_{n_l-1}$ *in* $\mathfrak{A}$ *if there exists a* Φ–$(l-1)$ *sequence* $a_0, \cdots, a_{n_0-1}, b_0, \cdots, a_{n_{l-1}-1}, b_{l-1}$ *such that*

$$(\mathbf{x}_{n_l}) \cdots (\mathbf{x}_{n_{l+1}-1})(\exists \mathbf{y}_{l+1}) \cdots (\exists \mathbf{y}_k)(\mathbf{x}_{n_k}) \cdots (\mathbf{x}_{n-1})$$
$$\Psi(a_0, \cdots, a_{n_0-1}, b_0, \cdots, a_{n_{l-1}-1}, b_{l-1}, a_{n_{l-1}}, \cdots, a_{n_l-1}, b_l,$$
$$\mathbf{x}_{n_l}, \cdots, \mathbf{x}_{n_{l+1}-1}, \mathbf{y}_{l+1}, \cdots, \mathbf{y}_k, \mathbf{x}_{n_k}, \cdots, \mathbf{x}_{n-1})$$

holds in $\mathfrak{A}$. *Then,* $a_0, \cdots, a_{n_0-1}, b_0, \cdots, a_{n_{l-1}-1}, b_{l-1}, a_{n_{l-1}}, \cdots, a_{n_l-1}, b_l$ *is a* Φ–l *sequence.*

Φ-inverse *will mean* Φ–l *inverse for some* $l < e(\Phi)$ *and* Σ-inverse *will mean* Φ*-inverse for some* $\Phi \in \Sigma$.

Note that in (ii) b_l is a Φ–i inverse of $a_0, \cdots, a_{n_i-1}$, for $0 < i \leq l$.

Most proofs of statements on inverses can be carried out only by induction on l as in Definition 1, which is sometimes technically involved. Therefore, we will work out the proofs only for sentences of the forms (1) and (2) and leave the details of formal proofs to the reader.

Lemma 1. *For every* $\Phi \in \Sigma$ *and* $l < e(\Phi)$ *there exists a formula* $\Phi^{[l]}(\mathbf{x}_0, \cdots, \mathbf{y})$ *in* $L(\tau)$ *such that for a* Σ*-structure* $\mathfrak{A}$ *and* $a_0, \cdots, b \in A$, b *is a* Φ–l *inverse of* $a_0, \cdots$ *if and only if* $\Phi^{[l]}(a_0, \cdots, b)$. *Furthermore, there exists a formula* $\Phi^{(l)}(\mathbf{x}_0, \cdots, \mathbf{x}_{n_0-1}, \mathbf{y}_0, \cdots, \mathbf{x}_{n_l-1}, \mathbf{y}_l)$ *in* $L(\tau)$ *such that if* $\mathfrak{A}$ *is a* Σ*-structure and* $a_0, \cdots, a_{n_0-1}, b_0, \cdots, a_{n_l-1}, b_l \in A$, *then* $a_0, \cdots, a_{n_0-1}, b_0, \cdots, a_{n_l-1}, b_l$ *is a* Φ–l *sequence in* $\mathfrak{A}$ *if and only if*

$$\Phi^{(l)}(a_0, \cdots, a_{n_0-1}, b_0, \cdots, a_{n_l-1}, b_l)$$

in $\mathfrak{A}$.

Proof. If Φ is of the form (1) then $\Phi^{[0]}(\mathbf{x}, \mathbf{y}) \equiv \Psi_1(\mathbf{x}, \mathbf{y}) \equiv \Phi^{(0)}(\mathbf{x}, \mathbf{y})$. If Φ is of the form (2), then $\Phi^{[0]}(\mathbf{x}, \mathbf{y}) \equiv (\mathbf{u})(\exists \mathbf{v})\Psi_2(\mathbf{x}, \mathbf{y}, \mathbf{u}, \mathbf{v})$ and

$$\Phi^{[1]}(\mathbf{x}_0, \mathbf{x}_1, \mathbf{y}) \equiv (\exists \mathbf{z})(\Phi^{[0]}(\mathbf{x}_0, \mathbf{z}) \wedge \Psi_2(\mathbf{x}_0, \mathbf{z}, \mathbf{x}_1, \mathbf{y})).$$

Furthermore, $\Phi^{(0)}(\mathbf{x}, \mathbf{y}) = \Phi^{[0]}(\mathbf{x}, \mathbf{y})$ and

$$\Phi^{(1)}(\mathbf{x}_0, \mathbf{y}_0, \mathbf{x}_1, \mathbf{y}_1) \equiv \Phi^{(0)}(\mathbf{x}_0, \mathbf{y}_0) \wedge \Psi_2(\mathbf{x}_0, \mathbf{y}_0, \mathbf{x}_1, \mathbf{y}_1).$$

Now, we are ready to define our substructure concept.

Definition 2. *Let $\mathfrak{A}$ be a Σ-structure and let $\mathfrak{B}$ be a substructure of $\mathfrak{A}$. Then $\mathfrak{B}$ is a* Σ-substructure of $\mathfrak{A}$ *if whenever $a_0, \cdots, a_t \in B, b \in A$ and b is a Σ-inverse of $a_0, \cdots, a_t$ in $\mathfrak{A}$, then $b \in B$.*

The most important property of Σ-substructures is the following:

Theorem 1. *Let $\mathfrak{A}$ be a Σ-structure, and let $\mathfrak{B}$ be a Σ-substructure of $\mathfrak{A}$; let $\Phi \in \Sigma$ and $l < e(\Phi)$. If $a_0, \cdots, b_0, \cdots \in B$ and*

$$\Phi^{(l)}(a_0, \cdots, b_0, \cdots, a_{n_l-1}, b_l)$$

in $\mathfrak{A}$, then it also holds in $\mathfrak{B}$.

Proof. This is trivial if Φ is of the form (1). Let Φ be of the form (2) and $\Phi^{(0)}(a_0, b_0)$ in $\mathfrak{A}$; that is $(\mathbf{u})(\exists \mathbf{v})\Psi_2(a_0, b_0, \mathbf{u}, \mathbf{v})$ holds in $\mathfrak{A}$. If $c \in B$ there exists a $d \in A$ with $\Psi_2(a_0, b_0, c, d)$. Now $\Phi^{(0)}(a_0, b_0)$ implies that d is a Φ–1 inverse of a and c, whence $d \in B$. This proves that $\Phi^{(0)}(a_0, b_0)$ in $\mathfrak{B}$. $\Phi^{(1)}(a_0, b_0, a_1, b_1)$ can be handled similarly.

Corollary 1. *Under the same conditions as in Theorem 1, if*

$$\Phi^{[l]}(a_0, \cdots, a_{n_l-1}, b)$$

in $\mathfrak{A}$, then it also holds in $\mathfrak{B}$.

Corollary 2. *A Σ-substructure of a Σ-structure is again a Σ-structure.*

Lemma 2. *Let $\mathfrak{A}$ be a Σ-structure and let $\mathfrak{B}$ be a Σ-substructure of $\mathfrak{A}$. Let $\mathfrak{C}$ be a Σ-substructure of $\mathfrak{B}$. Then $\mathfrak{C}$ is a Σ-substructure of $\mathfrak{A}$.*

Remark. By Corollary 2 to Theorem 1, $\mathfrak{B}$ is a Σ-structure.

Proof of Lemma 2. Let $a_0, \cdots, a_t \in C$, and $b \in A$ and let b be a Σ-inverse of $a_0, \cdots, a_t$. Then $b \in B$, since $\mathfrak{B}$ is a Σ-substructure. By Corollary 1 to Theorem 1, b is a Σ-inverse of $a_0, \cdots, a_t$ in $\mathfrak{B}$, whence $b \in C$, since $\mathfrak{C}$ is a Σ-substructure of $\mathfrak{B}$.

Lemma 3. *Let $\mathfrak{A}$ be a Σ-structure and $\varnothing \neq H \subseteq A$. Then there exists a smallest Σ-substructure $\mathfrak{B}$ with $H \subseteq B$.*

Proof. Obvious, since the intersection of Σ-substructures is again a Σ-substructure, provided it is not void.

We will set $B = [H]_\Sigma$ and we will say that H *Σ-generates* $\mathfrak{B}$ or H is a *Σ-generating set* of $\mathfrak{B}$.

Lemma 4. *Let $\mathfrak{A}$ be a Σ-structure, $\varnothing \neq H \subseteq A$. Set $H_0 = H$, $\overline{H}_{n-1} = \{a \mid a \in A$ and $a = p(a_0, \cdots, a_{k-1})$, where p is a polynomial and $a_0, \cdots, a_{k-1} \in H_{n-1}\}$, $H_n = \overline{H}_{n-1} \cup \{a \mid a \in A$ and there exist $b_0, \cdots, b_t \in \overline{H}_{n-1}$ such that a is a Σ-inverse of $b_0, \cdots, b_t$ in $\mathfrak{A}\}$. Then*

$$[H]_\Sigma = \bigcup (H_i \mid 0 \leqq i < \omega).$$

Proof. $H_n \subseteq [H]_\Sigma$ can be proved by induction on n, so we get

$$\bigcup (H_i \mid 0 \leqq i < \omega) \subseteq [H]_\Sigma.$$

It is routine to check that $\langle \bigcup (H_i \mid 0 \leqq i < \omega); F, R \rangle$ is a Σ-substructure, so we get equality.

A useful criterion for $a \in [H]_\Sigma$ can be given in terms of Σ-polynomials.

Definition 3. *Let n be a positive integer. The set $\mathbf{P}_n(\Sigma)$ of n-ary* Σ-polynomial symbols *is defined by rules* (i)–(iv) *below.*

(i) $\mathbf{x}_i \in \mathbf{P}_n(\Sigma)$, $\quad i = 0, \cdots, n-1$;
(ii) *if* $\mathbf{P}_0, \cdots, \mathbf{P}_{n_\gamma - 1} \in \mathbf{P}_n(\Sigma)$, *then* $\mathbf{f}_\gamma(\mathbf{P}_0, \cdots, \mathbf{P}_{n_\gamma - 1}) \in \mathbf{P}_n(\Sigma)$, *for* $\gamma < o_0(\tau)$;
(iii) *if* $\Phi \in \Sigma$, $l < e(\Phi)$, n_l *universal quantifiers precede* $\exists \mathbf{y}_l$ *and* $\mathbf{P}_0, \cdots, \mathbf{P}_{n_l - 1} \in \mathbf{P}_n(\Sigma)$, *then* $\Phi^{(l)}(\mathbf{P}_0, \cdots, \mathbf{P}_{n_l - 1}) \in \mathbf{P}_n(\Sigma)$;
(iv) $\mathbf{P}_n(\Sigma)$ *is the smallest set satisfying* (i)–(iii).

Definition 4. *Let $\mathbf{P} \in \mathbf{P}_n(\Sigma)$, let $\mathfrak{A}$ be a Σ-structure, and let $a_0, \cdots, a_{n-1} \in A$. Then $\mathbf{P}_\mathfrak{A}(a_0, \cdots, a_{n-1})$* (or, simply *$P(a_0, \cdots, a_{n-1})$*) *is a subset of A defined as follows:*

(i) *if* $\mathbf{P} = \mathbf{x}_i$, *then* $P(a_0, \cdots, a_{n-1}) = \{a_i\}$;
(ii) *if* $\mathbf{P} = \mathbf{f}_\gamma(\mathbf{P}_0, \cdots, \mathbf{P}_{n_\gamma - 1})$, *then*

$$P(a_0, \cdots, a_{n-1}) = \{a \mid a = f_\gamma(b_0, \cdots, b_{n_\gamma - 1}) \text{ for some } b_i \in P_i(a_0, \cdots, a_{n-1}),\ i = 0, \cdots, n_\gamma - 1\};$$

(iii) *if* $\mathbf{P} = \Phi^{(l)}(\mathbf{P}_0, \cdots, \mathbf{P}_{n_l - 1})$, *then*

$$P(a_0, \cdots, a_{n-1}) = \{a \mid a \text{ is a } \Phi\text{–}l \text{ inverse of some } b_0, \cdots, b_{n_l - 1} \text{ with } b_i \in P_i(a_0, \cdots, a_{n-1}),\ i = 0, \cdots, n_l - 1\}.$$

$\mathbf{P}_\mathfrak{A}$ *is called a* Σ-polynomial *(over $\mathfrak{A}$).*

Lemma 5. *Let $\mathfrak{A}$ be a Σ-structure, $\varnothing \neq H \subseteq A$. Then $a \in [H]_\Sigma$ if and only if for some positive integer n, $\mathbf{P} \in \mathbf{P}_n(\Sigma)$, and $h_0, \cdots, h_{n-1} \in H$ we have $a \in P(h_0, \cdots, h_{n-1})$.*

Proof. If $a \in [H]_\Sigma$, then by Lemma 4, $a \in H_i$ for some $i < \omega$ and then the proof of $a \in P(h_0, \cdots, h_{n-1})$ proceeds by an easy induction on i. Conversely, if $a \in P(h_0, \cdots, h_{n-1})$, then we can prove that $a \in H_i$ for some i, by induction on the "rank" of $\mathbf{P}$.

Corollary. *Let $\mathfrak{A}$ be a Σ-structure, and let $\mathfrak{B}$ be a Σ-substructure of $\mathfrak{A}$. Let $\mathbf{P} \in \mathbf{P}_n(\Sigma)$, $b_0, \cdots, b_{n-1} \in B$. Then $\mathbf{P}_{\mathfrak{A}}(b_0, \cdots, b_{n-1}) \subseteq B$.*

The following two lemmas will be used frequently.

Lemma 6. *Let $\mathbf{P} \in \mathbf{P}_n(\Sigma)$. Then there exists a formula $\mathbf{r}_{\mathbf{P}}(\mathbf{x}_0, \cdots, \mathbf{x}_{n-1}, \mathbf{y})$ in $L(\tau)$ such that if $\mathfrak{A}$ is a Σ-structure and $a_0, \cdots, a_{n-1}, b \in A$, then $b \in P(a_0, \cdots, a_{n-1})$ if and only if $r_{\mathbf{P}}(a_0, \cdots, a_{n-1}, b)$* in *$\mathfrak{A}$.*

Lemma 7. *Let $\mathfrak{A}$ be a Σ-structure, let $\mathfrak{B}$ be a Σ-substructure of $\mathfrak{A}$ and let $\mathbf{P} \in \mathbf{P}_n(\Sigma)$. If $a_0, \cdots, a_{n-1}, b \in B$ and $b \in \mathbf{P}_{\mathfrak{A}}(a_0, \cdots, a_{n-1})$, then $b \in \mathbf{P}_{\mathfrak{B}}(a_0, \cdots, a_{n-1})$. In other words, if $r_{\mathbf{P}}(a_0, \cdots, a_{n-1}, b)$ in $\mathfrak{A}$, then $r_{\mathbf{P}}(a_0, \cdots, a_{n-1}, b)$ in $\mathfrak{B}$.*

Lemma 6 and Lemma 7 follow from Lemma 1 by an easy induction.

§50. Σ-HOMOMORPHISMS AND SLENDER Σ-SUBSTRUCTURES

The example of lattices as partially ordered sets (see §49) shows that the usual concept of homomorphism may not preserve algebraic properties, e.g., the homomorphic image of a lattice may not be a lattice or the homomorphic image of a distributive lattice may be nondistributive. Therefore, we need a homomorphism concept which preserves the inverses.

Definition 1. *Let $\mathfrak{A}$ and $\mathfrak{B}$ be Σ-structures and let φ be a mapping of A into B. Then φ is called a* Σ-homomorphism *if φ is a homomorphism, and if for any positive integer n, $\mathbf{P} \in \mathbf{P}_n(\Sigma)$, and $a_0, \cdots, a_{n-1} \in A$ we have*

$$P(a_0, \cdots, a_{n-1})\varphi = P(a_0\varphi, \cdots, a_{n-1}\varphi).$$

It should be emphasized that a Σ-isomorphism is simply an isomorphism. Of course, one can give an equivalent definition without the use of Σ-polynomials.

Lemma 1. *Let $\mathfrak{A}$ and $\mathfrak{B}$ be Σ-structures and let φ be a mapping of A into B. Then φ is a Σ-homomorphism if and only if the following conditions are satisfied:*

(i) *φ is a homomorphism;*

(ii) *if $\Phi \in \Sigma$, $l < e(\Phi)$, $b, a_0, \cdots, a_t \in A$ and b is a Φ–l inverse of $a_0, \cdots, a_t$ in $\mathfrak{A}$ then $b\varphi$ is a Φ–l inverse of $a_0\varphi, \cdots, a_t\varphi$ in $\mathfrak{B}$;*

(iii) *if $\Phi \in \Sigma$, $l < e(\Phi)$, $a_0, \cdots, a_t \in A$, $\bar{b} \in B$ and $\bar{b}$ is a Φ–l inverse of*

$a_0\varphi, \cdots, a_t\varphi$ in $\mathfrak{B}$, then there exists a $b \in A$ such that b is a Φ–l inverse of $a_0, \cdots, a_t$ and $b\varphi = \bar{b}$.

Proof. Let φ be a Σ-homomorphism. Then (i) is satisfied by definition. Condition (ii) and (iii) follow easily by taking $\mathbf{P} = \Phi^{(l)}(\mathbf{x}_0, \cdots, \mathbf{x}_t)$ and applying the definition of Σ-homomorphism. Conversely, if (i)–(iii) are satisfied, then we prove $P(a_0, \cdots, a_{n-1})\varphi = P(a_0\varphi, \cdots, a_{n-1}\varphi)$ by induction. If $\mathbf{P} = \mathbf{x}_i$, the statement is trivial. If $\mathbf{P} = \mathbf{f}_\gamma(\mathbf{P}_0, \cdots, \mathbf{P}_{n_\gamma - 1})$, then it follows from (i). If $\mathbf{P} = \Phi^{(l)}(\mathbf{P}_0, \cdots, \mathbf{P}_{n_l - 1})$, it follows from (ii) and (iii).

Some obvious properties of Σ-homomorphisms are given in the following lemmas.

Lemma 2. *Let $\mathfrak{A}$ and $\mathfrak{B}$ be Σ-structures, and let φ be a Σ-homomorphism of $\mathfrak{A}$ into $\mathfrak{B}$; set $C = A\varphi$. Then $\mathfrak{C}$ is a Σ-substructure of $\mathfrak{B}$.*

Lemma 3. *Let $\mathfrak{A}$, $\mathfrak{B}$, and $\mathfrak{C}$ be Σ-structures, let φ be a Σ-homomorphism of $\mathfrak{A}$ into $\mathfrak{B}$, and let ψ be a Σ-homomorphism of $\mathfrak{B}$ into $\mathfrak{C}$. Then $\varphi\psi$ is a Σ-homomorphism of $\mathfrak{A}$ into $\mathfrak{C}$.*

A property of homomorphisms (which is very important in proofs concerning free algebras) fails to hold for Σ-homomorphisms. Namely, if φ is a Σ-homomorphism of $\mathfrak{A}$ into $\mathfrak{C}$ and $\mathfrak{B}$ is a Σ-substructure of $\mathfrak{A}$, then φ_B is not necessarily a Σ-homomorphism of $\mathfrak{B}$ into $\mathfrak{C}$. Let $b_0, \cdots, b_{n-1} \in B$, $\mathbf{P} \in \mathbf{P}_n(\Sigma)$; it then follows from the corollary to Lemma 49.5 that $\mathbf{P}_{\mathfrak{A}}(b_0, \cdots, b_{n-1}) \subseteq B$ and from Corollary 1 to Theorem 49.1 that $\mathbf{P}_{\mathfrak{A}}(b_0, \cdots, b_{n-1}) \subseteq \mathbf{P}_{\mathfrak{B}}(b_0, \cdots, b_{n-1})$. Whenever $\mathbf{P}_{\mathfrak{A}}(b_0, \cdots, b_{n-1}) \neq \mathbf{P}_{\mathfrak{B}}(b_0, \cdots, b_{n-1})$, we find that φ_B is not necessarily a Σ-homomorphism. This leads us to the definition of slender Σ-substructures.

Definition 2. *Let $\mathfrak{B}$ be a Σ-substructure of the Σ-structure $\mathfrak{A}$. Then $\mathfrak{B}$ is called a* slender Σ-substructure *if for any positive integer n, $\mathbf{P} \in \mathbf{P}_n(\Sigma)$ and $a_0, \cdots, a_{n-1} \in B$, we have that $\mathbf{P}_{\mathfrak{A}}(a_0, \cdots, a_{n-1}) = \mathbf{P}_{\mathfrak{B}}(a_0, \cdots, a_{n-1})$.*

Lemma 4. *Let $\mathfrak{B}$ be a Σ-substructure of a Σ-structure $\mathfrak{A}$. Then $\mathfrak{B}$ is slender if and only if for $\Phi \in \Sigma$, $l < e(\Phi)$ and $b, a_0, \cdots, a_t \in B$ we have that b is a Φ–l inverse of $a_0, \cdots, a_t$ in $\mathfrak{B}$ implies that b is a Φ–l inverse of $a_0, \cdots, a_t$ in $\mathfrak{A}$.*

The proof is again a simple induction based on Definition 49.3.

Lemma 5. *Let $\mathfrak{B}$ be a Σ-substructure of a Σ-structure $\mathfrak{A}$. The following conditions on $\mathfrak{B}$ are equivalent:*

(i) *$\mathfrak{B}$ is slender;*

(ii) *if* $\mathfrak{C}$ *is any* Σ*-structure and* φ *is a* Σ*-homomorphism of* $\mathfrak{A}$ *into* $\mathfrak{C}$, *then* φ_B *is a* Σ*-homomorphism of* $\mathfrak{B}$ *into* $\mathfrak{C}$;

(iii) *if* $\mathfrak{C}$ *is any* Σ*-structure and* φ *is a* Σ*-homomorphism of* $\mathfrak{C}$ *into* $\mathfrak{B}$, *then* φ *is a* Σ*-homomorphism of* $\mathfrak{C}$ *into* $\mathfrak{A}$;

(iv) *if* $\mathfrak{C}$ *is any* Σ*-structure and* φ *is a* Σ*-homomorphism of* $\mathfrak{C}$ *onto* $\mathfrak{B}$, *then* φ *is a* Σ*-homomorphism of* $\mathfrak{C}$ *into* $\mathfrak{A}$;

(v) $\varphi\colon x \to x$ *is a* Σ*-homomorphism of* $\mathfrak{B}$ *into* $\mathfrak{A}$.

Proof. The following implications are obvious: (i) implies (ii), (iii), (iv) and (v); (iii) implies (iv); (iv) implies (v) ($\mathfrak{B}=\mathfrak{C}$); (ii) implies (v) ($\mathfrak{A}=\mathfrak{C}$). Thus it suffices to prove that (v) implies (i); indeed (v) implies that $\mathbf{P}_{\mathfrak{B}}(b_0, \cdots, b_{n-1})\varphi = \mathbf{P}_{\mathfrak{A}}(b_0, \cdots, b_{n-1})$ $(b_0, \cdots, b_{n-1} \in B)$, that is, $\mathfrak{B}$ is slender.

Lemma 6. *Let* $\mathfrak{B}$ *be a slender* Σ*-substructure of a* Σ*-structure* $\mathfrak{A}$. *Then the following conditions hold:*

(i) *Let* $\mathfrak{C}$ *be a* Σ*-structure, and let* φ *be a* Σ*-homomorphism of* $\mathfrak{C}$ *into* $\mathfrak{A}$ *with* $C\varphi \subseteq B$; *then* φ *is a* Σ*-homomorphism of* $\mathfrak{C}$ *into* $\mathfrak{B}$;

(ii) *let* $\mathfrak{C}$ *be a* Σ*-substructure of* $\mathfrak{A}$ *with* $C \subseteq B$; *then* $\mathfrak{C}$ *is a* Σ*-substructure of* $\mathfrak{B}$;

(iii) *let* $H \subseteq B$; *then* $[H]_\Sigma$ *in* $\mathfrak{A}$ *equals* $[H]_\Sigma$ *in* $\mathfrak{B}$.

The proofs are trivial.

§51. FREE Σ-STRUCTURES AND THE UNIQUENESS THEOREM

Now we are ready to define free Σ-structures.

Definition 1. *Let* α *be an ordinal.* $\mathfrak{F}_\Sigma(\alpha)$ *is the* free Σ-structure *with* α Σ*-generators, if the following conditions are satisfied:*

(i) $\mathfrak{F}_\Sigma(\alpha)$ *is a* Σ*-structure;*

(ii) $\mathfrak{F}_\Sigma(\alpha)$ *is* Σ*-generated by the elements* $x_0, \cdots, x_\gamma, \cdots$ $(\gamma < \alpha)$;

(iii) *if* $\mathfrak{A}$ *is a* Σ*-structure and* $a_0, \cdots, a_\gamma, \cdots \in A$ *for* $\gamma < \alpha$, *then the mapping* $\varphi\colon x_\gamma \to a_\gamma, \gamma < \alpha$ *can be extended to a* Σ*-homomorphism* $\bar{\varphi}$ *of* $\mathfrak{F}_\Sigma(\alpha)$ *into* $\mathfrak{A}$.

Remark. The Σ-homomorphism $\bar{\varphi}$ in (iii) need not be unique. Indeed, let $o(\tau)=0$ and let Σ consists of the following two axioms:

$$(\mathbf{x})(\mathbf{y})(\mathbf{z})(\mathbf{u})(\mathbf{x} = \mathbf{y} \vee \mathbf{x} = \mathbf{z} \vee \mathbf{x} = \mathbf{u} \vee \mathbf{y} = \mathbf{z} \vee \mathbf{y} = \mathbf{u} \vee \mathbf{z} = \mathbf{u})$$
$$(\mathbf{x})(\exists \mathbf{y})(\exists \mathbf{z})(\mathbf{x} \neq \mathbf{y} \wedge \mathbf{x} \neq \mathbf{z} \wedge \mathbf{y} \neq \mathbf{z}).$$

Then a Σ-structure is a 3 element set. Let $A=\{a_0, a_1, a_2\}$, $B=\{b_0, b_1, b_2\}$. Then $\mathfrak{A}=\mathfrak{F}_\Sigma(1)$, e.g., a_0 is a free Σ-generator. The mapping $\varphi: a_0 \to b_0$ has two extensions to a Σ-homomorphism of $\mathfrak{A}$ onto $\mathfrak{B}$, namely, $a_0 \to b_0$, $a_1 \to b_1$, $a_2 \to b_2$ and $a_0 \to b_0$, $a_1 \to b_2$, $a_2 \to b_1$.

Most of the difficulties in the theory of free Σ-structures come from this fact.

The theory of free Σ-structures is based on the following result which, in a certain sense, is a substitute for the uniqueness of $\bar{\varphi}$.

Theorem 1. *Let us assume that $\mathfrak{F}_\Sigma(n)$ exists. Then every $\mathbf{P} \in \mathbf{P}_m(\Sigma)$ with $m \leqq n$ is "bounded", that is there exists a least positive integer $k_{\mathbf{P}}$ such that if $\mathfrak{A}$ is a Σ-structure, $a_0, \cdots, a_{m-1} \in A$, then*

$$|P(a_0, \cdots, a_{m-1})| \leqq k_{\mathbf{P}}.$$

Proof. Let us assume that Theorem 1 is not true. Then there exist $\mathbf{P} \in \mathbf{P}_m(\Sigma)$ with $m \leqq n$, Σ-structures $\mathfrak{A}_1, \mathfrak{A}_2, \cdots$ and $a_0^t, \cdots, a_{m-1}^t \in A_t$ $(t=1, 2, \cdots)$ such that

$$|P(a_0^t, \cdots, a_{m-1}^t)| \geqq t \qquad (t = 1, 2, \cdots).$$

Statement. *Under these conditions, for every cardinal $\mathfrak{m}$, there exists a Σ-structure $\mathfrak{A}$ and there exist $a_0, \cdots, a_{m-1} \in A$ such that*

$$|P(a_0, \cdots, a_{m-1})| \geqq \mathfrak{m}.$$

Proof of the Statement. Let α be the initial ordinal of cardinality $\mathfrak{m}$ and $\tau' = \tau \oplus (\alpha+m)$; that is, we get the type τ' by adjoining the constants $\mathbf{k}_{o(\tau)}, \cdots, \mathbf{k}_{o(\tau)+\gamma}, \cdots, \gamma<\alpha+m$ to τ. Set $\mathbf{l}_0=\mathbf{k}_{o(\tau)+\alpha}, \cdots, \mathbf{l}_{m-1}=\mathbf{k}_{o(\tau)+\alpha+m-1}$ and let us write $\mathbf{k}_\gamma$ for $\mathbf{k}_{o(\tau)+\gamma}$, $\gamma<\alpha$.

Let H be a finite set of ordinals $<\alpha$; we define a sentence Φ_H of $L(\tau')$ as follows:

$$\Phi_H = \bigwedge (\mathbf{r}_{\mathbf{P}}(\mathbf{l}_0, \cdots, \mathbf{l}_{m-1}, \mathbf{k}_\gamma) \mid \gamma \in H) \wedge \bigwedge (\mathbf{k}_\gamma \neq \mathbf{k}_\delta \mid \gamma, \delta \in H, \gamma \neq \delta),$$

where $\mathbf{r}_{\mathbf{P}}$ is the formula in $L(\tau)$ which was defined in Lemma 49.6.

Let Ω be the set of all Φ_H. We claim that there exists a structure $\mathfrak{A}'$ satisfying $\Sigma \cup \Omega$. By the compactness theorem (Theorem 39.2), it suffices to show that $\Sigma \cup \Omega_1$ has a model for all finite $\Omega_1 \subseteq \Omega$. Let

$$\Omega_1 = \{\Phi_{H_0}, \cdots, \Phi_{H_{t-1}}\}$$

and set $H=H_0 \cup \cdots \cup H_{t-1}$. Since $\Phi_H \models \Phi_{H_i}$, $i=0, \cdots, t-1$, it is sufficient to show that $\Sigma \cup \{\Phi_H\}$ has a model. Let $H=\{\gamma_0, \cdots, \gamma_{s-1}\}$. Let $\mathfrak{A}_s'$ be the structure that we get from $\mathfrak{A}_s$ by interpreting $\mathbf{l}_i$ as a_i^s $(i=0, \cdots, m-1)$ and $\mathbf{k}_{\gamma_0}, \cdots, \mathbf{k}_{\gamma_{s-1}}$ as distinct elements of $P(a_0^s, \cdots, a_{m-1}^s)$; we can do that since $|P(a_0^s, \cdots, a_{m-1}^s)| \geqq s$; let us interpret $\mathbf{k}_\gamma$, $\gamma \neq \gamma_0, \cdots, \gamma_{s-1}$ in an arbitrary manner. It is obvious then that $\mathfrak{A}_s'$ satisfies $\Sigma \cup \{\Phi_H\}$.

Now let $\mathfrak{A}'$ be a model of $\Sigma \cup \Omega$. Let $a_0, \cdots, a_{m-1}$ be the interpretations of $\mathbf{l}_0, \cdots, \mathbf{l}_{m-1}$ and let $b_0, \cdots, b_\gamma, \cdots$ be the interpretations of $\mathbf{k}_0, \cdots, \mathbf{k}_\gamma, \cdots$ for $\gamma < \alpha$. Then

$$b_0, \cdots, b_\gamma, \cdots \in P(a_0, \cdots, a_{m-1})$$

and $b_\gamma \neq b_\delta$ if $\gamma, \delta < \alpha$ and $\gamma \neq \delta$. Thus $|P(a_0, \cdots, a_{m-1})| \geqq \mathfrak{m}$. Therefore the τ-reduct $\mathfrak{A}$ of $\mathfrak{A}'$ satisfies the requirements, concluding the proof of the statement.

Let $x_0, \cdots, x_{n-1}$ be the Σ-generators of $\mathfrak{F}_\Sigma(n)$. Set $|P(x_0, \cdots, x_{m-1})| = \mathfrak{n}$. If $\mathfrak{A}$ is any Σ-structure and $a_0, \cdots, a_{m-1} \in A$, then there exists a Σ-homomorphism φ of $\mathfrak{F}_\Sigma(n)$ into $\mathfrak{A}$ with $x_0\varphi = a_0, \cdots, x_{m-1}\varphi = a_{m-1}$; thus

$$P(x_0, \cdots, x_{m-1})\varphi = P(a_0, \cdots, a_{m-1})$$

and therefore

$$|P(a_0, \cdots, a_{m-1})| \leqq |P(x_0, \cdots, x_{m-1})| = \mathfrak{n}.$$

Take any cardinal $\mathfrak{m}$ with $\mathfrak{n} < \mathfrak{m}$ and apply the Statement with $\mathfrak{m}$. The arising contradiction concludes the proof of Theorem 1.

Corollary. *Let us assume that $\mathfrak{F}_\Sigma(n)$ exists. Let $\mathbf{P} \in \mathbf{P}_k(\Sigma)$, and let $\mathfrak{A}$ be a Σ-structure, $a_0, \cdots, a_{k-1} \in A$. If there exist $b_0, \cdots, b_{m-1} \in A$ with $m \leqq n$ such that*

$$a_0, \cdots, a_{k-1} \in [b_0, \cdots, b_{m-1}]_\Sigma,$$

then $P(a_0, \cdots, a_{k-1})$ is finite.

Proof. Since $a_0, \cdots, a_{k-1} \in [b_0, \cdots, b_{m-1}]_\Sigma$, there exist

$$\mathbf{P}_0, \cdots, \mathbf{P}_{k-1} \in \mathbf{P}_m(\Sigma)$$

such that

$$a_i \in P_i(b_0, \cdots, b_{m-1}), \qquad i = 0, \cdots, k-1.$$

Thus

$$P(a_0, \cdots, a_{k-1}) \subseteq P(P_0(b_0, \cdots, b_{m-1}), \cdots, P_{k-1}(b_0, \cdots, b_{m-1}))$$

(where the right-hand side has its usual meaning) and the right-hand side is finite by Theorem 1.

Theorem 2 (The Uniqueness Theorem). *If the free Σ-structure on α generators, $\mathfrak{F}_\Sigma(\alpha)$ exists, then it is unique up to isomorphism.*

We will prove the following stronger version of Theorem 2.

Theorem 2′. *Let $\mathfrak{F}_\Sigma(\alpha)$ and $\mathfrak{F}_\Sigma'(\alpha)$ be free Σ-structures, with Σ-generators $x_0, \cdots, x_\gamma, \cdots$ and $x_0', \cdots, x_\gamma', \cdots, \gamma < \alpha$, respectively. Let φ be a Σ-homomorphism of $\mathfrak{F}_\Sigma(\alpha)$ into $\mathfrak{F}_\Sigma'(\alpha)$ with $x_\gamma\varphi = x_\gamma'$, for $\gamma < \alpha$. Then φ is an isomorphism.*

Since $\mathfrak{F}_\Sigma(\alpha)$ is free and $\mathfrak{F}_\Sigma'(\alpha)$ is a Σ-structure, it follows that such a φ exists; thus Theorem 2′ implies Theorem 2.

Proof. Let $a \in F_\Sigma'(\alpha)$; then there exist $n < \omega$, $\gamma_0, \cdots, \gamma_{n-1} < \alpha$, and $\mathbf{P} \in \mathbf{P}_n(\Sigma)$ such that $a \in P(x'_{\gamma_0}, \cdots, x'_{\gamma_{n-1}})$. Thus

$$F_\Sigma(\alpha)\varphi \supseteq P(x_{\gamma_0}, \cdots, x_{\gamma_{n-1}})\varphi = P(x'_{\gamma_0}, \cdots, x'_{\gamma_{n-1}}) \ni a,$$

which means that φ is onto.

Let φ' be a Σ-homomorphism of $\mathfrak{F}_\Sigma'(\alpha)$ into $\mathfrak{F}_\Sigma(\alpha)$ for which $x_\gamma'\varphi' = x_\gamma$ $(\gamma < \alpha)$. Let $\mathbf{P}'$ and $\mathbf{P}'' \in \mathbf{P}_n(\Sigma)$ and $\gamma_0, \cdots, \gamma_{n-1} < \alpha$. Then

$$\begin{aligned}(P'(x_{\gamma_0}, \cdots, x_{\gamma_{n-1}}) \cup P''(x_{\gamma_0}, \cdots, x_{\gamma_{n-1}}))\varphi\varphi' \\ &= (P'(x'_{\gamma_0}, \cdots, x'_{\gamma_{n-1}}) \cup P''(x'_{\gamma_0}, \cdots, x'_{\gamma_{n-1}}))\varphi' \\ &= P'(x_{\gamma_0}, \cdots, x_{\gamma_{n-1}}) \cup P''(x_{\gamma_0}, \cdots, x_{\gamma_{n-1}}).\end{aligned}$$

Since $P'(x_{\gamma_0}, \cdots, x_{\gamma_{n-1}}) \cup P''(x_{\gamma_0}, \cdots, x_{\gamma_{n-1}})$ is finite by Theorem 1, this implies that φ is 1–1 on this set. Since any two elements of $F_\Sigma(n)$ belong to a set of this form, we get that φ (and similarly φ') is a 1–1 and onto homomorphism. To prove that φ is an isomorphism we have to show that

$$r_\gamma(a_0\varphi, \cdots, a_{m_\gamma-1}\varphi) \text{ implies } r_\gamma(a_0, \cdots, a_{m_\gamma-1}).$$

(We can use this condition since φ is 1–1.)

Let $a_i \in P_i(x_{\gamma_0}, \cdots, x_{\gamma_{n-1}})$, $0 \leqq i < m_\gamma$ and form the sets

$$A = \prod (P_i(x_{\gamma_0}, \cdots, x_{\gamma_{n-1}}) \mid 0 \leqq i < m_\gamma)$$

and

$$A' = \prod (P_i(x'_{\gamma_0}, \cdots, x'_{\gamma_{n-1}}) \mid 0 \leqq i < m_\gamma).$$

Let $\varphi^{m_\gamma}: A \to A'$ and $(\varphi')^{m_\gamma}: A' \to A$ be the maps induced by φ and φ', respectively. Finally, let

$$B = \{\langle b_0, \cdots, b_{m_\gamma-1}\rangle \mid \langle b_0, \cdots, b_{m_\gamma-1}\rangle \in A \quad \text{and} \quad r_\gamma(b_0, \cdots, b_{m_\gamma-1})\}$$

and

$$B' = \{\langle b_0, \cdots, b_{m_\gamma-1}\rangle \mid \langle b_0, \cdots, b_{m_\gamma-1}\rangle \in A' \quad \text{and} \quad r_\gamma(b_0, \cdots, b_{m_\gamma-1})\}.$$

Then A and A' are finite sets, φ^{m_γ}, $(\varphi')^{m_\gamma}$ are 1–1 and onto maps. Furthermore, $B\varphi^{m_\gamma} \subseteq B'$ and $B'(\varphi')^{m_\gamma} \subseteq B$, thus φ^{m_γ} is a 1–1 and onto map between

B and B', showing that φ is an isomorphism. This completes the proof of Theorem 2.

Corollary. *Let α and β be ordinals with $\bar{\alpha}=\bar{\beta}$. Then if $\mathfrak{F}_\Sigma(\alpha)$ exists, $\mathfrak{F}_\Sigma(\beta)$ also exists and they are isomorphic.*

§52. ON THE FAMILY OF FREE Σ-STRUCTURES

Let $E(\Sigma)$ denote the class of all ordinals α for which $\mathfrak{F}_\Sigma(\alpha)$ exists. In this section we will characterize $E(\Sigma)$. The characterization theorem is based on the following result.

Theorem 1. *Assume that $\mathfrak{F}_\Sigma(\alpha)$ exists; let $x_0, \cdots, x_\gamma, \cdots (\gamma<\alpha)$ be a Σ-generating system of $\mathfrak{F}_\Sigma(\alpha)$. Let β be an ordinal, let $\gamma_\delta<\alpha$ for $\delta<\beta$ such that if $\delta\neq\delta'$, then $\gamma_\delta\neq\gamma_{\delta'}$, and set*

$$B = [\{x_{\gamma_\delta} \mid \delta < \beta\}]_\Sigma.$$

Then $\mathfrak{B}$ is a slender Σ-substructure of $\mathfrak{F}_\Sigma(\alpha)$. Therefore, $\mathfrak{F}_\Sigma(\beta)$ exists and it is isomorphic to $\mathfrak{B}$.

Proof. The second statement follows immediately from the first one and from Lemma 50.5(ii), as in Lemma 24.1.

In order to simplify our notations, let $\alpha=n<\omega$, $\beta=m(<\omega)$ and $\gamma_i=i$, $i<m$. Thus, we will prove that if $\mathfrak{A}=\mathfrak{F}_\Sigma(n)$ exists and

$$B = [x_0, \cdots, x_{m-1}]_\Sigma,$$

then $\mathfrak{B}$ is slender.

First, we make a few observations. Let φ be a Σ-homomorphism of $\mathfrak{A}$ into $\mathfrak{B}$ with $x_0\varphi=x_0, \cdots, x_{m-1}\varphi=x_{m-1}$. ($x_m\varphi, \cdots, x_{n-1}\varphi$ are arbitrary elements of B.)

(i) If $\mathbf{P} \in \mathbf{P}_m(\Sigma)$, then $\mathbf{P}_\mathfrak{A}(x_0, \cdots, x_{m-1})=\mathbf{P}_\mathfrak{B}(x_0, \cdots, x_{m-1})$.

Indeed, $\mathbf{P}_\mathfrak{A}(x_0, \cdots, x_{m-1})\subseteq\mathbf{P}_\mathfrak{B}(x_0, \cdots, x_{m-1})$ by Lemma 49.7. On the other hand,

$$\mathbf{P}_\mathfrak{A}(x_0, \cdots, x_{m-1})\varphi = \mathbf{P}_\mathfrak{B}(x_0, \cdots, x_{m-1}),$$

so that

$$|\mathbf{P}_\mathfrak{A}(x_0, \cdots, x_{m-1})| \geqq |\mathbf{P}_\mathfrak{B}(x_0, \cdots, x_{m-1})|.$$

Since by Theorem 51.1, $\mathbf{P}_\mathfrak{A}(x_0, \cdots, x_{m-1})$ is finite, we get the equality.

(ii) φ is onto.

Let $b \in B$; then $b \in \mathbf{P}_{\mathfrak{B}}(x_0, \cdots, x_{m-1})$ for some $\mathbf{P} \in \mathbf{P}_m(\Sigma)$. Thus

$$b \in \mathbf{P}_{\mathfrak{B}}(x_0, \cdots, x_{m-1}) = \mathbf{P}_{\mathfrak{A}}(x_0, \cdots, x_{m-1})\varphi \subseteq A\varphi.$$

(iii) φ_B is 1–1.

For $\mathbf{P}'$ and $\mathbf{P}'' \in P_m(\Sigma)$,

$$(\mathbf{P}'_{\mathfrak{A}}(x_0, \cdots, x_{m-1}) \cup \mathbf{P}''_{\mathfrak{A}}(x_0, \cdots, x_{m-1}))\varphi = \mathbf{P}'_{\mathfrak{B}}(x_0, \cdots, x_{m-1}) \cup \mathbf{P}''_{\mathfrak{B}}(x_0, \cdots, x_{m-1}).$$

Combining this with (i), we can argue as in the proof of Theorem 51.2′.

(iv) φ_B is an automorphism of $\mathfrak{B}$.

φ_B is a homomorphism; by (ii) and (iii) it is 1–1 and onto. Thus to prove that it is an automorphism it remains to show that

$$r_\gamma(a_0\varphi, \cdots, a_{m_\gamma-1}\varphi) \text{ implies } r_\gamma(a_0, \cdots, a_{m_\gamma-1}), \text{ for } a_0, \cdots, a_{m_\gamma-1} \in B.$$

Let $a_i \in P_i(x_0, \cdots, x_{m-1})$, $0 \leq i < m_\gamma$ and set

$$C = \prod (P_i(x_0, \cdots, x_{m-1}) \mid 0 \leq i < m_\gamma)$$

and

$$D = \{\langle b_0, \cdots, b_{m_\gamma-1}\rangle \mid \langle b_0, \cdots, b_{m_\gamma-1}\rangle \in C \text{ and } r_\gamma(b_0, \cdots, b_{m_\gamma-1})\}.$$

Then by (i)–(iii) and Theorem 51.1 the map $\varphi^{m_\gamma}: C \to C$, induced by φ, is 1–1 and onto on C, and C is a finite set. Furthermore, φ is a homomorphism, thus $D\varphi^{m_\gamma} \subseteq D$. Since φ^{m_γ} is 1–1 and D is finite, we get $D\varphi^{m_\gamma} = D$, a statement equivalent to the one that is to be proved.

Now† let $a_0, \cdots, a_t \in B$, $\Phi \in \Sigma$, $l < e(\Phi)$ and let $b_0, \cdots, b_{s-1}$ be the Φ–l inverses of $a_0, \cdots, a_t$ in $\mathfrak{B}$ (s is finite by the corollary to Theorem 51.1). Since φ_B is an automorphism of $\mathfrak{B}$, $b_0\varphi, \cdots, b_{s-1}\varphi$ are the Φ–l inverses of $a_0\varphi, \cdots, a_t\varphi$ in $\mathfrak{B}$. But φ is a Σ-homomorphism; thus by Lemma 50.1(iii), there are s Φ–l inverses $c_0, \cdots, c_{s-1}$ of $a_0, \cdots, a_t$ in $\mathfrak{A}$ such that $c_0\varphi = b_0\varphi, \cdots, c_{s-1}\varphi = b_{s-1}\varphi$. We get that $\{c_0, \cdots, c_{s-1}\} \subseteq B$, since $\mathfrak{B}$ is a Σ-substructure. Thus (iii) implies $c_0 = b_0, \cdots, c_{s-1} = b_{s-1}$. This means that every Φ–l inverse in $\mathfrak{B}$ is also a Φ–l inverse in $\mathfrak{A}$, completing the proof of Theorem 1.

Theorem 2. *If $\mathfrak{F}_\Sigma(n)$ exists for all $n < \omega$, then $\mathfrak{F}_\Sigma(\omega)$ also exists. In other words, if $n \in E(\Sigma)$, for all $n < \omega$, then $\omega \in E(\Sigma)$.*

Proof. Let $\mathfrak{F}_\Sigma(n)$ be Σ-generated by $x_0^n, \cdots, x_{n-1}^n$ $(n = 1, 2, \cdots)$. We can assume that $\mathfrak{F}_\Sigma(n)$ is disjoint from $\mathfrak{F}_\Sigma(m)$ if $n \neq m$.

† The original proof was continued using a rather long argument. This simplified version is due to G. H. Wenzel.

Let φ_n be a 1–1 Σ-homomorphism of $\mathfrak{F}_\Sigma(n)$ into $\mathfrak{F}_\Sigma(n+1)$ with $x_i^n\varphi_n = x_i^{n+1}$, $i = 0, \cdots, n-1$. For $n \leqq m$, set φ_{nn} = the identity map on $F_\Sigma(n)$, and

$$\varphi_{nm} = \varphi_n \cdots \varphi_{m-1}.$$

Then the Σ-structures $\mathfrak{F}_\Sigma(n)$ and the Σ-homomorphisms φ_{nm} form a direct family. Let $\mathfrak{A}$ denote its direct limit. If $\mathsf{x} \in A$, $\mathsf{x} = \langle x_n, x_{n+1}, \cdots\rangle$ (considering the equivalence class as a sequence), then the mapping

$$\varphi^n : x_n \to \mathsf{x}$$

is an embedding of $\mathfrak{F}_\Sigma(n)$ into $\mathfrak{A}$. Set $A_n = F_\Sigma(n)\varphi^n$. Then

$$A = \bigcup (A_n \mid n < \omega),$$
$$A_1 \subseteq A_2 \subseteq \cdots$$

and

$$\mathfrak{A}_n \cong \mathfrak{F}_\Sigma(n), \qquad n = 1, 2, \cdots.$$

First we prove that $\mathfrak{A}$ is a Σ-structure. We will verify only that if $\Phi = (\mathbf{x})(\exists \mathbf{y})(\mathbf{u})(\exists \mathbf{v})\Psi(\mathbf{x}, \mathbf{y}, \mathbf{u}, \mathbf{v}) \in \Sigma$, then Φ holds in $\mathfrak{A}$.

Let $a \in A$; then $a \in A_n$ for some $n < \omega$. Since $\mathfrak{A}_n$ is a Σ-structure, there exists a $b \in A_n$ such that $(\mathbf{u})(\exists \mathbf{v})\Psi(a, b, \mathbf{u}, \mathbf{v})$ holds in $\mathfrak{A}_n$. To prove that it also holds in $\mathfrak{A}$, take a $c \in A$ and an $m < \omega$, $n \leqq m$, with $a, b, c \in A_m$. Since $\mathfrak{F}_\Sigma(n)\varphi_{nm}\varphi^m$ is a slender Σ-substructure of $\mathfrak{F}_\Sigma(m)$ by Theorem 1, and $A_n = F_\Sigma(n)\varphi_{nm}\varphi^m$, $A_m = F_\Sigma(m)\varphi^m$, we get that $\mathfrak{A}_n$ is a slender Σ-substructure of $\mathfrak{A}_m$. Thus $(\mathbf{u})(\exists \mathbf{v})\Psi(a, b, \mathbf{u}, \mathbf{v})$ in $\mathfrak{A}_m$, hence there exists a $d \in A_m$ with $\Psi(a, b, c, d)$ in A_m. Therefore, $\Psi(a, b, c, d)$ in $\mathfrak{A}$, so $(\mathbf{u})(\exists \mathbf{v})\Psi(a, b, \mathbf{u}, \mathbf{v})$ in $\mathfrak{A}$, which was to be proved. A similar (but simpler) argument shows that if $a, c \in A$, then also a Φ–1 inverse exists.

Set

$$\mathsf{x}_i = \langle x_i^{i+1}, x_i^{i+2}, \cdots\rangle \quad \text{for} \quad i = 0, 1, 2, \cdots.$$

Then $\mathsf{x}_i \in A$, $A = [\mathsf{x}_0, \mathsf{x}_1, \cdots]_\Sigma$ and $A_n = [\mathsf{x}_0, \cdots, \mathsf{x}_{n-1}]_\Sigma$. Thus $\mathfrak{A}$ is Σ-generated by ω elements.

It remains to show that $\mathfrak{A}$ satisfies (iii) of Definition 51.1. Let $\mathfrak{B}$ be a Σ-structure, $b_0, b_1, \cdots \in B$. We can assume that A is disjoint from B. We want to construct a Σ-homomorphism φ of $\mathfrak{A}$ into $\mathfrak{B}$ with $\mathsf{x}_i\varphi = b_i$, $i = 0, 1, 2, \cdots$.

Set $C = A \cup B$; we define a structure $\mathfrak{C}$ on C:

(i) for every $a \in A$, there is a constant $\mathbf{k}_a$ and $(\mathbf{k}_a)_\mathfrak{C} = a$;
(ii) for every $d \in B$, there is a constant $\mathbf{l}_d$ and $(\mathbf{l}_d)_\mathfrak{C} = d$;
(iii) for $\gamma < o_1(\tau)$, $\mathbf{r}_\gamma$ is defined on A and B as it was in $\mathfrak{A}$ and $\mathfrak{B}$;
(iv) for $0 < n < \omega$, $\mathbf{P} \in \mathbf{P}_n(\Sigma)$, $\mathbf{r}_\mathbf{P}$ is defined on A and B as it was in $\mathfrak{A}$ and $\mathfrak{B}$ ($\mathbf{r}_\mathbf{P}$ was defined in Lemma 49.6).
(v) for $c_0, \cdots, c_{n-1}$ in A or in B and $\mathbf{P} \in \mathbf{P}_n(\Sigma)$ we define the constants

$\mathbf{c}(c_0, \cdots, c_{n-1}, \mathbf{P}, i)$ for $0 \leq i < k_{\mathbf{P}}$ (of Theorem 51.1); these are interpreted in $\mathfrak{C}$ such that every element of $P(c_0, \cdots, c_{n-1})$ is the interpretation of one of them and conversely.

Let $\mathfrak{C}$ denote the relational system defined by (i)–(v) (consider the constants as unary relations), and let τ^0 be the type of $\mathfrak{C}$. We want to define an additional relation $R(x, y)$ on $\mathfrak{C}$, satisfying the following universal sentences:

$$\mathbf{R}(\mathbf{k}_{x_i}, \mathbf{l}_{b_i}), \qquad i = 0, 1, \cdots \tag{1}$$

$$(\mathbf{r}(\mathbf{k}_{a_0}, \cdots, \mathbf{k}_{a_m}) \wedge \mathbf{R}(\mathbf{k}_{a_0}, \mathbf{l}_{d_0}) \wedge \cdots \wedge \mathbf{R}(\mathbf{k}_{a_m}, \mathbf{l}_{d_m})) \rightarrow \mathbf{r}(\mathbf{l}_{d_0}, \cdots, \mathbf{l}_{d_m}), \tag{2}$$

where $\mathbf{r}$ is some $\mathbf{r}_\gamma$ or $\mathbf{r}_{\mathbf{P}}$;

$$\begin{aligned}(\mathbf{R}(\mathbf{k}_{a_0}, \mathbf{l}_{d_0}) \wedge \cdots \wedge \mathbf{R}(\mathbf{k}_{a_{n-1}}, \mathbf{l}_{d_{n-1}}) \wedge \mathbf{R}(\mathbf{k}_a, \mathbf{l}_b) \wedge \mathbf{r}_{\mathbf{P}}(\mathbf{l}_{d_0}, \cdots, \mathbf{l}_{d_{n-1}}, \mathbf{l}_b))\\ \rightarrow (\mathbf{R}(\mathbf{c}(a_0, \cdots, a_{n-1}, \mathbf{P}, 0), \mathbf{l}_b) \vee \cdots\\ \vee \mathbf{R}(\mathbf{c}(a_0, \cdots, a_{n-1}, \mathbf{P}, k_{\mathbf{P}}-1), \mathbf{l}_b))\end{aligned} \tag{3}$$

$$(\mathbf{R}(\mathbf{k}_a, \mathbf{l}_d) \wedge \mathbf{R}(\mathbf{k}_a, \mathbf{l}_{d_1})) \rightarrow \mathbf{l}_d = \mathbf{l}_{d_1} \tag{4}$$

$$\begin{aligned}\mathbf{r}_{\mathbf{P}}(\mathbf{k}_{x_0}, \cdots, \mathbf{k}_{x_{n-1}}, \mathbf{k}_a)\\ \rightarrow (\mathbf{R}(\mathbf{k}_a, \mathbf{c}(b_0, \cdots, b_{n-1}, \mathbf{P}, 0)) \vee \cdots\\ \vee \mathbf{R}(\mathbf{k}_a, \mathbf{c}(b_0, \cdots, b_{n-1}, \mathbf{P}, k_{\mathbf{P}}-1)).\end{aligned} \tag{5}$$

If R can be defined so as to satisfy (1)–(5), then we can define a mapping φ of A into B by setting $a\varphi = d$ ($a \in A$, $b \in B$) if $R(a, d)$.

By (4), φ is well defined and by (5), φ is defined on the whole of A; (2) and (3) mean that φ is a Σ-homomorphism and by (1), $x_i\varphi = b_i$.

By Theorem 39.5, it is sufficient to prove that R can be defined on every finite subset of C. However, this is trivial, since if H is a finite subset of C, then for some n,

$$H \subseteq [x_0, \cdots, x_{n-1}]_\Sigma \cup [b_0, \cdots, b_{n-1}]_\Sigma \cup H',$$

where $H' = H \cap (B - [b_0, b_1, \cdots]_\Sigma)$. It follows from (4) and (5) that no element of H' occurs in (1)–(5); thus it suffices to consider $H'' = H - H'$. Since $\mathfrak{A}_n$ is the free Σ-structure on n Σ-generators, there is a homomorphism ψ of $\mathfrak{A}_n$ into $\mathfrak{B}$ for which $x_i\psi = b_i$, $i = 0, \cdots, n-1$. Define R on $A_n \cup B$ by $R(a, d)$ if $a\psi = d$. Obviously, R satisfies (1)–(5). This completes the proof of Theorem 2.

The following result is a more complicated version of Theorem 2.

Theorem 3. *Let α be a limit ordinal. If $\mathfrak{F}_\Sigma(\beta)$ exists for all $\beta < \alpha$, then also $\mathfrak{F}_\Sigma(\alpha)$ exists.*

Sketch of Proof. The proof of Theorem 2 started with the construction of a direct family. There we had no problem with $\varphi_{ln}\varphi_{nm} = \varphi_{lm}$ (for $l \leqq n \leqq m$) since we defined φ_{ln} as $\varphi_l \cdots \varphi_{n-1}$. However, we cannot do this now. In order to construct the direct family, we set

$$C = \bigcup (F_\Sigma(\beta) \mid \beta < \alpha),$$

where the $F_\Sigma(\beta)$ are assumed to be pairwise disjoint. We want to define on C a relation R such that $\varphi_{\beta\gamma}$ for $\beta \leqq \gamma < \alpha$ can be defined by $a\varphi_{\beta\gamma} = b$ for $a \in F_\Sigma(\beta)$ and $b \in F_\Sigma(\gamma)$, if $R(a, b)$. As in the proof of Theorem 2, we can do that by introducing sufficiently many constants and relations, which satisfy the analogues of (1)–(5), and

$$R(a, b) \wedge R(b, c) \rightarrow R(a, c). \tag{6}$$

We leave the obvious details to the reader. Then we form the direct limit $\mathfrak{A}$, and we proceed as in the proof of Theorem 2.

Now we are ready to characterize $E(\Sigma)$.

Theorem 4. *Either there exists a positive integer n such that $\mathfrak{F}_\Sigma(\alpha)$ exists if and only if $\alpha < n$, or $\mathfrak{F}_\Sigma(\alpha)$ exists for every α.*

In other words, either $E(\Sigma) = \{\alpha \mid \alpha < n\}$ or $E(\Sigma)$ is the class of all ordinals.

Proof. Let us assume that there is no n with $E(\Sigma) = \{\alpha \mid \alpha < n\}$. Then for every n there exists an $m \geqq n$ with $m \in E(\Sigma)$. By Theorem 1 this implies $n \in E(\Sigma)$; therefore by Theorem 2, $\omega \in E(\Sigma)$. Let us further assume that for some ordinal δ, $\delta \notin E(\Sigma)$. If δ is the smallest ordinal with $\delta \notin E(\Sigma)$, then by the corollary to Theorem 51.2, δ is an initial ordinal. Since $\omega < \delta$, δ is a limit ordinal, and if $\gamma < \delta$ then $\gamma \in E(\Sigma)$. Thus by Theorem 3, $\delta \in E(\Sigma)$. This contradiction and Theorem 1 prove Theorem 4.

§53. ON THE EXISTENCE OF FREE Σ-STRUCTURES

A necessary and sufficient condition for the existence of free Σ-structures is given in the following result:

Theorem 1. $\mathfrak{F}_\Sigma(n)$ *exists if and only if the following two conditions are satisfied:*

(B_n) *every* $\mathbf{P} \in \mathbf{P}_n(\Sigma)$ *is bounded;*

(C_n) *let $\mathfrak{A}$ and $\mathfrak{B}$ be Σ-structures, let $a_0, \cdots, a_{n-1} \in A$ and $b_0, \cdots, b_{n-1} \in B$. If $A = [a_0, \cdots, a_{n-1}]_\Sigma$, then there exists a Σ-structure $\mathfrak{C}$, Σ-generated by $c_0, \cdots, c_{n-1}$, and there exist Σ-homomorphisms $\varphi: C \rightarrow A$ and $\psi: C \rightarrow B$ such that $c_i\varphi = a_i$ and $c_i\psi = b_i$ for $0 \leqq i < n$.*

Proof. (B_n) is necessary by Theorem 51.1. It is obvious that (C_n) is also necessary, since we can always set $\mathfrak{C} = \mathfrak{F}_\Sigma(n)$.

Let us assume that (B_n) and (C_n) are satisfied. Let $\mathbf{P} \in \mathbf{P}_n(\Sigma)$; (B_n) implies that there exists a Σ-structure $\mathfrak{E}_\mathbf{P}$, Σ-generated by $a_0^\mathbf{P}, \cdots, a_{n-1}^\mathbf{P}$, such that

$$|P(a_0^\mathbf{P}, \cdots, a_{n-1}^\mathbf{P})| = k_\mathbf{P}.$$

Let $\mathfrak{E}_{\mathbf{P}'}$ be a Σ-structure which corresponds to $\mathbf{P}' \in \mathbf{P}_n(\Sigma)$, and let us apply (C_n) for $\mathfrak{E}_\mathbf{P}$ and $\mathfrak{E}_{\mathbf{P}'}$, obtaining a structure $\mathfrak{C}$ Σ-generated by $c_0, \cdots, c_{n-1}$. $|P(c_0, \cdots, c_{n-1})|$ and $|P'(c_0, \cdots, c_{n-1})|$ are maximal.

If $\mathbf{P}_0, \cdots, \mathbf{P}_{k-1} \in \mathbf{P}_n(\Sigma)$, then we can always find a minimal upper bound $k_{\mathbf{P}_0, \ldots, \mathbf{P}_{k-1}}$ for $\mathbf{P}_0 \cup \cdots \cup \mathbf{P}_{k-1}$. An obvious induction, combined with the argument given above, yields the following result:

Let H be a nonvoid finite subset of $\mathbf{P}_n(\Sigma)$; then there exists a least natural number k_H such that for every Σ-structure $\mathfrak{A}$ and $a_0, \cdots, a_{n-1} \in A$ we have

$$|\bigcup (P(a_0, \cdots, a_{n-1}) \mid \mathbf{P} \in H)| \leqq k_H.$$

Furthermore, there exists a Σ-structure $\mathfrak{A}_H$ and $a_0^H, \cdots, a_{n-1}^H \in A_H$ such that $A_H = [a_0^H, \cdots, a_{n-1}^H]_\Sigma$ and if $H' \subseteq H$, $H' \neq \varnothing$, then

$$|\bigcup (P(a_0^H, \cdots, a_{n-1}^H) \mid P \in H')| = k_{H'}.$$

Set $\mathbf{T} = \{H \mid H \text{ is finite}, \varnothing \neq H, \text{ and } H \subseteq \mathbf{P}_n(\Sigma)\}$ and for $H \in \mathbf{T}$ let

$$\mathbf{T}_H = \{K \mid K \in \mathbf{T} \text{ and } H \subseteq K\}.$$

Then $\mathbf{T}_{H_0} \cap \mathbf{T}_{H_1} = \mathbf{T}_{H_0 \cup H_1}$ and $\mathbf{T}_H \neq \varnothing$, and thus there exists a prime dual ideal $\mathscr{D}$ over $\mathbf{T}$ containing all the $\mathbf{T}_H$. Set

$$\mathfrak{A} = \prod_\mathscr{D} (\mathfrak{A}_H \mid H \in \mathbf{T}).$$

By Theorem 39.1, $\mathfrak{A}$ is a Σ-structure. Let f_i be the function for which $f_i(H) = a_i^H$ for all $H \in \mathbf{T}$, $i = 0, \cdots, n-1$. Then

$$\mathbf{T}'_H = \{K \mid |\bigcup (P(a_0^K, \cdots, a_{n-1}^K) \mid \mathbf{P} \in H)| = k_H\} \supseteq \mathbf{T}_H,$$

so $\mathbf{T}'_H \in \mathscr{D}$. Since there is a formula in our language which can express the statement

$$|\bigcup (P(a_0^K, \cdots, a_{n-1}^K) \mid \mathbf{P} \in H)| = k_H,$$

by Theorem 39.1, we conclude that

$$|\bigcup (P(f_0^\mathrm{v}, \cdots, f_{n-1}^\mathrm{v}) \mid \mathbf{P} \in H)| = k_H$$

for all $H \in \mathbf{T}$.

Let $\mathfrak{F}$ be the Σ-substructure of $\mathfrak{A}$, Σ-generated by $f_0^\mathrm{v}, \cdots, f_{n-1}^\mathrm{v}$. It is obvious that the above equality holds in $\mathfrak{F}$ as well.

Let $\mathfrak{B}$ be any Σ-structure and $b_0, \cdots, b_{n-1} \in B$. By (C_n), there exists a

Σ-structure $\mathfrak{C}$, Σ-generated by $c_0, \cdots, c_{n-1}$, and there exist Σ-homomorphisms $\varphi: C \to F$ and $\psi: C \to B$ with $c_i\varphi = f_i^{\mathrm{v}}$ and $c_i\psi = b_i$, $0 \leqq i < n$.

The mapping φ is obviously onto. Let $c, d \in C$ and let us choose $\mathbf{P}'$ and $\mathbf{P}'' \in \mathbf{P}_n(\Sigma)$ with $c \in P'(c_0, \cdots, c_{n-1})$ and $d \in P''(c_0, \cdots, c_{n-1})$ and set $H = \{\mathbf{P}', \mathbf{P}''\}$. By definition,

$$|P'(c_0, \cdots, c_{n-1}) \cup P''(c_0, \cdots, c_{n-1})| \leqq k_H.$$

On the other hand,

$$(P'(c_0, \cdots, c_{n-1}) \cup P''(c_0, \cdots, c_{n-1}))\varphi = P'(f_0^{\mathrm{v}}, \cdots, f_{n-1}^{\mathrm{v}}) \cup P''(f_0^{\mathrm{v}}, \cdots, f_{n-1}^{\mathrm{v}})$$

and

$$|P'(f_0^{\mathrm{v}}, \cdots, f_{n-1}^{\mathrm{v}}) \cup P''(f_0^{\mathrm{v}}, \cdots, f_{n-1}^{\mathrm{v}})| = k_H.$$

Thus φ is 1–1. Therefore, if we deal with Σ-algebras, then we can claim that φ is an isomorphism. This implies that $\varphi^{-1}\psi$ is a Σ-homomorphism of $\mathfrak{F}$ into $\mathfrak{B}$ with $f_i^{\mathrm{v}}\varphi^{-1}\psi = b_i$, for $0 \leqq i < n$, establishing that $\mathfrak{F}$ is the free Σ-algebra on n Σ-generators. However, in the general case φ need not be an isomorphism since φ^{-1} need not preserve relations. Let $\mathfrak{F} = \langle A; F, R \rangle$. Using ($\mathrm{B}_n$) and ($\mathrm{C}_n$) and some transfinite method, for instance Theorem 21.1 it can be verified that there exists a "smallest" Σ-structure $\mathfrak{A} = \langle A; F, R' \rangle$, such that for all Σ-polynomial symbols $\mathbf{P}$ we have $\mathbf{P}_{\mathfrak{F}} = \mathbf{P}_{\mathfrak{A}}$, for all $\gamma < o_0(\tau)$ we have $(\mathbf{f}_\gamma)_{\mathfrak{F}} = (\mathbf{f}_\gamma)_{\mathfrak{A}}$ and for all $\gamma < o_1(\tau)$, $(\mathbf{r}_\gamma)_{\mathfrak{A}}$ is smallest for all Σ-structures having these properties. For this $\mathfrak{A}$, in place of $\mathfrak{F}$, it is obvious that φ^{-1} is also a homomorphism, completing the proof of Theorem 1.

Corollary 1. *All free Σ-structures exist if and only if the following two conditions are satisfied:*

(B) *all Σ-polynomials are bounded;*

(C) *let $\mathfrak{A}$ and $\mathfrak{B}$ be Σ-structures, let $a_0, a_1, \cdots, a_n, \cdots \in A$, $b_0, b_1, \cdots, b_n, \cdots \in B$ and $A = [a_0, a_1, \cdots, a_n, \cdots]_\Sigma$; then there exists a Σ-structure $\mathfrak{C}$ with $C = [c_0, c_1, \cdots, c_n, \cdots]_\Sigma$ and there exist Σ-homomorphisms $\varphi: C \to A$ and $\psi: C \to B$ such that $c_i\varphi = a_i$ and $c_i\psi = b_i$, $i = 0, 1, 2, \cdots$.*

Corollary 1 is an obvious combination of Theorem 1 and Theorem 52.4.

Corollary 2. *Let Σ be universal. Then $\mathfrak{F}_\Sigma(n)$ (that is, the free structure on n generators over Σ) exists if and only if* (C_n) *holds. All free structures exist if and only if* (C) *holds.*

Indeed, if Σ is universal, then all Σ-polynomials are of bound 1, and thus (B_n) is always satisfied.

Corollary 2 is an analogue of Corollary 2 to Theorem 46.1.

It is interesting to compare Corollary 2 with Theorem 25.2. Is Corollary 2 true because we can apply this result to get a covering family and (C_n) implies that the members of the covering family are isomorphic? This is indeed the case and not only if Σ is universal but in a number of other cases as well. These cases are the ones which are described in the following definition.

Definition 1. *Σ is said to have property* (P) *if for every Φ in Σ either Φ is universal or $\Phi \equiv (\mathbf{x}_0), \cdots, (\mathbf{x}_{n-1})(\exists \mathbf{y})\Psi(\mathbf{x}_0, \cdots, \mathbf{x}_{n-1}, \mathbf{y})$ (where Ψ is the matrix of Φ), or Φ is positive.*

Let $\mathscr{A}$ be a well-ordered inverse family of the Σ-structures $\mathfrak{A}_\gamma$, $\gamma < \alpha$; let $\mathfrak{A}_\gamma$ be Σ-generated by $a_0^\gamma, \cdots, a_{n-1}^\gamma$; let the homomorphisms $\varphi_\delta^\gamma (\delta \leqq \gamma < \alpha)$ be Σ-homomorphisms and suppose

$$a_i^\gamma \varphi_\delta^\gamma = a_i^\delta, \quad \text{for} \quad \delta \leqq \gamma < \alpha, \qquad i = 0, \cdots, n-1.$$

Let $\mathfrak{A}$ be the inverse limit structure of $\mathscr{A}$.

Lemma 1. *Let $\gamma < \alpha$ and $a \in A_\gamma$. If* (B_n) *is satisfied, then there exists an* $\mathbf{a} \in A$ *with $a(\gamma) = a$.*

Proof. Choose $\mathbf{P} \in \mathbf{P}_n(\Sigma)$ such that

$$a \in P(a_0^\gamma, \cdots, a_{n-1}^\gamma).$$

For $\delta \geqq \gamma$, set

$$U_\delta = \{b \mid b \in P(a_0^\delta, \cdots, a_{n-1}^\delta) \text{ and } b\varphi_\gamma^\delta = a\}.$$

Since $a \in P(a_0^\delta, \cdots, a_{n-1}^\delta)\varphi_\gamma^\delta$ and φ_γ^δ is a Σ-homomorphism, U_δ is not void. By (B_n), U_δ is finite. Furthermore, $U_\delta \varphi_{\delta'}^\delta \subseteq U_{\delta'}$, if $\gamma \leqq \delta' \leqq \delta < \alpha$. Since the inverse limit of nonvoid finite sets is not void (Theorem 21.1), there exists $a(\delta) \in U_\delta$ for $\delta > \gamma$ such that $a(\delta)\varphi_{\delta'}^\delta = a(\delta')$ if $\gamma \leqq \delta' \leqq \delta < \alpha$. Set $a(\delta) = a\varphi_\delta^\gamma$ if $\delta \leqq \gamma$. Then for $\mathbf{a} = \langle a(\gamma) \mid \gamma < \alpha \rangle$ we have that $\mathbf{a} \in A$ and $a(\gamma) = a$.

Theorem 2. *If we assume* (P) *and* (B_n), *then $\mathfrak{A}$ is a Σ-structure.*

Proof. We first verify that if $\Phi = (\mathbf{x})(\exists \mathbf{y})(\mathbf{u})(\exists \mathbf{v})\Psi(\mathbf{x}, \mathbf{y}, \mathbf{u}, \mathbf{v}) \in \Sigma$, and Φ is positive, then Φ holds in $\mathfrak{A}$. Let $\mathbf{a} \in A$ and set

$$T_\gamma = \{b \mid b \in A_\gamma \text{ and } b \text{ is a } \Phi\text{–}0 \text{ inverse of } a(\gamma)\}.$$

It follows from (B_n) and from the corollary to Theorem 51.1 that T_γ is finite for all $\gamma < \alpha$ and $T_\gamma \neq \varnothing$. It is obvious that $T_\gamma \varphi_\delta^\gamma \subseteq T_\delta$ if $\delta \leqq \gamma < \alpha$. Thus there exists a $\mathbf{b} \in A$ with $b(\gamma) \in T_\gamma$ for all $\gamma < \alpha$, that is,

$$(\mathbf{u})(\exists \mathbf{v})\Psi(a(\gamma), b(\gamma), \mathbf{u}, \mathbf{v}) \text{ holds in } \mathfrak{A}_\gamma.$$

We want to prove that $(\mathbf{u})(\exists\mathbf{v})\Psi(\mathsf{a}, \mathsf{b}, \mathbf{u}, \mathbf{v})$ holds in $\mathfrak{A}$. Let $\mathsf{c} \in A$ and set

$$U_\gamma = \{d \mid \Psi(a(\gamma), b(\gamma), c(\gamma), d)\} \qquad \text{for } \gamma < \alpha.$$

Then

$$U_\gamma \subseteq \{d \mid d \text{ is a } \Phi\text{–1 inverse of } a(\gamma) \text{ and } c(\gamma)\}.$$

Since the right-hand side is finite, U_γ is finite for all $\gamma < \alpha$. Now let $d \in U_\gamma$ and $\delta < \gamma < \alpha$. Then $\Psi(a(\gamma), b(\gamma), c(\gamma), d)$ and since Ψ is positive, $\Psi(a(\delta), b(\delta), c(\delta), d\varphi_\delta{}^\gamma)$. Thus $U_\gamma\varphi_\delta{}^\gamma \subseteq U_\delta$. So we can choose $\mathsf{d} \in A$ with $d(\gamma) \in U_\gamma$. Therefore, $\Psi(a(\gamma), b(\gamma), c(\gamma), d(\gamma))$ for all $\gamma < \alpha$, which implies $\Psi(\mathsf{a}, \mathsf{b}, \mathsf{c}, \mathsf{d})$. The existence of Φ–1 inverses is proved by a similar argument.

Now let Φ be universal, $\Phi = (\mathbf{x}_0) \cdots (\mathbf{x}_{m-1})\Psi(\mathbf{x}_0, \cdots, \mathbf{x}_{m-1})$. Let $\mathsf{a}_0, \cdots, \mathsf{a}_{m-1} \in A$; then $\Psi(a_0(\gamma), \cdots, a_{m-1}(\gamma))$ for all $\gamma < \alpha$, whence $\Psi(\mathsf{a}_0, \cdots, \mathsf{a}_{m-1})$.

Finally, let $\Phi = (\mathbf{x}_0) \cdots (\mathbf{x}_{m-1})(\exists\mathbf{y})\Psi(\mathbf{x}_0, \cdots, \mathbf{x}_{m-1}, \mathbf{y})$ and let

$$\mathsf{a}_0, \cdots, \mathsf{a}_{m-1} \in A.$$

Set

$$T_\gamma = \{b \mid \Psi(a_0(\gamma), \cdots, a_{m-1}(\gamma), b)\}.$$

By (B_n) and from the corollary to Theorem 51.1, T_γ is finite. Since $T_\gamma\varphi_\delta{}^\gamma \subseteq T_\delta$ is obvious for $\delta \leqq \gamma < \alpha$, there exists a $b \in A$ with $b(\gamma) \in T_\gamma$ for $\gamma < \alpha$. Thus $\Psi(a_0(\gamma), \cdots, a_{m-1}(\gamma), b(\gamma))$ for $\gamma < \alpha$, which implies that $\Psi(\mathsf{a}_0, \cdots, \mathsf{a}_{m-1}, \mathsf{b})$, completing the proof of Theorem 2.

It is easy to see that the proof of Theorem 2 yields the following result.

Corollary. *For* $\mathsf{a}_0, \cdots, \mathsf{a}_{m-1}, \mathsf{b} \in A$ *and* $\mathbf{P} \in \mathbf{P}_m(\Sigma)$, *if* $b(\gamma) \in P(a_0(\gamma), \cdots, a_{m-1}(\gamma))$ *for all* $\gamma < \alpha$, *then* $\mathsf{b} \in P(\mathsf{a}_0, \cdots, \mathsf{a}_{m-1})$.

The converse of this corollary is also true.

Lemma 2. *Under the conditions of Theorem 2 and its corollary, if* $\mathsf{b} \in P(\mathsf{a}_0, \cdots, \mathsf{a}_{m-1})$, *then* $b(\gamma) \in P(a_0(\gamma), \cdots, a_{m-1}(\gamma))$ *for all* $\gamma < \alpha$.

Proof. It is sufficient to prove that if b is a Φ-inverse of $\mathsf{a}_0, \cdots, \mathsf{a}_{m-1}$, then $b(\gamma)$ is a Φ-inverse of $a_0(\gamma), \cdots, a_{m-1}(\gamma)$ for all $\gamma < \alpha$. If Φ is universal, there is nothing to be proved, so let

$$\Phi \equiv (\mathbf{x})(\exists\mathbf{y})(\mathbf{u})(\exists\mathbf{v})\Psi(\mathbf{x}, \mathbf{y}, \mathbf{u}, \mathbf{v}) \in \Sigma,$$

let Ψ be positive, and let b be a Φ–0 inverse of a. We have to prove that

$$(\mathbf{u})(\exists\mathbf{v})\Psi(a(\gamma), b(\gamma), \mathbf{u}, \mathbf{v}) \text{ holds in } \mathfrak{A}_\gamma \text{ for all } \gamma < \alpha.$$

Let $c \in A_\gamma$; by Lemma 1 there exists a $\mathsf{c} \in A$ with $c(\gamma) = c$. Since $(\mathbf{u})(\exists \mathbf{v})\Psi(\mathsf{a}, \mathsf{b}, \mathbf{u}, \mathbf{v})$ holds in $\mathfrak{A}$, there exists a $\mathsf{d} \in A$ with $\Psi(\mathsf{a}, \mathsf{b}, \mathsf{c}, \mathsf{d})$. Hence $\Psi(a(\delta), b(\delta), c(\delta), d(\delta))$ for all $\delta \geqq \delta_0$, where $\delta_0 < \alpha$. Since Ψ is positive, we get $\Psi(a(\gamma), b(\gamma), c, d(\gamma))$, completing the proof.

The same statement for Φ–1 inverses is even simpler to prove.

Now let $\Phi = (\mathbf{x}_0) \cdots (\mathbf{x}_{m-1})(\exists \mathbf{y})\Psi(\mathbf{x}_0, \cdots, \mathbf{x}_{m-1}, \mathbf{y})$. Let b be an inverse of $\mathsf{a}_0, \cdots, \mathsf{a}_{m-1}$. Then $\Psi(\mathsf{a}_0, \cdots, \mathsf{a}_{m-1}, \mathsf{b})$, so $\Psi(a_0(\delta), \cdots, a_{m-1}(\delta), b(\delta))$ holds for all $\delta \geqq \delta_0$, for some $\delta_0 < \alpha$. Choose δ such that $\delta > \max\{\gamma, \delta_0\}$. Since $b(\delta)$ is a Φ–0 inverse of $a_0(\delta), \cdots, a_{m-1}(\delta)$, it follows that $b(\gamma) = b(\delta)\varphi_\gamma{}^\delta$ is a Φ–0 inverse of $a_0(\gamma) = a_0(\delta)\varphi_\gamma{}^\delta, \cdots, a_{m-1}(\gamma) = a_{m-1}(\delta)\varphi_\gamma{}^\delta$; that is, $\Psi(a_0(\gamma), \cdots, a_{m-1}(\gamma), b(\gamma))$ for all $\gamma < \alpha$, which was to be proved.

Set $\mathsf{a}_0 = \langle a_0{}^\gamma \mid \gamma < \alpha \rangle, \cdots, \mathsf{a}_{n-1} = \langle a_{n-1}^\gamma \mid \gamma < \alpha \rangle$ and let $\bar{\mathfrak{A}}$ denote the Σ-substructure of $\mathfrak{A}$, Σ-generated by $\mathsf{a}_0, \cdots, \mathsf{a}_{n-1}$.

Lemma 3. *$\bar{\mathfrak{A}}$ is a slender Σ-substructure of $\mathfrak{A}$.*

Proof. We should note that the a of Lemma 1 is in $\bar{\mathfrak{A}}$. Thus, by repeating the proof of Lemma 2, and restricting a, c to $\bar{\mathfrak{A}}$, we get that the conclusion of Lemma 2 holds for $\bar{\mathfrak{A}}$, that is, if b is a Φ-inverse of $\mathsf{c}_0, \cdots, \mathsf{c}_{m-1}$ in $\bar{\mathfrak{A}}$, then $b(\gamma)$ is a Φ-inverse of $c_0(\gamma), \cdots, c_{m-1}(\gamma)$ in $\mathfrak{A}_\gamma$ for all $\gamma < \alpha$. Thus the corollary to Theorem 2 implies that b is a Φ-inverse of $\mathsf{c}_0, \cdots, \mathsf{c}_{m-1}$ in $\mathfrak{A}$, which was to be proved.

Corollary. *The mapping ψ_γ: $\mathsf{c} \to c(\gamma)$ is a Σ-homomorphism of $\bar{\mathfrak{A}}$ onto $\mathfrak{A}_\gamma$.*

Now we are ready to prove the main result.

Theorem 3. *Let us assume* (P) *and* (B_n). *Let $\mathfrak{A}_\gamma$ be Σ-structures and $A_\gamma = [a_0{}^\gamma, \cdots, a_{n-1}^\gamma]_\Sigma$ for $\gamma < \alpha$. Let us assume for all $\gamma \leqq \delta < \alpha$ that there exists a Σ-homomorphism $\varphi_\gamma{}^\delta$ such that $a_i{}^\delta\varphi_\gamma{}^\delta = a_i{}^\gamma$, $0 \leqq i < n$. Then there exists a Σ-structure $\mathfrak{A}$ and there exist $a_0, \cdots, a_{n-1} \in A$ such that $A = [a_0, \cdots, a_{n-1}]_\Sigma$ and for each $\gamma < \alpha$ there exists a Σ-homomorphism ψ_γ of $\mathfrak{A}$ onto $\mathfrak{A}_\gamma$, with $a_i\psi_\gamma = a_i{}^\gamma$, $0 \leqq i < n$.*

Proof. If we have that $\varphi_\gamma{}^\delta\varphi_\beta{}^\gamma = \varphi_\beta{}^\delta$, whenever $\beta \leqq \gamma \leqq \delta < \alpha$, the $\mathfrak{A}_\gamma$ form an inverse family and we can take the $\bar{\mathfrak{A}}$ as in Lemma 3; then by the corollary to Lemma 3, we have the ψ_γ for $\gamma < \alpha$. However $\varphi_\gamma{}^\delta\varphi_\beta{}^\gamma = \varphi_\beta{}^\delta$ need not hold. We are going to prove that the $\varphi_\gamma{}^\delta$ can be replaced by $\psi_\gamma{}^\delta$ in such a way that we still have $a_i{}^\delta\psi_\gamma{}^\delta = a_i{}^\gamma$ and also

$$\psi_\gamma{}^\delta\psi_\beta{}^\gamma = \psi_\beta{}^\delta \quad \text{for} \quad \beta \leqq \gamma \leqq \delta < \alpha.$$

Let us assume that A_γ and A_δ are disjoint if $\gamma \neq \delta$ and let us form

$$C = \bigcup (A_\gamma \mid \gamma < \alpha).$$

We will think of the required family of ψ_γ^δ as a single binary relation $R(x, y)$ on C, where $x\psi_\gamma^\delta = y$ means $x \in A_\delta$, $y \in A_\gamma$ and $R(x, y)$. Using the same tricks as in the second part of the proof of Theorem 52.2, we can introduce unary relations $R_\gamma(x)$ for $x \in A_\gamma$ and we can introduce sufficiently many relations and constants such that a set of universal sentences Ω will express that $R(x, y) \wedge R_\delta(x) \wedge R_\gamma(y)$ defines a Σ-homomorphism ψ_γ^δ of A_δ onto A_γ with $a_i^\delta\psi_\gamma^\delta = a_i^\gamma$, $0 \leqq i < n$. Let Ω^* be Ω along with the sentence

$$(\mathbf{x})(\mathbf{y})(\mathbf{z})((\mathbf{R}(\mathbf{x}, \mathbf{y}) \wedge \mathbf{R}(\mathbf{y}, \mathbf{z})) \rightarrow \mathbf{R}(\mathbf{x}, \mathbf{z})).$$

Let us observe that on every finite subset of C we can define R so as to satisfy Ω^*. Indeed, if H is finite, $H \subseteq C$, then there exist $\gamma_0 < \gamma_1 < \cdots < \gamma_{k-1} < \alpha$ such that $H \subseteq \bigcup (A_{\gamma_i} \mid 0 \leqq i < k)$. Now set

$$\psi_{\gamma_i}^{\gamma_l} = \varphi_{\gamma_{l-1}}^{\gamma_l} \varphi_{\gamma_{l-2}}^{\gamma_{l-1}} \cdots \varphi_{\gamma_i}^{\gamma_{i+1}}$$

for $i < l$ and let $R(x, y)$ mean that $x \in A_{\gamma_l}$, $y \in A_{\gamma_i}$ and $x\psi_{\gamma_i}^{\gamma_l} = y$ for some $0 \leqq i < l < k$. Then R obviously satisfies Ω^*. Thus by Theorem 39.5, R can be defined on C so as to satisfy Ω^*, completing the proof of Theorem 3.

Definition 2. *Let $\mathfrak{A}$ be a Σ-structure, Σ-generated by $a_0, \cdots, a_\gamma, \cdots, \gamma < \alpha$. Then $\mathfrak{A}$ is called a* maximally free Σ-structure, *in notation,* $\mathfrak{MF}_\Sigma(\alpha)$, *with respect to the Σ-generating set $\{a_\gamma \mid \gamma < \alpha\}$, if whenever $\mathfrak{B}$ is a Σ-structure Σ-generated by $b_0, \cdots, b_\gamma, \cdots, \gamma < \alpha$ and φ is a Σ-homomorphism of $\mathfrak{B}$ into $\mathfrak{A}$ with $b_\gamma\varphi = a_\gamma$, for $\gamma < \alpha$, then φ is an isomorphism.*

Definition 3. *Let K be a set of maximally free Σ-structures on α Σ-generators. K is called a* (Σ, α)-covering family *if for any Σ-structure $\mathfrak{B}$, Σ-generated by $b_0, \cdots, b_\gamma, \cdots, \gamma < \alpha$, there exists an $\mathfrak{A} \in K$ (with the Σ-generating family $a_0, \cdots, a_\gamma, \cdots, \gamma < \alpha$) and a Σ-homomorphism φ of $\mathfrak{A}$ onto $\mathfrak{B}$ with $a_\gamma\varphi = b\gamma$, for $\gamma < \alpha$.*

Theorem 4. *Let us assume* (P) *and* (B_n). *Then there exists a* (Σ, n)*-covering family.*

Proof. Let $\mathfrak{A}$ be a Σ-structure, $A = [h_0, \cdots, h_{n-1}]_\Sigma$. Consider the class of all pairs $\langle \mathfrak{A}_1, H_1 \rangle$, where $\mathfrak{A}_1$ is a Σ-structure, $H_1 = \langle h_0^1, \cdots, h_{n-1}^1 \rangle$, $A_1 = [h_0^1, \cdots, h_{n-1}^1]_\Sigma$ with the property that there exists a Σ-homomorphism φ of $\mathfrak{A}_1$ into $\mathfrak{A}$ with $h_0^1\varphi = h_0, \cdots, h_{n-1}^1\varphi = h_{n-1}$. Let us say that $\langle \mathfrak{A}_1, H_1 \rangle$ is isomorphic to $\langle \mathfrak{A}_2, H_2 \rangle$ if there exists an isomorphism φ of $\mathfrak{A}_1$ with $\mathfrak{A}_2$ satisfying $h_i^1\varphi = h_i^2$, for $i = 0, \cdots, n-1$, where $H_1 = \langle h_0^1, \cdots, h_{n-1}^1 \rangle$ and $H_2 = \langle h_0^2, \cdots, h_{n-1}^1 \rangle$. Let P be a class of such pairs, such that every pair has an isomorphic copy in P and there are no two isomorphic pairs in P. Using (B_n) it is easy to give an upper bound for the cardinality of P, so

P is a set. We introduce a binary relation $\leqq$ on P: $\langle \mathfrak{A}_1, H_1\rangle \leqq \langle \mathfrak{A}_2, H_2\rangle$ if there exists a Σ-homomorphism φ of $\mathfrak{A}_2$ into $\mathfrak{A}_1$ such that $h_i{}^2\varphi = h_i{}^1$ for $i=0, \cdots, n-1$. Then $\mathfrak{P} = \langle P; \leqq\rangle$ is a partially ordered set. The only nontrivial part in checking this is to prove that $\leqq$ is antisymmetric; this is an easy modification of the argument of Theorem 51.2′ (the freeness of the structures involved was used there only to prove (B_n); now we have (B_n) by assumption). Theorem 3 states that Zorn's Lemma can be applied to $\mathfrak{P}$. Any maximal element of $\mathfrak{P}$ will be maximally free (again use (B_n) and the argument of Theorem 51.2′).

It follows from (B_n) that a maximal class of nonisomorphic pairs of the $\langle \mathfrak{A}, H\rangle$ is a set. Using the above construction, we choose for each $\langle \mathfrak{A}, H\rangle$ an $\mathfrak{M}\mathfrak{F}_\Sigma(n)$ containing $\langle \mathfrak{A}, H\rangle$ in $\mathfrak{P}$ and thus we get a (Σ, n)-covering family.

Necessary and sufficient conditions for the existence of $\mathfrak{F}_\Sigma(n)$ for a universal Σ will be given in the exercises.

§54. STRONG FREE Σ-STRUCTURES AND THE INVERSE PRESERVING PROPERTY

If K is the class of all groups $\langle G; \cdot, 1\rangle$ defined in the usual way by Σ, then all free Σ-structures exist and they are the free groups in the usual sense. However, no one would use the theory of free Σ-structures to prove the existence of free groups. The most convenient way of proving the existence of free groups is the introduction of x^{-1} as an operation, because then in Σ the existential quantifiers are eliminated, and in this richer language, Σ is equivalent to a universal $\bar{\Sigma}$, to which the simple methods of Chapter 4 apply. In this section we will discuss the problem of when it is possible to eliminate the existential quantifiers in some Σ such that the resulting $\bar{\Sigma}$ can be used to construct free Σ-structures.

First, we introduce a property of a set Σ of sentences.

Definition 1. *Σ is said to have the* Inverse Preserving Property (IP) *if every Σ-substructure is slender.*

Theorem 1. *The following conditions on Σ are equivalent:*

(i) *Σ has* IP;

(ii) *if $\mathfrak{A}$, $\mathfrak{B}$, $\mathfrak{C}$ are Σ-structures, $\mathfrak{B}$ is a Σ-substructure of $\mathfrak{A}$ and φ is a Σ-homomorphism of $\mathfrak{A}$ into $\mathfrak{C}$, then φ_B is a Σ-homomorphism of $\mathfrak{B}$ into $\mathfrak{C}$;*

(iii) *if $\mathfrak{A}$, $\mathfrak{B}$, $\mathfrak{C}$ are Σ-structures, $\mathfrak{B}$ is a Σ-substructure of $\mathfrak{A}$ and φ is a Σ-homomorphism of $\mathfrak{C}$ into $\mathfrak{B}$, then φ is a Σ-homomorphism of $\mathfrak{C}$ into $\mathfrak{A}$;*

(iv) *if* $\mathfrak{B}$ *is a* Σ*-substructure of* $\mathfrak{A}$, *then* $\varphi: x \to x$ *is a* Σ*-homomorphism of* $\mathfrak{B}$ *into* $\mathfrak{A}$.

Corollary. *If* Σ *has* IP, *then every* Σ*-homomorphism* φ *can be written in the form* $\varphi = \psi\chi$, *where* ψ *is an onto* Σ*-homomorphism and* χ *is a* 1–1 Σ*-homomorphism.*

The proofs are trivial consequences of Lemmas 50.5 and 50.6.

We will also need a property of free Σ-structures.

Definition 2. *A free* Σ*-structure is* strong *if the* $\bar{\varphi}$ *of Definition* 51.1 *is always unique.*

That is, any mapping of the Σ-generators into a Σ-structure can be *uniquely* extended to a Σ-homomorphism.

Corollary. *Let* $\mathfrak{A}$ *be a free* Σ*-structure on* α Σ*-generators and let* $\mathfrak{B}$ *be a free* Σ*-structure on* β Σ*-generators. If* $\bar{\alpha} = \beta$, *then* $\mathfrak{A}$ *is strong if and only if* $\mathfrak{B}$ *is strong.*

This is trivial from the Uniqueness Theorem.

Theorem 2. *If the free* Σ*-structure* $\mathfrak{F}_\Sigma(\omega)$ *exists and is strong, then all free* Σ*-structures exist and all are strong.*

Proof. The existence of free Σ-structures follows from Theorem 52.4. It is obvious that if $\alpha = \lim \beta_i$ and each $\mathfrak{F}_\Sigma(\beta_i)$ is strong, then so is $\mathfrak{F}_\Sigma(\alpha)$. It remains to prove that if $\mathfrak{F}_\Sigma(\alpha)$ is strong and $\beta < \alpha$, then $\mathfrak{F}_\Sigma(\beta)$ is strong. Let $x_0, \cdots, x_\gamma, \cdots, \gamma < \alpha$ be the Σ-generators of $\mathfrak{F}_\Sigma(\alpha)$. By Theorem 52.1 and the corollary to Definition 2, we can assume that

$$F_\Sigma(\beta) = [x_0, \cdots, x_\gamma, \cdots]_\Sigma, \ \gamma < \beta.$$

Let χ be a Σ-homomorphism of $\mathfrak{F}_\Sigma(\alpha)$ onto $\mathfrak{F}_\Sigma(\beta)$ with $x_\gamma\chi = x_\gamma$ for $\gamma < \beta$ and $x_\gamma\chi = x_0$ for $\beta \leqq \gamma$. If $\mathfrak{F}_\Sigma(\beta)$ is not strong then there exists a Σ-structure $\mathfrak{B}$ and there exist $b_0, \cdots, b_\gamma, \cdots, \gamma < \beta$ elements of B such that $x_\gamma \to b_\gamma$ ($\gamma < \beta$) has two extensions to Σ-homomorphisms φ and ψ. Then the mapping $x_\gamma \to b_\gamma$ for $\gamma < \beta$ and $x_\gamma \to b_0$ for $\gamma \geqq \beta$ has two extensions to Σ-homomorphisms, namely $\chi\varphi$ and $\chi\psi$, contradicting the assumption that $\mathfrak{F}_\Sigma(\alpha)$ is strong.

Now we are ready to state and to prove the main result.

Theorem 3. *Let us assume that Σ has* IP *and that $\mathfrak{F}_\Sigma(\omega)$ exists and is strong. Then on every Σ-structure $\mathfrak{A}=\langle A; F, R\rangle$ we can define additional operations $f \in \bar{F}-F$, such that the correspondence*

$$\mathfrak{A} = \langle A; F, R\rangle \to \bar{\mathfrak{A}} = \langle A; \bar{F}, R\rangle$$

has the following properties:

(i) *$\mathfrak{A}$ is a Σ-substructure of $\mathfrak{B}$ if and only if $\bar{\mathfrak{A}}$ is a substructure of $\bar{\mathfrak{B}}$;*

(ii) *let ψ map A into B; then ψ is a Σ-homomorphism of $\mathfrak{A}$ into $\mathfrak{B}$ if and only if ψ is a homomorphism of $\bar{\mathfrak{A}}$ into $\bar{\mathfrak{B}}$;*

(iii) *let K denote the class of all $\bar{\mathfrak{A}}$; then $\mathfrak{F}_K(\alpha)$ exists for all α.*

Proof. For every $1 \leqq n < \omega$ and $\mathbf{P} \in \mathbf{P}_n(\Sigma)$ we introduce $k_\mathbf{P}$ n-ary operations, $f_0^\mathbf{P}, \cdots, f_{k_\mathbf{P}-1}^\mathbf{P}$ as follows:

Take $\mathfrak{F}_\Sigma(n)$ with the Σ-generators $x_0, \cdots, x_{n-1}$; define $f_i^\mathbf{P}(x_0, \cdots, x_{n-1})$, $i<k_\mathbf{P}$, such that $P(x_0, \cdots, x_{n-1})=\{f_i^\mathbf{P}(x_0, \cdots, x_{n-1}) \mid i<k_\mathbf{P}\}$; let $\mathfrak{A}$ be an arbitrary Σ-structure, $a_0, \cdots, a_{n-1} \in A$ and φ a Σ-homomorphism of $\mathfrak{F}_\Sigma(n)$ into $\mathfrak{A}$ with $x_0\varphi=a_0, \cdots, x_{n-1}\varphi=a_{n-1}$. Set

$$f_i^\mathbf{P}(a_0, \cdots, a_{n-1}) = f_i^\mathbf{P}(x_0, \cdots, x_{n-1})\varphi \quad \text{for} \quad i < k_\mathbf{P}.$$

Set

$$\bar{F} = F \cup \bigcup (\{f_i^\mathbf{P} \mid \mathbf{P} \in \mathbf{P}_n(\Sigma),\, i < k_\mathbf{P}\} \mid 1 \leqq n < \omega);$$

and $K=\{\langle A, \bar{F}, R\rangle \mid \langle A; F, R\rangle$ is a Σ-structure$\}$.

It is obvious that $\langle A; \bar{F}, R\rangle$ is well defined, since, by Theorem 2, φ is unique.

Now we will verify (i)–(iii).

Ad (i). Let $\mathfrak{A}$ be a Σ-substructure of $\mathfrak{B}$, $a_0, \cdots, a_{n-1} \in A$ and let $f \in \bar{F}$ be an n-ary operation. If $f \in F$, then $f(a_0, \cdots, a_{n-1}) \in A$. If $f \notin F$, then $f=f_i^\mathbf{P}$ for some $\mathbf{P} \in \mathbf{P}_n(\Sigma)$ and $i<k_\mathbf{P}$. Then

$$f_i^\mathbf{P}(x_0, \cdots, x_{n-1}) \in P(x_0, \cdots, x_{n-1})$$

in $\mathfrak{F}_\Sigma(n)$, so

$$\begin{aligned} f_i^\mathbf{P}(a_0, \cdots, a_{n-1}) &= f_i^\mathbf{P}(x_0, \cdots, x_{n-1})\varphi \in P(x_0, \cdots, x_{n-1})\varphi \\ &= P(a_0, \cdots, a_{n-1}) \subseteq A, \end{aligned}$$

since φ is an Σ-homomorphism. Thus $\bar{\mathfrak{A}}$ is a substructure of $\bar{\mathfrak{B}}$.

Let $\bar{\mathfrak{A}}$ be a substructure of $\bar{\mathfrak{B}}$; then $\mathfrak{A}$ is a substructure of $\mathfrak{B}$. To prove that it is a Σ-substructure, let $a_0, \cdots, a_{n-1} \in A$ and $\mathbf{P} \in \mathbf{P}_n(\Sigma)$. If $b \in P_\mathfrak{B}(a_0, \cdots, a_{n-1})$, then $b=f(a_0, \cdots, a_{n-1})$ for some $f \in \bar{F}$. Thus $b \in A$.

Ad (ii). Let ψ be a Σ-homomorphism of $\mathfrak{A}$ into $\mathfrak{B}$, $a_0, \cdots, a_{n-1} \in A$ and $f \in \bar{F}$. We want to prove that

$$f(a_0, \cdots, a_{n-1})\psi = f(a_0\psi, \cdots, a_{n-1}\psi).$$

This is obvious if $f \in F$. Let $f \notin F$, that is, $f = f_i^{\mathbf{P}}$ for some $\mathbf{P} \in \mathbf{P}_n(\Sigma)$ and $i < k_{\mathbf{P}}$.

Let φ and χ be the Σ-homomorphisms of $\mathfrak{F}_\Sigma(n)$ into $\mathfrak{A}$ and $\mathfrak{B}$, respectively, with $x_i\varphi = a_i$ and $x_i\chi = a_i\psi$, for $i < n$. Since χ is unique (Theorem 2) we get $\chi = \varphi\psi$. Thus

$$\begin{aligned} f(a_0, \cdots, a_{n-1})\psi &= f(x_0, \cdots, x_{n-1})\varphi\psi = f(x_0, \cdots, x_{n-1})\chi \\ &= f(a_0\psi, \cdots, a_{n-1}\psi), \end{aligned}$$

which was to be proved.

Let ψ be a homomorphism of $\bar{\mathfrak{A}}$ into $\bar{\mathfrak{B}}$; then ψ is a homomorphism of $\mathfrak{A}$ into $\mathfrak{B}$. To prove that ψ is a Σ-homomorphism, take $\mathbf{P} \in \mathbf{P}_n(\Sigma)$ and $a_0, \cdots, a_{n-1} \in A$. Let us define φ and χ as above. Let $b \in P(a_0, \cdots, a_{n-1})$. Since φ is a Σ-homomorphism, there exists a $u \in P(x_0, \cdots, x_{n-1})$ with $u\varphi = b$. Then $u = f_i^{\mathbf{P}}(x_0, \cdots, x_{n-1})$ for some $i < k_{\mathbf{P}}$. By the definition of $f_i^{\mathbf{P}}$ we have that $b = f_i^{\mathbf{P}}(a_0, \cdots, a_{n-1})$. Since ψ is a homomorphism, we get that $b\psi = f_i^{\mathbf{P}}(a_0\psi, \cdots, a_{n-1}\psi)$. Again, by the definition of $f_i^{\mathbf{P}}$, there exists a $v \in F_\Sigma(n)$ with $v = f_i^{\mathbf{P}}(x_0, \cdots, x_{n-1})$ and $v\chi = b\psi$. Since

$$v \in P(x_0, \cdots, x_{n-1})$$

and χ is a Σ-homomorphism, we get that

$$b \in P(a_0, \cdots, a_{n-1})$$

implies that

$$b\psi \in P(a_0\psi, \cdots, a_{n-1}\psi).$$

The converse of this statement can be proved similarly; thus

$$P(a_0, \cdots, a_{n-1})\psi = P(a_0\psi, \cdots, a_{n-1}\psi),$$

which was to be proved.

Ad (iii). It follows from the assumption that $\mathfrak{F}_\Sigma(\alpha)$ exists for all α. (i) and (ii) imply that $\bar{\mathfrak{F}}_\Sigma(\alpha)$ is $\mathfrak{F}_K(\alpha)$.

This completes the proof of Theorem 3.

Theorem 3 is the best possible result, since the following holds:

Theorem 4. *Let us assume that the conclusions of Theorem 3 hold for* Σ. *Then* Σ *has* IP, *and* $\mathfrak{F}_\Sigma(\omega)$ *exists and is strong.*

Proof. $\mathfrak{F}_K(\omega)$ exists by (iii) so, by (i) and (ii), $\mathfrak{F}_\Sigma(\omega)$ exists. Since a free algebra over K is always strong, $\mathfrak{F}_\Sigma(\omega)$ is also strong by (ii). Using (i) and (ii), condition (iv) of Theorem 1 can easily be verified; thus by Theorem 1, Σ has IP.

EXERCISES

1. Give an example to show that $\Psi(a, b, c, d)$ does not imply that d is an $(\mathbf{x})(\exists \mathbf{y})(\mathbf{u})(\exists \mathbf{v})\Psi(\mathbf{x}, \mathbf{y}, \mathbf{u}, \mathbf{v})$-inverse of a and c.
2. Let $\Phi \equiv (\mathbf{x})(\exists \mathbf{y})(\mathbf{u})(\exists \mathbf{v})\Psi(\mathbf{x}, \mathbf{y}, \mathbf{u}, \mathbf{v})$. Show that b is a Φ-inverse of a, and d is a Φ-inverse of a, c do not imply $\Psi(a, b, c, d)$.
3. Give a formal proof of Lemma 49.1.
4. Find a Σ-structure $\mathfrak{A}$, a Σ-substructure $\mathfrak{B}$ of $\mathfrak{A}$, and $a_0, \cdots, a_{n-1}, b \in B$ such that b is an inverse of $a_0, \cdots, a_{n-1}$ in $\mathfrak{B}$ but not in $\mathfrak{A}$.
5. Find Σ-structures $\mathfrak{A}$, $\mathfrak{B}$, $\mathfrak{C}$ such that $\mathfrak{B}$ and $\mathfrak{C}$ are Σ-substructures of $\mathfrak{A}$, $C \subseteq B$, but $\mathfrak{C}$ is not a Σ-substructure of $\mathfrak{B}$.
6. Prove Lemmas 49.6 and 49.7.
7. Find Σ-structures $\mathfrak{A}$, $\mathfrak{B}$, a Σ-substructure $\mathfrak{C}$ of $\mathfrak{B}$ and a Σ-homomorphism φ of $\mathfrak{A}$ into $\mathfrak{B}$, such that $A\varphi = C$, but φ is not a Σ-homomorphism of $\mathfrak{A}$ into $\mathfrak{C}$.
8. If $\mathfrak{A}$ and $\mathfrak{B}$ are Σ-algebras, then in Definition 50.1 one can omit "if φ is a homomorphism". Why can this not be omitted in general?
9. Show that conditions (i)–(iii) of Lemma 50.6 do not characterize slender Σ-substructures.
10. Describe $K(\tau)$, the class of all algebras of type τ, as relational systems: the class of all models of some Σ such that Σ-homomorphism and Σ-substructure be equivalent to homomorphism and subalgebra, respectively, and every Σ-substructure be slender.
11. Prove Theorem 51.1 using prime products.
12. Let $\Phi(\mathbf{x})$ be a formula free at most in $\mathbf{x}$ and let Σ be a set of sentences. Φ is *bounded in* Σ if for some positive integer n the following condition holds: if $\mathfrak{A}$ is a Σ-structure, $a_0, \cdots, a_{n-1} \in A$, $a_i \neq a_j$ if $i \neq j$ and $\Phi(a_i)$ for $0 \leqq i < n$, then $\Phi(a)$, $a \in A$ implies that $a = a_i$ for some $0 \leqq i < n$. Prove the following form of the statement in the proof of Theorem 51.1:

   If $\Phi(\mathbf{x})$ is not bounded in Σ, then for every infinite cardinal $\mathfrak{m}$ there exists a Σ-structure $\mathfrak{A}$ such that $|\{a \mid a \in A \text{ and } \Phi(a)\}| \geqq \mathfrak{m}$.
13. (A. Robinson [2]) $\Phi(\mathbf{x})$ is bounded in Σ if and only if for every ascending chain of Σ-structures $\mathfrak{A}_0, \mathfrak{A}_1, \cdots$ (where $\mathfrak{A}_i$ is a substructure of $\mathfrak{A}_{i+1}$) and every sequence $a_0, a_1, \cdots$ such that $a_i \in A_i$ and $\Phi(a_i)$ holds in $\mathfrak{A}_j$ for all $i \leqq j$, there exists a natural number n, such that

$$\{a_0, \cdots, a_n\} = \{a_0, \cdots, a_n, \cdots\}.$$

14. Formulate Theorem 51.1 using the concept of Ex.12.
15. There are many statements in Chapter 8 which are proved by referring to the proof of Theorem 51.2′ rather than to the statement of Theorem 51.2′. Formulate and prove a more general version of Theorem 51.2′ which is applicable in all these situations.
16. Can Theorem 52.2 be proved using prime products to construct $\mathfrak{F}_\Sigma(\omega)$?
17. Show that the existence of a $(\Sigma\text{-}n)$-covering family implies condition (B_n).

* * *

In Ex. 18-26 (which are based on G. Grätzer [11]) Σ is a set of universal sentences given in reduced normal form as in §46 (see Definitions 46.1, 46.2, and Lemmas 46.1 and 46.2). We assume that every $\Phi \in \Sigma$ is of the form $\Theta_0 \vee \cdots \vee \Theta_{m-1}$, where the Θ_i are negations of atomic formulas for $i < s(\Phi)$ and atomic formulas for $i \geqq s(\Phi)$.

Let $\Phi \in \Sigma$, $\Phi \equiv \Theta_0 \vee \cdots \vee \Theta_{m-1}$, where Φ is free at most in $\mathbf{x}_0, \cdots, \mathbf{x}_{k-1}$. Φ has property (P_n) means: if for the n-ary polynomials $\mathbf{p}_0, \cdots, \mathbf{p}_{k-1}$ we have $\Sigma \models (\mathbf{x}_0) \cdots (\mathbf{x}_{n-1})\Theta_i(\mathbf{p}_0, \cdots, \mathbf{p}_{k-1})$ for $i < s(\Phi)$, then there exists an i with $s(\Phi) \leqq i < m$ such that $\Sigma \models (\mathbf{x}_0) \cdots (\mathbf{x}_{n-1})\Theta_i(\mathbf{p}_0, \cdots, \mathbf{p}_{k-1})$. Φ has (P_ω) if it has (P_n) for all $n < \omega$. Σ has (P_α) $(\alpha \leqq \omega)$ if each $\Phi \in \Sigma$ has (P_α).

18. If for $\Phi \in \Sigma$ we have $m = 1$, then Φ has (P_ω).
19. If for $\Phi \in \Sigma$, $s(\Phi) \geqq m - 1$, then Φ has (P_ω).
20. If for $\Phi \in \Sigma$, $s(\Phi) = 0$ and Φ is free at most in $\mathbf{x}_0, \cdots, \mathbf{x}_{n-1}$, then Φ does not have (P_n).
21. Set $\Sigma^* = K$, $n < \omega$. $\mathfrak{F}_K(n)$ exists if and only if Σ has (P_n).
22. $\mathfrak{F}_K(\omega)$ exists if and only if Σ has (P_ω).
23. All free algebras over K exist if and only if Σ has (P_ω).
24. For a given $n < \omega$, find a Σ which has (P_k) if and only if $k \leqq n$.
25. (W. Peremans [2].) If every $\Phi \in \Sigma$ contains at most one atomic formula, then all free algebras over Σ^* exist.
26. Show that the sufficient condition of Ex. 26 is not necessary.

* * *

27. Prove Theorem 52.4 directly for strong free Σ-structures.
28. Let Σ satisfy (B) (of §53). A Σ-*chain* $\mathfrak{A}_\gamma$, $\gamma < \alpha$ is a chain of Σ-structures such that if $\gamma < \delta < \alpha$, then $\mathfrak{A}_\gamma$ is a Σ-substructure of $\mathfrak{A}_\delta$. Prove that the union of a Σ-chain is a Σ-structure.
29. We say that Σ is *of bound* 1 if for every Σ-polynomial $\mathbf{P}$, Σ-structure $\mathfrak{A}$, and $a_0, \cdots, a_{n-1} \in A$, $|P(a_0, \cdots, a_{n-1})| \leqq 1$. If Σ is of bound 1, then Σ has IP.
30. If Σ is of bound 1, and $\mathfrak{F}_\Sigma(\alpha)$ exists, then $\mathfrak{F}_\Sigma(\alpha)$ is strong.
31. If Σ is of bound 1, then there exists a set of universal sentences Σ^+ over a richer language such that Σ^* and $(\Sigma^+)^*$ are equivalent in the sense of Theorem 54.3. In particular, $\mathfrak{F}_\Sigma(\alpha)$ exists if and only if $\mathfrak{F}_{(\Sigma^+)^*}(\alpha)$ exists.
32. For an axiomatic class K, let $A(K)$ denote the set of all sets of sentences Σ with $\Sigma^* = K$. For $\Sigma, \Sigma' \in A(K)$ write $\Sigma \leqq \Sigma'$ if for every n-ary Σ-polynomial $\mathbf{P}$, there exists an n-ary Σ'-polynomial $\mathbf{P}'$ such that $P(a_0, \cdots, a_{n-1}) = P'(a_0, \cdots, a_{n-1})$ for any $\mathfrak{A} \in K$, $a_0, \cdots, a_{n-1} \in A$. Set

$$\hat{\Sigma} = \{\Sigma_1 \mid \Sigma \leqq \Sigma_1 \text{ and } \Sigma_1 \leqq \Sigma\} \text{ and } \hat{A}(K) = \{\hat{\Sigma} \mid \Sigma \in A(K)\}.$$

Define $\hat{\Sigma} \leqq \hat{\Sigma}_1$ if and only if $\Sigma \leqq \Sigma_1$. Prove that $\langle \hat{A}(K); \leqq \rangle$ is a partially ordered set.
33. For any $\varnothing \neq H \subseteq \hat{A}(K)$, l.u.b.$(H)$ exists.
34. If K is a universal class, then $\langle \hat{A}(K); \leqq \rangle$ is a complete lattice.
35. Let K be the class of all groups $\langle G; \cdot \rangle$. Is $\langle \hat{A}(K); \leqq \rangle$ a complete lattice? (No. C. R. Platt.)
36. Prove Theorem 52.2 using Theorem 21.1.

PROBLEMS

80. Does Theorem 51.1 (or the statement in the proof of Theorem 51.1) imply the compactness theorem?
81. For what axiomatic classes K does $\langle \hat{A}(K); \leqq \rangle$ (see Ex. 32) have a least element?
82. What is the structure of $\langle \hat{A}(K); \leqq \rangle$?
83. Give necessary and sufficient conditions for the existence of a (Σ, α)-covering family.
84. Characterize the automorphism groups of free Σ-structures.
85. Define a *Σ-inverse (direct) family* as an inverse (direct) family of Σ-structures such that the homomorphisms are Σ-homomorphisms. When is the inverse (direct) limit of a Σ-inverse (direct) family a Σ-structure?
86. Under what conditions is the class $\{\bar{\mathfrak{A}} \mid \mathfrak{A} \in \Sigma^*\}$ of Theorem 54.3 an axiomatic (universal) class?
87. Give a natural construction of a structure $\mathfrak{P}_\Sigma^{(\alpha)}$ such that whenever $\mathfrak{F}_\Sigma(\alpha)$ exists, $\mathfrak{P}_\Sigma^{(\alpha)} \cong \mathfrak{F}_\Sigma(\alpha)$. (If Σ is universal, the polynomial algebra with a natural definition of relations is always good.)
88. Let K be a universal class. Under what condition does K have a finite covering family? Generalize the result to the class of Σ-algebras, where Σ has property (P) of §53.
89. Characterize the abstract category, and also the concrete category of Σ-algebras. What restrictions are imposed by IP or (B)?
90. Develop a theory of homogeneous and universal Σ-structures (compare to B. Jónsson [3], M. Morley and R. Vaught [1]).
91.† In the theory of Σ-structures the set Σ plays three roles: it defines the class to be considered (Σ-structures), the maps that are homomorphisms (Σ-homomorphisms), and the substructures (Σ-substructures). Could Σ be replaced by $\langle \Sigma_0, \Sigma_1, \Sigma_2 \rangle$ one Σ_i for each role of Σ, such that some of the theory of Σ-structures generalizes to the $\langle \Sigma_0, \Sigma_1, \Sigma_2 \rangle$-theory?
92. Can one find for all pairs of ordinals $\alpha < \beta$ a Σ-homomorphism $\varphi_{\alpha\beta}$ of $\mathfrak{F}_\Sigma(\alpha)$ into $\mathfrak{F}_\Sigma(\beta)$ such that $x_i^\alpha \varphi_{\alpha\beta} = x_i^\beta$, and for all $\alpha < \beta < \gamma$, $\varphi_{\alpha\beta}\varphi_{\beta\gamma} = \varphi_{\alpha\gamma}$?
93. Find a concept of "weak-adjointness" of functors such that the "free Σ-structure" functor be weak-adjoint to the underlying set functor and such that weak adjoints be unique up to (nonunique!) natural equivalence.
94. Define free Σ-product of Σ-structures as in Definition 29.2 by requiring that the $\mathfrak{A}_i$ are Σ-structures, the ψ_i are Σ-homomorphisms, $\mathfrak{A}$ is Σ-generated by $\bigcup (A_i\psi_i \mid i \in I)$ and $\mathfrak{B}$ is a Σ-structure, the φ_i and φ Σ-homomorphisms. Under what conditions are free Σ-products unique up to isomorphism?
95. Give a sufficient condition for the existence of free Σ-products. (See e.g. Corollary 1 to Theorem 29.2.)
96. Is $\mathfrak{F}_\Sigma(\mathfrak{m})$ the free Σ-product of $\mathfrak{m}$ copies of $\mathfrak{F}_\Sigma(1)$?

† Problems 91–93 were suggested by F. W. Lawvere.

97. Define the Σ-structure Σ-freely generated by a partial structure, and prove a result analogous to Theorem 29.2.
98. Determine the group of those automorphisms of $\mathfrak{F}_\Sigma(\mathfrak{m})$ that permute a fixed set of Σ-generators.
99. When is $\langle \hat{A}(K); \leqq \rangle$ finite? When is it distributive? (A semilattice is *distributive* if $t \leqq a \vee b$ implies $t = a' \vee b'$ for some $a' \leqq a$ and $b' \leqq b$.)
100. Define $T_n \subseteq \hat{A}(K)$: $\hat{\Sigma} \in T_n$ if $\mathfrak{F}_\Sigma(n)$ exists. Then $T_1 \supseteq T_2 \supseteq \cdots \supseteq T_n \supseteq \cdots$. When is $T_n \neq T_{n+1}$ for all $n < \omega$? What is T_n in $\hat{A}(K)$?
101. Assume that $\mathfrak{F}_\Sigma(\omega)$ exists and for some $n < \omega$ $|F_\Sigma(\omega)| = |F_\Sigma(n)| = \aleph_0$. Is it true that every countable Σ-structure can be embedded in some $\mathfrak{F}_\Sigma(m)$, $m < \omega$?
102. Is it possible to give a direct proof of Theorem 51.2 (i.e. a proof that avoids Theorem 51.1 and preferably avoids the Compactness Theorem in any form)?

APPENDIX 1

GENERAL SURVEY

The present appendix attempts to survey recent developments not covered in the other appendices. In §55, many sections of the book are updated. Some corrections are also mentioned. §56 surveys related structures, one of the most interesting fields of universal algebra. §57 surveys some more topics of recent interest.

§55. A SURVEY BY SECTIONS

Polynomial algebras (§8) or absolutely free algebras are considered in a number of papers by J. Schmidt. For characterizations of polynomial algebras see also K. H. Diener [2] and [1969] and K. H. Diener and G. Grätzer [1]. Most of these papers also treat the infinitary case. See also J. Mycielski and W. Taylor [1976].

Congruence lattices (§10, §17, §18). See §56. Note also L. A. Skornjakov [1971] which deals with algebras whose congruence lattices are complemented.

Basic notions of partial algebras (§13). See the discussion of Problems 13, 14, 18, 24, and 25 in Appendix 2, the papers of J. Schmidt and J. Słomiński, and H. Andréka and I. Németi [a], [b]. Weak and strong congruences are discussed in H. Höft [1973], V. T. Kulik [1971], B. Wojdylo [1973]. A special type of partial algebra is introduced in G. Birkhoff and J. D. Lipson [1970] and [1974].

Free partial algebras (§14) are discussed in P. Burmeister and J. Schmidt [1967], P. Burmeister [1970], H. Andréka and I. Németi [a], [b].

Extension of congruence relations (§15). A variant of the main result of this section is treated simply in I. Fleischer [1975].

Direct products (§19). The investigations of C. C. Chang, B. Jónsson, and A. Tarski concerning the refinement properties of direct products of finite algebras were continued in C. C. Chang [1967]. This topic was taken up in R. McKenzie [1968] where the concept of "prime" structure is introduced (see also R. J. Seifert [1971]) and R. McKenzie [1971] and [1972] where a method is worked out to obtain refinement theorems; these results are applied to groups, idempotent semigroups, and so on.

L. Lovász [1967] proved two elegant cancellation theorems: (1) If $\mathfrak{A}$, $\mathfrak{B}$, $\mathfrak{C}$ are finite algebras, $\mathfrak{C}$ contains a one-element subalgebra, and $\mathfrak{A} \times \mathfrak{C} \cong \mathfrak{B} \times \mathfrak{C}$, then $\mathfrak{A} \cong \mathfrak{B}$. (2) If $\mathfrak{A}$ and $\mathfrak{B}$ are finite algebras and $\mathfrak{A}^n \cong \mathfrak{B}^n$, then $\mathfrak{A} \cong \mathfrak{B}$. If we drop the assumption that $\mathfrak{C}$ contains a one-element subalgebra one can still obtain "isotopy", see H. P. Gumm [b]. For some related results, see L. Lovász [1971], [1972], R. R. Appleson [1976], and R. R. Appleson and L. Lovász [1975]. See also A. F. Bravcev [1967].

The second result of Lovász suggests the investigation of the semigroup $S(\mathfrak{A})$, generated by $\{a\}$, subject to the defining relation $a^n = a^m$ whenever $\mathfrak{A}^n \cong \mathfrak{A}^m$, where $\mathfrak{A}$ is an infinite algebra. Extending this to a set of algebras $A = \{\mathfrak{A}_i \mid i \in I\}$ we obtain the semigroup $S(A)$ generated by $\{a_i \mid i \in I\}$ subject to the relations ${a_i}^{n_i} {a_j}^{n_j} \cdots = {a_k}^{n_k} {a_m}^{n_m} \cdots$ whenever ${\mathfrak{A}_i}^{n_i} \times {\mathfrak{A}_j}^{n_j} \times \cdots \cong {\mathfrak{A}_k}^{n_k} \times {\mathfrak{A}_m}^{n_m} \times \cdots$. V. Trnková [1975] obtained the astonishing converse: every commutative semigroup can be so represented. See also V. Trnková [1975 a], [1976], [a] and V. Koubek and V. Trnková [a].

Direct product of algebras of not necessarily the same type was introduced in the Marczewski seminar in Wrocław in the nineteen sixties, see also G. Grätzer [1970] and A. Goetz [1971]; this is applied in characterizations of Mal'cev-type conditions.

Subdirect products (§20). Theorem 20.4 is generalized to more than two factors in G. H. Wenzel [1]. See B. A. Davey [1973] for the connection between subdirect representations and sheaf representations; see also S. D. Comer [1971], [1974], [1974 a], W. H. Cornish [a], B. A. Davey [a], H. Draškovičová [1972], K. Keimel [1970], U. M. Swamy [1974], A. Wolf [1974]. R. Wille [1969] finds a new way of looking at subdirect products; this approach proved to be useful for finite lattices.

W. Taylor [1972] considers equational classes in which the subdirectly irreducible algebras form a set (up to isomorphism). For more detail on this important topic see item 8 of §71.

R. W. Quackenbush [1971 b] found a peculiar property of subdirectly irreducible algebras for equational classes generated by a finite algebra: if there are only finitely many finite subdirectly irreducible algebras, then there are no infinite subdirectly irreducible algebras. It is still open whether for such classes the existence of infinitely many finite subdirectly irreducible algebras implies the existence of an infinite subdirectly irreducible algebra.

Direct limits (§21). J. Płonka [1967] and [1967 b] introduced a new construction which is now known as a Płonka sum; it resembles direct limits where the underlying poset is a semilattice. This is investigated in a number of papers by J. Płonka [1967 c], [1967 d], [1968 a], [1973 b], [1974 b], H. Lakser, R. Padmanabhan, C. R. Platt [1972], A. Mitschke [1973]. A generalization is given in G. Grätzer and J. Sichler [1974]; see also E. Graczyńska [a], E. Graczyńska and A. Wroński [1975] and [1975 a].

Products associated with direct products (§22). M. Armbrust [1966], [1967] defines a concept of almost-direct product related to Theorem 22.1. V. N. Saliĭ [1969] extends the construction given in Definition 22.7. A surprising number of papers deal with extensions of Theorems 22.3 and 22.4; see also Appendix 5.

Operators (§23). See the references to Problems 24 and 25 in Appendix 2.

Identities (§26). Variants of Birkhoff's characterization of equational classes can be found in S. L. Bloom [1976] (ordered algebras), N. R. Brumberg [1969] (poly-varieties), S. Fajtlowicz [1969], H. Andréka and I. Németi [a], [c], L. Polák [1976] (nonindexed algebras), P. Hájek [1965], R. John [1978], R. Kerkhoff [1970], B. Schepull [1976] (for partial algebras), G. Matthiessen [1976] (heterogeneous algebras), J. M. Movsisjan [1974] (algebras of degree 2), V. S. Poythress [1973] (with partial morphisms), J. Tiuryn [a] (algebras defined on certain types of trees).

Identities in partial algebras are considered in many papers; see, e.g., H. Andréka and I. Németi [a], [b], P. Burmeister [1970], [1973], P. Burmeister and J. Schmidt [1967], and H. Höft [1973 a].

An interesting concept of attainability of congruences is introduced in T. Tamura [1966]; see also T. Tamura and F. M. Yaqub [1965]. L. N. Ševrin and L. M. Martynov [1971] contains some interesting results on attainability and lists some unsolved problems.

For some applications of Mal'cev's result (Theorem 26.4), see Appendices 3 and 5. For generalizations, see Appendices 3 and 4. See also T. Fujiwara [1965].

Equational completeness (§27). Many topics mentioned in this section are discussed in Appendices 4 and 5; see also the references given for Problems 32 and 33 in §58 and §70.

Free products (§29). The paper of T. M. Baranovič [3] is continued in V. P. Matus [1970] and A. I. Pilatovskaja [1968] and in T. M. Baranovič [1966]. The common refinement property for free products is established for regular equational classes in B. Jónsson and E. Nelson [1974]. For a general existence theorem, see A. I. Čeremisin [1969].

Word problem (§30). For new developments, see the papers of T. Evans, especially [1978] and his survey article [a]. For connections among residual finiteness, finite embeddability, and the word problem, see T. Evans [1969], [1972] and W. Taylor's survey referred to as EL in Appendix 4 (and also B. Banaschewski and E. Nelson [1972 a]).

Independence (§31 and §32). For a survey article, see E. Marczewski [11]. See also papers of S. Fajtlowicz, E. Marczewski and J. Płonka in the Bibliography and the references in Appendix 2 to Problems 50, 52, 53, and 56.

Invariants (§33). See the references in §58 to Problem 50; see also the papers of J. Dudek and J. Płonka in the Bibliography.

Independent sets (§34). See the references in §58 to Problems 52 and 53.

Preservation theorems (Chapter 7). This field has become too large to be covered here. The reader is referred to C. C. Chang and H. J. Keisler [1973] and to M. Makkai [1969] for an up-to-date treatment and new developments. M. Makkai and G. E. Reyes [1977] contains a categorical approach to first order theories.

The open question mentioned at the end of §41 was answered in S. Shelah [1971]; see also C. C. Chang and H. J. Keisler [1973] and H. Andréka, B. Dahn, I. Németi [1976].

Subdirect products (§48). In his review of G. Grätzer [5] (on which §48 is based) in the Journal of Symbolic Logic, F. Galvin points out that the concept of "implication" in Definition 48.1 is too vague. If we make it more precise by defining "Ψ implies Ψ'" to mean that "Ψ' is obtained from Ψ by deleting all atomic formulae that include variables not among $\mathbf{x}_{i_0}, \cdots, \mathbf{x}_{i_{k-1}}$", then the results of the section remain true but the examples mentioned in Exercises 42–45 will no longer fall under Theorem 48.2.

Thus it remains open how to modify Definition 48.1 so as to keep Theorem 48.2 true and yield an affirmative answer to Exercises 42–45.

Free Σ-structures (Chapter 8). The introduction (and the text) should have made the goal of this investigation clear. Here is some clarification, even if it comes a few years too late.

For a class K of algebras or structures, the basic interest of logicians seems to be whether K is axiomatic, that is, K is the class of all models of a set Σ of (first order) sentences. Given such a Σ we can introduce concepts of subalgebras and homomorphisms that reduce to the usual concept in case Σ is a set of universal sentences for algebras.

If we are given K and we find a Σ axiomatizing K, then we are provided with the two basic tools of algebra to investigate the properties of the class K. It is clear, however, that for a given K many axiom systems Σ can be found. Obviously, an axiom system giving a rich structure is to be preferred over one that gives a more meager structure.

A simple example should illustrate this point. Let K be the class of all rings with unit and with at least two elements. If Ω is the usual axiom system for rings with a unit, then both

$$\Sigma_1 = \Omega \cup \{\mathbf{0} \neq \mathbf{1}\},$$
$$\Sigma_2 = \Omega \cup \{(\mathbf{x})(\exists \mathbf{y})(\mathbf{x} \neq \mathbf{y})\}$$

axiomatize K. It is obvious that, in K, Σ_1-homomorphism and Σ_1-subalgebra mean ring homomorphism and subring, respectively, while a Σ_2-homomorphism is an isomorphism and a Σ_2-subalgebra is the whole ring.

Thus, in a sense, Σ_1 is a best axiomatization of K and Σ_2 is a worst axiomatization.

Now we can state the purpose of this line of inquiry: Firstly, to investigate the concept of Σ-subalgebras and Σ-homomorphisms and, secondly, for a given axiomatic class K, to search for axiom systems Σ that provide K with the richest algebraic structure.

A formalization of this second line is started in Exercises 32–35 and in Problems 81 and 82.

§56. RELATED STRUCTURES

There are a number of lattices, semigroups, groups, etc., associated with an algebra: the subalgebra lattice, the congruence lattice, the automorphism group, and the endomorphism monoid are the best known examples. The topic: related structures (this name was coined in 1963 as the title of an NSF research proposal) deals with the abstract and concrete representation problems of these structures and with their interdependence.

(a) *Subalgebra lattice.* The concrete characterization of $\mathscr{S}(\mathfrak{A})$ is given in Theorem 9.2 and the abstract characterization (that is, the characterization of $\mathscr{S}(\mathfrak{A})$ as a lattice) describes $\mathscr{S}(\mathfrak{A})$ as an algebraic lattice by Theorem 6.5.

The problem becomes more interesting if the type of $\mathfrak{A}$ is restricted. If the type of $\mathfrak{A}$ is countable, then every compact element of the subalgebra lattice contains at most countably many compact elements, and the converse holds even for groupoids (that is, algebras of type $\langle 2 \rangle$) by W. Hanf [1956].

For subalgebra lattices of unary algebras, see W. A. Lampe [1969], [1974] and G. Piegari [1972], [1975].

For subalgebra systems a complete solution is obtained in the papers of M. Gould, see the references for Problem 1 in Appendix 2.

Infinitary algebras pose no special difficulties; see G. Grätzer [8] and H. Andréka and I. Németi [1974].

The subalgebra systems of two algebras $\mathfrak{A}$, $\mathfrak{B}$ connected with a homomorphism are described in M. Gould and C. R. Platt [1971 a] and S. Burris [1970].

Subalgebra lattices of algebras of the form $\mathfrak{A}^2$ are characterized in A. A. Iskander [1]; see also G. Grätzer and W. A. Lampe [1]. Related more recent papers are A. A. Iskander [1971] and [1972 b].

(b) *Automorphism group.* The abstract problem was solved in Corollary 12.1; up to isomorphism every group is the automorphism group of a suitable (unary) algebra. The concrete problem was solved in B. Jónsson [1968]; see also E. Płonka [1968]. For algebras of finite type (or, equivalently, with a single operation), the concrete problem was solved in M. Gould [1972 a].

For automorphism groups of algebras of a restricted type and the abstract problem, see (c); for finite groupoids, see M. Gould [1972 c]. A case of special unary algebras is considered in J. Hyman [1974], J. Hyman and J. B. Nation [1974] and G. H. Wenzel [1969].

Automorphism groups of algebras of the form $\mathfrak{A}^2$ are characterized in M. Gould [1975 a] as groups containing an element $x \neq 1$ of order 2. In M. Gould [a] it is proved that automorphism groups of algebras of the form $\mathfrak{A}^n$ are the same as automorphism groups of (infinitary) free algebras on n generators. See also M. Gould and H. H. James [a].

Variations on this theme are the groups of inner automorphisms (B. Csákány [4]), splitting automorphisms (J. Płonka [1973 c]), weak automorphisms (A. Goetz [1]), and the semigroup of partial automorphisms (O. I. Domanov [1971]; see also T. Tichý and J. Vinárek [1972], D. A. Bredikhin [1976], [1976 a]). The only representation theorem here is due to J. Sichler [1973]; he represents an arbitrary group G and normal subgroup N of G as the group of weak automorphisms of an algebra such that the members of N are the automorphisms.

(c) *Endomorphism monoids.* Every monoid is isomorphic to the endomorphism monoid of some algebra by Corollary 12.2. This settles the abstract problem; for references to the concrete problem, see Problem 3 in §58.

One of the most fruitful problems of the last decade was the abstract problem considered in a categorical framework for some special equational classes of algebras. Here are some typical results. Every monoid is isomorphic to the endomorphism monoid of:

(1) a groupoid or an algebra with two unary operations (Z. Hedrlín and A. Pultr [1]);

(2) a semigroup (Z. Hedrlín and J. Lambek [1969]);

(3) an algebra with two unary operations and only five polynomials (J. Sichler, unpublished);

(4) bounded lattices—endomorphisms preserve the bounds by definition (G. Grätzer and J. Sichler [1971]);

(5) integral domains—endomorphisms preserve the unit element by definition (E. Fried and J. Sichler [1977]).

There are many results of this type and all these results have categorical versions. The book A. Pultr and V. Trnková [a] gives a full accounting of the methods and the results.

There are some recent references: M. E. Adams and J. Sichler [a], [b], [c], [d], [e], E. Fried [1977] and [b], E. Fried and J. Kollár [a], [b], E. Fried and J. Sichler [1973], [1977], J. Kollár [a], [b], [c], [d].

The endomorphism semigroups of algebras of the form $\mathfrak{A}^2$ are characterized in M. Gould [1975 a].

The endomorphism monoid of an algebra is independent from the endomorphism monoid of a subalgebra; see Z. Hedrlín and E. Mendelsohn [1969] and C. R. Platt [1970].

(d) *Congruence lattices.* For references to the abstract problem, see Problems 1 and 2 in Appendix 2.

The abstract problem was solved in Theorem 18.2. Variants and simplifications of this proof appear in W. A. Lampe [1969], [1972 a], [1973], E. Nelson [1974 b], E. T. Schmidt [1969]. The proof of P. Pudlák [1976] is considerably simpler than the previous ones.

All the proofs (except Pudlák's) yield a congruence lattice with a type 3 join (that is, if $a \equiv b(\Theta \vee \Phi)$, then there exist x_1, x_2, x_3 such that $a \equiv x_1(\Theta)$, $x_1 \equiv x_2(\Phi)$, $x_2 \equiv x_3(\Theta)$, $x_3 \equiv b(\Phi)$). By a result of B. Jónsson [1953] if all joins are type 2 (that is, if $a \equiv b(\Theta \vee \Phi)$, then there exist x_1, x_2 such that $a \equiv x_1(\Theta)$, $x_1 \equiv x_2(\Phi)$, $x_2 \equiv b(\Theta)$), then the congruence lattice is modular. The converse of this statement, that every modular algebraic lattice can be represented as a congruence lattice with type 2 joins, was announced in G. Grätzer and W. A. Lampe [1971/1972]; for a proof, see Appendix 7.

See also the references for Problems 13 and 17 in §58.

(e) *Interdependence.* The major result is due to W. A. Lampe [1972 a]: given algebraic lattices $\mathfrak{L}_a$, $\mathfrak{L}_c$ with more than one element each and a group $\mathfrak{G}$, there exists an algebra whose subalgebra lattice is isomorphic to $\mathfrak{L}_a$, whose congruence lattice is isomorphic to $\mathfrak{L}_c$, and whose automorphism group is isomorphic to $\mathfrak{G}$. (See Appendix 7.)

A special case of this result, namely the independence of the subalgebra lattice and the automorphism group was proved in E. T. Schmidt [1]. The concrete version of this case is treated in M. G. Stone [1969], [1972] and M. Gould [1972 b]. The most recent reference is L. Szabó [1978].

For any automorphism α of an algebra $\mathfrak{A}$, the map $\varphi_\alpha \colon X \to X\alpha$ is an automorphism of the subalgebra lattice of $\mathfrak{A}$. The map $\alpha \to \varphi_\alpha$ is a homomorphism of the automorphism group of $\mathfrak{A}$ into the automorphism group of the subalgebra lattice of $\mathfrak{A}$. The converse, namely, that any homomorphism of a group into the automorphism group of an algebraic lattice can be represented this way, is proved in E. Fried and G. Grätzer [1975]. For the connections between the automorphism group of an algebra and those of its various subalgebras, see E. Fried and J. Sichler [a] and J. Kollár [e].

If α is an endomorphism the induced map is considered in M. Gould [1972] and [1974]. No results are known about the induced map of the congruence lattice.

For the interdependence of the endomorphism monoid and the congruence lattice, see the references to Problem 7 in §58. For a related result, see M. Gould and C. R. Platt [1971].

§57. MISCELLANY

We mention here a few topics that, in our view, will prove to be important in the next few years.

Applications to Computer Science. In the last decade, the basic concepts of universal algebra and category theory found more and more widespread use in computer science. A partial list of papers relevant to this topic compiled by H. Andréka and I. Németi is 20 pages long. So we cannot hope to do justice to this new field; it is hoped, however, that the following few references will give the reader a starting point.

The Springer-Verlag Lecture Notes on Computer Science, volume 25, 1975, Category Theory Applied to Computation and Control, is a good introduction, with a fairly complete list of references. See also volume 56, 1977, Fundamentals of Computer Science, where Section B is mostly based on universal algebra and lattice theory.

For recent developments in the theory of automata, see the books S. Eilenberg [1974] and [1976] and the survey article J. Adámek and V. Trnková [a].

Some current references to the algebraic treatment of flow-chart schemes are B. Courcelle and M. Nivat [1976], J. W. Thatcher, E. G. Wagner, J. B. Wright [1977], J. A. Goguen, J. W. Thatcher, E. G. Wagner, J. B. Wright [1977 a], G. D. Plotkin [1976], J. Tiuryn [a], W. Damm, E. Fehr, K. Indermark [1978].

The basic reference to the algebraic theory of abstract data types is D. Scott [1976]; see also J. A. Goguen, J. W. Thatcher, E. G. Wagner, J. B. Wright [1977] and D. J. Lehman and M. B. Smyth [1977].

Continuous algebras arise naturally in the topics mentioned in the last two paragraphs. B. Courcelle and J. C. Raoult [a] and G. Markowsky [1977] are two additional references.

Universal algebra is used in the semantics of programming languages and in the methodology of proving properties of programs. See H. Andréka, T. Gergely, I. Németi [1977], A. Arnold and M. Nivat [1977], R. M. Burstall and P. J. Landin [1969], B. Courcelle, I. Guessarian and M. Nivat [a]; P. van Emde Boas and T. M. V. Janssen [1977], J. A. Goguen, J. W. Thatcher, E. G. Wagner, J. B. Wright [1976], R. Milner and R. Weyrauch [1972].

Congruence schemes. If for a class K of algebras, for all $\mathfrak{A} \in K$, and for all $a, b, c, d \in A$, $c \equiv d(\Theta(a, b))$ can be described as in Theorem 10.3 by the same sequence of polynomials and by the same "switching function" $f: \{0, 1, \cdots, n-1\} \to \{0, 1\}$ (that determines whether $p_i(a) = z_i$ or $p_i(a) = z_{i+1}$), then we say that K has a congruence scheme. The congruence scheme itself is the sequence of polynomials and f.

This concept occurs in the papers of R. Magari who investigates problems of the following type: let K have a congruence scheme; under what conditions on K can we conclude that the equational class generated by K also has the same congruence scheme.

E. Fried, G. Grätzer, and R. W. Quackenbush [1976], [a], [b] find conditions for an equational class to have a congruence scheme and they apply the results to weakly associative lattices.

J. Berman and G. Grätzer [1976] find for a congruence scheme S not containing constants, a necessary and sufficient condition for the existence of a nontrivial equational class K having S as a congruence scheme. (The condition is that no polynomial in S be unary.) J. Berman and G. Grätzer [1976 a] investigate embedding algebras into members of an equational class having a fixed S as a congruence scheme; there is an improvement of this result by C. R. Platt.

Quasivarieties. A *quasivariety* is a class of algebras closed under ultraproducts, subalgebras, products, and isomorphic copies, or equivalently, a class defined by equational implications (see §63); other names used are implicational classes, universal strict Horn classes, etc.

Mal'cev's papers (see A. I. Mal'cev [1971] and [1973]) provide a good background for work in this field. Many results parallel those for varieties: in a quasivariety K, every algebra is a subdirect product of subdirectly irreducible algebras; but "subdirectly irreducible" is not an absolute concept, it depends on K.

For an equational class K and for quasivarieties V and W in K define $\mathfrak{A} \in V \circ W$ (with respect to K) iff $\mathfrak{A} \in K$ and there is a congruence relation Θ such that $\mathfrak{A}/\Theta \in W$ and for all $a \in A$, if $[a]\Theta$ is a subalgebra of $\mathfrak{A}$, then $[a]\Theta \in V$. This concept was first studied for groups (see H. Neumann [1967]) and then for algebras in general (A. I. Mal'cev [1967 a]). The subject of products of varieties has recently been investigated for special types of algebras: Brouwerian semilattices (P. Köhler [a]), generalized interior algebras (W. J. Blok and P. Köhler [a]), lattice ordered groups (A. M. W. Glass, W. C. Holland, S. H. McCleary [a]), lattices (G. Grätzer and D. Kelly [1977]). Here are some typical results for lattices: $D \circ D$ and $M \circ D$ are varieties where D and M are the class of all distributive and modular lattices, respectively; there are $2^{\aleph_0}$ equational classes contained in $D \circ D$; every variety of lattices is a product of irreducible quasivarieties; apart from some trivial cases, $X \circ Y \neq X \vee Y$. (The class $D \circ D$ plays an important role in S. V. Polin [1977].)

Other papers on quasivarieties deal with the lattice of subquasivarieties of a quasivariety (which is much richer than the lattice of subvarieties), the atoms of this lattice, finite bases for equational implications, categoricity in some infinite power. The following is a partial list of references on quasivarieties: A. I. Abakumov, E. A. Paljutin, M. A. Taĭclin and Ju. E. Šišmarev [1972], M. E. Adams [1976], H. Andréka and I. Németi [a], M. Armbrust and K. Kaiser [1974], K. A. Baker [1974], J. T. Baldwin and A. H. Lachlan [1973], B. Banaschewski and H. Herrlich [1976], V. P. Belkin [1976], [1977], V. P. Belkin and V. A. Gorbunov [1975], W. J. Blok and P. Köhler [a], A. I. Budkin and V. A. Gorbunov [1973] and [1975], S. Burris [a], A. I. Čeremisin [1969], T. Fujiwara [1971 a], F. Galvin [1970], V. A. Gorbunov [1976], G. Grätzer and D. Kelly [1977], G. Grätzer and H. Lakser [1973], [1978], [a], V. I. Igošin [1971], [1974], H. Lakser [a], A. I. Mal'cev [1967 a], A. Yu. Ol'shanskiĭ [1974], E. A. Paljutin [1972], [1973], and [1975], A. Shafaat [1969], [1970], [1970 a], [1970 b], [1971], [1973 a], [1974 a], and [1975], H. Tabata [1969], [1971], A. A. Vinogradov [1965] and [1965 a].

Combinatorial algebra. The application of universal algebraic methods to combinatorial problems was pioneered by T. Evans and N. S. Mendelsohn. For a review of this field see the survey articles: T. Evans [1975], [b] and N. S. Mendelsohn [1975]. Here are some recent references: T. Evans and C. C. Lindner [1977], B. Ganter [a] and [b], B. Ganter and H. Werner [1975] and [1975 a], N. K. Pukharev [1966], R. W. Quackenbush [1974 c], [1975], [1976], [1977].

Infinitary algebras. Most results are infinitary analogues of finitary results. Many such references are mentioned in Appendix 2.

Appendix 7 is another illustration. Apart from the switch to closest elements in the range (as opposed to the domain), the proof follows the path of the finitary case.

P. Burmeister [1968] and [1970 a] (announced independently in G. Grätzer [1966]) characterized the cardinalities of bases of an infinitary free algebra. The result is strikingly different from the finitary case (mentioned at the end of §31).

Some interesting infinitary equational classes can be found in E. Nelson [1974] and [1974 a]. Example: while every finitary nontrivial equational class contains a simple algebra (R. Magari [1969]), this does not hold for infinitary equational classes.

B. Banaschewski and E. Nelson [1973] and G. A. Edgar [1974] provide easy examples of infinitary algebras that are not subdirect products of subdirectly irreducible algebras. (For earlier examples, see G. Grätzer and W. A. Lampe [1971/1972] and Appendix 7.)

Surveys. T. M. Baranovič [1968] and L. M. Gluskin [1970] are useful surveys of universal algebra from a Soviet point of view.

The polynomial algebras over an algebra give rise to many interesting problems. This field is surveyed in H. Lausch and W. Nöbauer [1973].

The contributions of A. Tarski and his students to equational logic are surveyed in A. Tarski [1968] and the finite basis problem for finite algebras in S. O. MacDonald [1973]. (See also §67.)

Filtrality of congruences is surveyed in R. Magari [1971 a] and R. Franci [1976].

Some approaches to linear dependence are described in R. Rado [1966].

Results on the p_n-sequence (size of free algebras) are surveyed up to 1969 in G. Grätzer [1970 a]. (See also item 5 of §71.)

A. F. Pixley [c] is an up-to-date survey of primality and related topics. (See also Appendix 5.)

Various topics discussed in these Appendices are surveyed up to 1971 in B. Jónsson [1972] and B. Jónsson [1974].

APPENDIX 2

THE PROBLEMS

§58. SOLUTIONS AND PARTIAL SOLUTIONS OF 46 PROBLEMS

Solutions and partial solutions were first reported in G. Grätzer [14] which, in fact, was only published in 1970.

Reading this section the reader should keep in mind that some papers contributing to a problem were not written as a result of the problem having been proposed in this book. This is especially clear with Tarski's solution of Problems 34 and 35, see the footnote on p. 194. On the other hand, some papers appeared solving problems not specifically proposed here, which, however, were obviously influenced by a development that started from a problem in this book.

Problem 1. The concept of the multiplicity type and the quasi-ordering by coordinatization may have been more important than the problem itself.

The original problem was solved in M. Gould [1968] (part (a)), M. Gould [1971] (part (b)); see also S. Burris [1968].

The related quasi-ordering for permutation groups on a set to be represented as the automorphism group of an algebra is considered in M. Gould [1972 a], [1972 b].

For lattices of equivalence relations to be represented as congruences of an algebra the quasi-ordering is completely described in T. P. Whaley [1971] and B. Jónsson and T. P. Whaley [1974].

For automorphisms and endomorphisms a description of the quasi-ordering appears still to be open. Similarly, descriptions for combinations, such as congruences and subalgebras, are also open.

Problem 2. It is not clear whether this problem will ever have a solution. Some general results are stated in B. Jónsson [1972], §4.4 and in H. Werner [1976]. Special cases were approached in M. Armbrust [1970], [1973], [1973 a] and H. J. Bandelt [1975].

A related interesting question is considered in R. W. Quackenbush and B. Wolk [1971], this problem is resolved in P. Pudlák [1977].

Problem 3. If the members of E are all permutations, the problem is settled in B. Jónsson [1968]; see also E. Płonka [1968]. See also the references given above for Problem 1. The general case is considered in M. G. Stone [1969]. An interesting approach is taken by J. Ježek [1975].

Two recent papers propose complete solutions: N. Sauer and M. G. Stone [1977] and L. Szabó [1978]; see also N. Sauer and M. G. Stone [1977 a] and [a].

Problem 5. A negative solution is given in A. Pasini [1971]. See also M. Gould [1].

Problem 6. The origin of this problem is the result in Exercise 1.27 concerning Boolean algebras. This was extended to any equational class generated by finitely many independent primal algebras by Tah-kai Hu [1971 a]. This was further extended by R. A. Knoebel [1974 a]. The extension to p-rings was given in A. Iskander [1972] and to finite arithmetical algebras in A. F. Pixley [1972 a].

The hemi-primal algebras of A. L. Foster are closely related to this problem. See A. L. Foster [1970], [1970 a], A. F. Pixley [1972], [1972 a] and for the related concept of affine completeness, see H. Werner [1974 a]. See also H. Werner [1971] and K. Keimel and H. Werner [1974].

In a recent paper, A. F. Pixley [a] and, independently, J. Hagemann and Ch. Herrmann [a] show that a local version of the property defined in this problem holds for every algebra of an equational class iff the equational class is arithmetical (i.e., the algebras are congruence distributive and permutable).

Problem 7. The best result is W. A. Lampe [1972 a] showing that such an algebra always exists if $\mathfrak{G}$ is a group and $\mathfrak{L}$ is an algebraic lattice with more than one element, solving the independence problem of automorphism groups and congruence lattices. The proof in Appendix 7 contains this result. Other independence results are discussed in §56.

For some partial results on the general case, see P. Burmeister [1971 a] and W. A. Lampe [1969].

Problem 9. W. A. Lampe [1969] has some necessary and also some sufficient conditions for the case of 3-element chain as a congruence lattice.

The related question: which monoids are endomorphism semigroups of algebras with a two-element subalgebra lattice, is settled in M. Gould and C. R. Platt [1971]. The characterization is the dual of the result in G. Grätzer [7]. Then they prove that such monoids can be paired with an arbitrary algebraic lattice as a subalgebra lattice. The corresponding result for congruence lattices is not known.

Problem 10. For a set of algebras K, the problem is solved by C. R. Platt [1971]. The conditions closely correspond to the one-object case. Platt also shows that these conditions are not sufficient for a class K; see C. R. Platt [1971 b].

Problem 12. Solved independently by P. Burmeister [1971], J. S. Johnson [1971], and A. Pasini [1971 a]. The condition is: $\mathfrak{L}_1$ is algebraic and $\mathfrak{L}_2$ is isomorphic to a principal ideal of $\mathfrak{L}_1$.

Problem 13. $L_0 = L_1$ has recently been proved in P. Pudlák and J. Tůma [a]. $L_2 = L_3$ was proved in J. Berman [1971]. P. Pudlák and J. Tůma [1976] found a large class of finite lattices contained in L_3. It is interesting to note that their condition (P) is equivalent to condition $(T_\vee)$ of H. Gaskill, G. Grätzer, and C. R. Platt [1975] as shown in A. Day [b].

In P. P. Pálfy and P. Pudlák [a], it is shown that $L_0 = L_3$ iff every finite lattice is isomorphic to an interval in the subgroup lattice of some finite group.

Problem 14. An affirmative solution is given in J. Słomiński [1968], G. H. Wenzel [1970], and G. Grätzer and H. Wenzel [1]. See also I. Fleischer [1975].

Problem 15. Nothing is known about the first question. $\mathfrak{C}(\mathfrak{A})$ is characterized as a complete lattice in Appendix 7; see also G. Grätzer and W. A. Lampe [1971/1972] and E. Nelson [1974 b].

Problem 16. T. P. Whaley [1969 a] points out that $\mathfrak{C}_F(\mathfrak{A})$ is not a subsemigroup and characterizes the pair $\langle \mathfrak{C}(\mathfrak{A}), \mathfrak{C}_S(\mathfrak{A}) \rangle$.

Problem 17. Two contributions to this problem were made by W. A. Lampe. In [a] he proved that every algebraic lattice with a compact unit element is isomorphic to the congruence lattice of an algebra with a single binary operation. In [1977], for any infinite cardinal $\mathfrak{n}$, he constructs an algebraic lattice L such that L is not isomorphic to the congruence lattice of an algebra with less than $\mathfrak{n}$ operations. This was improved in W. Taylor [1977 a] and by R. Freese (independently) by constructing a modular lattice L with this property; in fact, L can be chosen to be the subspace lattice of an infinite dimensional vector space over any field of cardinality $\geqq \mathfrak{n}$.

Problem 18. For partial algebras this is solved in A. A. Iskander [1971].

Problem 19. This is reported as solved in Part II of L. Szabó [1978].

Problem 20. This problem was considered for semigroups in L. Babai and F. Pastijn [a]. For instance, they show that if the automorphism group of a semigroup S contains a transitive torsion subgroup, then S is a rectangular band. For references to related papers by L. Babai and a discussion of current unsolved problems concerning automorphisms and endomorphisms of algebras, graphs, semigroups, and so on, see L. Babai [d].

Problem 22. A partial solution is reported by P. Zlatoš [a].

Problems 24 and 25. Partial solutions can be found in E. Nelson [1967]. See also H. Höft [1972], [1974], A. Iwanik [1974]. For the semigroup generated by **H**, **S**, and **P**, see D. Pigozzi [1972], S. D. Comer and J. S. Johnson [1972], and P. M. Neumann [1970].

Problems 26 and 27. C. R. Platt [1971 a] answers in the affirmative the first question both for direct and inverse limits. He also shows that if there are no arbitrarily large measurable cardinals, then the answer to the second question is in the negative both for direct and inverse limits. C. R. Platt [1974] constructs classes of lattices with the same properties.

Problem 29. It is easily seen that $K_{\mathfrak{A}}$ is an equational class iff $\mathfrak{A}$ is subdirectly irreducible and is weakly projective (that is, $\mathfrak{A} \in \mathbf{H}(\mathfrak{B})$ and $\mathfrak{B} \in K$ implies that $\mathfrak{A} \in \mathbf{IS}(\mathfrak{B})$); thus $\mathfrak{A}$ is a subalgebra of $\mathfrak{F}_K(\aleph_0)$. This is interesting for the case of finite lattices. If $\mathfrak{A}$ is a finite lattice, then $\mathfrak{A}$ is weakly projective iff $\mathfrak{A}$ is projective iff $\mathfrak{A}$ is a sublattice of a free lattice by a result of R. McKenzie [1972 a].

Problem 32. J. Ježek [1969] shows that $|L(\tau)| \leqq \aleph_0$ iff the type τ consists of finitely many nullary and at most one unary operations. He also characterizes all such countable lattices.

Various properties of the lattice $L(\tau)$ are given in J. Ježek [1968 b], [1971 a], [1976 a] and Ju. K. Rebane [1967]. The most striking result is in J. Ježek [1976 a]: every algebraic lattice with countably many compact elements is isomorphic to an interval of some $L(\tau)$.

Problem 33. Solved independently by S. Burris [1971 b] and J. Ježek [1970]. See also A. D. Bol'bot [1970].

Problems 34 and 35. Solved in A. Tarski [1968] (this paper includes some joint work with T. C. Green). A. Tarski characterizes $D(K)$ as a convex set of integers. For groups or rings, $D(K) = [1)$ or $[2)$.

Problem 38. Solved in R. McKenzie [1975].

Problem 40. Solved independently by G. A. Fraser and A. Horn [1970] and Tah-kai Hu [1970].

Problem 41. S. Fajtlowicz pointed out that part (β) is trivially false. (Example: pointed sets.)

Problem 42. Let p_n denote the number of essentially n-ary polynomials of K. Then a simple combinatorial formula expresses c_n by the p_i, $i \leqq n$ and vice versa. This yields a necessary condition for the c_n: p_n (as expressed by the c_n) $\geqq 0$; and reduces the problem of characterizing $\langle c_n \rangle$ to characterizing $\langle p_n \rangle$.

A review of this problem up to 1969 is given in G. Grätzer [1970 a], see also item 5 of §71.

A very interesting contribution has been made by A. Kisielewicz [a]. Verifying a conjecture of G. Grätzer and J. Płonka [1970 b] and [1973], he proved that, with the exception of a few types of algebras (such as semilattices), for every idempotent algebra $\mathfrak{A}$ there exists an integer m such that $p_n(\mathfrak{A})+1 \leqq p_{n+1}(\mathfrak{A})$ for all $n \geqq m$.

Problem 44. A special kind of amalgamation property for L was found in R. Freese and B. Jónsson [1976]. Recently, quite a bit of work has been done on the equational class generated by L, see Appendix 3.

Problem 45. A negative solution to this problem was given by R. McKenzie [1970].

Problem 46. This problem has just been solved in the affirmative in D. Pigozzi [c].

Problem 47. In the extremely special case when $\mathfrak{A}$ and $\mathfrak{B}$ are independent, the answer is in the affirmative (R. A. Knoebel [1973]). The general case is still open.

Problem 50. All pairs and some triples are characterized in G. H. Wenzel [1966] (see also G. H. Wenzel [3]). The six-tuple is characterized for finite free algebras in G. H. Wenzel [2].

Problem 52. Many papers give partial solutions. See, e.g., S. Fajtlowicz [1], [1969 a], and J. S. Johnson [1969].

Problem 53. This was solved in R. M. Vancko [1969] (see also R. M. Vancko [1972]) and, independently, in K. Truöl [1969]. The former appears to be a much better solution.

Problem 55. S. Fajtlowicz described in a letter to the author how to construct an algebra of finite type in which every nonempty subset is a weak basis, giving an affirmative answer to the first question raised in this problem and also answering the second question.

Problem 56. It turns out that K is equivalent to the class of all R-modules for some semiring R. This was proved by B. Csákány [1975], J. S. Johnson and E. G. Manes [1970], and R. M. Vancko [1974]. To get R-modules for a ring R one needs some additional hypothesis, such as congruence modularity.

Problem 58. For some related work, see A. F. Pixley [1974].

Problem 63. Answered in the affirmative in R. McKenzie [1975].

Problem 75. For a partial solution, see K. A. Baker and A. W. Hales [1974].

Problem 76. Such lattices are now called finite transferable lattices. H. Gaskill [1972] and H. Gaskill, G. Grätzer, and C. R. Platt [1975] led to the characterization of sharply transferable lattices. Some very recent papers are: A. Day [b], G. Grätzer and C. R. Platt [a], G. Grätzer, C. R. Platt, and B. Sands [1977], C. R. Platt [1977]. These together with the older results, show that for a finite lattice the following are equivalent: transferability, sharp transferability, $(T_\vee)$, projectivity, being a sublattice of a free lattice, (P), etc.

Problem 86. Always, see F. Gécseg [1970].

Problem 101. Trivially, no.

APPENDIX 3

CONGRUENCE VARIETIES

By Bjarni Jónsson

§59. ALGEBRAS AND THEIR CONGRUENCE LATTICES

According to a theorem by G. Birkhoff and O. Frink (Theorem 10.2), the congruence relations on an arbitrary algebra $\mathfrak{A}$ (with operations of finite rank) form an algebraic lattice $\mathfrak{C}(\mathfrak{A})$. The converse of this theorem, due to G. Grätzer and E. T. Schmidt (Theorem 18.3), states that every algebraic lattice is isomorphic to the congruence lattice of some algebra. The significance of this result is obvious, for it shows that unless something more is known about the algebra $\mathfrak{A}$, nothing can be said about its congruence lattice that does not follow from the fact that it is algebraic. This, however, raises the question of what happens when additional conditions are imposed on $\mathfrak{A}$. The most obvious restriction would be to specify the similarity type of the algebra. The algebra in the proof of the Grätzer-Schmidt theorem has a large number of operations, and we now know that this cannot be avoided; by a recent result of W. A. Lampe [1977] no upper bound can be placed on the number of operations. A stronger version of Lampe's theorem, obtained independently by R. Freese and by W. Taylor, shows that this holds even when the lattice is assumed to be modular.

Theorem 1 (*R. Freese, W. A. Lampe, W. Taylor*). *For every similarity type* τ *there exists a modular algebraic lattice* L *such that* $L \not\cong \mathfrak{C}(\mathfrak{A})$ *for any algebra* $\mathfrak{A}$ *of type* τ.

Thus the similarity type of an algebra does give some information about its congruence lattice, but our present knowledge of this subject is very limited. Some fragmentary results can be found in B. Jónsson [1972], Section 4.7. If we consider some of the most familiar classes of algebras, we have some information about the congruence lattices of their members. E.g., the congruence lattice of a group $\mathfrak{G}$, or the lattice of all normal subgroups of $\mathfrak{G}$, is always modular. This is one of the major reasons for the importance of the notion of a modular lattice. G. Birkhoff showed that, more generally, if $\mathfrak{A}$ is any algebra whose congruences permute, then $\mathfrak{C}(\mathfrak{A})$ is modular, and in B. Jónsson [1953] it is shown that under the same assumptions $\mathfrak{C}(\mathfrak{A})$ satisfies an identity that is stronger than the modular law, the so-called Arguesian identity. Since the geometric background of this identity will motivate some of our later arguments, we pause to recall the relevant geometric facts.

The subspaces of a projective geometry form a complemented modular lattice that is algebraic and atomic, with the points being the atoms. The geometry is said to satisfy *Desargues' Law* if any two triangles that are centrally perspective are also axially perspective. M. P. Schützenberger [1945] first showed that this property can be expressed as a lattice identity. We shall describe here a slightly different, but equivalent, identity that can be found in B. Jónsson [1953]. Consider six lattice elements, a_i, b_i, $(i=0, 1, 2)$, form the elements

$$c_0 = (a_1 \vee a_2) \wedge (b_1 \vee b_2)$$

and cyclically, and let

$$c_0' = c_0 \wedge (c_1 \vee c_2).$$

The inclusion

$$(a_0 \vee b_0) \wedge (a_1 \vee b_1) \wedge (a_2 \vee b_2) \leqq (a_1 \wedge (c_0' \vee a_2)) \vee b_1$$

is called the *Arguesian identity*, and a lattice in which this identity holds is said to be *Arguesian*. (Recall that any lattice inclusion is equivalent to a lattice identity, since $u \leqq v$ iff $u \vee v = v$.) It is easy to see that this identity implies Desargues' Law, for if a_i and b_i, $i=0, 1, 2$, are corresponding vertices of two triangles that are centrally perspective but not axially perspective, then c_0, c_1 and c_2 are the meets of corresponding sides, and since, by hypothesis, these are not collinear, c_0' will be 0. Hence the right-hand side of the inclusion will be b_1, and the inclusion fails since the left-hand side is the center of perspectivity. It is not hard to give a direct proof of the converse implication, but we shall not need it, and in any case, it will follow from the theorem that we are about to prove, together with the classical coordinatization theorem. We therefore return to the congruence relations.

Theorem 2 (*B. Jónsson* [1953]). *If $\mathfrak{A}$ is an algebra whose congruences permute, then $\mathfrak{C}(\mathfrak{A})$ is Arguesian.*

Proof. Suppose Θ_i, $\Phi_i \in \mathfrak{C}(\mathfrak{A})$ for $i=0, 1, 2$, and let $\Psi_0 = (\Theta_1 \vee \Theta_2) \wedge (\Phi_1 \vee \Phi_2)$ and cyclically, and $\Psi_0' = \Psi_0 \wedge (\Psi_1 \vee \Psi_2)$. If $\langle a, b\rangle \in \Theta_i \vee \Phi_i$ for $i=0, 1, 2$, then there exist c_0, c_1, c_2 with $\langle a, c_i\rangle \in \Theta_i$ and $\langle c_i, b\rangle \in \Phi_i$. It follows that $\langle c_1, c_2\rangle \in \Psi_0$ and cyclically, therefore $\langle c_1, c_2\rangle \in \Psi_0'$. We therefore have $\langle a, c_1\rangle \in \Theta_1 \wedge (\Theta_2 \vee \Psi_0')$, hence, $\langle a, b\rangle \in (\Theta_1 \wedge (\Theta_2 \vee \Psi_0')) \vee \Phi_1$, showing that the Arguesian identity holds.

Incidentally, it is easily seen that the Arguesian identity does imply the modular law.

Theorem 3 (*B. Jónsson* [1954]). *Every Arguesian lattice is modular.*

We shall later need the fact that the Arguesian identity is equivalent to certain lattice implications. To bring out the motivation for these implications, we adopt a geometric terminology. We refer to an ordered triple $a = \langle a_0, a_1, a_2 \rangle$ of lattice elements as a *triangle*, and we say that two triangles a and b are *centrally perspective* if

$$(a_0 \vee b_0) \wedge (a_1 \vee b_1) \leqq a_2 \vee b_2,$$

and we then call the left-hand side the *center of perspectivity*. Letting

$$c_0 = (a_1 \vee a_2) \wedge (b_1 \vee b_2)$$

and cyclically, we say that a and b are *axially perspective* if $c_2 \leqq c_0 \vee c_1$.

Theorem 4. *For any modular lattice L, the following conditions are equivalent:*

(i) *L is Arguesian.*

(ii) *Any two triangles in L that are centrally perspective are also axially perspective.*

(iii) *For any triangles a and b in L that are centrally perspective, if $a_0 \wedge a_1 \leqq a_2$ and $b_0 \wedge b_1 \leqq b_2$, and if the center of perspectivity, p, satisfies the conditions*

$$a_i \vee p = b_i \vee p = a_i \vee b_i \quad \textit{for } i = 0, 1, 2,$$

then a and b are axially perspective.

The proof of this theorem, which is rather computational in nature, will not be given here. It was shown in B. Jónsson and G. S. Monk [1969] that (i) implies (ii), and in G. Grätzer, B. Jónsson and H. Lakser [1973] that (iii) implies (i). The third condition is clearly a special case of the second.

It is a consequence of this theorem that the Arguesian identity is equivalent to its own dual.

Corollary (*B. Jónsson* [1972 a]). *The class of all Arguesian lattices is self dual.*

Lattices do not in general have permutable congruence relations, the simplest counter-example being a three-element chain. Nonetheless, the congruence lattice of a lattice is not only modular, but even distributive. We shall later prove a more general result, but it is instructive to look at this special case.

Theorem 5 (*N. Funayama and T. Nakayama* [1942]). *For any lattice L, $\mathfrak{C}(L)$ is distributive.*

Proof. Suppose $\Theta, \Phi, \Psi \in \mathfrak{C}(L)$, and suppose $\Theta \wedge (\Phi \vee \Psi)$ identifies x and y. The problem is to show that $(\Theta \wedge \Phi) \vee (\Theta \wedge \Psi)$ also identifies x and y. By the hypothesis, $x \Theta y$ and $x = z_0 \Phi z_1 \Psi z_2 \Phi z_3 \Psi \cdots z_n = y$ for some elements $z_i \in L$. We need to replace the elements z_i by elements z_i', all of which belong to the same Θ-class. There are many ways in which this can be done, e.g., we can take $z_i' = (z_i \wedge x) \vee (z_i \wedge y) \vee (x \wedge y)$.

It is an open question whether every distributive algebraic lattice is isomorphic to the congruence lattice of some lattice. In general, there are very few classes of algebras for which we have a complete description of the congruence lattices. Three such classes are: Boolean algebras, generalized Boolean algebras, and distributive lattices (cf. B. Jónsson [1972], pp. 210–211).

§60. MAL'CEV CLASSES

From the fact that the congruence lattice of a group is modular it follows that if an algebra $\mathfrak{A}$ is a group with respect to some of its polynomial (or, more generally, algebraic) operations, then $\mathfrak{C}(\mathfrak{A})$ is modular. Similarly, if $\mathfrak{A}$ is a lattice with respect to some of its polynomial operations, then $\mathfrak{C}(\mathfrak{A})$ is distributive. These simple observations form the basis for the idea of a Mal'cev class of varieties.

A variety V of algebras is said to be *finitely based* if there exists a finite set Σ of identities such that V is the class of all models of Σ. More generally, a class K of algebras, or a single algebra $\mathfrak{A}$, is said to be *finitely based* if the variety generated by K, or by $\mathfrak{A}$, is finitely based. A finitely based variety whose similarity type is finite is said to be *finitely presented*. Suppose now that $m = \langle m_i \mid i \in I \rangle$ and $n = \langle n_j \mid j \in J \rangle$ are two similarity types and $p = \langle p_i \mid i \in I \rangle$ is a system of polynomial symbols in the language of n, with no variable other than $x_0, x_1, \cdots, x_{m_i - 1}$ occurring in p_i. With any algebra $\mathfrak{A} = \langle A; \{g_j \mid j \in J\} \rangle$ of the type n we can then associate an algebra $\mathfrak{A}^p = \langle A; \{f_i \mid i \in I\} \rangle$ of type m, where f_i is the m_i-ary operation on A that is induced by p_i. If U and V are varieties of type m and n, respectively, we say that p is a *representation* of U in V provided $\mathfrak{A}^p \in U$ whenever $\mathfrak{A} \in V$. We say that U and V are (*definitionally*) *equivalent* if there exist a representation p of U in V and a representation q of V in U such that, $(\mathfrak{A}^p)^q = \mathfrak{A}$ for all $\mathfrak{A} \in V$ and $(\mathfrak{B}^q)^p = \mathfrak{B}$ for all $\mathfrak{B} \in U$. Examples of equivalent varieties arise whenever we have different choices for the basic operations of the "same" variety. E.g., in defining groups one can start with the group multiplication, inverses, and the identity element, or with the operation xy^{-1} and the identity element, or even with the ternary operation $xy^{-1}z$ and the identity element. Another familiar example is the equivalence of Boolean algebras and Boolean rings. We shall however be more concerned with the relation of representability. This relation is obviously reflexive and transitive, but it is not symmetric. If $V \subseteq U$, then U is representable in V, and if U is a reduct of V, then U is representable in V. If V is a finitely based variety, say with an equational basis Σ, and U is the reduct of V obtained by dropping all those operations that do not occur in Σ, then U is finitely presented, and U and V are each representable in the other (although they need not be equivalent).

Consider a class $\mathscr{K}$ of varieties. We say that $\mathscr{K}$ is a *strong Mal'cev class* (*is definable by a strong Mal'cev condition*) if there exists a finitely presented variety U such that $\mathscr{K}$ is precisely the class of those varieties in which U has a representation. We say that $\mathscr{K}$ is a *Mal'cev class* (*is definable by a Mal'cev condition*) if $\mathscr{K}$ is the union of a non-decreasing sequence of strong Mal'cev classes. Equivalently, $\mathscr{K}$ is a Mal'cev class iff there exist finitely presented varieties U_i, $i=0, 1, \cdots$, such that U_{i+1} has a representation in U_i for each $i \in \omega$, and $\mathscr{K}$ is precisely the class of those varieties in which some U_i has a representation. Finally, we say that $\mathscr{K}$ is a *weak Mal'cev class* if it is the intersection of countably many Mal'cev classes.

According to Theorem 26.4, the class of all varieties $\mathscr{V}$ whose algebras have permutable congruences is a strong Mal'cev class. Here we can take U to be the class of all algebras $\mathfrak{A} = \langle A; f \rangle$ with f a ternary operation such that, for all $a, b, c \in A$,

$$f(a, b, b) = a \quad \text{and} \quad f(a, a, b) = b.$$

Historically, this was the first example of a strong Mal'cev class. The second one, varieties with permutable and distributive congruences, can be found in A. F. Pixley [1963]. The first two examples of Mal'cev classes that are not strong Mal'cev classes are congruence distributive varieties in B. Jónsson [1967] and congruence modular varieties in A. Day [1969]. The phrase "Mal'cev-type condition" is first used in G. Grätzer [1970], and since then the examples of such classes have proliferated. General studies of Mal'cev classes can be found in R. Wille [1970], A. F. Pixley [1972 b], W. Taylor [1973], W. D. Neumann [1974] and J. T. Baldwin and J. Berman [a]. The classes treated in the next three theorems have been chosen because they will be needed in the next section.

An algebra $\mathfrak{A}$ is said to have *n-permutable congruences* if, for all Θ, $\Phi \in \mathfrak{C}(\mathfrak{A})$,

$$\Theta \vee \Phi = \Theta \circ \Phi \circ \Theta \circ \Phi \cdots \text{ (n factors)}$$

or, equivalently, if

$$\Theta \circ \Phi \circ \Theta \circ \Phi \cdots \subseteq \Phi \circ \Theta \circ \Phi \circ \Theta \cdots,$$

where the number of factors on each side is n. Thus 2-permutable means the same as permutable. The obvious generalization of Mal'cev's theorem to varieties with n-permutable congruences yields polynomials in $n+1$ variables, but one can get by with just three.

Theorem 1 (*J. Hagemann and A. Mitschke* [1973]). *For any variety V of algebras and any integer $n > 1$, the following conditions are equivalent:*

(i) *Every $\mathfrak{A} \in V$ has n-permutable congruences.*
(ii) *$\mathfrak{F}_V(n+1)$ has n-permutable congruences.*
(iii) *There exist ternary polynomials $p_0, p_1, \cdots, p_n$ such that the identities*

$$p_0(x, y, z) = x, \qquad p_n(x, y, z) = z,$$
$$p_i(x, x, z) = p_{i+1}(x, z, z) \qquad (i < n)$$

hold in V.

Proof. For notational convenience we assume that n is even, say $n=2m$. The alternative case can be treated similarly.

Assume (ii), and let the generators of $\mathfrak{F}_V(n+1)$ be $a_0, a_1, \cdots, a_n$. Let

$$\Theta = \bigvee_{i<m} \Theta(a_{2i}, a_{2i+1}), \qquad \Phi = \bigvee_{i<m} \Theta(a_{2i+1}, a_{2i+2}).$$

Then

$$a_0 \Theta a_1 \Phi a_2 \Theta a_3 \Phi \cdots \Theta a_{n-1} \Phi a_n. \tag{1}$$

Hence there exist elements b_i $(i \leqq n)$ such that

$$a_0 = b_0 \Phi b_1 \Theta b_2 \Phi b_3 \Theta \cdots \Phi b_{n-1} \Theta b_n = a_n. \tag{2}$$

Each b_i is of the form $b_i = q_i(a_0, a_1, \cdots, a_n)$ for some polynomial q_i. For i even,

$$q_i(a_0, a_1, \cdots, a_n) \Phi q_{i+1}(a_0, a_1, \cdots, a_n).$$

Since Φ identifies a_{2j+1} with a_{2j+2}, it follows that Φ identifies $q_i(a_0, a_1, a_1, a_3, a_3, \cdots)$ with $q_{i+1}(a_0, a_1, a_1, a_3, a_3, \cdots)$. But this implies that these two elements must be equal, for the restriction of Φ to the subalgebra generated by a_0 and the elements a_{2j+1} $(j<m)$ is trivial. Consequently, the identities

$$q_i(x_0, x_1, x_1, x_3, x_3, \cdots) = q_{i+1}(x_0, x_1, x_1, x_3, x_3, \cdots)$$

with i even hold in V. Similarly, for i odd,

$$q_i(x_0, x_0, x_2, x_2, \cdots) = q_{i+1}(x_0, x_0, x_2, x_2, \cdots)$$

holds. We now let $p_0(x, y, z) = x$ and $p_n(x, y, z) = z$, and for $0 < i < n$,

$$p_i(x, y, z) = q_i(x, x, \cdots, x, y, z, z, \cdots, z)$$

with x occurring i times and z $n-i$ times, and the identities in (iii) are easily seen to hold.

Now suppose (iii) holds, and suppose $\mathfrak{A} \in V$, Θ, $\Phi \in \mathfrak{C}(\mathfrak{A})$, and (1) holds. Letting $b_0 = a_0$, $b_n = a_n$ and $b_i = p_i(a_{i-1}, a_i, a_{i+1})$ for $0 < i < n$, one easily checks that (2) holds. E.g., if $0 < i < n$ and i is even, then

$$\begin{aligned} b_i &= p_i(a_{i-1}, a_i, a_{i+1}) \Phi p_i(a_i, a_i, a_{i+1}) \\ &= p_{i+1}(a_i, a_{i+1}, a_{i+1}) \Phi p_{i+1}(a_i, a_{i+1}, a_{i+2}) = b_{i+1}. \end{aligned}$$

Thus (i) holds. Since (ii) is a special case of (i), the theorem follows.

Theorem 2 (*B. Jónsson* [1967]). *For any variety V of algebras, the following conditions are equivalent:*

(i) *V is congruence distributive.*

(ii) *$\mathfrak{F}_V(3)$ is congruence distributive.*

(iii) *For any $\mathfrak{A} \in V$ and any $a, b, c \in A$,*

$$\langle a, c\rangle \in (\Theta(a, b) \wedge \Theta(a, c)) \vee (\Theta(b, c) \wedge \Theta(a, c)).$$

(iv) *For some integer $n \geqq 1$, there exist ternary polynomials $p_0, p_1, \cdots, p_n$ such that for $i = 0, 1, \cdots, n-1$ the identities*

$$\begin{aligned} &p_0(x, y, z) = x, \qquad p_n(x, y, z) = z, \qquad p_i(x, y, x) = x, \\ &p_i(x, x, z) = p_{i+1}(x, x, z) \quad \text{for } i < n,\ i \text{ even}, \\ &p_i(x, z, z) = p_{i+1}(x, z, z) \quad \text{for } i < n,\ i \text{ odd} \end{aligned}$$

hold in V.

Proof. Obviously (i) implies (ii). If (ii) holds, then every member of V that is generated by a three-element set must be congruence distributive, and from this (iii) easily follows. Assuming (iii), take $\mathfrak{A} = \mathfrak{F}_V(3)$, with a, b and c the free generators. Let $\Theta = \Theta(a, c)$, $\Phi = \Theta(a, b)$, $\Psi = \Theta(b, c)$. Then there exist finitely many elements $d_0, d_1, \cdots, d_n$, all in the same Θ-class, such that

$$a = d_0 \Phi d_1 \Psi d_2 \Phi \cdots d_n = c.$$

Each d_i is of the form $d_i = p_i(a, b, c)$ for some ternary polynomial p_i. It is now a simple matter to show that the equations in (iv) are satisfied for $x = a$, $y = b$, $z = c$. E.g., for i even, $p_i(a, b, c) \Phi p_{i+1}(a, b, c)$ and $a \Phi b$, hence $p_i(a, a, c) \Phi p_{i+1}(a, a, c)$. Since the restriction of Φ to the subalgebra generated by a and c is trivial, this yields $p_i(a, a, c) = p_{i+1}(a, a, c)$.

Now assume that (iv) holds. Consider any $\mathfrak{A} \in V$ and $\Theta, \Phi, \Psi \in \mathfrak{C}(\mathfrak{A})$, and let $\alpha_k = \Phi \circ \Psi \circ \Phi \circ \Psi \cdots$ (k factors). Then $\Theta \wedge (\Phi \vee \Psi)$ is the union of the relations $\Theta \cap \alpha_k$, $k = 1, 2, \cdots$, and to prove the distributivity of $\mathfrak{C}(\mathfrak{A})$ it therefore suffices to show that $\Theta \cap \alpha_k \subseteq (\Theta \wedge \Phi) \vee (\Theta \wedge \Psi)$. For $k = 1$ this is obvious, so we assume that the inclusion holds for a given value of k, and show that the formula also holds with k replaced by $k+1$. Now $\alpha_{k+1}=\alpha_k \circ \beta$ where β is either Φ or Ψ, and we claim that

$$\Theta \cap \alpha_{k+1} \subseteq (\Theta \cap \alpha_k{}^{-1}) \circ (\Theta \cap \alpha_k) \circ (\Theta \cap \beta) \circ (\Theta \cap \beta^{-1}) \circ (\Theta \cap \alpha_k{}^{-1}) \cdots \tag{1}$$

with $2n$ factors on the right. In fact, assume that $a\Theta c$ and $a\alpha_k b\beta c$. Let $d_i=p_i(a, b, c)$. Then $a=d_0$, $c=d_n$ and, for $i=0, 1, \cdots, n$,

$$d_i = p_i(a, b, c)\Theta p_i(a, b, a) = a,$$
$$p_i(a, a, c)\Theta p_i(a, a, a) = a,$$

so that all the elements d_i and also $p_i(a, a, c)$ belong to the same Θ-class. For i even,

$$d_i = p_i(a, b, c)\alpha_k^{-1}p_i(a, a, c) = p_{i+1}(a, a, c)\alpha_k p_{i+1}(a, b, c) = d_{i+1},$$

consequently

$$d_i(\Theta \cap \alpha_k^{-1})p_i(a, a, c)(\Theta \cap \alpha_k)d_{i+1}.$$

A similar argument shows that if i is odd, then

$$d_i(\Theta \cap \beta)p_i(a, c, c)(\Theta \cap \beta^{-1})d_{i+1}.$$

Hence (1) follows.

The analogue of this result for the modular case is due to A. Day. We present it here without proof.

Theorem 3 (*A. Day* [1969]). *For any variety V of algebras, the following conditions are equivalent:*

(i) *V is congruence modular.*
(ii) *$\mathfrak{F}_V(4)$ is congruence modular.*
(iii) *For any $\mathfrak{A} \in V$ and $a, b, c, d \in A$,*

$$\langle a, d\rangle \in \Theta(b, c) \vee (\Theta(a, d) \vee \Theta(b, c)) \wedge (\Theta(a, b) \vee \Theta(c, d)).$$

(iv) *For some integer $n \geqq 1$ there exist polynomials $p_0, p_1, \cdots, p_n$ in four variables such that for $i=0, 1, \cdots, n-1$ the following identities hold in V:*

$$p_0(x, y, z, u) = x, \quad p_n(x, y, z, u) = u, \quad p_i(x, y, y, x) = x,$$
$$p_i(x, y, y, u) = p_{i+1}(x, y, y, u) \quad \textit{for } i \textit{ odd}.$$
$$p_i(x, x, u, u) = p_{i+1}(x, x, u, u) \quad \textit{for } i \textit{ even}.$$

We consider some examples.

Example 1. *Quasigroups* are algebras $\langle G; \cdot, /, \backslash\rangle$ with three binary operations, satisfying the identities

$$(x/y)y = x, \quad (xy)/y = x,$$
$$x(x\backslash y) = y, \quad x\backslash(xy) = y.$$

Quasigroups have permutable congruences: take $p(x, y, z) = (x/(y\backslash y))(y\backslash z)$.

Example 2 (*A. Mitschke* [1971], *J. Hagemann and A. Mitschke* [1973]). *Implication algebras* are groupoids $\langle G; \rightarrow\rangle$ satisfying the identities

$$(x \rightarrow y) \rightarrow x = x, \qquad (x \rightarrow y) \rightarrow y = (y \rightarrow x) \rightarrow x,$$
$$x \rightarrow (y \rightarrow z) = y \rightarrow (x \rightarrow z).$$

Implication algebras have 3-permutable congruences: Take $p_1(x, y, z) = (z \rightarrow y) \rightarrow x$, $p_2(x, y, z) = (x \rightarrow y) \rightarrow z$. From the 3-permutability it follows that the congruence lattices are modular, but they are in fact distributive: apply Theorem 2 with $n=3$ and

$$p_1(x, y, z) = (y \rightarrow (z \rightarrow x)) \rightarrow x, \qquad p_2(x, y, z) = (x \rightarrow y) \rightarrow z.$$

Example 3 (*J. Hagemann and A. Mitschke* [1973]). A *right complemented semigroup* is an algebra $\langle A; \cdot, *\rangle$ with two binary operations, satisfying the identities

$$x \cdot (x * y) = y \cdot (y * x), \qquad (x \cdot y) * z = y * (x * z), \qquad x \cdot (y * y) = x.$$

Right complemented semigroups have 3-permutable congruences; Theorem 1 applies with $n=3$.

Example 4. A *median algebra* is an algebra $\langle A; m\rangle$ with a ternary operation m that satisfies the identities

$$m(x, x, y) = m(x, y, x) = m(y, x, x) = x.$$

Median algebras have distributive congruences; Theorem 2 applies with $n=2$ and $p_1(x, y, z) = m(x, y, z)$.

Example 5. Lattices have distributive congruences. This was proved directly in Theorem 59.5, but it also follows from Theorem 2 by taking $n=2$, and taking p_1 to be the median polynomial,

$$p_1(x, y, z) = ((x \wedge y) \vee (y \wedge z)) \vee (z \wedge x).$$

Example 6 (*E. Fried* [1970] *and H. L. Skala* [1971]). A *weakly associative lattice*, or a *trellis*, is an algebra $\langle A; \wedge, \vee\rangle$ with two binary operations, that satisfies the identities

$$x \wedge y = y \wedge x, \qquad x \wedge (x \vee y) = x, \qquad ((x \wedge z) \vee (y \wedge z)) \vee z = z$$

and their duals. Weakly associative lattices have distributive congruences; the same polynomial works here as in the preceding example.

Example 7 (*K. A. Baker* [b], [c], *K. Fichtner* [1972], *A. Mitschke* [1971], *D. Kelly* [1973]). For $n > 1$, let Δ_n be the condition that there exist ternary polynomials $p_0, p_1, \cdots, p_n$ satisfying the identities in Theorem 2(v). Median algebras satisfy Δ_2, and hence so do lattices and, more generally, weakly associative lattices. Implication algebras satisfy Δ_3 but not Δ_2 (A. Mitschke). In a nontrivial lattice L, take $f(x, y, z) = x \wedge (y \vee z)$. Then the variety generated by $\langle L; f \rangle$ satisfies Δ_4 but not Δ_3 (K. A. Baker). For $n > 4$, there are no known natural examples of varieties that satisfy Δ_n but not Δ_{n-1}. However, we can use the identities in Theorem 2(iv) to define a variety V_n, taking the polynomials p_i to be basic operations, and it turns out (K. Fichtner, D. Kelly) that V_n does not satisfy Δ_{n-1}. Similarly, the conditions in Theorem 3(iv) become weaker as n grows larger, and $(n+1)$-permutability does not imply n-permutability.

The next two theorems characterize strong Mal'cev classes and Mal'cev classes, respectively. First, a definition is needed. By the *nonindexed product* $\bigotimes(\mathfrak{A}_i \mid i \in I)$ of a system of algebras $\mathfrak{A}_i$ $(i \in I)$ we mean the algebra whose universe is the Cartesian product of the sets A_i, and which has an n-ary operation f corresponding to each sequence of n-ary polynomials $\langle p_i \mid i \in I \rangle$, where f is defined by

$$f(a_0, a_1, \cdots)(i) = p_i(a_0(i), a_1(i), \cdots).$$

By the *nonindexed product* $\bigotimes(V_i \mid i \in I)$ of a system of varieties V_i we mean the variety generated by the algebras $\bigotimes(\mathfrak{A}_i \mid i \in I)$ with $\mathfrak{A}_i \in V_i$. It is clear that the product $W = \bigotimes(V_i \mid i \in I)$ has a representation in each factor V_i. In fact, the operation symbols f_k $(k \in K)$ in W are associated with systems of polynomials $\langle p_{k,i} \mid i \in I \rangle$, and we represent W by the system $\langle p_{k,i} \mid k \in K \rangle$. The nonindexed product of two algebras $\mathfrak{A}'$ and $\mathfrak{A}''$ is written $\mathfrak{A}' \otimes \mathfrak{A}''$, and the product of two varieties V' and V'', $V' \otimes V''$. The following statements are easily verified:

If $\mathfrak{A} = \mathfrak{A}' \otimes \mathfrak{A}''$, then the subalgebras of $\mathfrak{A}$ are precisely the algebras $\mathfrak{B}' \otimes \mathfrak{B}''$ with $\mathfrak{B}'$ a subalgebra of $\mathfrak{A}'$ and $\mathfrak{B}''$ a subalgebra of $\mathfrak{A}''$.

If $\mathfrak{A} = \mathfrak{A}' \otimes \mathfrak{A}''$, then the congruence relations on $\mathfrak{A}$ are precisely the relations $\Theta' \times \Theta''$ with $\Theta' \in \mathfrak{C}(\mathfrak{A}')$ and $\Theta'' \in \mathfrak{C}(\mathfrak{A}'')$ (where $\langle a', a'' \rangle (\Theta' \times \Theta'') \langle b', b'' \rangle$ means that $a' \Theta' b'$ and $a'' \Theta'' b''$).

For two varieties V' and V'', $\mathfrak{A} \in V' \otimes V''$ iff $\mathfrak{A} \cong \mathfrak{A}' \otimes \mathfrak{A}''$ for some $\mathfrak{A}' \in V'$ and $\mathfrak{A}'' \in V''$.

These statements can of course be extended to products with an arbitrary finite number of factors, but the corresponding statements for infinite products are false.

Theorem 4 (*W. Taylor* [1973], *W. D. Neumann* [1974]). *A class $\mathcal{K}$ of varieties is a strong Mal'cev class iff the following conditions hold:*

(i) *Every variety in which some member of $\mathcal{K}$ has a representation belongs to $\mathcal{K}$.*

(ii) *The nonindexed product of countably many members of $\mathscr{K}$ always belongs to $\mathscr{K}$.*

(iii) *Every member of $\mathscr{K}$ is contained in a finitely based variety that also belongs to $\mathscr{K}$.*

Proof. These conditions are obviously necessary. Assuming that (i)–(iii) hold, let W be the nonindexed product of all the finitely presented members of $\mathscr{K}$. (Apart from the indexing, there are only countably many finitely presented varieties.) Then $W \in \mathscr{K}$, and hence there exists a finitely based variety $W' \in \mathscr{K}$ with $W \subseteq W'$. There exists a finitely presented variety U such that U and W' are representable in each other. Therefore $U \in \mathscr{K}$, and U is representable in W, hence in all the finitely based members of $\mathscr{K}$, and therefore by (iii) in every member of $\mathscr{K}$. Conversely, by (i) every variety in which U can be represented belongs to $\mathscr{K}$. Therefore $\mathscr{K}$ is a strong Mal'cev class.

Theorem 5 (*W. Taylor* [1973], *W. D. Neumann* [1974]). *A class $\mathscr{K}$ of varieties is a Mal'cev class iff the following conditions hold:*

(i) *Every variety in which some member of $\mathscr{K}$ has a representation belongs to $\mathscr{K}$.*

(ii) *The nonindexed product of two members of $\mathscr{K}$ always belongs to $\mathscr{K}$.*

(iii) *Every member of $\mathscr{K}$ is contained in a finitely based variety that also belongs to $\mathscr{K}$.*

Proof. Since every Mal'cev class is the union of a chain of strong Mal'cev classes, the necessity of (i)–(iii) follows from the preceding theorem. Assuming (i)–(iii), arrange the finitely presented members of $\mathscr{K}$ into a sequence $V_0, V_1, V_2, \cdots$. Let $U_0 = V_0$ and, assuming that $U_k \in \mathscr{K}$ has been chosen and is finitely presented, choose a finitely based variety $U'_{k+1} \in \mathscr{K}$ with $U_k \otimes V_{k+1} \subseteq U'_{k+1}$, and choose a finitely presented variety U_{k+1} such that U_{k+1} and U'_{k+1} are representable in each other. Then U_{k+1} is representable in U_k for each $k \in \omega$, and it is easy to see that $\mathscr{K}$ is precisely the class of those varieties in which some U_k is representable. Hence $\mathscr{K}$ is a Mal'cev class.

We have seen three examples of properties of the congruence lattices of algebras that give rise to Mal'cev classes of varieties. These conditions involve the lattice operations in the congruence lattice and the relative multiplication. Since the relative product of two congruence relations is not in general a congruence relation, it is useful in the general study of such phenomena to replace the congruence lattice $\mathfrak{C}(\mathfrak{A})$ by the larger structure $\mathfrak{R}(\mathfrak{A}) = \langle R(\mathfrak{A}); \wedge, \cdot, ^{-1}, C\rangle$ where $R(\mathfrak{A})$ is the set of all reflexive binary relations on A that are compatible with the operations of $\mathfrak{A}$, $\wedge$ is set intersection, $\cdot$ denotes relative products, Θ^{-1} is the inverse of Θ, and $C(\Theta)$ is the transitive closure of Θ; $\leqq$ will denote $\subseteq$. If V is a variety of algebras, we let $\mathfrak{R}(V)$ be the class of all $\mathfrak{R}(\mathfrak{A})$ with $\mathfrak{A} \in V$. Observe that if $\Theta \in R(\mathfrak{A})$, then $C(\Theta \cdot \Theta^{-1})$ is the congruence relation generated by Θ. Hence, if Θ and Φ are congruence relations on $\mathfrak{A}$, then $\Theta \vee \Phi = C(\Theta \cdot \Phi)$ Observe also that

$$\mathfrak{R}(\mathfrak{A} \otimes \mathfrak{B}) \cong \mathfrak{R}(\mathfrak{A}) \times \mathfrak{R}(\mathfrak{B}).$$

Theorem 6 (*A. F. Pixley* [1972 b] *and R. Wille* [1970]). *Suppose ε is an implication*

$$\bigwedge_{i<m} q_i \leqq r_i \Rightarrow q \leqq r$$

where q_i, r_i, q and r are polynomials in $\wedge$, $\cdot$, $^{-1}$ and C, and let $\mathscr{K}$ be the class of all varieties V such that $\mathfrak{R}(V) \models \varepsilon$. Then

(i) *$\mathscr{K}$ satisfies Conditions* (i) *and* (ii) *of Theorem* 5.
(ii) *If C does not occur in q_i and r_i, then $\mathscr{K}$ is a weak Mal'cev class.*
(iii) *If C does not occur in q_i, r_i and q, then $\mathscr{K}$ is a Mal'cev class.*

Proof. Suppose $U \in \mathscr{K}$, and suppose there exists a representation p of U in V. Then, for any $\mathfrak{A} \in V$, $\mathfrak{A}^p$ belongs to U, hence $\mathfrak{R}(\mathfrak{A}^p) \models \varepsilon$. Since $\mathfrak{R}(\mathfrak{A})$ is a substructure of $\mathfrak{R}(\mathfrak{A}^p)$, it follows that $\mathfrak{R}(\mathfrak{A}) \models \varepsilon$. The class $\mathscr{K}$ therefore satisfies Condition (i) of Theorem 5.

Suppose $V = V' \otimes V''$, $V', V'' \in \mathscr{K}$. For any $\mathfrak{A} \in V$ we then have $\mathfrak{A} \cong \mathfrak{A}' \otimes \mathfrak{A}''$ with $\mathfrak{A}' \in V'$ and $\mathfrak{A}'' \in V''$, and it follows that $\mathfrak{R}(\mathfrak{A}) \cong \mathfrak{R}(\mathfrak{A}') \times \mathfrak{R}(\mathfrak{A}'')$. Since ε holds in $\mathfrak{R}(\mathfrak{A}')$ and $\mathfrak{R}(\mathfrak{A}'')$, we infer that it also holds in $\mathfrak{R}(\mathfrak{A})$. Thus $\mathscr{K}$ satisfies Condition (ii) of Theorem 5.

We next prove (iii). For $k \in \omega$ let r^k be the polynomial obtained from r by replacing C by the operation C_k, where $C_k(\Theta) = \Theta \circ \Theta \circ \cdots \circ \Theta$ with k factors. In particular, $C_0(\Theta) = \omega$. It is clear that if $x_0, x_1, \cdots, x_{n-1}$ are the variables that occur in ε, then $r(\Theta_0, \Theta_1, \cdots, \Theta_{n-1})$ is the union of the relations $r^k(\Theta_0, \Theta_1, \cdots, \Theta_{n-1})$ for $k \in \omega$. Let ε_k be the formula obtained from ε by replacing r by r^k.

Suppose V is a variety such that $\mathfrak{R}(V) \models \varepsilon$, and let W be the class of all structures $\mathfrak{A}' = \langle \mathfrak{A}, \Theta_0, \Theta_1, \cdots, \Theta_{n-1}, a, b\rangle$ such that $\mathfrak{A}$ is an algebra of the similarity type of V, $\Theta_0, \Theta_1, \cdots, \Theta_{n-1} \in R(\mathfrak{A})$, and $a, b \in A$. For each $k \in \omega$ there exists a first order formula ε_k' in the language of W such that for any structure $\mathfrak{A}' \in W$, $\mathfrak{A}' \models \varepsilon_k'$ iff $q_i(\bar{\Theta}) \leqq r_i(\bar{\Theta})$ for $i < m$, $\langle a, b\rangle \in q(\bar{\Theta})$ and $\langle a, b\rangle \notin r^k(\bar{\Theta})$, where $\bar{\Theta} = \langle \Theta_0, \Theta_1, \cdots, \Theta_{n-1}\rangle$. From this it is clear that one of the formulas ε_k must hold in $\mathfrak{R}(V)$, for we could otherwise find for each $k \in \omega$ an algebra $\mathfrak{A}_k \in V$ such that one of the associated structures

$\mathfrak{A}_k'$ satisfies ε_k', and by the compactness theorem we could therefore find $\mathfrak{A} \in V$ such that one of the associated structures $\mathfrak{A}'$ satisfies all the formulas ε_k', contrary to the fact that ε holds in $\mathfrak{R}(\mathfrak{A})$. Thus Condition (iii) of Theorem 5 is also satisfied, and $\mathscr{K}$ is a Mal'cev class.

Finally, under the hypothesis of (ii), let q^k be the polynomial obtained from q by replacing the operation C everywhere by C_k. Then ε holds in $\mathfrak{R}(\mathfrak{A})$ iff, for each $k \in \omega$, the formula obtained from it by replacing q by q^k holds. Applying (iii) to each of these formulas, we conclude that $\mathscr{K}$ is a weak Mal'cev class.

There obviously should be a fourth part to Theorem 6, stating that under certain conditions $\mathscr{K}$ is a strong Mal'cev class. However, no general result of this type is known, and in particular, we do not know whether $\mathscr{K}$ is necessarily a strong Mal'cev class whenever C does not occur in ε.

In Part (ii), the antecedent of ε is not allowed to contain the symbol C, but we are nevertheless able to formulate conditions of the form "if $\Theta_0, \Theta_1, \cdots$ are congruence relations, then $\cdots$," for Θ_i is a congruence relation iff $\Theta_i \circ \Theta_i^{-1} = \Theta_i$. Thus the class of varieties whose congruence lattices satisfy some given identity is always a weak Mal'cev class. It is not known whether it is always a Mal'cev class. If the antecedent of ε contains C, then we do not even know whether $\mathscr{K}$ is a weak Mal'cev class. An interesting example would be the semi-distributive law

$$x \vee y = x \vee z \rightarrow x \vee y = x \vee (y \wedge z).$$

The dual of this law

$$x \wedge y = x \wedge z \rightarrow x \wedge y = x \wedge (y \vee z),$$

does yield a weak Mal'cev class.

§61. CONGRUENCE VARIETIES

With any variety V of algebras we associate the lattice variety $\mathrm{Con}(V)$ generated by the class of all congruence lattices $\mathfrak{C}(\mathfrak{A})$ of the algebras $\mathfrak{A} \in V$. We shall here be primarily concerned with the question, which varieties K of lattices are equal to $\mathrm{Con}(V)$ for some variety V of algebras. We call such varieties K *congruence varieties*, and, more specifically, we call $\mathrm{Con}(V)$ the congruence variety of V. Most of the known results are of a negative nature, in that they show that certain varieties of lattices are not congruence varieties. In order to facilitate the discussion of this phenomenon we introduce a new consequence relation: If Γ is a set of lattice identities and ε is a lattice identity, $\Gamma \models_c \varepsilon$ will mean that, for every variety V of algebras, $\mathrm{Con}(V) \models \Gamma$ implies $\mathrm{Con}(V) \models \varepsilon$. The distributive, modular and Arguesian identities will be abbreviated dist, mod, and arg, respectively.

One can classify lattice inclusions according to the *complexity* of the polynomials involved. Assign to each polynomial a weight, 0 for variables, 1 for joins and meets of variables, 2 for joins of meets of variables and meets of joins of variables, etc., and call $p \leqq q$ an (m, n)-inclusion if p has weight m and q has weight n. E.g., the distributive law, $x \wedge (y \vee z) \leqq (x \wedge y) \vee (x \wedge z)$, is a (2, 2)-inclusion, the modular law, $x \wedge ((x \wedge y) \vee z) \leqq (x \wedge y) \vee (x \wedge z)$, is a (3, 2)-inclusion, and the Arguesian law is a (2, 5)-inclusion. We can now state one of the major results to be established in this section, although its proof will require some preliminaries.

Theorem 1. *If* ε *is a nontrivial* (3, 3)*-inclusion, then* $\varepsilon \models_c$ mod.

It was first discovered by J. B. Nation that $\varepsilon \models_c$ mod does not imply $\varepsilon \models$ mod. In fact, he showed in J. B. Nation [1974] that if ε is a nontrivial inclusion of the form

$$\sigma_0 \wedge w \leqq \bigvee_{1 \leqq i \leqq n} \sigma_0 \wedge \sigma_i$$

where the polynomials σ_i are joins of variables, then $\varepsilon \models_c$ mod. Other inclusions with the same property can be found in A. Day [1973 a] and [1975], B. Jónsson [1976 a] and P. Mederly [1975]. A. Day [a] considers inclusions of the form

$$\bigwedge_{1 \leqq i \leqq n} (x_i \vee y) \leqq \bigvee_{1 \leqq j \leqq n} \Big((y_j' \vee x) \wedge \bigwedge_{1 \leqq i \leqq m} (x_i' \vee y)\Big) \qquad (\delta_{m,n})$$

where x is the join of all the variables x_i while x_i' is the join of the variables x_k with $k \neq i$, and y and y_j' are defined similarly. (These inclusions arise naturally as the conjugate equations of certain splitting lattices.) Finally, R. Freese and J. B. Nation [1977] shows that if ε is a nontrivial (3,3)-inclusion, then $\varepsilon \models \delta_{m,n}$ for some m and n; by A. Day, $\delta_{m,n} \models_c$ mod, thereby obtaining the above theorem. The reduction of the general theorem to the special case makes use of Day's technique of splitting an interval or a quotient. If $I = u/v$ is an interval in a lattice L, we form a new lattice $L[I]$ by replacing each member a of u/v by the two ordered pairs $\langle a, 0\rangle$ and $\langle a, 1\rangle$. The inclusion relation in $L[I]$ is defined in an obvious manner: for $a, b \in L - I$, $c, d \in I$ and $i, j = 0, 1$,

$$\begin{aligned} a &\leqq b \text{ in } L[I] \quad \text{iff} \quad a \leqq b \text{ in } L, \\ a &\leqq \langle d, j\rangle \quad \text{iff} \quad a \leqq d, \\ \langle c, i\rangle &\leqq b \quad \text{iff} \quad c \leqq b, \\ \langle c, i\rangle &\leqq \langle d, j\rangle \quad \text{iff} \quad c < d \text{ or } c = d \text{ and } i \leqq j. \end{aligned}$$

It is a routine matter to check that $L[I]$ is a lattice, and that there is a homomorphism of $L[I]$ onto L that maps each member of $L - I$ onto itself and maps $\langle c, i\rangle$ onto c for $c \in I$. If I consists of a single element, i.e., if $u = v$, we write $L[u]$ for $L[I]$.

Lemma 1 (*R. Freese and J. B. Nation* [1977]). *Any nontrivial* (3, 3)-*inclusion fails in* $D[a]$ *for some finite distributive lattice* D *and some* $a \in D$.

Proof. The given inclusion, $\pi \leqq \sigma$, may be assumed to have the form

$$\bigwedge_{i \in I} \bigvee_{j \in J_i} \pi_{i,j} \leqq \bigvee_{r \in R} \bigwedge_{s \in S_r} \sigma_{r,s}$$

where each $\pi_{i,j}$ is a meet of variables and each $\sigma_{r,s}$ is a join of variables. We may assume that the inclusion holds in every distributive lattice, for otherwise it fails in every nontrivial lattice, and D may be taken to be a one-element lattice. Let X be the set of all variables that occur in the inclusion. We identify the polynomials in X with the members of the free lattice $FL(X)$, and for $\lambda \in FL(X)$ we let $\tilde{\lambda}$ be the image of λ under the canonical homomorphism of $FL(X)$ onto $FD(X)$. Let Θ be the smallest congruence relation on $FD(X)$ that identifies $\tilde{\pi}$ and $\tilde{\sigma}$, and let $D = FD(X)/\Theta$. Finally, let f be the canonical homomorphism of $FL(X)$ onto D, and let $a = \pi f = \sigma f$.

Let g be the homomorphism of $FL(X)$ into $D[a]$ that takes an element $x \in X$ into $\langle a, 1\rangle$ if $xf = a$, but into xf otherwise. Observe that $f = gh$, where h is the obvious homomorphism of $D[a]$ onto D. The proof of the lemma will be completed by showing that

$$\sigma g \leqq \langle a, 0\rangle < \langle a, 1\rangle \leqq \pi g.$$

We need some simple facts from lattice theory: Since $FD(X)$ is distributive, the principal congruence $\Theta = \Theta(\tilde{\sigma}, \tilde{\pi})$ is characterized by the condition that, for all $u, v \in FD(X)$,

$$u \Theta v \quad \text{iff} \quad u \vee \tilde{\sigma} = v \vee \tilde{\sigma} \text{ and } u \wedge \tilde{\pi} = v \wedge \tilde{\pi}.$$

In particular, the Θ-class containing $\tilde{\sigma}$ and $\tilde{\pi}$ is precisely the quotient $\tilde{\sigma}/\tilde{\pi}$.

For $i \in I$ and $r \in R$ let

$$\pi_i = \bigvee_{j \in J_i} \pi_{i,j}, \qquad \sigma_r = \bigwedge_{s \in S_r} \sigma_{r,s}.$$

We want to show that $\pi_i g > \langle a, 1\rangle$ or, in other words, that Θ does not identify $\tilde{\pi}_i$ and $\tilde{\pi}$. Since $\tilde{\pi}_i \geqq \tilde{\pi}$, this is equivalent to the assertion that $\tilde{\pi}_i \nleqq \tilde{\sigma}$. Certainly $\pi_i \nleqq \sigma$, and hence for some $j \in J_i$, $\pi_{i,j} \nleqq \sigma$. From this it follows that $\pi_{i,j} \nleqq \sigma_r$ for all $r \in R$. For each $r \in R$ we can therefore choose $r\varphi \in S_r$ such that $\pi_{i,j} \nleqq \sigma_{r,r\varphi}$ or, in other words, such that $\pi_{i,j}$ and $\sigma_{r,r\varphi}$ have no variable in common. Now

$$\tilde{\sigma} = \bigwedge_{\psi \in F} \bigvee_{r \in R} \tilde{\sigma}_{r,r\psi}$$

where F is the Cartesian product of the sets S_r, and we clearly have

$$\tilde{\pi}_{i,j} \nleqq \bigvee_{r \in R} \tilde{\sigma}_{r,r\varphi},$$

so that $\bar{\pi}_{i,j} \nleqq \bar{\sigma}$, hence $\bar{\pi}_i \nleqq \bar{\sigma}$ as was to be shown. We thus have $\pi_i g = \pi_i f > \langle a, 1 \rangle$ for all $i \in I$, hence $\bar{\pi} g \geqq \langle a, 1 \rangle$. Similarly, $\bar{\sigma} g \leqq \langle a, 0 \rangle$, and the proof is complete.

By a *critical quotient* in a subdirectly irreducible lattice S we mean a nontrivial quotient that is collapsed by every nontrivial congruence relation on S. A quotient u/v in any lattice is said to be *prime* (or *atomic*) if u covers v.

Lemma 2 (*R. McKenzie* [1972 a]). *Suppose S is a subdirectly irreducible lattice, u/v is a critical prime quotient in S, and f is a homomorphism of $FL(X)$ onto S. For any $p, q \in FL(X)$, the lattice inclusion $p \leqq q$ fails in S iff there exist $r, s \in FL(X)$ such that $p \leqq q \vDash r \leqq s$, $rf = u$ and $sf = v$.*

Proof. If $p \leqq q$ holds in S, and if f takes r to u and s to v, then we cannot have $p \leqq q \vDash r \leqq s$, for $r \leqq s$ fails in S. Conversely, suppose $p \leqq q$ fails in S. Let V be the variety of all lattices that satisfy $p \leqq q$, and let f' be the canonical homomorphism of $FL(X)$ onto $F_V(X)$, the V-free lattice generated by X. If Θ and Θ' are the kernels of f and f', then $\Theta' \nsubseteq \Theta$, because $S \notin V$. Therefore, if g and g' are the canonical homomorphisms of S and $F_V(X)$ onto $FL(X)/\Theta \vee \Theta'$, then g identifies u and v. Taking any elements $r, s \in FL(X)$ with $rf = u$ and $sf = v$, we therefore have $r(\Theta \vee \Theta')s$. Thus there exist $r = t_0, t_1, \cdots, t_n = s$ in $FL(X)$ such that for $i = 0, 1, \cdots, n-1$ we have alternatingly $t_i \Theta t_{i+1}$ and $t_i \Theta' t_{i+1}$. We can choose r and s with $r \geqq s$, and the elements t_i can then be so chosen that $t_0 \geqq t_1 \geqq \cdots \geqq t_n$. Since u covers v, there must exist $i < n$ with $t_i f = u$ and $t_{i+1} f = v$. For this value of i we do not have $t_i \Theta t_{i+1}$ and must therefore have $t_i \Theta' t_{i+1}$, i.e., $p \leqq q \vDash t_i \leqq t_{i+1}$. With an obvious change in notation, this is the desired conclusion.

An element a of a lattice L is said to be *join-reducible* if a is the join of two smaller elements, and a is said to be *meet-reducible* if a is the meet of two larger elements. If a is both join-reducible and meet-reducible, then it is said to be *doubly reducible*. An element that is not join-reducible is said to be *join-irreducible*, and an element that is not meet-reducible is said to be *meet-irreducible*. We say that a is *join-prime* if $a \leqq b \vee c$ always implies that $a \leqq b$ or $a \leqq c$, and we say that a is *meet-prime* if $a \geqq b \wedge c$ always implies that $a \geqq b$ or $a \geqq c$. In a distributive lattice, join-prime is equivalent to join-irreducible, and meet-prime is equivalent to meet-irreducible.

Lemma 3 (*A. Day* [a]). *Let a be a doubly reducible element of a finite Boolean algebra B, and let B have m atoms that are not contained in a and n atoms that are contained in a. If $p \leqq q$ is any lattice inclusion that fails in $B[a]$, then $p \leqq q \vDash \delta_{m,n}$.*

Proof. Let $X=\{x_1, x_2, \cdots, x_m, y_1, y_2, \cdots, y_n\}$, and let f be a homomorphism of $FL(X)$ onto $B[a]$ that maps $y_1, y_2, \cdots, y_n$ onto the atoms contained in a and $x_1, x_2, \cdots, x_m$ onto the remaining atoms. Let x, y, x_i' and y_j' have the same meaning as in $\delta_{m,n}$, and let π and σ be the left and right sides of $\delta_{m,n}$.

It is not hard to show that $B[a]$ is subdirectly irreducible, with $\langle a, 1\rangle / \langle a, 0\rangle$ a critical prime quotient, and Lemma 2 therefore applies. Thus there exist $r, s \in FL(X)$ with $p \leqq q \models r \leqq s$, $rf=\langle a, 1\rangle$ and $sf=\langle a, 0\rangle$. Clearly $\pi f=\langle a, 1\rangle$ and $\sigma f=\langle a, 0\rangle$. We claim that π and σ are, respectively, the smallest member of $FL(X)$ that is mapped onto $\langle a, 1\rangle$ and the largest member that is mapped onto $\langle a, 0\rangle$. This will yield $\pi \leqq r$ and $s \leqq \sigma$, hence $p \leqq q \models \pi \leqq \sigma$.

Let g be the obvious homomorphism of $B[u]$ onto B. We claim that for $u \in B$ the set $u(fg)^{-1}$ has a smallest element $\alpha(u)$ and a largest element $\beta(u)$. We first look at the atoms $b_i=x_i f=x_i fg$ and $c_j=y_j f=y_j fg$. For any $z \in FL(X)$ we have either $x_i \leqq z$ or $z \leqq x_i' \vee y$. This is obvious if $z \in X$, and the set of elements z with this property is easily seen to form a sublattice of $FL(X)$. It follows that if $z < x_i$, then $z \leqq x_i \wedge (x_i' \vee y)$, and therefore $zfg=0$. Therefore $\alpha(b_i)=x_i$. Similarly, $\alpha(c_j)=y_j$, and for the coatoms b_i' and c_j' we have $\beta(b_i')=x_i' \vee y$ and $\beta(c_j')=y_j' \vee x$. The functions α and β are then extended to all of B in such a way that α preserves joins and β preserves meets. Clearly $\alpha(u) \geqq \pi$ whenever $u > a$, and since $\langle a, 1\rangle$ is join prime, it follows that the meet of the elements $\alpha(u)$ with $u > a$ is the smallest element of $FL(X)$ that is mapped to $\langle a, 1\rangle$ by f. But this meet is precisely π. Similarly, σ is the largest element of $FL(X)$ that is mapped to $\langle a, 0\rangle$ by f.

Remark. In R. McKenzie [1972 a], a lattice homomorphism $f: L_1 \to L_2$ is said to be *bounded* if, for each $u \in L_2$, the set of all $v \in L_1$ with $vf \leqq u$ is either empty or else has a largest element, and if the dual condition also holds. Our proofs would be better motivated if they were preceded by a systematic study of bounded homomorphisms, but this would require more space than could be justified by the limited use made of this concept here. For a more complete account we refer the reader to R. McKenzie [1972 a] and A. Day [a].

Corollary (*R. Freese and J. B. Nation* [1977]). *If* ε *is a nontrivial* (3, 3)-*inclusion, then* $\varepsilon \models \delta_{m,n}$ *for some* m *and* n.

Proof. By Lemma 1, ε fails in $D[a]$ for some finite distributive lattice D and some $a \in D$. We can embed D in a finite Boolean algebra B in such a way that a is doubly reducible in B, and the conclusion then follows from Lemma 3.

Proof of Theorem 1. By Corollary 2, ε implies $\delta_{m,n}$ for some positive integers m and n. Suppose V is a variety of algebras whose congruence variety satisfies ε, and hence $\delta_{m,n}$. Call the left-hand side of $\delta_{m,n}$ π and the right-hand side σ, and let $X=\{x_1, x_2, \cdots, x_m, y_1, y_2, \cdots, y_n\}$. Consider a V-free algebra $\mathfrak{F}(T)$ where $T=\{s, t, u_{i,j} \mid 1 \leqq i \leqq m, 1 \leqq j \leqq n+1\}$, and so define the homomorphism h of $FL(X)$ to $\mathfrak{C}(\mathfrak{F}(T))$

$$x_i h = \Theta(s, u_{i,1}) \vee \Theta(t, u_{i,n+1}),$$
$$y_j h = \bigvee_{1 \leqq i \leqq n} \Theta(u_{i,j}, u_{i,j+1}).$$

Then $\langle s, t\rangle \in \pi h$, hence $\langle s, t\rangle \in \sigma h$, and there exist $s=v_0, v_1, \cdots, v_r=t$, all in the same equivalence class $\operatorname{mod}(x_i' \vee y)h$ for $1 \leqq i \leqq n$, such that $\langle v_k, v_{k+1}\rangle \in \langle y_j' \vee x)h$ whenever $k \equiv (j-1) \bmod m$.

Consider the V-free algebra $\mathfrak{F}(S)$, where $S=\{a, b, c, d\}$, and let

$$\alpha = \Theta(c, d), \qquad \beta = \Theta(a, c) \vee \Theta(b, d), \qquad \gamma = \Theta(a, b) \vee \Theta(c, d).$$

We want to consider certain homomorphisms ξ of $\mathfrak{F}(T)$ into $\mathfrak{F}(S)$ that map s to a and t to b, and take the relation σh into γ. Such a homomorphism is determined by its values on T, and may therefore be visualized as a matrix with entries $u_{i,j}\xi$. Now each $(x_i' \vee y)h$ is determined by its restriction to T, and it is easily seen to partition T into two classes, the ith row of the matrix $(u_{i,j})$, and the remaining members of T. We therefore consider homomorphisms whose matrices have the property that one row contains only c's and d's, and the remaining rows are all a's and b's. Let H be the set of all such homomorphisms. Clearly each $\xi \in H$ sends some $(x_i' \vee y)h$ into γ, and therefore takes σh into γ.

We want to show that, for $0 \leqq k < r$, there exists $\xi_k \in H$ with $\langle v_k\xi_k, v_{k+1}\xi_k\rangle \in \beta$. Recall that $\langle v_k, v_{k+1}\rangle \in (y_j' \vee x)h$ for some j. Now $(y_j' \vee x)h$ is generated by its restriction R_j to T, and R_j partitions T into two classes, the s-class, which also contains the first j columns of $(u_{i,j})$, and the t-class, which contains the remaining columns. We therefore let ξ_k be the homomorphism in H that takes $u_{1,\kappa}$ to c when $\kappa \leqq j$ and to d when $\kappa > j$, and for $i > 1$ takes $u_{i,\kappa}$ to a when $\kappa \leqq j$ and to b when $\kappa > j$. Then ξ_k maps the s-class of $(y_j' \vee x)h$ into $\{a, c\}$ and the t-class into $\{b, d\}$, and thus maps R_j into β. It follows that ξ_k maps $(y_j' \vee x)h$ into β, and in particular $\langle v_k\xi_k, v_{k+1}\xi_k\rangle \in \beta$.

We wish to show that $v_{k+1}\xi_k$ and $v_{k+1}\xi_{k+1}$ can be connected by a sequence of elements such that any two successive terms are identified by either α or $\beta \wedge \gamma$. We actually prove a stronger statement. For ξ, $\eta \in H$, write $\xi\bar{\alpha}\eta$ if α identifies $w\xi$ and $w\eta$ for all $w \in F(T)$, and define $\bar{\beta}$ similarly. We show that any two members ξ and η of H can be connected by a sequence of members of H such that any two successive terms are identified by either $\bar{\alpha}$ or $\bar{\beta}$. It suffices to consider the case in which the c's and d's do not occur in the same row in the matrix for ξ as in the matrix for η, and we may therefore assume that they occur in the first row of the first matrix and in the second row of the second matrix. We consider two types of transformations of H: the first type replaces each member of H by another member in the same $\bar{\alpha}$-class and the second type replaces each member of H by another member in the same $\bar{\beta}$-class. The first type of transformation consists of replacing some of the c's by d's and replacing some of the d's by c's. For the second type of transformation we replace all the c's by a's and all the d's by b's, and in one of the rows containing a's and b's replace all the a's by c's and all the b's by d's. We now describe how one can get from ξ to η by a succession of such transformations.

First, to adjust the ith row, where $i>2$, we apply a transformation of the second kind, replacing the a's and b's in the ith row by c's and d's, and, of course, replacing the c's and d's in the first row by a's and b's. We then apply a transformation of the first kind, putting c's in those places in the ith row where η has an a, and d's where η has a b. By another transformation of the second kind we put the c's and d's back in the first row, and a's and b's in the ith.

Thus we obtain a homomorphism in H whose matrix agrees with the one for η except in the first two rows. Next we apply a transformation of the first kind, putting c's in those places in the first row where η has a's, and d's where η has b's. A transformation of the second kind, putting c's and d's in the second row, yields a homomorphism whose matrix agrees with the one for η except possibly in the second row, and a final transformation of the first kind yields η.

Thus $a=s\xi_0$ and $b=t\xi_{r-1}$ can be connected by a sequence of elements in $F(S)$ such that any two successive terms are identified by either α or β. Furthermore, all the terms are of the form $v_k\eta$ for some $\eta \in H$, and they are therefore all in the same γ-class. Consequently, $\langle a, b\rangle \in \alpha \vee (\beta \wedge \gamma)$, and we conclude by Theorem 60.3 that V is congruence modular.

Nation's original result on the relation $\models_c$ is easily derived from the preceding theorem.

Corollary (*J. B. Nation* [1974]). *Suppose ε is a nontrivial lattice inclusion of the form*

$$\sigma_0 \wedge w \leqq \bigvee_{1 \leqq i \leqq n} \sigma_0 \wedge \sigma_i$$

where $\sigma_0, \sigma_1, \cdots, \sigma_n$ are joins of variables. Then $\varepsilon \models_c$ mod.

Proof. Let Y be the set of all variables that occur in w, and let Y_i be the set of all variables that occur in w but not in σ_i. For $i \geqq 1$, Y_i must be nonempty, for otherwise the inclusion $w \leqq \sigma_i$ would hold in every lattice, and ε would be trivial. Let $w_i = \bigwedge Y_i$. For each $u \in FL(Y)$ we have either $u \leqq \sigma_i$ or $w_i \leqq u$. (Note that, by definition, σ_i is in the lattice generated by all the variables that occur in the formula.) This is obviously the case for $u \in Y$, and the set of elements for which it holds is readily seen to be closed under the lattice operations. In particular, $w_i \leqq w$. Letting $w' = w_1 \vee w_2 \vee \cdots \vee w_n$, we infer that the inclusion

$$\sigma_0 \wedge w' \leqq \bigvee_{1 \leqq i \leqq n} \sigma_0 \wedge \sigma_i$$

is a consequence of ε. This inclusion, call it ε', is easily seen to be non-trivial. Since ε' is a (3,3)-inclusion, we conclude by Theorem 1 that $\varepsilon' \models_c$ mod, hence $\varepsilon \models_c$ mod.

By imposing additional conditions on the inclusion ε in the preceding corollary one obtains the stronger conclusion that $\varepsilon \models_c$ dist.

Theorem 2 (*J. B. Nation* [1974]). *Suppose ε is a nontrivial lattice inclusion of the form*

$$\sigma_0 \wedge w \leqq \bigvee_{1 \leqq i \leqq n} \sigma_0 \wedge \sigma_i,$$

where $\sigma_0, \sigma_1, \cdots, \sigma_n$ are joins of variables and σ_0 and w have no variable in common. Then $\varepsilon \models_c$ dist.

Proof. Let X, Y and Z_i be the sets consisting of all those variables that occur in ε, in w, and in σ_i, respectively, and let $Y_i = Y - Z_i$. Since ε is nontrivial, Z_0 must contain some variable x_0 that is in no other Z_i, and all the sets Y_i must be nonempty. Letting $w_i = \bigwedge Y_i$, we have for each $u \in FL(Y)$ either $u \leqq \sigma_i$ or $w_i \leqq u$, hence in particular $w_i \leqq w$. Letting $w' = w_1 \vee w_2 \vee \cdots \vee w_n$, we infer that the inclusion

$$\sigma_0 \wedge w' \leqq \bigvee_{1 \leqq i \leqq n} \sigma_0 \wedge \sigma_i$$

is a consequence of ε. Calling this inclusion ε', we complete the proof by showing that $\varepsilon' \models_c$ dist.

Consider any variety V with $\mathrm{Con}(V) \models \varepsilon'$. Let $\mathfrak{F}(T)$ be the V-free algebra generated by the set $T = \{u_0, u_1, \cdots, u_n\}$, and let $s = u_0$, $t = u_n$. Let $x_0\Phi$ be the equivalence relation on T that identifies no two distinct elements except s and t, and for $x_0 \neq x \in X$ let $x\Phi$ be the equivalence relation on T such that, for $0 \leqq i < j \leqq n$,

$$\langle u_i, u_j \rangle \in x\Phi \quad \text{iff} \quad x \in Y_k \quad \text{whenever } i < k \leqq j.$$

Let ψ be the homomorphism of $FL(X)$ into $\mathfrak{C}(\mathfrak{F}(T))$ with $x\psi = \Theta(x\Phi)$ for $x \in X$. Clearly $\sigma_0\psi$ identifies s and t, and $x\psi$ identifies u_{i-1} and u_i whenever $x \in Y_i$. Therefore $\langle s, t\rangle \in (\sigma_0 \wedge w')\psi$. It follows that there exist $s = v_0, v_1, v_2, \cdots, v_m = t$ in $F(T)$, all in the same $\sigma_0\psi$-class, such that for each i with $0 < i \leqq m$ there is an integer $j(i)$ with $1 \leqq j(i) \leqq n$ for which $\langle v_{i-1}, v_i\rangle \in \sigma_{j(i)}\psi$.

We now consider the V-free algebra $\mathfrak{F}(S)$ generated by a 3-element set $S = \{a, b, c\}$, and let

$$\alpha = \Theta(a, b), \qquad \beta = \Theta(a, c) \qquad \gamma = \Theta(b, c).$$

Let H be the set of all homomorphisms of $\mathfrak{F}(T)$ into $\mathfrak{F}(S)$ that map T into S, s onto a and t onto b. Observe that $x\psi$ is trivial whenever $x_0 \neq x \in Z_0$, and therefore $\sigma_0\psi = x_0\psi = \Theta(s, t)$. Hence each member of H maps $\sigma_0\psi$ into α. In particular, all the elements $v_i\xi$ with $0 \leqq i \leqq m$ and $\xi \in H$ belong to the same α-class, $[a]\alpha = [b]\alpha$.

We want to find $\xi_i \in H$ such that $v_{i-1}\xi_i = v_i\xi_i$. Since $\langle v_{i-1}, v_i\rangle \in \sigma_{j(i)}\psi$, it suffices to choose ξ_i so that it maps $\sigma_{j(i)}\psi$ onto the identity relation in $\mathfrak{F}(S)$, and since $\sigma_{j(i)}\psi$ is generated by its restriction to T, it suffices to choose ξ_i so that any two $\sigma_{j(i)}\psi$-equivalent members of T are mapped onto the same member of S. We therefore define

$$u_k\xi_i = \begin{cases} a & \text{if } u_k \equiv s \bmod \sigma_{j(i)}\psi, \\ b & \text{if } u_k \equiv t \bmod \sigma_{j(i)}\psi, \\ c & \text{otherwise.} \end{cases}$$

It is essential here that $\sigma_{j(i)}\psi$ does not identify s and t. This is so because if $x \in Z_{j(i)}$, then $x \notin Y_{j(i)}$, and hence $x\Phi$ is contained in the equivalence relation that partitions T into the two sets $\{u_k \mid k < j(i)\}$ and $\{u_k \mid k \geqq j(i)\}$.

For $\xi, \eta \in H$ write $\xi\bar{\beta}\eta$ if $\langle w\xi, w\eta\rangle \in \beta$ for all $w \in F(T)$, and define $\bar{\gamma}$ similarly. We next show that for any $\xi, \eta \in H$ there exist $\zeta, \zeta' \in H$ such that $\xi\bar{\beta}\zeta\bar{\gamma}\zeta'\bar{\beta}\eta$. The images of the elements u_k under ζ and ζ' will depend on the images under ξ and η. Since the possible values for $u_k\xi$ and $u_k\eta$ are a, b and c, this gives rise to nine cases. The values of $u_k\zeta$ and $u_k\zeta'$ in each case are determined according to the following table:

$u_k\xi$	a	a	a	b	b	b	c	c	c
$u_k\zeta$	a	c	c	b	b	b	a	c	c
$u_k\zeta'$	a	b	c	c	b	c	a	b	c
$u_k\eta$	a	b	c	a	b	c	a	b	c

We have shown that for $0 < i \leqq m$, $v_{i-1}\xi_{i-1}$ can be connected to $v_{i-1}\xi_i$ by a sequence whose successive terms are always identified by either β or γ, and $v_{i-1}\xi_i$ and $v_i\xi_i$ can also be connected by such a sequence. Hence, so can the elements $a = v_0\xi_0 = s\xi_0$ and $b = v_m\xi_m = t\xi_m$. Furthermore, all the terms of these sequences are of the form $v_k\zeta$, and they are therefore all in the same α-class. Therefore $\langle a, b\rangle \in (\alpha \wedge \beta) \vee (\alpha \wedge \gamma)$, and we conclude by Theorem 60.2 that V is congruence distributive.

Theorem 3 (*B. Jónsson* [1975]). *If ε is a nontrivial* (2, 2)*-inclusion, then* $\varepsilon \models_c$ dist.

This can be derived from the preceding theorem with the aid of the following technical lemma, which will be stated here without a proof. A lattice L is said to be *n-distributive* if, for all $a, b_1, b_2, \cdots, b_n \in L$,

$$a \wedge \bigvee_{1 \leqq i \leqq n} b_i = \bigvee_{1 \leqq k \leqq n} (a \wedge \bigvee_{i \neq k} b_i).$$

By M_m we mean the lattice of length 2 and order $m+2$, i.e., the lattice consisting of a zero element 0, a unit element 1, and m elements that cover 0 and are covered by 1.

Lemma 4 (*J. Doyen and Ch. Herrmann* [1976]). *For any positive integer m there exists a positive integer n such that if L is any modular lattice that does not contain M_m as a sublattice, then L is n-distributive.*

Proof of Theorem 3. We shall show that ε fails in M_m for some m. Since, by Theorem 1, $\varepsilon \models_c$ mod, this implies that if V is any variety with $\mathrm{Con}(V) \models \varepsilon$, then $\mathrm{Con}(V)$ is n-distributive for some n, and by Theorem 2 this implies that $\mathrm{Con}(V)$ is distributive.

The given inclusion can be written in the form

$$\bigwedge_{j \in J} (\bigvee Y_j) \leq \bigvee_{k \in K} (\bigwedge Z_k)$$

where Y_j and Z_k are nonempty finite sets of variables. Let J' be the set of all $j \in J$ such that Y_j is a one-element set $\{y_j\}$, and let K' be the set of all $k \in K$ such that Z_k is a one-element set $\{z_k\}$. Also let $J''=J-J'$ and $K''=K-K'$. We may assume that no two of the sets Y_j are comparable, and similarly for the sets Z_k, for if they were then ε could be replaced by a simpler but equivalent inclusion. We may also assume that J' and K' are nonempty, for we could replace the left side by its meet with some variable that does not occur in ε, and replace the right side by its join with another such variable.

Observe that $y_j \neq z_k$ for $j \in J'$ and $k \in K'$, for otherwise ε would be trivial. We can therefore replace all the variables y_j with $j \in J'$, wherever they occur, by a single variable y_0, and all the variables z_k by a single variable z_0. The resulting inclusion ε',

$$y_0 \wedge \bigwedge_{j \in J''} (\bigvee U_j) \leqq z_0 \vee \bigvee_{k \in K''} (\bigwedge V_k)$$

is clearly a consequence of ε. Observe that none of the variables y_i with $i \in J'$ belongs to Y_j for $j \in J''$, for then Y_i and Y_j would be comparable. Also, Y_j cannot be contained in $\{z_k \mid k \in K'\}$, for then ε would be trivial. Hence each U_j contains at least two distinct elements, and the same is true of the sets V_k. It is now clear that ε' fails in M_m if we assign distinct atoms to all the variables that occur in ε', for the value of the left side will be the value assigned to y_0, and the value of the right-hand side will be the value assigned to z_0.

We have seen several examples of lattice identities ε such that $\varepsilon \models_c$ mod but not $\varepsilon \models$ mod. There also exist nontrivial identities ε for which $\varepsilon \models_c$ mod fails. This was first proved by S. V. Polin [1977] and A. Day used his technique to produce uncountably many nonmodular congruence varieties. The particular identity used below was also first considered by Day. Considering three variables x, y, z, let $y_1 = y$, $z_1 = z$, $y_{n+1} = y \vee (x \wedge z_n)$, $z_{n+1} = z \vee (x \wedge y_n)$, and let δ_n be the inclusion

$$x \wedge (y \vee z) \leqq (x \wedge y_n) \vee (x \wedge z_n).$$

The next lemma is due to A. Day, who credits the idea of the proof to J. B. Nation.

Lemma 5 (*A. Day*). *Suppose L is a sublattice of the congruence lattice of a semi-lattice S. If L is $2n$-permutable, then $L \models \delta_{2n}$.*

Proof. Suppose $\alpha, \beta, \gamma \in L$, let $\beta_1 = \beta$ and $\gamma_1 = \gamma$, and for $k = 1, 2, \cdots$ let $\beta_{k+1} = \beta \vee (\alpha \wedge \gamma_k)$ and $\gamma_{k+1} = \gamma \vee (\alpha \wedge \beta_k)$. Assuming that $\langle a, b\rangle \in \alpha \wedge (\beta \vee \gamma)$, we need to show that $\langle a, b\rangle \in (\alpha \wedge \beta_{2n}) \vee (\alpha \wedge \gamma_{2n})$.

By the $2n$-permutability of L, there exist $a = c_0, c_1, \cdots, c_{2n} = b$ such that, for $0 \leqq i < n$, $c_{2i}\beta c_{2i+1}\gamma c_{2i+2}$, and of course we have $a\alpha b$. For $1 \leqq i \leqq 2n$, the three elements $a \wedge c_i$, $a \wedge b \wedge c_i$ and $b \wedge c_i$ of S are obviously in the same α-class. We claim that

$$(a \wedge b \wedge c_i)\gamma_i(b \wedge c_i) \quad \text{for } i \text{ even},$$
$$(a \wedge b \wedge c_i)\beta_i(b \wedge c_i) \quad \text{for } i \text{ odd}.$$

For $i = 1$ this follows from the fact that $\beta_1 = \beta$ and $a\beta c_1$. Assuming that it holds for a given i, consider the index $i+1$. If i is odd, then

$$(a \wedge b \wedge c_{i+1})\gamma(a \wedge b \wedge c_i)(\alpha \wedge \beta_i)(b \wedge c_i)\gamma(b \wedge c_{i+1}),$$

and hence $a \wedge b \wedge c_{i+1}$ and $b \wedge c_{i+1}$ are identified by $\gamma_{i+1} = \gamma \vee (\alpha \wedge \beta_i)$. The argument for i even is similar. Since $c_{2n} = b$, we infer that $(a \wedge b)\gamma_{2n}b$. Similarly, for $0 \leqq i \leqq 2n - 1$,

$$(a \wedge c_i)\gamma_{2n-i}(a \wedge b \wedge c_i) \quad \text{for } i \text{ odd},$$
$$(a \wedge c_i)\beta_{2n-i}(a \wedge b \wedge c_i) \quad \text{for } i \text{ even}.$$

In this case we start with $i = 2n - 1$ and work our way down to $i = 0$. The case $i = 0$ yields $a\beta_{2n}(a \wedge b)$. Thus

$$a(\alpha \wedge \beta_{2n})(a \wedge b)(\alpha \wedge \gamma_{2n})b.$$

Theorem 4 (*A. Day*). *It is not the case that* $\delta_4 \models_c$ mod.

Proof. We are going to construct an algebra $\mathfrak{A} = \langle A; \wedge, f_1, f_2, f_3\rangle$ such that $\langle A; \wedge\rangle$ is a semi-lattice and f_1, f_2, f_3 are ternary operations that satisfy the identities in Theorem 60.1(iii) with $n=4$. This will guarantee that the algebras in the variety V generated by $\mathfrak{A}$ have 4-permutable congruences, and hence by the preceding lemma that $\mathrm{Con}(V)$ satisfies δ_4. On the other hand, $\mathfrak{C}(\mathfrak{A})$ will be shown to be nonmodular.

Take two nontrivial Boolean algebras $\mathfrak{B}_1$ and $\mathfrak{B}_2$ and let $\mathfrak{A} = \langle A; \wedge, f_1, f_2, f_3\rangle$ be the algebra with one binary operation and three ternary operations such that $A = B_1 \times B_2$ and, for $\langle a, s\rangle, \langle b, t\rangle, \langle c, u\rangle \in A$,

$$\begin{aligned}
\langle a, s\rangle \wedge \langle b, t\rangle &= \langle a \wedge b, s \wedge t\rangle\\
f_1(\langle a, s\rangle, \langle b, t\rangle, \langle c, u\rangle) &= \langle a \wedge (b' \vee c), s\rangle,\\
f_2(\langle a, s\rangle, \langle b, t\rangle, \langle c, u\rangle) &= \langle a \wedge b \wedge c, s \oplus t \oplus u\rangle,\\
f_3(\langle a, s\rangle, \langle b, t\rangle, \langle c, u\rangle) &= \langle (a \vee b') \wedge c, u\rangle,
\end{aligned}$$

where the prime denotes complements, and $\oplus$ is the symmetric difference operation, $s \oplus t = (s \wedge t') \vee (s' \wedge t)$. It is a simple matter to check that the identities

$$\begin{aligned}
f_1(x, z, z) &= x, & f_1(x, x, z) &= f_2(x, z, z)\\
f_2(x, x, z) &= f_3(x, z, z), & f_3(x, x, z) &= z
\end{aligned}$$

hold in $\mathfrak{A}$ and, as we already observed, this implies that $\mathrm{Con}(V) \models \delta_4$. Finally, defining

$$\begin{aligned}
\langle a, s\rangle \alpha \langle b, t\rangle &\text{ iff } a = b \text{ and } (a = 0 \text{ or } s = t),\\
\langle a, s\rangle \beta \langle b, t\rangle &\text{ iff } s = t,\\
\langle a, s\rangle \gamma \langle b, t\rangle &\text{ iff } a = b,
\end{aligned}$$

one easily checks that $\alpha, \beta, \gamma \in \mathfrak{C}(\mathfrak{A})$, $\alpha < \gamma$, $\alpha \vee \beta = \gamma \vee \beta = \iota$ and $\alpha \wedge \beta = \gamma \wedge \beta = \omega$, so that $\mathfrak{C}(\mathfrak{A})$ is not modular.

The final result of this section will show that the variety of all modular lattices is not a congruence variety.

Theorem 5 (*R. Freese and B. Jónsson* [1976]). mod $\models_c$ arg.

The proof will be modeled on the classical proof that a projective plane that can be embedded in a projective 3-space satisfies Desargues' Law. We begin with two lemmas. The first one is a lattice-theoretic version of the geometric theorem just stated, and the second shows how to "raise the dimension" of a congruence lattice. We state the first without proof.

Lemma 6 (*G. Grätzer, B. Jónsson and H. Lakser* [1973]). *Suppose L is a modular lattice and $a=\langle a_0, a_1, a_2\rangle$ and $b=\langle b_0, b_1, b_2\rangle$ are centrally perspective triangles in L, whose center of perspectivity p satisfies the conditions*

$$p \vee a_i = p \vee b_i = a_i \vee b_i \quad \text{for } i = 0, 1, 2.$$

Let $u=a_0 \vee a_1 \vee a_2 \vee b_0 \vee b_1 \vee b_2$. If there exist $q, r \in L$ such that

$$p \vee q = q \vee r = p \vee r, \qquad u \wedge q = p \wedge a_2, \qquad u \wedge r = p \wedge b_2,$$

then a and b are axially perspective.

Lemma 7 (*R. Freese and B. Jónsson* [1976]). *Suppose V is a variety of algebras, and let K be the class of all lattices L such that L is embeddable in the dual of $\mathfrak{C}(\mathfrak{A})$ for some $\mathfrak{A} \in V$. For any $L \in K$, and any $p, s, t, u \in L$, if*

$$p \vee s = p \vee t = s \vee t \leqq u,$$

then L has an extension L' in K such that, for some $q, r \in L'$,

$$p \vee q = p \vee r = q \vee r, \qquad q \wedge u = p \wedge s, \qquad r \wedge u = p \wedge t.$$

Proof. It may help to keep a geometric picture in mind. We are thinking of p, q and r as collinear points in a plane u, and we want to embed this plane in a larger space in which there is a line meeting u in p, with q and r two other points on that line.

There exists a dual embedding f of L into $\mathfrak{C}(\mathfrak{A})$ for some $\mathfrak{A} \in V$. Let the images of p, q, r, and u be α, β, γ, and μ, respectively. Consider a subdirect product $\mathfrak{B}$ of $\mathfrak{A}$ and $\mathfrak{A}$. The two projections $\langle a_0, a_1\rangle \to a_0$ and $\langle a_0, a_1\rangle \to a_1$ of $\mathfrak{B}$ onto $\mathfrak{A}$ induce embeddings $\lambda \to \lambda_i$ of $\mathfrak{C}(\mathfrak{A})$ into $\mathfrak{C}(\mathfrak{B})$. Thus we have two copies μ_0 and μ_1 of the "plane" u in the dual of $\mathfrak{C}(\mathfrak{B})$, and we want to choose $\mathfrak{B}$ so that α_0 and α_1, the two images of p, coincide. This means that, for elements $\langle a_0, a_1\rangle$ and $\langle b_0, b_1\rangle$ of $\mathfrak{B}$, the conditions $a_0 \alpha b_0$ and $a_1 \alpha b_1$ should be equivalent. Thinking of B as a binary relation on $A \times A$, we see that this holds iff $B \subseteq \alpha$. We take $B=\alpha$.

Identifying the members x of L with their images under the map $x \to (xf)_0$, we are going to take for L' the dual of $\mathfrak{C}(\mathfrak{B})$, and for q and r the elements β_1 and γ_1. To complete the proof we need to show that $\beta_1 \vee \mu_0 = \alpha_0 \vee \beta_0$ and $\gamma_1 \vee \mu_0 = \alpha_0 \vee \gamma_0$. This follows readily from the fact that

$$\alpha_0 = \alpha_1 = \omega_0 \vee \omega_1 \qquad \xi_0 \vee \omega_1 = \omega_0 \vee \xi_1,$$

where ω is the zero element of $\mathfrak{C}(\mathfrak{A})$, and ξ is an arbitrary member of $\mathfrak{C}(\mathfrak{A})$. To prove this, note that if $\langle a_0, a_1\rangle \alpha_i \langle b_0, b_1\rangle$, then $\langle a_0, b_1\rangle$ is in B, and hence

$$\langle a_0, a_1\rangle \omega_1 \langle a_0, b_1\rangle \omega_1 \langle b_0, b_1\rangle,$$

and that if $\langle a_0, a_1\rangle\xi_0\langle b_0, b_1\rangle$, then $a_0\xi b_0$, hence

$$\langle a_0, a_1\rangle\omega_0\langle a_0, a_0\rangle\xi_1\langle b_0, b_0\rangle\omega_0\langle b_0, b_1\rangle.$$

Proof of Theorem 5. Suppose V is a congruence modular variety and $\mathfrak{A} \in V$. Making use of Theorem 59.4 and its corollary, we consider two triangles $a = \langle a_0, a_1, a_2\rangle$ and $b = \langle b_0, b_1, b_2\rangle$ in the dual L of $\mathfrak{C}(\mathfrak{A})$ that are centrally perspective, and whose center of perspectivity p satisfies the condition $p \vee a_2 = p \vee b_2 = a_2 \vee b_2$. We then apply the preceding lemma with $s = a_2$ and $t = b_2$ to embed L in a modular lattice L' that satisfies the hypothesis of Lemma 7, and we infer by that lemma that a and b are axially perspective. Consequently, by Theorem 59.4, L is Arguesian, and therefore by the corollary to Theorem 59.4, so is $\mathfrak{C}(\mathfrak{A})$.

According to a recent result by R. Freese, the variety of all Arguesian lattices is not a congruence variety. In fact, he shows that there are identities stronger than the Arguesian identity that hold in every modular congruence variety. He has also shown that the join of two congruence varieties is a congruence variety, but their meet need not be. Of the many open questions related to the results treated in this section, we mention only a few. Is the join of arbitrarily many congruence varieties a congruence variety? Is there a largest modular congruence variety? (A possible candidate is the congruence variety of the variety of all groups.) Does any congruence variety other than the variety of all lattices contain all the modular lattices? Is the consequence relation $\models_c$ compact? I.e., given a set Γ of lattice identities and a lattice identity ε with $\Gamma \models_c \varepsilon$, does there always exist a finite subset Γ' of Γ with $\Gamma' \models_c \varepsilon$?

§62. CONGRUENCE DISTRIBUTIVITY AND FINITE BASES

R. C. Lyndon [2] made the rather surprising discovery that a finite algebra of a finite similarity type need not be finitely based (Theorem 27.4), and since then many such examples have been found. However, there have also been some important positive results. In particular, S. Oates and M. B. Powell [1964] showed that every finite group is finitely based, and R. McKenzie [1970] showed that every finite lattice with finitely many additional operations is finitely based. Shortly afterward, K. A. Baker made the remarkable discovery that the only property of lattices that is needed is the fact that their congruence lattices are distributive, i.e., he showed that every finite algebra of finite type that generates a congruence distributive variety is finitely based. He also generalized McKenzie's result in other ways. His result was announced in 1972, and the details of his proof were first published in K. A. Baker [a]. For other proofs, see S. Burris [b], B. Jónsson [a], M. Makkai [1973] and W. Taylor [1978]. Ch. Herrmann [1973] is also relevant, although it only deals with varieties of lattices, for it contains concepts and techniques that were adapted by Baker to congruence distributive varieties in general, and were used by him in the final version of his proof.

Our presentation will follow B. Jónsson [a], which contains the most general version of the theorem. We start with a finite basis criterion that does not involve congruence distributivity, and then borrow four lemmas from K. A. Baker [a] to show that this result applies to the situation in Baker's theorem. By the compactness theorem, Theorem 39.2, a variety V is finitely based iff it is an elementary class. A standard model-theoretic technique for proving that an axiomatic class is elementary is to show that its complement is closed under ultraproducts. By an obvious generalization of Theorem 42.4, this complement may be taken with respect to an arbitrary elementary class that contains the given class.

If B is any class of algebras, we let B_{SI} be the class of all subdirectly irreducible members of B, and B_{FSI} the class of all finitely subdirectly irreducible members. (An algebra $\mathfrak{A}$ is *finitely subdirectly irreducible* if in any representation of $\mathfrak{A}$ as a subdirect product of finitely many algebras, the projection onto at least one of the factors is one-to-one.)

Theorem 1 (*B. Jónsson* [a]). *Suppose V is a variety of algebras and B is an elementary class that contains V. If there exists an axiomatic class C such that $B_{\mathrm{SI}} \subseteq C$ and $V \cap C$ is elementary, then V is finitely based.*

Proof. If V is not finitely based, then there exists an algebra $\mathfrak{A} \in V$ that is an ultraproduct of algebras $\mathfrak{B}_i \in B - V$ $(i \in I)$ modulo some ultrafilter U on I. Each $\mathfrak{B}_i$ has as a homomorphic image a subdirectly irreducible algebra $\mathfrak{B}_i'$ that does not belong to V, and the ultraproduct $\mathfrak{A}'$ of the algebras $\mathfrak{B}_i'$ modulo U is in V, since it is a homomorphic image of $\mathfrak{A}$. The algebras $\mathfrak{B}_i'$ need not all be in B, but $\mathfrak{A}'$ is in B, and since B is elementary it follows that the set $\{i \in I \mid \mathfrak{B}_i' \in B\}$ belongs to U. We may therefore assume that all the algebras $\mathfrak{B}_i'$ belong to B, and therefore also to C. Consequently $\mathfrak{A}' \in V \cap C$. However, this is a contradiction, for $V \cap C$ is an elementary class, and $\mathfrak{A}'$ is an ultraproduct of algebras that do not belong to $V \cap C$.

Theorem 2 (*K. A. Baker* [a], *B. Jónsson* [a]). *If V is a congruence distributive variety of a finite type, and if V_{FSI} is an elementary class, then V is finitely based.*

The four lemmas on which the proof is based require some additional terminology. Given an algebra $\mathfrak{A}=\langle A; F\rangle$, a map g from A to A is called a *0-translation* if g is either constant or the identity map, and g is called a *1-translation* if it is a 0-translation or is obtained from one of the basic operations $f \in F$ by freezing all but one of the variables. A *k-translation*, for $k > 1$, is a map that is a composition of k 1-translations. We say that g is a *translation* if it is a k-translation for some k. For $a, b \in A$, and for $k \in \omega$, let $\Gamma_k(a, b)$ be the set of all ordered pairs $\langle c, d\rangle$ such that $\{c, d\}=\{g(a), g(b)\}$ for some k-translation g, and let $\Gamma(a, b)$ be the union of the relations $\Gamma_k(a, b)$ for $k=0, 1, 2, \cdots$. Then $\Theta(a, b)$ is the transitive closure of $\Gamma(a, b)$. We say that $\langle a, b\rangle$ and $\langle a', b'\rangle$ are bounded if $\Gamma(a, b) \cap \Gamma(a', b') \neq \omega$, and we say that $\langle a, b\rangle$ and $\langle a', b'\rangle$ are k-bounded if $\Gamma_k(a, b) \cap \Gamma_k(a', b') \neq \omega$. If there exists a natural number k such that, for all $a, b, a', b' \in A$, $\Theta(a, b) \cap \Theta(a', b') \neq \omega$ implies $\Gamma_k(a, b) \cap \Gamma_k(a', b') \neq \omega$, then the smallest such integer k is called the *radius* of $\mathfrak{A}$—in symbols $R(\mathfrak{A})$. If no such k exists, then we set $R(\mathfrak{A})=\infty$. For a class K of algebras, we let $R(K)$ be the sup of $R(\mathfrak{A})$ for $\mathfrak{A} \in K$.

Suppose V is a congruence distributive variety of finite similarity type. By Theorem 60.2 there exist ternary polynomials $p_0, p_1, \cdots, p_n$ such that the following identities hold in V:

$$\begin{aligned} p_0(x, y, z) &= x, \qquad p_n(x, y, z) = z, \qquad p_i(x, y, x) = x, \\ p_i(x, x, z) &= p_{i+1}(x, x, z) \quad \text{for } i < n, i \text{ even}, \\ p_i(x, z, z) &= p_{i+1}(x, z, z) \quad \text{for } i < n, i \text{ odd}. \end{aligned}$$

Let V_0 be the class of all algebras of the same type as in V that satisfy these identities. Then V_0 is a finitely based congruence distributive variety that contains V. We may assume that the p_i's are among the operation symbols for the variety V. (Alternatively, we could modify the definition of a 1-translation.) The above notation and assumptions will be in effect throughout the next four lemmas.

Lemma 1. If $\mathfrak{A} \in V_0$, $e_0, e_1, \cdots, e_m \in A$, *and* $e_0 \neq e_m$, *then there exists* $i < m$ *such that* $\langle e_0, e_m\rangle$ *and* $\langle e_i, e_{i+1}\rangle$ *are* 1*-bounded.*

Proof. Let $e_{i,j}=p_j(e_0, e_i, e_m)$ for $i \leqq m$ and $j \leqq n$, and let q be the smallest index such that the elements $e_{i,q}$ are not all equal to e_0. Such q exists because $e_{i,n}=e_m$, and we have $q>0$ because $e_{i,0}=e_0$. If q is odd, then $e_{0,q}=e_{0,q-1}=e_0$, and we can therefore choose $r<m$ such that $e_{r,q}=e_0 \neq e_{r+1,q}$. In this case take $c=e_0$ and $d=e_{r+1,q}$, and consider the 1-translations $g(x)=p_q(e_0, e_{r+1}, x)$, $h(x)=p_q(e_0, x, e_m)$. In the alternative case, when q is even, and therefore $e_{m,q}=e_{m,q-1}=e_0$, choose $r<m$ so that $e_{r,q} \neq e_0=e_{r+1,q}$, and let $c=e_{r,q}$, $d=e_0$, $g(x)=p_q(e_0, e_r, x)$ and $h(x)=p_q(e_0, x, e_m)$. In either case,

$$\{c, d\} = \{g(e_0), g(e_m)\} = \{h(e_r), h(e_{r+1})\}$$

and hence $\langle c, d\rangle \in \Gamma_1(e_0, e_m) \cap \Gamma_1(e_r, e_{r+1})$. Therefore $\langle e_0, e_m\rangle$ and $\langle e_r, e_{r+1}\rangle$ are 1-bounded.

Lemma 2. *For all* $\mathfrak{A} \in V_0$ *and* $a, b, a', b' \in A$, *if* $\Theta(a, b) \cap \Theta(a', b') \neq \omega$, *then* $\Gamma(a, b) \cap \Gamma(a', b') \neq \omega$.

Proof. Suppose $\langle c, d\rangle \in \Theta(a, b) \cap \Theta(a', b')$ and $c \neq d$. Then there exist $c = e_0, e_1, \cdots, e_m = d$ such that $\langle e_i, e_{i+1}\rangle \in \Gamma(a, b)$ for $i < m$. As in the preceding proof, let $e_{i,j} = p_j(e_0, e_i, e_m)$, and choose $r < m$ and $q < n$ so that the elements $c' = p_q(e_0, e_r, e_m)$ and $d' = p_q(e_0, e_{r+1}, e_m)$ are distinct. Then $\langle c', d'\rangle \in \Gamma(a, b)$, and we also have $\langle c', d'\rangle \in \Theta(a', b')$ because $\Theta(a', b')$ identifies every one of the elements $e_{i,j}$ with $p_j(e_0, e_i, e_0) = e_0$. Thus there exist $c' = e_0', e_1', \cdots, e'_{m'} = d'$ with $\langle e_i', e'_{i+1}\rangle \in \Gamma(a', b')$ for $i < m'$. By the preceding lemma, there exists $i < m'$ such that $\langle c', d'\rangle$ and $\langle e_i', e'_{i+1}\rangle$ are 1-bounded, and we conclude that $\langle a, b\rangle$ and $\langle a', b'\rangle$ are bounded, as was to be shown.

Lemma 3. *For any axiomatic subclass* C *of* V_0, C_{FSI} *is axiomatic iff* $R(C_{\mathrm{FSI}}) < \infty$.

Proof. For each $k \in \omega$ we can construct a formula $\varphi_k(x, y, x', y')$ such that, for all $\mathfrak{A} \in V_0$ and $a, b, a', b' \in A$, $\mathfrak{A} \models \varphi_k(a, b, a', b')$ iff $\langle a, b\rangle$ and $\langle a', b'\rangle$ are k-bounded. (It is essential here that the similarity type of V_0 is finite.) By the preceding lemma, an algebra $\mathfrak{A} \in V_0$ is finitely subdirectly irreducible iff it satisfies the infinite formula that is the disjunction of the equations $x = y$ and $x' = y'$ and of all the formulas $\varphi_k(x, y, x', y')$. It follows that if C_{FSI} is axiomatic, then it must satisfy the disjunction of finitely many of these formulas. In fact, since the formulas φ_k decrease in strength as k grows large, C_{FSI} must satisfy the disjunction of $x = y$, $x' = y'$ and one of the formulas φ_k. The smallest such k is clearly $R(C_{\mathrm{FSI}})$.

Conversely, if $R(C_{\mathrm{FSI}}) = k < \infty$, then C_{FSI} is precisely the class of all algebras $\mathfrak{A} \in C$ that satisfy the disjunction of $x = y$, $x' = y'$ and $\varphi_k(x, y, x', y')$.

Lemma 4. *If* $R(V_{\mathrm{FSI}}) = k < \infty$, *then* $R(V) \leqq k + 2$.

Proof. Consider any $\mathfrak{A} \in V$ and $a_0, b_0, a_1, b_1 \in A$, and suppose $\langle c, d\rangle \in \Theta(a_0, b_0) \cap \Theta(a_1, b_1)$ and $c \neq d$. There exists a homomorphism of $\mathfrak{A}$ onto a subdirectly irreducible algebra $\mathfrak{A}'$ that maps c and d onto distinct elements. It follows that $a_0' \neq b_0'$ and $a_1' \neq b_1'$, where the primes denote images in A'. Therefore $\langle a_0', b_0'\rangle$ and $\langle a_1', b_1'\rangle$ are k-bounded, say $\langle u, v\rangle \in \Gamma_k(a_0', b_0') \cap \Gamma_k(a_1', b_1')$. It is easy to see that there exist $\langle u_i, v_i\rangle \in \Gamma_k(a_i, b_i)$ such that

$u_i'=u$ and $v_i'=v$. In fact, if g' is a k-translation for $\mathfrak{A}'$ with $u=g'(a_i')$ and $v=g'(b_i')$, then a corresponding k-translation g for $\mathfrak{A}$ is obtained by replacing each element of A' that was used in the construction of g' by one of its counterimages in A, and we let $u_i=g(a_i)$ and $v_i=g(b_i)$.

We now choose $j<n$ so that the elements $u^*=p_j(u_0, u_1, v_0)$ and $v^*=p_j(u_0, v_1, v_0)$ are distinct. This can be done because in $\mathfrak{A}'$ we cannot have $p_j(u, u, v)=p_j(u, v, v)$ for all $j<n$, since this would give $u=p_0(u, u, v)=p_1(u, u, v)=p_1(u, v, v)=p_2(u, v, v)=\cdots=p_n(u, v, v)=v$. Observe that $\langle u^*, v^*\rangle \in \Gamma_1(u_1, v_1)$, and that $\langle u^*, u_0\rangle$, $\langle u_0, v^*\rangle \in \Gamma_1(u_0, v_0)$, as is easily seen by considering the 1-translations $p_j(u_0, x, v_0)$, $p_j(u_0, u_1, x)$ and $p_j(u_0, v_1, x)$. According to Lemma 1, either $\langle u^*, v^*\rangle$ and $\langle u^*, u_0\rangle$ are 1-bounded, or else $\langle u^*, v^*\rangle$ and $\langle u_0, v^*\rangle$ are 1-bounded. In either case, $\langle u_0, v_0\rangle$ and $\langle u_1, v_1\rangle$ are 2-bounded, and $\langle a_0, b_0\rangle$ and $\langle a_1, b_1\rangle$ are therefore $(k+2)$-bounded.

Proof of Theorem 2. By Lemma 3, $R(V_{\mathrm{FSI}})=k$ is finite and hence $R(V)\leqq k+2$ by Lemma 4. We let B be the class of all $\mathfrak{A}\in V_0$ such that $R(\mathfrak{A})\leqq k+2$. Since V_0 is elementary and the condition $R(\mathfrak{A})\leqq k+2$ can be expressed by a first order formula, B is an elementary class. Obviously $R(B_{\mathrm{FSI}})\leqq k+2$, and B_{FSI} is therefore elementary by Lemma 3. Since $V\cap B_{\mathrm{FSI}}=V_{\mathrm{FSI}}$ is elementary by hypothesis, we can apply Theorem 1 with $C=B_{\mathrm{FSI}}$ to conclude that V is finitely based.

The proof of the Oates–Powell finite basis theorem for groups is based on quite technical investigations in group theory. The question naturally arose whether this result could be extended to the universal algebra level in the same way as Baker had extended McKenzie's finite basis theorem for lattices. A common generalization would of course be particularly attractive. The most obvious common feature of the two results is that the varieties are congruence modular. However, any hope that congruence distributivity could be replaced by congruence modularity was dashed by the result in S. V. Polin [1976] that there exists a finite algebra of a finite type that generates a non-finitely based variety whose algebras have permutable congruences. A simpler example is given in M. R. Vaughan-Lee [a]. At the present we do not even have a conjecture as to what the common generalization should be, and it does not appear that we are close to a solution of this important problem.

APPENDIX 4

EQUATIONAL LOGIC

by Walter Taylor

This is an abridged version of a survey which appears in full in the Houston Journal of Mathematics. Some material which is not of immediate interest to universal algebra is absent from this version. The reader is advised to consult the original survey (henceforth referred to as EL) for many interesting examples and also for an almost 800-item bibliography.

§63. EQUATIONALLY DEFINED CLASSES

An *equationally defined* class of algebras, alias a *variety*, is a class V for which there exists a set Σ of equations with

$$V = \operatorname{mod} \Sigma = \{\mathfrak{A} \mid \forall e \in \Sigma,\ \mathfrak{A} \models e\}.$$

(Here "$\operatorname{mod} \Sigma$" abbreviates "the class of all models of Σ" and $\mathfrak{A} \models e$ signifies that the equation (identity) holds in $\mathfrak{A}$.)

The basic result is Birkhoff's Theorem 26.3.

This theorem has been followed over the years by many others of a similar format—sometimes called "preservation theorems" since Birkhoff's theorem (together with compactness) has the corollary that if a sentence φ is preserved under formation of homomorphic images, subalgebras and products, then φ is equivalent to a conjunction of equations. For instance H. J. Keisler and S. Shelah proved that *a class L of structures is definable by a set of first order sentences iff L is closed under the formation of isomorphic structures, ultraproducts and ultraroots.*† (See Theorem 41.2 and Corollary. Keisler proved this assuming the G.C.H. and S. Shelah [1971] without. See also H. Andréka, B. Dahn and I. Németi [1976].) For many other preservation theorems, see e.g., M. Makkai [1969]. More in keeping with algebraic results are the following three theorems.

Theorem 1 (*B. Jónsson and E. Nelson* [1974]; *see also J. Płonka* [1975 a]). *V is definable by regular equations if and only if V is closed under the formation of products, subalgebras, homomorphic images and sup-algebras.*

† Inverse of "ultrapower".

(An equation is *regular* if and only if exactly the same variables appear on both sides. The *sup-algebra* of type $\langle n_t \mid t \in T\rangle$ (unique within isomorphism) is the algebra $\langle\{0, 1\}; \{F_t \mid t \in T\}\rangle$ where for each t,

$$F_t(a_1, \cdots, a_{n_t}) = \begin{cases} 0 & \text{if } a_1 = \cdots = a_{n_t} = 0 \\ 1 & \text{otherwise.} \end{cases}$$

Theorem 2 (*N. D. Gautam* [1]; *M. N. Bleicher, H. Schneider and R. L. Wilson* [1973]; *see also A. Shafaat* [1974 b], *G. Grätzer and S. Whitney* [1978]). *V is definable by linear equations if and only if V is closed under the formation of products, subalgebras, homomorphic images and complex algebras.*

(An equation is *linear* iff each side has at most one occurrence of every variable. If $\mathfrak{A} = \langle A; \{F_t \mid t \in T\}\rangle$ is any algebra, the *complex algebra* of $\mathfrak{A}$ is $\mathfrak{B} = \langle B; \{G_t \mid t \in T\}\rangle$, where B is the set of nonempty subsets of A, and

$$G_t(u_1, \cdots, u_{n_t}) = \{F_t(a_1, \cdots, a_{n_t}) \mid a_i \in u_i\ (1 \leqq i \leqq n_t)\}.)$$

Definability of V by unary equations is equivalent to V being closed under *covers* (that is, if an algebra $\mathfrak{A}$ is the set union of subalgebras from V, then $\mathfrak{A} \in V$) by G. Grätzer [9].

The next important result really goes back to J. C. C. McKinsey [1] in 1943 (he proved a theorem which, in combination with the above theorem of Keisler and Shelah, immediately yields our statement). The present formulation was probably first given by S. R. Kogalovskiĭ [5]; see also A. I. Mal'cev [1973], p. 214, [1971], p. 29. Many other proofs have been independently given: G. Grätzer and H. Lakser [1973], G. McNulty [a], A. Shafaat [1969], T. Fujiwara [1971 a], B. Banaschewski [1972]—although the precise formulation differs from author to author. See also C. C. Chang and H. J. Keisler [1973], Theorem 6.2.8, p. 337, and for related results B. Banaschewski and H. Herrlich [1976], O. Keane [1975], W. Hodges [a], [b], H. Andréka and I. Németi [b] and R. John [1977].

Theorem 3. *V is definable by equational implications iff V is closed under the formation of isomorphic images, products, subalgebras, and direct limits.*

An *equational implication* is a formula of the form

$$(e_1 \mathbin{\&} e_2 \mathbin{\&} \cdots \mathbin{\&} e_n) \rightarrow e,$$

where $e, e_1, \cdots, e_n$ are equations, for example, the formula

$$(xy = xz \rightarrow y = z)$$

defining *left-cancellative* semigroups among all semigroups. For some interesting classes defined by equational implication, see M. A. Taylor

[1975] and G. M. Bergman [1975]. For some infinitary analogs of Birkhoff's theorem see J. Słomiński [3], and of McKinsey's theorem, see W. Hodges [b].

In the next result, infinitary formulas are in a sense forced upon one, even though it is a result about ordinary finitary algebras. A *generalized equational implication* is a formula

$$\bigwedge_{i \in I} e_i \to e,$$

where e, e_i $(i \in I)$ are equations (possibly infinitely many). The next theorem was perhaps first stated in B. Banaschewski and H. Herrlich [1976], although maybe some other people knew of it.

Theorem 4. *V is definable by a class of generalized equational implications iff V is closed under the formation of isomorphic images, products, and subalgebras.*

E. R. Fisher [1977] has in fact shown that we can always take this class of formulas to be a *set* iff *Vopěnka's principle* holds. (This is one of the proposed "higher" axioms of set theory.) Some less conclusive results about classes closed under the formation of products and subalgebras occur in J. R. Isbell [1], W. S. Hatcher [1970], and W. S. Hatcher and A. Shafaat [1975].

For some infinitary (in this case, topological) analogs of Birkhoff's theorem, see G. Edgar [1973], P. G. Dixon [1976], and W. Taylor [a]. (A unified treatment appears in P. G. Dixon [1977].) For instance, the condition

$$n!\, x \to 0 \qquad (*)$$

defines a class of topological Abelian groups (here $\to$ means "converges to"), which contains all finite discrete groups but not the circle group. In W. Taylor [a] there is a theory of classes defined by conditions similar to (*); these classes are called "varieties" of topological algebras.

For some other analogs of Birkhoff's theorem which go beyond pure algebra, see S. L. Bloom [1976], H. Andréka and I. Németi [a] and G. Matthiessen [1976].

§64. EQUATIONAL THEORIES

Birkhoff's Theorem 26.3 sets up a one-one correspondence between varieties V and certain sets Σ of equations

$$V \leftrightarrow \Sigma$$

via

$$\begin{aligned} V &\to \operatorname{Eq} V \\ \operatorname{mod} \Sigma &\leftarrow \Sigma. \end{aligned}$$

The sets Σ appearing here (as Eq V) are called *equational theories*. One easily sees that Σ is an equational theory iff $e \in \Sigma$ whenever $\Sigma \models e$; i.e., e is true in every model of Σ, i.e., e is a *consequence* of Σ. Birkhoff's next result was to *axiomatize* the consequence relation, as follows:

(1) $\sigma = \sigma$ is always an axiom.
(2) From $\sigma = \tau$, deduce $\tau = \sigma$.
(3) From $\rho = \sigma$ and $\sigma = \tau$, deduce $\rho = \tau$.
(4) From $\sigma_i = \tau_i$ $(1 \leq i \leq n_t)$, deduce $F_t(\sigma_1, \cdots, \sigma_{n_t}) = F_t(\tau_1, \cdots, \tau_{n_t})$.
(5) From $\sigma(x_1, \cdots, x_n) = \tau(x_1, \cdots, x_n)$, deduce
$\sigma(\rho_1, \cdots, \rho_n) = \tau(\rho_1, \cdots, \rho_n)$.

We write $\Sigma \vdash e$ if there exists a (finite) proof of e starting from Σ and using only the rules (1)–(5).

Theorem 1 (*G. Birkhoff* [2]). $\Sigma \models e$ *iff* $\Sigma \vdash e$.

(This is essentially the same as Theorem 26.2.)

It is sometimes useful to know refined versions of this "completeness" theorem, which state a similar result for different (usually more restrictive) variations on the notion of $\vdash$. See for instance S. Burris [1971 b], p. 40 for one; similar methods go back to A. Tarski.

Theorems parallel to this completeness theorem of Birkhoff are not numerous. There is of course Gödel's complete set of rules of proof for first order logic. A. Selman [1972] has given a set of rules for equational implications (independently discovered by D. Kelly (unpublished)), and some different rules were discovered by H. Andréka and I. Németi (unpublished). J. Słomiński [3] gave an infinitary analog of Birkhoff's theorem.

§65. EQUIVALENT VARIETIES

We mention two of the many possible ways of axiomatizing group theory equationally (not to mention non-equational forms such as $\forall x \exists y (xy = e)$).

$$\Gamma_1: \quad x(yz) = (xy)z$$
$$u \cdot (xx^{-1}) = (y^{-1} \cdot y) \cdot u = u$$

$$\Gamma_2: \quad (xy)z = x(yz) \qquad ex = xe = x$$
$$x/x = e \qquad x/y = x(e/y)$$
$$u(e/u) = (e/u)u = e$$

(where e is the group unit and / denotes "division"). Clearly Γ_1 and Γ_2 do not define the *same* variety, for they are of different types—$\langle 2,1 \rangle$ and $\langle 2, 2, 0 \rangle$. But examination of the models of Γ_1 and the models of Γ_2 will convince one that there is no *essential* difference between a Γ_1-group and a Γ_2-group.† To make this sameness precise we introduce equations which will serve as *definitions*:

$$\begin{aligned} \Delta_1: \quad & x/y = x \cdot y^{-1} \\ & e = x \cdot x^{-1} \\ \Delta_2: \quad & x^{-1} = e/x. \end{aligned}$$

Now one may check that

$$\Gamma_1, \Delta_1 \vdash \Gamma_2 \quad \text{and} \quad \Gamma_2, \Delta_2 \vdash \Gamma_1. \tag{*}$$

One more point is important. If we take one of the Δ_1 definitions of an operation F, i.e., $F=\alpha$, and substitute into α all the Δ_2 definitions, we get $F=\alpha[\Delta_2]$; then one should have

$$\Gamma_2 \vdash F = \alpha[\Delta_2] \tag{**}$$

and likewise with the rôles of Γ_1, Γ_2 and Δ_1, Δ_2 reversed. (E.g., Δ_1 says $x/y=x\cdot y^{-1}$. Upon substituting the Δ_2 definitions, we get $x/y=x\cdot(e/y)$, and this is indeed provable from Γ_2.) Now generally, equational theories Γ_1, Γ_2 are said to be *equivalent* iff there exist sets of definitions Δ_1, Δ_2 such that (*) and (**) hold.

Equivalence has its model-theoretic aspect, too. *Varieties V_1 and V_2 are equivalent* (i.e., Eq V_1 and Eq V_2 are equivalent in the above sense) *iff there exists an isomorphism of categories Φ: $V_1 \to V_2$ which commutes with the forgetful functor to sets* (i.e., $\Phi\mathfrak{A}$ has the same universe as $\mathfrak{A}$, and a similar fact holds for homomorphisms). (For various references and remarks on this theorem of A. I. Mal'cev, see W. Taylor [1973], p. 355.) Perhaps the first historical example of an equivalence of varieties is the well-known natural correspondence between *Boolean algebras* and *Boolean rings* (with unit). Also consider the correspondence between the varieties of *Abelian groups* and *Z-modules*—here the equivalence is so easy that some people write as if it were an equality. Some other interesting examples of equivalence may be found in B. Csákány and L. Megyesi [1975].

Very close to the idea of equivalence (in its model theoretic form) is the idea of *weak isomorphism* as developed in Wrocław. This together with an emphasis on *independent sets* over free algebras gave equational logic a somewhat different direction and flavor in that school. See E. Marczewski [11] for an introduction to these ideas. Briefly, algebras $\mathfrak{A}$ and $\mathfrak{B}$ are weakly isomorphic iff there is a bijection φ: $A \to B$ such that the algebraic operations of $\mathfrak{A}$ are exactly the same as the operations

† There is one intrinsic difference: Γ_1 has an empty model, but Γ_2 does not. Nonetheless Γ_1 and Γ_2 are generally regarded as equivalent. To this extent, empty algebras, do not matter; in fact this book, like many others, completely avoids this question by taking all algebras to be nonempty.

$\varphi^{-1}F(\varphi x_1, \cdots, \varphi x_n)$ where $F(x_1, \cdots, x_n)$ is an algebraic operation of $\mathfrak{B}$. (Here, by the family of algebraic operations, we mean the closure under composition of the family of all operations F_t together with all projection functions.) Then two varieties are equivalent iff they can be generated by weakly isomorphic algebras.

Properties of varieties seem more natural and interesting if they are equivalence-invariant, if only because then they do not force us to make any "unnatural" choice between, say, Γ_1 and Γ_2 above. For example, the similarity type $\langle 2, 1\rangle$ is obviously not intrinsic to the idea of a group. Many of the properties considered below are (obviously) equivalence-invariant, but a few, such as being "one-based" are not (compare 68.1 with 68.7, or 68.3 with 68.9).

It is possible to define equational classes so as to make all expressable properties automatically equivalence-invariant, i.e., to give no preference to any of the possible equivalent forms of a given variety. This amounts to considering the *set* (rather than a sequence $\langle F_t \mid t \in T\rangle$) of *all* possible operations defined by V-terms. This idea goes back to P. Hall (see P. M. Cohn [1], pp. 126–132), and has been worked out independently in detail by W. D. Neumann [1970] and F. W. Lawvere [1]; for more details see, e.g., G. C. Wraith [1970] or W. Felscher [1968], [1969]. See p. 362 of W. Taylor [1973] for more detailed historical remarks, and pp. 390–392 for a proof—also found by W. Felscher [1972]—of the equivalence of these two approaches. We refer the reader to W. Felscher [1972], for additional details and historical remarks.

§66. BASES AND GENERIC ALGEBRAS

As we have seen, if Σ_0 is any set of equations, the smallest equational theory $\supseteq \Sigma_0$ is

$$\text{Eq mod } \Sigma_0 = \{e \mid \Sigma_0 \vdash e\},$$

and in this case we say that Σ_0 is a *set of axioms*, or an *equational base* for Σ. Several of the next sections are concerned with the problem of finding (various sorts of) bases Σ_0.

Here we consider what amounts to some concrete examples of Birkhoff's Theorem 26.3, namely, we look for a base Σ_0 for a single algebra $\mathfrak{A}$. I.e., we want

$$\text{mod } \Sigma_0 = \mathbf{HSP}\ \mathfrak{A}.$$

Actually, given Σ_0, $\mathfrak{A}$ may be regarded as unknown. Here we refer to $\mathfrak{A}$ as *generic* for the variety mod Σ_0 or for the theory $\Sigma = \text{Eq mod } \Sigma_0$. Using $\mathbf{P}$ one can easily see that every variety V has a generic algebra. One such

$\mathfrak{A}$ is the V-free algebra on $\aleph_0$ generators; see also A. Tarski [1]. We mention here a few examples of such $\mathfrak{A}$ and Σ_0.

1. The ring Z of integers is generic for the theory of commutative rings.

2. The 2-element Boolean ring (with unit) is generic for the theory of Boolean rings (with unit), given by the laws for rings (with unit) together with the law $x^2 = x$. (Similarly for Boolean algebras, by remarks in §65; this fact may be interpreted as a completeness theorem for propositional logic.)

3. The algebra $\langle A; \cap, \cup, -, {}^{-}\rangle$, where A is the family of all subsets of the real line and ${}^{-}$ denotes topological closure, is a generic *closure algebra* (J. C. C. McKinsey and A. Tarski [1]; see also A. Tarski [1]). This fact may be regarded as a completeness theorem for the intuitionistic propositional logic.

4. The group of all monotone permutations of $\langle R; \leqq\rangle$ is a generic *lattice-ordered group* (W. C. Holland [1976]). (Here R denotes the set of real numbers, and $\leqq$ its usual ordering.)

5. For fixed $p \in R$, the algebra

$$\langle R; px+(1-p)y\rangle$$

is generic for the laws

$$\begin{aligned} xx &= x \\ (xy)(zw) &= (xz)(yw) \end{aligned}$$

iff p is transcendental (S. Fajtlowicz and J. Mycielski [1974]).

6. The algebra $\langle \omega; x^y\rangle$ is generic for the law

$$(x^y)^z = (x^z)^y$$

(C. F. Martin [1973], p. 56). (Here $\omega = \{0, 1, 2, \cdots\}$ and x^y denotes ordinary exponentiation with $0^0 = 1$.) The proof is surprisingly long.

7. The algebras $\langle \omega; \{A_n \mid n \geqq 3\}\rangle$ and $\langle \omega; \{O_n \mid n \geqq 4\}\rangle$ are each generic for the variety of *all* algebras of type $\langle 2, 2, 2, \cdots\rangle$ (i.e., for $\Sigma_0 = \varnothing$). (C. F. Martin [1973], p. 131, p. 134.) Here A_n ($n \geqq 3$) are the Ackermann operations beyond exponentiation, and the O_n are some related operations invented by J. Doner and A. Tarksi [1969], who conjectured a somewhat stronger statement.

For more examples and more references, see EL.

In place of finding a basis Σ_0 of a given $\mathfrak{A}$, one can often be content with the knowledge that a *finite* Σ_0 exists or does not exist, as the case may be: this is the idea of the next section.

§67. FINITELY BASED THEORIES

We say that an equational theory Σ is ***finitely based*** iff there exists a finite set Σ_0 of axioms for Σ. (This is an equivalence-invariant property of all Σ which have finitely many operations.) Evidently many familiar theories are finitely based—groups, Boolean algebras, rings, lattices, etc.; so also the various examples in §66. Here we list some algebras $\mathfrak{A}$ with Eq $\mathfrak{A}$ known to be ***finitely based:***

1. Any two-element algebra (R. C. Lyndon [1]).

2. Any finite group (S. Oates and M. B. Powell [1964]).

3. Any commutative semigroup (P. Perkins [1969]). (In other words, every variety of commutative semigroups is finitely based. This is also proved in T. Evans [1971].) Also, any 3-element semigroup, see P. Perkins [1969] and any 4-element semigroup, see A. D. Bol'bot [1971]. Also, any idempotent semigroup (A. P. Birjukov [1970], C. F. Fennemore [1971] and J. A. Gerhard [1970]).

4. Any finite ring (R. L. Kruse [1973], I. V. Lvov [1973]).

5. Any finite lattice (possibly with operators) (R. McKenzie [1970]). (This answers Problem 45 of this book.) More generally:

6. Any finite algebra which generates a congruence-distributive variety (K. A. Baker [a]—see also M. Makkai [1973], W. Taylor [1978] and §62). The special case (of 5 and 6) of primal algebras was known much earlier (P. C. Rosenbloom [1], A. Yaqub [1]), see also Appendix 5.

7. Any finite simple algebra with no proper subalgebras except one-element subalgebras, which generates a congruence-permutable variety (R. McKenzie [a]).

8. If V has only finitely many subdirectly irreducible algebras, all of them are finite, and V has definable principal congruence relations, then V is finitely based. As a corollary, if V is a locally finite variety and there

exist $\mathfrak{A}_1, \cdots, \mathfrak{A}_k \in V$ so that every finite $\mathfrak{A} \in V$ is isomorphic to some $\mathfrak{A}_1^{n_1} \times \mathfrak{A}_2^{n_2} \times \cdots \times \mathfrak{A}_k^{n_k}$, then V is finitely based (R. McKenzie [1978 a]). Thus the *para-primal* varieties of D. M. Clark and P. H. Krauss [1976] are finitely based.

9. Any finite $\otimes$-product of finitely based theories is finitely based (see W. Taylor [1973], pp. 357–358 and W. Taylor [1975], pp. 266–267).

10. Recently V. L. Murskiĭ [1975] has proved that "almost all" finite algebras have a finite basis for their identities (i.e., for fixed type, the fraction of such algebras among all algebras of power k approaches 1 as $k \to \infty$—or even, for fixed k, as the number of operations approaches ∞.)

We now turn to equational theories which are not finitely based. Of course it is almost trivial to construct such theories using infinitely many operations F_t $(t \in T)$, even some which are equivalent to the (finitely based!) theory with no operations. As G. M. Bergman pointed out, non-finitely based theories with *finite* T arise almost automatically if we consider a semigroup S which is finitely generated (say by $F \subseteq S$), but not finitely related. Our theory can be taken to have unary operations f for $f \in F$ and laws $f_1 f_2 \cdots f_k x = f_{k+1} \cdots f_s x$ whenever $f_1 \cdots f_k = f_{k+1} \cdots f_s$ in S. Some more interesting research has centered on finding less obvious, but more important, examples of theories and algebras which have T finite and are still ***not finitely based:***

11. The algebra with universe $\{0, 1, 2\}$ and binary operation

	0	1	2
0	0	0	0
1	0	0	1
2	0	2	2

(V. L. Murskiĭ [1], following R. C. Lyndon [2]).

12. The six-element semigroup

$$\left\{\begin{bmatrix}0 & 0\\0 & 0\end{bmatrix}, \begin{bmatrix}1 & 0\\0 & 1\end{bmatrix}, \begin{bmatrix}1 & 0\\0 & 0\end{bmatrix}, \begin{bmatrix}0 & 1\\0 & 0\end{bmatrix}, \begin{bmatrix}0 & 0\\1 & 0\end{bmatrix}, \begin{bmatrix}0 & 0\\0 & 1\end{bmatrix}\right\}$$

with ordinary matrix multiplication, see P. Perkins [1969]. Earlier A. K. Austin [1966] gave some other varieties of semigroups which are not finitely based.

13. There is a 64-element finite nonassociative ring with no finite equational basis—see S. V. Polin [1976].

14. The algebra with universe {0, 1, 2, 3} and binary operation

	0	1	2	3
0	0	1	3	3
1	1	1	2	3
2	3	2	2	3
3	3	3	3	3

(R. E. Park [1976]). This is the only known commutative idempotent groupoid which is not finitely based.

15. A certain *equationally complete* finite algebra, all of whose elements are algebraic constants (D. Pigozzi [c]). This answers Problem 46 of this book.

§68. ONE-BASED THEORIES

Taking Σ_0 and Σ as in §67 we say that Σ (or $V = \text{mod}\,\Sigma$) is *one-based* iff there exists a set of axioms Σ_0 with $|\Sigma_0| = 1$. Here are some algebras or ***theories which are one-based***:

1. The variety of all lattices (R. McKenzie [1970]). McKenzie's original proof yields a single equation of length about 300,000 with 34 variables. R. Padmanabhan [1976] has reduced it to a length of about 300, with 7 variables. Here we mean lattices formulated as usual with meet and join.

2. Any variety which has a polynomial m obeying

$$m(x, x, y) = m(x, y, x) = m(y, x, x) = x$$

(a *majority polynomial*) and is defined by *absorption identities*, i.e., equations of the form $x = p(x, y, \cdots)$. (R. McKenzie [1970]; see also R. Padmanabhan [1976].)

3. Any finitely based variety V of Γ_1-groups (see the beginning of §65) (G. Higman and B. H. Neumann [1952]).

4. Boolean algebras. (G. Grätzer, R. McKenzie and A. Tarski—see G. Grätzer [1971], p. 63.) (Cf. F. M. Sioson [1964].)

5. Any two-element binary algebra except (within isomorphism) as in 10 below (D. H. Potts [1965]).

6. Every finitely based variety with permutable and distributive congruences (R. McKenzie [1975]; R. Padmanabhan and R. W. Quackenbush [1973]). By 67.6 this applies to any finite algebra which generates a variety with permutable and distributive congruences, e.g., a *quasi-primal* algebra (see A. F. Pixley [1971], R. W. Quackenbush [1974 b]). This result for primal algebras was already known to G. Grätzer and R. McKenzie [1967]. (R. McKenzie [1975] contains some very interesting special one-based varieties.)

Here are some theories (and algebras) which are 2-based but ***not 1-based***:

7. The variety of all lattices given in terms of the single quaternary operation $D(x, y, z, w) = (x \vee y) \wedge (z \vee w)$ (R. McKenzie [1970]). (Cf. 1.)

8. Any finitely based variety of lattices other than the variety of all lattices and the trivial variety defined by $x = y$ (R. McKenzie [1970]). (Here again we mean the usual lattice operations. The fact that they are 2-based was discovered by R. Padmanabhan [1969].)

9. Any nontrivial finitely based variety of Γ_2-groups (defined at the beginning of §65) (T. C. Green and A. Tarski [1970], A. Tarski [1968]). (Cf. 3 above.)

10. $\mathfrak{A} = \langle\{0, 1\}; \vee\rangle$ and $\mathfrak{A} = \langle\{0, 1\}; \rightarrow\rangle$ with

$\vee$	0	1
0	0	1
1	1	1

$\rightarrow$	0	1
0	1	1
1	0	1

(D. H. Potts [1965]).

11. If Σ is a finitely based theory of type $\langle m_1, m_2\rangle$ with $m_1, m_2 \geqq 2$ in which F_1 and F_2 are each idempotent, i.e.,

$$\Sigma \vdash F_i(x, x, \cdots, x) = x \qquad (i = 1, 2),$$

then Σ is 2-based (and may also be 1-based) (R. Padmanabhan [1972]).

12. If Σ is a finitely based theory with a majority polynomial (as in 2 above), then Σ is 2-based (and sometimes 1-based) (R. Padmanabhan and R. W. Quackenbush [1973]). (R. McKenzie [1970] had this result for varieties in which lattices are definable.)

§69. IRREDUNDANT BASES

Σ_0 is an *irredundant base* for Σ iff Σ_0 is a base for Σ but no proper subset of Σ_0 is a base. A. Tarski [1968] has defined

$$\nabla(\Sigma) = \{|\Sigma_0| \mid \Sigma_0 \text{ is an irredundant base for } \Sigma\}.$$

Tarski's interpolation theorem (A. Tarski [1968] and [1975]) states that $\nabla(\Sigma)$ is always an interval (see G. McNulty and W. Taylor [1975] for a connection between this and some other interpolation theorems, especially in graph theory; see also S. Givant [1975]). One easily checks that (at least for a type $\langle n_t \mid t \in T\rangle$ with T finite), either $\nabla(\Sigma)=\varnothing$, $\nabla(\Sigma)=\{\aleph_0\}$ or $\nabla(\Sigma)$ is an interval of natural numbers. All these cases can occur.

R. McKenzie proved that $\nabla(\Sigma)$ can be any interval, and J. Ng showed that Σ can be found with one binary operation (see A. Tarski [1968]). For example, if

$$\Sigma = \{F^6x_1\cdots x_6 = F^6x_2\cdots x_6x_1\},$$

then $\nabla(\Sigma)=\{1, 2\}$, essentially because the cyclic group C_6 has both a single generator and an irredundant set of two generators. $\nabla(\Sigma)$ is an unbounded interval if $\Sigma \vdash \tau = x$, where τ contains x at least twice (A. Tarski [1968]), strengthened by G. McNulty [1976] to the case where τ has at least one operation of rank $\geqq 2$. On the other hand, if Σ is defined by balanced equations, and Σ is finitely based, then $\nabla(\Sigma)$ is a bounded interval (G. McNulty [1976]). (An equation $\sigma = \tau$ is *balanced* iff each variable, each nullary operation symbol and each unary operation symbol occurs equally often in σ and τ.) T. C. Green obtained irredundant bases of power n (any $n \in \omega$) for groups (T. C. Green and A. Tarski [1970], see also A. Tarski [1968]). $\nabla(\Sigma)$ is also defined on pp. 194–195. Finally, note that ∇ is *not* an equivalence invariant.

§70. THE LATTICE OF EQUATIONAL THEORIES

Dually to §26 of Chapter 4, for a fixed type $\langle n_t \mid t \in T\rangle$ ordering the family Λ of all equational theories by inclusion we obtain a *complete lattice*. More specifically,

$$\bigvee_{i\in I} \Sigma_i = \{e \mid \bigcup_{i\in I} \Sigma_i \vdash e\} = \mathrm{Eq}(\bigcap_{i\in I} \operatorname{mod} \Sigma_i).$$

From the proof-theoretic characterization of $\bigvee$ it follows that Λ is an algebraic closure system and hence an *algebraic lattice*. Specifically, the compact elements of Λ are the finitely based theories of §67 above, and every element is the join (actually the union) of all its finitely based subtheories. Obviously the join of two finitely based theories is finitely based (this holds for compact elements in any lattice); but the meet (i.e., intersection) of two finitely based theories can fail to be finitely based. We present an example of J. Karnofsky (unpublished—see D. Pigozzi [1970]) (here and below we will sometimes express a theory by one of its finite bases without further mention):

$$\begin{aligned}\Sigma_1: \quad x(yz) &= (xy)z\\ (xyz)^2 &= x^2y^2z^2\\ x^3y^3z^2w^3 &= y^3x^3z^2w^3\end{aligned}$$

$$\begin{aligned}\Sigma_2: \quad x(yz) &= (xy)z\\ x^3y^3 &= y^3x^3.\end{aligned}$$

The theory $\Sigma_1 \cap \Sigma_2$ is not finitely based, for every basis must contain equations essentially the same as

$$x^3y^3v_0{}^2\cdots v_{2k}^2w^3 = y^3x^3v_0{}^2\cdots v_{2k}^2w^3 \qquad (k = 1, 2, 3, \cdots).$$

B. Jónsson [1976] found two finitely based equational theories of lattices whose meet is not finitely based (also found by K. A. Baker—unpublished, but see D. Pigozzi [1970]). Whether there exist such theories of groups is unknown.

It is of interest to know what the lattices Λ look like. It has become clear that they are very complicated, as we will see. S. Burris [1971 b] and J. Ježek [1970] have proved that if the type $\langle n_t \mid t \in T\rangle$ has some $n_t \geqq 2$ or if $n_t \geqq 1$ for two values of t, then Λ contains an infinite partition lattice, and hence obeys no special lattice laws at all.

Thus, it has proved fruitful to proceed by studying some (often simpler) sublattices of Λ, namely for fixed Σ, the lattice $\Lambda(\Sigma)$ of all equational theories $\supseteq \Sigma$. (Dually, the lattice of all subvarieties of mod Σ.) There is only one Σ with $|\Lambda(\Sigma)| = 1$, namely $\Sigma = \{x = y\}$. Theories Σ with $|\Lambda(\Sigma)| = 2$, i.e.,

$$\Lambda(\Sigma) = \begin{array}{l}\bullet\ x = y\\ \big|\\ \bullet\ \Sigma\end{array}$$

are called *equationally complete* (see §27 of Chapter 4). Since every Λ is an algebraic closure system and $\{x = y\}$ is finitely based, every theory has an equationally complete extension, and thus the top of Λ consists wholly of replicas of the above picture. An algebra $\mathfrak{A}$ is *equationally complete* iff Eq $\mathfrak{A}$ is equationally complete. It has been determined that there exist many equationally complete theories (and algebras), in two senses. First, J. Kalicki [2] (Theorem 27.2) proved that in a type with one binary operation there exist $2^{\aleph_0}$ distinct equationally complete theories (and the corresponding number has been evaluated for all types by S. Burris [1971 b] and J. Ježek [1970], answering Problem 33 of this book). Second, A. D. Bol'bot [1970] and J. Ježek [1970] proved that (given at least two unary operations or one operation of rank $\geqq 2$) Λ is *dually pseudo-atomic*, i.e., the zero of Λ (i.e., $\Sigma_0 = \varnothing$) is the meet of all dual atoms (i.e., equationally complete theories). But see some of the examples below for varying numbers of equationally complete theories in various $\Lambda(\Sigma)$.

J. Kalicki and D. Scott [2] found all equationally complete semigroups; there are only $\aleph_0$ of them. All equationally complete rings were found by A. Tarski [1956]; again, there are $\aleph_0$ of them. We cannot begin to cover all the information presently known on equational completeness. For further information, see Chapter 4 and D. Pigozzi [1970], Chapter 2. Here we sample just a few very recent results.

Theorem 1 (*D. Pigozzi* [a]). *There exists an equationally complete variety which does not have the amalgamation property.*

(This answers a question of S. Fajtlowicz.) (See 71.6 below for the amalgamation property.)

Theorem 2 (*D. M. Clark and P. H. Krauss* [1976]). *If V is a locally finite congruence-permutable equationally complete variety, then V has a plain para-primal direct Stone generator.*

(See D. M. Clark and P. H. Krauss [1976] for the meaning of these terms—roughly speaking, this means that V is generated in the manner either of Boolean algebras or of primary Abelian groups of exponent p. Cf. Appendix 5.)

For $\Sigma = \varnothing$ in a type with just one unary operation, E. Jacobs and R. Schwabauer [1964] have a complete description of Λ. For unary operations and constants, see J. Ježek [1969].

If $\Sigma = \mathrm{Eq}\,\mathfrak{A}$ for a finite algebra $\mathfrak{A}$ in a finite similarity type, then D. Scott [1] showed that $\Lambda(\Sigma)$ has only finitely many co-atoms (i.e., equationally complete varieties) (Theorem 27.3). If $\mathfrak{A}$ generates a congruence-distributive variety, then Jónsson's lemma (Theorem 39.6) easily implies that $\Lambda(\Sigma)$ is finite. If $\mathfrak{A}$ is quasi-primal (see R. W. Quackenbush [1974 b]), then $\Lambda(\Sigma)$ is a finite distributive lattice with a unique atom ($= \mathbf{HSP}\{\mathfrak{B} \mid \mathfrak{B} \subset \mathfrak{A}\}$), and, conversely, every finite distributive lattice with unique atom can be represented in this way (H. P. Gumm, unpublished).

There have been extensive studies on the lattice $\Lambda(\Sigma)$ in the setting of "intermediate" (intuitionistic propositional) logics, and in the setting of modal logics. For references and discussion, the reader is referred to W. J. Blok [1976].

For more examples of $\Lambda(\Sigma)$, see EL.

We close this section with some remarkable general results of R. McKenzie on the full lattice $\Lambda = \Lambda(\langle n_t \mid t \in T \rangle)$, where, temporarily, we make the type explicit.

Theorem 3 (*R. McKenzie* [1971 a]). *From the isomorphism type of* $\Lambda(\langle n_t \mid t \in T \rangle)$ *one can recover the multiplicity type.*

Theorem 4 (*R. McKenzie* [1971 a]). (*Appropriate* $\langle n_t \mid t \in T \rangle$.) *There exists a first order formula* $\varphi(x)$ *with one free variable in the language of lattice theory such that the unique element of* Λ *satisfying* $\varphi(x)$ *is the equational theory of groups.* (*Respectively, semigroups, lattices, distributive lattices, commutative semigroups, Boolean algebras.*)

This last theorem had a precursor in J. Ježek [1971 a]: the variety of commutative semigroups obeying $x^2y = xy$ is definable (in a similar fashion).

J. Ježek [1976 a] has proved that L is isomorphic to an interval in $\Lambda(\langle 2 \rangle)$ iff L is algebraic and has only countably many compact elements.

§71. SOME FURTHER INVARIANTS OF THE EQUIVALENCE CLASS OF A VARIETY

1. The *spectrum* of V is defined as

$$\text{spec } V = \{n \in \omega \mid (\exists \mathfrak{A} \in V)\ |A| = n\}.$$

Clearly $1 \in \text{spec } V$, and since V is closed under the formation of products, spec V is multiplicatively closed. G. Grätzer [12] proved that, conversely, any multiplicatively closed set containing 1 is the spectrum of some variety (see also T. Evans [7]). J. Froemke and R. W. Quackenbush [1975] showed that this variety need have only one binary operation. The characterization of sets spec V for V *finitely based* seems to be much more difficult. R. McKenzie [1975] proved that if $K \subseteq \omega$ is the spectrum of any first order sentence, then there exists a single identity $\sigma = \tau$ such that the multiplicative closure of $K \cup \{1\}$ is the spectrum of $(\sigma = \tau)$. (See also B. H. Neumann [3].) Characterizations of first order spectra are known in terms of time-bounded machine recognizability—see R. Fagin [1974] for detailed statements and further references. Note that the definition of spec can be extended to mean the image of any forgetful functor (or any pseudo-elementary class)—see R. Fagin [1974] and [1975] for more details. For example, we can consider

$$T(V) = \{A \mid A \text{ is a topological space and there exists } \langle A; F \rangle \in V \text{ with all } F_t \in F \text{ continuous}\}.$$

(Some preliminary investigations on $T(V)$ appear in W. Taylor [b].) Of course, many descriptions of individual varieties in the literature yield spec V. A certain amount of attention has focused on the condition spec $V=\{1\}$. (See, e.g., B. Jónsson and A. Tarski [2], A. K. Austin [1965] and remarks and references given in W. Taylor [1973], p. 382.) For instance, this equation of A. K. Austin [1965]

$$((y^2 \cdot y) \cdot x)((y^2 \cdot (y^2 \cdot y)) \cdot z) = x$$

has infinite models but no nontrivial finite models; the same holds for any variety which has a binary congruence scheme (see §57), that is, all the polynomials occurring in the scheme are binary (see J. Berman and G. Grätzer [1976]).

N. S. Mendelsohn [1977] has shown that if V is an idempotent binary variety given by 2-variable equations, then spec V is ultimately periodic.

2. The *fine spectrum* of V is the function

$$f_V(n) = \text{the number of nonisomorphic algebras of power } n \text{ in } V.$$

Characterization of such functions seems hopeless. A typical theorem is that of S. Fajtlowicz [1973] (see also W. Taylor [1975], pp. 299–300 for a proof): if $f_V(\mathfrak{n})=1$ for all cardinals $\mathfrak{n} \geqq 1$, then V must be (within equivalence) one of two varieties: "sets" (no operations at all) or "pointed sets" (one unary operation f which obeys the law $fx=fy$). For some related results see W. Taylor [1975], R. W. Quackenbush [a], R. McKenzie [1978 a] and D. M. Clark and P. H. Krauss [1977].

3. *Categoricity in power.* Varieties obeying the condition $f_V(\mathfrak{n})=1$ for all infinite $\mathfrak{n} \geqq$ the cardinality of the similarity type of V have been characterized (within equivalence) by S. Givant [1975 a] and E. A. Paljutin [1975]. For a detailed statement, also see, e.g., W. Taylor [1975], p. 299.

4. *Varietal chains.* For any variety V we may define

$$V_1 \subseteq V_2 \subseteq \cdots \subseteq V \subseteq \cdots \subseteq V^2 \subseteq V^1$$

with

$$\bigcup V_n = \bigcap V^n = V$$

as follows. $V_n = \mathbf{HSP}(\mathfrak{F}_V(n))$ and V^n is the variety defined by all n-variable identities holding in V, i.e., $\mathfrak{A} \in V^n$ iff every n-generated subalgebra of $\mathfrak{A}$ is in V (see the Exercises on p. 191 and Problem 37). As an equivalence invariant, one may take the set of proper inclusions in either of these two chains. B. Jónsson, G. McNulty and R. W. Quackenbush [1975] prove that, with a few possible exceptions, almost any sequences of proper inclusions can occur.

5. The *size of free algebras* is a subject with a long history: precisely, define the invariant

$$\begin{aligned}\omega = \omega(V) &= \langle \omega_n(V) \mid n = 0, 1, 2, \cdots \rangle \\ &= \langle |\mathfrak{F}_V(n)| \mid n = 0, 1, 2, \cdots \rangle\end{aligned}$$

(i.e., the cardinalities of V-free algebras). (This notation was introduced in E. Marczewski [11] and the first extensive description of $\omega(V)$ occurred in G. Grätzer [1970 a]; see also S. Fajtlowicz [1970 a], G. Grätzer and J. Płonka [1970], [1970 b], [1973], J. Płonka [1971 c] and [1971 d], A. Sekaninová [1973].) In Problem 42 of this book, Grätzer asked for a complete characterization of the functions $\omega(V)$—our references represent only a partial solution.

This invariant has been explicitly evaluated for only a few of the better known varieties: vector spaces over a q-element field ($\omega_n = q^n$), Boolean algebras ($\omega_n = 2^{2^n}$), semilattices ($\omega_n = 2^n - 1$), the variety of groups given by the law $x^3 = 1$ ($\omega_n = 3^{(1/6)[n(n^2+5)]}$) (F. W. Levi and B. L. van der Waerden [1933]; see also pp. 320–324 of M. Hall [1959]), and the variety of Heyting algebras defined by "Stone's identity" (see A. Horn [1969]). The quasi-primal varieties of Pixley *et al.* often have very easily calculated ω (see, e.g., A. F. Pixley [1971] or R. W. Quackenbush [1974 b] for quasi-primal varieties); see the chart on page 291 of W. Taylor [1975] for some explicit calculations.

But for most garden varieties, the invariant $\omega(V)$ is either trivial (because infinite) or hopelessly complicated. Sometimes special cases can be calculated. Dedekind found in 1900 that the free modular lattice on 3 generators has 28 elements ("free algebra" had not yet been defined); see G. Birkhoff [7], p. 63. For some other special calculations (distributive lattices, etc.), see G. Birkhoff [7], p. 63, A. G. Waterman [1965], J. Berman, A. J. Burger and P. Köhler [1975] and J. Berman and B. Wolk [1976]. The class of all finite $\omega(V)$ is closed under (coordinatewise) multiplication (see, e.g., W. Taylor [1975], p. 266), and it forms a closed set in the space ω^ω (S. Świerczkowski—see E. Marczewski [11], p. 181).

In general algebra, a typical theorem is that of J. Płonka [1971 c] and [1971 d]: if $\omega_n(V) = n \cdot 2^{n-1}$, then V must be equivalent to one of four varieties, namely those given by Σ_1–Σ_4:

$$\begin{aligned}\Sigma_1: \qquad xx &= x \\ (xy)z &= x(yz) \\ x(yz) &= x(zy)\end{aligned}$$

$$\begin{aligned}
\Sigma_2: \quad & xx = x \\
& (xy)z = (xz)y \\
& x(yz) = xy \\
& (xy)y = xy \\
\Sigma_3: \quad & xx = x \\
& (xy)z = (xz)y \\
& x(yz) = xy \\
& (xy)y = x \\
\Sigma_4: \quad & (xyz)uv = x(yzu)v = xy(zuv) \\
& xyy = x \\
& xyz = xzy
\end{aligned}$$

where Σ_4 has a single ternary operation, denoted by juxtaposition.

6. The *amalgamation property* (AP) for V generalizes the existence (due to O. Schreier) of amalgamated free products in group theory. (One form of) the AP states that given $\mathfrak{A}$, $\mathfrak{B}$, $\mathfrak{C} \in V$ and embeddings $f\colon \mathfrak{A} \to \mathfrak{B}$, $g\colon \mathfrak{A} \to \mathfrak{C}$, there exists $\mathfrak{D} \in V$ and embeddings $f'\colon \mathfrak{B} \to \mathfrak{D}$, $g'\colon \mathfrak{C} \to \mathfrak{D}$ such that $ff' = gg'$. A general investigation of AP began with B. Jónsson [3], [6] and now there is an extensive literature—e.g., see D. Pigozzi [1971], P. D. Bacsich [1975], [1975 a], J. T. Baldwin [1973], P. Dwinger [1970], H. Hule and W. B. Müller [1976], W. K. Forrest [a], J. L. MacDonald [1974], H. Simmons [1972], P. E. Schupp [a], M. Yasuhara [1974], P. D. Bacsich and D. Rowlands-Hughes [1974]; see also G. Grätzer [1978], §V.4. Varieties known to have AP are relatively rare, but include groups, lattices, distributive lattices and semilattices (see 9 below). No other nontrivial variety of modular lattices has AP, see G. Grätzer, B. Jónsson and H. Lakser [1973]; a lattice variety having AP is join-irreducible in the lattice of all varieties of lattices, see G. Grätzer [1976]; and AP fails for semigroups—J. M. Howie [1962], N. Kimura [1957]. We cannot begin to mention all results on the AP, but one representative theorem comes from D. A. Bryars [1973] (also see P. D. Bacsich [1975]): V has the AP iff for any universal formulas $\alpha_1(x_1, x_2, \cdots)$, $\alpha_2(x_1, x_2, \cdots)$ such that $V \models \alpha_1 \vee \alpha_2$, there exist existential formulas β_1, β_2 such that $V \models \beta_i \to \alpha_i$ $(i = 1, 2)$ and $V \models \beta_1 \vee \beta_2$.

For a related property, see J. Ježek and T. Kepka [1975].

7. A variety V has the *congruence extension property* (CEP) iff every congruence Θ on a subalgebra $\mathfrak{B}$ of $\mathfrak{A} \in V$ can be extended to all of $\mathfrak{A}$, i.e., there exists a congruence Ψ on $\mathfrak{A}$ such that $\Theta = \Psi \cap B^2$. Abelian groups and distributive lattices have CEP, but groups and lattices do not. See, e.g., B. Banaschewski [1970], D. Pigozzi [1971], B. A. Davey [1977], A. Day [1971] and [1973], E. Fried, G. Grätzer and R. W. Quackenbush [a] and [b], G. Grätzer and H. Lakser [1972], P. D. Bacsich and D. Row-

lands-Hughes [1974], R. Magari [1973 a], G. Mazzanti [1974] (where one will find some other references to the Italian school—in Italian usage, "regolare" means "having the CEP"). In P. D. Bacsich [1975] there is a syntactic characterization of CEP in the style of that for AP in 6 above, although these two properties are really rather different. This characterization is closely related to A. Day [1971]. Notice that in Boolean algebras the CEP can be checked rather easily because every algebra $\mathfrak{B}$ is a subalgebra of some power $\mathfrak{A}^I$, where $\mathfrak{A}$ is the two-element algebra and every congruence Θ on $\mathfrak{B}$ is of the form

$$\langle a_i \rangle \Theta \langle b_i \rangle \quad \text{iff} \quad \{i \mid a_i = b_i\} \in \mathfrak{F}$$

for some filter $\mathfrak{F}$ of subsets of I. (Of course, the same filter $\mathfrak{F}$ may be used to extend Θ to larger algebras.) A variety in which congruences can be described by filters in this manner is called *filtral*—see, e.g., E. Fried, G. Grätzer and R. W. Quackenbush [1976], [a], and [b], G. Mazzanti [1974], R. Magari [1973 a] and especially G. M. Bergman [1972]. But, e.g., semilattices form a nonfiltral variety which has CEP. It is open whether filtrality implies congruence-distributivity. For CEP see also A. R. Stralka [1971].

In G. Grätzer [1971], p. 192, the question is raised: if V satisfies

$$\mathbf{HS}X = \mathbf{SH}X, \quad \text{for all } X \subseteq V$$

then does V have the CEP? This is obviously true for lattice varieties.

8. A variety V is *residually small* iff V contains only a set of subdirectly irreducible (s.i.) algebras, i.e., the s.i. algebras do not form a proper class, equivalently, there is a bound on their cardinality. It turns out that this bound, if it exists, may be taken as $2^{\mathfrak{n}}$, where $\mathfrak{n} = \aleph_0 +$ the number of operations in V (see W. Taylor [1972]). For some conditions equivalent to residual smallness, see W. Taylor [1972] and B. Banaschewski and E. Nelson [1972]; also see J. T. Baldwin and J. Berman [1975] where, e.g., finite bounds on s.i. algebras are considered. R. McKenzie and S. Shelah [1974] consider bounds on the size of *simple* algebras in V and obtain a result analogous to the bound of $2^{\mathfrak{n}}$ just above.

Some residually small varieties: Abelian groups, commutative rings with a law $x^m = x$, semilattices, distributive lattices, various "linear" varieties (as in 3 above); also if $V = \mathbf{HSP}\,\mathfrak{A}$ for $\mathfrak{A}$ finite and V has distributive congruences (e.g., if $\mathfrak{A}$ is any finite lattice), then V is residually small by Jónsson's lemma (Theorem 39.6). Also any $\otimes$-product of two residually small varieties is residually small. (Similarly for AP and CEP.) Some varieties which are not residually small: groups, rings, pseudocomplemented semilattices, modular lattices, and $\mathbf{HSP}\,\mathfrak{A}$ for $\mathfrak{A}$ either 8-element non-Abelian group (both generate the same variety).

A variety V is residually small iff every $\mathfrak{A}$ in V can be embedded in an equationally compact algebra $\mathfrak{B}$ (W. Taylor [1972]). J. Mycielski [3] defined $\mathfrak{B}$ to be *equationally compact* iff every set Γ of equations with constants from $\mathfrak{B}$ is satisfiable in $\mathfrak{B}$ if every finite subset of Γ is satisfiable in $\mathfrak{B}$. Equational compactness is implied by topological compactness, but not conversely. For a detailed treatment and examples, consult Appendix 6; see also W. Taylor [1975 a]. Thus we are led to the following question (W. Taylor [1972]): if V is residually small, can every algebra in V be embedded in a compact Hausdorff topological algebra? An affirmative answer has been given for many varieties.

Among the equationally compact $\mathfrak{B} \supseteq \mathfrak{A}$ there is one which is "smallest", i.e., a "compactification of $\mathfrak{A}$"—see W. Taylor [1972], p. 40 or B. Banaschewski [1974 a]. B. Węglorz [2] proved that this compactification is always in **HSP** $\mathfrak{A}$.

9. The conjunction of AP, CEP and residual smallness (6–8) is equivalent to the purely category-theoretic property of *injective completeness* (see B. Banaschewski [1970]). (R. S. Pierce [1968] noticed that injective completeness implies AP.) V is *injectively complete* (or, "has enough injectives") iff every algebra in V is embeddable in a V-injective, i.e., an algebra $\mathfrak{A} \in V$ such that whenever $\mathfrak{B} \subseteq \mathfrak{C} \in V$ and $f: \mathfrak{B} \to \mathfrak{A}$ is a homomorphism, there exists an extension of f to $g: \mathfrak{C} \to \mathfrak{A}$. (See, e.g., P. Freyd [1964], §4.9 but remember that most varieties are not Abelian categories.) The variety of semilattices has enough injectives (G. Bruns and H. Lakser [1970], A. Horn and N. Kimura [1971]) and so does that of distributive lattices (B. Banaschewski and G. Bruns [1968], R. Balbes [1967]). For a theory of injective hulls in varieties, see W. Taylor [1971 a], p. 411. We know examples to show that AP, CEP and residual smallness are completely independent properties, except for one case. It is not known whether AP and residual smallness imply CEP. For further information on varieties with enough injectives, see A. Day [1972] and O. D. García [1974]. In 10 just below we will mention another category-theoretic property of varieties. Yet another one is that of being a *binding* category, investigated for varieties in J. Sichler [1973 a] and Z. Hedrlín and J. Sichler [1971].

10. *Unique factorization of finite algebras* (UFF) in V (i.e., if $\mathfrak{A}_1 \times \cdots \times \mathfrak{A}_n \cong \mathfrak{B}_1 \times \cdots \times \mathfrak{B}_s \in V$ is finite and no $\mathfrak{A}_i$ or $\mathfrak{B}_j$ can be further decomposed as a product of smaller factors, then $n = s$ and after suitably renumbering, $\mathfrak{A}_1 \cong \mathfrak{B}_1$, $\mathfrak{A}_2 \cong \mathfrak{B}_2, \cdots, \mathfrak{A}_n \cong \mathfrak{B}_n$). Historically this problem has been approached independently of any mention of V but the results obtained often have an equational character. G. Birkhoff [7], p. 169 proved that V has UFF if V has a constant term a such that

$$V \models F(a, \cdots, a) = a \quad \text{for all operations } F \text{ of } V \qquad (*)$$

and V has permutable congruences, and B. Jónsson [6] improved "permutable" to "modular". B. Jónsson and A. Tarski [1] proved that V has UFF if V has a constant term a obeying (*) and a binary operation $+$ such that

$$V \models x+a = x = a+x.$$

R. McKenzie [1972] showed that the variety of idempotent semigroups has UFF, but the variety of commutative semigroups does not (see G. Birkhoff [7], p. 170). UFF has an influence on the fine spectrum (defined in 2)—see W. Taylor [1975], pp. 285–286. For the closely related subject of "cancellation", see L. Lovász [1971] and R. R. Appleson [1976].

11. *Universal varieties.* We must refer the reader to the papers of D. Pigozzi [1976] and [b] for this relatively new notion which promises to be quite important. V is *universal* iff for every similarity type there exist V-terms α_t corresponding to the operations F_t of this type, such that for each $\mathfrak{A}$ of this type there exists $\mathfrak{B} \in V$ such that $\mathfrak{A}$ is a subalgebra of $\langle B; \{\alpha_t^{\mathfrak{B}} \mid t \in T\}\rangle$, which obeys exactly the same laws as $\mathfrak{A}$. (E.g., the variety of quasigroups is universal.) Many undecidability and lattice-theoretic results extend to universal varieties.

12. The *Schreier* property (all subalgebras of free algebras are free) is investigated in S. Meskin [1969], P. Kelenson [1972], C. Aust [1974], A. I. Budkin [1974], J. Ježek [1976].

§72. MAL'CEV CONDITIONS AND CONGRUENCE IDENTITIES

Mal'cev's Theorem 26.4 and B. Jónsson's result given in Exercise 5.70 are two early examples of "Mal'cev conditions", as named in G. Grätzer [1970]; see also W. Taylor [1973], W. D. Neumann [1974] or J. T. Baldwin and J. Berman [a] for a precise definition. The number of properties known to be Mal'cev-definable has been growing rapidly—see W. Taylor [1973] for a summary of those known up to 1973 and S. Bulman-Fleming and W. Taylor [1976] for a partial updating; also see K. A. Baker and A. F. Pixley [1975] and G. M. Bergman [1977]. They include the following:

$|A|$ divides $|B|$ whenever $\mathfrak{A} \subseteq \mathfrak{B} \in V$, $\mathfrak{B}$ finite;
V has no topological algebras with noncommutative homotopy;
no $\mathfrak{A} \in V$ is a union of two proper subalgebras;
V has no nontrivial finite algebras.

The first three of these hold for all groups. The last one has the distinction that there is no way to enumerate recursively a Mal'cev condition for it, as was observed by J. Malitz—see W. Taylor [1973], p. 383. See W. Taylor [1973], W. D. Neumann [1974], or J. T. Baldwin and J. Berman [a] for a necessary and sufficient condition for a property of varieties to be Mal'cev-definable which easily entails all the above examples (and many more) except the second.

Permutability and modularity of congruences have been important from earliest times in universal algebra (cf. §71.10 above and for recent examples, see H. P. Gumm [b] and [d], W. Taylor [a] and the recent work of J. D. H. Smith [1976]). But the condition which has been most important recently is distributivity; this importance stems from Jónsson's lemma (Theorem 39.6). (Also see K. A. Baker [1974].) For the finite case, see also R. W. Quackenbush [1974] for a somewhat simpler proof; a similar argument had earlier been known to A. F. Pixley [1970]. Among many uses of this result has been the investigation of congruence lattices, and the "internal" model theory of many individual varieties whose algebras have the operations of lattice theory among their operations. See, e.g., J. Berman [1974], B. A. Davey [1977], and references given there. A very important kind of algebra generating a congruence-distributive variety is a *quasi-primal algebra* (see §75), i.e., within equivalence, an algebra $\mathfrak{A}=\langle A; T, F_1, F_2, \cdots\rangle$ where A is a finite set and

$$T(x, y, z) = \begin{cases} x & \text{if } x \neq y, \\ z & \text{if } x = y. \end{cases}$$

Every finite algebra in **HSP** $\mathfrak{A}$ is uniquely a product of subalgebras of $\mathfrak{A}$. Many of the equivalence-invariants of these varieties (e.g., the fine spectrum, CEP, AP, $\omega_n(V)$—see §71) are relatively easy to evaluate. See A. F. Pixley [1971] and R. W. Quackenbush [1974 b] for details and further references—the notion goes back essentially to A. F. Pixley, building on work of A. L. Foster and P. C. Rosenbloom; see also M. I. Gould and G. Grätzer [1] and §23. For infinite analogs of primal algebras, see S. Tulipani [1972] and A. Iwanik [1974 a].

D. M. Clark and P. H. Krauss [1976], [1977] have developed a remarkable theory of *para-primal* algebras, a kind of nondistributive generalization of quasi-primal algebras, combining ideas of quasi-primality and linear algebra—see §77. Also see R. W. Quackenbush [a], R. McKenzie [1978 a]; and §67.8 above.

A. F. Pixley and R. Wille established an algorithm (A. F. Pixley [1972 b], R. Wille [1970]; also see W. Taylor [1973], Theorem 5.1) to convert every congruence identity in $\wedge$, $\vee$, and $\circ$ into a weaker form of Mal'cev condition. Which of these conditions are "new" remains an open question. For more information, consult §60.

Another Mal'cev-definable property of varieties V which has received wide attention is that $\mathfrak{F}_V(n) \cong \mathfrak{F}_V(m)$. (See, e.g., E. Marczewski [11].) For fixed n_0, the set of numbers

$$\{n \in \omega \mid \mathfrak{F}_V(n) \cong \mathfrak{F}_V(n_0)\}$$

is always an arithmetic progression, and any progression can occur (see §31). If $\mathfrak{F}_V(n) \cong \mathfrak{F}_V(m)$ with $m \neq n$, then V has no nontrivial finite algebras (B. Jónsson and A. Tarski [2]). (Also see D. M. Clark [1969].)

See B. Csákány [1976] for a collection of properties of varieties resembling, but more general than, Mal'cev conditions. A nice special example is in L. Klukovits [1975].

APPENDIX 5

PRIMALITY: THE INFLUENCE OF BOOLEAN ALGEBRAS IN UNIVERSAL ALGEBRA

By Robert W. Quackenbush

§73. INTRODUCTION

The equational class of Boolean algebras is one of the most thoroughly studied classes of algebras even though important questions remain unanswered and important discoveries are still being made (such as the characterization of projective Boolean algebras by R. Freese and J. B. Nation). We shall review some of the important ways in which classes of algebras mimic the behavior of Boolean algebras, based on the ideas of A. L. Foster.

We first concentrate on $\mathfrak{B}_1$, the 2-element Boolean algebra. It is the smallest nontrivial Boolean algebra and every Boolean algebra can be built up from $\mathfrak{B}_1$ using only direct products and subalgebras.

Let $\mathfrak{A}$ be a 2-element algebra; when is $\mathfrak{A}$ Boolean up to† *polynomial equivalence* (i.e., when do $\mathfrak{A}$ and $\mathfrak{B}_1$ have the same set of functions as polynomials; for a precise description of polynomial equivalence, see §8 and §60; see also §65)?

One of the most important properties of $\mathfrak{B}_1$ is that every function on $\mathfrak{B}_1$ is a polynomial; in fact, this characterizes $\mathfrak{B}_1$. Thus we have: $\mathfrak{A}$ is $\mathfrak{B}_1$ (up to polynomial equivalence) if and only if every function on $\mathfrak{A}$ is a polynomial on $\mathfrak{A}$.

This characterization is of great interest to logicians in their study of truth functions and to computer scientists in their study of logic circuits. Is there a characterization of $\mathfrak{B}_1$ that would be of more interest to algebraists? Indeed there is, and it is due to A. L. Foster and A. F. Pixley [1]: $\mathfrak{A}$ is $\mathfrak{B}_1$ if and only if (a) $\mathfrak{A}$ is simple, (b) $\mathfrak{A}$ has no proper subalgebras, (c) $\mathfrak{A}$ has no proper automorphisms, (d) $\mathfrak{A}$ generates a congruence permutable equational class, (e) $\mathfrak{A}$ generates a congruence distributive equational class.

Clearly $\mathfrak{B}_1$ satisfies (a)–(e). For the converse, it suffices to prove that $\mathfrak{F}_{\mathfrak{A}}(n)$ is $\mathfrak{A}^{|A|^n}$. Of course, $\mathfrak{F}_{\mathfrak{A}}(n)$ is a subalgebra of $\mathfrak{A}^{|A|^n}$. By (b), $\mathfrak{F}_{\mathfrak{A}}(n)$ is a

† Called definitional equivalence in Appendix 3 and equivalence in Appendix 4.

subdirect power of $\mathfrak{A}$. Then (a) and (d) say that $\mathfrak{F}_{\mathfrak{A}}(n)$ is $\mathfrak{A}^m$ for some $m \leqq |A|^n$. Next, (a) and (e) say that m is the number of dual atoms in $\mathfrak{C}(\mathfrak{F}_{\mathfrak{A}}(n))$. Finally, (a), (b) and (c) say that $\mathfrak{C}(\mathfrak{F}_{\mathfrak{A}}(n))$ has $|A|^n$ dual atoms.

A third characterization of $\mathfrak{B}_1$ involves the subalgebras of finite powers of $\mathfrak{B}_1$; this characterization is due to P. H. Krauss [1972 a]. Consider $\mathfrak{B} \subseteq (\mathfrak{B}_1)^n$ and $1 \leqq i, j \leqq n$. Set $i \sim j$ if for every $b \in B$, $b_i = b_j$; clearly $\sim$ is an equivalence relation. Let $J \subseteq \{1, \cdots, n\}$; J is *$\mathfrak{B}$-reduced* if $\sim$ restricted to J is the identity relation. Let π_J be the projection of $(\mathfrak{B}_1)^n$ onto $(\mathfrak{B}_1)^J$; then $(\pi_J)_B$ is onto $(\mathfrak{B}_1)^J$ if and only if J is $\mathfrak{B}$-reduced. Thus $\mathfrak{A}$ is $\mathfrak{B}_1$ if and only if for every finite I, every $\mathfrak{B} \subseteq \mathfrak{A}^I$ and every $\mathfrak{B}$-reduced $J \subseteq I$, $(\pi_J)_B$ is onto $\mathfrak{A}^J$. The proof is quite easy: For any finite algebra $\mathfrak{A}$, $\mathfrak{F}_{\mathfrak{A}}(n) \subseteq \mathfrak{A}^{|A|^n}$ and $|A|^n$ is always $\mathfrak{F}_{\mathfrak{A}}(n)$-reduced.

§74. PRIMAL ALGEBRAS

Notice that none of the three characterizations of $\mathfrak{B}_1$ make use of the fact that $\mathfrak{A}$ is a 2-element algebra. Thus, as with A. L. Foster [3], we define a finite nontrivial algebra $\mathfrak{A}$ to be *primal* if every function on $\mathfrak{A}$ is a polynomial of $\mathfrak{A}$. Moreover, the second and third characterizations of $\mathfrak{B}_1$ carry over verbatim to primal algebras.

Example (*D. Webb* [1936]). $\langle \{0, 1, \cdots, n-1\}; * \rangle$ is primal, where $x*y = \max(x+1, y+1)$ with addition mod n.

How closely to $\mathfrak{B}_1$ does a primal algebra behave? On the basis of the three characterizations, we expect quite similar behavior. In some sense the behavior is as close as possible: The equational class generated by a primal algebra is isomorphic as a category to the category of Boolean algebras (where morphisms are all homomorphisms); moreover, any equational class isomorphic to Boolean algebras is generated by a primal algebra (see T.-K. Hu [1969]). This isomorphism shows that many important properties of Boolean algebras transfer to equational classes generated by a primal algebra. One interesting property which transfers but is not implied by the category isomorphism is that the equational class generated by a primal algebra (of finite type) is defined by a single identity (see G. Grätzer and R. McKenzie [1967]).

It seems clear that type plays no significant role in the study of primal algebras since the characterizations involve polynomials rather than operations. However, this is not the case. The five conditions (a)–(e) of the algebraic characterization of primal algebras are independent; yet if we restrict ourselves to only one operation (which must be at least binary), then conditions (d) and (e) are redundant. This unexpected result is due to

G. Rousseau [1967] and is based upon an important but extremely difficult theorem of I. Rosenberg [1970 b] (its proof covers some 80 pages). Rosenberg's theorem describes all maximal proper sets of functions closed under composition on a finite set. Thus a finite nontrivial algebra $\mathfrak{A}$ is primal if and only if for each such maximal set there is an operation of $\mathfrak{A}$ not belonging to the set.

Since primal algebras have so many special properties, they must be very sparse among all finite algebras of fixed type. Let us restrict our attention to groupoids. If $N(n)$ is the number (of isomorphism classes) of groupoids of cardinality n and $P(n)$ is the number of primal algebras of cardinality n, then we are interested in the behavior of $P(n)/N(n)$ as $n \to \infty$. For instance, for a random finite groupoid, what is the probability that it has no 1-element subalgebras? If $|\mathfrak{A}| = n$, it is $[(n-1)/n]^n$ since $\mathfrak{A}$ has no 1-element subalgebras if and only if $a \cdot a \neq a$ for every $a \in \mathfrak{A}$. This ratio tends to e^{-1}; thus the limiting value of $P(n)/N(n)$ (if it exists) is $\leq e^{-1}$. Remarkably, the limit does exist and is equal to e^{-1}. I first saw this result in an unpublished manuscript of R. O. Davies. The proof uses both Rousseau's theorem and Rosenberg's theorem and shows that almost every finite groupoid is simple, has no proper automorphisms and has no proper nontrivial subalgebras.

§75. QUASI-PRIMAL ALGEBRAS

We now turn our attention to the generalizations of primal algebras, whose hyphenated names make quite a cacophony. The most important of these generalizations is the concept of a quasi-primal algebra. For this we need the *ternary discriminator function*, $t(x, y, z)$ which is defined on any set A by: $t(a, a, b) = b$ and $t(a, b, c) = a$ if $a \neq b$. A finite nontrivial algebra, $\mathfrak{A}$, is *quasi-primal* if $t(x, y, z)$ is a polynomial of $\mathfrak{A}$. Quasi-primal algebras were introduced by A. F. Pixley [1970] under the name "simple algebraic algebra" and were given their current name by him in A. F. Pixley [1971].

We can now characterize quasi-primal algebras in a manner similar to the three characterizations of primal algebras:

(1) A finite nontrivial algebra $\mathfrak{A}$ is quasi-primal if and only if for every $n \geq 1$, every function $f\colon A^n \to A$ which preserves the subalgebras and internal isomorphisms of $\mathfrak{A}$ is a polynomial of $\mathfrak{A}$. (We say *f preserves the subalgebras* of $\mathfrak{A}$ if for every $a_1, \cdots, a_n \in A$, $f(a_1, \cdots, a_n)$ is contained in the subalgebra of $\mathfrak{A}$ generated by $\{a_1, \cdots, a_n\}$. By an *internal isomorphism*, φ, of $\mathfrak{A}$ is meant an isomorphism between (not necessarily distinct) subalgebras of $\mathfrak{A}$; *f preserves* $\varphi\colon \mathfrak{A}_1 \to \mathfrak{A}_2$ if for all $a_1, \cdots, a_n \in \mathfrak{A}_1, f(\varphi(a_1), \cdots,$

$\varphi(a_n))=\varphi(f(a_1, \cdots, a_n))$.) This characterization is in actuality Pixley's original definition; in A. F. Pixley [1971] he proved that this condition is equivalent to $t(x, y, z)$ being a polynomial of $\mathfrak{A}$.

(2) A finite nontrivial algebra $\mathfrak{A}$ is quasi-primal if and only if (a′) every subalgebra of $\mathfrak{A}$ is simple, (d), (e). Notice that condition (a) is replaced by (a′) since $t(x, y, z)$ being a polynomial of $\mathfrak{A}$ forces all subalgebras to be simple. This characterization is due to A. F. Pixley [1970].

(3) A finite nontrivial algebra $\mathfrak{A}$ is quasi-primal if and only if for each finite I, each $\mathfrak{B} \subseteq \mathfrak{A}^I$, each $\mathfrak{B}$-irredundant subset $J \subseteq I$, $\pi_J(\mathfrak{B}) = \prod(\pi_j(\mathfrak{B}) \mid j \in J)$. ($J$ is *$\mathfrak{B}$-irredundant* if the restriction of the equivalence relation $\sim$ to J is the identity relation, where $\sim$ is defined by: $i \sim j$ if and only if there is an isomorphism φ from $\pi_i(\mathfrak{B})$ onto $\pi_j(\mathfrak{B})$ such that for all $b \in \mathfrak{B}$, $\varphi(b_i) = b_j$.) This characterization is due to P. H. Krauss [1973].

A quasi-primal algebra $\mathfrak{A}$ is *demi-semi-primal* (R. W. Quackenbush [1971]) if every internal isomorphism of $\mathfrak{A}$ can be extended to an automorphism of $\mathfrak{A}$; $\mathfrak{A}$ is *demi-primal* (R. W. Quackenbush [1971]) if $\mathfrak{A}$ contains no proper subalgebras, and $\mathfrak{A}$ is *semi-primal* (A. L. Foster and A. F. Pixley [1]) if the only internal isomorphisms on $\mathfrak{A}$ are the identity maps on the subalgebras. Note that both demi-primal and semi-primal algebras are demi-semi-primal. Properly between demi-semi-primal and semi-primal algebras are the *infra-primal algebras* (A. L. Foster [1971]): $\mathfrak{A}$ is infra-primal if it is demi-semi-primal and if each internal isomorphism is an automorphism on its domain.

Example 1. GF(q), the Galois field on q elements, is quasi-primal. In fact, for p prime, GF(p) is semi-primal. For $n \geqq 2$, GF(p^n) is infra-primal but not semi-primal.

Example 2. A simple monadic algebra is a Boolean algebra with an added unary operation C such that $C(1)=1$ and for $a<1$, $C(a)=0$. Finite simple monadic algebras are quasi-primal. The 2-element simple monadic algebra is Boolean and hence primal while the 4-element one is demi-semi-primal. No others are demi-semi-primal.

Example 3. Build your own quasi-primal algebra: (i) Take a set A with $1<|A|<\omega$. (ii) Take a collection $S(A)$ of subsets of A (with $A \in S(A)$) closed under set intersection. (iii) Take a collection $I(A)$ of bijections with domains and ranges being members of $S(A)$ closed under composition, inverse and restriction (i.e., if $\varphi: A_1 \to A_2 \in I(A)$, $A_1 \supseteq A_3 \in S(A)$, then $\varphi_{A_3} \in I(A)$). (iv) Make sure that if $\varphi \in I(A)$ then $\{a \mid a = \varphi(a)\} \in S(A)$. (v) Let F be the set of all functions on A which preserve $S(A)$ and $I(A)$.

Then $\langle A; F\rangle$ is quasi-primal, $S(A)$ is the set of subalgebras of $\mathfrak{A}$ and those $\varphi \in I(A)$ with nontrivial domain are the internal isomorphisms of $\mathfrak{A}$. This recipe can be found in M. G. Stone [1972].

Example 4. Almost every finite algebra is semi-primal. Just as the proportion of primal groupoids is e^{-1}, V. L. Murskiĭ [1975] shows that the proportion of semi-primal groupoids all of whose proper subalgebras are trivial is 1. This obviously implies the same result for any fixed type (containing at least one at least binary operation).

Using the method of Example 3 it is easy to show that the special cases of quasi-primality so far introduced have no inclusions among themselves other than the obvious ones. The reader is invited to modify the three characterizations of quasi-primal algebras for each of the special cases introduced above.

§76. ARITHMETICAL ALGEBRAS

Let us return to the second characterization of quasi-primal algebras. There three conditions are listed. What happens if we delete one? If we delete condition (a′), then we merely have that $\mathfrak{A}$ generates a congruence permutable and congruence distributive equational class. By a well-known result of A. F. Pixley [1] this is equivalent to the existence of a polynomial $p(x, y, z)$ satisfying:

$$p(x, y, x) = p(x, y, y) = p(y, y, x) = x. \tag{*}$$

This is a rather weak condition for a single algebra (although it is a very strong and important condition for the equational class it generates); in particular, it is true for every algebra in the equational class generated by it. Thus, we want to add a condition. One such condition is to require that every function $F: A^n \to A$ which preserves the congruences of $\mathfrak{A}$ be a polynomial of $\mathfrak{A}$ (*f preserves* $\Theta \in C(\mathfrak{A})$ if for every $a_1, \cdots, a_n, b_1, \cdots, b_n \in \mathfrak{A}$ with $a_i \equiv b_i(\Theta)$ for $1 \leqq i \leqq n$, we have $f(a_1, \cdots, a_n) \equiv f(b_1, \cdots, b_n)(\Theta)$. See Problem 6 and Exercise 27 of Chapter 1.) In particular, this means that every element of $\mathfrak{A}$ is the value of a constant polynomial. Thus, following A. F. Pixley [1972 a], we call an algebra $\mathfrak{A}$ *arithmetical* if $\mathfrak{C}(\mathfrak{A})$ is distributive and permutable. Following A. L. Foster [1970], we call a finite nontrivial algebra $\mathfrak{A}$ *hemi-primal* if for every $n \geqq 1$, every function $f: A^n \to A$ which preserves $\mathfrak{C}(\mathfrak{A})$ is a polynomial of $\mathfrak{A}$. Then A. F. Pixley [1972 a] proves the following characterization theorem for arithmetical hemi-primal algebras.

A finite nontrivial algebra $\mathfrak{A}$ is arithmetical hemi-primal if and only if (b), (c′) if $\Theta_1, \Theta_2 \in C(\mathfrak{A})$ and $g: \mathfrak{A}/\Theta_1 \to \mathfrak{A}/\Theta_2$ is an isomorphism, then g is the identity function (and so $\Theta_1 = \Theta_2$), (d), (e).

Arithmetical hemi-primal algebras are as yet poorly understood; there is a dearth of natural examples and there is no characterization in terms of subalgebras of $\mathfrak{A}^I$. However, as in the quasi-primal case, we can build our own arithmetical hemi-primal algebras: (i) Take a set A with $1<|A|<\omega$. (ii) Take a distributive, permuting $\{0, 1\}$-sublattice $\mathfrak{C}(A)$ of $E(A)$, the lattice of equivalence relations on A (in particular, any chain containing ω and ι is such). (iii) Take F to be all functions on A which preserve $\mathfrak{C}(A)$. By a result of R. W. Quackenbush and B. Wolk [1971], $\mathfrak{C}(A)$ is the congruence lattice of $\langle A; F\rangle$. By A. F. Pixley [1972 a], F contains a function $p(x, y, z)$ satisfying (*) and hence $\langle A; F\rangle$ is arithmetical hemi-primal. In the case that $\mathfrak{C}(A)$ is a chain, $\langle A; F\rangle$ is called *linear hemi-primal*.

§77. PARA-PRIMAL ALGEBRAS

Let us consider conditions (a') and (d), thus dropping congruence distributivity. Conditions (a') and (d) are still strong enough to guarantee that every subalgebra of a finite power of $\mathfrak{A}$ is a direct product of subalgebras of $\mathfrak{A}$. However, the third characterization of quasi-primal algebras does not carry over to this case. For instance, if $Z_2=\langle\{0, 1\}; +\rangle$ is the 2-element group, then the congruence lattice of $(Z_2)^2$ is not distributive. Let $\mathfrak{B}\subseteq(Z_2)^4$ be $\{\langle 0, 0, 0, 0\rangle, \langle 0, 1, 0, 1\rangle, \langle 1, 0, 1, 0\rangle, \langle 1, 1, 1, 1\rangle, \langle 0, 0, 1, 1\rangle, \langle 1, 1, 0, 0\rangle, \langle 0, 1, 1, 0\rangle, \langle 1, 0, 0, 1\rangle\}$; as a subset of $(Z_2)^2\times(Z_2)^2$, $\mathfrak{B}$ is the congruence on $(Z_2)^2$ induced by the subgroup $\{\langle 0, 0\rangle, \langle 1, 1\rangle\}$. Now note that $\{1, 2, 3, 4\}$ is $\mathfrak{B}$-irredundant but that $\mathfrak{B}\neq(Z_2)^4$.

Also notice that if we project $(Z_2)^4$ onto its first 3 components, then the restriction to $\mathfrak{B}$ is 1-1. Thus while $\{1, 2, 3, 4\}$ is $\mathfrak{B}$-irredundant, it is in an obvious sense not minimal. We formalize this as follows: a $\mathfrak{B}$-irredundant subset $J\subseteq I$ is *$\mathfrak{B}$-minimal* if $(\pi_J)_B$ is 1-1 but for no proper subset $J'\subset J$ is $(\pi_{J'})_B$ 1-1. Since a maximal $\mathfrak{B}$-irredundant set J has the property that $(\pi_J)_B$ is 1-1, $\mathfrak{B}$-minimal subsets of I always exist. Following D. M. Clark and P. H. Krauss [1976] we define a finite nontrivial algebra $\mathfrak{A}$ to be *para-primal* if for every finite I, every $\mathfrak{B}\subseteq\mathfrak{A}^I$, and every $\mathfrak{B}$-minimal subset $J\subseteq I$, we have that $\pi_J(\mathfrak{B})=\prod(\pi_j(\mathfrak{B})\mid j\in J)$. Their fundamental result is:

A finite nontrivial algebra $\mathfrak{A}$ is para-primal if and only if (a') and (d).

The biggest difficulty in the proof is to show that a para-primal algebra generates a congruence permutable equational class. This was first done in R. W. Quackenbush [a] in a round-about manner. This result has a nice generalization which gives a remarkable converse to the following well-known proposition:

If K is a locally finite congruence permutable equational class in which each finite subdirectly irreducible algebra is simple, then each finite algebra in K is a direct product of simple algebras.

Notice that K is generated by $\mathbf{Sim}_f(K)$, the set of finite simple algebras in K and that the proposition remains true if we delete "locally finite". However, in order to prove the converse, some condition like local finiteness is needed. If K is locally finite, then $\mathbf{Sim}_f(K)$ contains only finitely many n-generated algebras for each n; call such a set of finite algebras *generically finite*. Notice that any finite subdirect product of algebras from $\mathbf{Sim}_f(K)$ is isomorphic to a direct product of some subset of the factors. Call any such set of finite algebras a *strong direct factor set*. Now we can state the converse:

Let K' be a generically finite strong direct factor set of finite algebras with $\mathbf{S}(K')\subseteq\mathbf{P}(K')$. Then K, the equational class generated by K', has permutable congruences.

Interestingly, $K'\subseteq\mathbf{Sim}_f(K)$, but equality need not hold. However, $\mathbf{Sim}_f(K)$ does contain all finite subdirectly irreducible algebras in K, $\mathbf{Sim}_f(K)$ is a generically finite strong direct factor set, and the set of cardinalities of members of $\mathbf{Sim}_f(K)$ is the same as that of K'.

There is as yet no characterization of para-primal algebras in terms of polynomials. Any nice solution is likely to make use of the following result (H. Werner [1974]): In a congruence permutable equational class, a direct product of simple algebras, $\mathfrak{A}_1\times\cdots\times\mathfrak{A}_n$, has a distributive congruence lattice if and only if for each $1\leqq i<j\leqq n$, $\mathfrak{A}_i\times\mathfrak{A}_j$ has a distributive congruence lattice. Thus if $\mathfrak{A}$ is para-primal but not quasi-primal, then $\mathfrak{C}(\mathfrak{F}_{\mathfrak{A}}(3))$ is not distributive. Since $\mathfrak{F}_{\mathfrak{A}}(3)$ is a direct product of subalgebras of $\mathfrak{A}$, we have that for some subalgebras $\mathfrak{A}_0$, $\mathfrak{A}_1$, of $\mathfrak{A}$, $\mathfrak{C}(\mathfrak{A}_0\times\mathfrak{A}_1)$ is not distributive. But then it can be shown that $\mathfrak{C}(\mathfrak{A}_0^2)$ is not distributive. Call such an $\mathfrak{A}_0$ *affine*. We really only need to characterize the polynomials of affine para-primal algebras since nonaffine para-primal algebras behave very much like quasi-primal algebras as far as their contribution to free algebras is concerned.

Fortunately, R. McKenzie [1978 a] has characterized affine para-primal algebras. He calls a nontrivial finite algebra $\langle A;F\rangle$ *affine* if there are p and n such that $A=(Z_p)^n$ ($Z_p=\langle\{0,1,\cdots,p-1\};+\ (\mathrm{mod}\ p)\rangle$ for p prime) and each m-ary $f\in F$ is of the form $\varphi+c$ where $\varphi\colon (Z_p)^{mn}\to(Z_p)^n$ is a homomorphism and $c\in(Z_p)^n$. In particular, each finite abelian group of prime power order is affine. Then McKenzie's characterization asserts that the simple affine algebras are just the para-primal affine algebras (and in particular, they have prime power order).

Finally, the very nice paper of H. P. Gumm [d] must be mentioned. It gives very elegant proofs for many of the results mentioned in this section.

§78. DUAL-DISCRIMINATOR ALGEBRAS

Now we turn to considering conditions (a′) and (e). This area is virtually unexplored. If we add condition (b), then these algebras have been studied by P. D. Bacsich [1973] under the name "crypto-primal algebras". However, his study is restricted to model theoretic properties of crypto-primal algebras, and he does not address himself to the questions we are investigating here.

A very interesting and important paper of E. Fried and A. F. Pixley [a] does take a big step in investigating conditions (a′) and (e). They study the *dual-discriminator function*, $d(x, y, z)$, which is defined on any set A by: $d(a, a, b)=a$ and $d(a, b, c)=c$ if $a\neq b$. The dual-discriminator can always be obtained from the discriminator: $d(x, y, z)=t(x, t(x, y, z), z)$. But the 2-element lattice (in which $d(x, y, z)=(x\vee y)\wedge(x\vee z)\wedge(y\vee z)$) shows that $t(x, y, z)$ cannot always be obtained from $d(x, y, z)$.

Call a finite nontrivial algebra $\mathfrak{A}$ a *dual-discriminator algebra* if $d(x, y, z)$ is a polynomial of $\mathfrak{A}$. Since $d(x, x, y)=d(x, y, x)=d(y, x, x)=x$, $\mathfrak{A}$ generates a congruence distributive equational class; moreover, using $d(x, y, z)$, it is easy to prove that each subalgebra of $\mathfrak{A}$ is simple. Thus a dual-discriminator algebra satisfies (a′) and (e).

E. Fried and A. F. Pixley characterize dual-discriminator algebras in terms of their polynomials. A subalgebra, $\mathfrak{B}$, of $\mathfrak{A}_0\times\mathfrak{A}_1$ is *rectangular* if for each $a_0\in A_0$ and $a_1\in A_1$, $|\{b_0\in A_0\mid\langle b_0, a_1\rangle\in B\}|$ is 0, 1, or $|\pi_0(B)|$ and $|\{b_1\in A_1\mid\langle a_0, b_1\rangle\in B\}|$ is 0, 1, or $|\pi_1(B)|$. The characterization is:

A finite nontrivial algebra, $\mathfrak{A}$, is a dual-discriminator algebra if and only if every function $f\colon A^n\to A$ which preserves the rectangular subalgebras of $\mathfrak{A}^2$ is a polynomial of $\mathfrak{A}$ (f *preserves* $\mathfrak{B}$, a subalgebra of $\mathfrak{A}^2$, if $\langle f, f\rangle\colon A^{2n}\to A^2$ preserves $\mathfrak{B}$).

Notice that there is no explicit reference to f preserving the subalgebras and internal isomorphisms. This has been taken care of by the rectangular subalgebras of $\mathfrak{A}^2$: if $\mathfrak{A}_0\subseteq\mathfrak{A}$, then $(\mathfrak{A}_0)^2$ is a rectangular subalgebra of $\mathfrak{A}^2$ while if $\varphi\colon A_0\to A_1$ is an internal isomorphism of $\mathfrak{A}$, then $\{\langle a, \varphi(a)\rangle\mid a\in A_0\}$ is a rectangular subalgebra of $\mathfrak{A}^2$.

There is as yet no characterization of dual-discriminator algebras in terms of subalgebras of powers of $\mathfrak{A}$. Such a characterization can likely be obtained by making use of a result in G. M. Bergman [1977]: In any equational class with a *majority polynomial* (i.e., a polynomial $m(x, y, z)$ satisfying $m(x, x, y)=m(x, y, x)=m(y, x, x)=x$) any subalgebra of $\mathfrak{A}_1\times\cdots\times\mathfrak{A}_n$ is determined by its projections onto $\mathfrak{A}_i\times\mathfrak{A}_j$ for $1\leqq i<j\leqq n$.

Example. Any finite nontrivial weakly associative lattice with the unique bound property is a dual-discriminator algebra. ($\langle A;\vee,\wedge\rangle$ is such an algebra if $\vee$ and $\wedge$ are idempotent, commutative, absorptive and for any two distinct elements $a, b \in A$, $a \vee b$ is the unique element $x \in A$ such that $a \vee x = x = b \vee x$ and dually.)

§79. FUNCTIONAL COMPLETENESS

A concept closely related to primality is functional completeness. A finite nontrivial algebra $\mathfrak{A}$ is *functionally complete* if and only if for every $n \geqq 1$, every function $f: A^n \to A$ is an algebraic function on $\mathfrak{A}$. Thus $\mathfrak{A}$ is functionally complete if and only if $\mathfrak{A}^*$, the algebra obtained from $\mathfrak{A}$ by making each element of $\mathfrak{A}$ the value of a new nullary operation, is primal. Thus $\mathfrak{A}$ is functionally complete if and only if $t(x, y, z)$ is an algebraic function on $\mathfrak{A}$; this was first shown by H. Werner [1970]. E. Fried and A. F. Pixley [a] show that if $|A| \geqq 3$, then $\mathfrak{A}$ is functionally complete if and only if $d(x, y, z)$ is an algebraic function. Notice that the 2-element lattice has $d(x, y, z)$ as a polynomial, but that it is not functionally complete.

Let $\mathfrak{A}$ be a finite nontrivial algebra. An easy necessary condition for $\mathfrak{A}$ to be functionally complete is that $\mathfrak{A}$ be simple; thus let us also assume that $\mathfrak{A}$ is simple. When is $\mathfrak{A}$ functionally complete? If $\mathfrak{A}$ generates a congruence distributive and congruence permutable equational class, then $\mathfrak{A}$ is functionally complete. R. McKenzie [1978 a] has shown that if $\mathfrak{A}$ generates a congruence permutable equational class, then $\mathfrak{A}$ is functionally complete if and only if $\mathfrak{A}$ is not affine. For $\mathfrak{A}$ to generate a congruence distributive equational class, we have the obvious result, via Fried and Pixley, that a dual-discriminator algebra with at least three elements is functionally complete. However, it is easy to construct a functionally complete $\mathfrak{A}$ generating a congruence distributive equational class with $\mathfrak{A}$ not a dual-discriminator algebra. What is needed is an analog of McKenzie's result.

A completeness concept generalizing functional completeness but not requiring simplicity is affine completeness. A finite nontrivial algebra $\mathfrak{A}$ is *affine complete* if for every $n \geqq 1$, every function $f: A^n \to A$ which preserves $\mathfrak{C}(\mathfrak{A})$ is an algebraic function (see Problem 6 and its discussion in Appendix 2). Of course, if $\mathfrak{A}$ is simple, then $\mathfrak{A}$ is affine complete if and only if it is functionally complete. A. F. Pixley [1972 a] has shown that if $\mathfrak{A}$ is a finite nontrivial algebra generating an arithmetical equational class, then $\mathfrak{A}$ is affine complete. G. Grätzer has shown that every Boolean algebra, finite or infinite, is affine complete. However, A. A. Iskander [1972] shows that in the equational class generated by a finite prime field, not every infinite algebra is affine complete.

§80. REPRESENTATION THEORY

The class of Boolean algebras, B, is generated by $\mathfrak{B}_1$: $B = \mathbf{HSP}(\mathfrak{B}_1)$. But this is not a particularly exciting result; every equational class is generated by some algebra. The important representation theorem for Boolean algebras is that $B = \mathbf{SP}(\mathfrak{B}_1)$. Since $\mathbf{SP}(K) = \mathbf{P}_s\mathbf{S}(K)$, this says that all subdirectly irreducible algebras in B are subalgebras of $\mathfrak{B}_1$. But we can say more since $\mathfrak{B}_1$ has no proper subalgebras: $B = \mathbf{P}_s(\mathfrak{B}_1)$; that is, $\mathfrak{B}_1$ is the only subdirectly irreducible algebra in B. We can say even more; the finite Boolean algebras are just the direct powers of $\mathfrak{B}_1$.

Let us now give these properties some names. But first, some notation: $\mathbf{V}(K)$ is the equational class generated by K; $\mathbf{V}_f(K)$ is the class of finite members of $\mathbf{V}(K)$, and $\mathbf{P}_f(K)$ is the class of direct products of finitely many copies of members of K. Note that if $\mathbf{V}(K)$ is locally finite (e.g., if K is a finite set of finite algebras) then $\mathbf{V}_f(K) = \mathbf{HSP}_f(K)$. In general, $\mathbf{HSP}_f(K)$ is smaller than $\mathbf{V}_f(K)$: Take $K = \{Z_q \mid q \text{ is a power of } 2\}$; then $\mathbf{HSP}_f(K)$ is the class of all abelian groups whose cardinality is a power of 2 while $\mathbf{V}_f(K)$ is the class of all finite abelian groups. Let $\mathfrak{A}$ be a finite nontrivial algebra; we say that $\mathfrak{A}$ is an **SP**-*algebra* ($\mathbf{P}_s$-*algebra*) if $\mathbf{V}(\mathfrak{A}) = \mathbf{SP}(\mathfrak{A})$ ($\mathbf{V}(\mathfrak{A}) = \mathbf{P}_s(\mathfrak{A})$). An **SP**-algebra ($\mathbf{P}_s$-algebra) $\mathfrak{A}$ is *direct* if $\mathbf{V}_f(\mathfrak{A}) = \mathbf{P}_f\mathbf{S}(\mathfrak{A})$ ($\mathbf{V}_f(\mathfrak{A}) = \mathbf{P}_f(\mathfrak{A})$). This terminology is new; there are two alternate terminologies in the literature, one established by A. L. Foster and A. F. Pixley, and the other by D. M. Clark and P. H. Krauss. For instance, an **SP**-algebra is called semi-categorical by A. L. Foster and A. F. Pixley, and a subdirect Stone generator by D. M. Clark and P. H. Krauss.

The crucial result for proving representation theorems for equational classes generated by the algebras we have considered is a corollary of B. Jónsson's lemma (Theorem 39.6): If $\mathfrak{A}$ is a finite algebra in a congruence distributive equational class then $\mathbf{V}(\mathfrak{A}) = \mathbf{P}_s\mathbf{HS}(\mathfrak{A})$. Jónsson's proof uses ultraproducts. A direct proof is due to A. F. Pixley [1970]. Using congruence distributivity, it is readily shown that all finite subdirectly irreducible algebras in $\mathbf{V}(\mathfrak{A})$ are in $\mathbf{HS}(\mathfrak{A})$. Then A. F. Pixley uses a generalization of a result of A. Astromoff [1]: If every finitely generated subalgebra of $\mathfrak{B}$ can be embedded in a power of $\mathfrak{A}$, then so can $\mathfrak{B}$ itself.

If all subalgebras of $\mathfrak{A}$ are simple, then, modulo the trivial algebra, $\mathbf{HS}(\mathfrak{A}) = \mathbf{S}(\mathfrak{A})$ so that if $\mathfrak{A}$ generates a congruence distributive equational class, then $\mathfrak{A}$ is an **SP**-algebra. Thus every dual-discriminator algebra (and hence every quasi-primal algebra) is an **SP**-algebra. What if $\mathfrak{A}$ is para-primal but not quasi-primal? Then $Z_2{}^* = \langle\{0, 1\}; +, 1\rangle$ shows that $\mathfrak{A}$ need not be an **SP**-algebra since $(Z_2{}^*)^2$ has $Z_2 = \langle\{0, 1\}; +, 0\rangle$ as a homomorphic image. As D. M. Clark and P. H. Krauss [1977] tell us, this completely typifies when $\mathfrak{A}$ is not an **SP**-algebra: A para-primal algebra $\mathfrak{A}$ is an **SP**-algebra if and only if every affine subalgebra of $\mathfrak{A}$ contains a trivial subalgebra. Of course, a hemi-primal algebra which is not primal cannot be an **SP**-algebra since it has proper nontrivial homomorphic images but no proper subalgebras.

When is an **SP**-algebra a $\mathbf{P}_s$-algebra? Exactly when it has no proper nontrivial subalgebras. When is an **SP**-algebra $\mathfrak{A}$ direct? Certainly this is so if the subalgebras of $\mathfrak{A}$ are simple and if $\mathbf{V}(\mathfrak{A})$ is congruence permutable. Thus a para-primal **SP**-algebra is direct, and a dual-discriminator algebra is direct if and only if it is para-primal (and therefore quasi-primal). But every para-primal algebra $\mathfrak{A}$ is direct in the sense that every algebra in $\mathbf{V}_f(\mathfrak{A})$ is a direct product of simple algebras; the simple algebras in $\mathbf{V}(\mathfrak{A})$ all have cardinality equal to some subalgebra of $\mathfrak{A}$, and there are only finitely many simple algebras in $\mathbf{V}(\mathfrak{A})$.

§81. CONGRUENCES

Since $B = \mathbf{SP}(\mathfrak{B}_1)$, we really do not need homomorphisms to form Boolean algebras; however, to understand Boolean algebras, we need to know how congruences and homomorphisms behave. Congruences of Boolean algebras are as well-behaved as possible. Besides being distributive and permutable, Boolean congruences correspond to ideals (or filters) and, more generally, any congruence is uniquely determined by any of its congruence classes (i.e., congruences are *regular*); any two congruence classes of the same congruence have the same cardinality (congruences are *uniform*). Except for distributivity, these properties carry over to $\mathbf{V}(\mathfrak{A})$ for $\mathfrak{A}$ para-primal (D. M. Clark and P. H. Krauss [1976]). Neither regularity nor uniformity carry over to $\mathbf{V}(\mathfrak{A})$ for $\mathfrak{A}$ an arithmetical hemi-primal algebra or a dual-discriminator algebra.

Since every congruence is a join of principal congruences, it is very useful to have a nice description of principal congruences. The usual criterion for niceness is definability. A class K of algebras has *definable principal congruences* if there is a first order sentence $\varphi(x, y, u, v)$ such that for all $\mathfrak{A} \in K$ and all $a, b, c, d \in A$, $c \equiv d(\Theta(a, b))$ if and only if $\varphi(a, b, c, d)$. For B we have $c \equiv d(\Theta(a, b))$ if and only if $c \oplus d \leq a \oplus b$ where $\oplus$ is symmetric difference. R. McKenzie [1978 a] proves that for a para-primal algebra $\mathfrak{A}$, $\mathbf{V}(\mathfrak{A})$ has definable principal congruences. E. Fried and A. F. Pixley [a] prove that for a dual-discriminator algebra $\mathfrak{A}$, $\mathbf{V}(\mathfrak{A})$ has definable principal congruences. In fact, they prove more: $\mathbf{V}(\mathfrak{A})$ has *equationally definable* principal congruences in the sense that $\varphi(x, y, u, v)$ can be chosen to be a conjunction of equations. The concept of equationally definable principal congruences is introduced and discussed in E. Fried, G. Grätzer and R. W. Quackenbush [1976], [a], and [b]. In particular, it is shown there that if a congruence distributive equational class has definable principal congruences, then it has equationally definable principal congruences. It is easily seen that the equational class generated by an affine para-primal algebra does not have equationally definable principal congruences. It is not known whether an arithmetical hemi-primal algebra generates an equational class with definable principal congruences.

One of the most important congruence properties a class of algebras can have is the *congruence extension property*: an algebra $\mathfrak{A}$ has the congruence extension property if for every subalgebra, $\mathfrak{B}$, of $\mathfrak{A}$, the restriction mapping from $\mathfrak{C}(\mathfrak{A})$ to $\mathfrak{C}(\mathfrak{B})$ is onto (i.e., every $\Theta \in C(\mathfrak{B})$ extends to $\bar{\Theta} \in C(\mathfrak{A})$ such that if $a, b \in B$ and $a \equiv b(\bar{\Theta})$, then $a \equiv b(\Theta)$); a class has the congruence extension property if every algebra in it does. By a result of A. Day [1971], we can restrict our attention to 4-generated subalgebras of $\mathfrak{A}$. For instance, for commutative rings with unit we have $c \equiv d(\Theta(a, b))$ if and only if $(\exists z)\ (c-d=z(a-b))$ (i.e., $c-d$ is contained in the principal ideal generated by $a-b$). Since commutative rings with 1 do not have congruence extension, we cannot eliminate the z. Of course, B has the congruence extension property.

E. Fried and A. F. Pixley [a] show that for any dual-discriminator algebra $\mathfrak{A}$, $\mathbf{V}(\mathfrak{A})$ has the congruence extension property. D. M. Clark and P. H. Krauss [1977] show that for a para-primal algebra $\mathfrak{A}$, $\mathbf{V}(\mathfrak{A})$ has the congruence extension property if and only if $\mathfrak{A}$ is quasi-primal or affine. The result of E. Fried and A. F. Pixley follows from a very general result of B. A. Davey [1977] which says that for a congruence distributive equational class K in which $\mathbf{S}_i(K)$, the class of subdirectly irreducible algebras in K, is axiomatic, K has the congruence extension property if and only if $\mathbf{S}_i(K)$ does. This means that if $\mathfrak{A}$ is arithmetical hemi-primal, then $\mathbf{V}(\mathfrak{A})$ has the congruence extension property.

§82. INJECTIVITY AND PROJECTIVITY

An algebra $\mathfrak{A} \in K$ is *injective* in K if for any $\mathfrak{A}_1, \mathfrak{A}_2 \in K$, any homomorphism $h: A_1 \to A$ and any monomorphism $g: A_1 \to A_2$, there is a homomorphism $f: A_2 \to A$ such that $fg=h$, i.e., any homomorphism from $\mathfrak{A}_1$ into $\mathfrak{A}$ can be extended to a homomorphism from any extension $\mathfrak{A}_2$ of $\mathfrak{A}_1$ into $\mathfrak{A}$. P. R. Halmos [1961] showed that a Boolean algebra is injective if and only if it is complete. A good reference for general results on injectivity is B. Banaschewski [1970]. That $\mathfrak{B}_1$ is injective is the prime ideal theorem. In any equational class, a direct product of injective algebras is injective and a retract of an injective algebra is injective ($\mathfrak{B}$ is a *retract* of $\mathfrak{A}$ if there is a monomorphism $\chi: B \to A$ and an epimorphism $\varepsilon: A \to B$ such that $\varepsilon\chi: B \to B$ is the identity). Thus to prove the Halmos result, one needs to show that the complete Boolean algebras are just retracts of direct powers of $\mathfrak{B}_1$.

For $\mathbf{V}(\mathfrak{A})$ where $\mathfrak{A}$ is quasi-primal, the injective algebras were described in R. W. Quackenbush [1971]. If $\mathfrak{A}$ is not demi-semi-primal, then $\mathbf{V}(\mathfrak{A})$ has no nontrivial injective algebras. If $\mathfrak{A}$ is demi-semi-primal, then the injective algebras in $\mathbf{V}(\mathfrak{A})$ are just the extensions of $\mathfrak{A}$ by complete Boolean algebras.

If $\mathfrak{A}$ is quasi-primal but not demi-semi-primal, then $\mathfrak{A}$ is not injective because $\mathfrak{A}$ has an internal isomorphism which does not extend to an automorphism of $\mathfrak{A}$. However, if we consider only homomorphisms onto $\mathfrak{A}$, then such homomorphisms can always be extended. With G. Grätzer and H. Lakser [1971] and [1972 a], we define $\mathfrak{A}$ to be *weak injective* in K by modifying the definition of injectivity to require h to be onto. Closely related is the concept of an absolute subretract; $\mathfrak{A}$ is an *absolute subretract* in K if $\mathfrak{A} \in K$ and $\mathfrak{A}$ is a retract of every extension of $\mathfrak{A}$ in K. Trivially, if $\mathfrak{A}$ is injective, then it is weak injective, and if $\mathfrak{A}$ is weak injective, then it is an absolute subretract. If $\mathfrak{A}$ is a *maximal subdirectly irreducible algebra* in K (i.e., $\mathfrak{A}$ is maximal in the class of subdirectly irreducible algebras of K), then $\mathfrak{A}$ is an absolute retract in $\mathfrak{A}$. By G. Grätzer and H. Lakser [1971] and [1972 a], if K is a class of algebras with the congruence extension property and $\mathfrak{A} \in K$ is an absolute subretract, then $\mathfrak{A}$ is weak injective in K. Thus if $\mathfrak{A}$ is quasi-primal then $\mathfrak{A}$ is weak injective in $\mathbf{V}(\mathfrak{A})$. In R. W. Quackenbush [1974], it is shown that the weak injectives in $\mathbf{V}(\mathfrak{A})$ are just the extensions of $\mathfrak{A}$ by complete Boolean algebras.

If $\mathfrak{A}$ is para-primal but not quasi-primal, then $\mathbf{V}(\mathfrak{A})$ does not have the congruence extension property; hence we cannot reason as above. It is easy to show that $\mathfrak{A}$ is weak injective in $\mathbf{V}_f(\mathfrak{A})$. Now one can mimic the inverse limit argument of R. W. Quackenbush [1976] and show that $\mathfrak{A}$ is weak injective in $\mathbf{V}(\mathfrak{A})$. Finally, essentially the same argument as for the quasi-primal case shows that the weak injectives in $\mathbf{V}(\mathfrak{A})$ are just the extensions of $\mathfrak{A}$ by complete Boolean algebras.

If $\mathfrak{A}$ is arithmetical hemi-primal or is a dual-discriminator algebra, then $\mathbf{V}(\mathfrak{A})$ has the congruence extension property. If $\mathfrak{A}$ is a dual-discriminator algebra, then $\mathfrak{A}$ is weak injective in $\mathbf{V}(\mathfrak{A})$; thus each extension of $\mathfrak{A}$ by a complete Boolean algebra is weak injective in $\mathbf{V}(\mathfrak{A})$. It is not yet known whether these are all the weak injectives in $\mathbf{V}(\mathfrak{A})$. If each internal isomorphism of $\mathfrak{A}$ extends to an automorphism of $\mathfrak{A}$, then $\mathfrak{A}$ is injective $\mathbf{V}(\mathfrak{A})$.

If $\mathfrak{A}$ is arithmetical hemi-primal, let $\{\mathfrak{A}_1, \cdots, \mathfrak{A}_n\}$ be the subdirectly irreducible homomorphic images of $\mathfrak{A}$; each $\mathfrak{A}_i$ is a maximal subdirectly irreducible algebra in $\mathbf{V}(\mathfrak{A})$ and so is weak injective. Since any homomorphism into $\mathfrak{A}$ is onto $\mathfrak{A}$, the same is true for each $\mathfrak{A}_i$. Thus each $\mathfrak{A}_i$ is injective, and $\prod_{i=1}^{n} \mathfrak{A}_i[\mathfrak{C}_i]$ is injective in $V(\mathfrak{A})$, where each $\mathfrak{C}_i$ is a complete Boolean algebra. It is not yet known whether these are all the injectives in $\mathbf{V}(\mathfrak{A})$.

A property closely related to injectivity and congruence extension is the amalgamation property. Let $\mathfrak{A}$, $\mathfrak{B}$, $\mathfrak{C} \in K$ and let $\beta: A \to B$, $\gamma: A \to C$ be monomorphisms. Then $\langle \mathfrak{A}, \beta, \mathfrak{B}, \gamma, \mathfrak{C} \rangle$ is an *amalgam* in K. We say that K has the *amalgamation property* if for every amalgam $\langle \mathfrak{A}, \beta, \mathfrak{B}, \gamma, \mathfrak{C} \rangle$ in K there is a $D \in K$ and monomorphisms $\beta': B \to D$ and $\gamma': C \to D$ such that $\beta'\beta = \gamma'\gamma$. By B. Banaschewski [1970], every algebra in an equational class K can be embedded in an injective algebra in K if and only if K has the congruence extension property, the amalgamation property, and every $\mathfrak{A} \in K$ has a set of essential extensions ($\mathfrak{D}$, an extension of $\mathfrak{B}$, is *essential* if for any $\Theta \in C(\mathfrak{D})$, $\Theta_B = \omega_B$ implies $\Theta = \omega_D$). It is not difficult to show that for the algebras we have been considering, every $\mathfrak{B} \in \mathbf{V}(\mathfrak{A})$ has only a set of essential extensions. Thus if $\mathfrak{A}$ is demi-semi-primal or arithmetical hemi-primal, then $\mathbf{V}(\mathfrak{A})$ has the amalgamation property.

If $\mathfrak{A}$ is para-primal and $\mathbf{V}(\mathfrak{A})$ does not have the congruence extension property, then, of course, we cannot apply Banaschewski's result. For a para-primal algebra $\mathfrak{A}$, it is easy to show that if $\mathbf{V}(\mathfrak{A})$ does not have the congruence extension property then it does not have the amalgamation property either. Following G. Grätzer and H. Lakser [1971], let **Amal**(K) be the class of all algebras $\mathfrak{A} \in K$ such that every amalgam $\langle \mathfrak{A}, \beta, \mathfrak{B}, \gamma, \mathfrak{C} \rangle$ in K can always be amalgamated in K. What can we say about **Amal**($\mathbf{V}(\mathfrak{A})$)? As shown in R. W. Quackenbush [1974], it appears hopeless to find a nice characterization of **Amal**($\mathbf{V}(\mathfrak{A})$). On the other hand, M. Yasuhara [1974] shows that for any equational class K, $K = \mathbf{S}(\mathbf{Amal}(K))$. This is somewhat disconcerting since, for instance, for the equational class, M, of modular lattices, no nontrivial member of **Amal**(M) is known (for more details, see G. Grätzer, B. Jónsson, H. Lakser [1973]).

Let $\mathfrak{A}$ be quasi-primal and $\mathfrak{B} \in \mathbf{V}(\mathfrak{A})$. Let $\mathfrak{C}$ be a maximal essential extension of $\mathfrak{B}$ in $\mathbf{V}(\mathfrak{A})$ (since $\mathfrak{B}$ has only a set of essential extensions, such exists by Zorn's lemma). Since a maximal essential extension is an absolute subretract and since $\mathbf{V}(\mathfrak{A})$ has the congruence extension property, $\mathfrak{C}$ is weak injective in $\mathbf{V}(\mathfrak{A})$. Any such $\mathfrak{C}$ will be called a *weak injective hull* of $\mathfrak{B}$; if $\mathfrak{C}$ is injective, then it is an *injective hull* of $\mathfrak{B}$. Obvious diagram chasing shows that injective hulls are unique up to isomorphism. With respect to weak injective hulls, this diagram chasing shows that if $\mathfrak{B} \in \mathbf{Amal}(\mathbf{V}(\mathfrak{A}))$, then any two weak injective hulls of $\mathfrak{B}$ are isomorphic. However, it is easy to construct an equational class in which weak injective hulls are not unique. But in these examples, weak injective hulls in the class of subdirectly irreducible algebras are not unique. Thus the following question remains unanswered: Let K be an equational class in which each algebra has a weak injective hull and such that in the class of subdirectly irreducible algebras of K weak injective hulls are unique; when are weak injective hulls unique in K?

Our last topic for this section is projectivity, the dual of injectivity. An algebra $\mathfrak{A} \in K$ is *projective* if for any $\mathfrak{B}$, $\mathfrak{C} \in K$, any epimorphism $\beta: B \to C$

and any homomorphism $\gamma: A \to C$, there is a homomorphism $\alpha: A \to B$ such that $\beta\alpha = \gamma$. It is well-known that in an equational class the projective algebras are just the retracts of the free algebras. One would like an internal characterization of projective algebras. As mentioned earlier, this has only recently been done for Boolean algebras by R. Freese and J. B. Nation. Thus the problem of characterizing the projective algebras in the various equational classes we have considered is largely unexplored. If $\mathfrak{A}$ is quasi-primal, then the finite projective algebras in $\mathbf{V}(\mathfrak{A})$ were characterized in R. W. Quackenbush [1971]; they are just the finite algebras in $\mathbf{V}(\mathfrak{A})$ which have each demi-primal subalgebra of $\mathfrak{A}$ as a direct factor.

§83. FURTHER REFERENCES AND COMMENTS

In this last section we give further references for topics discussed in this appendix, and we mention and give references for some topics which could not be included here due to space limitations.

We first mention the survey article by A. F. Pixley [c]. This gives an alternate viewpoint to much of the material in this appendix. Next we mention H. Werner [1975]. It contains discussions of many important topics not included in this appendix. Thirdly, I. Rosenberg [1977] gives a survey of results on composition of functions and has a bibliography of 464 items.

Functional completeness is a topic where algebraic functions rather than polynomials play a fundamental role. In the book by H. Lausch and W. Nöbauer [1973], universal algebra is developed from the point of view of algebraic functions rather than polynomials.

In the section on representation theory no mention is made of sheaf representations for equational classes generated by quasi-primal algebras. This very important topic got its start in S. D. Comer [1971] and has been developed in B. A. Davey [1973] and S. Burris and H. Werner [a]. Closely related is duality theory, generalizing Stone duality for Boolean algebras. Two good references for this are K. Keimel and H. Werner [1974] and B. A. Davey [a]. These two topics are closely related to Boolean extensions; see S. Burris [1976].

No mention has been made of the theories of ideal classes and filtrality which have been developed by R. Magari and his students; this topic is closely related to equationally definable principal congruences. A partial summary of results in this area can be found in R. Magari [1973], and an extensive discussion can be found in R. Franci [1976].

One important topic not mentioned in this appendix is independence for equational classes. References are G. Grätzer, H. Lakser and J. Płonka

[1969], J. Froemke [1971 a], T.-K. Hu and P. Kelenson [1971], A. L. Foster and A. F. Pixley [1971], [1972].

Another important topic not mentioned is the extension of primal algebra theory to infinite algebras. An algebra $\mathfrak{A}$ is *locally primal* if every partial function on A with a finite domain agrees with some polynomial on that domain. References are A. L. Foster [11], A. F. Pixley [1972 b], [1972 c], T.-K. Hu [1973], P. Kelenson [1973], [1973 a]. This topic is very much underdeveloped and potentially of great importance. However, it seems likely that further significant progress awaits a proper blending of local primality with the theory of topological universal algebra (a good reference for this latter topic is W. Taylor [a]). W. Taylor [d] has taken a first step in this direction.

Finally, we mention an application of quasi-primal algebras. This application is to the problem of the spectrum of an equational class, i.e., the set of cardinalities of finite members of the equational class. This set is a monoid of positive integers under multiplication. G. Grätzer [12] proved the converse by using an equational class generated by a set of primal algebras which had the ternary discriminator as an operation. In J. Froemke and R. W. Quackenbush [1975] this result was extended to the class of groupoids (where, of course, the ternary discriminator cannot be an operation).

APPENDIX 6

EQUATIONAL COMPACTNESS

By Günter H. Wenzel

In the beginning there were the algebraically compact Abelian groups as introduced by I. Kaplansky in the 1954 edition of his monograph Infinite Abelian groups (I. Kaplansky [1954]): An Abelian group is *algebraically compact* if it has the form $G=C\oplus D$ where C is divisible and D is a complete direct sum of groups D_p, one for each prime p; D_p is a module over the p-adic integers with no elements of infinite height and it is complete in its p-adic topology. The 1969 edition of the same book gives a quite different definition, reflecting a development that originated in Poland. S. Balcerzyk [1957] and J. Łoś [1957] discovered that Kaplansky's rather involved structural definition is equivalent to either one of the following two: (1) Every finitely solvable system of algebraic equations over the given group G is solvable (in G). (2) G is a direct summand of a compact topological group.

J. Mycielski [3] gave the concept its proper universal algebraic setting and initiated a series of investigations dealing with the new topic. We shall restrict ourselves to what we believe are the central results from a universal algebraic standpoint. We cannot even attempt to cover all aspects or all results. The proofs will be concise, some will be omitted; nevertheless we hope to pass on the flavor of the subject.

§84. EQUATIONAL AND ATOMIC COMPACTNESS—FIRST EXAMPLES

Equational compactness represents a successful attempt to carry over properties and methods of the theory of compact topological spaces to certain algebraic model theoretic questions. Some fascinating behavior of classical arithmetical structures on the borderline of algebra and naive set theory on the one hand, some surprising interplays between topology, equational compactness, algebra and model theory in important classes of algebraic systems on the other hand have created increasing interest in the field. The Zariski topology in affine or projective spaces of algebraic geometry can be considered as possibly the first realization of this phenomenon: Solution sets of polynomial equations are made the subbase of closed sets of a topology—in spite of the fact that this topology becomes very weak and atypical from a topologist's point of view. Of course, the analogy ends quickly. There we have quasi-compactness immediately (the Zariski topologies over algebraically closed fields are Noetherian with respect to open sets). With regard to the algebraic questions it is here that the problem begins: When do the solution sets of algebraic equations "behave quasi-compact"? What does such behavior mean for individual algebras? What does it mean for classes or—of primary interest—for equational classes? The compactness theorem and its various generalizations offer interesting problems of their own.

To illustrate the ideas we begin with a few examples. We start with a precise definition of equational compactness.

Definition 1. (1) *An algebra* $\mathfrak{A}=\langle A; F\rangle$ *is* equationally compact *if every infinite system* Σ *of algebraic equations over* $\mathfrak{A}$ *in the variables* x_γ, $\gamma<\beta$, *is simultaneously solvable in* $\mathfrak{A}$ *provided that every finite subset has a simultaneous solution. An* algebraic equation over $\mathfrak{A}$ *is an equation* $p=q$ *where* p, q *are algebraic functions of* $\mathfrak{A}$ (*i.e., polynomials with constants from* A). *We consider the solution sets in* A^β (*or in some* A^γ *for* $\gamma\geqq\beta$).

(2) *A structure* $\mathfrak{A}=\langle A; F, R\rangle$ *is* atomic compact *if every infinite system of atomic formulas with constants in* A *and variables* x_γ, $\gamma<\beta$, *is simultaneously satisfiable in* $\mathfrak{A}$ *provided the conjunction of every finite subset is satisfiable.*

This appendix is concerned with *equational* compactness. Thus, we will adhere strictly to finitary universal algebras, although many or even most of the results presented can be given a broader setting within the framework of structures or of infinitary structures. In spite of this limitation we introduced atomic compactness to facilitate the presentation of some interesting algebraic constructions of W. Taylor (see §87).

Remark. An algebra $\mathfrak{A}=\langle A; F\rangle$ is equationally compact if and only if the associated relational system is atomic compact.

Now some examples: Finite algebras (structures) are, of course, equationally (atomic) compact. If an algebra (structure) carries a Hausdorff topology that makes the operations continuous (and has closed relations), then we speak of a *topological algebra* (*topological structure*). The next result was observed by J. Mycielski [3].

Theorem 1. *Compact topological algebras* (*structures*) *are equationally compact* (*atomic compact*).

Proof. Due to the Hausdorff property, the associated relational systems have only closed relations, so the very definition of atomic compactness shows that compactness implies atomic compactness.

Theorem 2. *Direct products and retracts* (*a* retract *is a homomorphic image under a homomorphism that is the identity on its image*) *of equationally compact algebras, respectively, atomic compact structures, are again equationally compact, respectively, atomic compact.*

Equationally compact Abelian groups, i.e., algebraically compact Abelian groups, are exactly the direct summands of compact topological groups (see, e.g., J. Łoś [1957]). The group $\langle Q; +, -\rangle$ of rational numbers is divisible, hence equationally compact, without being a compact topological group. The group $\langle Z; +, -\rangle$ of integers has created interesting problems. First of all, it is not equationally compact. This follows immediately from Kaplansky's characterization theorem. One may, however, also use the following explicit system of equations: $\{3x_0+x_1=1, x_1=2x_2, \cdots, x_n=2x_{n+1}, \cdots \mid n \in N\}$ (see J. Mycielski [3]). It is finitely solvable but not solvable. J. Mycielski [3] has also produced a system of $\aleph_1$ equations over the *ring* $\langle Z; +, -, \cdot\rangle$ which is not solvable, but every countable subsystem is. Such is

$$\Sigma_1 = \{x_{\xi,\eta}\cdot(5z_\xi+2)+y_{\xi,\eta}\cdot(5z_\eta+2) = 1 \mid 0 \leqq \xi \neq \eta < \omega_1\}.$$

The nonsolvability is obvious for $\xi\neq\eta$ implies $z_\xi\neq z_\eta$. The fact that every countable subsystem is solvable relies on Dirichlet's prime number theorem which allows for every countable system of indices $\xi_0, \xi_1, \xi_2, \cdots$ a choice of integers z_{ξ_n} such that all $5z_{\xi_n}+2$ are prime numbers. Mycielski's result was improved by R. McKenzie [1971 b] who produced a system of $\aleph_1$ equations over the *group* $\langle Z; +, -\rangle$ with the same property. To this end choose $\aleph_1$ infinite sets of prime numbers, say $P_0, P_1 \cdots, P_\mu, \cdots, \mu<\omega_1$, such that any two have finite intersection (Zorn's lemma). For every $0\leqq\xi<\eta<\omega_1$ choose a prime number $p(\xi, \eta)$ such that $\{p(\xi, \eta) \mid 0\leqq \xi<\eta<\omega_1\}$ satisfies the following two properties:

(a) $p(\xi_1, \eta)=p(\xi_2, \eta)$ implies $\xi_1=\xi_2$,
(b) $p(\xi, \eta) \in P_\eta-P_\xi$.

The system is then given by the system of congruences

$$\Sigma_2 = \{x_\eta \equiv x_\xi+1 \pmod{p(\xi, \eta)} \mid 0 \leqq \xi < \eta < \omega_1\}.$$

Again the nonsolvability follows from $x_\eta\neq x_\xi$ for $\eta\neq\xi$. The solvability of every countable subsystem follows from a skilful application of the Chinese remainder theorem. We summarize these examples in Theorem 3 and supplement them with a result of J. Łoś (implicit in J. Łoś [1959]).

Theorem 3. (1) *Let* $\mathfrak{m} \in \{\aleph_0, \aleph_1\}$. *There exists a system* Σ *of algebraic equations over* $\langle Z; +, -\rangle$ *with* $|\Sigma| = \mathfrak{m}$ *such that* Σ *is not solvable in* Z *while each subsystem with* $< \mathfrak{m}$ *equations is solvable.*

(2) *If* λ *is a nonmeasurable regular cardinal, then there exists a cardinal* $\mathfrak{m}$ *with* $\lambda \leqq \mathfrak{m} \leqq 2^\lambda$ *for which there is a system* Σ *with the same properties as in* (1).

Once we know that every fixed finitely solvable system of algebraic equations over an algebra $\mathfrak{A}$ can be solved in a suitable extension $\mathfrak{B}$ of $\mathfrak{A}$ we realize that all algebras that are injective in $K(\tau)$ are equationally compact. Thus equational compactness is a loosened version of injectivity. R-modules have injective hulls, so they can be *equationally compactified*, i.e., embedded into equationally compact algebras. Equational compactification is not always possible, for instance *algebraically closed groups* (i.e., groups that are pure subgroups of all extension groups) can never be equationally compactified. To see this, we shall need some theory. A more elementary example is the lattice $\mathfrak{L}(\mathfrak{n})$ (see J. Mycielski and C. Ryll-Nardzewski [1]): If $\mathfrak{n}$ is an infinite cardinal let $L(\mathfrak{n}) = \{0, 1\} \dot\cup S$ where $|S| = \mathfrak{n}$ and define $s \vee t = 1$, $s \wedge t = 0$ for s, $t \in S$ $(s \neq t)$. The system $\Sigma_\mathfrak{m} = \{x_i \vee x_j = 1, x_i \wedge x_j = 0 \mid i, j \in T, i \neq j\}$ for a set T of cardinality $\mathfrak{m}$ shows that $\mathfrak{L}(\mathfrak{n})$ cannot be equationally compactified if we choose $\mathfrak{m}$ large enough.

§85. RELATED COMPACTNESS CONCEPTS AND CHARACTERIZATIONS

Let τ be a fixed type of algebras. By changing (see §41 viz. §36) the set of variables from a countable set to $\{x_\gamma \mid \gamma < \alpha\}$ for an arbitrary ordinal $\alpha \geqq \omega$, we obtain a language $L^{(\alpha)}(\tau)$ for every such α. Clearly the language $L(\tau)$ of this book is our $L^{(\omega)}(\tau)$. We set $L^\infty(\tau) = \bigcup (L^{(\alpha)}(\tau) \mid \alpha \geqq \omega)$. The semantic side is modified accordingly by choosing *solutions* (i.e., "satisfying sequences") in A^α rather than in A^ω for every $\mathfrak{A} = \langle A; F\rangle \in K(\tau)$. If C is an arbitrary fixed set we obtain $L_C{}^{(\alpha)}(\tau)$ and $L_C{}^\infty(\tau)$ by allowing formal substitution of some variables by elements from C, considering these elements $c \in C$ as new nullary operational symbols. If $\mathfrak{A} \in K(\tau)$ then $L_A^{(\omega)}(\tau)$ is Robinson's diagram language (see Chapter 6, §39 and Exercise 73). If $C \subseteq A$, $\mathfrak{A} = \langle A; F\rangle \in K(\tau)$, and γ is a well-ordering of C, then $\mathfrak{A}_C$ denotes the algebra $\langle \mathfrak{A}, \gamma\rangle$ (see §39) in the canonically enriched type.

Definition 1. *Let* L *be a fixed subclass of* $L^\infty(\tau)$, $\mathfrak{A}$, $\mathfrak{B} \in K(\tau)$, $C \subseteq A$, B, *and* $\mathfrak{m}$ *an infinite cardinal number:*

(1) $\mathfrak{B}$ *is* L-$(C, \mathfrak{A}, \mathfrak{m})$-compact *if every system* Σ *of* $\mathfrak{m}$ *formulas from* L_C *is simultaneously solvable* (*i.e., satisfiable*) *in* $\mathfrak{B}$ *provided every finite subsystem of* Σ *is solvable in* $\mathfrak{A}$. *We will refer to this* finite solvability condition *of the system* Σ *as* $\mathfrak{A}$-consistency of Σ.

(2) $\mathfrak{B}$ *is* L-$(C, \mathfrak{A})$-compact *if it is* L-$(C, \mathfrak{A}, \mathfrak{m})$-*compact for all cardinals* $\mathfrak{m} \geqq \aleph_0$.

This definition subsumes all the compactness concepts with which we shall be dealing. It subsumes, in particular, the various compactness concepts introduced by J. Mycielski [3] (see also B. Węglorz [1], [2]). The next definition provides the linguistic link to the more familiar terminology.

Definition 2. *We use the following equivalent terminology for algebras* $\mathfrak{A}$ (At *and* Pos *denote the class of all atomic and positive formulas, respectively*)*:*

$\mathfrak{A}$ *is*

(1) *weakly equationally* $\mathfrak{m}$-*compact*	At-$(\varnothing, \mathfrak{A}, \mathfrak{m})$-*compact*
(1′) *weakly equationally compact*	At-$(\varnothing, \mathfrak{A})$-*compact*
(2) *equationally* $\mathfrak{m}$-*compact*	At-$(A, \mathfrak{A}, \mathfrak{m})$-*compact*
(2′) *equationally compact* (*see Definition 1*)	At-$(A, \mathfrak{A})$-*compact*
(3) *weakly positively* $\mathfrak{m}$-*compact*	Pos-$(\varnothing, \mathfrak{A}, \mathfrak{m})$-*compact*
(3′) *weakly positively compact*	Pos-$(\varnothing, \mathfrak{A})$-*compact*
(4) *positively* $\mathfrak{m}$-*compact*	Pos-$(A, \mathfrak{A}, \mathfrak{m})$-*compact*
(4′) *positively compact*	Pos-$(A, \mathfrak{A})$-*compact*
(5) *weakly elementarily* $\mathfrak{m}$-*compact*	$L^{\infty}(\tau)$-$(\varnothing, \mathfrak{A}, \mathfrak{m})$-*compact*
(5′) *weakly elementarily compact*	$L^{\infty}(\tau)$-$(\varnothing, \mathfrak{A})$-*compact*
(6) *elementarily* $\mathfrak{m}$-*compact*	$L^{\infty}(\tau)$-$(A, \mathfrak{A}, \mathfrak{m})$-*compact*
(6′) *elementarily compact*	$L^{\infty}(\tau)$-$(A, \mathfrak{A})$-*compact*

Remark. Properties (5′) and (6′) are without interest since they are satisfied by a finite $\mathfrak{A}$ only. Reason: If $\bar{\alpha} > |A| \geqq \aleph_0$, then the system $\{x_\gamma \neq x_\delta \mid 0 \leqq \gamma < \delta < \alpha\}$ is $\mathfrak{A}$-consistent but not solvable in $\mathfrak{A}$. Of primary interest to us are (1′), (2′) and (4′).

The introduction of these concepts by J. Mycielski in 1964 was followed by B. Węglorz's very useful characterization theorems for some of them (B. Węglorz [1]). We give first the simplest formulation that contains the basic idea; this will be followed by a less appealing formulation which has the advantage that it covers simultaneously all cases of interest to us. Purity plays a central role in these and other characterizations, so we shall precede the theorems with its definition.

Definition 3. *Let* $\mathfrak{A}$, $\mathfrak{B}$ *be algebras* (*or structures*), $\mathfrak{A} \in \mathbf{S}(\mathfrak{B})$, $C \subseteq A$. *Then* $\mathfrak{B}$ *is a* C-pure extension *of* $\mathfrak{A}$ *if every finite set of atomic formulas with constants in* C *is solvable in* $\mathfrak{A}$ *provided that it is solvable in* $\mathfrak{B}$. $\varnothing$-*pure extensions of* $\mathfrak{A}$ *are called* weakly pure *and* A-*pure extensions are called* pure.

Theorem 1. *Let $\mathfrak{A}$ be an algebra in $K(\tau)$. The following conditions are equivalent:*

(1) *$\mathfrak{A}$ is weakly equationally compact.*
(2) *$\mathfrak{A}$ contains a homomorphic image of every pure extension.*
(3) *$\mathfrak{A}$ contains a homomorphic image of every elementary extension.*
(4) *$\mathfrak{A}$ contains a homomorphic image of every ultrapower $\mathfrak{A}_D^I$.*

Theorem 1′. *Let $\mathfrak{A}$, $\mathfrak{B}$ be algebras in $K(\tau)$, $C \subseteq A$, B. The following conditions are equivalent:*

(1) *$\mathfrak{A}$ is* At-$(C, \mathfrak{B})$*-compact.*

(2) *If $\mathfrak{P}$ is a C-pure extension of $\mathfrak{B}$, then there exists a C-homomorphism $g\colon \mathfrak{P} \to \mathfrak{A}$ (i.e., $g|_C = \mathrm{id}$).*

(3) *$\mathfrak{A}$ contains a C-homomorphic image of every pure extension of $\mathfrak{B}$.*
(4) *$\mathfrak{A}$ contains a C-homomorphic image of every elementary extension of $\mathfrak{B}$.*
(5) *$\mathfrak{A}$ contains a C-homomorphic image of every ultrapower $\mathfrak{B}_D^I$.*

If we set $\mathfrak{B} = \mathfrak{A}$ and $C = \varnothing$ in Theorem 1′ then we obtain Theorem 1. If we set $C = A = B$, then the C-homomorphisms become retractions and Theorem 1′ yields B. Węglorz's characterization of equational compactness. Of course, since equational compactness of $\mathfrak{A}$ is exactly weak equational compactness of $\mathfrak{A}_A$ the latter is also contained in Theorem 1.

Proof of Theorem 1′. Only (1) implies (2) and (5) implies (1) require proof. (1) implies (2): In Chapter 6, Exercise 73 the diagram $D(\mathfrak{P})$ of $\mathfrak{P}$ is defined. We define the *positive diagram* $D_0(\mathfrak{P})$ in the same way but without allowing the negation of atomic formulas. If we replace in $D_0(\mathfrak{P})$ each occurrence of $p \in P - C$ by the variable x_p then $D_0(\mathfrak{P})$ turns into a $\mathfrak{B}$-consistent system Σ of algebraic equations with constants in C. Thus, Σ is solvable in $\mathfrak{A}$. If $x_p = a_p$ ($p \in P - C$) is a solution of Σ in A then $g\colon P \to A$ with $pg = a_p$ for $p \notin C$ and $cg = c$ for $c \in C$ is the required C-homomorphism. (5) implies (1) follows from Lemma 1 below.

Let $\mathfrak{m}$ be an infinite cardinal number. We choose a set J with $|J| = \mathfrak{m}$ and let I be the ideal of finite subsets of J. For arbitrary $x \in J$ we define $I_x = \{T \mid T \in I, x \in T\}$ and embed $\{I_x \mid x \in J\}$ into an ultrafilter D over I.

Definition 4. *Every ultraproduct $\prod_D (\mathfrak{A}_i \mid i \in I)$ (D, I as above) is called an* $\mathfrak{m}$-associated ultraproduct *and $\mathfrak{A}_D^I$, denoted by $\mathfrak{A}(\mathfrak{m})$, an* $\mathfrak{m}$-associated ultrapower.

Lemma 1. *If Σ is an $\mathfrak{A}$-consistent set of $\mathfrak{m}$ ($\geqq \aleph_0$) formulas in $L_C^{(\alpha)}(\tau)$ (with $C \subseteq A$) then Σ has a simultaneous solution in every $\mathfrak{A}(\mathfrak{m})$. Each $\mathfrak{A}(\mathfrak{m})$ satisfies $|A(\mathfrak{m})| = |A|^{\mathfrak{m}}$.*

Proof. Follow the pattern of the proof for the compactness theorem. For the cardinality statement see also J. Mycielski and C. Ryll-Nardzewski [1], Lemma 2.

Corollary. *Let* $\mathfrak{A}$, $\mathfrak{B}$ *be algebras in* $K(\tau)$, $\mathfrak{B} \in \mathbf{S}(\mathfrak{A})$, $\mathfrak{m} \geqq \aleph_0$, *and* $\mathfrak{B}(\mathfrak{m})$ *a fixed* $\mathfrak{m}$*-associated ultrapower. If* $\mathfrak{A}$ *contains a B-homomorphic image of* $\mathfrak{B}(\mathfrak{m})$, *then* $\mathfrak{A}$ *is* At-$(B, \mathfrak{B}, \mathfrak{m})$*-compact.*

The next result (G. H. Wenzel [1973]) is the universal algebraic counterpart to a theorem of J. Łoś [1959] that states that an Abelian group is algebraically compact if and only if generalized λ-limits exist for every ordinal λ. λ-limits can be viewed as an alternative for the ultrapowers that need to be considered in Theorem 1.

If $\mathfrak{A}$ is an algebra, then a *λ-limit over* $\mathfrak{A}$ is a retraction $\mathrm{Lim}_\lambda\colon \mathfrak{A}^{\omega_\lambda} \to \mathfrak{A}$ ($\mathfrak{A}$ is identified with the diagonal) with the property that $\langle x_\xi \rangle_{\xi < \omega_\lambda} \mathrm{Lim}_\lambda = \langle y_\xi \rangle_{\xi < \omega_\lambda} \mathrm{Lim}_\lambda$ holds if there exists $\xi_0 < \omega_\lambda$ with $x_\xi = y_\xi$ for all $\xi \geqq \xi_0$. If we take on $\{\xi \mid \xi < \omega_\lambda\}$ the filter E_λ generated by $\{\{\xi \mid \xi_0 \leqq \xi < \omega_\lambda\} \mid \xi_0 < \omega_\lambda\}$, then we can say equivalently that Lim_λ is a retraction from $\mathfrak{A}^{\omega_\lambda}_{E_\lambda}$ to $\mathfrak{A}$.

Theorem 2. *The algebra* $\mathfrak{A}$ *is equationally compact if and only if all* λ*-limits exist over* $\mathfrak{A}$.

Proof. Let $\mathfrak{A}$ be equationally compact. Since $\mathfrak{A}^{\omega_\lambda}_{E_\lambda}$ is a pure extension of $\mathfrak{A}$, there exists a retraction $g\colon \mathfrak{A}^{\omega_\lambda}_{E_\lambda} \to \mathfrak{A}$ (Theorem 1′). Conversely, we have to show that every system Σ of algebraic equations over $\mathfrak{A}$ with $\aleph_\lambda$ equations is solvable in $\mathfrak{A}$ provided that every subsystem with less than $\aleph_\lambda$ equations is solvable. To see this, we choose such a Σ in the variables x_γ ($\gamma < \beta$) and write $\Sigma = \bigcup (\Sigma_\gamma \mid \gamma < \omega_\lambda)$ where $|\Sigma_\lambda| < \aleph_\lambda$ and $\gamma_1 \leqq \gamma_2$ implies $\Sigma_{\gamma_1} \subseteq \Sigma_{\gamma_2}$. Each Σ_γ has a solution $\langle c_0{}^\gamma, \cdots, c_\delta{}^\gamma, \cdots \rangle_{\delta < \beta} \in A^\beta$. With $\xi^\delta = \langle c_\delta{}^0, c_\delta{}^1, \cdots, c_\delta{}^\gamma, \cdots \rangle_{\gamma < \omega_\lambda} \in A^{\omega_\lambda}$, $\delta < \beta$, we have that $\langle \hat{\xi}^0, \cdots, \hat{\xi}^\delta, \cdots \rangle_{\delta < \beta} \in (A^{\omega_\lambda}_{E_\lambda})^\beta$ is a solution of Σ in $\mathfrak{A}^{\omega_\lambda}_{E_\lambda}$. Then, by hypothesis, Lim_λ yields a solution of Σ in $\mathfrak{A}$.

Another very interesting characterization is due to B. Węglorz [1].

Theorem 3. *The algebra* $\mathfrak{A}$ *is equationally compact if and only if it is positively compact.*

Proof. Let $\mathfrak{A}$ be equationally compact and Σ be an $\mathfrak{A}$-consistent set of positive formulas with constants in A. By Lemma 1, Σ is solvable in some ultrapower $\mathfrak{A}_D{}^I$ that retracts onto $\mathfrak{A}$ (Theorem 1′). Since homomorphisms preserve positive formulas, Σ is solvable in $\mathfrak{A}$.

The analogous result does not hold for weak equational compactness as the following example shows (G. H. Wenzel [1970 b]): The ring $\langle Z; +, -, \cdot\rangle$ of integers is weakly equationally compact. However, the following system of positive formulas is finitely solvable in Z without being solvable: $\Sigma = \{(x_{\omega+1})(x_\omega \cdot x_{\omega+1} = x_{\omega+1})\} \bigcup \{3x_0 + x_1 = x_\omega, x_1 = 2x_2, \cdots, x_n = 2x_{n+1}, \cdots\}$. Another counterexample was given by G. Fuhrken and W. Taylor [1971] (Example 4.2, p. 139).

The next two theorems reveal once more the special role played by purity in the investigation of equationally compact algebras. The first was independently observed by many and it is in the spirit of B. Węglorz's original characterization. The second is based on the notion of "pure essential extensions" and was discovered by W. Taylor (see W. Taylor [1972], also B. Banaschewski and E. Nelson [1972]). W. Taylor pointed out that his results followed discussions with R. B. Warfield in the course of which Warfield conjectured "that all the things about purity which were known for Abelian groups would probably go through for general algebra".

Theorem 4. *The algebra $\mathfrak{A}$ is equationally compact if and only if it is pure-injective in $K(\tau)$ (i.e., every homomorphism $g\colon \mathfrak{B} \to \mathfrak{A}$ can be extended to a homomorphism $g_0\colon \mathfrak{C} \to \mathfrak{A}$ for every pure extension $\mathfrak{C}$ of $\mathfrak{B}$).*

Proof. We transform the positive diagram $D_0(\mathfrak{C})$ (see proof of Theorem 1′) to a system Σ of equations over $\mathfrak{A}$ by replacing each $c \in C - B$ by a variable x_c and each $b \in B$ by bg. Σ is solvable in $\mathfrak{A}$. If $x_c = a_c \in A$ is a solution, then we define $cg_0 = a_c$ for $c \in C - B$ and $g_0|_B = g$. The converse is obvious.

Definition 5. *Let $\mathfrak{A}$, $\mathfrak{B}$ be algebras and $\mathfrak{A} \in \mathbf{S}(\mathfrak{B})$. $\mathfrak{B}$ is an* essential extension *of $\mathfrak{A}$ if $\Theta \in C(\mathfrak{B})$, $\Theta \neq \omega$ implies $\Theta_A \neq \omega$. $\mathfrak{B}$ is a* pure-essential extension *of $\mathfrak{A}$ if $\mathfrak{B}$ is a pure extension of $\mathfrak{A}$ and whenever $\Theta \in C(\mathfrak{B})$, $\Theta_A = \omega$, and $\mathfrak{B}/\Theta$ is a pure extension of $\mathfrak{A}$, then $\Theta = \omega$.*

Lemma 2. *If $\mathfrak{B}$ is a pure extension of $\mathfrak{A}$, then there exists $\Theta \in C(\mathfrak{B})$ such that $\mathfrak{B}/\Theta$ is a pure-essential extension of $\mathfrak{A}$.*

Proof. $\{\Phi \mid \Phi \in C(\mathfrak{B}), \Phi_A = \omega, \mathfrak{B}/\Phi$ is a pure extension of $\mathfrak{A}\}$ has a maximal element Θ. We take $\mathfrak{C} = \mathfrak{B}/\Theta$.

Theorem 5. *The following four conditions are equivalent for an algebra $\mathfrak{A}$:*

(1) *$\mathfrak{A}$ is equationally compact.*
(2) *$\mathfrak{A}$ has no proper pure-essential extension in $K(\tau)$.*
(3) *$\mathfrak{A}$ is a maximal pure-essential extension of some $\mathfrak{B} \in K(\tau)$.*
(4) *$\mathfrak{A}$ has no proper pure-essential extension in* $\mathbf{HSP}(\mathfrak{A})$.

Proof. (1) implies (2). Every pure extension of $\mathfrak{A}$ can be retracted to $\mathfrak{A}$ by Theorem 1. (2) implies (3). Take $\mathfrak{B}=\mathfrak{A}$. (3) implies (4). Let $\mathfrak{C}$ [$\in$ **HSP**($\mathfrak{A}$)] be a pure-essential extension of $\mathfrak{A}$. Then $\mathfrak{C}$ is a pure extension of $\mathfrak{B}$. By Lemma 2 we find $\Theta \in C(\mathfrak{C})$ such that $\mathfrak{C}/\Theta$ is a pure-essential extension of $\mathfrak{B}$. Since then $\mathfrak{A}/\Theta_A$ is a pure extension of $\mathfrak{B}$ we get $\Theta_A=\omega$, thus $\mathfrak{A} \in$ **S**($\mathfrak{C}/\Theta$), i.e., $\mathfrak{A}=\mathfrak{C}/\Theta$. Hence, $\mathfrak{A}$ is a retract of $\mathfrak{C}$ which implies $\mathfrak{A}=\mathfrak{C}$. (We needed to argue this way since W. Taylor [1972 b] showed that pure-essentialness is not transitive.) (4) implies (1). Let $\mathfrak{B}$ be an ultrapower of $\mathfrak{A}$. By Lemma 2, $\mathfrak{B}/\Theta$ is a pure-essential extension of $\mathfrak{A}$ for suitable Θ and, of course, $\mathfrak{B}/\Theta \in$ **HSP**($\mathfrak{A}$). Thus, $\mathfrak{B}/\Theta=\mathfrak{A}$, i.e., $\mathfrak{B}$ retracts to $\mathfrak{A}$.

§86. CONNECTIONS WITH $|A|$

One can limit the size of the systems of equations that need be considered for testing equational compactness or weak equational compactness of an algebra. This is the content of two important results. Theorem 1 was discovered by J. Mycielski and C. Ryll-Nardzewski [1]; Theorem 2 was conjectured in the same paper and proved by G. Fuhrken and W. Taylor [1971]. The next lemma forms the basis of both results (J. Mycielski and C. Ryll-Nardzewski [1]).

Lemma 1. *The following two statements are equivalent for an infinite algebra $\mathfrak{A}$ and a cardinal number $\mathfrak{m} \geqq \aleph_0$:*

(1) *$\mathfrak{A}$ is equationally $\mathfrak{m}$-compact.*

(2) *Every $\mathfrak{A}$-consistent system Σ of formulas of the type $(\exists x_1) \cdots (\exists x_m)$ $(e_0 \wedge e_1 \wedge \cdots \wedge e_r)$ with algebraic equations e_i over $\mathfrak{A}$ and exactly one free variable x_0 is solvable in $\mathfrak{A}$ if $|\Sigma| \leqq \min(|A|, \mathfrak{m})$.*

Proof. (1) implies (2) holds evidently even for $\min(|A|, \mathfrak{m})$ replaced by $\mathfrak{m}$. So we prove that (2) implies (1). If $\Sigma=\Sigma(X)$ (with $X=\{x_\gamma\}_{\gamma<\alpha}$) is a system of algebraic equations in the variables x_γ and with $|\Sigma| \leqq \mathfrak{m}$ then we construct by transfinite induction a sequence $\langle a_0, a_1, \cdots, a_\gamma, \cdots\rangle_{\gamma<\alpha} \in A^\alpha$ such that for each $\tau \leqq \alpha$ the substitution $x_\gamma \rightarrow a_\gamma$ $(\gamma<\tau)$ yields a system Σ_τ that is still $\mathfrak{A}$-consistent. Set $\Sigma_0=\Sigma$. Assume that Σ_τ has been found for some $\tau<\alpha$ but there is no $a_\tau \in A$ such that the substitution $x_\tau \rightarrow a_\tau$ turns Σ_τ into an $\mathfrak{A}$-consistent system $\Sigma_{\tau+1}$. Then we find for each $a \in A$ a finite subsystem $\Sigma^{(a)}=\Sigma^{(a)}(x_{\gamma_0}, \cdots, x_{\gamma_t}, x_\tau)=\{e_1, \cdots, e_r\}$ of Σ_τ which has no solution with $x_\tau=a$. We set $\Phi^{(a)}=(\exists x_{\gamma_0}) \cdots (\exists x_{\gamma_t})(e_1 \wedge e_2 \wedge \cdots \wedge e_r)$. $\Lambda=\{\Phi^{(a)} \mid a \in A\}$ is then an $\mathfrak{A}$-consistent system of the type described in (2) with $|\Lambda| \leqq |A|$. $|\Sigma| \leqq \mathfrak{m}$ implies, of course, that $|\Lambda| \leqq \mathfrak{m}$, i.e., $|\Lambda| \leqq \min(|A|, \mathfrak{m})$. Since Λ is not solvable in $\mathfrak{A}$ we arrive at a contradiction. There is no problem in the remaining limit process.

Remark. D. K. Haley [1974 a] showed that we cannot assume in Lemma 1 (2) that every e_i actually depends on x_0.

Theorem 1. *The algebra $\mathfrak{A}$ is equationally compact if and only if it is equationally $|A|$-compact.*

Proof. Without loss of generality we can assume that $|A| \geqq \aleph_0$. If $\mathfrak{A}$ is equationally $|A|$-compact, then (Lemma 1, (2) implies (1)) $\mathfrak{A}$ is equationally $\mathfrak{m}$-compact for all infinite $\mathfrak{m}$, i.e., $\mathfrak{A}$ is equationally compact.

Corollary. *$\mathfrak{A}$ is equationally compact if and only if $\mathfrak{A}$ is a retract of any fixed $\mathfrak{A}(|A|)$.*

The following result was observed by G. Fuhrken and published without proof in G. Fuhrken and W. Taylor [1971]. The proof given below is due to W. Taylor.

Lemma 2. *If $\mathfrak{A} = \langle A; F\rangle$ is an algebra with $\mathfrak{m} \geqq |A| \geqq \aleph_0$, then $\mathfrak{A}$ is weakly equationally $\mathfrak{m}$-compact if and only if every $\mathfrak{A}$-consistent system of polynomial equations $\Sigma(X)$ with $|X| \leqq \mathfrak{m}$ is solvable in $\mathfrak{A}$.*

Proof. Only one direction requires proof. We assume weak equational $\mathfrak{m}$-compactness and fix an $\mathfrak{A}$-consistent system of polynomial equations $\Sigma = \Sigma(X)$ with the set of variables X and $|X| \leqq \mathfrak{m}$. We have to prove that Σ has a solution in A^X. For $Y \subseteq X$, Σ_Y denotes the subset of Σ that involves only the variables from Y. Every solution of Σ_Y is canonically an element of A^Y. For each $\Theta \in A^Y$ we have either that Θ solves Σ_Y or there is some $e_\Theta{}^Y \in \Sigma_Y$ which is not solved by Θ. We set, for every finite $Y \subseteq X$, $\Sigma_Y{}^0 = \{e_\Theta{}^Y \mid \Theta \in A^Y,\ \Theta \text{ is no solution of } \Sigma_Y\}$. We have $\Sigma_Y{}^0 \subseteq \Sigma_Y \subseteq \Sigma$ and $|\Sigma_Y{}^0| \leqq |A| \leqq \mathfrak{m}$. Finally, we set $\Sigma^0 = \bigcup (\Sigma_Y{}^0 \mid Y \subseteq X,\ |Y| < \aleph_0)$ and have $\Sigma^0 \subseteq \Sigma$ with $|\Sigma^0| \leqq \mathfrak{m}$. By assumption there is a solution $\Theta \in A^X$ of Σ^0. It is easily seen that Θ is a solution of Σ.

Theorem 2. *The algebra $\mathfrak{A}$ is weakly equationally compact if and only if it is weakly equationally $|A|^+$-compact.*

Proof. Of course, we assume $|A| \geqq \aleph_0$. Let $\Sigma = \Sigma(X)$ be any $\mathfrak{A}$-consistent system of polynomial equations with the set of variables X. Call $Z \subseteq X^2$ *Σ-compatible* if $\Sigma \cup \{x = y \mid \langle x, y\rangle \in Z\}$ is still $\mathfrak{A}$-consistent. By Zorn's lemma we find a maximal Σ-compatible Z_0. Z_0 is an equivalence on X and we choose Y as a complete system of representatives of X/Z_0. By replacing in $\Sigma(X)$ each x by its representative we obtain $\Sigma(Y)$ which is still $\mathfrak{A}$-consistent. Assume $|Y| > |A|$ and choose $Y_0 \subseteq Y$ with $|Y_0| = |A|^+$.

For all $x \neq y \in Y_0$ we can find $\Sigma_{x,y} \subseteq \Sigma(Y)$ such that $\Sigma_{x,y}$ is finite and $\sum_{x,y} \cup \{x=y\}$ is not solvable in $\mathfrak{A}$. Then $\Sigma_1 = \bigcup (\Sigma_{x,y} \mid x, y \in Y_0, x \neq y)$ is $\mathfrak{A}$-consistent and $|\Sigma_1| = |A|^+$. Weak equational $|A|^+$-compactness yields that Σ_1 is solvable in $\mathfrak{A}$. A solution $\{a_y \mid y \in Y\}$ satisfies $a_x \neq a_y$ for all $x, y \in Y_0, x \neq y$. Since the set of variables of Σ_1 contains Y_0 and $|Y_0| = |A|^+$ we arrive at a contradiction. Thus, $|Y| \leqq |A|$, i.e., $\Sigma(Y)$ is solvable. But a solution of $\Sigma(Y)$ leads naturally to a solution of Σ. A different proof (assuming GCH) is contained in G. H. Wenzel [1970 b].

Example 1. By Theorem 1, an Abelian group $\mathfrak{G}$ is equationally compact if and only if it is $|G|$-compact. This result can be improved: $\mathfrak{G}$ is equationally compact if and only if it is equationally $\aleph_0$-compact. This is a result of S. Balcerzyk [1956] which states that an Abelian group $\mathfrak{G}$ is equationally compact if and only if every $\mathfrak{G}$-consistent system of equations $\{x_0 - a_n = n!\, x_n \mid n \in N\}$ with constants a_n and variables x_n is solvable in $\mathfrak{G}$. J. Mycielski and C. Ryll-Nardzewski [1] generalized this result to $\mathfrak{R}$-modules: An $\mathfrak{R}$-module $\mathfrak{M}$ is equationally compact if and only if $\mathfrak{M}$ is $(|R| + \aleph_0)$-compact. This result is a direct application of Lemma 1 if we notice that there are exactly $|R| + \aleph_0$ systems Σ of linear equations of the type described in (2) of Lemma 1 and that the solution set of any $\Phi \in \Sigma$ is a coset of the group $\langle M; +, - \rangle$ with respect to a subgroup completely determined by Φ.

Example 2. Theorem 1 cannot be strengthened. If we add the elements 0, 1 to the type, then the lattice $\mathfrak{L}(\aleph_0)$ of §84 is weakly equationally $\aleph_0$-compact but not weakly equationally compact.

Example 3. Theorem 85.2 can be modified in the spirit of this section: $\mathfrak{A}$ with $|A| = \aleph_\alpha$ is equationally compact if and only if all λ-limits for $\lambda \leqq \alpha$ exist (G. H. Wenzel [1973]). The existence of the α-limit alone does not suffice for equational compactness (S. Bulman-Fleming and W. Taylor [1972]).

Example 4. Lemma 1 suggests a link to topology. Let $\mathfrak{A}$ be an algebra and E the class of all formulas of the type described in (2) of Lemma 1. Then $\{\{a_0 \mid a_0 \in A, \mathfrak{A} \models \Phi(a_0)\} \mid \Phi \in E\}$ is the subbase of closed sets of a topology $\mathscr{T}$ on A. By Lemma 1, $\mathfrak{A}$ is equationally compact if and only if the topology $\mathscr{T}$ is quasi-compact. This topology and its generalizations are used in W. Taylor [1969] and D. K. Haley [1977]. They all have the disadvantage that they are not Hausdorff and that they do not, in general, make the operations continuous. From a topologist's viewpoint the ideas presented in the next section are of greater interest.

§87. THE MYCIELSKI QUESTION: CHROMATIC NUMBERS AND TOPOLOGY

J. Mycielski [3] raised the question that generated much of the interest in the field: "Is every equationally compact algebra a retract of a compact topological algebra?" The class of Abelian groups yields again a strong motivation to pose the question. As has been noted before (J. Łoś [1957]), algebraically compact Abelian groups are exactly the direct summands of compact topological Abelian groups. The "Bohr compactification" of an Abelian group $\mathfrak{G}=\langle G; +, -\rangle$ is the Abelian group $\mathfrak{B}(\mathfrak{G})=\langle B(\mathfrak{G}); +, -\rangle$ with $B(\mathfrak{G})=\mathrm{Hom}(\mathrm{Hom}(\mathfrak{G}, \mathfrak{C}), \mathfrak{C})$. Here $\mathfrak{C}$ is the circle group in the complex plane, $\mathfrak{G}$ and $\mathrm{Hom}(\mathfrak{G}, \mathfrak{C})$ are considered as discrete topological groups and $\mathfrak{B}(\mathfrak{G})$ as a compact subgroup of the Tychonoff product $\mathfrak{C}^{\mathrm{Hom}(\mathfrak{G},\mathfrak{C})}$. Every homomorphism $g: \mathfrak{G} \to \mathfrak{D}$ into a compact topological Abelian group $\mathfrak{D}$ can be uniquely extended to a homomorphism $g_0: \mathfrak{B}(\mathfrak{G}) \to \mathfrak{D}$. In case $\mathfrak{G}$ is an $\mathfrak{R}$-module we can therefore extend the scalar multiplications $r: \mathfrak{G} \to \mathfrak{G}$ to scalar multiplications $r: \mathfrak{B}(\mathfrak{G}) \to \mathfrak{B}(\mathfrak{G})$. Thus we obtain the Bohr compactification of an $\mathfrak{R}$-module. R. B. Warfield [1969] showed that every equationally compact $\mathfrak{R}$-module is a retract of its Bohr compactification; so the answer to Mycielski's question is positive again. For non-Abelian groups not much is known to this day.

G. Grätzer and H. Lakser [1969] characterized equationally compact semilattices $\mathfrak{S}=\langle S; \vee\rangle$ as exactly those semilattices that satisfy the following three conditions: (1) every nonempty subset $T\subseteq S$ has a least upper bound $\bigvee T$; (2) every downward directed set $D\subseteq S$ has a greatest lower bound $\bigwedge D$; (3) if $a\in S$ and D is a downward directed set in S then $a\vee(\bigwedge D)=\bigwedge(a\vee d \mid d\in D)$. Using this characterization S. Bulman-Fleming [1972] proved that Mycielski's question has a positive answer in the class of semilattices. W. Taylor [1974] showed that every equationally compact semilattice is a retract of a product of finite semilattices. These results were improved by S. Bulman-Fleming, I. Fleischer and K. Keimel [1978] to *sendos*, i.e., semilattices to which an arbitrary set of semilattice endomorphisms is added as unary operations. They show that a semilattice endomorphism $g: \langle S; \vee\rangle \to \langle S; \vee\rangle$ can be extended to a semilattice endomorphism on $\mathfrak{F}(\mathfrak{S})=\langle F(\mathfrak{S}); \cup\rangle$ where $F(\mathfrak{S})$ denotes the filters of $\langle S; \vee\rangle$, and they produce a sendo-retraction from $\mathfrak{F}(\mathfrak{S})$ onto $\mathfrak{S}$. $\mathfrak{F}(\mathfrak{S})$, however, is in the natural way a compact topological sendo via the topology induced from 2^S.

B. Węglorz [1] shows that the equationally compact Boolean algebras are exactly the complete, i.e., the injective ones. Since every Boolean algebra $\mathfrak{A}$ can be embedded in the Boolean algebra $\mathfrak{A}'$ of all subsets of the Stone space $S(\mathfrak{A})$ of $\mathfrak{A}$, and since $\mathfrak{A}'$ (being the direct product of 2-element Boolean algebras) is compact, we have again a positive answer to Mycielski's question.

There is not much we know about lattices in general. D. Kelly [1972] characterized equationally compact lattices without infinite anti-chains but it is not known how they behave topologically. In particular there is neither a characterization of equationally compact distributive lattices nor an answer to Mycielski's question.

Abelian groups and Boolean algebras can be considered as special examples of rings (zero rings, Boolean rings). A characterization of equational compactness in general rings is very difficult and an unsolved problem. A rather deep study of rings with ascending chain condition on left ideals was carried out by D. K. Haley [1970], [1973], [1974], [1976], [1977] and produced a very strong positive answer to Mycielski's question: Equational and topological compactness coincide in that class of rings. This connection is used to show that compact topological rings in the class of rings with ascending chain condition carry a unique compact topology, namely the ideal topology determined by the Jacobson radical.

W. Taylor [1976] and S. Bulman-Fleming and H. Werner [1977] settle Mycielski's question in the case of quasi-primal varieties. Again we have an affirmative answer.

G. H. Wenzel [1970 a] characterizes equationally compact mono-unary algebras and proves that equationally compact mono-unary algebras are the retracts of their Stone–Čech compactification. Bi-unary algebras $\langle A; f, g\rangle$ furnish the first negative answer as W. Taylor [1971] proved. The construction is based on graph-theoretical considerations that were generalized (W. Taylor [1970] or [1971]) to reveal a surprising relationship between "chromatic numbers" and atomic compactness of relational systems. Since the first counterexample was a highlight in the short history of our subject, we discuss the relevant ideas and present both the first counterexample and also a subsequent one in the class of semigroups.

W. Taylor [1970] introduces a common generalization to relational systems of the classical chromatic number of a graph and the chromatic number of a uniform set system of P. Erdös and A. Hajnal [1966]. For any equivalence ρ of $\{1, 2, \cdots n\}$ and any set B we denote by $\mathfrak{B}_\rho = \langle B; S_\rho\rangle$ the relational system with one n-ary relation S_ρ defined as follows: $\langle x_1, \cdots, x_n\rangle \in S_\rho$ holds if and only if there are $i, j \in \{1, \cdots, n\}$, $i \neq j$, with $i \equiv j$ (ρ) but $x_i \neq x_j$. If $\mathfrak{A} = \langle A; r\rangle$ is any relational system with one n-ary relation r $(n \geqq 2)$ then the *ρ-chromatic number* $\chi_\rho(\mathfrak{A})$ is the least cardinal number $\mathfrak{m}$ such that there is a $\mathfrak{B}_\rho$ with $|B| = \mathfrak{m}$ and a homomorphism $g: \mathfrak{A} \to \mathfrak{B}_\rho$. If no such homomorphism exists, then we set $\chi_\rho(\mathfrak{A}) = \infty$.

Remarks. (1) The classical chromatic number $\chi(\mathfrak{G})$ of a graph $\mathfrak{G} = \langle G; r\rangle$ is $\rho_i(\mathfrak{G})$ for the equivalence i on $\{1, 2\}$. (2) If $\chi_\rho(\mathfrak{A}) \neq \infty$ then id_A is a homomorphism from $\mathfrak{A}$ to $\mathfrak{A}_\rho$ and vice versa. In particular $\chi_\rho(\mathfrak{A}) \leqq |A|$.

Theorem 1. *If the relational system $\mathfrak{A}=\langle A; r\rangle$ with one n-ary relation r $(n \geqq 2)$ is a retract of a compact topological relational system, then $\chi_\rho(\mathfrak{A}) \in N \cup \{\infty\}$ holds for all equivalences ρ on $\{1, 2, \cdots, n\}$.*

Proof. Assume that $\chi_\rho(\mathfrak{A}) \notin N$ for some ρ. We may assume that $\mathfrak{A}$ itself is a compact topological relational system. If $\Theta \in E(A)$ has only finitely many equivalence classes, then $\chi_\rho(\mathfrak{A}) \notin N$ implies the existence of $\langle a_{1\Theta}, \cdots, a_{n\Theta}\rangle \in r$ with $i \equiv j(\rho)$ only if $a_{i\Theta} \equiv a_{j\Theta}(\Theta)$. If $E'(A)$ denotes the set of equivalences on A with finite quotient sets, then $E'(A)$ is directed in the usual partial order, so for each $i \in \{1, 2, \cdots, n\}$ we have the net $\langle a_{i\Theta} \mid \Theta \in E'(A)\rangle$. We choose, for fixed i, a subnet $\langle a_{i\Theta} \mid \Theta \in E_i\rangle$ that converges, say, to $a_i \in A$. Obviously each $\langle a_{j\Theta} \mid \Theta \in E_i\rangle$ for $j \equiv i(\rho)$ converges to a_i, too. We continue this process by taking further subnets for all equivalence classes of ρ and obtain $\langle a_1, \cdots, a_n\rangle \in \bar{r} - S_\rho = r - S_\rho$. Hence, id_A is not a homomorphism from $\mathfrak{A}$ to $\mathfrak{A}_\rho$. In view of Remark (2), we are finished.

Corollary. *If the graph $\mathfrak{G}$ is a retract of a compact topological relational system then $\chi(\mathfrak{G}) < \aleph_0$.*

The next theorem is based on this corollary.

Theorem 2. *There exists an atomic compact graph $\mathfrak{G}=\langle G; r\rangle$ of infinite chromatic number. It is not a retract of a compact topological relational system.*

Proof. (1) If $\mathfrak{G}^{(n)}$, $n=2, 3, 4, \cdots$, are finite graphs with $\chi(\mathfrak{G}^{(n)}) \geqq n$ and without odd cycles of length $\leqq n$, then the graph $\mathfrak{G}=\dot{\bigcup}(\mathfrak{G}^{(n)} \mid n \in N)$ is atomic compact with $\chi(\mathfrak{G})=\aleph_0$. This is easily seen: $\chi(\mathfrak{G})=\aleph_0$ is clear. Atomic compactness follows from Theorem 85.1 (which holds verbatim for atomic compactness rather than equational compactness) since any elementary extension $\mathfrak{E}$ has the property that $\mathfrak{G}$ is a *full subgraph*, i.e., $x \in G$, $y \in E$, $\langle x, y\rangle \in r$ implies $y \in G$. Moreover, $E-G$ contains no cycles of odd length (since the number of odd cycles of length m is finite in G and can be fixed by a first-order sentence), so it can be mapped homomorphically onto an edge of $\mathfrak{G}$.

(2) Let $k, n \in N$, $e \in \mathbb{R}$ $(e>0)$ such that $2n \leqq (2-e)/\sqrt{4e-e^2}$. If B^k is the unit ball in $\mathbb{R}^k$ and the binary relation $r_{e,k,n}$ on B^k is defined by $\langle a, b\rangle \in r_{e,k,n}$ iff $|a-b|>2-e$, then $\langle B^k; r_{e,k,n}\rangle$ has no odd cycles of length $\leqq 2n+1$. Moreover, for every $l \in N$ there is some $k(l) \in N$ such that $\chi(\langle B^k; r_{e,k,n}\rangle) \geqq l$ for all $k > k(l)$.

(3) (1) and (2) yield the proof in view of a result of N. G. de Bruijn and P. Erdös [1961] which states that an infinite graph has finite chromatic number m only if some finite subgraph has chromatic number m.

We use the graph of Theorem 2 to define a bi-unary algebra $\langle A; f, g\rangle$ as follows:

$$A = G \dot\cup r \dot\cup \{e_1, e_2\}$$

$$f(x) = \begin{cases} a & \text{if } x = \langle a, b\rangle \in r \\ e_1 & \text{otherwise} \end{cases}$$

$$g(x) = \begin{cases} b & \text{if } x = \langle a, b\rangle \in r \\ e_2 & \text{otherwise} \end{cases}$$

This is W. Taylor's original example showing that the answer to Mycielski's question is not always in the affirmative.

Theorem 3. *If $\mathfrak{G}$ is the graph of Theorem 2 and $\mathfrak{A} = \langle A; f, g\rangle$ the bi-unary algebra constructed above, then $\mathfrak{A}$ is equationally compact but it is not a retract of a compact topological algebra.*

Later W. Taylor [1972 c] provided a negative answer to Mycielski's question within the class of semigroups. As S. Bulman-Fleming noted, this counterexample can easily be made into a commutative semigroup. Nothing is known about the situation for groups, rings, lattices, distributive lattices $\cdots$ and the problem appears to be hard in every instance. The semigroup is again rooted in graph theory. If $\mathfrak{A} = \langle A; F\rangle$ is a universal algebra, $G \subseteq A$ and $\mathfrak{G} = \langle G; r\rangle$ is a graph, then $\mathfrak{G}$ is called *equationally definable in* $\mathfrak{A}$ if there exist algebraic equations $p_1 = q_1, \cdots, p_m = q_m$ in two variables x, y over $\mathfrak{A}$ such that $\langle a, b\rangle \in r$ is equivalent to $p_i(a, b) = q_i(a, b)$ for $i = 1, 2, \cdots, m$ $(a, b \in A)$.

Theorem 4. *If the graph $\mathfrak{G}$ is equationally definable in the algebra $\mathfrak{A}$ and $\chi(\mathfrak{G}) \geqq \aleph_0$, then $\mathfrak{A}$ is not a retract of a compact topological algebra $\mathfrak{B}$.*

Proof. Assume that $h: \mathfrak{B} \to \mathfrak{A}$ is a retraction. If $\mathfrak{G}$ is defined by the equations $p_i(x, y) = q_i(x, y)$, $i = 1, 2, \cdots, m$, then we define r' on B by $\langle c, d\rangle \in r'$ iff $p_i(c, d) = q_i(c, d)$, $i = 1, 2, \cdots, m$, and let r^B be the symmetric hull of r'. Obviously $\langle \text{domain } (r^B); r^B\rangle$ is a compact topological graph that retracts onto $\mathfrak{G}$, contradicting Theorem 1.

Theorem 5. *There is an equationally compact commutative semigroup $\mathfrak{S} = \langle S; \cdot\rangle$ which is not a retract of a compact topological semigroup.*

Proof. We take an atomic compact graph $\mathfrak{G} = \langle G; r\rangle$ with $\chi(\mathfrak{G}) \geqq \aleph_0$ and let $\mathfrak{S}_0$ be the free commutative semigroup over the basis G in the variety K of all commutative semigroups that satisfy $xyz = uvw$. We define $a \equiv b(\Theta)$ $(a, b \in S_0)$ iff (i) $a = b$ or (ii) $a = c_1 \cdot c_2$, $b = d_1 \cdot d_2$ and $\langle c_1, c_2\rangle$, $\langle d_1, d_2\rangle \in r$. $\Theta \in C(\mathfrak{S}_0)$ and $\mathfrak{S} = \mathfrak{S}_0/\Theta \in K$ contains canonically G. If

$\langle a, b\rangle \in r$ then we denote $[a \cdot b]\Theta$ by t and obtain for $a', b' \in S$: $\langle a', b'\rangle \in r$ holds exactly if $a' \cdot b' = t$. So $\mathfrak{S}$ is equationally definable in $\mathfrak{S}$, i.e., $\mathfrak{S}$ is not a retract of a compact topological algebra. Using the ultrapower part of Theorem 85.1 one proves that $\mathfrak{S}$ is equationally compact.

§88. MINIMUM COMPACTNESS

There are interesting links between equational compactness and weak equational compactness. One such is based on the observation that weakly equationally compact algebras all of whose endomorphisms are monomorphisms ($\mathrm{End}(\mathfrak{A}) = \mathrm{Mon}(\mathfrak{A})$) are equationally compact and all endomorphisms are automorphisms ($\mathrm{End}(\mathfrak{A}) = \mathrm{Aut}(\mathfrak{A})$) (G. H. Wenzel [1971]). W. Taylor's discovery of minimum compact algebras goes in the same direction and proves fundamental for a theory of compactifications (W. Taylor [1971 a]).

Theorem 1. *Let $\mathfrak{A}$ be a weakly equationally compact algebra with* $\mathrm{End}(\mathfrak{A}) = \mathrm{Mon}(\mathfrak{A})$. *Then* $\mathrm{End}(\mathfrak{A}) = \mathrm{Aut}(\mathfrak{A})$.

Proof. If $g\colon \mathfrak{A} \to \mathfrak{A}$ is a proper monomorphism, then we get a chain $\mathfrak{A} = \mathfrak{A}_0 \subset \mathfrak{A}_1 \subset \cdots \subset \mathfrak{A}_n \subset \cdots$ of isomorphic algebras, and $\mathfrak{A}_\omega = \bigcup (\mathfrak{A}_n \mid n \in N)$ is a weakly pure extension of $\mathfrak{A}$. Thus there exists a homomorphism $r\colon \mathfrak{A}_\omega \to \mathfrak{A}$ which implies that $\mathrm{End}(\mathfrak{A}_\omega) = \mathrm{Mon}(\mathfrak{A}_\omega)$ and that $\mathfrak{A}_\omega$ is weakly equationally compact. In particular, r is a monomorphism. We now build by transfinite induction an increasing chain of arbitrary length $\mathfrak{A} = \mathfrak{A}_0 \subset \mathfrak{A}_1 \subset \cdots \subset \mathfrak{A}_\gamma \subset \cdots$ $(\gamma < \alpha)$ such that each $\mathfrak{A}_\gamma$ is weakly equationally compact, has only monic endomorphisms and admits a monomorphism $g\colon \mathfrak{A}_\gamma \to \mathfrak{A}$. This leads to a cardinality contradiction.

Theorem 1 has the following application to simple algebras (G. H. Wenzel [1971]).

Theorem 2. *A simple algebra $\mathfrak{A}$ without one-element subalgebras is weakly equationally compact if and only if it is equationally compact.*

Proof. $\mathfrak{A}$ contains a homomorphic image of every ultrapower $\mathfrak{A}_D{}^I$. The respective homomorphism, restricted to A, is an automorphism of $\mathfrak{A}$. Theorem 85.1 settles the matter.

(In W. Taylor [1971 a] it is shown that pseudosimplicity of $\mathfrak{A}$ suffices in Theorem 2. $\mathfrak{A}$ is *pseudosimple* if for all $\Theta \in C(\mathfrak{A})$ either $|A/\Theta| = 1$ or $\mathfrak{A}/\Theta \cong \mathfrak{A}$.)

These results lead naturally to a closer scrutiny of algebras that satisfy $\mathrm{End}(\mathfrak{A}) = \mathrm{Mon}(\mathfrak{A})$. Various authors contributed to the naming of the related phenomena.

Definition 1. (1) (*B. Banaschewski*) *If* $\mathfrak{A}$, $\mathfrak{B}$ *are algebras,* $\mathfrak{A} \in \mathbf{S}(\mathfrak{B})$, *then* $\mathfrak{B}$ *is a* firm extension of $\mathfrak{A}$ *if* $\mathrm{End}(\mathfrak{B}_A) = \mathrm{Mon}(\mathfrak{B}_A)$.

(2a) *Let* $\mathfrak{A}$, $\mathfrak{B}$ *be algebras:* $\mathfrak{B}$ *is* weakly pure with respect to $\mathfrak{A}$ *if every finite set of polynomial equations that is solvable in* $\mathfrak{B}$ *is also solvable in* $\mathfrak{A}$ *(see also W. Taylor* [1971], *Definition 2.2).*

(2b) (*J. Mycielski*) $\mathfrak{A}$ *is a* folded algebra *if every homomorphism* $h: \mathfrak{A} \to \mathfrak{B}$ *into an algebra* $\mathfrak{B}$ *which is weakly pure with respect to* $\mathfrak{A}$ *is a monomorphism.*

(3) (*W. Taylor*) *The algebra* $\mathfrak{A}$ *is* minimum compact *if it is folded and weakly equationally compact.*

The next remarks follow immediately from the above definition.

Remarks. (1) If $\mathfrak{A}$ is subalgebra of $\mathfrak{B}$, then $\mathfrak{B}$ is weakly pure with respect to $\mathfrak{A}$ if and only if $\mathfrak{B}$ is a weakly pure extension of $\mathfrak{A}$.

(2) $\mathfrak{B}$ is a weakly equationally compact firm extension of $\mathfrak{A}$ if and only if $\mathfrak{B}_A$ is minimum compact.

(3) $\mathfrak{A}$ is minimum compact if and only if $\mathfrak{A}$ is equationally compact and $\mathrm{End}(\mathfrak{A}) = \mathrm{Aut}(\mathfrak{A})$, i.e., if and only if $\mathfrak{A}$ is weakly equationally compact and $\mathrm{End}(\mathfrak{A}) = \mathrm{Aut}(\mathfrak{A})$.

Theorems 3 and 4 are taken from W. Taylor [1971 a].

Theorem 3. *Let* $\mathfrak{A}$ *be a weakly equationally compact algebra. Then there exists (up to isomorphism) a unique minimum compact algebra* $\mathfrak{C}$ *such that* $\mathfrak{A}$ *and* $\mathfrak{C}$ *are mutually weakly pure.* $\mathfrak{C}$ *can be chosen to be a retract of* $\mathfrak{A}$.

Corollary. *Let* $\mathfrak{A}$ *be a subalgebra of* $\mathfrak{B}$ *and assume that* $\mathfrak{B}_A$ *is weakly equationally compact. Then there exists (up to isomorphism) a unique firm extension* $\mathfrak{C}$ *of* $\mathfrak{A}$ *that satisfies* $\mathfrak{A} \subseteq \mathfrak{C} \subseteq \mathfrak{B}$ *and* $\mathfrak{C}$ *is retract of* $\mathfrak{B}$.

Proof of Theorem 3. The set of all congruences $\Theta \in C(\mathfrak{A})$ such that $\mathfrak{A}$ and $\mathfrak{A}/\Theta$ are mutually weakly pure has a maximal element Θ_0. Define $\mathfrak{C} = \mathfrak{A}/\Theta_0$. $\mathfrak{C}$ is evidently minimum compact and $\mathfrak{C}$ and $\mathfrak{A}$ are mutually weakly pure. By the standard "diagram argument" we can find a homomorphism $h: \mathfrak{C} \to \mathfrak{A}$ which is monic. $\mathfrak{C}h \subseteq \mathfrak{A}$ is a retract of $\mathfrak{A}$ since it contains a homomorphic image of $\mathfrak{A}$ and $\mathrm{End}(\mathfrak{A}h) = \mathrm{Aut}(\mathfrak{A}h)$.

Definition 2. *Following B. Banaschewski* [1974 a] *we define the* $\mathfrak{C}$ *of Theorem 3 as the* core of $\mathfrak{A}$ *and the* $\mathfrak{C}$ *of the Corollary as the* core of $\mathfrak{B}$ over $\mathfrak{A}$.

Theorem 4 (an application of minimum compact algebras) is a decomposition theorem for equationally compact algebras in terms of minimum compact algebras in a suitably enriched type.

Theorem 4. *Let* $\mathfrak{A}=\langle A; F\rangle$ *be an infinite algebra in* $K(\tau)$ *such that for each finite* $H\subseteq A$ *the algebra* $\mathfrak{A}_H$ *is weakly equationally compact (special case:* $\mathfrak{A}$ *is equationally compact). Further let I be the set of finite subsets of A. Then there exists a subdirect representation* $g\colon \mathfrak{A}\to \prod (\mathfrak{C}^{(i)} \mid i\in I)$ *with the following two properties;*

(1) $i\subseteq C^{(i)}$ *and* $\mathfrak{C}_i^{(i)}$ *is minimum compact.*

(2) *g is a pure embedding.*

Proof. For each i the algebra $\mathfrak{A}_i$ (in the enriched type) is weakly equationally compact. We choose $\mathfrak{C}_i^{(i)}$ as the core of $\mathfrak{A}_i$ and define $\mathfrak{C}^{(i)}$ as the τ-reduct of $\mathfrak{C}_i^{(i)}$. By Theorem 3 there exist retractions $p_i\colon \mathfrak{A}_i \to \mathfrak{C}_i^{(i)}$. We define $g\colon \mathfrak{A}\to\prod(\mathfrak{C}^{(i)} \mid i\in I)$ by $ag=\langle ap_i\rangle_{i\in I}$ and verify that g is a pure subdirect representation.

Some more applications of minimum compact algebras will be given later. We finish this section with two cardinality results. The first is due to W. Taylor [1971 a] and it plays an important role in subsequent sections. The second one follows from R. McKenzie and S. Shelah [1974] and points to a different direction of research.

Theorem 5. *Let* $\mathfrak{A}=\langle A; F\rangle$ *be a minimum compact algebra in* $K(\tau)$. *Then*

(1) $|A|\leqq 2^{\mathfrak{m}}$, *where* $\mathfrak{m}=\aleph_0+\mathrm{o}(\tau)$.

(2) *If* $\mathrm{o}(\tau)\leqq\aleph_0$, *then* $|A|\leqq\aleph_0$ *or* $|A|=2^{\aleph_0}$.

Proof. We outline the proof of (1). The proof is based on the following set-theoretical result of P. Erdös [1942]: If S, I are infinite sets with $|S|>2^{|I|}$, $S^{(2)}$ denotes the set of two-element subsets of S, and $S^{(2)}=\bigcup(C_i \mid i\in I)$, then there exists an $i_0\in I$ and a subset $T\subseteq S$ such that $T^{(2)}\subseteq C_{i_0}$ and $|T|>|I|$. Let I be the set of all formulas $\Phi(y_0, y_1)$ of the type $(\exists x_{\gamma_0})\cdots(\exists x_{\gamma_n})(a_0\wedge a_1\wedge\cdots\wedge a_t)$ with polynomial equations a_j ($j=0,1,\cdots,t$) and y_0, y_1 as the only free variables such that $\mathfrak{A}\models\neg(\exists x)(\Phi(x,x))$. For each $\Phi\in I$ set $B_\Phi=\{\{a,b\}\mid a,b\in A$ and $\mathfrak{A}\models\Phi(a,b)\}$. The following equality can be proved: $A^{(2)}=\bigcup(B_\Phi \mid \Phi\in I)$. Of course, $|I|=\mathfrak{m}$. If $|A|>2^{\mathfrak{m}}$, then the result of Erdös yields an infinite subset B of A (in fact, $|B|>|I|$)

and some $\Phi_0 \in I$ such that $B^{(2)} \subseteq B_{\Phi_0}$. For any set X of variables we define a system $\Sigma(X)$ of positive formulas as follows: $\Sigma(X) = \{\Phi_0(x, y) \vee \Phi_0(y, x) \mid x, y \in X, x \neq y\}$. $\Sigma(X)$ is $\mathfrak{A}$-consistent, hence solvable in $\mathfrak{A}$. If we choose X such that $|X| > |A|$, we deduce $\Phi_0(a, a)$ for some $a \in A$, a contradiction.

Corollary. *If* $\mathfrak{C}$ *is the core of* $\mathfrak{B}$ *over* $\mathfrak{A}$, *then* $|C| \leq 2^{\mathfrak{n}}$, *where* $\mathfrak{n} = \aleph_0 + \overline{o(\tau)} + |A|$.

Examples. All finite fields are minimum compact if 1 is added to the type. $\langle Q; +, -, c\rangle$ with Q the set of rational numbers, $c \neq 0$, is minimum compact. $\langle Z; f\rangle$ and $\langle Z/(n); f\rangle$ with $f(z) = z + 1$ are minimum compact. Simple weakly equationally compact algebras without one-element subalgebras are minimum compact. Hence, the group SO(3) of all rotations of the 2-sphere is minimum compact if a constant $c \neq \mathrm{id}$ is added to the type.

§89. COMPACTIFICATION OF ALGEBRAS

Let $\mathfrak{A}$, $\mathfrak{B}$ be algebras, $\mathfrak{A} \in \mathbf{S}(\mathfrak{B})$. If $\mathfrak{B}$ is not an equationally compact extension of $\mathfrak{A}$, then it may still satisfy one or both of the following two conditions:

(i) $\mathfrak{B}$ is At-$(A, \mathfrak{B})$-compact, i.e., every $\mathfrak{B}$-consistent system of algebraic equations with constants in A is solvable in $\mathfrak{B}$.

(ii) $\mathfrak{B}$ is At-$(A, \mathfrak{A})$-compact, i.e., every $\mathfrak{A}$-consistent system of algebraic equations with constants in A is solvable in $\mathfrak{B}$.

Definition 1. In case (i) we call $\mathfrak{B}$ a *closure* of $\mathfrak{A}$, in case (ii) a *quasi-compact extension.*

The interrelation of these concepts and some of their properties were investigated in B. Węglorz [2], W. Taylor [1971 a], G. H. Wenzel [1971], B. Banaschewski [1974 a].

Not every algebra can be embedded into an equationally compact algebra. The lattices $\mathfrak{L}(\mathfrak{n})$ and algebraically closed groups have been given as examples in §84. The reason for groups is twofold: (1) Every group can be embedded into an algebraically closed group, (2) every algebraically closed group is a simple group (see B. H. Neumann [1952]). Thus, assume that the algebraically closed group $\mathfrak{G}$ has an equationally compact extension group $\mathfrak{H}$. Then $\mathfrak{H}$ is At-$(G, \mathfrak{G})$-compact. Theorem 85.1′ yields a G-homomorphism $f: \mathfrak{G}_1 \to \mathfrak{H}$ for every pure extension $\mathfrak{G}_1$ of $\mathfrak{G}$. Since $\mathfrak{G}_1$ can be chosen to be an algebraically compact group containing $\mathfrak{H}$ with $|G_1| >$

$|H|$ we get a contradiction to the simplicity of $\mathfrak{G}_1$. Also the group of all linear substitutions of the form $f_{k,r}(z) = 2^k z + r$ ($k \in Z$, $r \in Q$) cannot be a subgroup of an equationally compact group. This was proved by J. Mycielski and C. Ryll-Nardzewski [1] as a generalization of a result of H. Freudenthal stating that this group cannot be embedded into a compact topological group. The following example of B. Węglorz [3] is very illuminating for the complexity of the problem: The lattice

$$\mathfrak{L} = \bigoplus_{0<n<\aleph_0} \mathfrak{L}(n) \oplus \mathfrak{L}(0)$$

(i.e., the lattices $\mathfrak{L}(n)$ are put "above one another" and $\mathfrak{L}(0)$ on top of them all) is complete and even a compact topological lattice in the interval topology. However, no nontrivial countable ultrapower of $\mathfrak{L}$ has an equationally compact extension, for each such $\mathfrak{L}_D{}^I$ contains $\prod_D (\mathfrak{L}(n) \mid n \in N)$ which is isomorphic to $\mathfrak{L}(2^{\aleph_0})$. Of course, every nontrivial ultraproduct of finite fields of mutually different characteristics is another example of an ultraproduct of compact topological algebras that cannot be embedded into an equationally compact algebra. Can an algebra $\mathfrak{A}$ that has no equationally compact extension be embedded into a closure or at least a quasi-compact extension? As we shall see, the answer to the first question is in the negative while the answer to the second is still unknown. The next result shows that we can limit ourselves to a class very tightly described by $\mathfrak{A}$ in searching for compact extensions.

Theorem 1. *Let $\mathfrak{A}$ be an algebra with an equationally compact (quasi-compact) extension $\mathfrak{B}$. Then there exists an equationally compact (quasi-compact) extension $\mathfrak{C}$ of $\mathfrak{A}$ in* $\mathbf{S}(\mathfrak{B})$ *that satisfies every universal-existential sentence Φ with constants in A that is satisfied by $\mathfrak{A}$. In particular there exists such an algebra $\mathfrak{C}$ in* $\mathbf{HSP}(\mathfrak{A})$. *$\mathfrak{C}$ can be chosen as a retract of $\mathfrak{B}$.*

Proof. The set of algebras $\mathfrak{D}$ with $\mathfrak{A} \subseteq \mathfrak{D} \subseteq \mathfrak{B}$ that satisfy all universal-existential sentences of $\mathfrak{A}$ has a maximal element $\mathfrak{C}$ because of the Corollary to Theorem 45.2. If $\mathfrak{C}_D{}^I$ is an arbitrary ultrapower of $\mathfrak{C}$ and $\mathfrak{B}_D{}^I$ the corresponding ultrapower of $\mathfrak{B}$, then there exists a retraction $g\colon \mathfrak{B}_D{}^I \to \mathfrak{B}$ that maps $\mathfrak{C}_D{}^I$ onto $\mathfrak{C}$.

Of course, one cannot hope to get always elementary equationally compact extensions if there are equationally compact extensions. B. Banaschewski [1974 a] and W. Taylor [1971 a] gave examples of algebras with equationally compact extensions that do not even have pure equationally compact extension. W. Taylor [1971 a] cites an example of G. Fuhrken of an algebra which has an equationally compact pure extension, but no equationally compact elementary extension. The best positive result in that direction deals with elementary $\mathfrak{m}$-compactness and is due to J. Mycielski and C. Ryll-Nardzewski [1].

Theorem 2. *Every infinite algebra $\mathfrak{A}$ has an elementarily $\mathfrak{m}$-compact elementary extension $\mathfrak{B}$ with $|B| = |A|^{\mathfrak{m}}$ for every cardinal number $\mathfrak{m}$.*

Proof. For the terminology see Definition 85.4 and Lemma 85.1. We choose $\mathfrak{A}_0 = \mathfrak{A}$ and, for every ordinal $\xi > 0$, $\mathfrak{A}_\xi = \bigcup (\mathfrak{A}_\eta(\mathfrak{m}) \mid \eta < \xi)$. If α is the initial ordinal of $\mathfrak{m}^+$ then $\mathfrak{B} = \mathfrak{A}_\alpha$ is the desired extension.

Example (G. H. Wenzel [1971]). It is, in general, not true that the existence of a weak equationally compact extension of $\mathfrak{A}$ implies the existence of a weak equationally compact extension within **HSP**($\mathfrak{A}$). To show this we observe that an algebra without nullary operations is always weakly equationally compactifiable by the obvious one-point compactification. We choose $\mathfrak{A} = \langle A; \vee, \wedge, {}^* \rangle$ of type (2, 2, 1) with $A = \{a_n \mid n \in N\} \dot{\cup} \{0, 1\}$ such that $\langle A; \vee, \wedge \rangle \cong \mathfrak{L}(\aleph_0)$. Moreover $a_i{}^* = a_{i+1}$, $0^* = 1$, $1^* = 0$. $\mathfrak{A}$ satisfies the identities $x_0 \wedge x_0{}^* = x_1 \wedge x_1{}^*$ and $x_0 \vee x_0{}^* = x_1 \vee x_1{}^*$, but no weakly equationally compact algebra containing $\mathfrak{A}$ satisfies these.

Theorem 3 was independently proved by W. Taylor [1972] and G. H. Wenzel [1971]. Later B. Banaschewski [1974 a] added a third proof.

Theorem 3. *Let $\mathfrak{A}$ be an algebra with a closure $\mathfrak{B}$. Then there exists an equationally compact extension $\mathfrak{C}$ of $\mathfrak{A}$ that is a retract of $\mathfrak{B}$.*

Proof. See the Corollary to Theorem 88.3.

A study of compactifications either along the lines of topological compactifications that lead to the Stone-Čech compactification or along the lines of injective embeddings in category theory that lead to injective hulls, soon suggests a separate investigation of the following two successively stronger cases:

(A) (The pure case.) There exists a pure equationally compact extension of $\mathfrak{A}$.

(B) (The general case.) There does not necessarily exist a pure equationally compact extension, but there are equationally compact extensions.

Both cases were investigated initially by W. Taylor [1971 a] and [1972]. B. Banaschewski and E. Nelson [1972] gave new algebraic proofs that guide us in this presentation. The following lemma shows that pure-essential extensions of an algebra $\mathfrak{A}$ are "small" compared with pure equationally compact extensions.

Lemma 1. *Every pure-essential extension* $\mathfrak{C}$ *of* $\mathfrak{A}$ *can be embedded in every equationally compact pure extension* $\mathfrak{B}$ *of* $\mathfrak{A}$. *Thus, the isomorphism classes of pure-essential extensions form a set if there exists a pure equationally compact extension.*

Proof. The "diagram argument" yields an A-homomorphism $g: \mathfrak{C} \to \mathfrak{B}$. Consequently g is one-to-one.

Lemma 1 provides us with maximal pure-essential extensions of $\mathfrak{A}$, which are equationally compact by Theorem 85.5.

Definition 2. *An* equationally compact hull of $\mathfrak{A}$ *is an equationally compact pure-essential extension of* $\mathfrak{A}$.

The next theorem is based on Lemma 1 and the remark following it.

Theorem 4. *The following conditions are equivalent for an algebra* $\mathfrak{A}$:

(1) *There exists a pure equationally compact extension of* $\mathfrak{A}$.
(2) $\mathfrak{A}$ *has an equationally compact hull.*
(3) $\mathfrak{A}$ *has (up to isomorphism) only a set of pure-essential extensions.*

Theorem 5. *Equationally compact hulls* $\mathfrak{B}$ *of an algebra* $\mathfrak{A}$ *are (if they exist) exactly the maximal pure-essential extensions of* $\mathfrak{A}$ *and as such are mutually* A*-isomorphic. Moreover,* $|B| \leqq 2^{\mathfrak{n}}$, *where* $\mathfrak{n} = \aleph_0 + \overline{o(\tau)} + |A|$.

Proof. Surely $\mathfrak{A}$ has *one* equationally compact hull $\mathfrak{B}$ that is a maximal pure-essential extension of $\mathfrak{A}$. Every other equationally compact hull is mapped into $\mathfrak{B}$ by an A-monomorphism. Theorem 85.5 completes the proof. For the cardinality statement we refer to the Corollary to Theorem 88.5.

Let us point out one crucial difference between injective hulls and equationally compact hulls. The former depend essentially on the equational class within which they are considered, the latter not. As we shall see the existence of an equationally compact extension does not imply the existence of a pure such extension. However, the famed Löwenheim–Skolem–Tarski theorem yields immediately the following result (see also B. Banaschewski and E. Nelson [1972]).

Theorem 6. *For every pair of algebras* $\mathfrak{A}$, $\mathfrak{B}$ *with* $\mathfrak{A} \in \mathbf{S}(\mathfrak{B})$ *there is a* $\mathfrak{C} \in \mathbf{S}(\mathfrak{B})$ *with* $A \subseteq C \subseteq B$ *such that* $|C| \leqq \aleph_0 + \overline{o(\tau)} + |A|$ *and* $\mathfrak{B}$ *is a pure extension of* $\mathfrak{C}$.

Definition 3. *A* compactification $\mathfrak{B}$ *of the algebra* $\mathfrak{A}$ *is an equationally compact extension of* $\mathfrak{A}$ *which is its own core over* $\mathfrak{A}$.

The corollary to Theorem 88.3 yields immediately that every algebra that has an equationally compact extension has also a compactification. Compactifications are firm extensions. So "being its own core over $\mathfrak{A}$" can be viewed as some algebraic variant of topological denseness (B. Banaschewski [1974 a]). Injective hulls in equational classes of algebras are well-known examples of compactifications. So are the equationally compact hulls just introduced (Theorem 85.5 and Theorem 5). If F_p is the prime field of characteristic p, then all its finite extensions are compactifications. This simple example of B. Banaschewski makes two points immediately clear: (1) $\mathfrak{A}$ may have compactifications that are not in **HSP**($\mathfrak{A}$) although Theorem 1 guarantees that, if there are any, there is always one in **HSP**($\mathfrak{A}$). (2) Compactifications of a given $\mathfrak{A}$ are neither, in general, minimal or even smallest equationally compact extensions nor are they unique. Equationally compact hulls behave much better in this respect.

Remark. Equationally compact hulls are the smallest compactifications of $\mathfrak{A}$ with respect to set inclusion.

Theorem 7. *For every equationally compact extension* $\mathfrak{B}$ *of* $\mathfrak{A}$ *the following conditions are equivalent:*

(1) $\mathfrak{B}$ *is a compactification of* $\mathfrak{A}$.

(2) $\mathfrak{B}$ *is a firm extension of* $\mathfrak{A}$.

(3) *Whenever* $\mathfrak{B}$ *is a pure extension of some subalgebra* $\mathfrak{D}$ *with* $A \subseteq D$ *then it is a pure-essential extension (i.e., equationally compact hull) of* $\mathfrak{D}$.

Proof. The remarks following Definition 88.1 show the equivalence of (1) and (2). (2) implies (3). By Lemma 85.2 we find a pure-essential extension $\mathfrak{B}/\Theta$ of $\mathfrak{D}$ and a D-homomorphism $g: \mathfrak{B}/\Theta \to \mathfrak{B}$. Since $\pi \circ g$ (π is the natural projection) is an automorphism, we get $\Theta = \omega$. (3) implies (2). Let $\mathfrak{B}'$ be the core of $\mathfrak{B}$ over $\mathfrak{A}$. Then $\mathfrak{B}$ is a pure-essential extension of $\mathfrak{B}'$. Since $\mathfrak{B}'$ is a retract of $\mathfrak{B}$ we get $\mathfrak{B} = \mathfrak{B}'$.

Corollary 1. *Any algebra* $\mathfrak{A}$ *has only a set of compactifications (up to isomorphism).*

Proof. Put Theorems 6, 4(3) and 7(3) together and add Zorn's lemma.

Corollary 2. *If* $\mathfrak{B}$ *is a pure equationally compact extension of* $\mathfrak{A}$, *then* $\mathfrak{B}$ *is a compactification of* $\mathfrak{A}$ *if and only if it is the equationally compact hull. Thus there is (up to isomorphism) only one pure compactification of* $\mathfrak{A}$.

Definition 4. *The compactification* $\mathfrak{B}$ *of* $\mathfrak{A}$ *is called a* maximal compactification *if there exists an A-homomorphism* $g\colon \mathfrak{B} \to \mathfrak{C}$ *for every compactification* $\mathfrak{C}$ *of* $\mathfrak{A}$.

This "Stone–Čech-maximality condition" endows these maximal compactifications with the same special role amongst all compactifications as the Stone–Čech compactification of completely regular Hausdorff spaces enjoys amongst all topological compactifications. In particular one proves easily the following theorem (W. Taylor [1971 a], B. Banaschewski [1974 a]; the arguments in both papers are rooted in category theory).

Theorem 8. *Let* $\mathfrak{A}$ *be an algebra with equationally compact extensions.*

(1) *There is (up to A-isomorphism) exactly one maximal compactification of* $\mathfrak{A}$.

(2) *Any homomorphism* $h\colon \mathfrak{A} \to \mathfrak{D}$ *from* $\mathfrak{A}$ *to an equationally compact algebra* $\mathfrak{D}$ *extends to the maximal compactification of* $\mathfrak{A}$.

As the next and last theorem of this section states, the maximal compactifications of algebras $\mathfrak{A}$ with at least one pure equationally compact extension are characterized by purity. B. Banaschewski's question, whether there is a similar characterization for the maximal compactifications in general, remains unanswered. "Similar" means here a characterization that does not refer to the set of *all* compactifications.

Theorem 9. *If* $\mathfrak{A}$ *has a pure equationally compact extension then a compactification* $\mathfrak{B}$ *of* $\mathfrak{A}$ *is maximal if and only if* $\mathfrak{B}$ *is a pure extension, i.e., if and only if* $\mathfrak{B}$ *is an equationally compact hull of* $\mathfrak{A}$.

Proof. Let $\mathfrak{C}$ be an equationally compact hull and let $\mathfrak{D}$ be an arbitrary maximal compactification of $\mathfrak{A}$. There is an A-homomorphism $g\colon \mathfrak{D} \to \mathfrak{C}$, so $\mathfrak{D}$ is a pure extension. Corollary 2 to Theorem 7 finishes the proof.

Examples. (1) H. Numakura [1952] proved that every cancellative compact topological semigroup is in fact a group. B. Węglorz [2] generalized this to cancellative equationally compact semigroups. B. Banaschewski [1974 a] refined the result further by showing that every cancellative monoid with an equationally compact pure extension is a group. It is quite interesting that one can go one more step: Any cancellative semigroup with an equationally compact weakly pure extension is a monoid (not yet a group as $\langle N_0; +\rangle$ demonstrates). Thus, every cancellative semigroup with an equationally compact pure extension is a group. Hence (B. Banaschewski [1974 a]) any cancellative commutative monoid which is not a group has equationally compact extensions but none of these are pure.

(2) B. Banaschewski [1974 a] also shows that any free monoid or free commutative monoid $\mathfrak{A}$ has a one-point compactification $\mathfrak{B}$ (i.e., $|B-A|=1$). Compactifications of Peano algebras are considered in J. Mycielski and W. Taylor [1976]. Any Peano algebra of any type has compactifications.

(3) B. Banaschewski [1974] characterizes equationally compact $\mathfrak{G}$-sets for arbitrary groups $\mathfrak{G}$. $\mathfrak{A}=\langle A; \{f_g \mid g \in G\}\rangle$ of type $\langle 1, 1, 1, \cdots\rangle$ is a $\mathfrak{G}$-set if $f_{gh}=f_g \circ f_h$ and $f_1=\mathrm{id}_A$ hold. A $\mathfrak{G}$-set is equationally compact if and only if any subgroup $\mathfrak{H}$ of $\mathfrak{G}$ for which every finitely generated subgroup is contained in some stability subgroup of A is itself contained in a stability subgroup of A. Using this it can be shown that every $\mathfrak{G}$-set has an equationally compact hull. B. Banaschewski gives an explicit description of this hull. Mono-unary algebras have equationally compact hulls (W. Taylor [1974]); semilattices, in general, do not (E. Nelson [1975 b]).

The last example falls naturally into the circle of problems that we touch upon in the next and last section: When do *all* algebras of a given equational class have compactifications or equationally compact hulls?

§90. APPLICATION TO EQUATIONAL CLASSES OF ALGEBRAS

The results of this section are essentially due to W. Taylor [1972]; the approach has been altered from a model-theoretic to an algebraic one by B. Banaschewski and E. Nelson [1972]. The latter will be presented here.

Taylor's point of departure was the question of when an equational class V can be written as $V=\mathbf{ISP}(K)$ for some *set* K. In view of G. Birkhoff's subdirect representation theorem we always have $V=\mathbf{ISP}(K)$ for the *class* K of all subdirectly irreducible algebras in V. Thus the problem may be reformulated as to when the subdirectly irreducible members of an equational class V form (up to isomorphism, of course) a set. Such equational classes are called *residually small.* The following theorem reveals the problem as a relevant part of our topic.

Theorem 1. *If V is an equational class of algebras, then the following conditions are equivalent:*

(1) *Every $\mathfrak{A} \in V$ has an equationally compact extension.*

(2) *If $\mathfrak{B}$ is an essential extension of $\mathfrak{A} \in V$, then $|B| \leq 2^{\mathfrak{n}}$, where $\mathfrak{n}=\aleph_0+\overline{o(\tau)}+|A|$.*

(3) *If $\mathfrak{A} \in V$ is subdirectly irreducible, then $|A| \leq 2^{\mathfrak{m}}$, where $\mathfrak{m}=\aleph_0+\overline{o(\tau)}$.*

(4) *V is residually small.*

(5) *Every $\mathfrak{A} \in V$ can be embedded into an absolute retract in V.*

Proof. (1) implies (2). We take a compactification $\mathfrak{B}'$ of $\mathfrak{B}$ and choose $\mathfrak{C}$ with $A \subseteq C \subseteq B'$ such that $\mathfrak{B}'$ is a pure extension of $\mathfrak{C}$ and $|C| \leq \mathfrak{n}$ (Theorem 89.6). By Theorem 89.7(3) the algebra $\mathfrak{B}'$ is an equationally compact hull of $\mathfrak{C}$ and, of course, $|B| \leq |B'| \leq 2^{\mathfrak{n}}$. (2) implies (3). Every subdirectly irreducible algebra is an essential extension of a suitable algebra generated by two elements. (4) implies (5). Essential extensions of subdirectly irreducible algebras are again subdirectly irreducible. Thus Zorn's lemma assures maximal essential extensions of any given subdirectly irreducible algebra. These are absolute retracts in V. If (5) holds for subdirectly irreducible algebras, then it holds in general.

Examples. (1) Abelian groups, $\mathbf{HSP}(\mathfrak{S}_3)$ (i.e., the equational class generated by the symmetric group on 3 letters), Boolean algebras, distributive lattices, semilattices, rings with 1 satisfying $x^n = x$, mono-unary and bi-unary algebras, $\mathfrak{G}$-sets, all quasi-primal varieties are residually small equational classes.

(2) The varieties of groups, modular lattices, lattices, commutative rings with 1 are not residually small.

(3) In W. Taylor [1976] we find the following interesting examples: The group SO(3) generates a non-residually small equational class, although it is equationally compact. The same holds for the 8-element quaternion group (due to S. O. MacDonald and H. Groves) or the 3-element semigroup $\langle\{0, 1, 2\}; \cdot\rangle$ defined by the algebraic equations $0 \cdot x = 0$, $1 \cdot x = 1$ and $2 \cdot x = x$ (due to J. A. Gerhard).

The attempts to characterize equational classes of algebras with enough equationally compact hulls leads to a nice analog to the role of subdirectly irreducible algebras in the "pure situation" that takes the place of G. Birkhoff's famed result.

Definition 1. *The algebra $\mathfrak{A}$ is called* pure-irreducible *if for any embedding $g\colon \mathfrak{A} \to \prod(\mathfrak{B}_i \mid i \in I)$ such that $\mathfrak{A}g$ is a pure subalgebra of the direct product (we call such g a* pure embedding*) the composite of g with some projection $\pi_{i_0}\colon \prod(\mathfrak{B}_i \mid i \in I) \to \mathfrak{B}_{i_0}$ is a monomorphism.*

One easily proves the following different description of this concept.

Remark. $\mathfrak{A}$ is pure-irreducible if and only if there is a finite set of algebraic equations $\{p_1 = q_1, \cdots, p_n = q_n\}$ with constants in A such that this system is solvable in $\mathfrak{A}$ modulo every congruence $\Theta \neq \omega$, while the system itself is not solvable.

Remark. Subdirectly irreducible algebras are pure-irreducible. The converse is false as, e.g., the three-element lattice shows. W. Taylor [1972] observed the following analog of Birkhoff's theorem.

Theorem 2. *Any algebra* $\mathfrak{A}$ *is the subdirect product of pure-irreducible algebras* $\mathfrak{B}_i$, $i \in I$, *in such a way that* $\mathfrak{A}$ *is a pure subalgebra of* $\prod(\mathfrak{B}_i \mid i \in I)$.

Pure-irreducible algebras allow the following improvement over Theorem 89.6 and a generalization of the fact that every subdirectly irreducible algebra is an essential extension of a subalgebra generated by two elements.

Lemma 1. *If* $\mathfrak{A}$ *is a pure-irreducible algebra, then there exists a subalgebra* $\mathfrak{B}$ *of* $\mathfrak{A}$ *with* $|B| \leqq \aleph_0 + \overline{o(\tau)}$ *such that* $\mathfrak{A}$ *is a pure-essential extension of* $\mathfrak{B}$. $\mathfrak{B}$ *may be chosen in such a way that it contains any given finite set* $K \subseteq A$.

Proof. Following B. Banaschewski and E. Nelson [1972] we take $S = \{p_i = q_i \mid i = 1, \cdots, n\}$ as in the remark following Definition 1. If $K \subseteq A$ is a given finite set and L is the set of constants appearing in S then there is a finitely generated subalgebra $\mathfrak{A}_0$ of $\mathfrak{A}$ with $K \cup L \subseteq A_0$. We choose a pure subalgebra $\mathfrak{B}$ of $\mathfrak{A}$ with $A_0 \subseteq B$ and $|B| \leqq \aleph_0 + \overline{o(\tau)}$ (Theorem 89.6). If Θ is an arbitrary congruence of $\mathfrak{A}$, $\Theta \neq \omega$, then $p_i \equiv q_i \pmod{\Theta}$, $i = 1, \cdots, n$, is solvable in $\mathfrak{A}$ but S is not solvable in $\mathfrak{B}$. We conclude that no $\mathfrak{A}/\Theta$ is a pure extension of $\mathfrak{B}$, i.e., $\mathfrak{A}$ is a pure-essential extension of $\mathfrak{B}$.

We have all the pieces to prove Taylor's "pure analog" of Theorem 2.

Theorem 3. *If* V *is an equational class of algebras, then the following conditions are equivalent:*

(1) *Every* $\mathfrak{A} \in V$ *has a pure equationally compact extension.*
(2) *If* $\mathfrak{B}$ *is a pure-essential extension of* $\mathfrak{A} \in V$, *then* $|B| \leqq 2^{\mathfrak{n}}$, *where* $\mathfrak{n} = \aleph_0 + \overline{o(\tau)} + |A|$.
(3) *If* $\mathfrak{A} \in V$ *is pure-irreducible, then* $|A| \leqq 2^{\mathfrak{m}}$, *where* $\mathfrak{m} = \aleph_0 + \overline{o(\tau)}$.
(4) *The isomorphism classes of pure-irreducible algebras form a set.*

Proof. (1) implies (2). If $\mathfrak{C}$ is an equationally compact hull of $\mathfrak{A}$, then it is a maximal pure-essential extension and $\mathfrak{B}$ can be embedded in $\mathfrak{C}$ (Lemma 89.1). By Theorem 89.5 we have $|B| \leqq |C| \leqq 2^{\mathfrak{n}}$. (2) implies (3). Let $\mathfrak{A}$ be pure-irreducible. By Lemma 1 there exists some $\mathfrak{B} \in \mathbf{S}(\mathfrak{A})$ with $|B| \leqq \mathfrak{m}$ such that $\mathfrak{A}$ is a pure-essential extension of $\mathfrak{B}$. We conclude $|A| \leqq 2^{\mathfrak{m} + |B|} = 2^{\mathfrak{m}}$. (3) implies (4) is trivial; (4) implies (1) by Theorem 89.4.

Examples. Abelian groups, mono-unary algebras, $\mathfrak{G}$-sets, and Boolean algebras constitute residually small equational classes in which every algebra has even a *pure* equationally compact extension. The last example is a quasi-primal variety. Equationally compact and compact topological algebras in such varieties were characterized by S. Bulman-Fleming and H. Werner [1977] via Boolean extensions. W. Taylor [1976] showed that every algebra in such a variety is a pure subalgebra of a direct product of finite simple algebras. So quasi-primal varieties have always *pure* equationally compact extensions. Bi-unary algebras, semilattices (mentioned before), and distributive lattices are residually small equational classes *without* this property. The last example is due to R. McKenzie (see W. Taylor [1972]) who constructed pure-irreducible distributive lattices of arbitrarily large cardinality.

One last remark may be of interest: it is an open problem whether the existence of quasi-compact extensions implies the existence of equationally compact extensions for a fixed algebra $\mathfrak{A}$. A slight modification of the original Banaschewski–Nelson proof for Theorem 1 (mimeographed notes, 1971) reveals that one can give a positive answer if one asks the question for *all* algebras of a given equational class. A proof can also be based on the following lemma of W. Scott [1951].

Lemma 2. *Every algebra $\mathfrak{A}$ can be embedded into an absolutely pure algebra $\mathfrak{B} \in$ **HSP**($\mathfrak{A}$) (i.e., $\mathfrak{B}$ is a pure subalgebra of every extension $\mathfrak{C} \in$ **HSP** ($\mathfrak{A}$)).*

Theorem 4. *Let V be an equational class of algebras. The following two statements are equivalent:*

(1) *Every $\mathfrak{A} \in V$ has an equationally compact extension.*
(2) *Every $\mathfrak{A} \in V$ has a quasi-compact extension.*

Proof. We choose a quasi-compact extension $\mathfrak{B}$ of an absolutely pure extension $\mathfrak{C}$ of $\mathfrak{A}$ in **HSP**($\mathfrak{A}$). $\mathfrak{B}$ is a closure of $\mathfrak{C}$ and Theorem 89.3 settles the matter.

§91. CONCLUDING REMARKS

We have restricted our attention to universal algebras. E. Nelson [1975 a] investigated the natural functor Γ from the class of all algebras of a fixed type to a suitable class of relational systems which preserves underlying sets and replaces every n_γ-ary operation by its $(n_\gamma + 1)$-ary "graph" (associated relational system). For relational systems we can define atomic compactness and atomic hulls in complete analogy to equational compactness and equationally compact hulls. Γ preserves and reflects purity, atomic compactness, pure-essentialness, hence the existence of atomic hulls, although it does not reflect the existence of atomic compact extensions in general. In another paper E. Nelson [1975] produces a second functor Δ from the class of all relational system (really structures) of a fixed type to the class of all groupoids that preserves and reflects purity and atomic compactness, preserves pure-essential extensions, hence atomic-compact hulls, and reflects the existence of atomic compact hulls. $\Gamma \circ \Delta$ and Δ are therefore functors from the class of all algebras of a fixed type to the class of all groupoids that have the above properties. The two papers treat the topic (in parts at least) within the more general framework of infinitary algebras. In E. Nelson [1974] equational compactness for infinitary algebras is investigated.

We have not reported on any detailed investigation in special classes of algebras nor on problems concerned with limiting the number of variables (rather than the number of equations), in the systems of equations (or formulas) that occur in the basic definitions. A. Abian's [1970] result that a Boolean algebra is equationally compact if and only if every consistent system of equations of the type $a \wedge x = b$ is solvable and similar results for semilattices by G. Grätzer and H. Lakser [1969] or for certain lattices by D. Kelly [1972] indicate that such considerations can bring forth interesting results.

The answer to Mycielski's question is not known for many classes of algebras—most notably for distributive lattices, groups and rings. The best result for rings seems to be an approximation of a negative answer by D. K. Haley [1977] that reads as follows: There exists a commutative ring $\mathfrak{R}$ and a structure $\mathfrak{A} = \langle R; +, r \rangle$ such that $\langle R; + \rangle$ is the group-reduct of the ring $\mathfrak{R}$, the binary relation r is "equationally definable in $\mathfrak{R}$" and such that $\mathfrak{A}$ is atomic compact without being a retract of a compact topological structure.

There are interesting purely model-theoretic results that we had to skip. The cardinality result of R. McKenzie and S. Shelah (Theorem 88.5), for example, points into an area of problems that goes beyond the scope of our discussion here. Another class of results can be exemplified by the following theorem that has been proved by various mathematicians (B. Węglorz, A. I. Omarov, W. Taylor, J. T. Baldwin): All models of an $\aleph_1$-categorical Horn theory are atomic compact. A separate survey of such questions (put into another framework) would surely be of interest.

In conclusion the following quotation from the 1969 edition of Kaplansky's Infinite Abelian groups [1969] seems particularly fitting: "I realized that I was in a position to tell my colleague Paul Halmos exactly what were the possibilities for the algebraic structure of a compact Abelian group (modulo some final fiddling with cardinal numbers). It was in order to be able to state a precise theorem without this annoying investigation of cardinal numbers that I broadened the class of compact groups slightly and called the generalized object 'algebraically compact'. There was a lucky accident here, for the concept had ramifications considerably beyond anything I envisaged."

§92. SOME PROBLEMS

In this section we collect some open problems on equational compactness. Naturally the selection is biased. However we have tried to concentrate on those problems that have been around in the literature for a while. They are partly folklore, partly re-appearing in various publications by the authors who have been frequently mentioned in this report; so we will state them without attempting to trace their origins.

1. Can every equationally compact algebra be embedded in a compact topological algebra? If not, does such an embedding exist in interesting classes of algebras? (W. Taylor [1972] mentions specifically residually small equational classes.)
2. Is the algebra $\mathfrak{A}$ a retract of a compact topological algebra if its associated relational system is a retract of a compact topological relational system? (The known counterexamples to Mycielski's question remain counterexamples if we switch to the associated relational systems.)
3. Does an algebra $\mathfrak{A}$ have an equationally compact extension if it has a quasi-compact extension? (See B. Węglorz [2].)
4. Can every equationally compact partial algebra be embedded as a relative subalgebra into an equationally compact algebra?
5. Is there a description of the maximal compactifications of an algebra that does not refer to the set of *all* compactifications? (See B. Banaschewski [1974 a]). Can one give a "construction" of the core of a weakly equationally compact algebra? (W. Taylor [1971 a].)
6. Characterize minimum compact algebras in interesting equational classes with finitely many new constants.
7. Can the representation in Theorem 88.4 be made in any sense "canonical" (i.e., unique or irreducible)? (See W. Taylor [1971 a].)
8. Study equationally compact lattices. More specifically:
   (a) Is a distributive lattice equationally compact if and only if it is complete and fully distributive? (This is a conjecture by G. Grätzer that was independently made by W. Taylor in his Ph.D. Thesis.) What about Mycielski's question in the class of distributive lattices?
   (b) Does Mycielski's question have a positive answer in the class of all lattices without infinite anti-chains? (See D. Kelly [1972].)

9. Is there a system of $\aleph_2$ equations over $\langle Z; +, -\rangle$ or $\langle Z; +, -, \cdot\rangle$ which is unsolvable in Z although every subsystem with $\leqq \aleph_1$ equations has a solution? (See Theorem 84.3 and R. McKenzie [1971 b].)
10. Investigate equational compactness in classes of groups other than the Abelian groups. FC-groups, for instance, seem to be more accessible to model-theoretic methods than just arbitrary groups (*FC-groups* are groups with finite conjugacy classes). What about locally finite FC-groups? J. Mycielski [3] suggests an investigation of non-Abelian, connected, locally compact topological groups.
11. S. Balcerzyk [1956] proved that an Abelian group is equationally compact if and only if every set of equations of the particular form $\{x_0 - a_n = n!\, x_n \mid n = 1, 2, \cdots\}$ is solvable provided it is finitely solvable. Do such particular forms exist for arbitrary $\mathfrak{R}$-modules?
12. What is the answer to Mycielski's question in the class of groups or the class of (commutative) rings?
13. Are simple equationally compact rings (with 1) finite? (A corresponding result holds for compact topological rings.) Are there simple non-equationally compact rings that allow a quasi-compact extension?
14. (W. Taylor [1974 a]; Problem 2.7.) Does there exist a residually small variety K that is congruence-regular but not congruence-uniform? (I.e., there are blocks B_1, B_2 of a congruence Θ on some $\mathfrak{A} \in K$ with $|B_2| > |B_1| \geqq \aleph_0$.)

APPENDIX 7
THE INDEPENDENCE PROOF

By G. Grätzer and W. A. Lampe

§93. STATEMENT OF THE MAIN RESULTS

In this appendix we shall prove the independence of the congruence lattice, the subalgebra lattice, and the automorphism group of an (infinitary) algebra and characterize type-2 congruence lattices.

Theorem 1. *Let* $\mathfrak{m}$ *be an infinite regular cardinal. Let* $\mathfrak{G}$ *be a group and let* $\mathfrak{L}_c$ *and* $\mathfrak{L}_a$ *be* $\mathfrak{m}$*-algebraic lattices with more than one element. Then there exists an algebra* $\mathfrak{A}$ *of characteristic* $\mathfrak{m}$ *satisfying the following conditions:*

(i) *The congruence lattice, the subalgebra lattice, and the automorphism group of* $\mathfrak{A}$ *are isomorphic to* $\mathfrak{L}_c$, $\mathfrak{L}_a$, *and* $\mathfrak{G}$, *respectively.*

(ii) *The congruence lattice of* $\mathfrak{A}$ *is a sublattice of the partition lattice (i.e., the lattice of equivalence relations) of* A*; in fact, the congruences have type-3 joins.*

(iii) *If* $\mathfrak{L}_a$ *is modular and* $\mathfrak{G}$ *has only one element, then the congruence lattice of* $\mathfrak{A}$ *has type-2 joins.*

In this theorem, type-3 (respectively, type-2) joins for congruences means that $\Theta \vee \Phi = \Theta\Phi\Theta\Phi$ (respectively, $\Theta \vee \Phi = \Theta\Phi\Theta$).

In §94 we discuss some preliminaries and define two crucial concepts: closest elements and invertible operations. In §95 we axiomatize the systems we wish to deal with (Cls-quintuples) and the appropriate concept of expansion (Cls-expansions) and prove that $\mathfrak{A}[f]$ gives us such an expansion.

The construction $\mathfrak{A}^*$ corresponds to $\mathfrak{A}'$ of §17, but without g_1, g_2, g_3. Under suitable hypotheses, this again yields a Cls-expansion. If g_1, g_2, g_3 are also considered, then stronger hypotheses are needed to get a Cls-expansion; this gives the $\mathfrak{A}^{**}$ construction. These two constructions are discussed in §96.

Two intermediate constructions needed for subalgebras and automorphisms are introduced in §97. Finally, in §98, we give the initial construction. Starting with this construction, the proof of Theorem 1 is then an easy transfinite series of applications of the constructions introduced in §§95–97.

§94. PRELIMINARIES

We need some notation, terminology, and preliminary observations. In this appendix, all algebras or partial algebras will be of characteristic $\leqq \mathfrak{m}$, where $\mathfrak{m}$ is a fixed infinite regular cardinal. A number of results of Chapters 0–2 are valid for $\mathfrak{m}$-complete semilattices and for partial algebras and algebras of characteristic $\leqq \mathfrak{m}$ (see, in particular, Exercises 82–84 of Chapter 0, Exercises 88–96 of Chapter 1, Exercise 45 of Chapter 2).

If α is an ordinal, A is a set, $\mathbf{a} \in A^{\alpha}$, and $\gamma < \alpha$, then a_{γ} denotes the γth component of $\mathbf{a}$. If σ maps A into itself, $\mathbf{a}\sigma$ is the sequence $\langle a_{\gamma}\sigma \rangle$. If $\mathbf{a}, \mathbf{b} \in A^{\alpha}$ and Θ is a binary relation on A, then $\mathbf{a} \equiv \mathbf{b}(\Theta)$ means that $a_{\gamma} \equiv b_{\gamma}(\Theta)$ for all $\gamma < \alpha$. If C is a closure system on A, we write $[\mathbf{a}]_C$ or $[a_0, a_1, \cdots]_C$ for $[\{a_0, a_1, \cdots\}]_C$. For X, $Y \in C$ we set

$$X \vee_C Y = [X \cup Y]_C.$$

The following two observations will be useful:

Lemma 1. *Let C be a set of subsets of the set S. If $\langle C; \subseteq \rangle$ is a complete lattice and for each $s \in S$ there is an $[s] \in C$ satisfying:*

(i) $s \in [s]$;

(ii) *for $D \in C$, $s \in D$ implies that $[s] \subseteq D$,*

then C is a closure system.

Lemma 2. *A closure system C on the set A is $\mathfrak{m}$-algebraic iff a subset X of A is closed whenever X contains the closure of Y for every $Y \subseteq X$ satisfying $|Y| < \mathfrak{m}$.*

The easy proofs are left to the reader.

In an $\mathfrak{m}$-algebraic closure system C on the set A, *$\mathfrak{m}$-compact equals principal* iff for any $B \subseteq A$ with $|B| < \mathfrak{m}$ there is a $b \in B$ such that $[B]_C = [b]_C$.

For a partial algebra $\mathfrak{A} = \langle A; F \rangle$ and $f \in F$, $R(f, \mathfrak{A})$ denotes the range of f.

Let $\mathfrak{A}$ and $\mathfrak{B}$ be partial algebras and $A \subseteq B$. $\mathfrak{B}$ is an *expansion* of $\mathfrak{A}$ iff the following hold: (i) each partial operation f of $\mathfrak{A}$ is a partial operation of $\mathfrak{B}$ and $D(f, \mathfrak{A}) \subseteq D(f, \mathfrak{B})$; (ii) for any partial operation f of $\mathfrak{A}$ and any $\mathbf{a} \in D(f, \mathfrak{A})$, the value $f(\mathbf{a})$ is the same in $\mathfrak{A}$ as it is in $\mathfrak{B}$. If $\mathfrak{A}$ and $\mathfrak{B}$ have the same partial operations and $\mathfrak{A}$ is a subalgebra of $\mathfrak{B}$, then $\mathfrak{B}$ is an *extension* of $\mathfrak{A}$.

Let $\mathfrak{B}$ be an expansion of $\mathfrak{A}=\langle A; F\rangle$, let $E \in \mathcal{S}(\mathfrak{B})$, $\Phi \in C(\mathfrak{B})$, $\tau \in G(\mathfrak{B})$, $D \in \mathcal{S}(\mathfrak{A})$, $\Theta \in C(\mathfrak{A})$, and $\sigma \in G(\mathfrak{A})$. E *is an extension of* D iff $E \cap A = D$. Φ *is an extension of* Θ iff Φ restricted to A is Θ. τ *is an extension of* σ iff τ and σ agree on A.

Repeated expansions lead to limits. Let α be a limit ordinal and let $\mathfrak{A}_\gamma$ be a partial algebra, for all $\gamma < \alpha$, with the property that $\mathfrak{A}_\delta$ is an expansion of $\mathfrak{A}_\gamma$ for all $\gamma \leqq \delta < \alpha$. Then we can form $A = \bigcup (A_\delta \mid \delta < \alpha)$ and $F = \bigcup (F_\gamma \mid \gamma < \alpha)$ and call $\mathfrak{A} = \langle A; F\rangle$ the *limit of* $\mathfrak{A}_\gamma$, $\gamma < \alpha$.

As in the finitary case, for the partial algebra $\mathfrak{A}=\langle A; F\rangle$ and $f \in F$, we define $\mathfrak{A}[f]$; $\mathfrak{A}[F]$ is the union of $\mathfrak{A}[f]$ as f ranges over F. It is crucial that Lemma 15.3 holds for infinitary partial algebras with no change in the proof. We denote by $\Theta[f]$ the extension of $\Theta \in C(\mathfrak{A})$ to $\mathfrak{A}[f]$ as given by Lemma 15.3.

For $H \subseteq C(\mathfrak{A})$, set $H[h] = \{\Theta[h] \mid \Theta \in H\}$.

Let D be a subalgebra of $\mathfrak{A}$. Denote by $D[h]$ the subalgebra generated by D in $\mathfrak{A}[h]$. Then $a \in D[h]$ iff $a \in D$ or $a = h(\mathbf{d})$ for a unique $\mathbf{d} \in D^\gamma - D(h, \mathfrak{A})$, where γ is the arity of h. So $D[h] \cap A = D$. For $K \subseteq \mathcal{S}(\mathfrak{A})$, we set $K[h] = \{D[h] \mid D \in K\}$.

It is clear that an automorphism σ of $\mathfrak{A}$ has a unique extension $\sigma[h] \in G(\mathfrak{A}[h])$. Set $H[h] = \{\sigma[h] \mid \sigma \in H\}$ for any $H \subseteq G(\mathfrak{A})$ and $\mathfrak{H}[h] = \langle H[h]; \circ\rangle$ for any subgroup $\mathfrak{H} = \langle H; \circ\rangle$ of $\mathfrak{G}(\mathfrak{A})$.

Let $\mathfrak{A}=\langle A; F\rangle$ be a partial algebra of characteristic $\mathfrak{m}$. Set $\mathfrak{A} = \mathfrak{A}[F]_0$. For any ordinal α, define $\mathfrak{A}[F]_{\alpha+1} = \langle A[F]_{\alpha+1}; F\rangle = (\mathfrak{A}[F]_\alpha)[F]$. If α is a limit ordinal, define $\mathfrak{A}[F]_\alpha$ as the limit of $\mathfrak{A}[F]_\gamma$, $\gamma < \alpha$. It is easy to see that $\mathfrak{A}[F]_\mathfrak{m}$ is an algebra and that $\mathfrak{A}[F]_\alpha = \mathfrak{A}[F]_\mathfrak{m}$ for any $\alpha \geqq \mathfrak{m}$. In particular, if $\mathfrak{A}$ is finitary $\mathfrak{A}[F]_\alpha = \mathfrak{A}[F]_\omega$ for any $\alpha \geqq \omega$. We shall use the notation $\mathbf{Fr}(\mathfrak{A}) = \langle Fr(\mathfrak{A}); F\rangle$ for $\mathfrak{A}[F]_\mathfrak{m}$, and we call $\mathbf{Fr}(\mathfrak{A})$ *the algebra (absolutely) freely generated by* $\mathfrak{A}$.

Now we introduce the first new concept, the concept of a closest element, modelled after the situation shown on the diagram of p. 101: for instance, for any element of $g_1(A)$, c is the closest element in A.

Let H be a set of equivalence relations on the set A. For an ordinal α and $\mathbf{a}, \mathbf{b} \in A^\alpha$ we set

$$\Theta_H(\mathbf{a}, \mathbf{b}) = \bigcap (\Phi \mid \Phi \in H \text{ and } \mathbf{a} \equiv \mathbf{b}(\Phi)).$$

If $\mathbf{a} = \langle a\rangle$ and $\mathbf{b} = \langle b\rangle$, we write $\Theta_H(a, b)$ for $\Theta_H(\langle a\rangle, \langle b\rangle)$. Observe that if $H = C(\mathfrak{A})$, then $\Theta_H(a, b) = \Theta(a, b)$. Intuitively, we think of $\Theta_H(a, b)$ as the distance, measured in H, between a and b. If H is a closure system, then $\Theta_H(\mathbf{a}, \mathbf{b}) \in H$.

Similarly, for $\mathbf{a} \in A^\alpha$ and $D \subseteq A^\alpha$ we set

$$H\text{-Dist}(\mathbf{a}, D) = \bigcap (\Theta_H(\mathbf{a}, \mathbf{b}) \mid \mathbf{b} \in D),$$

and call this the *H-distance of* $\mathbf{a}$ *from* D. If for some $\mathbf{b} \in D$,

$$\Theta_H(\mathbf{a}, \mathbf{b}) = H\text{-Dist}(\mathbf{a}, D),$$

we say that **b** *is an H-closest element in D to* **a**, in symbols,

$$\mathbf{b} \text{ Cls } \mathbf{a} \text{ (in } D, \text{mod } H).$$

Observe that this holds iff $\Theta_H(\mathbf{a}, \mathbf{b}) \leqq \Theta_H(\mathbf{a}, \mathbf{x})$ for any $\mathbf{x} \in D$. (Note that such a **b** is not unique. For instance, in the diagram of p. 101, any element of $g_2(A)$ has at least two closest elements in A.)

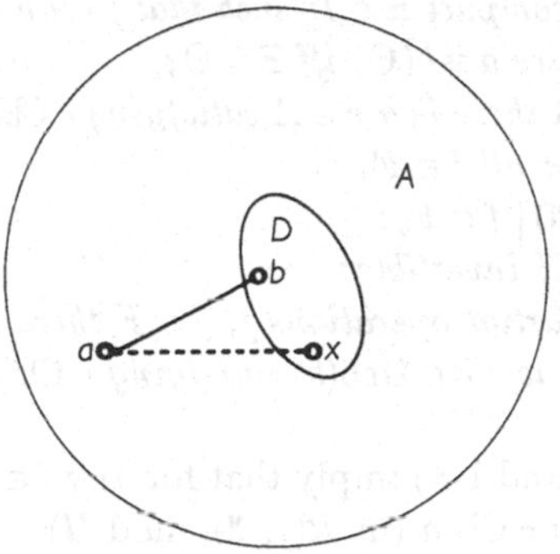

b is closest to *a* in *D* modulo *H*

$\Theta_H(x, a) \geqq \Theta_H(b, a)$

Fig. 1

We need two more concepts.

Let H be a set of equivalence relations on the set A. We call H a *unary-algebraic closure system* iff there is some family F of *unary* partial operations on A so that $H = C(\langle A; F\rangle)$. Equivalently, H is unary-algebraic iff (i) H is a closure system (on $A \times A$), (ii) $\omega \in H$, and (iii) for any *equivalence relation* Φ on A, $\Phi \in H$ iff $\Phi \supseteq \Theta_H(a, b)$ for every $\langle a, b\rangle \in \Phi$.

Again, let H be a set of equivalence relations on the set H and let f be an α-ary partial operation on A. Then f is called *H-invertible* iff for every $\Theta \in H$, $\mathbf{a}, \mathbf{b} \in D(f)$, $f(\mathbf{a}) \equiv f(\mathbf{b})(\Theta)$ implies that $\mathbf{a} \equiv \mathbf{b}(\Theta)$. This is just the converse of what is required for Θ to be a congruence relation. Observe that f is one-to-one iff it is $\{\omega\}$-invertible.

§95. Cls-EXPANSIONS AND FREE EXTENSIONS

The proof of Theorem 93.1 is based on two complicated definitions, motivated by §§17 and 18 of Chapter 2.

Definition 1. *Let* $\mathfrak{A}$ *be a partial algebra,* $H \subseteq C(\mathfrak{A})$, $K \subseteq \mathscr{S}(\mathfrak{A})$, $G \subseteq G(\mathfrak{A})$, *and let* $\mathfrak{m}$ *be an infinite regular cardinal.* $\langle \mathfrak{A}, H, K, G, \mathfrak{m} \rangle$ *is a* Cls-quintuple *iff the following conditions are satisfied:*

(i) $\mathfrak{A}$ *is of characteristic* $\leqq \mathfrak{m}$; *H and K are* $\mathfrak{m}$*-algebraic closure systems;* $\mathfrak{G} = \langle G; \circ \rangle$ *is a regular permutation group (that is, if* $a\sigma = a$ *for some* $a \in A$ *and* $\sigma \in G$, *then* σ *is the identity of* $\mathfrak{G}$*);*

(ii) $\omega \in H$ *and* $\varnothing \in K$;

(iii) $\mathrm{Orb}(a) = \{a\sigma \mid \sigma \in G\} \subseteq [a]_K$, *for* $a \in A$;

(iv) $\Theta_H(a, b) = \Theta_H(a\sigma, b\sigma)$, *for* $a, b \in A$, $\sigma \in G$;

(v) *there is an* $\mathfrak{m}$*-compact* $\Xi \in H$ *such that for any* $\Theta \in H, a, b \in A, a \neq b$, *and* $b \in \mathrm{Orb}(a)$, *we have* $a \equiv b(\Theta)$ *iff* $\Xi \subseteq \Theta$;

(vi) *for all* $a, b \in A$ *there is a* $c \in A$ *satisfying* c Cls a *(in* $\mathrm{Orb}(b)$, mod H*);*

(vii) $R(f, \mathfrak{A}) \neq \varnothing$ *for all* $f \in F$;

(viii) $A = \bigcup (R(f, \mathfrak{A}) \mid f \in F)$;

(ix) *each* $f \in F$ *is H-invertible;*

(x) *for any two partial operations* $f, g \in F$ *there is a* $b \in A$ *such that for any* $a \in R(g, \mathfrak{A})$ *there is a* $c \in \mathrm{Orb}(b)$ *satisfying* c Cls a (in $R(f, \mathfrak{A})$, mod H).

Remark 1. 1(viii) and 1(x) imply that for any $f \in F$ and $a \in A$ there is a $c \in R(f, \mathfrak{A})$ satisfying c Cls a (in $R(f, \mathfrak{A})$, mod H).

Remark 2. It follows from 1(v) and 1(vi) that 1(vi) holds also for sequences. Namely, if $\mathbf{a}, \mathbf{b} \in A^\alpha$, then there is a $\mathbf{c} \in A^\alpha$ satisfying $\mathbf{c}$ Cls $\mathbf{a}$ (in $\mathrm{Orb}(\mathbf{b})$, mod H). Proof: By 1(vi), there is a $c = b_0\tau$ ($\tau \in G$) satisfying $b_0\tau$ Cls a_0 (in $\mathrm{Orb}(b_0)$, mod H). Set $\mathbf{c} = \mathbf{b}\tau$ and let $\mathbf{b}\sigma \equiv \mathbf{a}(\Theta)$ for some $\sigma \in G$. We wish to show that $\mathbf{b}\tau \equiv \mathbf{a}(\Theta)$. If $\sigma = \tau$, there is nothing to prove. So let $\sigma \neq \tau$. Since $\mathbf{b}\sigma \equiv \mathbf{a}(\Theta)$, then, in particular, $b_0\sigma \equiv a_0(\Theta)$. Hence $b_0\tau \equiv a(\Theta)$. Since $\sigma \neq \tau$ and G is regular, $b_0\sigma \neq b_0\tau$ and $b_0\sigma \equiv b_0\tau(\Theta)$. Applying 1(v), we conclude that $\Xi \subseteq \Theta$ and so $b_\gamma\sigma \equiv b_\gamma\tau(\Theta)$ for all $\gamma < \alpha$, hence $\mathbf{b}\sigma \equiv \mathbf{b}\tau(\Theta)$. Thus $\mathbf{a} \equiv \mathbf{b}\sigma \equiv \mathbf{b}\tau(\Theta)$, as claimed.

Definition 2. $\langle \mathfrak{A}', H', K', G', \mathfrak{m} \rangle$ *is a* Cls-expansion *of* $\langle \mathfrak{A}, H, K, G, \mathfrak{m} \rangle$ *(where* $\mathfrak{A} = \langle A; F \rangle$ *and* $\mathfrak{A}' = \langle A'; F' \rangle$*) iff the following conditions are satisfied:*

(i) $\langle \mathfrak{A}, H, K, G, \mathfrak{m} \rangle$ *and* $\langle \mathfrak{A}', H', K', G', \mathfrak{m}' \rangle$ *are* Cls*-quintuples and* $\mathfrak{m} = \mathfrak{m}'$;

(ii) $\mathfrak{A}'$ *is an expansion of* $\mathfrak{A}$;

(iii) *for* $\Theta \in H$, $[\Theta]_{H'}$ *is an extension of* Θ; *for* $D \in K$, $[D]_{K'}$ *is an extension of* D; *for any* $\sigma \in G$ *there is a unique extension* $\sigma' \in G'$;

(iv) $H' = \{[\Theta]_{H'} \mid \Theta \in H\}$, $K' = \{[D]_{K'} \mid D \in K\}$, $G' = \{\sigma' \mid \sigma \in G\}$;

(v) *let* $f, g \in F$, $f \neq g$; *if* b *satisfies* 1(x) *for* f *and* g *in* $\mathfrak{A}$, *then* b *satisfies* 1(x) *for* f *and* g *in* $\mathfrak{A}'$.

Remark 1. The relation defined by Cls-expansion is transitive. We shall use this repeatedly without reference.

Remark 2. It follows from Definitions 1 and 2, that the Ξ' of H' given by 1(v) is in fact $[\Xi]_{H'}$.

Now we come to the first important statement:

Lemma 1. *Let* $\langle \mathfrak{A}, H, K, G, \mathfrak{m}\rangle$ *be a* Cls-*quintuple*, $\mathfrak{A} = \langle A; F\rangle$ *and* $h \in F$. *Then* $\langle \mathfrak{A}[h], H[h], K[h], G[h], \mathfrak{m}\rangle$ *is a* Cls-*expansion of* $\langle \mathfrak{A}, H, K, G, \mathfrak{m}\rangle$.

Proof. We start by verifying 1(i)–1(x) for the expanded quintuple. First, a convention: throughout this appendix, for $\Theta, \Phi \in H$ we write $\Theta \vee \Phi$ for $\Theta \vee_H \Phi (= [\Theta \cup \Phi]_H)$ and if $Y \subseteq H$ we set $\bigvee Y$ for $\bigvee_H Y\ (= [\bigcup Y]_H)$.

1(i). We know that $H[h] \subseteq C(\mathfrak{A}[h])$, $K[h] \subseteq \mathscr{S}(\mathfrak{A}[h])$, and $G[h] \subseteq G(\mathfrak{A}[h])$. Firstly, we verify that $H[h]$ is an $\mathfrak{m}$-algebraic closure system.

By Lemma 15.3(i), $\Theta \to \Theta[h]$ is an isomorphism between H and $H[h]$, hence $\langle H[h]; \subseteq\rangle$ is a complete lattice. To show that $H[h]$ is a closure system we verify that the conditions of Lemma 94.1 apply.

For $a, b \in A[h]$ we define $\Phi(a, b) \in H$:

(i) For $a, b \in A$, set $\Phi(a, b) = \Theta_H(a, b)$.

(ii) For $a \in A$ and $b = h(\mathbf{y}) \notin A$, let $c = h(\mathbf{w})$ be any element satisfying c Cls a (in $R(h, \mathfrak{A})$, mod H). Set $\Phi(a, b) = \Theta_H(\mathbf{w}, \mathbf{y}) \vee \Theta_H(a, c)$.

(iii) For $a \notin A$ and $b \in A$, set $\Phi(a, b) = \Phi(b, a)$ as defined in (ii).

(iv) For $a = h(\mathbf{x}) \notin A$ and $b = h(\mathbf{y}) \notin A$, set $\Phi(a, b) = \Theta_H(\mathbf{x}, \mathbf{y})$.

Now set

$$[\langle a, b\rangle] = \Phi(a, b)[h].$$

We claim that $[\langle a, b\rangle]$ satisfies the conditions of Lemma 94.1.

Comparing (i)–(iv) with the condition of Lemma 15.3, we see immediately that $\langle a, b\rangle \in [\langle a, b\rangle] \in H[h]$.

Now let $\Theta \in H$ and $a \equiv b(\Theta[h])$. We wish to show that $[\langle a, b\rangle] = \Phi(a, b)[h] \subseteq \Theta[h]$, or, equivalently, that $\Phi(a, b) \subseteq \Theta$.

If $a, b \in A$, then by 15.3(i) $a \equiv b(\Theta)$. Thus $\Phi(a, b) = \Theta_H(a, b) \subseteq \Theta$.

Let $a \in A$ and $b = h(\mathbf{y}) \notin A$. Let $c = h(\mathbf{w})$ be any element satisfying c Cls a (in $R(h, \mathfrak{A})$, mod H). By 15.3(ii), there is an $h(\mathbf{v}) \in A$ such that $a \equiv h(\mathbf{v})(\Theta)$ and $\mathbf{v} \equiv \mathbf{y}(\Theta)$. Since c is closest to a, $c \equiv a(\Theta)$ and so $h(\mathbf{w}) \equiv h(\mathbf{v})(\Theta)$. By 1(ix), h is H-invertible, and so $\mathbf{v} \equiv \mathbf{w}(\Theta)$. Thus $\mathbf{w} \equiv \mathbf{y}(\Theta)$. Hence $\Phi(a, b) = \Theta_H(\mathbf{w}, \mathbf{y}) \vee \Theta_H(a, c) \subseteq \Theta$.

Let $a \notin A$ and $b \in A$. Then $\Phi(a, b) = \Phi(b, a) \subseteq \Theta$.

Let $a = h(\mathbf{x}) \notin A$ and $b = h(\mathbf{y}) \notin A$. Then there exist by Lemma 15.3(iii),

$h(\mathbf{u}) \in A$ and $h(\mathbf{v}) \in A$ with $\mathbf{x} \equiv \mathbf{u}(\Theta)$, $h(\mathbf{u}) \equiv h(\mathbf{y})(\Theta)$, and $\mathbf{v} \equiv \mathbf{y}(\Theta)$. Since h is H-invertible, $\mathbf{u} \equiv \mathbf{v}(\Theta)$, and so $\mathbf{x} \equiv \mathbf{y}(\Theta)$. Thus 15.3(iii) implies that $\mathbf{x} \equiv \mathbf{y}(\Theta)$, and so $\Phi(a, b) \subseteq \Theta$.

Thus, by Lemma 94.1, $H[h]$ is a closure system.

Let $(\Theta_i \mid i \in I)$ be a family of members of H, and let $(\Theta_i[h] \mid i \in I)$ be an $\mathfrak{m}$-directed family. Then $(\Theta_i \mid i \in I)$ is an $\mathfrak{m}$-directed family. Hence $\bigcup (\Theta_i \mid i \in I) = \Phi \in H$. Since $\Theta_i \subseteq \Phi$ we have that $\Theta_i[h] \subseteq \Phi[h]$, for all $i \in I$. Thus $\bigcup (\Theta_i[h] \mid i \in I) \subseteq \Phi[h]$. Also $\Phi \subseteq \bigcup (\Theta_i \mid i \in I) \subseteq \bigcup (\Theta_i[h] \mid i \in I) \in C(\mathfrak{A}[h])$ containing Φ, and so $\Phi[h] \subseteq \bigcup (\Theta_i[h] \mid i \in I)$. Thus $\bigcup (\Theta_i[h] \mid i \in I) = \Phi[h] \in H[h]$. Hence H is an $\mathfrak{m}$-algebraic closure system. (Note that we utilized Lemma 94.2 and the condition that the characteristic of $\mathfrak{A}$ is at most $\mathfrak{m}$.)

It is much simpler to prove that $K[h]$ is an $\mathfrak{m}$-algebraic closure system. Let $a \in A[h]$. For $a \in A$, set $[a] = [a]_K[h]$. If $a \notin A$, then $a = h(\mathbf{x})$ for a unique $\mathbf{x}$; set $[a] = [\mathbf{x}]_K[h]$. Clearly, $[a]$ satisfies the conditions of Lemma 94.1 and so $K[h]$ is a closure system. The $\mathfrak{m}$-algebraic property can be verified as we did for $H[h]$.

Let $\tau = \sigma[h] \in G[h]$, and let us suppose that $h(\mathbf{a})\tau = h(\mathbf{a})$ for some $h(\mathbf{a}) \notin$ A. Since $h(\mathbf{a})\tau = h(\mathbf{a}\sigma)$, we conclude that $h(\mathbf{a}\sigma) = h(\mathbf{a})$ and so $\mathbf{a}\sigma = \mathbf{a}$. In particular, $a_0\sigma = a_0$. Since $\mathfrak{G}$ is a regular permutation group, it follows that σ is the identity and so $\mathfrak{G}[h]$ also is a regular permutation group.

We have now established that (i) of Definition 1 holds for $\langle \mathfrak{A}[h], H[h], K[h], G[h], \mathfrak{m} \rangle$.

1(ii). Condition 1(ii) trivially holds.

1(iii). Let $\mathbf{a} \in A^\alpha$ and $\sigma \in G$. Since 1(iii) holds for G and K, we have $a_\beta \sigma \in [\mathbf{a}]_K$ for any $\beta < \alpha$. Now let $h(\mathbf{a}) \notin A$. Then $h(\mathbf{a})\sigma[h] = h(\mathbf{a}\sigma) \in [\mathbf{a}]_K[h] = [h(\mathbf{a})]_{K[h]}$. It follows that 1(iii) holds for $G[h]$ and $K[h]$.

1(iv). Let $a, b \in A[h]$ and let $\tau[h] \in G[h]$. Let $\mathbf{x}, \mathbf{y} \in A^\alpha$ and $\alpha > 0$. Since 1(iv) holds for G and H, we have $\Theta_H(\mathbf{x}, \mathbf{y}) = \Theta_H(\mathbf{x}\tau, \mathbf{y}\tau)$. If $a, b \in A$, then $\Theta_{H[h]}(a, b) = [\langle a, b \rangle] = \Theta_H(a, b)[h] = \Theta_H(a\tau, b\tau)[h] = \Theta_{H[h]}(a\tau[h], b\tau[h])$. If $a \in A$ and $b = h(\mathbf{y}) \notin A$, then choose $c = h(\mathbf{w})$ satisfying c Cls a (in $R(h, \mathfrak{A})$, mod H). Then $c\tau \in R(h, \mathfrak{A})$ and 1(iv) implies that $c\tau$ Cls $a\tau$ (in $R(h, \mathfrak{A})$, mod H). Thus $\Theta_{H[h]}(a, b) = (\Theta_H(a, c) \vee \Theta_H(\mathbf{w}, \mathbf{y}))[h] = (\Theta_H(a\tau, c\tau) \vee \Theta_H(\mathbf{w}\tau, \mathbf{y}\tau))[h] = \Theta_{H[h]}(a\tau[h], b\tau[h])$.

If $a = h(\mathbf{x}) \notin A$ and $b = h(\mathbf{y}) \notin A$, then $\Phi(a, b) = \Theta_H(\mathbf{x}, \mathbf{y}) = \Theta_H(\mathbf{x}\tau, \mathbf{y}\tau) = \Phi(a\tau[h], b\tau[h])$, implying that $\Theta_{H[h]}(a, b) = \Theta_{H[h]}(a\tau[h], b\tau[h])$, completing the proof of 1(iv).

1(v). Set $\Xi' = \Xi[h]$. Let $a \in A[h]$ and $\tau = \sigma[h] \in G[h]$, $b = a\tau$. Let $\Theta = \Phi[h] \in H[h]$ and $a \equiv a\tau\ (\Phi[h])$. If $a \in A$, then $a \equiv a\sigma\ (\Phi)$, hence $\Xi \subseteq \Phi$ and $\Xi' \subseteq \Theta$. If $a = h(\mathbf{x}) \notin A$, then $a = h(\mathbf{x}) \equiv h(\mathbf{x})\sigma[h] = h(\mathbf{x}\sigma)(\Theta)$ and so $\mathbf{x} \equiv \mathbf{x}\sigma(\Phi)$, in particular, $x_0 \equiv x_0\sigma\ (\Phi)$. Since G is regular, $x_0 \neq x_0\sigma$, implying $\Xi \subseteq \Phi$ and $\Xi' \subseteq \Theta$.

Conversely, if $\Theta = \Phi[h] \in H[h]$ and $\Xi' \subseteq \Theta$, then we have to prove that for any $a \neq a\sigma[h]$ ($a \in A[h]$, $\sigma[h] \in G[h]$) the congruence $a \equiv a\sigma[h](\Theta)$ holds. In case $a \in A$, this holds because $\Xi \subseteq \Phi$ thus $a \equiv a\sigma(\Phi)$ while if $a \notin A$, then $a = h(\mathbf{x})$ and we conclude $\mathbf{x} \equiv \mathbf{x}\sigma(\Phi)$ and so $h(\mathbf{x}) \equiv h(\mathbf{x}\sigma) = h(\mathbf{x})\sigma[h](\Theta)$.

1(vi). Let $a, b \in A[h]$; we have to find $c \in \mathrm{Orb}(b)$ such that c Cls a (in $\mathrm{Orb}(b)$, mod $H[h]$). If $a, b \in A$, this is obvious.

Let $a = h(\mathbf{x}) \notin A$, $b \in A$. By 1(x) (and Remark 1) applied to $\mathfrak{A}$, we can find a $d = h(\mathbf{y}) \in A$ such that d Cls b (in $R(h, \mathfrak{A})$, mod H). By Remark 2 to Definition 1, there is a $\tau \in G$ such that $\mathbf{y}\tau$ Cls $\mathbf{x}$ (in $\mathrm{Orb}(\mathbf{y})$, mod H). By 1(vi) applied to b and $d\tau$ in $\mathfrak{A}$, there is a $c \in \mathrm{Orb}(b)$ satisfying c Cls $d\tau$ (in $\mathrm{Orb}(b)$, mod H). We claim that c Cls a (in $\mathrm{Orb}(b)$, mod $H[h]$). Indeed, let $\sigma \in G$, $\Theta \in H$, and $b\sigma \equiv a\ (\Theta[h])$. By Lemma 15.3, there exists an $h(\mathbf{w}) \in A$ satisfying $b\sigma \equiv h(\mathbf{w})(\Theta)$ and $\mathbf{x} \equiv \mathbf{w}(\Theta)$. By 1(iv), $b \equiv h(\mathbf{w}\sigma^{-1})(\Theta)$ and so, by the choice of d, $b \equiv d(\Theta)$. Thus $h(\mathbf{w}\sigma^{-1}) \equiv h(\mathbf{y})(\Theta)$ and by H-invertibility, $\mathbf{w}\sigma^{-1} \equiv \mathbf{y}(\Theta)$. Again by 1(iv), $\mathbf{w} \equiv \mathbf{y}\sigma(\Theta)$ which, together with $\mathbf{x} \equiv \mathbf{w}(\Theta)$, yields $\mathbf{x} \equiv \mathbf{y}\sigma(\Theta)$, that is, $a = h(\mathbf{x}) \equiv h(\mathbf{y}\sigma) = d\sigma(\Theta[h])$. By the choice of τ, $a \equiv d\tau(\Theta)$. From $b \equiv d(\Theta)$ we also conclude that $b\tau \equiv d\tau(\Theta)$, hence by the choice of c, $c \equiv d\tau(\Theta)$. Thus $a \equiv d\tau \equiv c(\Theta[h])$, which was to be proved.

Next, let $a \in A$, $b = h(\mathbf{x}) \notin A$. By 1(x), we can choose $d = h(\mathbf{y}) \in A$ such that d Cls a (in $R(h, \mathfrak{A})$, mod H). By Remark 2 to Definition 1, we can find $\sigma \in G$ such that $\mathbf{x}\sigma$ Cls $\mathbf{y}$ (in $\mathrm{Orb}(\mathbf{x})$, mod H). Set $c = h(\mathbf{x}\sigma)$; we claim that c Cls a (in Orb(b), mod $H[h]$). Indeed, let $e = b\tau \in \mathrm{Orb}(b)$ ($\tau \in G$), $\Theta \in H$, and $a \equiv e(\Theta[h])$. By Lemma 15.3, there is an $h(\mathbf{w}) \in A$ satisfying $a \equiv h(\mathbf{w})(\Theta)$ and $\mathbf{w} \equiv \mathbf{x}\tau(\Theta)$. By the choice of d, we obtain $d \equiv a(\Theta)$. Moreover, $e \equiv h(\mathbf{w}) \equiv a \equiv d(\Theta[h])$, hence by the choice of σ, we get $c \equiv d(\Theta[h])$, yielding $a \equiv c(\Theta[h])$, as desired.

For $a, b \notin A$, the statement follows from Remark 2 to Definition 1 and formula (iv) in the proof of 1(i).

1(vii). This is trivial.

1(viii). Any $a \in A$ is in some $R(f, \mathfrak{A}) \subseteq R(f, \mathfrak{A}[h])$. All $a \notin A$ are in $R(h, \mathfrak{A}[h])$.

1(ix). This statement is trivial for all $f \in F - \{h\}$. To prove it for h, let $h(\mathbf{x}) \equiv h(\mathbf{y})(\Theta[h])$ for some $\Theta \in H$. If $h(\mathbf{x}), h(\mathbf{y}) \in A$, then $\mathbf{x} \equiv \mathbf{y}(\Theta)$ since h is H-invertible in $\mathfrak{A}$. If $h(\mathbf{x}) \notin A$ or $h(\mathbf{y}) \notin A$, then by the formulas in (ii), (iii), and (iv) in the proof of 1(i), we obtain $\Phi(h(\mathbf{x}), h(\mathbf{y})) = \Theta_H(\mathbf{x}, \mathbf{y}) \subseteq \Theta$, that is, $\mathbf{x} \equiv \mathbf{y}(\Theta)$, which was to be proved.

1(x). Since $\mathfrak{A}[h] = \langle A[h]; F\rangle$, it is sufficient to verify 2(v). Choose $f, g \in F$, $f \neq g$. If $f \neq h$ and $g \neq h$, then $R(f, \mathfrak{A}[h])$, $R(g, \mathfrak{A}[h]) \subseteq A$ and the statement is immediate since any $\Theta[h]$ ($\Theta \in H$) restricted to A is Θ.

Thus we have two cases to consider: f, h and h, f, where $f \in F$ and $f \neq h$.

First pick a $b \in R(f, \mathfrak{A})$ such that for any $a \in R(h, \mathfrak{A})$ there is a $c \in \mathrm{Orb}(b)$ satisfying c Cls a (in $R(f, \mathfrak{A})$, mod H). This obviously implies that c Cls a (in $R(f, \mathfrak{A}[h])$, mod $H[h]$).

Now take an $a \in R(h, \mathfrak{A}[h])$, $a \notin A$. Then $a = h(\mathbf{x})$. Applying 1(x) to h and f in $\mathfrak{A}$, we find a $d = h(\mathbf{w}) \in A$ satisfying d Cls b (in $R(h, \mathfrak{A})$, mod H). Applying Remark 2 to Definition 1 to $\mathbf{x}$ and $\mathbf{w}$, we find a $\sigma \in G$ such that $\mathbf{w}\sigma$ Cls $\mathbf{x}$ (in $\mathrm{Orb}(\mathbf{w})$, mod H). Finally, by the choice of b, there is a $c \in \mathrm{Orb}(b)$ satisfying c Cls $d\sigma$ (in $R(f, \mathfrak{A})$, mod H). We claim that this c works also for $a = h(\mathbf{x})$, that is, c Cls a (in $R(f, \mathfrak{A}[h])$, mod $H[h]$). To see this, take an arbitrary $e \in R(f, \mathfrak{A}[h]) = R(f, \mathfrak{A})$ and a $\Theta \in H$ satisfying $e \equiv a(\Theta[h])$. Then, by Lemma 15.3, there is an $h(\mathbf{y}) \in A$ satisfying $e \equiv h(\mathbf{y})(\Theta)$ and $\mathbf{y} \equiv \mathbf{x}(\Theta)$. By the choice of b, there is a $b' = b\tau$ ($\tau \in G$) such that b' Cls $h(\mathbf{y})$ (in $R(f, \mathfrak{A})$, mod H). In particular, since $e \equiv h(\mathbf{y})(\Theta)$, we have $b' \equiv h(\mathbf{y})(\Theta)$. So by 1(iv), $\mathbf{b} \equiv h(\mathbf{y}\tau^{-1})(\Theta)$. By the choice of d, this implies that $b \equiv d(\Theta)$, and so $d = h(\mathbf{w}) \equiv b \equiv h(\mathbf{y}\tau^{-1})(\Theta)$. By H-invertibility, $\mathbf{w} \equiv \mathbf{y}\tau^{-1}(\Theta)$, and again by 1(iv), $\mathbf{y} \equiv \mathbf{w}\tau(\Theta)$. Recall that $\mathbf{x} \equiv \mathbf{y}(\Theta)$, thus $\mathbf{x} \equiv \mathbf{w}\tau(\Theta)$. By the choice of σ, $\mathbf{w}\sigma \equiv \mathbf{x}(\Theta)$ and so $d\sigma \equiv a(\Theta[h])$. Thus $b' \equiv h(\mathbf{y}) \equiv h(\mathbf{x}) \equiv h(\mathbf{w}\sigma) = d\sigma(\Theta[h])$ and $b' \equiv d\sigma(\Theta)$, and so, by the choice of c, $c \equiv d\sigma(\Theta)$. Therefore, $c \equiv a(\Theta[h])$, which was to be proved.

Secondly, we consider h and f. Applying 1(x) to $\mathfrak{A}$ we obtain a $b \in R(h, \mathfrak{A})$ such that for every $a \in R(f, \mathfrak{A}) = R(f, \mathfrak{A}[h])$ there exists a $c \in \mathrm{Orb}(b)$ satisfying c Cls a (in $R(h, \mathfrak{A})$, mod H). We claim that c Cls a (in $(R(h, \mathfrak{A}[h])$, mod $H[h]$). To see this, take an $h(\mathbf{x}) \in A[h]$ and $\Theta \in H$ satisfying $h(\mathbf{x}) \equiv a(\Theta[h])$. Then, by Lemma 15.3, there exists an $h(\mathbf{y}) \in A$ such that $\mathbf{x} \equiv \mathbf{y}(\Theta)$ and $h(\mathbf{y}) \equiv a(\Theta)$. By the choice of c, we obtain $c \equiv a(\Theta)$ and so $c \equiv a(\Theta[h])$, which was to be proved.

This concludes the verification of 1(i)–(x) for the expanded quintuple. We have also verified 2(v). Thus we also have 2(i). 2(ii) is trivial. 2(iii) and (iv) are trivial from Lemma 15.3. This completes the proof of Lemma 1.

Let us be given for all $\gamma < \alpha$ a Cls-quintuple $\langle \mathfrak{A}_\gamma, H_\gamma, K_\gamma, G_\gamma, \mathfrak{m} \rangle$ with the property that whenever $\gamma \leqq \delta < \alpha$, then $\langle \mathfrak{A}_\delta, H_\delta, K_\delta, G_\delta, \mathfrak{m} \rangle$ is a Cls-expansion of $\langle \mathfrak{A}_\gamma, H_\gamma, K_\gamma, G_\gamma, \mathfrak{m} \rangle$. Let $\mathfrak{A}_\alpha$ be the limit of $\mathfrak{A}_\gamma, \gamma < \alpha$. For $\Theta \in H_0$ let $\Theta_\gamma = [\Theta]_{H_\gamma}$ and $\Theta_\alpha = \bigcup (\Theta_\gamma \mid \gamma < \alpha)$; for $D \in K_0$, let $D_\gamma = [D]_{K_\gamma}$ and $D_\alpha = \bigcup (D_\gamma \mid \gamma < \alpha)$; for $\sigma \in G_0$, let σ_γ be the unique member of G_γ extending σ and let σ_α be the unique map of A_α extending all $\sigma_\gamma, \gamma < \alpha$. Set $H_\alpha = \{\Theta_\alpha \mid \Theta \in H_0\}$, $K_\alpha = \{D_\alpha \mid D \in K_0\}$, $G_\alpha = \{\sigma_\alpha \mid \sigma \in G_0\}$. Then we call $\langle \mathfrak{A}_\alpha, H_\alpha, K_\alpha, G_\alpha, \mathfrak{m} \rangle$ the *limit of* $\langle \mathfrak{A}_\gamma, H_\gamma, K_\gamma, G_\gamma, \mathfrak{m} \rangle, \gamma < \alpha$.

Lemma 2. *Let* $\langle \mathfrak{A}_\alpha, H_\alpha, K_\alpha, G_\alpha, \mathfrak{m} \rangle$ *be the limit of* $\langle \mathfrak{A}_\gamma, H_\gamma, K_\gamma, G_\gamma, \mathfrak{m} \rangle$. *Then* $\langle \mathfrak{A}_\alpha, H_\alpha, K_\alpha, G_\alpha, \mathfrak{m} \rangle$ *is a* Cls-*quintuple and it is a* Cls-*expansion of* $\langle \mathfrak{A}_\gamma, H_\gamma, K_\gamma, G_\gamma, \mathfrak{m} \rangle$ *for any* $\gamma < \alpha$.

Proof. For $a, b \in A_\alpha$ find a $\gamma<\alpha$ such that $a, b \in A_\gamma$ and define $[\langle a, b\rangle]=[\Theta_{H_\gamma}(a, b)]_{H_\alpha}$. This evidently satisfies the hypotheses of Lemma 94.1. Since $\langle H_\alpha; \subseteq\rangle \cong \langle H_0; \subseteq\rangle$, it follows that H_α is a closure system. To check that it is $\mathfrak{m}$-algebraic, let $X \subseteq H_\alpha$ be $\mathfrak{m}$-directed and $\Theta=\bigcup X$. Let $X_\gamma=\{\Phi_\gamma \mid \Phi \in X\}$, where Φ_γ is the restriction of Φ to A_γ. Then X_γ is $\mathfrak{m}$-directed and, since H_γ is $\mathfrak{m}$-algebraic, $\Theta_\gamma=\bigcup X_\gamma \in H_\gamma$. Thus $\Theta_\alpha=\bigcup(\Theta_\gamma \mid \gamma<\alpha) \in H_\alpha$. Obviously, $\Theta_\alpha=\Theta$ and so $\Theta \in H_\alpha$.

We can prove similarly that K_α is an $\mathfrak{m}$-algebraic closure system. G_α is obviously regular, verifying 1(i). The remainder of the conditions of Definitions 1 and 2 are trivial to check; they always deal with elements that are in some A_γ, $\gamma<\alpha$. Only 1(ix) deserves one more comment: if $f(\mathbf{a}) \equiv f(\mathbf{b})(\Theta)$ for $\Theta \in H_\alpha$, then $\mathbf{a}, \mathbf{b} \in D(f, \mathfrak{A}_\alpha)=\bigcup(D(f, \mathfrak{A}_\gamma) \mid \gamma<\alpha)$. Hence, again we conclude that $\mathbf{a}, \mathbf{b} \in D(f, \mathfrak{A}_\gamma)$ for some $\gamma<\alpha$ and we proceed in $\mathfrak{A}_\gamma$.

§96. Cls-EXPANSIONS WITH TWO ORBITS

The following observations on orbits are trivial. (In this lemma—and for the rest of the appendix—Ξ refers to the congruence relation given by Definition 95.1(v).)

Lemma 1. *Let* $\langle\mathfrak{A}, H, K, G, \mathfrak{m}\rangle$ *be a* Cls-*quintuple and let* $\mathbf{a}, \mathbf{b} \in A^\alpha$ *for some ordinal* $\alpha>0$.

(i) *If* $\mathbf{a}$ Cls $\mathbf{b}$ (in Orb($\mathbf{a}$), mod H), *then* $\mathbf{b}$ Cls $\mathbf{a}$ (*in* Orb($\mathbf{b}$), mod H).

(ii) *If* $\mathbf{a}$ Cls $\mathbf{b}$ (in Orb($\mathbf{a}$), mod H), *then, for all* $\mathbf{a}' \in$ Orb($\mathbf{a}$), $\Theta_H(\mathbf{a}', \mathbf{b})=\Theta_H(\mathbf{a}, \mathbf{b}) \vee \Xi$.

(iii) *If* $a_\gamma=b_\gamma$ *for some* $\gamma<\alpha$, *then* $\mathbf{a}$ Cls $\mathbf{b}$ (in Orb($\mathbf{a}$), mod H).

Now we come to our basic construction which corresponds to the construction of $\mathfrak{A}'$ in §17 without g_1, g_2, g_3. This construction is derived in part from those in B. Jónsson [1951] and W. A. Lampe [1972 a].

In what follows, let $\langle\mathfrak{A}, H, K, G, \mathfrak{m}\rangle$ be a Cls-quintuple. We choose $a_0, b_0 \in A$, an ordinal α, $0<\alpha<\mathfrak{m}$, $\mathbf{a}_i, \mathbf{b}_i \in A^\alpha$, $i=1,2,3$, and three ($\mathfrak{m}$-compact) congruence relations, $\Theta_i=\Theta_H(\mathbf{a}_i, \mathbf{b}_i)$, $i=1,2,3$. We set $\Theta_0=\Theta_H(a_0, b_0)$ and $\Theta_0'=H$-Dist(a_0, Orb(b_0)).

Take two elements x, y such that $\{x\} \times G$, $\{y\} \times G$, and A are pairwise disjoint. We set $r=\langle x, 1\rangle$, $s=\langle y, 1\rangle$ and for $\sigma \in G$ we define $r\sigma=\langle x, \sigma\rangle$, $s\sigma=\langle y, \sigma\rangle$, and

$$A^* = A \cup \{r\sigma \mid \sigma \in G\} \cup \{s\sigma \mid \sigma \in G\}.$$

We have already defined the action of G on A^* so

$$A^* = A \cup \text{Orb}(r) \cup \text{Orb}(s).$$

Under the hypotheses of Lemma 2, the definitions that follow define, for every $\Theta \in H$, an equivalence relation Θ^* on A^*. (Figures 2–6 illustrate Θ^*; the dotted areas are congruence classes of Θ^* that are not congruence classes of Θ. Since Θ^* restricted to A is Θ, this is the same as the congruence classes of Θ^* not completely contained in A.)

$(C, \varnothing)$. Let $\Theta \nsupseteq \Theta_1, \Theta_2, \Theta_3$.

(i) If $\Theta \nsupseteq \Xi$, then $[x]\Theta = [x]\Theta^*$ for $x \in A$ and $[x]\Theta^* = \{x\}$ for $x \in A^* - A$.

(ii) If $\Theta \supseteq \Xi$, then $[x]\Theta = [x]\Theta^*$ for $x \in A$ and $[x]\Theta^* = \mathrm{Orb}(x)$ for $x \in A^* - A$.

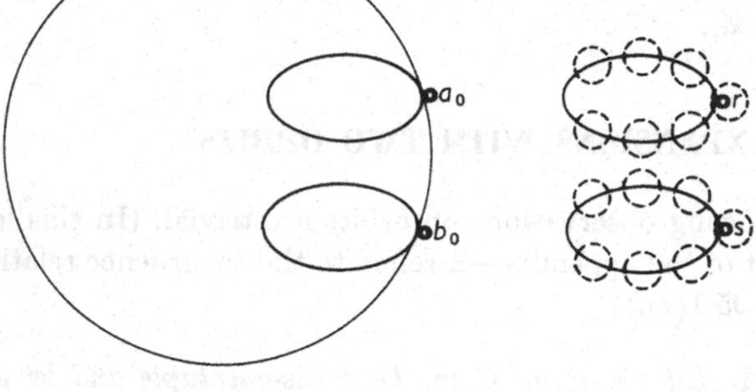

(i) $\Theta \nsupseteq \Xi$

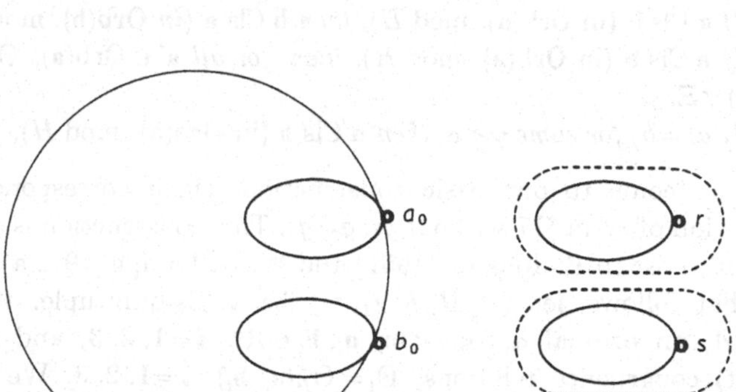

(ii) $\Theta \supseteq \Xi$

$(C, \varnothing)$

Fig. 2

$(C, 1)$. Let $\Theta \supseteq \Theta_1$, $\Theta \not\supseteq \Theta_2$, Θ_3.

(i) If $\Theta \not\supseteq \Xi$, then $[x]\Theta = [x]\Theta^*$ for $x \in A - [\mathrm{Orb}(a_0)]\Theta$; for $x = a_0\sigma$ $(\sigma \in G)$, $[a_0\sigma]\Theta^* = [a_0\sigma]\Theta \cup \{r\sigma\}$; for $x \in \mathrm{Orb}(s)$, $[x]\Theta^* = \{x\}$.

(ii) If $\Theta \supseteq \Xi$, then $[x]\Theta = [x]\Theta^*$ for $x \in A - [\mathrm{Orb}(a_0)]\Theta$; $[a_0]\Theta^* = [a_0]\Theta \cup \mathrm{Orb}(r)$; $[s]\Theta^* = \mathrm{Orb}(s)$.

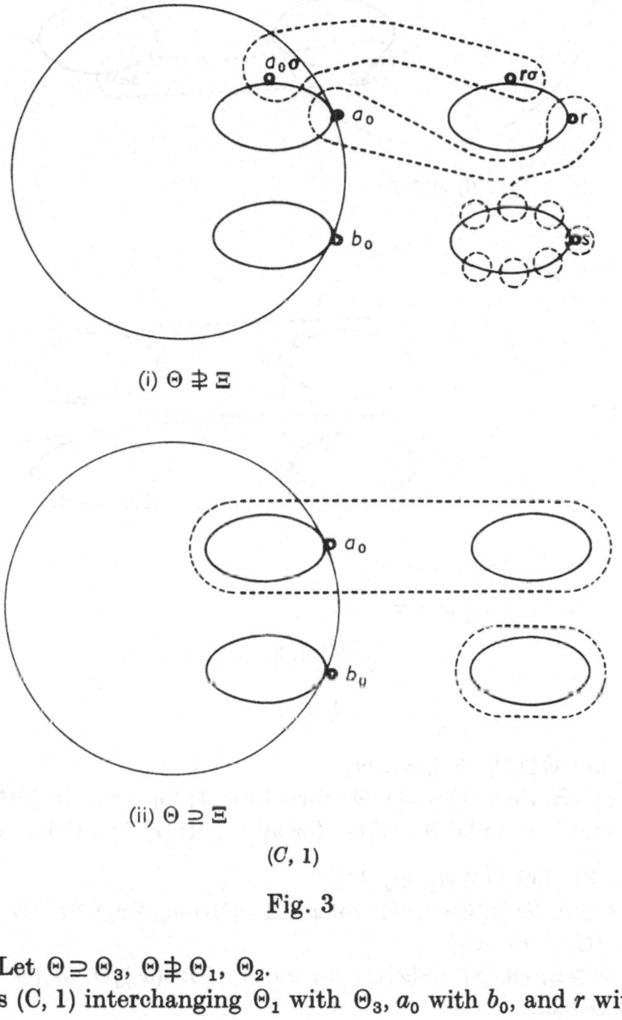

(i) $\Theta \not\supseteq \Xi$

(ii) $\Theta \supseteq \Xi$

$(C, 1)$

Fig. 3

$(C, 3)$. Let $\Theta \supseteq \Theta_3$, $\Theta \not\supseteq \Theta_1$, Θ_2.
Same as (C, 1) interchanging Θ_1 with Θ_3, a_0 with b_0, and r with s.

$(C, 1, 3)$. Let $\Theta \supseteq \Theta_1$, Θ_3, $\Theta \not\supseteq \Theta_2$.

(i) If $\Theta \not\supseteq \Xi$, then $[x]\Theta = [x]\Theta^*$ for $x \in A - [\mathrm{Orb}(a_0) \cup \mathrm{Orb}(b_0)]\Theta$; for $x = a_0\sigma$ $(\sigma \in G)$, $[x]\Theta^* = [x]\Theta \cup \{r\sigma\}$; for $x = b_0\sigma$ $(\sigma \in G)$, $[x]\Theta^* = [x]\Theta \cup \{s\sigma\}$.

(ii) If $\Theta \supseteq \Xi$, then $[x]\Theta = [x]\Theta^*$ for $x \in A - [\mathrm{Orb}(a_0) \cup \mathrm{Orb}(b_0)]\Theta$; $[a_0]\Theta^* = [a_0]\Theta \cup \mathrm{Orb}(r)$; $[b_0]\Theta^* = [b_0]\Theta \cup \mathrm{Orb}(s)$.

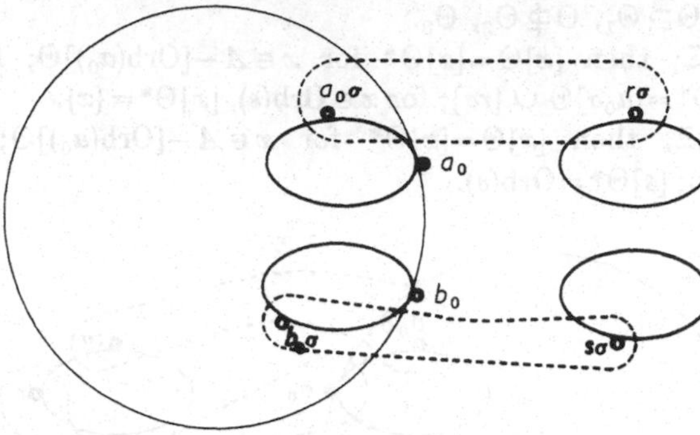

(i) $\Theta \not\supseteq \Xi$

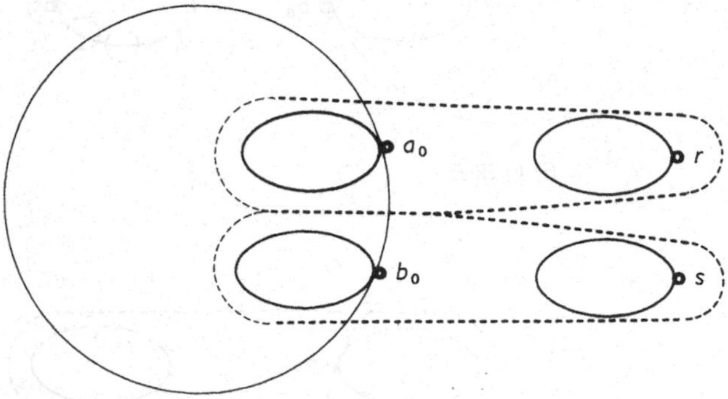

(ii) $\Theta \supseteq \Xi$

(*C*, 1, 3)

Fig. 4

(C, 2). Let $\Theta \supseteq \Theta_2$, $\Theta \not\supseteq \Theta_1$, Θ_3.

(i) If $\Theta \not\supseteq \Xi$, then $[x]\Theta = [x]\Theta^*$ for all $x \in A$; for $\sigma \in G$, $[r\sigma]\Theta^* = \{r\sigma, s\sigma\}$.

(ii) If $\Theta \supseteq \Xi$, then $[x]\Theta = [x]\Theta^*$ for all $x \in A$; $[r]\Theta^* = \mathrm{Orb}(r) \cup \mathrm{Orb}(s)$.

(C, 1, 2, 3). Let $\Theta \supseteq \Theta_1$, Θ_2, Θ_3.

(i) If $\Theta \not\supseteq \Xi$, let $[x]\Theta = [x]\Theta^*$ for $x \in A - [\mathrm{Orb}(a_0)]\Theta$; for $\sigma \in G$, $[a_0\sigma]\Theta^* = [\{a_0\sigma, b_0\sigma\}]\Theta \cup \{r\sigma, s\sigma\}$;

(ii) If $\Theta \supseteq \Xi$, let $[x]\Theta = [x]\Theta^*$ for $x \in A - [\mathrm{Orb}(a_0)]\Theta$; $[a_0]\Theta^* = [a_0]\Theta \cup \mathrm{Orb}(r) \cup \mathrm{Orb}(s)$.

Set $\Sigma = \Theta_0 \vee \Theta_1 \vee \Theta_2 \vee \Theta_3$.

Lemma 2. *Let us assume that*

$$\Sigma = \Theta_1 \vee \Theta_2 = \Theta_2 \vee \Theta_3 \qquad (\mathrm{P}_1)$$

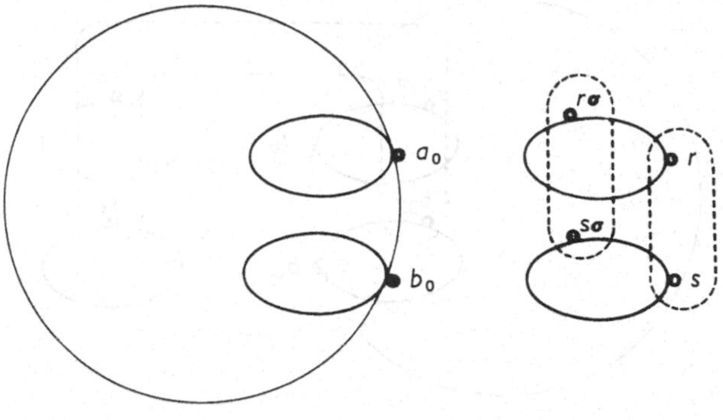

(i) $\Theta \not\supseteq \Xi$

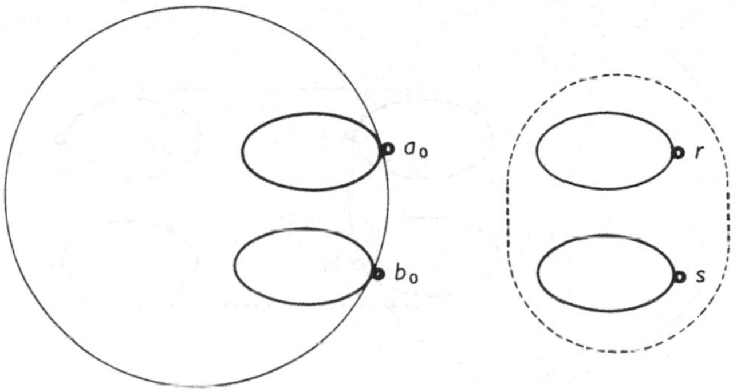

(ii) $\Theta \supseteq \Xi$

$(C, 2)$

Fig. 5

and

$$\Sigma = \Theta_1 \vee \Theta_0{}' \vee \Theta_3. \qquad (\mathrm{P}_2)$$

Then each $\Theta \in H$ *satisfies exactly one of the hypotheses of* $(C, \varnothing)$, $(C, 1)$, $(C, 3)$, $(C, 1, 3)$, $(C, 2)$, $(C, 1, 2, 3)$ *and the one condition that applies defines an extension* Θ^* *of* Θ.

Proof. If Θ contains no Θ_i, $i = 1, 2, 3$, Θ^* defined by $(C, \varnothing)$ is an extension of Θ. If Θ contains exactly one of Θ_1, Θ_2, and Θ_3, or all of them, then $(C, 1)$, $(C, 2)$, $(C, 3)$, and $(C, 1, 2, 3)$ will define an extension Θ^* of Θ. (To see this for $(C, 1, 2, 3)$ we have to use that $\Theta_0 \subseteq \Theta_1 \vee \Theta_2 \vee \Theta_3$, which follows from (P_1).)

So let Θ contain exactly two of Θ_1, Θ_2, Θ_3. By (P_1), the hypothesis of $(C, 1, 3)$ is the only possibility, that is $\Theta \supseteq \Theta_1$, Θ_3 and $\Theta \not\supseteq \Theta_2$. Since, by

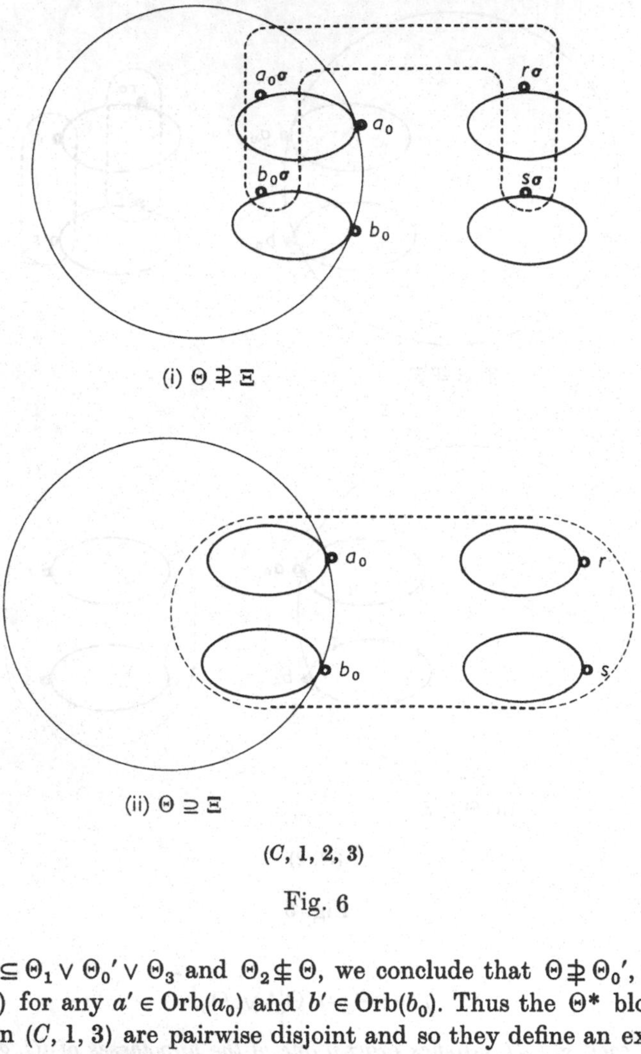

(C, 1, 2, 3)

Fig. 6

(P_2), $\Theta_2 \subseteq \Theta_1 \vee \Theta_0' \vee \Theta_3$ and $\Theta_2 \nsubseteq \Theta$, we conclude that $\Theta \not\supseteq \Theta_0'$, that is, $a' \not\equiv b'(\Theta)$ for any $a' \in \mathrm{Orb}(a_0)$ and $b' \in \mathrm{Orb}(b_0)$. Thus the Θ^* blocks described in $(C, 1, 3)$ are pairwise disjoint and so they define an extension Θ^* of Θ, completing the proof of Lemma 2.

Our next task is to show that $H^* = \{\Theta^* \mid \Theta \in H\}$ is an $\mathfrak{m}$-algebraic closure system. As usual, the crucial step is the proof of existence of the $\Phi(a, b)$:

Lemma 3. *Let us assume* (P_1) *and* (P_2). *For all* $a, b \in A^*$ *we define an* $\mathfrak{m}$*-compact* $\Phi(a, b) = \Phi(b, a) \in H$ *by the following rules:*

(i) *if* $a, b \in A$, $\Phi(a, b) = \Theta_H(a, b)$;
(ii) *if* $a \in A$, $b = r\sigma$ $(\sigma \in G)$, *then* $\Phi(a, b) = \Theta_H(a, a_0\sigma) \vee \Theta_1$;

(iii) *if* $a \in A$, $b = s\sigma$ $(\sigma \in G)$, *then* $\Phi(a, b) = \Theta_H(a, b_0\sigma) \vee \Theta_3$;

(iv) *if* $a = r\sigma$, $b = r\tau$ $(\sigma, \tau \in G)$, *then* $\Phi(a, b) = \omega$ *for* $\sigma = \tau$ *and* $\Phi(a, b) = \Xi$ *for* $\sigma \neq \tau$;

(v) *if* $a = r\sigma$, $b = s\tau$ $(\sigma, \tau \in G)$, *then* $\Phi(a, b) = \Theta_2$ *for* $\sigma = \tau$ *and* $\Phi(a, b) = \Theta_2 \vee \Xi$ *for* $\sigma \neq \tau$.

Then, for $\Theta \in H$, $a \equiv b(\Theta^*)$ *iff* $\Phi(a, b) \subseteq \Theta$.

Proof. Firstly, observe that $\Phi(a, b)$, as defined in (i)–(v), is indeed an $\mathfrak{m}$-compact member of H.

Now let $\Theta \in H$ and $\Phi(a, b) \subseteq \Theta$. Then we conclude $a \equiv b(\Theta^*)$: in case (i), by Lemma 2; in case (ii), $(C, 1)$, or $(C, 1, 3)$, or $(C, 1, 2, 3)$ apply and each one gives $a_0\sigma \equiv r\sigma(\Theta^*)$ and so, using $\Theta_H(a, a_0\sigma) \subseteq \Theta$, we obtain $a \equiv a_0\sigma \equiv r\sigma(\Theta^*)$; in case (iii), we proceed similarly; in case (iv), we can assume $\sigma \neq \tau$, and observe that $r\sigma \equiv r\tau(\Theta^*)$ whenever $\Theta \supseteq \Xi$, regardless of which $(C, \ )$ condition applies; in case (v), $(C, 2)$ or $(C, 1, 2, 3)$ apply and in either case $a \equiv b(\Theta^*)$.

Finally, let $\Theta \in H$ and $a \equiv b(\Theta^*)$. We take the five cases separately.

(i) Then $\Phi(a, b) \subseteq \Theta$ follows from Lemma 2.

(ii) $r\sigma$ is congruent to some element of A modulo Θ^* iff $(C, 1)$, $(C, 1, 3)$, or $(C, 1, 2, 3)$ applies, that is, iff $\Theta_1 \subseteq \Theta$. If $\Theta_1 \subseteq \Theta$, then in all three cases, $a_0\sigma \in [r\sigma]\Theta^*$, hence $a \equiv a_0\sigma(\Theta)$, implying that $\Phi(a, b) \subseteq \Theta$.

(iii) Proceed as under (ii).

(iv) We can assume that $\sigma \neq \tau$. Under any $(C, \ )$ condition, $r\sigma \equiv r\tau(\Theta^*)$ iff $\Theta \supseteq \Xi$, hence $r\sigma \equiv r\tau(\Theta^*)$ implies that $\Theta \supseteq \Phi(a, b)$.

(v) $[r\sigma]\Theta^* \cap \mathrm{Orb}(s) \neq \varnothing$ iff $(C, 2)$ or $(C, 1, 2, 3)$ applies, that is, iff $\Theta \supseteq \Theta_2$. If $\sigma = \tau$, then $\Phi(a, b) = \Theta_2 \subseteq \Theta$, as required. If $\sigma \neq \tau$, then both in $(C, 2)$ and $(C, 1, 2, 3)$ only in subcase (ii) do we have $r\sigma \equiv s\tau(\Theta^*)$, thus we must have $\Theta \supseteq \Xi$. Therefore, $\Theta \supseteq \Theta_2 \vee \Xi = \Phi(a, b)$. This completes the proof of Lemma 3.

Lemma 4. H^* is an $\mathfrak{m}$-algebraic closure system.

Proof. By the definition of Θ and by Lemma 2, $\Theta \to \Theta^*$ is an isomorphism between $\langle H; \subseteq \rangle$ and $\langle H^*; \subseteq \rangle$ and so $\langle H^*; \subseteq \rangle$ is a complete lattice. For $a, b \in A$, set $[\langle a, b \rangle] = (\Phi(a, b))^*$. Then, by Lemma 3, $[\langle a, b \rangle]$ satisfies 94.1(i) and 94.1(ii) and so H^* is a closure system.

To show that H^* is $\mathfrak{m}$-algebraic, let $X \subseteq H^*$ be $\mathfrak{m}$-directed. Set $Y = \{\Theta \mid \Theta \in H \text{ and } \Theta^* \in X^*\}$. Then Y is $\mathfrak{m}$-directed, hence $\Theta = \bigcup Y \in H$. We claim that $\Theta^* = \bigcup X$. Indeed, if $a \equiv b(\bigcup X)$, then $a \equiv b(\Phi^*)$ for some $\Phi^* \in X$ and so $\Phi(a, b) \subseteq \Phi \subseteq \Theta$. Thus $a \equiv b(\Theta^*)$. Conversely, let $a \equiv b(\Theta^*)$. Then $\Phi(a, b) \subseteq \Theta = \bigcup Y = \bigvee Y$. Since $\Phi(a, b)$ is $\mathfrak{m}$-compact, there exists a $Y_1 \subseteq Y$ such that $\Phi(a, b) \leqq \bigvee Y_1$ and $|Y_1| < \mathfrak{m}$. But Y is $\mathfrak{m}$-directed, hence there exists a $\Phi \in Y$ such that $\Psi \subseteq \Phi$ for all $\Psi \in Y_1$. Thus $\Phi(a, b) \leqq \bigvee Y_1 \leqq \Phi \in Y$ and so $a \equiv b(\Phi^*)$, implying that $a \equiv b(\bigcup X)$, since $\Phi^* \in X$.

This verifies the claim. The claim implies that $\bigcup X \in H^*$, proving that H^* is $\mathfrak{m}$-algebraic.

For every $D \in K$ we define a $D^* \subseteq A^*$ as follows (the motivation for this definition should be clear from §17 of Chapter 2 or from the definition of $\mathfrak{A}^{**}$ given below):

$$D^* = \begin{cases} D, & \text{if } \mathbf{b}_1, \mathbf{b}_2 \notin D^\alpha; \\ D \cup \mathrm{Orb}(r), & \text{if } \mathbf{b}_1 \in D^\alpha, \mathbf{b}_2 \notin D^\alpha; \\ D \cup \mathrm{Orb}(s), & \text{if } \mathbf{b}_1 \notin D^\alpha, \mathbf{b}_2 \in D^\alpha; \\ D \cup \mathrm{Orb}(s) \cup \mathrm{Orb}(r), & \text{if } \mathbf{b}_1, \mathbf{b}_2 \in D^\alpha. \end{cases}$$

Lemma 5. *For $D \in K$, D^* is an extension of D and $K^* = \{D^* \mid D \in K\}$ is an $\mathfrak{m}$-algebraic closure system.*

Proof. We claim that for $X \subseteq K$,

$$(\bigcap X)^* = \bigcap X^*,$$

where $X^* = \{D^* \mid D \in X\}$. Indeed, $\bigcap X \subseteq D$ for all $D \in X$, hence $(\bigcap X)^* \subseteq D^*$ and so $(\bigcap X)^* \subseteq \bigcap X^*$. Now let $a \in \bigcap X^*$. If $a \in A$, then $a \in \bigcap X$ and so $a \in (\bigcap X)^*$; so let $a \notin A$, say $a = r\sigma$, $\sigma \in G$ ($a = s\sigma$ can be handled similarly). Observe that $a \in D^*$ for $D \in H$ iff $[\mathbf{b}_1]_K \subseteq D$. Thus $a \in \bigcap X^*$ iff $[\mathbf{b}_1]_K \subseteq D$ for all $D \in K$, which in turn is equivalent to $[\mathbf{b}_1]_K \subseteq \bigcap X$; from this we conclude that $a \in (\bigcap X)^*$. This proves the claim. This claim obviously implies that K^* is a closure system.

To prove that K^* is $\mathfrak{m}$-algebraic, take an $\mathfrak{m}$-directed $X \subseteq K^*$; we have to show that $\bigcup X \in K^*$. Define $Y = \{D \mid D^* \in X\}$. Then Y is $\mathfrak{m}$-directed and so $\bigcup Y \in K$. Now $(\bigcup Y)^* = \bigcup X$ can be verified using the argument of Lemma 4 since $[\mathbf{b}_1]_K$ and $[\mathbf{b}_2]_K$ are $\mathfrak{m}$-compact. This completes the proof of Lemma 5.

Let $\mathfrak{A}^* = \langle A^*; F^* \rangle$ be defined as follows: $F^* = F \cup \{g_r, g_s\}$; $\mathfrak{A}$ is a subalgebra of $\mathfrak{A}^*$ and $D(f, \mathfrak{A}) = D(f, \mathfrak{A}^*)$ for all $f \in F$; $D(g_r, \mathfrak{A}^*) = \mathrm{Orb}(r)$ and $g_r(x) = x$ for $x \in \mathrm{Orb}(r)$; $D(g_s, \mathfrak{A}^*) = \mathrm{Orb}(s)$ and $g_s(x) = x$ for $x \in \mathrm{Orb}(s)$. Let G^* be the action of G on A^*, that is, $G^* = \{\sigma^* \mid \sigma \in G\}$ and $a\sigma^* = a\sigma$ for $a \in A$ and $(r\tau)\sigma^* = r(\tau\sigma)$, $(s\tau)\sigma^* = s(\tau\sigma)$.

Lemma 6. *If the congruences satisfy conditions* (P_1) *and* (P_2) *of Lemma* 2, *then* $\langle \mathfrak{A}^*, H^*, K^*, G^*, \mathfrak{m} \rangle$ *is a* Cls-*expansion of* $\langle \mathfrak{A}, H, K, G, \mathfrak{m} \rangle$.

Proof. We verify that $\langle \mathfrak{A}^*, H^*, K^*, G^*, \mathfrak{m} \rangle$ satisfies 95.1(i)–95.1(x). In view of the definition of g_r and g_s, evidently, $H^* \subseteq C(\mathfrak{A}^*)$ and $K^* \subseteq \mathscr{S}(\mathfrak{A}^*)$.

Thus, by Lemmas 4 and 5, 95.1(i) is satisfied. 95.1(ii) is trivial. 95.1(iii) is evident for $a \in A$, while $D \cap (A^* - A) = \varnothing$ or $\text{Orb}(r)$ or $\text{Orb}(r) \cup \text{Orb}(s)$ for all $D \in K^*$, so 95.1(iii) holds.

To verify 95.1(iv), take $a, b \in A^*$ and $\xi^* \in G^*$. Observe that if a and b are situated as described by 3(i)–3(v), then $a\xi^*$ and $b\xi^*$ belong to the same classification. Hence, listed according to 3(i)–3(v),

(i) $\Phi(a, b) = \Theta_H(a, b)$ and $\Phi(a\xi^*, b\xi^*) = \Theta_H(a\xi^*, b\xi^*)$;
(ii) $\Phi(a, b) = \Theta_H(a, a_0\sigma) \vee \Theta_1$ and $\Phi(a\xi^*, b\xi^*) = \Theta_H(a\xi^*, a_0\sigma\xi^*) \vee \Theta_1$;
(iii) $\Phi(a, b) = \Theta_H(a, b_0\sigma) \vee \Theta_3$ and $\Phi(a\xi^*, b\xi^*) = \Theta_H(a\xi^*, b_0\sigma\xi^*) \vee \Theta_1$;
(iv) if $\sigma = \tau$, then $\Phi(a, b) = \omega$ and $\Phi(a\xi^*, b\xi^*) = \omega$; if $\sigma \neq \tau$, then $\Phi(a, b) = \Xi$ and $\Phi(a\xi^*, b\xi^*) = \Xi$, since G^* is regular;
(v) if $\sigma = \tau$, then $\Phi(a, b) = \Theta_2$ and $\Phi(a\xi^*, b\xi^*) = \Theta_2$; if $\sigma \neq \tau$, then $\Phi(a, b) = \Theta_2 \vee \Xi$ and $\Phi(a\xi^*, b\xi^*) = \Theta_2 \vee \Xi$.

Since 95.1(iv) holds for $\langle \mathfrak{A}, H, K, G, \mathfrak{m} \rangle$, $\Phi(a, b) = \Phi(a\xi^*, b\xi^*)$ for all $a, b \in A$ and $\xi^* \in G$, hence

$$\Theta_{H^*}(a, b) = (\Phi(a, b))^* = (\Phi(a\xi^*, b\xi^*))^* = \Theta_{H^*}(a\xi^*, b\xi^*),$$

verifying 95.1(iv).

95.1(v) is obvious with Ξ^*.

To verify 95.1(vi) take $a, b \in A$. If $a, b \in A$, take $c \in \text{Orb}(b)$ satisfying c Cls a (in $\text{Orb}(b)$, mod H). Obviously, c Cls a (in $\text{Orb}(b)$, mod H^*).

Let $a \in A$, $b \notin A$, say $b = r\sigma$, $\sigma \in G$ ($b = s\sigma$ can be handled similarly). There is a $\tau \in G$ satisfying $a_0\tau$ Cls a (in $\text{Orb}(a_0)$, mod H). We claim that $r\tau$ Cls a (in $\text{Orb}(r)$, mod H^*). Indeed, for any $\xi \in G$, $\Phi(a, r\xi) = \Theta_H(a, a_0\xi) \vee \Theta_1$ and $a_0\tau \equiv a(\Theta_H(a, a_0\xi))$. Hence $\Theta_{H^*}(r\tau, a) = (\Phi(r\tau, a))^* = (\Theta_H(a, a_0\tau) \vee \Theta_1)^* \leqq (\Theta_H(a, a_0\xi) \vee \Theta_1)^* = (\Phi(r\xi, a))^* = \Theta_{H^*}(r\xi, a)$.

Let $a \notin A$, $b \in A$, say $a = r\sigma$, $\sigma \in G$. Then there exists a $c \in \text{Orb}(b)$, such that c Cls $a_0\sigma$ (in $\text{Orb}(b)$, mod H). We claim that c Cls $r\sigma$ (in $\text{Orb}(b)$, mod H^*). Indeed, if $c' \in \text{Orb}(b)$, then $\Theta_{H^*}(c', r\sigma) = (\Phi(c', r\sigma))^* = (\Theta_H(c', a_0\sigma) \vee \Theta_1)^* \geqq$ (since $\Theta_H(c', a_0\sigma) \geqq \Theta_H(c, a_0\sigma)$) $\geqq (\Theta_H(c, a_0\sigma) \vee \Theta_1)^* = (\Phi(c, r\sigma))^* = \Theta_{H^*}(c, r\sigma)$.

Let $a, b \notin A$. If $a = r\tau$, $b = r\sigma$ ($\tau, \sigma \in G$), then we can choose $c = a$. So let $a = r\tau$ and $b = s\sigma$. Then $s\tau$ Cls a (in $\text{Orb}(b)$, mod H^*), since $\Theta_{H^*}(s\tau, r\tau) = \Theta_2$ and, for any $\xi \neq \tau$, $\Theta_{H^*}(s\xi, r\tau) = (\Theta_2 \vee \Xi)^* \geqq \Theta_{H^*}(s\tau, r\tau)$. This concludes the proof of 95.1(vi).

95.1(vii)–95.1(ix) are trivial.

95.1(x) (and 95.2(v)) is clear for $f, g \in F$ since $R(f, \mathfrak{A}^*)$, $R(g, \mathfrak{A}^*) \subseteq A$.

Let $f \in F$ and $g \notin F$, say $g = g_r$ ($g = g_s$ can be handled similarly). There is a $b \in R(f, \mathfrak{A})$ satisfying b Cls a_0 (in $R(f, \mathfrak{A})$, mod H). Take an $a \in R(g_r, \mathfrak{A}^*)$, that is, $a = r\sigma$ for some $\sigma \in G$. Then we claim that $c = b\sigma$ Cls a (in $R(f, \mathfrak{A}^*)$, mod H^*). Indeed, let $d \in R(f, \mathfrak{A}^*) = R(f, \mathfrak{A})$, then

$$\begin{aligned}\Theta_{H^*}(d, a) &= (\Phi(d, a))^* = (\Theta_H(d, a_0\sigma) \vee \Theta_1)^* = (\text{by } 95.1(\text{iv}))\\ &= (\Theta_H(d\sigma^{-1}, a_0) \vee \Theta_1)^* \geqq (\text{since } d\sigma^{-1} \in R(f, \mathfrak{A}))\\ &\geqq (\Theta_H(b, a_0) \vee \Theta_1)^* = (\text{by } 95.1(\text{iv})) = (\Theta_H(b\sigma, a_0\sigma) \vee \Theta_1)^*\\ &= (\Phi(c, a))^* = \Theta_{H^*}(c, a).\end{aligned}$$

Let $f \notin F$. Then $R(f, \mathfrak{A}^*)$ is one orbit and so 95.1(x) follows from 95.1(vi) applied to that orbit.

This completes the proof of 95.1(i)–95.1(x) and, therefore, that of 95.2(i). 95.2(ii)–95.2(iv) hold by the definition of H^*, K^*, and G^*, while 95.2(v) has been verified while proving 95.1(x).

To proceed to the analogue of the $\mathfrak{A}'$ construction of §17 with the partial operations g_1, g_2, g_3, we have to impose some more stringent conditions on $\Theta_1, \Theta_2, \Theta_3$.

Now set $A^{**} = A^*$, $F^{**} = F \cup \{f_1, f_2, f_3\}$, $\mathfrak{A}^{**} = \langle A^{**}; F^{**}\rangle$. Define

$$\begin{aligned} D(f_i, \mathfrak{A}^{**}) &= \text{Orb}(\mathbf{a}_i) \cup \text{Orb}(\mathbf{b}_i), & i &= 1, 2, 3,\\ f_1(\mathbf{a}_1\sigma) &= a_0\sigma, & f_1(\mathbf{b}_1\sigma) &= r\sigma,\\ f_2(\mathbf{a}_2\sigma) &= r\sigma, & f_2(\mathbf{b}_2\sigma) &= s\sigma,\\ f_3(\mathbf{a}_3\sigma) &= s\sigma, & f_3(\mathbf{b}_3\sigma) &= b_0\sigma, \quad \text{for} \quad \sigma \in G. \end{aligned}$$

We define $H^{**} = H^*$, $K^{**} = K^*$, $G^{**} = G^*$.

Lemma 7. *Let us assume that*

(i) $\Theta_0 \leqq \Theta_1 = \Theta_2 = \Theta_3$;
(ii) $\text{Orb}(\mathbf{a}_i) \neq \text{Orb}(\mathbf{b}_i)$, $i = 1, 2, 3$;
(iii) $\mathbf{a}_i$ Cls $\mathbf{b}_i$ (in $\text{Orb}(\mathbf{a}_i)$, mod H), $i = 1, 2, 3$;
(iv) $a_0 \in [\mathbf{a}_1]_K$, $[\mathbf{b}_1]_K = [\mathbf{a}_2]_K$, $[\mathbf{b}_2]_K = [\mathbf{a}_3]_K$, $b_0 \in [\mathbf{b}_3]_K$.

Then $\langle \mathfrak{A}^{**}, H^{**}, K^{**}, G^{**}, \mathfrak{m}\rangle$ *is a* Cls-*expansion of* $\langle \mathfrak{A}, H, K, G, \mathfrak{m}\rangle$.

Remark. We shall see in the proof of Theorem 93.1 that 7(ii)–7(iv) are easily satisfied by a judicious choice of the $\mathbf{a}_i$, $\mathbf{b}_i$ representing the Θ_i.

Proof. First observe that 7(ii) implies that f_1, f_2, and f_3 are well-defined partial operations. Now let $\Theta \in H$; we claim that $\Theta^* = \Theta^{**}$ is a congruence relation of $\mathfrak{A}^{**}$.

7(i) implies conditions (P_1) and (P_2) of Lemma 96.2, hence $\Theta^* = \Theta^{**}$ is well-defined and Lemmas 2–4 apply. It is sufficient to prove the Substitution Property for f_1, f_2, f_3; so let $\mathbf{x}, \mathbf{y} \in D(f_i, \mathfrak{A}^{**})$ and $\mathbf{x} \equiv \mathbf{y}(\Theta)$. If $\mathbf{x} = \mathbf{a}_i\sigma$, $\mathbf{y} = \mathbf{b}_i\tau$ (or symmetrically) for some $\sigma, \tau \in G$, then $\mathbf{a}_i\sigma \equiv \mathbf{b}_i\tau(\Theta)$ which implies by 7(iii) (see Lemma 1) that $\mathbf{a}_i \equiv \mathbf{b}_i(\Theta)$, that is, $\Theta_i \subseteq \Theta$. If $\sigma = \tau$, this implies that $f_i(\mathbf{x}) \equiv f_i(\mathbf{y})(\Theta^*)$. If $\sigma \neq \tau$ then we conclude that $\mathbf{a}_i\sigma \equiv \mathbf{a}_i\tau(\Theta)$, hence by 95.1(v), $\Xi \subseteq \Theta$ and again we obtain $f_i(\mathbf{x}) \equiv f_i(\mathbf{y})(\Theta^*)$. If $\mathbf{x} = \mathbf{a}_i\sigma$ and $\mathbf{y} = \mathbf{a}_i\tau$ for $\sigma, \tau \in G$, then we can assume $\sigma \neq \tau$, implying that $\Xi \subseteq \Theta$ and so $f_i(\mathbf{a}_i\sigma) \equiv f_i(\mathbf{a}_i\tau)(\Theta^{**})$.

For $D \in K$ let $D^{**} = D^*$. We claim that D^{**} is a subalgebra of $\mathfrak{A}^{**}$. Indeed, if $\mathbf{x} \in (D^{**})^\alpha$ and $\mathbf{x} \in D(f_1, \mathfrak{A}^{**})$, then $\mathbf{x} = \mathbf{a}_1\sigma$ or $\mathbf{x} = \mathbf{b}_1\sigma$ for some $\sigma \in G$. Then $\mathbf{a}_1 \in D^\alpha$ or $\mathbf{b}_1 \in D^\alpha$. If $\mathbf{a}_1 \in D^\alpha$, then, by 7(iv), $a_0 \in [\mathbf{a}_1]_K \subseteq D$, and so $f_1(\mathbf{x}) = a_0\sigma \in D^{**}$; if $\mathbf{b}_1 \in D^\alpha$, then $\mathrm{Orb}(r) \subseteq D^{**}$ by the definition of $D^* = D^{**}$ hence $f_1(\mathbf{x}) = r\sigma \in D^{**}$. The proof for f_3 is the same, while for f_2 we have to utilize that $[\mathbf{b}_1]_K = [\mathbf{a}_2]_K$ and $[\mathbf{b}_2]_K = [\mathbf{a}_3]_K$ to obtain the result.

Observe, that we have proved that, in fact, D^{**} is the subalgebra of $\mathfrak{A}^{**}$ generated by D.

Thus $\mathfrak{A}^{**}$ is a partial algebra, $H^{**} \subseteq C(\mathfrak{A}^{**})$, $K^{**} \subseteq \mathscr{S}(\mathfrak{A}^{**})$; $G^{**} \subseteq G(\mathfrak{A}^{**})$ follows from $G^* \subseteq G(\mathfrak{A}^*)$ and from the definitions of f_1, f_2, f_3. Now we shall verify 95.1(i)–95.1(x) for $\langle \mathfrak{A}^{**}, H^{**}, K^{**}, G^{**}, \mathfrak{m} \rangle$.

95.1(i). $\mathfrak{A}^{**}$ is of characteristic $\leqq \mathfrak{m}$ since so is $\mathfrak{A}$ and $\alpha < \mathfrak{m}$. $H^{**} = H^*$ is an $\mathfrak{m}$-algebraic closure system by Lemma 6 and $K^{**} = K^*$ is an $\mathfrak{m}$-algebraic closure system by Lemma 5.

95.1(ii) is trivial.

95.1(iii), (iv), (v), and (vi) hold by Lemma 6 since $A^{**} = A^*$, $H^{**} = H^*$, $G^{**} = G^*$.

95.1(vii) and (viii) are obvious.

95.1(ix). Each $f \in F$ is H-invertible so they are H^{**}-invertible. To show that each f_1 is H^{**}-invertible let $f_1(\mathbf{x}) \equiv f_1(\mathbf{y})(\Theta^{**})$, $f_1(\mathbf{x}) \neq f_1(\mathbf{y})$, and $\Theta \in H$. If $f_1(\mathbf{x}), f_1(\mathbf{y}) \in \mathrm{Orb}(a_0)$ or $\mathrm{Orb}(r)$, then $\Xi \subseteq \Theta$ and therefore $\mathbf{x} \equiv \mathbf{y}(\Theta)$. If $f_1(\mathbf{x}) = a_0\sigma, f_2(\mathbf{y}) = r\tau$, $\sigma, \tau \in G$ (or symmetrically), then $\Phi(f_1(\mathbf{x}), f_1(\mathbf{y})) \subseteq \Theta$. Now if $\sigma = \tau$, then $\Phi(f_1(\mathbf{x}), f_1(\mathbf{y})) = \Theta_1$, hence $\mathbf{a}_1 \equiv \mathbf{b}_1(\Theta)$ and, by 95.1(iv), $\mathbf{x} = \mathbf{a}_1\sigma \equiv \mathbf{b}_1\sigma = \mathbf{y}(\Theta)$. Finally, let $\sigma \neq \tau$. Then $\Phi(f_1(\mathbf{x}), f_1(\mathbf{y})) = \Theta_1 \vee \Xi$ and so $\mathbf{x} = \mathbf{a}_1\sigma \equiv \mathbf{b}_1\tau = \mathbf{y}(\Theta)$. The proofs for f_2 and f_3 are the same, mutatis mutandis.

95.1(x). This is known to be true by Lemma 6 for $f, g \in F$. Let $f \in F$ and $g = f_1$. Then choose $b \in R(f, \mathfrak{A})$ satisfying b Cls a_0 (in $R(f, \mathfrak{A})$, mod H). Pick an $a \in R(f_1, \mathfrak{A}^{**}) = \mathrm{Orb}(a_0) \cup \mathrm{Orb}(r)$. Then $a = a_0\sigma$ or $a = r\sigma$ for some $\sigma \in G$. We define $c = b\sigma$. If $a = a_0\sigma$, then obviously $\Theta_{H^{**}}(c, a) = \Theta_{H^{**}}(b, a_0)$ and so c Cls a (in $R(f, \mathfrak{A}^{**})$, mod H^{**}). If $a = r\sigma$, then (by Lemma 3)

$$\Theta_{H^{**}}(a, c) = (\Theta_H(c, a_0\sigma) \vee \Theta_1)^{**}$$

and, for any $d \in R(f, \mathfrak{A})$,

$$\Theta_{H^{**}}(a, d) = (\Theta_H(d, a_0\sigma) \vee \Theta_1)^{**},$$

and so c Cls a (in $R(f, \mathfrak{A}^{**})$, mod H^{**}).

Let $f \in F$ and $g = f_2$. We choose again $b \in R(f, \mathfrak{A})$ to satisfy b Cls a_0 (in $R(f, \mathfrak{A})$, mod H) and take an $a \in R(f_2, \mathfrak{A}^{**}) = \mathrm{Orb}(r) \cup \mathrm{Orb}(s)$. Then $a = r\sigma$ or $a = s\sigma$ and define $c = b\sigma$. If $a = r\sigma$ we proceed as in the previous paragraph. If $a = s\sigma$ and $d \in R(f, \mathfrak{A})$, then (by Lemma 3)

$$\Theta_{H^{**}}(c, a) = (\Theta_H(c, b_0\sigma) \vee \Theta_3)^{**}$$

and

$$\Theta_{H^{**}}(d, a) = (\Theta_H(d, b_0\sigma) \vee \Theta_3)^{**}.$$

By 7(i), $\Theta_0 \leqq \Theta_3$, hence $a_0 \equiv b_0(\Theta_{H^{**}}(d, a))$ and so

$$d \equiv b_0\sigma \equiv a_0\sigma(\Theta_{H^{**}}(d, a)).$$

Since b Cls a_0 (in $R(f, \mathfrak{A})$, mod H), by 95.1(iv), $c = b\sigma$ Cls $a_0\sigma$ (in $R(f, \mathfrak{A})$, mod H), hence the last congruence implies that $c \equiv a_0\sigma(\Theta_{H^{**}}(d, a))$ and so $c \equiv b_0\sigma(\Theta_{H^{**}}(d, a))$. This implies that

$$\Theta_{H^{**}}(d, a) \geqq \Theta_{H^{**}}(c, a),$$

that is, c Cls a (in $R(f, \mathfrak{A}^{**})$, mod H^{**}).

$f \in F$ and $g = f_3$ can be handled just like $g = f_1$.

Now let $f = f_1$ and $g \in F$. Then we choose $b = a_0$. For $a \in R(g, \mathfrak{A})$ there is by 95.1(vi), a $c \in \mathrm{Orb}(a_0)$ such that c Cls a (in $\mathrm{Orb}(a_0)$, mod H). Let $d \in R(f_1, \mathfrak{A}^{**}) = \mathrm{Orb}(a_0) \cup \mathrm{Orb}(r)$. Then $d = a_0\sigma$ or $d = r\sigma$, $\sigma \in G$. If $d = a_0\sigma$, then $\Theta_{H^{**}}(d, a) = (\Theta_H(d, a))^{**} \geqq (\Theta_H(c, a))^{**} = \Theta_{H^{**}}(c, a)$ by the choice of c. If $d = r\sigma$, then, by Lemma 3,

$$\Theta_{H^{**}}(d, a) = (\Theta_H(a, a_0\sigma) \vee \Theta_1)^{**} \geqq (\Theta_H(c, a) \vee \Theta_1)^{**} \geqq \Theta_{H^{**}}(c, a),$$

hence c Cls a (in $R(f_1, \mathfrak{A}^{**})$, mod H^{**}).

Let $f = f_2$ and $g \in F$. We can pick any element of $R(f, \mathfrak{A}^{**})$ as b; let us choose, say, $b = r$. Let $a \in R(g, \mathfrak{A})$. By 95.1(vi), there is a $\sigma \in G$ such that $a_0\sigma$ Cls a (in $\mathrm{Orb}(a_0)$, mod H). Set $c = b\sigma$. Then

$$\Theta_{H^{**}}(a, c) = (\Theta_H(a, a_0\sigma) \vee \Theta_1)^{**}.$$

Let $d \in R(f_2, \mathfrak{A}^{**}) = \mathrm{Orb}(r) \cup \mathrm{Orb}(s)$, that is $d = r\tau$ or $d = s\tau$ for some $\tau \in G$. If $d = r\tau$, then

$$\Theta_{H^{**}}(d, a) = (\Theta_H(a, a_0\tau) \vee \Theta_1)^{**} \geqq \Theta_{H^{**}}(a, c)$$

by the choice of c. If $d = s\tau$, then, by Lemma 3,

$$\Theta_{H^{**}}(a, d) = (\Theta_H(a, b_0\tau) \vee \Theta_3)^{**}.$$

By 7(i), $\Theta_0 \leqq \Theta_1 = \Theta_3 \leqq \Theta_{H^{**}}(a, d)$, hence

$$a \equiv b_0\tau \equiv a_0\tau(\Theta_{H^{**}}(a, d))$$

and so, by the choice of $a_0\sigma$,

$$a \equiv a_0\sigma(\Theta_{H^{**}}(a, d)).$$

This implies that $\Theta_{H^{**}}(a, d) \geqq \Theta_H(a, a_0\sigma)$, Θ_1, and therefore

$$\Theta_{H^{**}}(a, d) \geqq \Theta_{H^{**}}(a, c),$$

proving that c Cls a (in $R(f_2, \mathfrak{A}^{**})$, mod H^{**}).

$f = f_3$ and $g \in F$ can be handled just as $f = f_1$, $g \in F$.

So finally, let $f, g \in \{f_1, f_2, f_3\}$. Since $R(f_1, \mathfrak{A}^{**}) \cap R(f_2, \mathfrak{A}^{**}) \neq \varnothing$ and $R(f_2, \mathfrak{A}^{**}) \cap R(f_3, \mathfrak{A}^{**}) \neq \varnothing$, for f_1, f_2 and f_2, f_3, b can always be chosen as any element in the intersection. We are left with $f=f_1$ and $g=f_3$ or $f=f_3$ and $g=f_1$. The two cases are symmetric so it is sufficient to consider $f=f_1$ and $g=f_3$. Then we define $b=a_0$. Let $a \in R(f_3, \mathfrak{A}^{**}) = \mathrm{Orb}(b_0) \cup \mathrm{Orb}(s)$; then $a=b_0\sigma$ or $a=s\sigma$, for some $\sigma \in G$. We define $c=b\sigma=a_0\sigma$. Let $d \in R(f_1, \mathfrak{A}^{**})$; then $d=a_0\tau$ or $r\tau$ for some $\tau \in G$. We have to verify that $\Theta_{H^{**}}(d, a) \geqq \Theta_{H^{**}}(c, a)$.

Case 1. $a=b_0\sigma$ and $d=a_0\tau$. If $\sigma=\tau$, then $c=d$ so we have nothing to prove. If $\sigma \neq \tau$, then

$$\Theta_{H^{**}}(d, a) = (\Theta_H(a_0\sigma, b_0\sigma) \vee \Xi)^{**} \geqq (\Theta_H(a_0\sigma, b_0\sigma))^{**} = \Theta_{H^{**}}(c, a).$$

Case 2. $a=b_0\sigma$ and $d=r\tau$. Then $\Theta_{H^{**}}(d, a) = (\Theta_1 \vee \Theta_H(a_0\tau, a))^{**} \geqq$ (by Case 1) $\geqq \Theta_{H^{**}}(c, a)$.

Case 3. $a=s\sigma$ and $d=a_0\tau$. Then $\Theta_{H^{**}}(d, a) = (\Theta_H(d, b_0\sigma) \vee \Theta_3)^{**} \geqq$ (since $\Theta_0 \leqq \Theta_3$ and so $a_0 \equiv b_0(\Theta_3)$ and $a_0\sigma \equiv b_0\sigma(\Theta_3)) \geqq (\Theta_H(a_0\sigma, b_0\sigma) \vee \Theta_3)^{**} = \Theta_{H^{**}}(c, a)$.

Case 4. $a=s\sigma$ and $d=r\tau$. Then

$$\Theta_{H^{**}}(d, a) \geqq \Theta_{H^{**}}(r\sigma, s\sigma) = \Theta_2 \geqq (\Theta_H(a_0\sigma, b_0\sigma) \vee \Theta_3)^{**} = \Theta_{H^{**}}(c, a),$$

since $\Theta_2 = \Theta_3$ and $\Theta_2 \geqq \Theta_0$ and so, by 95.1(iv), $a_0\sigma \equiv b_0\sigma(\Theta_2)$.

This completes the proof of 95.1(x). 95.2(i)–95.2(v) require no further proof. Thus the proof of Lemma 7 is complete.

§97. THREE MORE CONSTRUCTIONS

The following two extension theorems for $\mathfrak{m}$-algebraic closure systems will be needed. The first result is trivial while the second can be proved just as Lemma 96.4 is proved.

Lemma 1. (i) *Let C be an $\mathfrak{m}$-algebraic closure system on the set A. Then $\{X \times B \mid X \in C\}$ is an algebraic closure system on $A \times B$.*

(ii) *Let C be an $\mathfrak{m}$-algebraic closure system of equivalence relations on the set A and let $Z \in C$ be $\mathfrak{m}$-compact. Then*

$$\{X \times \omega_B \mid X \in C,\ Z \nsubseteq X\} \cup \{X \times \iota_B \mid X \in C,\ Z \subseteq X\}$$

is an $\mathfrak{m}$-algebraic closure system on $A \times B$.

Now let $\langle \mathfrak{A}, H, K, G, \mathfrak{m}\rangle$ be a Cls-quintuple. Let $a_0 \equiv b_0(\Theta_H(\mathbf{a}, \mathbf{b}))$, where $a_0, b_0 \in A$, $a_0 \neq b_0$, $\mathbf{a}, \mathbf{b} \in A^\alpha$ and $0 < \alpha < \mathfrak{m}$.

Let $\mathbf{c}$ be the concatenation of $\langle a_0, b_0\rangle$, $\mathbf{a}$, and $\mathbf{b}$ and let $\mathbf{d}$ be the concatenation of $\langle a_0, b_0\rangle$, $\mathbf{b}$, and $\mathbf{a}$ and let β be the ordinal $2+\alpha+\alpha$.

Lemma 2. $\mathbf{c}, \mathbf{d} \in A^\beta$, $0 < \beta < \mathfrak{m}$, $\Theta_H(\mathbf{a}, \mathbf{b}) = \Theta_H(\mathbf{c}, \mathbf{d})$, *and*

$$\begin{gathered} a_0, b_0 \in [\mathbf{c}]_K = [\mathbf{d}]_K, \\ \mathrm{Orb}(\mathbf{c}) \neq \mathrm{Orb}(\mathbf{d}), \\ \mathbf{c} \text{ Cls } \mathbf{d} \text{ (in Orb } (\mathbf{c}), \text{ mod } H). \end{gathered}$$

Proof. All but the last two statements are evident. The last statement follows from Lemma 96.1(iii). If $\mathrm{Orb}(\mathbf{c}) = \mathrm{Orb}(\mathbf{d})$, then $\mathbf{d} = \mathbf{c}\sigma$ for some $\sigma \in G$. Then $c_0\sigma = d_0$, that is, $a_0\sigma = a_0$, and so σ is the identity of G since G is regular. But then we obtain $\mathbf{c} = \mathbf{d}$ and $\mathbf{a} = \mathbf{b}$, contradicting $a_0 \neq b_0$.

Let $\langle \mathfrak{A}, H, K, G, \mathfrak{m}\rangle$ be a Cls-quintuple, $\mathfrak{A} = \langle A; F\rangle$, let F' be a family of partial operations, and let $\mathfrak{A}' = \langle \mathfrak{A}; F \cup F'\rangle$ satisfy the following conditions:

$$H \subseteq C(\mathfrak{A}'), \qquad K \subseteq \mathscr{S}(\mathfrak{A}'), \qquad G \subseteq G(\mathfrak{A}'),$$

the arity of each $f \in F'$ is $< \mathfrak{m}$, and for all $f \in F'$ there is an $e \in A$ such that $R(f, \mathfrak{A}') = \mathrm{Orb}(e)$, and any $f \in F'$ is H-invertible.

Lemma 3. $\langle \mathfrak{A}', H, K, G, \mathfrak{m}\rangle$ *is a* Cls-*expansion of* $\langle \mathfrak{A}, H, K, G, \mathfrak{m}\rangle$.

Proof. Only 95.1(x) needs verification. Let $f, g \in F \cup F'$. If $f, g \in F$, the statement is trivial. Let $f \in F$ and $g \in F'$, $R(g, \mathfrak{A}') = \mathrm{Orb}(e)$. Let $b \in A$ satisfy b Cls e (in $R(f, \mathfrak{A})$, mod H). Then for every $a \in R(g, \mathfrak{A})$, $a = e\sigma$ for some $\sigma \in G$, hence $c = b\sigma$ Cls a (in $R(f, \mathfrak{A})$, mod H). Let $f \notin F$, $R(f, \mathfrak{A}') = \mathrm{Orb}(e)$, $g \in F$. Then we choose $b = e$ and the rest follows from 95.1(vi). The case $f, g \in F'$ also follows from 95.1(vi).

We need two more types of expansions for the proof; one that will fix up the subalgebra system, and one that will adjust the automorphisms. First, we handle the subalgebras.

Let $\langle \mathfrak{A}, H, K, G, \mathfrak{m}\rangle$ be a Cls-quintuple, $\mathfrak{A} = \langle A; F\rangle$. For every ordinal α, $0 < \alpha < \mathfrak{m}$, every $\mathbf{a} \in A^\alpha$, and $a \in [\mathbf{a}]_K$, define an α-ary partial operation $f_{\mathbf{a}, a}$ by

$$D(f_{\mathbf{a}, a}, \mathfrak{A}^s) = \mathrm{Orb}(\mathbf{a})$$

and for $\sigma \in G$, let

$$f_{\mathbf{a}, a}(\mathbf{a}\sigma) = a\sigma.$$

Let F_s be the set of all such partial operations and let $\mathfrak{A}^s = \langle A; F \cup F_s\rangle$.

Lemma 4. *$\langle \mathfrak{A}^s, H, K, G, \mathfrak{m}\rangle$ is a* Cls-*expansion of $\langle \mathfrak{A}, H, K, G, \mathfrak{m}\rangle$ and $\mathscr{S}(\mathfrak{A}^s)=K$.*

Proof. Let $\Theta \in H$, $f=f_{\mathbf{a},a}$ be α-ary, $\mathbf{x}, \mathbf{y} \in D(f, \mathfrak{A}^s)$, $\mathbf{x}\neq\mathbf{y}$, and $\mathbf{x}\equiv\mathbf{y}(\Theta)$. Then $\mathbf{x}, \mathbf{y} \in \mathrm{Orb}(\mathbf{a})$ and so $\mathbf{x}=\mathbf{y}\sigma$ for some $\sigma \in G$. Since α is not the identity of G, $x_\gamma\neq y_\gamma\sigma$ for all $\gamma<\alpha$. Thus $\mathbf{x}\equiv\mathbf{y}(\Theta)$ iff $\Theta\supseteq\Xi$, which in turn is equivalent to $f(\mathbf{x})\equiv f(\mathbf{y})(\Theta)$, since $f(\mathbf{x})$ and $f(\mathbf{y})$ are two distinct elements in $\mathrm{Orb}(\mathbf{a})$. This proves that $H\subseteq C(\mathfrak{A}^s)$ and the H-invertibility of all $f\in F_s$, that is, 95.1(ix). Condition 95.1(x) follows from Lemma 3. All the other conditions are trivial.

Now by the definition of $f_{\mathbf{a},a}$, $K\subseteq\mathscr{S}(\mathfrak{A}^s)$. Conversely, let $D\in\mathscr{S}(\mathfrak{A}^s)$ and $a\in[D]_K$. Since K is $\mathfrak{m}$-algebraic, there is an $\mathbf{a}\in D^\alpha$, $\alpha<\mathfrak{m}$, such that $a\in[\mathbf{a}]_K$. Since $a=f_{\mathbf{a},a}(a)$, we conclude that $a\in D$, thus $[D]_K=D$, and $D\in K$.

Now we tackle the automorphisms. Let $\langle \mathfrak{A}, H, K, G, \mathfrak{m}\rangle$ be a Cls-quintuple, $\mathfrak{A}=\langle A; F\rangle$. For every pair $\langle b, c\rangle$ of A we introduce a partial operation $g_{b,c}$ by

$$D(g_{b,c}, \mathfrak{A}^a) = \mathrm{Orb}(\langle b, c\rangle).$$

and for $\langle x, y\rangle \in D(g_{b,c}, \mathfrak{A}')$, $g_{b,c}(x, y)=x$. Let F' be the set of all the $g_{b,c}$ and let $\mathfrak{A}=\langle A; F\cup F'\rangle$.

Lemma 5. *$\langle \mathfrak{A}^a, H, K, G, \mathfrak{m}\rangle$ is a* Cls-*expansion of $\langle \mathfrak{A}, H, K, G, \mathfrak{m}\rangle$ and $G(\mathfrak{A}^a)=G$.*

Proof. Again, by Lemma 3, to prove that we obtained a Cls-expansion it is sufficient to verify that $g_{b,c}$ is H-invertible. Indeed, if $g_{b,c}(x_0, y_0)\equiv g_{b,c}(x_1, y_1)(\Theta)$ for some $\langle x_0, y_0\rangle, \langle x_1, y_1\rangle \in D(g_{b,c}, \mathfrak{A}^a)$ and $\Theta\in H$, then $x_0\equiv x_1(\Theta)$. Now if $x_0=x_1$, then also $y_0=y_1$ and so $y_0\equiv y_1(\Theta)$. If $x_0\neq x_1$, then $x_1=x_0\sigma$ where $\sigma\in G$ and σ is not the identity of G. Thus by 95.1(v), and the regularity of G, $x_0\equiv x_1(\Theta)$ implies that $\Xi\subseteq\Theta$. By the definition of $D(g_{b,c}, \mathfrak{A}^a)$, $y_1=y_0\sigma$, therefore $y_0\equiv y_1(\Theta)$.

Obviously, $G\subseteq G(\mathfrak{A}^a)$. Conversely, let $\alpha\in G(\mathfrak{A}^a)$. For any $b, c\in A$, $\langle b, c\rangle \in D(g_{b,c}, \mathfrak{A}^a)=\{\langle b\sigma, c\sigma\rangle \mid \sigma\in G\}$, hence for any automorphism α, $\langle b, c\rangle\alpha \in D(g_{a,b}, \mathfrak{A}^a)$. Thus $\langle b, c\rangle\alpha=\langle b, c\rangle\tau$ for some $\tau\in G$, that is, $b\alpha=b\tau$ and $c\alpha=c\tau$. This τ does not depend on b and c since $b\alpha=b\tau$ uniquely determines τ by the regularity of G, hence $c\alpha=c\tau$ for all $c\in A$, that is, $\alpha=\tau$.

§98. PROOF OF THE MAIN THEOREM

Now we present the initial construction for the proof of Theorem 93.1. We are given an infinite regular cardinal $\mathfrak{m}$, $\mathfrak{m}$-algebraic lattices $\mathfrak{L}_c$ and $\mathfrak{L}_a$

with more than one element, and a group $\mathfrak{G}$. Let C_c and C_a be the set of $\mathfrak{m}$-compact elements of $\mathfrak{L}_c$ and $\mathfrak{L}_a$, respectively; $\mathfrak{C}_c = \langle C_c; \vee \rangle$ and $\mathfrak{C}_a = \langle C_a; \vee \rangle$ are the join semilattice on C_c and C_a, respectively. We set $C_a' = C_a - \{0\}$, where 0 is the zero of $\mathfrak{C}_a$. Since $|L_a| > 1$, $C_a' \neq \varnothing$.

We form the set $B = C_a' \times C_c \times G$.

For an $\mathfrak{m}$-ideal I of C_a, we define a subset of B:

$$D_I = (I - \{0\}) \times C_c \times G$$

and set

$$K = \{D_I \mid I \text{ is an } \mathfrak{m}\text{-ideal of } C_a\}.$$

Fix an arbitrary $z \in C_c$, $z \neq 0$. For any $\mathfrak{m}$-ideal J of C_c we define the equivalence relation Θ_J on B. Let $\mathbf{x}_i = \langle a_i, b_i, \sigma_i \rangle \in B$, $i = 0, 1$. If $z \notin J$, let $\mathbf{x}_0 \equiv \mathbf{x}_1(\Theta_J)$ iff $\langle a_0, \sigma_0 \rangle = \langle a_1, \sigma_1 \rangle$ and $b_0, b_1 \in J$ or $\mathbf{x}_0 = \mathbf{x}_1$; if $z \in I$, let $\mathbf{x}_0 \equiv \mathbf{x}_1(\Theta_J)$ iff $b_0, b_1 \in J$ or $b_0 = b_1$. Set

$$H = \{\Theta_J \mid J \text{ is a } \mathfrak{m}\text{-ideal of } C_c\}.$$

For $\tau \in G$, define the map $\rho_\tau \colon B \to B$, by

$$\langle a, b, \sigma \rangle \rho_\tau = \langle a, b, \sigma\tau \rangle$$

and $G' = \{\rho_\tau \mid \tau \in G\}$, $\mathfrak{G}' = \langle G'; \circ \rangle$.

Finally, we define $\mathfrak{B} = \langle B; \{\mathrm{id}\} \rangle$ where id is the identity function on B.

Lemma 1. $\langle \mathfrak{B}, H, K, G', \mathfrak{m} \rangle$ *is a* Cls-*quintuple and* $\langle H; \subseteq \rangle \cong \mathfrak{L}_c$, $\langle K; \subseteq \rangle \cong \mathfrak{L}_a$, $\mathfrak{G}' \cong \mathfrak{G}$.

Proof. H and K are $\mathfrak{m}$-algebraic closure systems by Lemma 97.1. G' is regular by definition. The isomorphisms of the lattices and of the groups are trivial, proving 95.1(i) and the isomorphisms. 95.1(ii) holds because we deleted 0 from C_c and $\Theta_{(0]} = \omega$. 95.1(iii) is clear.

Let $\mathbf{x}_i = \langle a_i, b_i, \sigma_i \rangle$, $i = 0, 1$. We describe $\Theta_H(\mathbf{x}_0, \mathbf{x}_1)$;

$$\Theta_H(\mathbf{x}_0, \mathbf{x}_1) = \begin{cases} \omega & \text{if } b_0 \neq b_1 \text{ and } \mathbf{x}_0 = \mathbf{x}_1, \\ \Theta_{(b_0 \vee b_1]} & \text{if } b_0 \neq b_1 \text{ and } \langle a_0, \sigma_0 \rangle = \langle a, \sigma_1 \rangle, \\ \Theta_{(b_0 \vee b_1 \vee z]} & \text{if } b_0 \neq b_1 \text{ and } \langle a_0, \sigma_0 \rangle \neq \langle a_1, \sigma_1 \rangle, \\ \Theta_{(z]} & \text{if } b_0 = b_1. \end{cases}$$

The hypotheses are invariant under ρ_τ, $\tau \in G$, hence 95.1(iv). Clearly, $\Xi = \Theta_{(z]}$ in 95.1(v). Again, from the formulas giving $\Theta_H(\mathbf{x}_0, \mathbf{x}_1)$, it is clear that $\langle a_0, b_0, \sigma_1 \rangle$ is closest to $\mathbf{x}_1$ in the orbit of x_0, verifying 95.1(vi). 95.1(vii)– 95.1(x) are trivial.

We construct the algebra $\mathfrak{A}$ of Theorem 93.1 as a limit of algebras $\mathfrak{A}_\gamma$, $\gamma < \mathfrak{m}$. We now describe the construction of the $\mathfrak{A}_\gamma$.

We set $\langle \mathfrak{A}_0, H_0, K_0, G_0, \mathfrak{m}\rangle = \langle \mathfrak{B}, H, K, G', \mathfrak{m}\rangle$ as given by Lemma 1. Now let $\delta < \mathfrak{m}$ and let us assume that $\langle \mathfrak{A}_\gamma, H_\gamma, K_\gamma, G_\gamma, \mathfrak{m}\rangle$ has been constructed for all $\gamma < \delta$ satisfying the following properties:

(C_1) For $\gamma < \delta$, $\mathfrak{A}_\gamma = \langle A_\gamma; F_\gamma\rangle$ is an algebra.

(C_2) For $\beta < \gamma < \delta$, $\langle \mathfrak{A}_\gamma, H_\gamma, K_\gamma, G_\gamma, \mathfrak{m}\rangle$ is a Cls-expansion of $\langle \mathfrak{A}_\beta, H_\beta, K_\beta, G_\beta, \mathfrak{m}\rangle$.

(C_3) For $\beta < \gamma < \delta$, and $a \in A_\beta$, $D \subseteq A_\beta$, if $a \in [D]_{K_\beta}$, then a is in the subalgebra of $\mathfrak{A}_\gamma$ generated by D.

(C_4) For $\beta < \gamma < \delta$, $a, b \in A_\beta$, $X \subseteq A_\beta \times A_\beta$, if $a \equiv b$ $([X]_H)$, then $a \equiv b(\Theta)$, where Θ is the congruence relation of $\mathfrak{A}_\gamma$ generated by X.

(C_5) For $\beta < \gamma < \delta$ and $\alpha < \mathfrak{m}$, let $a_0, b_0 \in A_\beta$, $\mathbf{a}_i, \mathbf{b}_i \in (A_\beta)^\alpha$, $i = 1, 2, 3$, $\Theta_i = \Theta_{H_\beta}(\mathbf{a}_i, \mathbf{b}_i)$, $\Theta_0 = \Theta_{H_\beta}(a_0, b_0)$, and $\Theta_0' = H_\beta - \mathrm{Dist}(a_0, \mathrm{Orb}(b_0))$; if Θ_0', Θ_0, Θ_1, Θ_2, Θ_3 satisfy conditions (P_1) and (P_2) of Lemma 96.2, then there exist $r, s \in A_\gamma$ satisfying $a_0 \equiv r([\Theta_1]_{H_\gamma})$, $r \equiv s([\Theta_2]_{H_\gamma})$, $s \equiv b_0([\Theta_3]_{H_\gamma})$.

(C_6) For $\beta < \gamma < \delta$ and $\alpha < \mathfrak{m}$, let $a_0, b_0 \in A_\beta$, $\mathbf{a}_i, \mathbf{b}_i \in (A_\beta)^\alpha$, $i = 1, 2, 3$, $\Theta_0 = \Theta_{H_\beta}(a_0, b_0)$, $\Theta_i = \Theta_{H_\beta}(\mathbf{a}_i, \mathbf{b}_i)$, $i = 1, 2, 3$, satisfy the hypotheses of Lemma 96.7 (with $H = H_\beta$ and $K = K_\beta$); then there exist $f_1, f_2, f_3 \in F_\gamma$ satisfying $a_0 = f_1(\mathbf{a}_1)$, $f_1(\mathbf{b}_1) = f_2(\mathbf{a}_2)$, $f_2(\mathbf{b}_2) = f_3(\mathbf{a}_3)$, $f_3(\mathbf{b}_3) = b_0$.

Let $\langle \mathfrak{A}_\delta', H_\delta', K_\delta', G_\delta', \mathfrak{m}\rangle$ be the limit of the $\langle \mathfrak{A}_\gamma, H_\gamma, K_\gamma, G_\gamma, \mathfrak{m}\rangle$, $\gamma < \delta$, and we set $\langle (\mathfrak{A}_\delta')^{s,a}, H_\delta', K_\delta', G_\delta', \mathfrak{m}\rangle = \langle \mathfrak{C}_0, H_{\delta,0}, K_{\delta,0}, G_{\delta,0}, \mathfrak{m}\rangle$, where $\mathfrak{C}_0 = \langle C_0; G_0\rangle$.

Now let $\varepsilon < \varphi$ be ordinals and let

$$\{\langle a_{0,\gamma}, b_{0,\gamma}, \mathbf{a}_{1,\gamma}, \mathbf{b}_{1,\gamma}, \mathbf{a}_{2,\gamma}, \mathbf{b}_{2,\gamma}, \mathbf{a}_{3,\gamma}, \mathbf{b}_{3,\gamma}\rangle \mid 0 < \gamma < \varphi\}$$

be the set of all sequences of elements of A_δ' satisfying conditions (i)–(iv) of Lemma 96.7 for $0 < \gamma < \varepsilon$ and conditions (P_1) and (P_2) for $\varepsilon \leqq \gamma < \varphi$ with $\Theta_i = \Theta_{H_\delta'}(\mathbf{a}_i, \mathbf{b}_i)$, $i = 1, 2, 3$, $\Theta_0 = \Theta_{H_\delta'}(a_0, b_0)$, and $\Theta_0' = H_\delta' - \mathrm{Dist}(a_0, \mathrm{Orb}(b_0))$. For $0 < \gamma < \varepsilon$, we define $\langle \mathfrak{C}_\gamma, H_{\delta,\gamma}, K_{\delta,\gamma}, G_{\delta,\gamma}, \mathfrak{m}\rangle$ as the limit $\langle \mathfrak{C}_\gamma', H_{\delta,\gamma}', K_{\delta,\gamma}', G_{\delta,\gamma}', \mathfrak{m}\rangle$ of $\langle \mathfrak{C}_\chi, H_{\delta,\chi}, K_{\delta,\chi}, G_{\delta,\chi}, \mathfrak{m}\rangle$, $\chi < \gamma$, on which we perform the **-construction if $0 < \gamma < \varepsilon$, and the *-construction if $\varepsilon \leqq \gamma$ with respect to $\langle a_{0,\gamma}, b_{0,\gamma}, \cdots, \mathbf{b}_{3,\gamma}\rangle$; let $\langle \mathfrak{C}_\gamma, H_{\delta,\gamma} K_{\delta,\gamma}, G_{\delta,\gamma}, \mathfrak{m}\rangle$ be the resulting Cls-expansion.

Finally, we obtain $\langle \mathfrak{C}_\varphi, H_{\delta,\varphi}, K_{\delta,\varphi}, G_{\delta,\varphi}, \mathfrak{m}\rangle$ as the limit of $\langle \mathfrak{C}_\gamma, H_{\delta,\gamma}, K_{\delta,\gamma}, G_{\delta,\gamma}, \mathfrak{m}\rangle$, $\gamma < \varphi$, and by $\mathfrak{m}$ repeated applications of the $\mathfrak{A}[F]$ expansions we get $\langle \mathbf{Fr}(\mathfrak{C}_\varphi), \mathrm{Fr}\, H_{\delta,\varphi}, \mathrm{Fr}\, K_{\delta,\varphi}, \mathrm{Fr}\, G_{\delta,\varphi}, \mathfrak{m}\rangle = \langle \mathfrak{A}_\delta, H_\delta, K_\delta, G_\delta, \mathfrak{m}\rangle$.

(C_1) is obvious. (C_2) follows from Lemma 95.2. Since we performed the construction of Lemma 97.3, this lemma implies (C_3). (C_5) and (C_6) follow from the construction and (C_4) follows from (C_6) in view of Lemma 97.2.

Now let $\mathfrak{A}$ be the limit of the $\mathfrak{A}_\gamma$, $\gamma < \mathfrak{m}$.

Theorem 1. *$\mathfrak{A}$ is an algebra of characteristic $\mathfrak{m}$, $\langle \mathfrak{A}, \mathfrak{C}(\mathfrak{A}), \mathscr{S}(\mathfrak{A}), \mathfrak{G}(\mathfrak{A}), \mathfrak{m}\rangle$ is a* Cls-*quintuple.* $\mathfrak{C}(\mathfrak{A}) \cong \mathfrak{L}_c$, $\mathscr{S}(\mathfrak{A}) \cong \mathfrak{L}_a$, $\mathfrak{G}(\mathfrak{A}) \cong \mathfrak{G}$, *and $\mathfrak{C}(\mathfrak{A})$ is a sublattice of*

Part A, *in fact,* $\mathfrak{C}(\mathfrak{A})$ *has type-3 joins. If* $a, b \in A$, $\Theta, \Phi \in C(\mathfrak{A})$, $a \equiv b(\Theta \vee \Phi)$, $\Phi \subseteq \Theta \vee \Theta(a, b)$, *and* a Cls b (in Orb(a), mod $C(\mathfrak{A})$), *then there exist* $r, s \in A$ *satisfying* $a \equiv r(\Theta)$, $r \equiv s(\Phi)$, $s \equiv b(\Theta)$. *In particular, if* $\mathfrak{L}_c$ *is modular and* $|G| = 1$, *then* $\mathfrak{C}(\mathfrak{A})$ *has type-2 joins.*

Remark. This is a slightly stronger version of the Theorem 93.1, in particular, we formulate how close $\mathfrak{A}$ is to being of type-2.

Proof. Since the ^{a}-construction was applied at least once (once would have been enough), $\mathfrak{G}(\mathfrak{A}) \cong \mathfrak{G}$ is clear. All the other statements but two are trivial from (C_1)–(C_6). One exception is the last one which follows from the construction. The other exception is the claim that $\mathfrak{C}(\mathfrak{A})$ has type-3 joins which we proceed to prove.

Let $a_0 \equiv b_0(\Theta \vee \Phi)$, where $a_0, b_0 \in A$, $\Theta, \Phi \in C(\mathfrak{A})$. Choose $\mathfrak{m}$-compact $\Theta' \leqq \Theta$ and $\Phi' \leqq \Phi$ such that $a \equiv b(\Theta' \vee \Phi')$. Then there exist $\mathfrak{m}$-compact ideals J_0 and J_1 of C_c (of the initial construction) such that Θ' and Φ' restricted to B is Θ_{J_0} and Θ_{J_1}, respectively. Since J_0 and J_1 are $\mathfrak{m}$-compact, they can be generated by $< \mathfrak{m}$ elements; let $J_0 = (X]$, $J_1 = (Y]$, $|X|$, $|Y| < \mathfrak{m}$. We can assume that $|X| = |Y|$ and let $\mathbf{a}$, $\mathbf{b}$ be well-orderings of the same type of X and Y, respectively. Let $\mathbf{0}$ denote a sequence of the same type all of whose components are the 0 of C_c. Then $\Theta' = \Theta(\mathbf{a}, \mathbf{0})$, $\Phi' = \Theta(\mathbf{b}, \mathbf{0})$, and $\Theta' \vee \Phi' = \Theta(\mathbf{a}, \mathbf{b})$. By applying Lemma 97.2, we can find sequences $\mathbf{a}'$ and $\mathbf{b}'$ and a sequence of 0-*s*, $\mathbf{0}'$, such that

$$\Theta' = \Theta(\mathbf{a}', \mathbf{0}'), \; \Phi' = \Theta(\mathbf{b}', \mathbf{0}'), \; \Theta' \vee \Phi' = \Theta(\mathbf{a}', \mathbf{b}'),$$

a_0, b_0; $\mathbf{a}', \mathbf{b}'$; $\mathbf{b}', \mathbf{a}'$; $\mathbf{a}', \mathbf{b}'$ and $\Theta_1 = \Theta_2 = \Theta_3 = \Theta' \vee \Phi'$ satisfy conditions (i)–(iv) of Lemma 96.7. Thus by (C_6), $\mathfrak{A}$ has operations f_1, f_2, f_3 satisfying $a_0 = f_1(\mathbf{a}')$, $f_1(\mathbf{b}') = f_2(\mathbf{b}')$, $f_2(\mathbf{a}') = f_3(\mathbf{a}')$, $f_3(\mathbf{b}') = b_0$. Then

$$\begin{aligned} a_0 &= f_1(\mathbf{a}') \equiv f_1(\mathbf{0}')(\Theta'),\\ f_1(\mathbf{0}') &\equiv f_1(\mathbf{b}') = f_2(\mathbf{b}') \equiv f_2(\mathbf{0}')(\Phi'),\\ f_2(\mathbf{0}') &\equiv f_2(\mathbf{a}') = f_3(\mathbf{a}') \equiv f_3(\mathbf{0}')(\Theta'),\\ f_3(\mathbf{0}') &\equiv f_3(\mathbf{b}') = b_0(\Phi'), \end{aligned}$$

obtaining the type-3 joins with the sequence $a_0, f_1(\mathbf{0}'), f_2(\mathbf{0}'), f_3(\mathbf{0}'), b_0$.

This completes the proof of Theorem 1 and, therefore, that of Theorem 93.1.

BIBLIOGRAPHY

The following list of papers and books is a bibliography of Universal Algebra rather than a bibliography of this book. An attempt was made to include in this bibliography every paper that undoubtedly qualifies as a paper on Universal Algebra. Many papers in this list do not qualify as such, but they have or may have some applications to Universal Algebra. No attempt has been made to make the list complete as far as related areas are concerned.

Several items in this bibliography are not mentioned in the book and also there are several references in the book that are not included in this bibliography. The standard abbreviations of the Mathematical Reviews are used. Complete references are given to the review in the Mathematical Reviews (abbreviated MR in the Bibliography), including the name of the reviewer.

A. Ádám

[1] A theorem on algebraic operators in the most general sense, Acta Sci. Math. (Szeged) 18 (1957), 205–206; MR 20, #14 (J. E. Whitesitt).

[2] On permutations of set products, Publ. Math. Debrecen 5 (1957), 147–149; MR 22, #1534 (P. J. Higgins).

[3] On the definitions of direct product in universal algebra, Publ. Math. Debrecen 6 (1959), 303–310; MR 22, #2571 (P. J. Higgins).

[4] On subsets of set products, Ann. Univ. Sci. Budapest. Eötvös Sect. Math., 2 (1959), 147–149; MR 22, #6719 (S. Ginsburg).

J. W. Addison

[1] The method of alternating chains, The theory of models, Proc. 1963 Int. Symp. at Berkeley, North-Holland, Amsterdam, 1965, 1–16.

Z. A. Almagambetov

[1] On classes of axioms closed with respect to given reduced products and powers (Russian), Algebra i Logika Sem. 4 (1965), no. 4, 71–77; MR 33, #7255 (J. Sonner).

K. I. Appel

[1] Horn sentences in identity theory, J. Symbolic Logic 24 (1959), 306–310; MR 24, #A2528 (C. C. Chang).

M. Armbrust and J. Schmidt

[1] Zum Cayleyschen Darstellungssatz, Math. Ann. 154 (1964), 70–72; MR 28, #3109 (N. Ito).

A. Astromoff

[1] Some structure theorems for primal and categorical algebras, Math. Z. 87 (1965), 365–377; MR 32, #2355 (B. M. Schein).

A. K. Austin

[1] Finite models for laws in two variables, Proc. Amer. Math. Soc. 17 (1966), 1410–1412; MR 33, #7284 (G. Grätzer).

P. C. Baayen

[1] Universal Morphisms, Mathematical Centre Tracts, 9, Mathematisch Centrum, Amsterdam, 1964, 182 pp; MR 30, #3044 (A. H. Copeland, Jr.).

R. Baer

[1] The homomorphism theorems for loops, Amer. J. Math. 67 (1945), 450–460; MR 7, 7 (H. H. Campaigne).

[2] Splitting endomorphisms, Trans. Amer. Math. Soc. 61 (1947), 508–516; MR 8, 563 (J. Kuntzmann).

[3] Direct decompositions, Trans. Amer. Math. Soc. 62 (1947), 62–98; MR 9, 134 (O. Borůvka).

[4] The role of the center in the theory of direct decompositions, Bull. Amer. Math. Soc. 54 (1948), 167–174; MR 9, 410 (D. Rees).

[5] Direct decompositions into infinitely many summands, Trans. Amer. Math. Soc. 64 (1948), 519–551; MR 10, 425 (P. Lorenzen).

T. M. Baranovič

[1] The equivalence of topological spaces in primitive algebra classes (Russian), Mat. Sb. (N.S.) 56 (98) (1962), 129–136; MR 25, #1464 (J. R. Isbell).

[2] On multi-identities in universal algebras (Russian), Sibirsk. Mat. Ž. 5 (1964), 976–986; MR 30, #47 (G. Grätzer).

[3] Free decompositions in the intersection of primitive classes of algebras (Russian), Mat. Sb. (N.S.) 67 (109) (1965), 135–153; MR 32, #68 (K. A. Hirsch).

D. Barbilian

[1] Neue Gesichtspunkte über klassische Sätze aus der axiomatischen Algebra, Disquist. Math. Phys. 6 (1948), 3–48; MR 10, 502 (O. Ore).

M. Barr

[1] Cohomology in tensored categories, Proceedings of the Conference on Categorical Algebra at La Jolla, 1965, Springer-Verlag, New York, N.Y., 1966, 344–354.

M. Barr and J. Beck

[1] Acyclic models and triples, Proceedings of the Conference on Categorical Algebra at La Jolla, 1965, Springer-Verlag, New York, N.Y., 1966, 336–343.

V. Baumann and J. Pfanzagl

[1] The closure operator in partial algebras with distributive operations, Math Z. 92 (1966), 416–424; MR 33, #2584 (P. M. Cohn).

J. Beck (*See* M. Barr and J. Beck.)

V. D. Belousov

[1] Grätzer algebras and S-systems of quasigroups (Russian), Mat. Issled. 1 (1966), 55–81; MR 34, #1242 (G. Grätzer).

J. Benabou

[1] Structures Algèbriques dans les Categories, Ph.D. Thesis, Univ. Paris, 1966.

M. Benado

[1] La notion de normalité et les théorémes de décomposition de l'algèbre, Acad. Repub. Pop. Romañe. Stud. Cerc. Mat. 1 (1950), 282–317; MR 16, 212 (P. M. Whitman).

[2] Sur les théorèmes de décomposition de l'algèbre, Acad. Repub. Pop. Romañe. Stud. Cerc. Mat. 3 (1952), 263–288; MR 16, 212 (P. M. Whitman).

G. Berman and R. J. Silverman

[1] Embedding of algebraic systems, Pacific J. Math. 10 (1960), 777–786; MR 22, #11060 (I. M. H. Etherington).

E. W. Beth

[1] On Padoa's method in the theory of definition, Nederl. Akad. Wetensch. Proc. Ser. A. 56 (1953), 330-339; MR 15, 385 (A. Robinson).

[2] Some consequences of the theorem of Löwenheim-Skolem-Gödel-Mal'cev, Nederl. Akad. Wetensch. Proc. Ser. A. 56 (1953), 66–71; MR 14, 714 (I. Novak Gál).

[3] Sur le parallélisme logico-mathématique. Les méthodes formelles en axiometique. Colloques Internationaux du Centre National de la Recherche Scientifique, No. 26, Paris, 1950, pp. 27–32; discussion, pp. 32–33; Centre National de la Recherche Scientifique, Paris, 1953; MR 15, 277 (I. Novak Gál).

K. Bing

[1] On arithmetical classes not closed under direct union, Proc. Amer. Math. Soc. 6 (1955), 836–846; MR 17, 226 (R. C. Lyndon).

G. Birkhoff

[1] On the combination of subalgebras, Proc. Cambridge Phil. Soc. 29 (1933), 441–464.

[2] On the structure of abstract algebras, Proc. Cambridge Phil. Soc. 31 (1935), 433–454.

[3] Subdirect unions in universal algebra, Bull. Amer. Math. Soc. 50 (1944), 764–768; MR 6, 33 (N. H. McCoy).

[4] Universal Algebra, Proc. First Canadian Math. Congress, Montreal (1945), 310–326, University of Toronto Press, Toronto, 1946; MR 8, 432 (N. H. McCoy).

[5] On groups of automorphisms (Spanish), Rev. Un. Math. Argentina 11 (1946), 155–157; MR 7, 411 (A. A. Bennett).

[6] Lattice Theory, Amer. Math. Soc. Colloq. Publ. vol. 25, revised edition, Amer. Math. Soc., New York, N.Y. (1948); MR 10, 673 (O. Frink).

[7] Lattice theory. Amer. Math. Soc. Colloq. Publ. vol. 25, third edition, Amer. Math. Soc., New York, N.Y. 1967.

G. Birkhoff and O. Frink

[1] Representations of lattices by sets, Trans. Amer. Math. Soc. 64 (1948), 299–316; MR 10, 279 (O. Borůvka).

M. N. Bleicher and H. Schneider

[1] Completions of partially ordered sets and universal algebras, Acta Math. Acad. Sci. Hungar. 17 (1966), 271–301; MR 34, #2503 (P. F. Conrad).

L. A. Bokut'

[1] Theorems of imbedding in the theory of algebras (Russian), Colloq. Math. 14 (1966), 349–353; MR 32, #1145 (O. Wyler).

N. Bourbaki

[1] Éléments de mathématique, 22. Première partie: les structures fondamentales de l'analyse. Livre 1: Théorie des ensembles. Chapitre 4: Structures. Actualités Sci. Ind. No. 1258, Hermann, Paris, 1957; MR 20, #3804 (B. Jónsson).

R. H. Bruck

[1] A survey of binary systems, Ergebnisse der Mathematik und ihrer Grenzgebiete, Neue Folge, Heft 20, Reihe: Gruppentheorie, Springer-Verlag, Berlin-Göttingen-Heidelberg, 1958; MR 20, #76 (L. J. Paige).

P. Brunovský

[1] On generalized algebraic systems (Czech), Acta Fac. Rerum. Natur. Univ. Comenian. 3 (1958), 41–54; MR 21, #3360 (J. Jakubík).

S. Bryant (*See* J. Marica and S. Bryant.)

J. R. Büchi

[1] Representation of complete lattices by sets, Portugal. Math. 11 (1952), 151–167; MR 14, 940 (M. Novotný).

J. R. Büchi and W. Craig

[1] Notes on the family PC_{Δ} of sets of models, J. Symbolic Logic 21 (1956), 222–223.

P. Burmeister and J. Schmidt

[1] Über die Dimension einer partiellen Algebra, mit endlichen oder unendlichen Operationen, II, Z. Math. Logik Grundlagen Math. 12 (1966), 311–315; MR 34, #1243 (G. Grätzer).

J. W. Butler

[1] On complete and independent sets of operations in finite algebras, Pacific J. Math. 10 (1960), 1169–1179; MR 27, #3543 (C. C. Chang).

M. S. Calenko

[1] Proper unions and special subdirect sums in categories (Russian), Math. Sb. (N.S.) 57 (99) (1962), 75–94; MR 25, #1123 (J. R. Isbell).

[2] Completion of categories by free and direct joins of objects (Russian), Mat. Sb. (N.S.) 60 (102) (1963), 235–256; MR 27, #1484 (H. J. Hoehnke).

C. C. Chang

[1] Some general theorems on direct products and their applications in the theory of models, Nederl. Akad. Wetensch. Proc. Ser. A. 57 (1954), 592–598; MR 16, 555 (A. Robinson).

[2] On arithmetical classes which are closed under homomorphisms, Bull. Amer. Math. Soc. 62 (1956), 599.

[3] On unions of chains of models, Proc. Amer. Math. Soc. 10 (1959), 120–127; MR 21, #2576 (J. D. Rutledge).

[4] Cardinal and ordinal multiplication of relation types, Proc. Sympos. Pure Math., Vol. II, pp. 123–128, Amer. Math. Soc., Providence, R.I., 1961; MR 24, #A50 (D. Kurepa).

[5] Some new results in definability, Bull. Amer. Math. Soc. 70 (1964), 808–813; MR 30, #1053 (E. Engeler).

[6] A note on two cardinal problems, Proc. Amer. Math. Soc. 16 (1965), 1148–1155; MR 34, #1238 (G. Fodor).

C. C. Chang, B. Jónsson, and A. Tarski

[1] Refinement properties for relational structures, Fund. Math. 55 (1964), 249–281; MR 30, #3029 (R. C. Lyndon).

C. C. Chang and H. J. Keisler

[1] Applications of ultraproducts of pairs of cardinals to the theory of models, Pacific J. Math. 12 (1962), 835–845; MR 26, #3606 (E. Engeler).

[2] Continuous model theory, Ann. of Math. Studies, No. 58, Princeton Univ. Press, 1966.

C. C. Chang and A. C. Morel

[1] On closure under direct product, J. Symbolic Logic 23 (1958), 149–154; MR 21, #3359 (R. C. Lyndon).

[2] Some cancellation theorems for ordinal products of relations, Duke Math. J. 27 (1960), 171–182; MR 22, #2553 (D. Kurepa).

C. C. Chen (*See* H. H. Teh and C. C. Chen.)

D. J. Christensen and R. S. Pierce

[1] Free products of α-distributive Boolean algebras, Math. Scand. 7 (1959), 81–105; MR 26, #6077 (G. Grätzer).

Al. Climescu

[1] Une définition unitaire des algèbres à opérations finitaires, Bul. Inst. Politehn. Iași (N.S.) 6 (10) (1960), 1–4; MR 24, #1859.

P. M. Cohn

[1] Universal algebra, Harper and Row, New York, N.Y., 1965; MR 31, #224 (J. R. Isbell).

W. Craig

[1] Three uses of the Herbrand-Gentzen theorem in relating model theory and proof theory, J. Symbolic Logic 22 (1957), 269–285; MR 21, #3318 (L. N. Gál). (*See also* J. R. Büchi and W. Craig.)

P. Crawley

[1] Direct decompositions with finite dimensional factors, Pacific J. Math. 12 (1962), 457–468; MR 26, #6086 (R. P. Dilworth).

P. Crawley and B. Jónsson

[1] Direct decompositions of algebraic systems, Bull. Amer. Math. Soc. 69 (1963), 541–547; MR 28, #52 (J. Guérindon).

[2] Refinements for infinite direct decompositions of algebraic systems, Pacific J. Math. 14 (1964), 797–855; MR 30, #49 (J. Wiegold).

B. Csákány

[1] On the equivalence of certain classes of algebraic systems (Russian), Acta Sci. Math. (Szeged) 23 (1962), 46–57; MR 25, #2991 (G. Grätzer).

[2] Primitive classes of algebras which are equivalent to classes of semi-modules and modules (Russian), Acta Sci. Math. (Szeged) 24 (1963), 157–164; MR 27, #2459 (P. M. Cohn).

[3] Abelian properties of primitive classes of universal algebras (Russian), Acta Sci. Math. (Szeged) 25 (1964), 202–208; MR 30, #48 (P. M. Cohn).
[4] Inner automorphisms of universal algebras, Publ. Math Debrecen 12 (1965), 331–333; MR 32, #2356 (M. A. Harrison).

G. Čupona
[1] Distributive systems of finitary operations (Macedonian), Fac. Sci. Univ. Skopje Annuaire 13 (1960), 5–15 (1963); MR 27, #5713 (St. Schwarz).
(*See also* B. Trpenowski and G. Čupona.)

A. Daigneault
[1] Tensor products of polyadic algebras, J. Symbolic Logic, 28 (1963), 177–200; MR 31, #1215 (H. J. Keisler).
[2] On automorphisms of polyadic algebras, Trans. Amer. Math. Soc. 112 (1964), 84–130; MR 29, #45 (H. J. Keisler).
[3] Freedom in polyadic algebras and two theorems of Beth and Craig, Michigan Math. J. 11 (1964), 129–135; MR 29, #3408 (H. J. Keisler).

R. L. Davis
[1] The number of structures of finite relations, Proc. Amer. Math. Soc. 4 (1953), 486–495; MR 14, 1053 (P. Lorenzen).

K. De Bouvère
[1] Synonymous theories, The theory of models, Proc. 1963 Int. Symp. at Berkeley North-Holland, Amsterdam, 1965, 402–406.

A. H. Diamond and J. C. C. McKinsey
[1] Algebras and their subalgebras, Bull. Amer. Math. Soc. 53 (1947), 959–962; MR 9, 324 (B. Jónsson).

K. H. Diener
[1] Über zwei Birkhoff-Frinksche Struktursätze der allgemeinen Algebra, Arch. Math. 7 (1956), 339–345; MR 18, 868 (G. Birkhoff).
[2] Order in absolutely free and related algebras, Colloq. Math. 14 (1966), 63–72; MR 32, #4059 (A. Yaqub).

K. H. Diener and G. Grätzer
[1] A note on absolutely free algebras, Proc. Amer. Math. Soc. 18 (1967), 551–553.

K. Dörge
[1] Bemerkung über Elimination in beliebigen Mengen mit Operationen, Math. Nachr. 4 (1951), 282–297; MR 12, 583 (D. Zelinsky).

K. Dörge and H. K. Schuff
[1] Über Elimination in beliebigen Mengen mit allgemeinsten Operationen, Math. Nachr. 10 (1953), 315–330; MR 15, 498 (D. Zelinsky).

P. Dubreil
[1] Algèbre, Tome I, Équivalences, Opérations, Groupes, Anneaux, Corps, Gauthier-Villars, Paris (1946); MR 8, 192 (G. Birkhoff). Third edition 1963; MR 28, #2145.
[2] Les relations d'équivalence et leurs principales applications, Les conférences du Palais de la Découverte, Série A, No. 194, Université de Paris (1954); MR 16, 667 (D. C. Murdoch).
[3] Endomorphismes, Séminaire Dubreil-Pisot, 18e année, 1964/65, No. 23.

R. J. Duffin and R. S. Pate
[1] An abstract theory of the Jordan-Hölder composition series, Duke Math. J. 10 (1943), 743-750; MR 5, 170 (D. C. Murdoch).

P. Dwinger
[1] Some theorems on universal algebras I, II, Nederl. Akad. Wetensch. Proc. Ser. A. 60 (1957), 182–189, 190–195; MR 19, 240 (R. P. Dilworth).
[2] Some theorems on universal algebras III, Nederl. Akad. Wetensch. Proc. Ser. A. 61 (1958), 70–76; MR 20, #5749 (R. P. Dilworth).

B. ECKMANN AND P. J. HILTON

[1] Structure maps in group theory, Fund. Math. 50 (1961/62) 207–221; MR 24 #A2612 (K. Gruenberg).

[2] Group-like structures in general categories. I. Multiplications and Comultiplications, Math. Ann. (1961/62), 227–255; MR 25, #108 (J. F. Adams).

[3] Group-like structures in general categories. II. Equalizers, limits, lengths, Math. Ann. 151 (1963), 150–186; MR 27, #3681 (J. F. Adams).

[4] Group-like structures in general categories. III. Primitive Categories, Math. Ann. 150 (1963), 165–187; MR 27, #3682 (J. F. Adams).

A. EHRENFEUCHT

[1] On theories categorical in power, Fund. Math. 44 (1957), 241–248; MR 20, #3089 (H. B. Curry).

[2] Elementary theories with models without automorphisms, The theory of models, Proc. 1963 Int. Symp. at Berkeley, North-Holland, Amsterdam, 1965, 70–76.

A. EHRENFEUCHT AND A. MOSTOWSKI

[1] Models of axiomatic theories admitting automorphisms, Fund. Math. 43 (1956), 50–68; MR 18, 863 (I. Novak Gál).

C. EHRESMANN

[1] Structures quotient et catégories quotient, C. R. Acad. Sci. Paris 256 (1963) 5031–5034; MR 27, #3684 (A. Heller).

[2] Catégories et Structures, Dunod, Paris, 1965.

S. EILENBERG AND G. M. KELLY

[1] Closed categories, Proceedings of the Conference on Categorical Algebra at La Jolla, 1965, Springer-Verlag, New York, N.Y., 1966, 421–562.

S. EILENBERG AND S. MACLANE

[1] The general theory of natural equivalence, Trans. Amer. Math. Soc. 58 (1945), 231–294; MR 7, 109 (P. A. Smith).

S. EILENBERG AND J. C. MOORE

[1] Adjoint functors and triples, Illinois J. Math. 9 (1965), 381–398, MR 32, #2455 (P. J. Huber).

E. ENGELER

[1] Eine Konstruktion von Modellerweiterungen, Z. Math. Logik Grundlagen Math. 5 (1959), 126–131; MR 22, #12 (P. Lorenzen).

[2] A characterization of theories with isomorphic denumerable models, Notices Amer. Math. Soc. 6 (1959), 161.

[3] Unendliche Formeln in der Modelltheorie, Z. Math. Logik Grundlagen Math. 7 (1961) 154–160; MR 27, #1371 (A. Levy).

[4] A reduction-principle for infinite formulas, Math. Ann. 151 (1963), 296–303; MR 28, # 2974 (A. Nerode).

[5] Combinatorial theorems for the constructions of models, The theory of models, Proc. 1963 Int. Symp. at Berkeley, North-Holland, Amsterdam, 1965, 77–88; MR 34, #1186 (M. Makkai).

[6] Categories of mapping filters, Proceedings of the Conference on Categorical Algebra at La Jolla, 1965, Springer-Verlag, New York, N.Y., 1966, 247–253.

P. ERDŐS AND A. HAJNAL

[1] On a problem of B. Jónsson, Bull. Acad. Polon. Sci. Sér. Sci. Math. Astronom. Phys. 14 (1966), 19–23.

JU. L. ERŠOV

[1] Axiomatizable classes of models with infinite signature (Russian), Algebra i Logika Sem. 1 (1962), no. 4, 32–44; MR 33, #3930 (M. Taĭclin).

T. EVANS

[1] The word problem for abstract algebras, J. London Math. Soc. 26 (1951), 64–71; MR 12, 475 (G. Bates).

[2] Embedding theorems for multiplicative systems and projective geometries, Proc. Amer. Math. Soc. 3 (1952), 614–620; MR 14, 347 (F. W. Levi).
[3] Embeddability and the word problem, J. London Math. Soc. 28 (1953), 76–80; MR 14, 839 (G. Bates).
[4] Endomorphisms of abstract algebras, Proc. Royal Soc. Edinburgh 66 (1961–62), 53–64.
[5] Properties of algebras almost equivalent to identities, J. London Math. Soc. 37 (1962), 53–59; MR 24, #3225 (W. E. Jenner).
[6] Products of points—some simple algebras and their identities, Amer. Math. Monthly 74 (1967), 362–372.
[7] The spectrum of a variety, Z. Math. Logik Grundlagen Math. 13 (1976), 213–218.

S. Fajtlowicz
[1] Properties of the family of independent subsets of a general algebra, Colloq. Math. 14 (1966), 225–231; MR 32, #1147 (F. M. Sioson).

S. Fajtlowicz, K. Głazek, and K. Urbanik
[1] Separable variables algebras, Colloq. Math. 15 (1966), 161–171; MR 34, #5723 (A. C. Choudhury).

S. Feferman and R. L. Vaught
[1] The first order properties of products of algebraic systems, Fund. Math. 47 (1959), 57–103; MR 21, #7171 (R. C. Lyndon).

S. Feigelstock
[1] A universal subalgebra theorem, Amer. Math. Monthly 72 (1965), 884–888; MR 32, #2357 (P. J. Higgins).

W. Felscher
[1] Zur Algebra unendlich langer Zeichenreihen, Z. Math. Logik Grundlagen Math. 11 (1965), 5–16; MR 32, #2358 (P. Abellanas).
[2] Adjungierte Funktoren und primitive Klassen, Sitzungsber. Heidelberger Akad. Wiss. Math.—nat Kl., 1965, 1–65.

J. M. G. Fell and A. Tarski
[1] On algebras whose factor algebras are Boolean, Pacific J. Math. 2 (1952), 297–318; MR 14, 130 (R. C. Lyndon).

G. D. Findlay
[1] Reflexive homomorphic relations, Canad. Math. Bull. 3 (1960), 131–132; MR 23, #A1565 (S. A. Amitsur).

I. Fleischer
[1] A note on subdirect products, Acta Math. Acad. Sci. Hungar. 6 (1955), 463–465; MR 17, 819 (R. P. Dilworth).
[2] Sur le probleme d'application universelle de M. Bourbaki, C. R. Acad. Sci. Paris 254 (1962), 3161–3163; MR 25, #1122 (P. J. Higgins).
[3] Elementary properties of inductive limits, Z. Math. Logik Grundlagen Math. 9 (1963), 347–350; MR 28, #1130 (C. C. Chang).

A. L. Foster
[1] On the permutational representation of general sets of operations by partition lattices, Trans. Amer. Math. Soc. 66 (1949), 366–388; MR 11, 75 (P. M. Whitman).
[2] Generalized "Boolean" theory of universal algebras, Part I: Subdirect sums and normal representation theorem, Math. Z. 58 (1953), 306–336; MR 15, 194 (B. Jónsson).
[3] Generalized "Boolean" theory of universal algebras, Part II: Identities and subdirect sums in functionally complete algebras, Math Z. 59 (1953), 191–199; MR 15, 194 (B. Jónsson).
[4] The identities of—and unique subdirect factorization within—classes of universal algebras, Math. Z. 62 (1955), 171–188; MR 17, 452 (B. Jónsson).
[5] On the finiteness of free (universal) algebras, Proc. Amer. Math. Soc. 7 (1956), 1011–1013; MR 18, 788 (B. Jónsson).

[6] Ideals and their structure in classes of operational algebras, Math Z. 65 (1956), 70–75; MR 18, 108 (B. Jónsson).
[7] The generalized Chinese remainder theorem for universal algebras; subdirect factorization, Math. Z. 66 (1956), 452–469; MR 18, 788 (B. Jónsson).
[8] An existence theorem for functionally complete universal algebras, Math. Z. 71 (1959), 69–82; MR 21, #7172 (B. Jónsson).
[9] On the imbeddability of universal algebras in relation to their identities I, Math. Ann. 138 (1959), 219–238; MR 22, #5603 (B. Jónsson).
[10] Functional completeness in the small. Algebraic structure theorems and identities, Math. Ann. 143 (1961), 29–53; MR 23, #A85 (B. Jónsson).
[11] Generalized equational maximality and primal-in-the-small algebras, Math. Z. 79 (1962), 127–146; MR 27, #4775 (O. Frink).
[12] Functional completeness in the small II. Algebraic cluster theorem, Math. Ann. 148 (1962), 173–191; MR 27; #5714 (H. Ribeiro).
[13] Families of algebras with unique (sub-) direct factorization. Equational characterization of factorization, Math. Ann. 166 (1966), 302–326; MR 33, #7285 (F. M. Sioson).

A. L. Foster and A. Pixley
[1] Semi-categorical algebras. I. Semi-primal algebras, Math. Z. 83 (1964), 147–169; MR 29, #1165 (H.-J. Hoehnke).
[2] Semi-categorical algebras. II, Math. Z. 85 (1964), 169–184; MR 29, #5771 (H.-J. Hoehnke).
[3] Algebraic and equational semi-maximality; equational spectra. I, Math. Z. 92 (1966), 30–50; MR 33, #3973 (H.-J. Hoehnke).

R. Fraïssé
[1] Sur une classification des systèmes de relations faisant intervenir les ordinaux transfinis, C. R. Acad. Sci. Paris 228 (1949), 1682–1684; MR 11, 17 (R. Arens).
[2] Sur les types de polyrelations et sur une hypothèse d'origine logistique, C. R. Acad. Sci. Paris 230 (1950), 1557–1559; MR 12, 14 (R. Arens).
[3] Sur une nouvelle classification des systèmes de relations, C. R. Acad. Sci. Paris 230 (1950), 1022–1024; MR 11, 585 (R. Arens).
[4] Sur la signification d'une hypothèse de la théorie des relations, du point de vue du calcul logique, C. R. Acad. Sci. Paris 232 (1951), 1793–1795; MR 13, 99 (R. Arens).
[5] Some elementary properties of universal classes, Bull. Amer. Math. Soc. 60 (1954), 396.
[6] Sur les rapports entre la théorie des relations et la sémantique au sens de A. Tarski, Applications scientifiques de la logique mathématique, Colloque de logique mathématique Paris 95 (1954).
[7] Sur l'extension aux relations de quelques propriétés des ordres, Ann. Sci. École Norm. Sup. 71 (1954), 363–388; MR 16, 1006 (G. Kurepa).
[8] Sur quelques classifications des systèmes de relations, Publ. Sci. Univ. Alger. Sér. A. 1 (1954), 35–182; MR 16, 1005.
[9] Sur quelques classifications des relations, basées sur des isomorphismes restreints, Publ. Sci. Univ. Alger. Sér. A. 2 (1955), 15–60 (1956); MR 18, 139 (G. Kurepa).
[10] Sur quelques classifications des relations, basées sur des isomorphismes restreints II. Application aux relations d'ordre, et construction d'exemples montrant que ces classifications sont distinctes, Publ. Sci. Univ. Alger. Sér. A. 2 (1955), 273–295 (1957); MR 19, 111 (G. Kurepa).
[11] Application des γ-opérateurs au calcul logique du premier echelon, Z. Math. Logik Grundlagen Math. 2 (1956), 76–92; MR 19, 829 (G. Kurepa).
[12] A hypothesis concerning the extension of finite relations and its verification for certain special cases, First Part, The theory of models, Proc. 1963 Int. Symp. at Berkeley, North-Holland, Amsterdam, 1965, 96–106; MR 33, #3942 (G. Sabidussi).

T. Frayne, A. C. Morel, and D. S. Scott
[1] Reduced direct products, Fund. Math. 51 (1962/63), 195-228; MR 26, #28 (C. C. Chang).

P. Freyd
[1] The theories of functors and models, The theory of models, Proc. 1963 Int. Symp. at Berkeley, North-Holland, Amsterdam, 1965, 107–120; MR 34, #1371 (R. L. Knighten).
[2] Algebra valued functors in general and tensor products in particular, Colloq. Math. 14 (1966), 89–106; MR 33, #4116 (F. E. J. Linton).

O.Frink (*See* G. Birkhoff and O. Frink.)

O. Frink and G. Grätzer
[1] The closed subalgebras of a topological algebra, Arch. Math. (Basel) 17 (1966), 154–158; MR 33, #7382 (K. G. Wolfson).

L. Fuchs
[1] On subdirect unions I, Acta Math. Acad. Sci. Hungar. 3 (1952), 103-120; MR 14, 612 (K. A. Hirsch).
[2] On partially ordered algebras, I, Colloq. Math. 14 (1966), 115–130; MR 32, #2359 (B. M. Schein).
[3] On partially ordered algebras. II, Acta Sci. Math. (Szeged) 26 (1965), 35–41; MR 31, #4749 (P. F. Conrad).

G. Fuhrken
[1] Bemerkung zu einer Arbeit E. Engelers, Z. Math. Logik Grundlagen Math. 8 (1962), 277–279; MR 26, #3600 (E. Engeler).
[2] Minimal- und Primmodelle, Z. Math. Logik Grundlagen Math. 9 (1966), 3–11.

T. Fujiwara
[1] On the structure of algebraic systems, Proc. Japan Acad. 30 (1954), 74–79; MR 16, 107 (O. Ore).
[2] Note on the isomorphism problem for free algebraic systems, Proc. Japan Acad. 31 (1955), 135-136; MR 17, 227 (R. C. Lyndon).
[3] Remarks on the Jordan-Hölder-Schreier theorem, Proc. Japan Acad. 31 (1955) 137–140; MR 17, 450 (B. Jónsson).
[4] On the existence of algebraically closed algebraic extensions, Osaka J. Math. 8 (1956), 23–33; MR 18, 11 (O. Ore).
[5] Note on free algebraic systems, Proc. Japan Acad. 32 (1956), 662–664; MR 18, 636 (O. Ore).
[6] Supplementary note on free algebraic systems, Proc. Japan Acad. 33 (1957), 633–635; MR 20, #5750a (O. Ore).
[7] Note on free products, Proc. Japan Acad. 33 (1957), 636–638; MR 20, #5750b (O. Ore).
[8] On mappings between algebraic systems, Osaka J. Math. 11 (1959), 153–172; MR 22, #1535 (G. B. Seligman).
[9] On mappings between algebraic systems II, Osaka J. Math. 12 (1960), 253–268; MR 22, #10933 (G. B. Seligman).

L. Novak Gál
[1] A note on direct products, J. Symbolic Logic 23 (1958), 1–6; MR 20, #5121 (E. J. Cogan).

N. D. Gautam
[1] The validity of equations of complex algebras, Arch. Math. Logik Grundlagenforsch. 3 (1957), 117–124; MR 19, 1152 (B. Jónsson).

F. Gécseg
[1] On certain classes of semi-modules and modules (Russian), Acta Sci. Math. (Szeged) 24 (1963), 165–172; MR 27, #2460 (P. M. Cohn).

H. Gericke
[1] Über den Begriff der algebraischen Struktur, Arch. Math. 4 (1953), 163–171; MR 15, 280 (O. Frink).

A. GHIKA

[1] Structures algébriques ternaires (Romanian), Com. Acad. R. P. Romîne 8 (1958), 257–261; MR 21, #1283a (M. M. Day).

[2] Structures algébrico-topologiques binaires-ternaires (Romanian), Com. Acad. R. P. Romîne 8 (1958), 447–450; MR 21, #1283b (M. M. Day).

P. C. GILMORE

[1] Some forms of completeness, J. Symbolic Logic 27 (1962), 344–352.

K. GŁAZEK (*See* S. FAJTLOWICZ, K. GŁAZEK, AND K. URBANIK.)

L. M. GLUSKIN

[1] Ideals of semigroups of transformations (Russian), Mat. Sb. (N.S.) 47 (89) (1959), 111–130; MR 21, #4197 (L. J. Paige).

[2] Ideals of semigroups (Russian), Mat. Sb. (N.S.) 55 (97) (1961), 421–448; MR 25, #3106 (Št. Schwarz).

[3] On dense embeddings (Russian), Mat. Sb. (N.S.) 61 (103) (1963), 175–206; MR 27, #2570 (K. A. Ross).

K. GÖDEL

[1] Die Vollständigkeit der Axiome des logischen Funktionenkalküls, Monatshefte für Mathematik und Physik 37 (1930), 349–360.

A. GOETZ

[1] On weak isomorphisms and weak homomorphisms of abstract algebras, Colloq. Math. 14 (1966), 163–167; MR 32, #2360 (H. Neumann).

A. GOETZ AND C. RYLL-NARDZEWSKI

[1] On bases of abstract algebras, Bull. Acad. Polon. Sci. Sér. Sci. Math. Astronom. Phys. 8 (1960), 157–161; MR 22, #6753 (O. Frink).

A. W. GOLDIE

[1] The Jordan-Hölder theorem for general abstract algebras, Proc. London Math. Soc. (2) 52 (1950), 107–131; MR 12, 238 (G. Birkhoff).

[2] The scope of the Jordan-Hölder theorem in abstract algebra, Proc. London Math. Soc. (3) 2 (1952), 349–368; MR 14, 129 (P. M. Whitman).

[3] On direct decompositions I, Proc. Cambridge Philos. Soc. 48 (1952), 1–22; MR 14, 9 (J. Riguet).

[4] On direct decompositions II, Proc. Cambridge Philos. Soc. 48 (1952), 23–34; MR 14, 10 (J. Riguet).

M. GOULD

[1] On extensions of Schreier's theorem to universal algebras, Studia Sci. Math. Hungar. 1 (1966), 369–377; MR 34, #2513 (G. Grätzer).

M. I. GOULD AND G. GRÄTZER

[1] Boolean extensions and normal subdirect powers of finite universal algebras, Math. Z. 99 (1967), 16–25.

G. GRÄTZER

[1] A representation theorem for multi-algebras, Arch. Math. 13 (1962), 452–456; MR 26, #3629 (A. Sade).

[2] A theorem on doubly transitive permutation groups with application to universal algebras, Fund. Math. 53 (1963), 25–41; MR 28, #53 (P. J. Higgins).

[3] Free algebras over first order axiom systems, Magyar Tud. Akad. Mat. Kutató Int. Közl. 8 (1963), 193–199; MR 29, #4716 (O. Frink).

[4] On the Jordan-Hölder theorem for universal algebras, Magyar. Tud. Akad. Mat. Kutató Int. Közl. 8 (1963), 397–406 (1964); MR 29, #4717 (O. Frink).

[5] On the class of subdirect powers of a finite algebra, Acta Sci. Math. (Szeged) 25 (1964), 160–168; MR 29, #5772 (P. J. Higgins).

[6] A new notion of independence in universal algebras, Colloq. Math. 17 (1967), 225–234.

[7] On the endomorphism semigroup of simple algebras, Math. Ann. 170 (1967), 334–338.

[8] On the family of certain subalgebras of a universal algebra, Nederl. Akad. Wetensch. Proc. Ser. A. 68 (1965), 790–802; MR 32; #7468 (E. T. Schmidt).

[9] On coverings of universal algebras, Arch. Math. 18 (1967), 113–117.

[10] Free Σ-structures, Trans. Amer. Math. Soc. 135 (1969), 517–542.

[11] On the existence of free structures over universal classes, Math. Nachr. 36 (1968), 135–140.

[12] On spectra of classes of algebras, Proc. Amer. Math. Soc. 18 (1967), 729–735.

[13] On polynomial algebras and free algebras, Canadian J. Math. 20 (1968), 575–581.

[14] Chapter on Universal algebra, in Trends in lattice theory, edited by J. C. Abbott, Van Nostrand Reinhold, New York, 1970.

(*See also* K. H. Diener and G. Grätzer; O. Frink and G. Grätzer; M. I. Gould and G. Grätzer.)

G. Grätzer and W. A. Lampe

[1] On the subalgebra lattice of universal algebras, J. Algebra 7 (1967), 263–270.

G. Grätzer and E. T. Schmidt

[1] On inaccessible and minimal congruence relations I, Acta. Sci. Math. (Szeged) 21 (1960), 337–342; MR 23, #A3697 (P. J. Higgins).

[2] Characterizations of congruence lattices of abstract algebras, Acta Sci. Math. (Szeged) 24 (1963), 34–59; MR 27, #1391 (P. M. Whitman).

G. Grätzer and G. H. Wenzel

[1] On the concept of congruence relation in partial algebras, Math. Scand. 20 (1967), 275–280.

J. A. Green

[1] A duality in abstract algebra, J. London Math. Soc. 27 (1952), 64–73; MR 14, 133 (R. D. Schafer).

A. Hajnal (*See* P. Erdös and A. Hajnal.)

P. Hall

[1] Some word-problems, J. London Math. Soc. 33 (1958), 482–496; MR 21, #1331 (R. H. Bruck).

P. R. Halmos

[1] Algebraic Logic, Chelsea Publishing Co., New York, N.Y. 1962, 271 pp; MR 24, #A1808.

E. Harzheim

[1] Über die Grundlagen der universellen Algebra, Math. Nachr. 31 (1966), 39–52; MR 33, #84 (P. M. Cohn).

J. Hashimoto

[1] Direct, subdirect decompositions and congruence relations, Osaka J. Math. 9 (1957), 87–112; MR 19, 935 (R. P. Dilworth).

[2] Congruence relations and congruence classes in lattices, Osaka J. Math. 15 (1963), 71–86; MR 27, #1394 (M. F. Smiley).

[3] On meromorphisms of algebraic systems, Nagoya Math. J. 27 (1966), 559–569; MR 33, #7286 (F. M. Sioson).

[4] On meromorphisms and congruence relations, J. Math. Soc. Japan 19 (1967), 70–81; MR 34, #4181 (F. M. Sioson).

Z. Hedrlin and A. Pultr

[1] On full embeddings of categories of algebras, Illinois J. Math. 10 (1966), 392–406; MR 33, #85 (J. R. Isbell).

L. A. Henkin

[1] The completeness of the first-order functional calculus, J. Symbolic Logic 14 (1949), 159–166; MR 11, 487 (J. C. C. McKinsey).

[2] Some interconnections between modern algebra and mathematical logic, Trans. Amer. Math. Soc. 74 (1953), 410–427; MR 14, 1052 (P. Lorenzen).

[3] A generalization of the concept of ω-consistency, J. Symbolic Logic 19 (1954) 183–196; MR 16, 103 (G. Kreisel).

[4] On a theorem of Vaught, Nederl. Akad. Wetensch. Proc. Ser. A. 58 (1955), 326-328; MR 16, 1080 (G. Kreisel).

[5] Two concepts from the theory of models, J. Symbolic Logic 21 (1956), 28–32; MR 17, 816 (G. Kreisel).

[6] A generalization of the concept of ω-completeness, J. Symbolic Logic 22 (1957), 1–14; MR 20, #1626 (A. Robinson).

[7] An extension of the Craig-Lyndon interpolation theorem, J. Symbolic Logic 28 (1963) 201–216; MR 30, #1927 (R. C. Lyndon).

G. C. Hewitt

[1] The existence of free unions in classes of abstract algebras, Proc. Amer. Math. Soc. 14 (1963), 417–422; MR 26, #6098 (G. Grätzer).

P. J. Higgins

[1] Algebras with a scheme of operators, Math. Nachr. 27 (1963), 115–132; MR 29, #1239 (A. Heller).

[2] Presentations of groupoids, with applications to groups, Proc. Cambridge Philos. Soc. 60 (1964), 7–20; MR 28, #1246 (H.-J. Hoehnke).

G. Higman

[1] Ordering by divisibility in abstract algebras, Proc. London Math. Soc. (3) 2 (1952), 326–336; MR 14, 238 (D. Zelinsky).

G. Higman and A. H. Stone

[1] On inverse systems with trivial limits, J. London Math. Soc. 29 (1954), 233–236; MR 15, 773 (D. Zelinsky).

P. J. Hilton

[1] Note on free and direct products in general categories, Bull. Soc. Math. Belg. 13 (1961), 38–49; MR 24, #A3189 (P. J. Higgins).

(*See also* B. Eckmann and P. J. Hilton.)

K. J. Hintikka

[1] An application of logic to algebra, Math. Scand. 2 (1954), 243–246; MR 17, 449 (A. Robinson).

Ja. V. Hion

[1] Ω-systems (Russian), Interuniv. Sci. Sympos. General Algebra, 123–129, Tartu Gos. Univ., Tartu, 1966; MR 33, #7288 (R. S. Pierce).

H.-J. Hoehnke

[1] Charakterisierung der Kongruenzklassen—Teilsysteme in binären partiellen Algebren, Math. Nachr. 25 (1963), 59–64; MR 28, #48 (Št. Schwarz).

[2] Einige neue Resultate über abstrakte Halbgruppen, Colloq. Math. 14 (1966), 329–348.

[3] Zur Strukturgleichheit axiomatischer Klassen, Z. Math. Logik Grundlagen Math. 12 (1966), 69–83; MR 32, #5499 (M. Makkai).

[4] Radikale in allgemeinen Algebren, Math. Nachr. 35 (1967), 347–383.

A. Horn

[1] On sentences which are true of direct unions of algebras, J. Symbolic Logic 16 (1951), 14–21; MR 12, 662 (O. Frink).

L. A. Hostinsky

[1] Direct decompositions in lattices, Amer. J. Math. 73 (1951), 741–755; MR 13, #525 (P. Lorenzen).

[2] Loewy chains and uniform splitting of lattices, Proc. Amer. Math. Soc. 5 (1954), 315–319; MR 15, #673 (P. M. Whitman).

Tah-kai Hu

[1] Distributive extensions and quasi-framal algebras, Canad. J. Math. 18 (1966), 265–281; MR 32, #7470 (D. Monk).

[2] On the bases of free algebras, Math. Scand. 16 (1965), 25–28; MR 32, #7469 (H. J. Keisler).

N. J. S. Hughes

[1] Refinement and uniqueness theorems for the decompositions of algebraic systems with a regularity condition, J. London Math. Soc. 30 (1955), 259–273; MR 16, 1083 (B. Jónsson).

[2] The Jordan-Hölder-Schreier Theorem for general algebraic systems, Compositio Math. 14 (1960), 228–236; MR 24, #A699 (J. Guerindon).

M. HUKUHARA

[1] Théorie des endomorphismes complètement continus, I, II, J. Fac. Sci. Univ. Tokyo 7 (1957/58), 511-525; MR 18, #909 (M. E. Shanks), MR 20, #2620 (J. H. Williamson).

A. HULANICKI, E. MARCZEWSKI, AND J. MYCIELSKI

[1] Exchange of independent sets in abstract algebras (I), Colloq. Math. 14 (1966), 203–215.

A. HULANICKI AND S. ŚWIERCZKOWSKI

[1] Number of algebras with a given set of elements, Bull. Acad. Polon, Sci. Sér. Sci. Math. Astronom. Phys. 8 (1960), 283–284; MR 24, #A3113 (Ph. Dwinger).

J. R. ISBELL

[1] Subobjects, adequacy, completeness and categories of algebras, Rozprawy Mat. 36 (1964) 33 pp; MR 29, #1238 (A. Heller).

[2] Epimorphisms and dominions, Proceedings of the Conference on Categorical Algebra at La Jolla, 1965, Springer-Verlag, New York, 1966, 232–246.

K. ISÉKI

[1] On endomorphism with fixed element on algebra, Proc. Japan Acad. 40 (1964), 403; MR 30, #1074 (A. Hulanicki).

A. A. ISKANDER

[1] Correspondence lattices of universal algebras (Russian), Izv. Akad. Nauk SSSR Ser. Mat. 29 (1965), 1357–1372; MR 33, #7289 (E. T. Schmidt).

[2] Universal algebras with relations and amalgams (Russian), Vestnik Moskov. Univ. Ser. I. Mat. Meh. 4 (1965), 22–28; MR 33, #86 (R. S. Pierce).

[3] Partial universal algebras with preassigned lattices of subalgebras and correspondences (Russian), Mat. Sb. (N.S) 70 (112) (1966), 438–456; MR 33, #5541 (R. S. Pierce).

O. A. IVANOVA

[1] On direct powers of unary algebras (Russian), Vestnik Moskov. Univ. Ser. I. Mat. Meh. 3 (1964), 31–38; MR 29, #2207 (R. S. Pierce).

J. JAKUBIK

[1] Congruence relations on abstract algebras (Russian), Czechoslovak Math. J. 4 (79) (1954), 314–317; MR 16, 787 (P. M. Whitman).

[2] Direct decompositions of the unity in modular lattices (Russian), Czechoslovak Math. J. 5 (80) (1955), 399–411; MR 18, 106 (I. Kaplansky).

[3] On the existence of algebras (Czech), Časopis Pěst. Mat. 81 (1956), 43–54; MR 18, 275 (G. Birkhoff).

B. JÓNSSON

[1] Universal relational systems, Math. Scand. 4 (1956), 193–208; MR 20, #3091 (S. Ginsburg).

[2] On isomorphism types of groups and other algebraic systems, Math. Scand. 5 (1957), 224–229; MR 21, #7169 (Št. Schwarz).

[3] Homogeneous universal relational systems, Math. Scand. 8 (1960), 137–142; MR 23, #A2328 (S. Ginsburg).

[4] Sublattices of a free lattice, Canad. J. Math., 13 (1961), 256–264; MR 23, #A818 (R. P. Dilworth).

[5] Algebraic extensions of relational systems, Math. Scand. 11 (1962), 179–205; MR 27, #4777 (I. G. Amemiya).

[6] Extensions of relational structures, Theory of Models, Proc. 1963 Int. Symp. at Berkeley, North-Holland, Amsterdam, 1965, 146–157; MR 34, #2463 (J. E. Rubin).

[7] The unique factorization problem for finite relational structures, Colloq. Math. 14 (1966), 1–32; MR 33, #87 (C. C. Chang)

[8] Algebras whose congruence lattices are distributive, Math. Scand. 21 (1967), 110–121.

(*See also* C. C. Chang, B. Jónsson, and A. Tarski; P. Crawley and B. Jónsson.)

B. Jónsson and A. Tarski

[1] Direct Decompositions of Finite Algebraic Systems, Notre Dame Mathematical Lectures No. 5, University of Notre Dame, Notre Dame, Ind., 1947; MR 8, 560 (G. Birkhoff).

[2] On two properties of free algebras, Math. Scand. 9 (1961), 95–101; MR 23, #A3695 (J. McLaughlin).

J. Kalicki

[1] On comparison of finite algebras, Proc. Amer. Math. Soc. 3 (1952), 36–40; MR 13, 898 (P. Lorenzen).

[2] The number of equationally complete classes of equations, Nederl. Akad. Wetensch. Proc. Ser. A. 58 (1955), 660–662; MR 17, 571 (A. Robinson).

J. Kalicki and D. Scott

[1] Some equationally complete algebras, Bull. Amer. Math. Soc. 59 (1953), 77–78.

[2] Equational completeness of abstract algebras, Nederl. Akad. Wetensch. Proc. Ser. A. 58 (1955), 650–659; MR 17, 571 (A. Robinson).

J. A. Kalman

[1] Equational completeness and families of sets closed under subtraction, Indag. Math. 22 (1960), 402–405; MR 25, #2992 (R. C. Lyndon).

L. N. Karolinskaja

[1] Direct decompositions of abstract algebras with distinguished subalgebras (Russian), Izv. Vysš. Učebn. Zaved. Matematika 17 (1960), 106–113; MR 28, #5021 (R. M. Baer).

H. J. Keisler

[1] Theory of models with generalized atomic formulas, J. Symbolic Logic 25 (1960), 1–26; MR 24, #A36 (L. N. Gál).

[2] Ultraproducts and elementary classes, Nederl. Akad. Wetensch. Proc. Ser. A. 64 (1961), 477–495; MR 25, #3816 (C. C. Chang).

[3] On some results of Jónsson and Tarski concerning free algebras, Math. Scand. 9 (1961), 102–106; MR 23, #A3696 (J. McLaughlin).

[4] Limit ultrapowers, Trans. Amer. Math. Soc. 107 (1963), 382–408; MR 26, #6054 (A. Mostowski).

[5] Good ideals in fields of sets, Ann. of Math. (2) 79 (1964), 338–359; MR 29, #3383 (A. Lévy).

[6] On cardinalities of ultraproducts, Bull. Amer. Math. Soc. 70 (1964), 644–647; MR 29, #3384 (C. C. Chang).

[7] Unions of relational systems, Proc. Amer. Math. Soc. 15 (1964), 540–545; MR 29, #2185 (J. Mycielski).

[8] Ultraproducts and saturated models, Nederl. Akad. Wetensch. Proc. Ser. A. 67 (1964), 178–186; MR 29 #5745 (A. Lévy).

[9] Reduced products and Horn classes, Trans. Amer. Math. Soc. 117 (1965), 307–328; MR 30, #1047 (C. C. Chang).

[10] Finite approximations of infinitely long formulas, The theory of models, Proc. 1963 Int. Symp. at Berkeley, North-Holland, Amsterdam, 1965, 158–169; MR 34, #2464 (D. Monk).

[11] Some applications of infinitely long formulas, J. Symbolic Logic 30 (1965), 339–349.

[12] Limit ultraproducts, J. Symbolic Logic 30 (1965), 212–234; MR 33, #46 (R. C. Lyndon).

[13] First order properties of pairs of cardinals, Bull. Amer. Math. Soc. 72 (1966), 141–144; MR 32, #1117 (M. Morley).

[14] A survey of ultraproducts, Proc. of the 1964 Int. Congress of Logic, Meth. and Phil. of Sci., Amsterdam, 1965, 112–126; MR 34, #5678 (A. Daigneault).

[15] Some model-theoretic results for ω-logic, Israel J. Math. 4 (1966), 249–261.

[16] Ultraproducts of finite sets, J. Symbolic Logic 32 (1967), 47–57.

[17] Ultraproducts which are not saturated, J. Symbolic Logic 32 (1967), 23–46. (*See also* C. C. Chang and H. J. Keisler.)

G. M. Kelly (*See* S. Eilenberg and G. M. Kelly.)

R. Kerkhoff

[1] Eine Konstruktion absolut freier Algebren, Math. Ann. 158 (1965), 109–112; MR 31, #2185 (H. F. J. Lowig).

S. Kochen

[1] Ultraproducts in the theory of models, Ann. of Math. (2) 74 (1961), 221–261; MR 25, #1992 (G. Kreisel).

[2] Topics in the theory of definition, The theory of models, Proc. 1963 Int. Symp. at Berkeley, North-Holland, Amsterdam, 1965, 170–176.

S. R. Kogalovskiĭ

[1] Universal classes of algebras (Russian), Dokl. Akad. Nauk SSSR 122 (1958), 759–761; MR 21 #1281 (R. A. Good).

[2] On universal classes of algebras, closed under direct products (Russian), Izv. Vysš, Učebn. Zaved. Mathematika 10 (1959), 88–96; MR 27, #84 (R. M. Baer).

[3] Universal classes of models (Russian), Dokl. Akad. Nauk SSSR 124 (1959), 260–263; MR 21, #1932 (E. J. Cogan).

[4] A general method of obtaining the structural characteristics of axiomatized classes (Russian), Dokl. Akad. Nauk SSSR 136 (1961), 1291–1294; MR 22, #12045 (E. Mendelson).

[5] Structural characteristics of universal classes (Russian), Sibirsk. Mat. Ž. 4 (1963), 97–119; MR 26, #4909 (E. J. Cogan).

[6] On the relation between finitely projective and finitely reductive classes of models (Russian), Dokl. Akad. Nauk SSSR 155 (1964), 1255–1257; MR 30, #1048 (H. Rasiowa).

[7] Some remarks on ultraproducts (Russian), Izv. Akad. Nauk SSSR, Ser. Mat. 29 (1965), 997–1004; MR 34, #4135 (E. Mendelson).

[8] Certain remarks on non-elementary axiomatizable classes of models (Russian) Ural Gos. Univ. Mat. Zap. 5 (1965), 54–60; MR 34, #1188 (M. Misicu).

[9] Generalized quasiuniversal classes of models (Russian), Izv. Akad. Nauk SSSR Ser. Mat. 29 (1965), 1273–1282; MR 33, #3932 (G. Asser).

[10] On the theorem of Birkhoff (Russian), Uspehi Mat. Nauk 20 (1965), 206–207; MR 34 #1249 (V. Dlab).

[11] On finitely reductive classes of models (Russian), Sibirsk. Mat. Ž. 6 (1965), 1021–1025.

[12] Two problems on finitely projective classes (Russian), Sibirsk. Mat. Ž. 6 (1965), 1429–1431; MR 34, #36 (E. Mendelson).

[13] On properties preserved under algebraic constructions (Russian), Interuniv. Sci. Sympos. General Algebra (Russian), 44–51. Tartu Gos. Univ., Tartu, 1966; MR 34, #5679 (K. Drbohlav).

M. Kolibiar

[1] Über direkte Produkte von Relativen, Acta Fac. Natur. Univ. Comenian, 10 (1965), 1–9; MR 34, #2514 (A. C. Choudhury).

V. S. Krishnan

[1] Homomorphisms and congruences in general algebra, Math. Student 13 (1945), 1–9; MR 7, 110 (G. Birkhoff).

A. H. Kruse

[1] Completion of mathematical systems, Pacific J. Math. 12 (1962), 589–605; MR 26, #1273 (E. Engeler).

[2] An abstract property P for groupoids such that locally locally P is weaker than locally P, J. London Math. Soc. 42 (1967), 81–85; MR 34, #4183 (M. Yamada).

A. G. Kuroš

[1] Free sums of multiple operator algebras (Russian), Sibirsk. Mat. Ž. 1 (1960), 62–70; correction 638; MR 24, #A1278 (L. Fuchs).

[2] Lectures on general algebra, Gosudarstv. Izdat. Fiz.-Mat. Lit., Moscow, 1962; MR 25, #5097 (G. Grätzer). (Two English translations are available.)

A. G. Kuroš, A. H. Livšic, and E. G. Šul'geĭfer

[1] Foundations of the theory of categories (Russian), Usephi Mat. Nauk 15 (1960), 3–52; MR 22, #9526 (T. M. Beck).

G. Ladner

[1] On two problems of J. Schmidt, Proc. Amer. Math. Soc. 6 (1955), 647–650; MR 17, 226 (P. Lorenzen).

J. Lambek

[1] Goursat's theorem and the Zassenhaus lemma, Canad. J. Math. 10 (1958), 45–46; MR 20, #4600 (O. Borůvka).

[2] Completions of categories, Springer lecture notes in mathematics, vol. 24, 1966.

W. A. Lampe (*See* G. Grätzer and W. A. Lampe.)

F. W. Lawvere

[1] Functorial semantics of algebraic theories, Proc. Nat. Acad. Sci., U.S.A. 50 (1963), 869–872; MR 28, #2143 (M. Artin).

[2] Algebraic theories, algebraic categories, and algebraic functors, The theory of models, Proc. 1963 Int. Symp. at Berkeley, North-Holland, Amsterdam, 1965, 413–418.

[3] The category of categories as a foundation for mathematics, Proceedings of the Conference on Categorical Algebra at La Jolla, 1965, Springer-Verlag, New York, N.Y., 1966, 1–20.

M. H. Lim (*See* H. H. Teh and M. H. Lim.)

F. E. J. Linton

[1] Some aspects of equational categories, Proceedings of the Conference on Categorical Algebra at La Jolla, 1965, Springer-Verlag, New York, N.Y., 1966, 84–94.

[2] Autonomous categories and duality of functors, J. Alg. 2 (1965), 315–349.

[3] Autonomous equational categories, J. Math. Mech. 15 (1966), 637–642; MR 32, #7619.

A. H. Livšic (*See* A. G. Kuroš, A. H. Livšic, and E. G. Šul'geĭfer.)

E. S. Ljapin

[1] Abstract characterization of the class of endomorphism semigroups of systems of a general type (Russian), Mat. Sb. (N.S.) 70 (112) (1966), 171–179.

P. Lorenzen

[1] Eine Bemerkung zum Schreierschen Verfeinerungssatz, Math. Z. 49 (1944), 647–653; MR 6, 143 (R. Baer).

[2] Über die Korrespondenzen einer Struktur, Math. Z. 60 (1954), 61–65; MR 16, 787 (O. Frink).

J. Łoś

[1] On the categoricity in power of elementary deductive systems and some related problems, Colloq. Math. 3 (1954), 58–62; MR 15, 845 (H. B. Curry).

[2] Quelques remarques théorèmes et problèmes sur les classes définissables d'algèbras, Mathematical interpretation of formal systems, North-Holland, Amsterdam, 1955, 98–113; MR 17, 700 (E. Mendelson).

[3] On the extending of models I, Fund. Math. 42 (1955), 38–54; MR 17, 224 (G Kreisel).

[4] Common extension in equational classes, Log. Method. and Phil. of Sci. (Proc 1960 Int. Congr.), Stanford Univ. Press, Stanford, Calif., 1962, 136–142; MR 27 #2400 (E. Engeler).

[5] Free product in general algebra, The theory of models, Proc. 1963 Int. Symp. at Berkeley, North-Holland, Amsterdam, 1965, 229–237; MR 33, #3976 (R. C. Lyndon).

J. ŁOŚ, J. SŁOMIŃSKI, AND R. SUSZKO

[1] On extending of models. V. Embedding theorems for relational models, Fund. Math. 48 (1959/60), 113–121; MR 22, #3676 (C. C. Chang).

J. ŁOŚ AND R. SUSZKO

[1] On the infinite sums of models, Bull. Acad. Polon. Sci. Sér. Sci. Math. Astronom. Phys. 3 (1955), 201–202; MR 17, 224 (G. Kreisel).

[2] On the extending of models II, Common extensions, Fund. Math. 42 (1955), 343–347; MR 17, 815 (G. Kreisel).

[3] On the extending of models IV, Fund. Math. 44 (1957), 52–60; MR 19, 724 (G. Kreisel).

H. F. J. LOWIG (LÖWIG)

[1] On the properties of freely generated algebras, J. Reine Angew. Math. 190 (1952), 65–74; MR 14, 443 (R. C. Lyndon).

[2] Gesetzrelationen über frei erzeugten Algebren, J. Reine Angew. Math. 193 (1954), 129–142; MR 16, 786 (H. A. Thurston).

[3] On the existence of freely generated algebras, Proc. Cambridge Philos. Soc. 53 (1957), 790–795; MR 19, 1153 (H. A. Thurston).

[4] On some representations of lattices of law relations, Osaka J. Math. 10 (1958), 159–180; MR 21, #1282 (H. A. Thurston).

[5] On the composition of some representations of lattices of law relations, Osaka J. Math. 13 (1961), 217–228; MR 25, #2012 (H. A. Thurston).

R. C. LYNDON

[1] Identities in two-valued calculi, Trans. Amer. Math. Soc. 71 (1951), 457–465; MR 13, 422 (A. Rose).

[2] Identities in finite algebras, Proc. Amer. Math. Soc. 5 (1954), 8–9; MR 15, 676 (B. Jónsson).

[3] An interpolation theorem in the predicate calculus, Pacific J. Math. 9 (1959), 129–142; MR 21, #5555 (E. W. Beth).

[4] Properties preserved under homomorphism, Pacific J. Math. 9 (1959) 143–154; MR 21, #7157 (J. Jakubik).

[5] Properties preserved in subdirect products, Pacific J. Math. 9 (1959), 155–164; MR 21, #6331 (J. Jakubik).

[6] Properties preserved under algebraic constructions, Bull. Amer. Math. Soc. 65 (1959), 287–299; MR 22, #2549 (H. Rasiowa).

[7] Existential Horn sentences, Proc. Amer. Math. Soc. 10 (1959), 994–998; MR 22, #6712 (B. Jónsson).

[8] Metamathematics and algebra: An example, Logic, Methodology and Philosophy of Science (Proc. 1960 Int. Congr.), Stanford Univ. Press, Stanford, Calif., 1962, 143–150; MR 28, #2052 (H. Rasiowa).

M. MACHOVER

[1] A note on sentences preserved under direct products and powers, Bull. Acad. Polon. Sci. Sér. Sci. Math. Astronom. Phys. 8 (1960), 519–523; MR 28, #4997 (B. van Rootselaar).

S. MACLANE

[1] Categorical algebra, Bull. Amer. Math. Soc. 71 (1965), 40–106; MR 30, #2053 (A. Dold).

(*See also* S. EILENBERG AND S. MACLANE.)

F. MAEDA

[1] A lattice formulation for algebraic and transcendental extensions in abstract algebras, J. Sci. Hiroshima Univ. Ser. A. 1 Math. 16 (1953), 383–397; MR 15, 675 (O. Frink).

M. MAKKAI

[1] Solution of a problem of G. Grätzer concerning endomorphism semigroups, Acta. Math. Acad. Sci. Hungar. 15 (1964), 297–307, MR 29, #3564 (B. M. Schein).

[2] On PC_Δ-classes in the theory of models, Magyar Tud. Akad. Mat. Kutató Int. Közl. 9 (1964), 159–194; MR 30, #1052 (C. C. Chang).
[3] On a generalization of a theorem of E. W. Beth, Acta Math. Acad. Sci. Hungar. 15 (1964), 227–235; MR 28, #5001 (H. J. Keisler).
[4] Remarks on my paper "On PC_Δ-classes in the theory of models", Magyar Tud. Akad. Mat. Kutató Int. Közl. 9 (1964), 601–602; MR 33, #48 (C. C. Chang).
[5] A compactness result concerning direct products of models, Fund. Math. 57 (1965), 313–325.

A. I. Mal'cev
[1] Untersuchungen aus dem Gebiete der mathematische Logik, Mat. Sb. 1 (43) (1936), 323–336.
[2] On a general method for obtaining local theorems in group theory (Russian), Ivanov. Gos. Ped. Inst. Uč. Zap. Fiz.-Mat. Fak. 1 (1941), 3–9; MR 17, 823 (I. Kaplansky).
[3] On the general theory of algebraic systems (Russian), Mat. Sb. (N.S.) 35 (77) (1954), 3–20; MR 16, 440 (I. Kaplansky).
[4] On representations of models (Russian), Dokl. Akad. Nauk SSSR 108 (1956), 27–29; MR 18, 370 (E. J. Cogan).
[5] Quasiprimitive classes of abstract algebras (Russian), Dokl. Akad. Nauk SSSR 108 (1956), 187–189; MR 18, 107 (R. A. Good).
[6] Subdirect products of models (Russian), Dokl. Akad. Nauk SSSR 109 (1956), 264–266; MR 19, 240 (V. E. Beneš).
[7] On derived operations and predicates (Russian), Dokl. Akad. Nauk SSSR 116 (1957), 24–27; MR 20, #1647 (A. Heyting).
[8] Free topological algebras (Russian), Izv. Akad. Nauk SSSR Ser. Mat. 21 (1957), 171–198; MR 20, #5249 (R. A. Good).
[9] On classes of models which possess the operation of generation (Russian), Dokl. Akad. Nauk SSSR 116 (1957), 738–741; MR 20, #2271 (E. J. Cogan).
[10] Defining relations in categories (Russian), Dokl. Akad. Nauk SSSR 119 (1958), 1095–1098; MR 20, #3805 (J. Isbell).
[11] The structure characteristic of some classes of algebras (Russian), Dokl. Akad. Nauk SSSR 120 (1958), 29–32; MR 20, #5154 (R. A. Good).
[12] On certain classes of models (Russian), Dokl. Akad. Nauk SSSR 120 (1958), 245–248; MR 20, #5155 (R. A. Good).
[13] On small models (Russian), Dokl. Akad. Nauk SSSR 127 (1959), 258–261; MR 21, #5553 (E. Mendelson).
[14] Model correspondences (Russian), Izv. Akad. Nauk SSSR Ser. Mat. 23 (1959), 313–336; MR 22, #10909 (E. Mendelson).
[15] Constructive algebras I. (Russian), Uspehi Mat. Nauk. 16 (1961), 3–60; MR 27, #1362 (H. Rasiowa).
[16] Elementary theories of locally free universal algebras (Russian), Dokl. Akad. Nauk SSSR 138 (1961), 1009–1012; MR 24, #A3055 (H. Rasiowa).
[17] Axiomatizable classes of locally free algebras of certain types (Russian), Sibirsk. Math. Ž. 3 (1962), 729–743; MR 26, #59 (P. M. Cohn).
[18] Certain questions of the modern theory of classes of models (Russian), Trudy 4 Vsesojuzn. Mat. S'ezda, Vol. 1, Leningrad 1963, 169–198.
[19] Several remarks on quasivarieties of algebraic systems (Russian), Algebra i Logika Sem., no. 3, 5 (1966), 3–9; MR 34, #5728 (R. S. Pierce).

J. M. Maranda
[1] Formal categories, Canad. J. Math. 17 (1965), 758–801; MR 32, #70 (P. J. Higgins).

E. Marczewski
[1] Sur les congruences et les propriétés positives d'algèbres abstraites, Colloq. Math. 2 (1951), 220–228; MR 14, 347 (J. Riguet).
[2] A general scheme of the notions of independence in mathematics, Bull. Acad. Polon. Sci. Sér. Sci. Math. Astronom. Phys. 6 (1958), 731–736; MR 21, #3363 (B. Jónsson).

[3] Independence in some abstract algebras, Bull. Acad. Polon. Sci. Sér. Sci. Math. Astronom. Phys. 7 (1959), 611–616; MR 22, #6751 (B. Jónsson).

[4] Independence in algebras of sets and Boolean algebras, Fund. Math. 48 (1959/60), 135–145; MR 22, #2569 (S. Ginsburg).

[5] Independence and homomorphisms in abstract algebras, Fund. Math. 50 (1961/62), 45–61; MR 25, #2016 (A. Goetz).

[6] Nombre d'éléments indépendants et nombre d'éléments générateurs dans les algèbras abstraites finies, Ann. Mat. Pura Appl. (4) 59 (1962), 1–9; MR 26, #2382 (G. Grätzer).

[7] Correction au travail "Nombre d'éléments indépendants et nombre d'éléments générateurs dans les algèbres abstraites finies", Ann. Mat. Pura Appl. (4) 61 (1963), 349; MR 27, #5718 (G. Grätzer).

[8] Homogeneous operations and homogeneous algebras, Fund. Math. 56 (1964), 81–103; MR 31, #1218 (G. Grätzer).

[9] The number of independent elements in abstract algebras with unary or binary operations, Bull. Acad. Polon. Sci. Sér. Sci. Math. Astronom. Phys. 12 (1964), 723–727; MR 31, #106 (G. Grätzer).

[10] Remarks on symmetrical and quasi-symmetrical operations, Bull. Acad. Polon. Sci. Sér. Sci. Math. Astronom. Phys. 12 (1964), 735–737; MR 31, #107 (G. Grätzer).

[11] Independence in abstract algebras. Results and problems, Colloq. Math. 14 (1966), 169–188; MR 32, #2361 (G. Grätzer).

(*See also* A. Hulanicki, E. Marczewski, and J. Mycielski.)

E. Marczewski and K. Urbanik

[1] Abstract algebras in which all elements are independent, Colloq. Math. 9 (1962), 199–207; MR 26, #60 (J. Jakubik).

J. Marica

[1] A note on unary algebras, Proc. Amer. Math. Soc. 14 (1963), 792; MR 27, #2454 (J. W. Andrushkin).

J. Marica and S. Bryant

[1] Unary algebras, Pacific J. Math. 10 (1960), 1347–1359; MR 22, #9463 (J. W. Andrushkin).

J. C. C. McKinsey

[1] The decision problem for some classes of sentences without quantifiers, J. Symbolic Logic 8 (1943), 61–76; MR 5, 85 (O. Frink).

(*See also* A. H. Diamond and J. C. C. McKinsey.)

J. C. C. McKinsey and A. Tarski

[1] The algebra of topology, Ann. of Math. (2) 45 (1944), 141–191; MR 5, 211 (G. Birkhoff).

[2] On closed elements in closure algebras, Ann. of Math. (2) 47 (1946), 122–162; MR 7, 359 (G. Birkhoff).

D. H. McLain

[1] Local theorems in universal algebras, J. London Math. Soc. 34 (1959), 177–184; MR 21, #3361 (B. Jónsson).

B. Mitchell

[1] Theory of categories, Academic Press, New York, N.Y., and London, 1965.

D. Monk

[1] On pseudo-simple universal algebras, Proc. Amer. Math. Soc. 13 (1962), 543–546; MR 26, #2381 (Ph. Dwinger).

J. C. Moore (*See* S. Eilenberg and J. C. Moore.)

A. C. Morel (*See* C. C. Chang and A. C. Morel; T. Frayne, A. C. Morel, and D. S. Scott.)

M. Morley

[1] Categoricity in power, Trans. Amer. Math. Soc. 114 (1965), 514–538; MR 31, #58 (H. B. Curry).

[2] Omitting classes of elements, The theory of models, Proc. 1963 Int. Symp. at Berkeley, North-Holland, Amsterdam, 1965, 265–273; MR 34, #1189 (J. R. Schoenfield).

M. Morley and R. Vaught

[1] Homogeneous universal models, Math. Scand. 11 (1962), 37–57; MR 27, #37 (B. Jónsson).

A. Mostowski

[1] On direct products of theories, J. Symbolic Logic 17 (1952), 1–31; MR 13, 897 (P. Lorenzen).

[2] Thirty years of foundational studies. Lectures on the development of mathematical logic and the study of the foundations of mathematics in 1930–1964, Acta Philos. Fenn. Fasc. 17 (1965), 1–180; MR 33, #18 (H. B. Curry).

(*See also* A. Ehrenfeucht and A. Mostowski.)

A. Mostowski and A. Tarski

[1] Arithmetical classes and types of well ordered systems, Bull. Amer. Math. Soc. 55 (1949), 65 (corrections: 1192).

V. L. Murskiĭ

[1] The existence in the three-valued logic of a closed class with a finite basis having no finite complete system of identities (Russian), Dokl. Akad. Nauk SSSR 163 (1965), 815–818; MR 32, #3998 (R. C. Lyndon).

J. Mycielski

[1] A characterization of arithmetical classes, Bull. Acad. Polon. Sci. Sér. Sci. Math. Astronom. Phys. 5 (1957), 1025–1027; MR 20, #5 (A. Robinson).

[2] Independent sets in topological algebras, Fund. Math. 55 (1964), 139–147; MR 30, #3855 (Th. J. Dekker).

[3] Some compactifications of general algebras, Colloq. Math. 13 (1964), 1–9;

[4] On unions of denumerable models, Algebra i Logika Sem., no. 4, 4 (1965), 57–58; MR 31, #5799 (H. J. Keisler).

(*See also* A. Hulanicki, E. Marczewski, and J. Mycielski.)

J. Mycielski and C. Ryll-Nardzewski

[1] Equationally compact algebras II, Fund. Math. 61 (1968), 271–281.

W. Narkiewicz

[1] Independence in a certain class of abstract algebras, Fund. Math 50 (1961/62), 333–340; MR 25, #36 (K. Urbanik).

[2] A note on v^*-algebras, Fund. Math. 52 (1963), 289–290; MR 27, #3575 (Ph. Dwinger).

[3] On a certain class of abstract algebras, Fund. Math. 54 (1964), 115–124; MR 28, #5020 (G. Grätzer). (Corrections: Fund. Math. 58 (1966), 111.)

[4] Remarks on abstract algebras having bases with different number of elements, Colloq. Math. 15 (1966), 11–17; MR 33, #3977 (G. Grätzer).

A. Nerode

[1] Composita, equations, and freely generated algebras, Trans. Amer. Math. Soc. 91 (1959), 139–151; MR 21, #3362 (R. C. Lyndon).

B. H. Neumann

[1] An embedding theorem for algebraic systems, Proc. London Math. Soc. (3) 4 (1954), 138–153; MR 17, 448 (K. A. Hirsch).

[2] Universal algebra, Lecture notes, Courant Institute of Math. Sci., New York University, 1962.

[3] On a problem of G. Grätzer, Publ. Math. Debrecen 14 (1967), 325–329.

W. Nitka

[1] Self-dependent elements in abstract algebras, Colloq. Math. 8 (1961), 15–17; MR 23, #A3107 (S. Ginsburg).

W. Nöbauer

[1] Über die Einfachheit von Funktionenalgebren, Monatsch. Math. 66 (1962), 441–452; MR 26, #3633 (A. Sade).

[2] Über die Darstellung von universellen Algebren durch Funktionenalgebren, Publ. Math. Debrecen 10 (1963), 151–154; MR 30, #1075 (B. Schweizer).
[3] Transformation von Teilalgebren und Kongruenzrelationen in allgemeinen Algebren, J. Reine Angew. Math. 214/215 (1964), 412–418; MR 29, #3412 (J. D. Swift).

W. Nöbauer and W. Philipp
[1] Die Einfachheit der mehrdimensionalen Funktionenalgebren, Arch. Math. 15 (1964), 1–5; MR 28, #3950 (A. Sade).

A. Oberschelp
[1] Über die Axiome produkt-abgeschlossener arithmetischer Klassen, Arch. Math. Logik Grundlagenforsch. 4 (1958), 95–123; MR 21, #6330 (R. C. Lyndon).
[2] Über die Axiome arithmetischer Klassen mit Abgeschlossenheitsbedingungen, Arch. Math. Logik Grundlagenforsch. 5 (1960) 26–36, MR 22, #6711 (R. C. Lyndon).

E. S. O'Keefe
[1] On the independence of primal algebras, Math. Z. 73 (1960), 79–94; MR 22, #12064 (Ph. Dwinger).
[2] Primal clusters of two-element algebras, Pacific J. Math. 11 (1961), 1505–1510; MR 25, #3881 (B. Jónsson).
[3] Independence in the small among universal algebras, Math. Ann. 154 (1964) 273–284; MR 29, #46 (O. Pretzel).

O. Ore
[1] On the foundation of abstract algebra I, II, Ann. of Math. (2) 36 (1935), 406–437; (2) 37 (1936), 265–292.
[2] On the decomposition theorems of algebra, C.R. Congress Internat. Math., Oslo, 1936, 297–307.
[3] Theory of equivalence relations, Duke Math. J. 9 (1942), 573–627; MR 4, 128 (L. W. Griffiths).

R. S. Pate (*See* R. J. Duffin and R. S. Pate.)

J. E. Penzov
[1] On the arithmetic of *n*-relations (Russian), Izv. Vysš. Učebn. Zaved. Matematika 23 (1961), 78–92; MR 25, #3835 (H. B. Curry).

W. Peremans
[1] A remark on free algebras (Dutch), Math. Centrum Amsterdam, Rapport ZW-1949-015 (1949); MR 11, 414 (H. A. Thurston).
[2] Some theorems on free algebras and on direct products of algebras, Simon Stevin 29 (1952), 51–59; MR 14, 347 (B. Jónsson).
[3] Free algebras with an empty set of generators, Nederl. Akad. Wetensch. Proc. Ser. A. 59 (1956), 565–570; MR 18, 788 (B. Jónsson).

J. Pfanzagl (*See* V. Baumann and J. Pfanzagl.)

W. Philipp (*See* W. Nöbauer and W. Philipp.)

G. Pickert
[1] Bemerkungen zum Homomorphiebegriff, Math. Z. 53 (1950), 375–386; MR 12, 583 (O. Ore).
[2] Direkte Zerlegungen von algebraischen Strukturen mit Relationen, Math. Z. 57 (1953), 395–404; MR 14, 718 (O. Ore).

H. E. Pickett
[1] Subdirect representations of relational systems, Fund. Math. 56 (1964), 223–240.

R. S. Pierce
[1] A note on free products of abstract algebras, Nederl. Akad. Wetensch. Proc. Ser. A. 66 (1963), 401–407; MR 27, #1400 (B. H. Neumann).
[2] A note on free algebras, Proc. Amer. Math. Soc. 14 (1963), 845–846; MR 27, #2455 (M. F. Newman).
(*See also* D. J. Christensen and R. S. Pierce.)

A. F. PIXLEY

[1] Distributivity and permutability of congruence relations in equational classes of algebras, Proc. Amer. Math. Soc. 14 (1963), 105–109; MR 26, #3630 (G. Birkhoff).

(*See also* A. FOSTER AND A. PIXLEY.)

J. PŁONKA

[1] Diagonal algebras and algebraic independence, Bull. Acad. Polon. Sci. Sér. Sci. Math. Astronom. Phys. 12 (1964), 729–733; MR 31, #108 (G. Grätzer).

[2] Exchange of independent sets in abstract algebras (II), Colloq. Math. 14 (1966), 217–224; MR 33, #2586 (A. C. Choudhury).

[3] On the number of independent elements in finite abstract algebras having a binary operation, Colloq. Math. 14 (1966), 189–201; MR 33, #88 (A. C. Choudhury).

[4] Remarks on diagonal and generalized diagonal algebras, Colloq. Math. 15 (1966), 19–23; MR 33, #2587 (G. Grätzer).

[5] Diagonal algebras, Fund. Math. 58 (1966), 309–321; MR 33, #2588 (B. M. Schein).

[6] Exchange of independent sets in abstract algebras (III), Colloq. Math. 15 (1966), 173–180; MR 34, #5729 (A. C. Choudhury).

[7] On functionally uniform symmetrical algebras, Colloq. Math. 15 (1966), 181–188; MR 34, #5730 (A. C. Choudhury).

B. I. PLOTKIN

[1] The automorphism groups of algebraic systems (Russian), Izd. 'Nauka", Moscow, 1966.

E. L. POST

[1] The Two-valued Iterative Systems of Mathematical Logic, Annals of Mathematics Studies No. 5, Princeton University Press, Princeton, N.J. 1941; MR 2, 337 (S. C. Kleene).

G. B. PRESTON

[1] The arithmetic of a lattice of sub-algebras of a general algebra, J. London Math. Soc. 29 (1954), 1–15; MR 15, 390 (D. C. Murdoch).

[2] Factorization of ideals in general algebras, J. London Math. Soc. 29 (1954), 363–368; MR 15, 927 (D. C. Murdoch).

A. PULTR (*See* Z. HEDRLIN AND A. PULTR.)

M. O. RABIN

[1] Arithmetical extensions with prescribed cardinality, Nederl. Akad. Wetensch. Proc. Ser. A. 62 (1959), 439–466; MR 21, #5564 (E. Mendelson).

[2] Computable algebra, general theory and theory of computable fields, Trans. Amer. Math. Soc. 95 (1960), 341–360; MR 22, #4639 (G. Kreisel).

[3] Classes of models and sets of sentences with the intersection property, Ann. Fac. Sci. Univ. Clermont-Ferrand 7 (1962), 39–53.

[4] Universal groups of automorphisms of models, The theory of models, Proc. 1963 Int. Symp. at Berkeley, North-Holland, Amsterdam, 1965, 274–284; MR 34, #1191 (H. J. Keisler).

H. RASIOWA

[1] A proof of the compactness theorem for arithmetical classes, Fund. Math. 39 (1952), 8–14 (1953); MR 14, 938 (I. Novak Gál).

H. RASIOWA AND R. SIKORSKI

[1] A proof of the completeness theorem of Gödel, Fund. Math. 37 (1950), 193–200; MR 12, 661 (I. L. Novak).

JU. K. REBANE

[1] Representation of universal algebras in commutative semigroups (Russian), Sibirsk. Mat. Ž. 7 (1966), 878–885; MR 33, #7290 (P. M. Cohn).

O. REIMER

[1] On the direct decompositions of algebras, Spisy Přírod. Fak. Univ. Brno 1962, 449–457; MR 27, #4780 (O. Pretzel).

H. Ribeiro
[1] The notion of universal completeness, Portugal. Math. 15 (1956), 83–86; MR 18, 785 (L. N. Gál).
[2] On the universal completeness of classes of relational systems, Arch. Math. Logik Grundlagenforsch. 5 (1961), 90–95; MR 30, #26 (K. Ono).

H. Ribeiro and R. Schwabauer
[1] A remark on equational completeness, Arch. Math. Logik Grundlagenforsch. 7 (1963), 122–123.

J. Riguet
[1] Sur les rapports entre les concepts de machine de multipole et de structure algébrique, C. R. Acad. Sci. Paris 237 (1953), 425–427; MR 15, 559 (G. Birkhoff).

A. Robinson
[1] On axiomatic systems which possess finite models, Methodos 3 (1951), 140–199.
[2] On the metamathematics of algebra, Studies in Logic and the Foundations of Mathematics, North-Holland, Amsterdam, 1951; MR 13, 715 (P. Lorenzen).
[3] On the application of symbolic logic to algebra, Proceedings of the International Congress of Mathematicians, Cambridge, Mass., 1950, vol. 1, 686–694, Amer. Math. Soc., Providence, R.I., 1952; MR 13, 716 (P. Lorenzen).
[4] L'application de la logique formelle aux mathématiques, Applications scientifiques de la logique mathématique (Actes du 2e Colloque International de Logique Mathématique, Paris, 1952, 51–63; discussion, 64. Gauthier-Villars, Paris; E. Nauwelaerts, Louvain, 1954; MR 16, 782 (I. Novak Gál).
[5] Note on an embedding theorem for algebraic systems, J. London Math. Soc. 30 (1955), 249–252; MR 17, 449 (P. Lorenzen).
[6] Ordered structures and related concepts, Mathematical interpretation of formal systems 51–56, North-Holland, Amsterdam, 1955; MR 17, 700 (E. Mendelson).
[7] Théorie métamathématique des idéaux, Gauthier-Villars, Paris; E. Nauwelaerts, Louvain, 1955; MR 16, 1080 (P. Lorenzen).
[8] A result on consistency and its application to the theory of definition, Nederl. Akad. Wetensch. Proc. Ser. A. 59 (1956), 47–58; MR 17, 1172 (P. Lorenzen).
[9] Complete theories, North-Holland, Amsterdam, 1956; MR 17, 817 (P. R. Halmos).
[10] Note on a problem of L. Henkin, J. Symbolic Logic 21 (1956), 33–35; MR 17, 817 (G. Kreisel).
[11] Some problems of definability in the lower predicate calculus, Fund. Math. 44 (1957), 309–329; MR 19, 1032 (E. J. Cogan).
[12] Relative model-completeness and the elimination of quantifiers, Dialecta 12 (1958), 394–406; MR 21, #1265 (E. J. Cogan).
[13] Obstructions to arithmetical extension and the theorem of Łoś and Suszko, Nederl. Akad. Wetensch. Proc. Ser. A. 62 (1959), 489–495; MR 22, #2544 (G. Kreisel).
[14] On the construction of models, Essays on the foundations of mathematics, 207–217, Magnes Press Hebrew Univ., Jerusalem, 1961; MR 29, #23 (S. Jaskowski).
[15] A note on embedding problems, Fund. Math. 50 (1961/62), 455–461; MR 25, #2946 (G. Grätzer).
[16] Recent developments in model theory, Logic, Methodology and Philosophy of Science (Proc. 1960 Internat. Congr.) 60–79, Stanford Univ. Press, Stanford, Calif., 1962; MR 29, #4668 (D. Monk).
[17] On the construction of models, Essays on the foundations of mathematics, Jerusalem, 1961, and Amsterdam, 1962, 207–217; MR 29, #23 (S. Jáskow).
[18] Introduction to model theory and to the metamathematics of algebra, North-Holland, Amsterdam, 1963; MR 27, #3533 (E. Engeler).

P. Rokos
[1] Generalization of theorems in "General Algebra" (Greek), Bull. Soc. Math. Gréce 28 (1954), 167–187; MR 15, 927 (L. A. Kokoris).

P. C. ROSENBLOOM

[1] Post algebras. I. Postulates and general theory, Amer. J. Math. 64 (1942), 167–188; MR 3, 262 (O. Frink).

C. RYLL-NARDZEWSKI

[1] On the categoricity in power $\leqq \aleph_0$, Bull. Acad. Polon. Sci. Sér. Sci. Math. Astronom. Phys. 7 (1959), 545–548; MR 22, #2543 (H. B. Curry).

(*See also* A. GOETZ AND C. RYLL-NARDZEWSKI; J. MYCIELSKI AND C. RYLL-NARDZEWSKI.)

M. SAADE

[1] On some classes of point algebras, Comment. Math. Univ. Carolinae 12 (1971), 33–36.

A. SALOMAA

[1] On the composition of functions of several variables ranging over a finite set, Ann. Univ. Turku. Ser. A.I. No. 41 (1960), 48 pp; MR 22, #6696 (A. Rose).

[2] On the number of simple bases of the set of functions over a finite domain, Ann. Univ. Turku. Ser. A.I. No. 52 (1962), 4 pp; MR 26, #3612 (C. C. Chang).

[3] Some completeness criteria for sets of functions over a finite domain. I, Ann. Univ. Turku. Ser. A.I. No. 53 (1962), 10 pp; MR 26, #3613 (C. C. Chang).

[4] Some completeness criteria for sets of functions over a finite domain. II, Ann. Univ. Turku. Ser. A.I. No. 63 (1963), 19 pp; MR 27, #3539 (C. C. Chang).

[5] On sequences of functions over an arbitrary domain, Ann. Univ. Turku. Ser. A.I. No. 62 (1963), 5 pp; MR 27, #3538 (C. C. Chang).

B. M. SCHEIN (Šain)

[1] Relation algebras, Bull. Acad. Polon. Sci. Sér. Sci. Math. Astronom. Phys. 13 (1965), 1–5; MR 31, #2188 (R. C. Lyndon).

[2] On the theorem of Birkhoff-Kogalovskiĭ (Russian), Uspehi Mat. Nauk. 20 (1965), 173–174; MR 34, #1250 (V. Dlab).

[3] Relation algebras (Russian), Scientific symposium on general algebra, Reports, Talks, Summaries, Tartu, 1966, 130–168.

E. T. SCHMIDT

[1] Universale Algebren mit gegebenen Automorphismengruppen und Unteralgebrenverbänden, Acta. Sci. Math. (Szeged) 24 (1963), 251–254; MR 28, #2987 (G. Grätzer).

[2] Universale Algebren mit gegebenen Automorphismengruppen und Kongruenzverbänden, Acta. Math. Acad. Sci. Hungar. 15 (1964), 37–45; MR 29, #3411 (W. Beuz).

[3] Über endliche Verbände, die in einen Zerlegungsverband einbettbar sind, Studia Sci. Math. Hungar. 1 (1966), 427–429; MR 34, #5714 (G. Grätzer).

(*See also* G. GRÄTZER AND E. T. SCHMIDT.)

J. SCHMIDT

[1] Über die Rolle der transfiniten Schlussweisen in einer allgemeinen Idealtheorie, Math. Nachr. 7 (1952), 165–182; MR 13, 904 (H. Curry).

[2] Einige grundlegende Begriffe und Sätze aus der Theorie der Hüllenoperatoren. Bericht über die Mathematiker-Tagung in Berlin, January 1953, 21–48, Deutscher Verlag der Wissenschaften, Berlin 1953; MR 16, 1083 (O. Ore).

[3] Peano-Räume, Z. Math. Logik Grundlagen Math. 6 (1960), 225–239.

[4] Einige algebraische Äquivalente zum Auswahlaxiom, Fund. Math. 50 (1961/62), 485–496; MR 25, #2990 (G. Grätzer).

[5] On the definition of algebraic operations in finitary algebras, Colloq. Math. 9 (1962), 189–197; MR 25, #5017 (G. Birkhoff).

[6] Die Charakteristik einer allgemeinen Algebra I, Arch. Math. 13 (1962), 457–470; MR 27, #80 (H. F. J. Lowig).

[7] Algebraic operations and algebraic independence in algebras with infinitary operations, Math. Japon. 6 (1961/62), 77–112; MR 28, #54 (B. Jónsson).

[8] Some properties of algebraically independent sets in algebras with infinitary operations, Fund. Math. 55 (1964), 123–137; MR 32, #74 (O. Pretzel).

[9] Concerning some theorems of Marczewski on algebraic independence, Colloq. Math. 13 (1964), 11–15; MR 31, #1219 (H. J. Keisler).

[10] Die Charakteristik einer allgemeinen Algebra. II, Arch. Math. 15 (1964), 286–301; MR 32, #73 (O. Pretzel).

[11] Über die Dimension einer partiellen Algebra mit endlichen oder unendlichen Operationen, Z. Math. Logik Grundlagen Math. 11 (1965), 227–239; MR 32, #76 (G. Grätzer).

[12] Die überinvarianten und verwandte Kongruenzrelationen einer allgemeinen Algebra, Math. Ann. 158 (1965), 131–157; MR 32, #75 (G. Grätzer).

[13] A general existence theorem on partial algebras and its special cases, Colloq. Math. 14 (1966), 73–87; MR 32, #2363 (R. Sikorski).

(*See also* M. Armbrust and J. Schmidt; P. Burmeister and J. Schmidt.)

H. Schneider (*See* M. N. Bleicher and H. Schneider.)

H. K. Schuff

[1] Über die Summation neutraler Zerschlagungen in beliebigen algebraischen Bereichen, Math. Nachr. 11 (1954), 295–301; MR 16, 119 (D. Kurepa).

[2] Polynome über allgemeinen algebraischen Systemen, Math. Nachr. 13 (1955), 343–366; MR 17, 571 (H. A. Thurston).

(*See also* K. Dörge and H. K. Schuff.)

M. P. Schützenberger

[1] Remarques sur la notion de clivage dans les structures algébriques et son application aux treillis, C. R. Acad. Sci. Paris 224 (1947), 512–514; MR 8, 366 (G. Birkhoff).

R. Schwabauer (*See* H. Ribeiro and R. Schwabauer.)

D. Scott

[1] Equationally complete extensions of finite algebras, Nederl. Akad. Wetensch. Proc. Ser. A. 59 (1956), 35–38; MR 18, 636 (R. C. Lyndon).

(*See also* T. Frayne, A. C. Morel, and D. S. Scott; J. Kalicki and D. Scott.)

L. Sedláček

[1] Universal algebras (Czech), Acta Univ. Palac. Olomucensis, Fac. Rer. Nat. 15 (1964), 39–68.

Z. Semadeni

[1] Free and direct objects, Bull. Amer. Math. Soc. 69 (1963), 63–66; MR 25, #5020 (J. R. Isbell).

K. Shoda

[1] Über die allgemeinen algebraischen Systeme I–VIII, Proc. Imp. Acad. Tokyo 17 (1941), 323–327; 18 (1942), 179–184, 227–232, 276–279; 19 (1943), 114–118, 259–263, 515–517; 20 (1944), 584–588; MR 7, 408–409 (O. Ore).

[2] Über die Schreiersche Erweiterungstheorie, Proc. Imp. Acad. Tokyo 19 (1943), 518–519; MR 7, 410 (O. Ore).

[3] General Algebra (Japanese), Kyôritsu-shuppan, Tokyo 1947; MR 12, 313 (T. Nakayama).

[4] Allgemeine Algebra, Osaka J. Math. 1 (1949), 182–225; MR 11, 308 (O. Ore).

[5] Zur Theorie der algebraischen Erweiterungen, Osaka J. Math. 4 (1952), 133–143; MR 14, 614 (O. Ore).

[6] Über die nicht algebraischen Erweiterungen algebraischer Systeme, Proc. Japan Acad. 30 (1954), 70–73; MR 16, 107 (O. Ore).

[7] Bemerkungen über die Existenz der algebraisch abgeschlossenen Erweiterung, Proc. Japan Acad. 31 (1955), 128–130; MR 17, 6 (O. Ore).

[8] Berichtigungen zu den Arbeiten über die Erweiterungen algebraischer Systeme, Osaka J. Math. 9 (1957), 239–240; MR 20, #4512 (O. Ore).

W. Sierpinski

[1] Sur les fonctions de plusieurs variables, Fund. Math. 33 (1945), 169–173; MR 8, 18 (J. Todd).

R. SIKORSKI

[1] Products of abstract algebras, Fund. Math. 39 (1952), 211–228 (1953); MR 14, 839 (B. Jónsson).

[2] Products of generalized algebras and products of realizations, Colloq. Math. 10 (1963), 1–13.

(*See also* H. RASIOWA AND R. SIKORSKI.)

R. J. SILVERMAN (*See* G. BERMAN AND R. J. SILVERMAN.)

F. M. SIOSON

[1] The structure lattice of a unary algebra, Natural and Applied Science Bulletin 18 (1960), 75–94.

[2] Some primal clusters, Math. Z. 75 (1960/61), 201–210; MR 25, #2018 (R. C. Lyndon).

[3] Free-algebraic characterizations of primal and independent algebras, Proc. Amer. Math. Soc. 12 (1961), 435–439; MR 23, #A3670 (A. Rose).

[4] Decompositions of generalized algebras I, II, Proc. Japan Acad. 923–932; MR 33, #5542, #5543 (G. Grätzer).

[5] On generalized algebras, Portugal. Math. 25 (1966), 67–90.

J. SŁOMIŃSKI

[1] On the extending of models III. Extensions in equationally definable classes of algebras, Fund. Math. 43 (1956), 69–76; MR 18, 2 (G. Kreisel).

[2] Theory of models with infinitary operations and relations, Bull. Acad. Polon. Sci. Sér. Sci. Math. Astronom. Phys. 6 (1958), 449–456; MR 20, #4479 (D. Kurepa).

[3] The theory of abstract algebras with infinitary operations, Rozprawy Mat. 18 (1959), 1–67; MR 21, #7173 (D. Kurepa).

[4] On the determining of the form of congruences in abstract algebras with equationally definable constant elements, Fund. Math. 48 (1959/60), 325–341; MR 23, #A84 (B. Jónsson).

[5] On the embedding of abstract quasi algebras into equationally definable classes of abstract algebras, Bull. Acad. Polon. Sci. Sér. Sci. Math. Astronom. Phys. 8 (1960), 11–17; MR 22, #9464 (R. C. Lyndon).

[6] On the common embedding of abstract quasi algebras into equationally definable classes of abstract algebras, Bull. Acad. Polon. Sci. Sér. Sci. Math. Astronom. Phys. 8 (1960), 277–282; MR 24, #A3111 (D. Kurepa).

[7] A theory of extensions of map-systems into equationally definable classes of abstract algebras, Bull. Acad. Polon. Sci. Sér. Sci. Math. Astronom. Phys. 10 (1962), 621–626; MR 27, #2456 (D. Kurepa).

[8] On the solving of systems of equations over quasi algebras and algebras, Bull. Acad. Polon. Sci. Sér. Sci. Math. Astronom. Phys. 10 (1962), 627–635; MR 27, #2457 (D. Kurepa).

[9] A theory of extensions of quasi-algebras to algebras, Rozprawy Mat. 40 (1964), 63 pp; MR 31, #111 (D. Kurepa).

[10] On certain existence theorems for models, Bull. Acad. Polon. Sci. Sér. Sci. Math. Astronom. Phys. 13 (1965), 769–775.

[11] A theory of *P*-homomorphisms, Colloq. Math. 14 (1966), 135–162; MR 32, #2364 (G. B. Seligman).

(*See also* J. ŁOŚ, J. SŁOMIŃSKI, AND R. SUSZKO.)

J. SONNER

[1] Canonical categories, Proceedings of the Conference on Categorical Algebra at La Jolla, 1965, Springer-Verlag, New York, N.Y., 1966, 272–294.

M. G. STANLEY

[1] Generation of full varieties, Michigan Math. J., 13 (1966), 127–128; MR 33, #90 (P. M. Cohn).

O. STEINFELD

[1] Über das Zassenhaussche Lemma in allgemeinen algebraischen Strukturen,

Ann. Univ. Sci. Budapest. Eötvös Sect. Math. 3–4 (1960/61), 309–314; MR 24, #A3110 (O. Frink).

A. H. STONE (*See* G. HIGMAN AND A. H. STONE.)

E. G. ŠUL'GEĬFER (*See* A. G. KUROŠ, A. H. LIVŠIC, AND E. G. ŠUL'GEĬFER.)

R. SUSZKO (*See* J. ŁOŚ, J. SLOMINSKI, AND R. SUSZKO; J. ŁOŚ AND R. SUSZKO.)

L. SVENONIUS

[1] A theorem on permutations in models, Theoria (Lund) 25 (1959), 173–178; MR 25, #1986b (R. C. Lyndon).

[2] $\aleph_0$-categoricity in first-order predicate calculus. Theoria (Lund) 25 (1959), 82–94; MR 25, #1986a (R. C. Lyndon).

[3] On minimal models of first order systems, Theoria (Lund) 26 (1960), 44–52; MR 25, #1986c (R. C. Lyndon).

[4] On the denumerable models of theories with extra predicates, The theory of models, Proc. 1963 Int. Symp. at Berkeley, North-Holland, Amsterdam, 1965, 376–389.

S. ŚWIERCZKOWSKI

[1] On independent elements in finitely generated algebras, Bull. Acad. Polon. Sci. Sér. Sci. Math. Astronom. Phys. 6 (1958), 749–752; MR 21, #3364 (B. Jónsson).

[2] Algebras independently generated by every *n* elements, Bull. Acad. Polon. Sci. Sér. Sci. Math. Astronom. Phys. 7 (1959), 501–502; MR 22, #6750 (B. Jónsson).

[3] On isomorphic free algebras, Fund. Math. 50 (1961), 35–44; MR 25, #2017 (A. Goetz).

[4] Algebras which are independently generated by every *n* elements, Fund. Math. 49 (1960/61), 93–104; MR 24, #A700 (K. Urbanik).

[5] A sufficient condition for independence, Colloq. Math. 9 (1962), 39–42; MR 25, #3880 (B. Jónsson).

[6] On two numerical constants associated with finite algebras, Ann. Mat. Pura Appl. (4) 62 (1963), 241–245; MR 28, #1151 (G. Grätzer).

[7] On the independence of continuous functions, Fund. Math. 52 (1963), 41–58; MR 26, #1398 (P. V. Reichelderfer).

[8] Topologies in free algebras, Proc. London Math. Soc. (3) 14 (1964), 566–576; MR 28, #5140 (K. H. Hofmann).

(*See also* A. HULANICKI AND S. ŚWIERCZKOWSKI.)

A. D. TAĬMANOV

[1] Class of models closed with respect to direct union (Russian), Dokl. Akad. Nauk SSSR 127 (1959), 1173–1175; MR 21, #5554 (R. A. Good).

[2] A class of models closed under direct products (Russian), Izv. Akad. Nauk SSSR Ser. Mat. 24 (1960), 493–510; MR 22, #6713 (R. A. Good).

[3] Characterization of finitely axiomatizable classes of models (Russian), Sibirsk. Mat. Ž. 2 (1961), 759–766; MR 26, #1257 (E. Mendelson).

[4] Characteristic properties of axiomatizable classes of models I (Russian), Izv. Akad. Nauk SSSR Ser. Mat. 25 (1961), 601–620; MR 26, #1255 (E. Mendelson).

[5] Characteristic properties of axiomatizable classes of models II (Russian), Izv. Akad. Nauk SSSR Ser. Mat. 25 (1961), 755–764; MR 26, #1256 (E. Mendelson).

[6] The characteristics of axiomatizable classes of models (Russian), Algebra i Logika Sem. 1 (1962), no. 4, 4–31; MR 33, #5490 (M. Taíclin).

[7] On a theorem of Beth and Kochen (Russian), Algebra i Logika Sem. 1 (1962/63), No. 6, 4–16; MR 27, #4752 (J. Mycielski).

K. TAKEUCHI

[1] The word problem for free algebraic systems (Japanese), Sûgaku 8 (1956/57), 218–229; MR 20, #897 (Y. Kawada).

T. TANAKA

[1] Canonical subdirect factorizations of lattices, J. Sci. Hiroshima Univ. Ser. A I. Math. 16 (1952), 239–246; MR 15, 674 (O. Frink).

A. TARSKI

[1] A remark on functionally free algebras, Ann. of Math. (2), 47 (1946), 163–165; MR 7, 360 (G. Birkhoff).

[2] Cardinal algebras, with an appendix: B. Jónsson and A. Tarski, Cardinal products of isomorphism types, Oxford University Press, New York, N.Y., 1949, xii + 326 pp; MR 10, 686 (S. MacLane).

[3] Some notions and methods on the borderline of algebra and metamathematics, Proceedings of the International Congress of Mathematicians, Cambridge, Mass., 1950, 705–720, Amer. Math. Soc., Providence, R.I., 1952; MR 13, 521 (P. Lorenzen).

[4] Contributions to the theory of models I, II, III, Nederl. Akad. Wetensch. Proc. Ser. A. 57 (1954), 572–581, 582–588; 58 (1955), 56–64; MR 16, 554 (A. Robinson).

[5] Ordinal algebras, with appendices by Chen-Chung Chang and Bjarni Jónsson, North-Holland, Amsterdam, 1956, i + 133 pp; MR 18, 632 (R. C. Lyndon).

(*See also* C. C. CHANG, B. JÓNSSON, AND A. TARSKI; J. M. G. FELL AND A. TARSKI; B. JÓNSSON AND A. TARSKI; J. C. C. MCKINSEY AND A. TARSKI.)

A. TARSKI AND R. L. VAUGHT

[1] Arithmetical extensions of relational systems, Compositio Math. 13 (1958), 81–102; MR 20, #1627 (A. Robinson).

H. H. TEH AND C. C. CHEN

[1] Some contributions to the study of universal algebras, Bull. Math. Soc. Nanyang Univ. 1964, 1–62; MR 32, #4063 (S. P. Bandyopadhyay).

H. H. TEH AND M. H. LIM

[1] The net concept in the study of abstract algebras and groups, Bull. Math. Soc. Nanyang Univ. 1964, 139–152; MR 32, #77 (W. E. Deskins).

A. A. TEREHOV

[1] On algebras in which the direct and free products coincide (Russian), Učen. Zap. Ivan. Ped. Inst. 18 (1958), 61–66.

F. B. THOMPSON

[1] A note on the unique factorization of abstract algebras, Bull. Amer. Math. Soc. 55 (1949), 1137–1141; MR 11, 309 (J. C. Moore).

H. A. THURSTON

[1] A note on continued products, J. London Math. Soc. 27 (1952), 239–241; MR 14, 238 (J. Riguet).

[2] Equivalences and mappings, Proc. London Math. Soc. (3) 2 (1952), 175–182; MR 14, 241 (O. Ore).

[3] The structure of an operation, J. London Math. Soc. 27 (1952), 271–279; MR 14, 239 (J. Riguet).

[4] Derived operations and congruences, Proc. London Math. Soc. (3) 8 (1958), 127–134; MR 19, 1033 (A. A. Grau).

B. TRPENOVSKI AND G. ČUPONA

[1] Finitary associative operations with neutral elements (Macedonian), Bull. Soc. Math. Phys. Macédoine 12 (1961), 15–24 (1963); MR 27, #224 (J. Wiegold).

K. URBANIK

[1] A representation theorem for Marczewski's algebras, Fund. Math. 48 (1959/60), 147–167; MR 22, #2570 (S. Ginsburg).

[2] A representation theorem for v^*-algebras, Fund. Math. 52 (1963), 291–317; MR 27, #3574 (Ph. Dwinger).

[3] Remarks on independence in finite algebras, Colloq. Math. 11 (1963), 1–12; MR 28, #2068 (R. S. Pierce).

[4] On algebraic operations in idempotent algebras, Colloq. Math. 13 (1965), 129–157.

[5] On a class of universal algebras, Fund. Math. 57 (1965), 327–350; MR 32, #4066 (G. Grätzer).

[6] A representation theorem for two-dimensional v^*-algebras, Fund. Math. 57 (1965), 215–236; MR 32, #4065 (Ph. Dwinger).
[7] Linear independence in abstract algebras, Colloq. Math. 14 (1966), 233–255; MR 32, #2365 (G. Grätzer).
[8] Remarks on symmetrical operations, Colloq. Math. 15 (1966), 1–9.
[9] Remarks on quasi-symmetrical operations, Bull. Acad. Polon. Sci. Sér. Sci. Math. Astronom. Phys. 13 (1965), 389–392; MR 32, #4064 (G. Grätzer).
[10] On some numerical constants associated with abstract algebras, Fund. Math. 59 (1966), 263–288.
(*See also* S. FAJTLOWICZ, K. GŁAZEK, AND K. URBANIK; E. MARCZEWSKI AND K. URBANIK.)

I. I. VALUCÈ
[1] Left ideals of the semigroup of endomorphisms of a free universal algebra (Russian), Mat. Sb. (N.S.) 62 (104) (1963), 371–384; MR 28, #47 (J. Jakubík).
[2] Universal algebras with proper, but not permutable, congruences (Russian), Uspehi Mat. Nauk 18 (1963), 145–148; MR 27, #5712 (M. Kolibiar).

R. L. VAUGHT
[1] On sentences holding in direct products of relational systems, Proceedings of the International Congress of Mathematicians, 2, Amsterdam 1954, 409.
[2] Remarks on universal classes of relational systems, Nederl. Akad. Wetensch. Proc. Ser. A. 57 (1954), 589–591; MR 16, 554 (A. Robinson).
[3] Applications of the Löwenheim-Skolem-Tarski theorem to problems of completeness and decidability, Nederl. Akad. Wetensch. Proc. Ser. A. 57 (1954), 467–472; MR 16, 208 (G. Kreisel).
[4] Denumerable models of complete theories, Infinitistic Methods, Proceedings of the Symposium on Foundations of Mathematics at Warsaw, 1959, Warszawa, 1961, 303–321.
[5] The elementary character of two notions from general algebra, Essays on the foundations of mathematics, 226–233, Magnes Press, Hebrew Univ., Jerusalem, 1961; MR 30, #1051 (H. Rasiowa).
[6] Models of complete theories, Bull. Amer. Math. Soc. 69 (1963), 299–313; MR 26, #4912 (E. Engeler).
[7] A Löwenheim–Skolem theorem for cardinals far apart, The theory of models, Proc. 1963 Int. Symp. at Berkeley, North-Holland, Amsterdam, 1965, 390–401.
[8] Elementary classes closed under descending intersections, Proc. Amer. Math. Soc. 17 (1966), 430–433; MR 33, #49 (M. Makkai).
(*See also* S. FEFERMAN AND R. VAUGHT; A. TARSKI AND R. L. VAUGHT.)

A. A. VINOGRADOV
[1] On the decomposability of algebras of a certain class into a direct product of simple algebras (Russian), Dokl. Akad. Nauk SSSR 163 (1965), 14–17; MR 33, #1266 (B. Vinograde).

V. V. VIŠIN
[1] Identity transformations in a four-valued logic (Russian), Dokl. Akad. Nauk SSSR 150 (1963), 719–721; MR 34, #1176 (D. H. Potts).

N. N. VOROB'EV
[1] On congruences of algebras (Russian), Dokl. Akad. Nauk SSSR 93 (1953), 607–608; MR 15, 595 (P. M. Whitman).

L. I. WADE
[1] Post algebras and rings, Duke Math. J. 12 (1945), 389–395; MR 7, 1 (N.H. McCoy).

B. WĘGLORZ
[1] Equationally compact algebras I, Fund. Math. 59 (1966), 289–298; MR 32, #7471 (J. Mycielski).
[2] Equationally compact algebras III, Fund. Math. 60 (1967), 89–93.
[3] Completeness and compactness in lattices, Colloq. Math. 16 (1967), 243–248.

G. H. Wenzel

[1] Note on a subdirect representation of universal algebras, Acta Math. Acad. Sci. Hungar. 18 (1967), 329–333.

[2] Konstanten in endlichen, freien universellen Algebren, Math. Z. 102 (1967), 205–215.

[3] On Marczewski's six-tuple of constants in finite universal algebras, Bull. Acad. Polon. Sci. Sér. Sci. Math. Astronom. Phys. 11 (1967), 759–764.

(*See also* G. Grätzer and G. H. Wenzel.)

A. N. Whitehead

[1] A treatise on universal algebra, Cambridge at the University Press, 1898.

O. Wyler

[1] Ein Isomorphiesatz, Arch. Math. 14 (1963), 13–15; MR 26, #6096 (Ph. Dwinger).

[2] Operational categories, Proceedings of the Conference on Categorical Algebra at La Jolla, 1965, Springer-Verlag, New York, N.Y., 1966, 295–316.

A. Yaqub

[1] On the identities of certain algebras, Proc. Amer. Math. Soc. 8 (1957), 522–524; MR 19, 831 (R. C. Lyndon).

[2] On the identities of direct products of certain algebras, Amer. Math. Monthly 68 (1961), 239–241; MR 23, #A1566 (J. McLaughlin).

[3] Primal clusters, Pacific J. Math. 16 (1966), 379–388, MR 32, #5568 (G. Grätzer).

[4] On certain classes of—and an existence theorem for—primal clusters, An. Scuola Norm. Sup. Pisa (3) 29 (1966), 1–13; MR 33, #7292 (F. M. Sioson).

[5] Semi-primal categorical independent algebras, Math. Z. 91 (1966), 395–403; MR 34, #121 (P. J. Higgins).

M. Yoeli

[1] Multivalued homomorphic mappings. Manuscript.

D. A. Zaharov

[1] On the theorem of Łoś and Suszko, Uspehi Mat. Nauk 16 (98) (1961), 200–201.

K. A. Zareckiĭ

[1] The semigroup of binary relations (Russian), Mat. Sb. (N.S.) 61 (103) (1963), 291–305; MR 27, #5719 (J. Wiegold).

R. G. Zubieta

[1] Arithmetical classes defined without equality, Bol. Soc. Mat. Mexicana 2 (1957), 45–53; MR 20, #2272 (E. J. Cogan).

ADDITIONAL BIBLIOGRAPHY

A. I. Abakumov, E. A. Paljutin, M. A. Taĭclin and Ju. E. Šišmarev

[1972] Categorical quasivarieties. (Russian) Algebra i Logika 11, 3–38, 121.

A. Abian

[1970] On the solvability of infinite systems of Boolean polynomial equations. Colloq. Math. 21, 27–30.

[1970 a] Generalized completeness theorem and solvability of systems of Boolean polynomial equations. Z. Math. Logik Grundlagen Math. 16, 263–264.

J. Adámek and J. Reiterman

[1974] Fixed-point property of unary algebras. Algebra Universalis 4, 163–165.

J. Adámek and V. Trnková

[a] Varietors and machines in a category. Algebra Universalis.

M. E. Adams

[1976] Implicational classes of pseudocomplemented distributive lattices. J. London Math. Soc. (2) 13, 381–384.

M. E. Adams and J. Sichler

[a] Homomorphisms of bounded lattices with a given sublattice. Arch. Math. (Basel) 30 (1978), 122–128.

[b] Bounded endomorphisms of lattices of finite height. Canad. J. Math. 29 (1977), 1254–1263.

[c] Endomorphism monoids of distributive double p-algebras. Glasgow Math. J.

[d] Subfunctors of full embeddings of algebraic categories.

[e] Cover set lattices.

[f] Quotients of rigid graphs.

A. A. Akataev

[1970] The varieties $\mathfrak{A}_{m,n}$. (Russian) Algebra i Logika 9, 127–136.

[1971] Axiomatic rank of the variety $\mathfrak{A}_{n-1,n}$. (Russian) Algebra i Logika 10, 125–134.

A. A. Akataev and D. M. Smirnov

[1968] Lattices of subvarieties of algebra varieties. (Russian) Algebra i Logika 7, no. 1, 5–25.

I. Sh. Aliev

[1966] On the minimal variety of a symmetric algebra. Algebra i Logika 5, no. 6, 5–14.

H. Andréka, B. Dahn and I. Németi

[1976] On a proof of Shelah. Bull. Acad. Polon. Sci. Sér. Sci. Math. Astronom. Phys. 24, no. 1, 1–7.

H. Andréka and I. Németi

[1974] Subalgebra systems of algebras with finite and infinite, regular and irregular arity. Ann. Univ. Sci. Budapest. Eötvös Sect. Math. 17, 103–118.

[1975] A simple, purely algebraic proof of the completeness of some first order logics. Algebra Universalis 5, 8–15.

[a] Generalisations of variety and quasivariety concepts to partial algebras through category theory.

[b] Formulas and ultraproducts in categories. Beiträge Algebra Geom.

[c] Formulas, filters and categories. Proceedings of the Colloquium held in Esztergom, 1977. Colloquia Mathematica Societatis János Bolyai. North-Holland Publishing Co., Amsterdam.

H. ANDRÉKA, T. GERGELY AND I. NÉMETI

[1977] On universal algebraic construction of logics. Studia Logica 36, 1–2, 9–47.

J. ANUSIAK

[1972] On transitive operations in abstract algebras. Colloq. Math. 25, 15–23.

J. ANUSIAK AND B. WĘGLORZ

[1971] Remarks on C-independence in Cartesian products of abstract algebras. Colloq. Math. 22, 161–165.

R. R. APPLESON

[1976] Zero divisors among finite structures of a fixed type. Algebra Universalis 6, 25–35.

R. R. APPLESON AND L. LOVÁSZ

[1975] A characterization of cancellable k-ary structures. Period. Math. Hungar. 6, 17–19.

M. ARMBRUST

[1966] Die fastdirekten Zerlegungen einer allgemeinen Algebra. I. Colloq. Math. 14, 39–62.

[1967] Die fastdirekten Zerlegungen einer allgemeinen Algebra. II. Colloq. Math. 17, 1–22.

[1970] On set-theoretic characterization of congruence lattices. Z. Math. Logik Grundlagen Math. 16, 417–419.

[1973] A first order approach to correspondence and congruence systems. Z. Math. Logik Grundlagen Math. 19, 215–222.

[1973 a] Direct limits of congruence systems. Colloq. Math. 27, 177–185.

M. ARMBRUST AND K. KAISER

[1974] On some properties a projective model class passes on to the generated axiomatic class. Arch. Math. Logik Grundlagenforsch. 16, 133–136.

A. ARNOLD AND M. NIVAT

[1977] Non-deterministic recursive program schemes. Fundamentals on Computer Science, Lecture Notes on Computer Science 56, 12–22. Springer Verlag, Berlin-New York.

V. A. ARTAMONOV

[1969] Clones of multilinear operations and multiple operator algebras. (Russian) Uspehi Mat. Nauk. 24, 47–59.

C. J. ASH

[1975] Reduced powers and Boolean extensions. J. London Math. Soc. (2) 9, 429–432.

C. AUST

[1974] Primitive elements and one relation algebras. Trans. Amer. Math. Soc. 193, 375–387.

A. K. AUSTIN

[1965] A note on models of identities. Proc. Amer. Math. Soc. 16, 522–523.

[1966] A closed set of laws which is not generated by a finite set of laws. Quart. J. Math. Oxford Ser. (2) 17, 11–13.

L. BABAI

[a] Infinite digraphs with given regular automorphism groups. J. Combinatorial Theory.

[b] Endomorphisms of sub- and factor semigroups. Proc. Conf. on Semigroup Theory, Szeged, 1976.

[c] Infinite graphs with given regular automorphism groups.

[d] Problems and results on automorphism groups and endomorphism monoids.

L. BABAI AND P. FRANKL

[a] Infinite quasigroups with given regular automorphism groups. Algebra Universalis 8 (1978), 310–319.

L. BABAI AND F. PASTIJN

[a] On semigroups with high symmetry. Simon Stevin.

P. D. Bacsich

[1972] Cofinal simplicity and algebraic closedness. Algebra Universalis 2, 354–360.

[1972 a] Injectivity in model theory. Colloq. Math. 25, 165–176.

[1973] Primality and model-completions. Algebra Universalis 3, 265–270.

[1975] Amalgamation properties and interpolation theorems for equational theories. Algebra Universalis 5, 45–55.

[1975 a] The strong amalgamation property. Colloq. Math. 33, no. 1, 13–23.

P. D. Bacsich and D. Rowlands-Hughes

[1974] Syntactic characterizations of amalgamation, convexity and related properties. J. Symbolic Logic 39, 433–451.

R. A. Baĭramov

[1968] On endomorphisms of certain algebraic systems. (Russian) Akad. Nauk Azerbaĭdžan. SSR Dokl. 24, 3–7.

[1971] The subalgebra lattices of certain algebras. (Russian) Izv. Akad. Nauk Azerbaĭdžan. SSR Ser. Fiz.-Tehn. Mat. Nauk, no. 2, 149–156.

[1974] On small and locally small varieties. (Russian) Vestnik Moskov. Univ. Ser. I Mat. Meh. 4, 133–134.

[1975] Interconnections between the subalgebra lattice and generating sets. (Russian) Vestnik Moskov. Univ. Ser. I Mat. Meh. 6, 115.

[1975 a] On generating sets and subalgebras of universal algebras. (Russian) XIIIth All-Union Algebra Symposium, (Russian) Gomel, 375–376.

R. A. Baĭramov and V. M. Džabbarzadeh

[1976] On generation of varieties and quasivarieties. (Russian) Materials of the Republican Conference of Young Scientists, Math. and Mech., (Russian) Baku, 5–9.

[1978] On the theory of varieties, quasivarieties and prevarieties of universal algebras. (Russian) Materials of the II-nd Republican Conference of Young Scientists, Math. and Mech., (Russian) Baku.

R. A. Baĭramov, R. B. Feĭzullaev and A. A. Mahmudov

[1977] On certain finiteness conditions and operators in the theory of varieties. (Russian) Izv. Akad. Nauk Azerbaĭdžan. SSR Ser. Fiz.-Tehn. Mat. Nauk 5, 93–98.

K. A. Baker

[1974] Primitive satisfaction and equational problems for lattices and other algebras. Trans. Amer. Math. Soc. 190, 125–150.

[a] Finite equational bases for finite algebras in a congruence-distributive equational class. Advances in Math. 24 (1977), 207–243.

[b] Congruence-distributive polynomial reducts of lattices. Algebra Universalis.

[c] Jónsson Δ_4-algebras from lattices. Algebra Universalis.

K. A. Baker and A. W. Hales

[1974] From a lattice to its ideal lattice. Algebra Universalis 4, 250–258.

K. A. Baker and A. F. Pixley

[1975] Polynomial interpolation and the Chinese remainder theorem for algebraic systems. Math. Z. 143, Heft 2, 165–174.

R. Balbes

[1967] Projective and injective distributive lattices. Pacific J. Math. 21, 405–420.

S. Balcerzyk

[1956] Remark on a paper of S. Gacsályi. Publ. Math. Debrecen 4, 357–358.

[1957] On algebraically compact groups of I. Kaplansky. Fund. Math. 44, 91–93.

J. T. Baldwin

[1973] A sufficient condition for a variety to have the amalgamation property. Colloq. Math. 28, 181–183, 329.

J. T. Baldwin and J. Berman

[1975] The number of subdirectly irreducible algebras in a variety. Algebra Universalis 5, 379–389.

[1976] Varieties and finite closure conditions. Colloq. Math. 35, no. 1, 15–20.
[a] A model-theoretic approach to Mal'cev conditions. J. Symbolic Logic 42 (1977), 277–288.

J. T. Baldwin and A. H. Lachlan
[1973] On universal Horn classes categorical in some infinite power. Algebra Universalis 3, 98–111.

B. Banaschewski
[1970] Injectivity and essential extensions in equational classes of algebras. Proc. Conf. on Universal Algebra (Queen's Univ., Kingston, Ont., 1969), pp. 131–147. Queen's Univ., Kingston, Ont.
[1972] On profinite universal algebras. General topology and its relations to modern analysis and algebra, III (Proc. Third Prague Topological Sympos., 1971), pp. 51–62. Academia, Prague.
[1972 a] An introduction to universal algebra. Manuscript.
[1974] Equational compactness of G-sets. Canad. Math. Bull. 17, 11–18.
[1974 a] On equationally compact extensions of algebras. Algebra Universalis 4, 20–35.

B. Banaschewski and G. Bruns
[1968] Injective hulls in the category of distributive lattices. J. Reine Angew. Math. 232, 102–109.

B. Banaschewski and H. Herrlich
[1976] Subcategories defined by implications. Houston J. Math. 2, no. 2, 149–171.

B. Banaschewski and E. Nelson
[1972] Equational compactness in equational classes of algebras. Algebra Universalis 2, 152–165.
[1972 a] On residual finiteness and finite embeddability. Algebra Universalis 2, 361–364.
[1973] Equational compactness in infinitary algebras. Colloq. Math. 27, 197–205.
[1977] Elementary properties of limit reduced powers with applications to Boolean powers. Contributions to universal algebra. Proceedings of the Colloquium held in Szeged, 1975. Colloquia Mathematica Societatis János Bolyai, Vol. 17. North-Holland Publishing Co., Amsterdam, pp. 21–25.
[a] Boolean powers as algebras of continuous functions.

H. J. Bandelt
[1975] Zur konkreten Charakterisierung von Kongruenzverbänden. Arch. Math. (Basel) 26, 8–13.
[1977] On congruence lattices of 2-valued algebras. Contributions to universal algebra. Proceedings of the Colloquium held in Szeged, 1975. Colloquia Mathematica Societatis János Bolyai, Vol. 17. North-Holland Publishing Co., Amsterdam, pp. 27–31.

T. M. Baranovič
[1966] Free decompositions in certain primitive classes of universal algebras. (Russian) Sibirsk. Mat. Ž. 7, 1230–1249.
[1968] Universal algebras. (Russian) Algebra. Topology. Geometry. 1966 (Russian), pp. 109–136. Akad. Nauk SSSR Inst. Naučn. Informacii, Moscow.
[1970] The categories that are structurally equivalent to certain categories of algebras. (Russian) Mat. Sb. (N.S.) 83 (125), 3–14.
[1972] Certain theorems in the theory of multioperator algebras. (Russian) Sibirsk. Mat. Ž. 13, 6–16.
[1973] Free decompositions of algebras with binary and arbitrary operations. (Russian) Trudy Moskov. Mat. Obšč. 29, 79–85.

T. M. Baranovič and M. S. Burgin
[1975] Linear Ω-algebras. (Russian) Uspehi Mat. Nauk 30, 61–106.

C. M. de Barros
[1967] Quelques structures algébriques définies par des lois de compositions partielles et associatives. C. R. Acad. Sci. Paris Sér. A–B, 265, A163–A166.

C. A. BATESON
[1977] Interplay between algebra and topology: groupoids in a variety. Ph. D. Thesis, University of Colorado.

W. BAUR
[1974] Rekursive Algebren mit Kettenbedingungen. Z. Math. Logik Grundlagen Math. 20, 37–46.

O. V. BELEGRADEK AND M. A. TAĬCLIN
[1972] Two remarks on the varieties $\mathfrak{A}_{m,n}$. (Russian) Algebra i Logika 11, 501–508, 614.

V. P. BELKIN
[1976] On certain lattices of quasivarieties in algebras. (Russian) Algebra i Logika 15, 12–21.
[1977] On quasi-identities in certain finite algebras. (Russian) Mat. Zametki 22, 335–338.

V. P. BELKIN AND V. A. GORBUNOV
[1975] Filters in lattices of quasivarieties of algebraic systems. (Russian) Algebra i Logika 14, no. 4, 373–392.

J. BÉNABOU
[1966] Structures algébriques dans les catégories. Cahiers de Topologie et de Géometrie Différentielle 10, pp. 1–126.

P. B. BENDIX
(*See* D. E. KNUTH AND P. B. BENDIX)

J. H. BENNETT
[1962] On spectra. (Abstract) J. Symbolic Logic 30, 264.

G. M. BERGMAN
[1972] Sulle classi filtrali di algebre. Ann. Univ. Ferrara Sez. VII (N.S.) 17, 35–42.
[1975] Some category-theoretic ideas in algebra. Proc. of Inter. Cong. of Mathematicians, Vancouver, B.C., 1974. Canadian Math. Congress, 285–296.
[1977] On the existence of subalgebras of direct products with prescribed d-fold projections. Algebra Universalis 7, 341–356.
[1978] Terms and cyclic permutations. Algebra Universalis 8, 129–130.

J. BERMAN
[1971] Strong congruence lattices of finite partial algebras. Algebra Universalis 1, 133–135.
[1972] On the congruence lattices of unary algebras. Proc. Amer. Math. Soc. 36, 34–38.
[1974] Notes on equational classes of algebras. Lecture Notes, University of Illinois at Chicago Circle.
[1975] Algebras with modular lattice reducts and simple subdirectly irreducibles. Discrete Math. 11, 1–8.
(*See also* J. T. BALDWIN AND J. BERMAN)

J. BERMAN, A. J. BURGER AND P. KÖHLER
[1975] The free distributive lattice on seven generators. (Abstract) Notices Amer. Math. Soc. 22, A-622.

J. BERMAN AND G. GRÄTZER
[1976] Uniform representations of congruence schemes. (Abstract) Notices Amer. Math. Soc. 23, A-430.
[1976 a] A testing algebra for congruence schemes. (Abstract) Notices Amer. Math. Soc. 23, A-574.

J. BERMAN AND B. WOLK
[1976] Free lattices in some locally finite varieties. (Abstract) Notices Amer. Math. Soc. 23, A-358.

C. BERNARDI
[1972] Su alcune condizioni necessarie per l'indipendenza di due varietà di algebre. Boll. Un. Mat. Ital. (4) 6, 410–421.

[1973] Idealità e indipendenza. Boll. Un. Mat. Ital. (4) 7, 94–101.
[1973 a] Sull'unione di classi filtrali. Ann. Univ. Ferrara Sez. VII (N.S.) 18, 1–14.

E. Beutler
[1977] The c-ideal lattice and subalgebra lattice are independent. Contributions to universal algebra. Proceedings of the Colloquium held in Szeged, 1975. Colloquia Mathematica Societatis János Bolyai, Vol. 17. North-Holland Publishing Co., Amsterdam, pp. 33–39.
[1978] An idealtheoretic characterization of varieties of abelian Ω-groups. Algebra Universalis 8, 91–100.

E. H. Bird
[1973] Automorphism groups of partial orders. Bull. Amer. Math. Soc. 79, 1011–1015.

G. Birkhoff
[1976]. Note on universal topological algebra. Algebra Universalis 6, 21–23.

G. Birkhoff and J. D. Lipson
[1970] Heterogeneous algebras. J. Combinatorial Theory 8, 115–133.
[1974] Universal algebra and automata. Proceedings of the Tarski Symposium (Proc. Sympos. Pure Math., Vol. XXV, Univ. of California, Berkeley, Calif., 1971), pp. 41–51. Amer. Math. Soc., Providence, R.I.

A. P. Birjukov
[1970] Varieties of idempotent semigroups. (Russian) Algebra i Logika 9, 255–273.

B. Biró, E. Kiss and P. P. Pálfy
[a] On the congruence extension property. Proceedings of the Colloquium held in Esztergom, 1977. Colloquia Mathematica Societatis János Bolyai. North-Holland Publishing Co., Amsterdam.

A. Blass and P. M. Neumann
[1974] An application of universal algebra in group theory. Michigan Math. J. 21, 167–169.

M. N. Bleicher, H. Schneider and R. L. Wilson
[1973] Permanence of identities on algebras. Algebra Universalis 3, 72–93.

W. J. Blok
[1976] Varieties of interior algebras. Ph.D. Thesis, University of Amsterdam.

W. J. Blok and P. Köhler
[a] The semigroup of varieties of generalized interior algebras. Houston J. Math.

S. L. Bloom
[1976] Varieties of ordered algebras. J. Comput. System Sci. 13, 200–212.

S. L. Bloom and C. C. Elgot
[1976] The existence and construction of free iterative theories. J. Comput. System Sci. 12, 305–318.

E. K. Blum and D. R. Estes
[1977] A generalization of the homomorphism concept. Algebra Universalis 7, 143–161.

A. D. Bol'bot
[1970] Varieties of Ω-algebras. (Russian) Algebra i Logika 9, 406–414.
[1971] The finiteness of the base for identities of four-element semigroups. (Russian) XIth All Union Algebra Colloquium (Russian), Kišinev, pp. 185–186.

A. F. Bravcev
[1967] Congruences of a direct product of algebras. (Russian) Algebra i Logika Sem. 6, no. 1, 39–43.

D. A. Bredikhin
[1976] Inverse semigroups of local automorphisms of universal algebras. (Russian) Sibirsk. Mat. Ž. 17, 499–507.
[1976 a] A concrete characterization of semigroups of local endomorphisms of hereditarily simple universal algebras. (Russian) In Theory of Functions, Differential Equations and Their Applications. Elista, pp. 40–45.

M. Brinzei
[1969] On generalized algebras. An. Sti. Univ. "Al. I. Cuza" Iași Secț. I a Mat. (N.S.) 15, 291–297.

N. G. de Bruijn and P. Erdös
[1961] A colour problem for infinite graphs and a problem in the theory of relations. Indag. Math. 13, 369–373.

N. R. Brumberg
[1969] Poly-varieties. (Russian) Mat. Zametki 5, 545–551.

G. Bruns
(*See* B. Banaschewski and G. Bruns)

G. Bruns and H. Lakser
[1970] Injective hulls of semilattices. Canad. Math. Bull. 13, 115–118.

D. A. Bryars
[1973] On the syntactic characterization of some model-theoretic relations. Ph.D. Thesis, London.

A. I. Budkin
[1974] Semivarieties and Schreier varieties of unary algebras. (Russian) Mat. Zametki 15, 263–270.

A. I. Budkin and V. A. Gorbunov
[1973] Implicative classes of algebras. (Russian) Algebra i Logika 12, no. 3, 249–268.
[1975] Quasivarieties of algebraic systems. Algebra i Logika 14, no. 2, 123–142.

S. Bulman-Fleming
[1971] Congruence topologies on universal algebras. Math. Z. 119, 287–289.
[1972] On equationally compact semilattices. Algebra Universalis 2, 146–151.
[1974] Algebraic compactness and its relations to topology. TOPO 72—general topology and its applications (Proc. Second Pittsburgh Internat. Conf., Carnegie-Mellon Univ. and Univ. of Pittsburgh, Pa., 1972; dedicated to the memory of Johannes H. de Groot), pp. 89–94. Lecture Notes in Math., Vol. 378, Springer, Berlin.
[1974 a] A note on equationally compact algebras. Algebra Universalis 4, 41–43.

S. Bulman-Fleming, A. Day and W. Taylor
[1974] Regularity and modularity of congruences. Algebra Universalis 4, 58–60.

S. Bulman-Fleming and I. Fleischer
[a] Equational compactness in semilattices with an additional unary operation.
[b] One-variable equational compactness in partially distributive semilattices with pseudocomplementation.

S. Bulman-Fleming, I. Fleischer and K. Keimel
[1978] The semilattices with distinguished endomorphisms which are equationally compact. Proc. Amer. Math. Soc.

S. Bulman-Fleming and W. Taylor
[1972] On a question of G. H. Wenzel. Algebra Universalis 2, 142–145.
[1976] Union-indecomposable varieties. Colloq. Math. 35, no. 2, 189–199.

S. Bulman-Fleming and H. Werner
[1977] Equational compactness in quasi-primal varieties. Algebra Universalis 7, 33–46.

A. J. Burger
(*See* J. Berman, A. J. Burger and P. Köhler)

M. S. Burgin
[1970] The groupoid of varieties of linear Ω-algebras. (Russian) Uspehi Mat. Nauk. 25, 263–264.
[1972] Free quotient algebras of free linear Ω-algebras. (Russian) Mat. Zametki 11, 537–544.
[1972 a] Free topological groups and universal algebras. (Russian) Dokl. Akad. Nauk SSSR 204, 9–11.

[1973] Topological algebras with continuous systems of operations. (Russian) Dokl. Akad. Nauk SSSR 213, 505–508.
[1973 a] Subalgebras of free products of linear Ω-algebras. (Russian) Trudy Moskov. Mat. 29, 101–117.
[1974] Schreier varieties of linear Ω-algebras. (Russian) Mat. Sb. 93, 554–572.
[1974 a] Cancellation law and accessible classes of linear Ω-algebras. (Russian) Mat. Zametki 16, 467–478.
(*See also* T. M. Baranovič and M. S. Burgin)

P. Burmeister
[1968] Über die Mächtigkeiten und Unabhängigkeitsgrade der Basen freier Algebren. I. Fund. Math. 62, 165–189.
[1970] Free partial algebras. J. Reine Angew. Math. 241, 75–86.
[1970 a] Über die Mächtigkeiten und Unabhängigkeitsgrade der Basen freier Algebren. II. Fund. Math. 67, 323–336.
[1971] On problem 12 in G. Grätzer's book "Universal Algebra". (Abstract) Notices Amer. Math. Soc. 18, 401.
[1971 a] On the kernels of endomorphisms of abstract algebras. (Abstract) Notices Amer. Math. Soc. 18, 633.
[1973] An embedding theorem for partial algebras and the free completion of a free partial algebra within a primitive class. Algebra Universalis 3, 271–279.

P. Burmeister and J. Schmidt
[1967] On the completion of partial algebras. Colloq. Math. 17, 235–245.

S. Burris
[1968] Representation theorems for closure spaces. Colloq. Math. 19, 187–193.
[1970] Closure homomorphisms. J. Algebra 15, 68–71.
[1971] A note on varieties of unary algebras. Colloq. Math. 22, 195–196.
[1971 a] The structure of closure congruences. Colloq. Math. 24, 3–5.
[1971 b] On the structure of the lattice of equational classes $\mathscr{L}(\tau)$. Algebra Universalis 1, 39–45.
[1971 c] Models in equational theories of unary algebras. Algebra Universalis 1, 386–392.
[1972] Embedding algebraic closure spaces in 2-ary closure spaces. Portugal. Math. 31, 183–185.
[1975] Separating sets in modular lattices with application to congruence lattices. Algebra Universalis 5, 213–223.
[1976] Boolean powers. Algebra Universalis 5, 341–360.
[1976 a] Subdirect representation in axiomatic classes. Colloq. Math. 34, 191–197.
[1977] An example concerning definable principal congruences. Algebra Universalis 7, 403–404.
[1978] Bounded Boolean powers and $\equiv_n$. Algebra Universalis 8, 137–138.
[1978 a] Rigid Boolean powers. Algebra Universalis 8, 264–265.
[a] Remarks on reducts of varieties. Proceedings of the Colloquium held in Esztergom, 1977. Colloquia Mathematica Societatis János Bolyai. North-Holland Publishing Co., Amsterdam.
[b] Kirby Baker's finite basis theorem for congruence distributive equational classes.

S. Burris and J. Lawrence
[a] Definable principal congruences in varieties of groups and rings. Algebra Universalis.

S. Burris and E. Nelson
[1971] Embedding the dual of Π_m in the lattice of equational classes of commutative semigroups. Proc. Amer. Math. Soc. 30, 37–39.
[1971 a] Embedding the dual of Π_∞ in the lattice of equational classes of semigroups. Algebra Universalis 1, 248–253.

S. Burris and H. P. Sankappanavar
[1975] Lattice-theoretic decision problems in universal algebra. Algebra Universalis 5, 163–177.

S. BURRIS AND H. WERNER
[a] Sheaf constructions and their elementary properties. Trans. Amer. Math. Soc.

R. M. BURSTALL AND P. J. LANDIN
[1969] Programs and their proofs: An algebraic approach. Machine Intelligence 4, pp. 17–43. American Elsevier, New York.

B. A. ČEPURNOV
[1973] Certain remarks on quasi-universal classes. (Russian) Učen. Zap. Ivanov. Gos. Ped. Inst. 125, 175–190.

B. A. ČEPURNOV AND S. R. KOGALOVSKIĬ
[1976] Certain criteria of hereditarity and locality for formulas of higher degrees. (Russian) In Studies in Set Theory and Nonclassic Logic. (Russian) Moscow, Izdat. Nauka, pp. 127–156.

A. I. ČEREMISIN
[1969] The existence of generalized free structures and generalized free unions of structures in categories. (Russian) Ivanov. Gos. Ped. Inst. Učen. Zap. 61, vyp. mat., 242–270.
[1969 a] Structural characterization of certain classes of models. (Russian) Ivanov. Gos. Ped. Inst. Učen. Zap. 61, vyp. mat., 271–286.

I. CHAJDA
[1973] Direct products of homomorphic mappings. Arch. Math. (Brno) 9, 61–65.
[1974] Direct products of homomorphic mappings. II. Arch. Math. (Brno) 10, no. 1, 1–8.

C. C. CHANG
[1967] Cardinal factorization of finite relational structures. Fund. Math. 60, 251–269.

C. C. CHANG AND H. J. KEISLER
[1973] Model theory. Studies in Logic and the Foundations of Mathematics, Vol. 73. North-Holland Publishing Co., Amsterdam-London; American Elsevier Publishing Co., Inc., New York.

C. C. CHEN
(*See* H. H. TEH AND C. C. CHEN)

M. R. CHIARO
[1975] A cluster theorem for polynomial complete algebras. Algebra Universalis 5, 197–202.

T. H. CHOE
[1977] Injective and projective zero-dimensional compact universal algebras. Algebra Universalis 7, 137–142.

J. P. CIRULIS
[1973] Complex systems instead of algebraic ones. (Russian) Latvian mathematical yearbook, 13 (Russian), pp. 169–184. Izdat. Zinatne, Riga.

D. M. CLARK
[1969] Varieties with isomorphic free algebras. Colloq. Math. 20, 181–187.
(*See also* P. H. KRAUSS AND D. M. CLARK)

D. M. CLARK AND P. H. KRAUSS
[1976] Para-primal algebras. Algebra Universalis 6, 165–192.
[1977] Varieties generated by para-primal algebras. Algebra Universalis 7, 93–114.

S. D. COMER
[1969] Classes without the amalgamation property. Pacific J. Math. 28, 309–318.
[1971] Representations by algebras of sections over Boolean spaces. Pacific J. Math. 38, 29–38.
[1973 a] Arithmetic properties of relatively free products. Proceedings of the University of Houston Lattice Theory Conference (Houston, Tex., 1973), pp. 180–193. Dept. Math., Univ. Houston, Houston, Tex.

[1974] Restricted direct products and sectional representations. Math. Nachr. 64, 333–344.
[1974 a] Elementary properties of structures of sections. Bol. Soc. Mat. Mexicana 19, 78–85.

S. D. Comer and J. S. Johnson
[1972] The standard semigroup of operators of a variety. Algebra Universalis 2, 77–79.

S. D. Comer and J. J. Le Tourneau
[1969] Isomorphism types of infinite algebras. Proc. Amer. Math. Soc. 21, 635–639.

W. H. Cornish
[a] The Chinese remainder theorem and sheaf representations. Fund. Math.
[b] On the Chinese remainder theorem of H. Draškovičová. Mat. Časopis Sloven. Akad. Vied.

W. H. Cornish and P. N. Stewart
[1976] Weakly regular algebras, Boolean orthogonalities and direct products of integral domains. Canad. J. Math. 28, 148–153.

B. Courcelle, I. Guessarian and M. Nivat
[a] The algebraic semantics of recursive program schemes.

B. Courcelle and M. Nivat
[1976] Algebraic families of interpretation. 17th Symposium on Foundations of Computer Science, Houston.

B. Courcelle and J. C. Raoult
[a] Completions of ordered magmas. Fund. Informat.

A. B. Cruse and M. F. Neff
[1975] Finite embeddability in a class of infinitary algebras. Algebra Universalis 5, 329–332.

B. Csákány
[1970] Characterizations of regular varieties. Acta Sci. Math. (Szeged) 31, 187–189.
[1975] Varieties of modules and affine modules. Acta Math. Acad. Sci. Hungar. 26, no. 3–4, 263–266.
[1975 a] Congruences and subalgebras. Ann. Univ. Sci. Budapest Eötvös Sect. Math. 18, 37–44.
[1975 b] Varieties of affine modules. Acta Sci. Math. (Szeged) 37, 3–10.
[1975 c] Varieties in which congruences and subalgebras are amicable. Acta Sci. Math. (Szeged) 37, 25–31.
[1975 d] On affine spaces over prime fields. Acta Sci. Math. (Szeged) 37, 33–36.
[1976] Conditions involving universally quantified function variables. Acta Sci. Math. (Szeged) 38, no. 1–2, 7–11.
[1976 a] Varieties whose algebras have no idempotent elements. Colloq. Math. 35, 201–203.

B. Csákány and L. Megyesi
[1975] Varieties of idempotent medial quasigroups. Acta Sci. Math. (Szeged) 37, 17–23.
[a] Varieties of idempotent medial n-quasigroups. Colloq. Math.

B. Csákány and E. T. Schmidt
[1970] Translations of regular algebras. Acta Sci. Math. (Szeged) 31, 157–160.

G. Čupona
[1966] On some primitive classes of universal algebras. Mat. Vesnik 3 (18), 105–108. (Correction: 6(21) (1969), 354.)
[1970] On a theorem of Cohn and Rebane. (Macedonian) Fac. Sci. Univ. Skopje Annuaire 20, 5–14.

G. Czédli
[a] An application of Mal'cev type theorems to congruence varieties. Proceedings of the Colloquium held in Esztergom, 1977. Colloquia Mathematica Societatis János Bolyai. North-Holland Publishing Co., Amsterdam.

G. Czédli and G. Hutchinson
[1978] A test for identities satisfied in lattices of submodules. Algebra Universalis 8, 269–309.

B. Dahn
(*See* H. Andréka, B. Dahn and I. Németi)

E. C. Dale
[1956] Semigroup and braid representations of varieties of algebras. Ph.D. Thesis, Manchester.

W. Damm, E. Fehr and K. Indermark
[1978] Higher type recursion and self-application as control structures. In Formal Description of Programming Concepts, pp. 461–489. North-Holland, Amsterdam.

G. Dantoni
[1969] Relazioni invarianti di un'algebra universale ed algebre con il sistema di operazioni completo rispetto ad una famiglia di relazioni invarianti. Matematiche (Catania) 24, 187–217.
[1969 a] Su certe famiglie di relazioni n-arie invarianti di un'algebra. Atti Accad. Naz. Lincei Rend. Cl. Sci. Fis. Mat. Natur. (8) 47, 456–464.
[1972] Relazioni n-arie invarianti di un'algebra universale. Struttura, lemma di Zassenhaus, teorema di Schreier. Matematiche (Catania) 27, 1–45.

B. A. Davey
[1973] Sheaf spaces and sheaves of universal algebras. Math. Z. 134, 275–290.
[1977] Weak injectivity and congruence extension in congruence-distributive equational classes. Canad. J. Math. 29, 449–459.
[a] Topological duality for prevarieties of universal algebras. Advances in Math.

B. A. Davey and H. Werner
[a] Injectivity and Boolean powers. Math. Z.

R. O. Davies
[1966] Two theorems on essential variables. J. London Math. Soc. 41, 333–335.

A. Day
[1969] A characterization of modularity for congruence lattices of algebras. Canad. Math. Bull. 12, 167–173.
[1971] A note on the congruence extension property. Algebra Universalis 1, 234–235.
[1972] Injectivity in equational classes of algebras. Canad. J. Math. 24, 209–220.
[1973] The congruence extension property and subdirectly irreducible algebras—an example. Algebra Universalis 3, 229–237.
[1973 a] p-modularity implies modularity in equational classes. Algebra Universalis 3, 398–399.
[1975] Splitting algebras and a weak notion of projectivity. Algebra Universalis 5, 153–162.
[1976] Lattice conditions implying congruence modularity. Algebra Universalis 6, 291–301.
[a] Splitting lattices and congruence-modularity. Contributions to universal algebra. Proceedings of the Colloquium held in Szeged, 1975. Colloquia Mathematica Societatis János Bolyai, Vol. 17. North-Holland Publishing Co., Amsterdam, pp. 57–71.
[b] Characterizations of finite lattices that are bounded-homomorphic images or sublattices of free lattices.
(*See also* S. Bulman-Fleming, A. Day and W. Taylor)

B. J. Day
[1977] Varieties of a closed category. Bull. Austral. Math. Soc. 16, 131–145.

K. H. Diener
[1969] A remark on equational classes generated by very small free algebras. Arch. Math. (Basel) 20, 491–494.

P. G. DIXON

[1976] Varieties of Banach algebras. Quart. J. Math. Oxford Ser. (2) 27, 481–487.

[1977] Classes of algebraic systems defined by universal Horn sentences. Algebra Universalis 7, 315–339.

V. DLAB

[1969] Lattice representation of general algebraic dependence. Math. Systems Theory 3, 289–299.

O. I. DOMANOV

[1971] Semigroups of all partial automorphisms of universal algebras. (Russian) Izv. Vysš. Učebn. Zaved. Matematika, no. 8 (111), 52–58.

J. DONER AND A. TARSKI

[1969] An extended arithmetic of ordinal numbers. Fund. Math. 65, 95–127.

J. DOYEN AND CH. HERRMANN

[1976] Projective lines and n-distributivity. Lattice theory. Proceedings of the Colloquium held in Szeged, 1974. Colloquia Mathematica Societatis János Bolyai, Vol. 14. North-Holland Publishing Co., Amsterdam, pp. 45–50.

H. DRAŠKOVIČOVÁ

[1972] On a generalization of permutable equivalence relations. Mat. Časopis Sloven. Akad. Vied. 22, 297–309.

[1972 a] External characterizations of subdirect representations of algebras. Acta Math. Acad. Sci. Hung. 23, 367–373.

[1973] Permutability, distributivity of equivalence relations and direct products. Mat. Časopis Sloven. Akad. Vied. 23, 69–87.

[1973 a] Independence of equational classes. Mat. Časopis Sloven. Akad. Vied 23, 125–135.

[1974] On a representation of lattices by congruence relations. Mat. Časopis Sloven. Akad. Vied 24, 69–75.

K. DRBOHLAV

[1965] A categorical generalization of a theorem of G. Birkhoff on primitive classes of universal algebras. Comment. Math. Univ. Carolinae 6, 21–41.

P. DUBREIL

[1969] Sur le demi-groupe des endomorphismes d'une algèbre abstraite. Atti Accad. Naz. Lincei Rend. Cl. Sci. Fis. Mat. Natur. (8) 46, 149–153.

J. DUDEK

[1970] Number of algebraic operations in idempotent groupoids. Colloq. Math. 21, 169–177.

[1971] Binary minimal algebras. Lectures on the theory of ordered sets and general algebra (Abstracts, Summer Session, Harmonia, 1970). Acta Fac. Rerum Natur. Univ. Comenian. Math. Mimoriadne Čislo, 21–22.

[1971 a] Number of polynomials in dependence preserving algebras. Colloq. Math. 22, 193–194.

[1971 b] Remarks on algebras having two bases of different cardinalities. Colloq. Math. 22, 197–200.

J. DUDEK AND K. GŁAZEK

[1977] Some remarks on weak automorphisms. Contributions to universal algebra. Proceedings of the Colloquium held in Szeged, 1975. Colloquia Mathematica Societatis János Bolyai, Vol. 17. North-Holland Publishing Co., Amsterdam, pp. 73–81.

J. DUDEK AND E. PŁONKA

[1971] Weak automorphisms of linear spaces and of some other abstract algebras. Colloq. Math. 22, 201–208.

P. DWINGER

[1970] The amalgamation problem from a categorical point of view. Proc. Conf. on Universal Algebra (Queen's Univ., Kingston, Ont., 1969), pp. 190–210. Queen's Univ., Kingston, Ont.

V. M. DŽABBARZADEH
(*See* R. A. BAĬRAMOV AND V. M. DŽABBARZADEH)

G. A. EDGAR
[1973] The class of topological spaces is equationally definable. Algebra Universalis 3, 139–146.
[1974] A completely divisible algebra. Algebra Universalis 4, 190–191.

A. EHRENFEUCHT, S. FAJTLOWICZ AND J. MYCIELSKI
[a] Homomorphisms of direct powers of algebras. Fund. Math.

G. EIGENTHALER
[1977] On polynomial algebras. Contributions to universal algebra. Proceedings of the Colloquium held in Szeged, 1975. Colloquia Mathematica Societatis János Bolyai, Vol. 17. North-Holland Publishing Co., Amsterdam, pp. 83–99.

S. EILENBERG
[1974] Automata, languages, and machines. A. Pure and Applied Mathematics, 59-A, Academic Press, New York.
[1976] Automata, languages, and machines. B. Pure and Applied Mathematics, 59-B, Academic Press, New York.

S. EILENBERG AND M. P. SCHÜTZENBERGER
[1976] On pseudovarieties. Advances in Math. 19, 413–418.

S. EILENBERG AND J. B. WRIGHT
[1967] Automata in general algebras. Information and Control 11, 452–470.

P. C. EKLOF
[1974] Algebraic closure operators and strong amalgamation bases. Algebra Universalis 4, 89–98.

C. C. ELGOT
(*See* S. L. BLOOM AND C. C. ELGOT)

P. VAN EMDE BOAS AND T. M. V. JANSSEN
[1977] The expressive power of intensional logic in the semantics of programming languages. Proc. 6th MFCS Symp., Lecture Notes on Computer Science 53, pp. 303–311. Springer Verlag, Berlin-New York.

P. ERDÖS
[1942] Some set-theoretical properties of graphs. Univ. Nac. Tucumán. Revista A.3, 363–367.
(*See also* N. G. DE BRUIJN AND P. ERDÖS)

P. ERDÖS AND S. FAJTLOWICZ
[1977] On composition of polynomials. Algebra Universalis 7, 357–360.

P. ERDÖS AND A. HAJNAL
[1966] On chromatic number of graphs and set-systems. Acta Math. Acad. Sci. Hungar. 17, 61–99.

P. ERDÖS AND R. RADO
[1956] A partition calculus in set theory. Bull. Amer. Math. Soc. 62, 427–489.

D. R. ESTES
(*See* E. K. BLUM AND D. R. ESTES)

T. EVANS
[1969] Some connections between residual finiteness, finite embeddability and the word problem. J. London Math. Soc. (2) 1, 399–403.
[1970] Residually finite semigroups of endomorphisms. J. London Math. Soc. (2) 2, 719–721.
[1971] The lattice of semigroup varieties. Semigroup Forum 2, 1–43.
[1972] Residual finiteness and finite embeddability. A remark on a paper by Banaschewski and Nelson. Algebra Universalis 2, 397.
[1975] Algebraic structures associated with latin squares and orthogonal arrays. Proc. Conf. on Algebraic Aspects of Combinatorics, University of Toronto, 31–52.

[1978] An algebra has a solvable word problem if and only if it is embeddable in a finitely generated simple algebra. Algebra Universalis.
[a] Some solvable word problems. Oxford Conference on Decision Problems, 1976. North-Holland, Amsterdam.
[b] Universal algebra and Euler's officer problem.

T. Evans and C. C. Lindner
[1977] Finite embedding theorems for partial designs and algebras. Les Presses de l'Université de Montréal.

T. Evans, K. I. Mandelberg and M. F. Neff
[1975] Embedding algebras with solvable word problems in simple algebras—some Boone-Higman type theorems. Logic Colloquium '73 (Bristol 1973), pp. 259–277. Studies in Logic and Foundations of Mathematics, Vol. 80, North-Holland, Amsterdam.

R. Fagin
[1974] Generalized first-order spectra and polynomial-time recognizable sets. Complexity of computation (Proc. Sympos., New York, 1973), pp. 43–73. SIAM-AMS Proc., Vol. VII, Amer. Math. Soc., Providence, R.I.
[1975] Monadic generalized spectra. Z. Math. Logik Grundlagen Math. 21, 89–96.

S. Fajtlowicz
[1967] On the exchange of independent sets in abstract algebras. Bull. Acad. Polon. Sci. Sér. Sci. Math. Astronom. Phys. 15, 765–767.
[1968] A remark on independence in projective spaces. Colloq. Math. 19, 23–25.
[1969] Birkhoff's theorem in the category of non-indexed algebras. Bull. Acad. Polon. Sci. Sér. Sci. Math. Astronom. Phys. 17, 273–275.
[1969 a] Families of independent sets in finite unary algebras. Colloq. Math. 20, 13–15.
[1970] Algebras of homomorphisms. Rend. Mat. (6) 3, 523–527.
[1970 a] On algebraic operations in binary algebras. Colloq. Math. 21, 23–26.
[1971] n-dimensional dice. Rend. Mat. (6) 4, 855–865.
[1973] Categoricity in varieties. (Abstract) Notices Amer. Math. Soc. 19, A-435.
[1977] Duality for algebras. Contributions to universal algebra. Proceedings of the Colloquium held in Szeged, 1975. Colloquia Mathematica Societatis János Bolyai, Vol. 17. North-Holland Publishing Co., Amsterdam, pp. 101–112.
(*See also* A. Ehrenfeucht, S. Fajtlowicz and J. Mycielski; P. Erdös and S. Fajtlowicz)

S. Fajtlowicz and K. Głazek
[1967] Independence in separable variables algebras. Colloq. Math. 17, 221–224.

S. Fajtlowicz, W. Holsztyński, J. Mycielski and B. Węglorz
[1968] On powers of bases in some compact algebras. Colloq. Math. 19, 43–46.

S. Fajtlowicz and E. Marczewski
[1969] On some properties of the family of independent sets in abstract algebras. Colloq. Math. 20, 189–195.

S. Fajtlowicz and J. Mycielski
[1974] On convex linear forms. Algebra Universalis 4, 244–249.

M. Fattorosi-Barnaba
[1969] Intorno ad alcuni operatori di chiusura sugli insiemi di leggi. Rend. Mat. 2, 355–368.
[1973] Un'osservazione sugli insiemi di identità che generano classi equazionali equazionalmente complete. Rend. Mat. 6, 131–137.

T. H. Fay
[1978] On categorical conditions for congruences to commute. Algebra Universalis 8, 173–179.

E. Fehr
(*See* W. Damm, E. Fehr and K. Indermark)

R. B. FEĬZULLAEV
(*See* R. A. BAĬRAMOV, R. B. FEĬZULLAEV AND A. A. MAHMUDOV)

W. FELSCHER
[1968] Equational maps. Contributions to Math. Logic (Colloquium, Hannover, 1966), pp. 121–161. North-Holland, Amsterdam.
[1969] Birkhoffsche und kategorische Algebra. Math. Ann. 180, 1–25.
[1972] Equational classes, clones, theories and triples. Manuscript.

W. FELSCHER AND G. JARFE
[1968] Free structures and categories. The Theory of Models, Proc. of the 1962 International Symposium at Berkeley, North-Holland Publishing Co., Amsterdam, pp. 427–428.

C. F. FENNEMORE
[1971] All varieties of bands. Semigroup Forum 1, 172–179.

K. FICHTNER
[1968] Varieties of universal algebras with ideals. (Russian) Mat. Sb. (N.S.) 75 (117), 445–453.
[1968 a] On the theory of universal algebras with ideals. (Russian) Mat. Sb. (N.S.) 77 (119), 125–135.
[1970] Eine Bemerkung über Mannigfaltigkeiten universeller Algebren mit Idealen. Monatsb. Deutsch. Akad. Wiss. Berlin 12, 21–25.
[1972] Distributivity and modularity in varieties of algebras. Acta Sci. Math. (Szeged) 33, 343–348.

E. R. FISHER
[1977] Vopěnka's principle, category theory, and universal algebra. (Abstract) Notices Amer. Math. Soc. 24, A-44.

I. FLEISCHER
[1966] A note on universal homogeneous models. Math. Scand. 19, 183–184.
[1975] On extending congruences from partial algebras. Fund. Math. 88, no. 1, 11–16.
[1976] Concerning my note on universal homogeneous models. Math. Scand. 38, 24.
[a] Extending a partial equivalence to a congruence and relative embedding in universal algebras. Fund. Math.
(*See also* S. BULMAN-FLEMING AND I. FLEISCHER; S. BULMAN-FLEMING, I. FLEISCHER AND K. KEIMEL)

I. FLEISCHER AND I. ROSENBERG
[a] The Galois connection between partial functions and relations. Pacific J. Math.

V. FLEISCHER
[1974] On endomorphisms of free polygons. (Russian) Acta Comm. Univ. Tartu 13, 189–205.
[1975] Determinability of a free polygon by its endomorphism semigroup. (Russian) Acta Comm. Univ. Tartu 16, 27–41.
[1977] Projective and injective varieties of abelian Ω-algebras. Contributions to universal algebra. Proceedings of the Colloquium held in Szeged, 1975. Colloquia Mathematica Societatis János Bolyai, Vol. 17. North-Holland Publishing Co., Amsterdam, pp. 113–132.

S. FOLDES
[a] Symmetries of directed graphs and the Chinese remainder theorem.

W. K. FORREST
[a] Model theory for universal classes with the amalgamation property: a study in the foundations of model theory and algebra. Ann. Math. Logic 11(1977), 263–366.

V. A. FORTUNATOV
[1974] Varieties of perfect algebras. (Russian) Studies in Algebra, No. 4 (Russian), pp. 110–114. Izdat. Saratov Univ., Saratov.

A. L. Foster
[1967] Semi-primal algebras: Characterization and normal-decomposition. Math. Z. 99, 105–116.
[1968] Algebraic function-spectra. Math. Z. 106, 225–244.
[1968 a] Pre-fields and universal algebraic extensions; equational precessions. Monatsh. Math. 72, 315–324.
[1969] Automorphisms and functional completeness in universal algebras. I. General automorphisms, structure theory and characterization. Math. Ann. 180, 138–169.
[1970] Congruence relations and functional completeness in universal algebras; structure theory of hemi-primals. I. Math. Z. 113, 293–308.
[1970 a] Homomorphisms and functional completeness. Hemi-primal algebras. II. Math. Z. 115, 23–32.
[1971] Functional completeness and automorphisms. General infra-primal theory and universal algebra "fields" of Galois-class. I. Monatsh. Math. 75, 303–315.
[1972] Functional completeness and automorphisms. General infra-primal theory and universal algebra "fields" of Galois-class. II. Monatsh. Math. 76, 226–238.

A. L. Foster and A. F. Pixley
[1966] Algebraic and equational semi-maximality. Equational Spectra. II. Math. Z. 94, 122–133.
[1971] Total algebras and weak independence. I. Math. Z. 123, 93–104.
[1972] Total algebras and weak independence. II. Math. Z. 125, 271–284.

R. Franci
[1971] Estensioni semigruppali. Ann. Univ. Ferrara Sez. VII (N.S.) 16, 45–53.
[1972] Estensioni semigruppali ed estensioni booleane. Boll. Un. Mat. Ital. (4) 5, 418–425.
[1973] Una nota sulle classi ideali. Boll. Un. Mat. Ital. (4) 7, 429–439.
[1976] Filtral and ideal classes of universal algebras. Quaderni dell'Istituto de Matematica dell'Università de Siena. Istituto de Matematica, Università degli Studi di Siena, Siena.

R. Franci and L. Toti-Rigatelli
[1969] Sulla varieta dei reticoli distributivi. Ann. Univ. Ferrara Sez. VII (N.S.) 14, 23–27.

P. Frankl
(*See* L. Babai and P. Frankl)

G. A. Fraser and A. Horn
[1970] Congruence relations in direct products. Proc. Amer. Math. Soc. 26, 390–394.

R. Freese
[1975] Congruence modularity. (Abstract) Notices Amer. Math. Soc. 22, A-301.

R. Freese and B. Jónsson
[1976] Congruence modularity implies the Arguesian identity. Algebra Universalis 6, 225–228.

R. Freese and J. B. Nation
[1977] 3-3 lattice inclusions imply congruence modularity. Algebra Universalis 7, 191–194.

P. Freyd
[1964] Abelian categories. An introduction to the theory of functors. Harper's Series in Modern Mathematics. Harper & Row, New York.

E. Fried
[1970] Tournaments and nonassociative lattices. Ann. Univ. Sci. Budapest Eötvös Sect. Math. 13, 151–164.
[1977] Automorphism group of integral domains fixing a given subring. Algebra Universalis 7, 373–387.

[a] A note on the congruence extension property. Acta. Sci. Math. (Szeged).
[b] The category of integral domains containing a given one. Acta. Math. Acad. Sci. Hungar.

E. Fried and G. Grätzer
[1975] On automorphisms of the subalgebra lattice induced by the automorphisms of the algebra. (Abstract) Notices Amer. Math. Soc. 22, A-380. (Acta Sci. Math. (Szeged) 40 (1978), 49–52.)

E. Fried, G. Grätzer and R. W. Quackenbush
[1976] Uniform Mal'cev sequences. (Abstract) Notices Amer. Math. Soc., 23, A-45.
[a] Uniform congruence schemes. Algebra Universalis.
[b] The equational class generated by weakly associative lattices with the Unique Bound Property. Ann. Univ. Sci. Budapest. Eötvös Sect. Math.

E. Fried and J. Kollár
[a] Automorphism groups of fields. Proceedings of the Colloquium held in Esztergom, 1977. Colloquia Mathematica Societatis János Bolyai. North-Holland Publishing Co., Amsterdam.
[b] Automorphism groups of algebraic number fields.

E. Fried and A. F. Pixley
[a] The dual discriminator function in universal algebra.

E. Fried and J. Sichler
[1973] Homomorphisms of commutative rings with unit element. Pacific J. Math. 45, 485–491.
[1977] Homomorphisms of integral domains of characteristic zero. Trans. Amer. Math. Soc. 225, 163–182.
[a] On automorphism groups of subalgebras of a universal algebra.

H. Friedman
[1976] On decidability of equational theories. J. Pure and Applied Alg. 7, 1–2.

O. Frink and R. S. Smith
[1972] On the distributivity of the lattice of filters of a groupoid. Pacific J. Math. 42, 313–322.

M. Froda-Schechter
[1974] Sur les homomorphismes des structures relationelles. I. Studia Univ. Babes-Bolyai Ser. Math.-Mech. 19, 20–25.
[1975] Sur les homomorphismes des structures relationelles. II. Studia Univ. Babes-Bolyai Ser. Math.-Mech. 20, 11–15.

J. Froemke
[1970] Independent factorizations of abstract algebras. J. Algebra 16, 311–325.
[1971] Maximal primal clusters are infinite. Math. Ann. 191, 99–120.
[1971 a] Pairwise and general independence of abstract algebras. Math. Z. 123, 1–17.
[1974] A note on functionally complete algebras with no non-trivial subalgebras. Math. Z. 136, 353–355.
[1975] Maximal pre-primal clusters. Canad. J. Math. 27, no. 4, 746–751.

J. Froemke and R. W. Quackenbush
[1975] The spectrum of an equational class of groupoids. Pacific J. Math. 58, no. 2, 381–386.

G. Fuhrken
[1973] On automorphisms of algebras with a single binary operation. Portugaliae Math. 32, 49–52.

G. Fuhrken and W. Taylor
[1971] Weakly atomic-compact relational structures. J. Symbolic Logic 36, 129–140.

T. Fujiwara
[1964] On the permutability of congruences on algebraic systems. Proc. Japan Acad. 40, 787–792.

[1965] Note on permutability of congruences on algebraic systems. Proc. Japan Acad. 41, 822–827.

[1971] Freely generable classes of structures. Proc. Japan Acad. 47, 761–764.

[1971 a] On the construction of the least universal Horn class containing a given class. Osaka J. Math. 8, 425–436.

T. FUJIWARA AND Y. NAKANO

[1970] On the relation between free structures and direct limits. Math. Japan. 15, 19–23.

N. FUNAYAMA AND T. NAKAYAMA

[1942] On the distributivity of a lattice of lattice congruences. Proc. Imp. Acad. Tokyo 18, 553–554.

S. GACSÁLYI

[1955] On pure subgroups and direct summands of Abelian groups. Publ. Math. Debrecen 4, 88–92.

F. GALVIN

[1970] Horn sentences. Ann. Math. Logic 1, 389–422.

F. GALVIN AND A. HORN

[1970] Operations preserving all equivalence relations. Proc. Amer. Math. Soc. 24, 521–523.

F. GALVIN AND K. PRIKRY

[1976] Infinitary Jónsson algebras and partition relations. Algebra Universalis 6, 367–376.

B. GANTER

[a] Combinatorial designs and algebras. Algebra Universalis.

[b] Kombinatorische Algebra: Optimale Geometrien.

B. GANTER, J. PŁONKA AND H. WERNER

[1973] Homogeneous algebras are simple. Fund. Math. 79, no. 3, 217–220.

B. GANTER AND H. WERNER

[1975] Equational classes of Steiner systems. Algebra Universalis 5, 125–140.

[1975 a] Equational classes of Steiner systems. II. Proc. Conf. on Algebraic Aspects of Combinatorics, University of Toronto, 283–285.

O. D. GARCÍA

[1974] Injectivity in categories of groups and algebras. An. Inst. Mat. Univ. Nac. Autonoma Mexico 14, 95–115.

B. J. GARDNER

[1975] Semi-simple radical classes of algebras and attainability of identities. Pacific J. Math. 61, 401–416.

V. S. GARVACKIĬ AND B. M. ŠAĬN

[1970] Injective and surjective morphisms in abstract categories. (Russian) Colloq. Math. 22, 51–57.

H. GASKILL

[1972] Transferability in lattices and semilattices. Ph.D. Thesis, Simon Fraser University.

H. GASKILL, G. GRÄTZER AND C. R. PLATT

[1975] Sharply transferable lattices. Canad. J. Math. 27, 1246–1262.

F. GÉCSEG

[1970] On certain classes of Σ-structures. Acta Sci. Math. (Szeged) 31, 191–195.

[1974] On subdirect representations of finite commutative unoids. Acta Sci. Math. (Szeged) 36, 33–38.

[1977] Universal algebras and tree automata. In Fundamentals of Computer Science Theory, Lecture Notes in Computer Science. Springer-Verlag, pp. 98–112.

F. GÉCSEG AND S. SZÉKELY

[1973] On equational classes of unoids. Acta Sci. Math. (Szeged) 34, 99–101.

E. Gedeonová

[1972] A characterization of p-modularity for congruence lattices of algebras. Acta Fac. Rerum Natur. Univ. Comenian. Math. Publ. 28, 99–106.

D. Geiger

[1968] Closed systems of functions and predicates. Pacific J. Math. 27, 95–100.

[1974] Coherent algebras. (Abstract) Notices Amer. Math. Soc. 21, A-436.

T. Gergely

(*See* H. Andréka, T. Gergely and I. Németi)

J. A. Gerhard

[1970] The lattice of equational classes of idempotent semigroups. J. Algebra 15, 195–224.

[1971] The number of polynomials of idempotent semigroups. J. Algebra 18, 366–376.

S. Givant

[1975] Possible cardinalities of irredundant bases for closure systems. Discrete Math. 12, 201–204.

[1975 a] Universal classes categorical or free in power. Ph. D. Thesis, Berkeley.

A. M. W. Glass, W. C. Holland and S. H. McCleary

[a] The structure of l-group varieties. Algebra Universalis.

K. Głazek

[1971] Independence with respect to family of mappings in abstract algebras. Dissertationes Math. Rozprawy Mat. 81.

[1971 a] On weak automorphisms of quasi-linear algebras. Colloq. Math. 23, 191–197.

[1971 b] Q-independence and various notions of independence in regular reducts of Boolean algebra. Lectures on the theory of ordered sets and general algebra (Abstracts, Summer Session, Harmonia, 1970). Acta Fac. Rerum. Natur. Univ. Comenian. Math. Mimoriadne Čislo, 25–37.

[1971 c] Weak automorphisms of integral domains and some other abstract algebras. Lectures on the theory of ordered sets and general algebra (Abstracts, Summer Session, Harmonia, 1970). Acta Fac. Rerum. Natur. Univ. Comenian. Math. Mimoriadne Čislo, 39–49.

[1975] Quasi-constants in universal algebras and independent subalgebras. Collection of papers on the theory of ordered sets and general algebra. Acta Fac. Rerum Natur. Univ. Comenian. Math., Special No., 9–16.

(*See also* J. Dudek and K. Głazek; S. Fajtlowicz and K. Głazek)

K. Głazek and A. Iwanik

[1974] Quasi-constants in general algebras. Colloq. Math. 29, 45–50, 159.

[1974 a] Independent subalgebras of a general algebra. Colloq. Math. 29, 189–194.

K. Głazek and J. Michalski

[1974] On weak homomorphisms of general non-indexed algebras. Bull. Acad. Polon. Sci. 22, 651–656.

[1977] Weak homomorphisms of general algebras. Ann. Soc. Math. Polon. 19, 211–228.

M. M. Gluhov

[1971] Free decompositions and algorithmic problems in R-varieties of universal algebras. (Russian) Mat. Sb. (N.S.) 85 (127), 307–338.

L. M. Gluskin

[1970] Studies in general algebra at Saratov. (Russian) Izv. Vysš. Učebn. Zaved. Matematika, no. 4 (95), 3–16.

G. Gnani

[1971] Sulle classi metaideali, Matematiche (Catania) 26, 368–380.

[1972] Un'osservazione sulle classi metaideali. Matematiche (Catania) 27, 105–110.

A. Goetz

[1967] Algebraic independence in an infinite Steiner triple system. Notre Dame J. Formal Logic 8, 51–55.

[1971] A generalization of the direct product of universal algebras. Colloq. Math. 22, 167–176.

A. Goetz and J. R. Senft

[1968] On the structure of normal subdirect powers. Bull. Acad. Polon. Sci. Sér. Sci. Math. Astronom. Phys. 16, 747–750.

J. A. Goguen, J. W. Thatcher, E. G. Wagner and J. B. Wright

[1976] Some fundamentals of order-algebraic semantics. IBM Research Report, RC 6020.

[1977] An initial algebra approach to the specification, correctness, and implementation of abstract data types. IBM Research Report, RC 6487.

[1977 a] Initial algebra semantics and continuous algebras. J. Assoc. Comput. Mach. 24, 68–95.

K. Golema

[1964] Free products of compact general algebras. Colloq. Math. 13, 165–166.

K. Golema-Hartman

[1973] Idempotent reducts of abelian groups and minimal algebras. Bull. Acad. Polon. Sci. Sér. Sci. Math. Astronom. Phys. 21, 809–812.

O. N. Golovin, A. I. Kostrikin, L. A. Skornjakov and A. L. Smel'kin

[1975] Research Seminar on General Algebra. Sessions of the Fall semester of the 1974/75 academic year. (Russian) Vestnik Moskov. Univ. Ser. I Mat. Meh. 30, no. 4, 114–117.

V. A. Gorbunov

[1976] On lattices of quasivarieties. (Russian) Algebra i Logika 15, 436–457.

(*See also* V. P. Belkin and V. A. Gorbunov; A. I. Budkin and V. A. Gorbunov)

M. Gould

[1968] Multiplicity type and subalgebra structure in universal algebras. Pacific J. Math. 26, 469–485.

[1971] Multiplicity type and subalgebra structure in infinitary universal algebras. Colloq. Math. 24, 109–116.

[1972] Subalgebra maps induced by endomorphisms and automorphisms of algebras. Algebra Universalis 2, 88–94.

[1972 a] Automorphism groups of algebras of finite type. Canad. J. Math. 24, 1065–1069.

[1972 b] Automorphism and subalgebra structure in algebras of finite type. Algebra Universalis 2, 369–374.

[1972 c] A note on automorphisms of groupoids. Algebra Universalis 2, 54–56.

[1974] Representable epimorphisms of monoids. Compositio Math. 29, 213–222.

[1975] An equational spectrum giving cardinalities of endomorphism monoids. Canad. Math. Bull. 18, no. 33, 427–429.

[1975 a] Endomorphism and automorphism structure of direct squares of universal algebras. Pacific J. Math. 59, 69–84.

[a] Automorphism groups of free algebras and direct powers. Proceedings of the Colloquium held in Esztergom, 1977. Colloquia Mathematica Societatis János Bolyai. North-Holland Publishing Co., Amsterdam.

[b] An easy proof of Ponizovski's theorem. Semigroup Forum.

M. Gould and H. H. James

[a] Automorphism groups retracting onto symmetric groups.

M. Gould and C. R. Platt

[1971] Versatile monoids and versatile categories. Algebra Universalis 1, 54–62.

[1971 a] Semilattice maps induced by homomorphisms of algebras. Algebra Universalis 1, 90–92.

E. Graczyńska

[a] On connections between sums of double systems of algebras and direct limits. Proceedings of the Colloquium held in Esztergom, 1977. Colloquia

Mathematica Societatis János Bolyai. North-Holland Publishing Co., Amsterdam.

E. GRACZYŃSKA AND A. WROŃSKI

[1975] On normal Agassiz systems of algebras. Polish Acad. Sci. Inst. Philos. Sociol. Bull. Sect. Logic 4, 143–149.

[1975 a] On weak Agassiz systems of algebras. Polish Acad. Sci. Inst. Philos. Sociol. Bull. Sect. Logic 4, 150–153.

G. GRÄTZER

[1966] On bases of infinitary algebras. (Abstract) Notices Amer. Math. Soc. 13, 592.

[1966 a] Equivalence relations of cardinals induced by equational classes of infinitary algebras. (Abstract) Notices Amer. Math. Soc. 13, 632–633.

[1968] Universal Algebra. The University Series in Higher Mathematics. D. Van Nostrand Co., Princeton, N.J.

[1969] Free Σ-structures. Trans. Amer. Math. Soc. 135, 517–542.

[1970] Two Mal'cev-type theorems in universal algebra. J. Combinatorial Theory 8, 334–342.

[1970 a] Composition of functions. Proc. Conf. on Universal Algebra (Queen's Univ., Kingston, Ont., 1969), pp. 1–106. Queen's Univ., Kingston, Ont.

[1971] Lattice theory. First concepts and distributive lattices. W. H. Freeman and Co., San Francisco, Calif.

[1976] On the Amalgamation Property. (Abstract) Notices Amer. Math. Soc. 23, A-268.

[1978] General lattice theory. Series on Pure and Applied Mathematics, Academic Press, New York, N.Y.; Mathematische Reihe, Band 52, Birkhäuser Verlag, Basel; Akademie Verlag, Berlin.

(*See also* J. BERMAN AND G. GRÄTZER; E. FRIED AND G. GRÄTZER; E. FRIED, G. GRÄTZER AND R. W. QUACKENBUSH; H. GASKILL, G. GRÄTZER AND C. R. PLATT)

G. GRÄTZER, B. JÓNSSON AND H. LAKSER

[1973] The amalgamation property in equational classes of modular lattices. Pacific J. Math. 45, 507–524.

G. GRÄTZER AND D. KELLY

[1977] On the product of lattice varieties. (Abstract) Notices Amer. Math. Soc. 24, A-526.

G. GRÄTZER AND H. LAKSER

[1969] Equationally compact semilattices. Colloq. Math. 20, 27–30.

[1971] The structure of pseudocomplemented distributive lattices. II. Congruence extension and amalgamation. Trans. Amer. Math. Soc. 156, 343–358.

[1972] Two observations on the congruence extension property. Proc. Amer. Math. Soc. 35, 63–64.

[1972 a] The structure of pseudocomplemented distributive lattices. III. Injectives and absolute subretracts. Trans. Amer. Math. Soc. 169, 475–487.

[1973] A note on the implicational class generated by a class of structures. Canad. Math. Bull. 16, 603–605.

[1978] A variety of lattices whose quasivarieties are varieties. Algebra Universalis 8, 135–136.

[a] The lattice of quasivarieties of lattices. Algebra Universalis 9 (1979), 102–115.

G. GRÄTZER, H. LAKSER AND J. PŁONKA

[1969] Joins and direct products of equational classes. Canad. Math. Bull. 12, 741–744.

G. GRÄTZER AND W. A. LAMPE

[1971/1972] Representations of complete lattices as congruence lattices of infinitary algebras. (Abstracts) Notices Amer. Math. Soc. 18, A-937, 19, A-683.

[1973] Modular algebraic lattices as congruence lattices of universal algebras. Manuscript.

G. GRÄTZER AND R. MCKENZIE
[1967] Equational spectra and reduction of identities. (Abstract) Notices Amer. Math. Soc. 20, A-505.

G. GRÄTZER AND C. R. PLATT
[1977] A characterization of sharply transferable lattices. (Abstract) Notices Amer. Math. Soc. 24, A-62.
[a] Two embedding theorems for lattices. Proc. Amer. Math. Soc. 69 (1978), 21–24.

G. GRÄTZER, C. R. PLATT AND B. SANDS
[1977] Lattice ideal embedding theorems. (Abstract) Notices Amer. Math. Soc. 24, A-527.

G. GRÄTZER AND J. PŁONKA
[1970] On the number of polynomials of a universal algebra. II. Colloq. Math. 22, 13–19.
[1970 a] A characterization of semilattices. Colloq. Math. 22, 21–24.
[1970 b] On the number of polynomials of an idempotent algebra. I. Pacific J. Math. 32, 697–709.
[1973] On the number of polynomials of an idempotent algebra. II. Pacific J. Math. 47, 99–113.

G. GRÄTZER, J. PŁONKA AND A. SEKANINOVÁ
[1970] On the number of polynomials of a universal algebra. I. Colloq. Math. 22, 9–11.

G. GRÄTZER AND J. SICHLER
[1971] Endomorphism semigroups (and categories) of bounded lattices. Pacific J. Math. 36, 639–647.
[1974] Agassiz sum of algebras. Colloq. Math. 30, 57–59.

G. GRÄTZER AND S. WHITNEY
[1978] Infinitary varieties of structures closed under the formation of complex structures. (Abstract) Notices Amer. Math. Soc. 25, A-224.

A. A. GRAU
[1947] Ternary Boolean algebra. Bull. Amer. Math. Soc. 53, 567–572.
[1951] A ternary operation related to the complete disjunction Boolean algebra. Univ. Nac. Tucuman Rev. Ser. A. 8, 121–126.

T. C. GREEN AND A. TARSKI
[1970] The minimum cardinality of equational bases for varieties of groups and rings. (Abstract) Notices Amer. Math. Soc. 17, 429–430.

G. GRIMEISEN
[1972] Extensions of topological partial algebras to topological algebras. Theory of sets and topology (in honour of Felix Hausdorff, 1868–1942), pp. 199–207. VEB Deutsch. Verlag Wissensch, Berlin.

I. GUESSARIAN
(*See* B. COURCELLE, I. GUESSARIAN AND M. NIVAT)

H. P. GUMM
[1975] Mal'cev conditions in sums of varieties and a new Mal'cev condition. Algebra Universalis 5, 56–64.
[1977] Congruence-equalities and Mal'cev conditions in regular equational classes. Acta Sci. Math. (Szeged) 39, 265–272.
[1978] Is there a Mal'cev theory for single algebras? Algebra Universalis 8, 320–329.
[a] Congruence equalities and Mal'cev conditions in regular equational classes. Acta. Sci. Math. (Szeged).

[b] A cancellation theorem for finite algebras. Proceedings of the Colloquium held in Esztergom, 1977. Colloquia Mathematica Societatis János Bolyai. North-Holland Publishing Co., Amsterdam.
[c] Über die Lösungsmengenen von Gleichungssystemen über allgemeinen Algebren.
[d] Algebras in congruence permutable varieties: Geometrical properties of affine algebras. Algebra Universalis 9 (1979), 8–34.

J. HAGEMANN
[a] Congruences on products and subdirect products of algebras. Algebra Universalis.
[b] On regular and weakly regular congruences.

J. HAGEMANN AND CH. HERRMANN
[a] A concrete ideal multiplication for algebraic systems and its relation with congruence distributivity.

J. HAGEMANN AND A. MITSCHKE
[1973] On n-permutable congruences. Algebra Universalis 3, 8–12.

P. HÁJEK
[1965] The concept of a primitive class of algebras (Birkhoff theorem). (Czech) Časopis Pěst. Mat. 90, 477–486.

A. HAJNAL
(*See* P. ERDÖS AND A. HAJNAL)

A. W. HALES
(*See* K. A. BAKER AND A. W. HALES)

D. K. HALEY
[1970] On compact commutative Noetherian rings. Math. Ann., 272–274.
[1973] Equationally compact Artinian rings. Canad. J. Math. 25, 273–283.
[1974] A note on equational compactness. Algebra Universalis 4, 36–40.
[1974 a] Note on compactifying Artinian rings. Canad. J. Math. 26, 580–582.
[1976] Equational compactness and compact topologies in rings satisfying A. C. C. Pacific J. Math. 62, 99–115.
[1977] Equational compactness in rings and applications to the theory of topological rings. Habilitationsschrift, Univ. Mannheim.

K. HALKOWSKA
[1975] Congruences and automorphisms of algebras belonging to equational classes defined by non-trivializing equalities. Bull. Soc. Roy. Sci. Liège 44, no. 1–2, 8–11.

M. HALL
[1959] The theory of groups. MacMillan, New York.

P. R. HALMOS
[1961] Injective and projective Boolean algebras. Lattice theory. Proc. Symp. Pure Math., Vol. II, pp. 114–122. Amer. Math. Soc., Providence, R.I.

W. HANF
[1956] Representations of lattices by subalgebras. (Abstract) Bull. Amer. Math. Soc. 62, 402.

M. A. HARRISON
[1966] Note on the number of finite algebras. J. Combinatorial Theory 1, 395–397.
[1966 a] The number of isomorphism types of finite algebras. Proc. Amer. Math. Soc. 17, 731–737.

W. S. HATCHER
[1970] Quasiprimitive subcategories. Math. Ann. 190, 93–96.

W. S. HATCHER AND A. SHAFAAT
[1975] Categorical languages for algebraic structures. Z. Math. Logik Grundlagen Math. 21, no. 5, 433–438.

Z. Hedrlín and J. Lambek
[1969] How comprehensive is the category of semigroups? J. Algebra 11, 195–212.

Z. Hedrlín and E. Mendelsohn
[1969] The category of graphs with a given subgraph—with applications to topology and algebra. Canad. J. Math. 21, 1506–1517.

Z. Hedrlín and A. Pultr
[1964] Relations (graphs) with given infinite semigroups. Monatsh. Math. 68. 421–425.

Z. Hedrlín and J. Sichler
[1971] Any boundable binding category contains a proper class of mutually disjoint copies of itself. Algebra Universalis 1, 97–103.

L. Henkin, J. D. Monk and A. Tarski
[1971] Cylindric algebras. Part I. Studies in Logic and the Foundations of Mathematics. North-Holland, Amsterdam.

H. Herrlich
(*See* B. Banaschewski and H. Herrlich)

Ch. Herrmann
[1973] Weak (projective) radius and finite equational bases for classes of lattices. Algebra Universalis 3, 51–58.
[a] Affine algebras in congruence modular varieties. Acta Sci. Math. (Szeged).
(*See also* J. Doyen and Ch. Herrmann; J. Hagemann and Ch. Herrmann)

K. K. Hickin
[1973] Countable type local theorems in algebra. J. of Algebra 27, 523–537.

D. Higgs
[1971] Remarks on residually small varieties. Algebra Universalis 1, 383–385.

G. Higman and B. H. Neumann
[1952] Groups as groupoids with one law. Publ. Math. Debrecen 2, 215–221.

W. Hodges
[a] Compactness and interpolation for Horn sentences.
[b] Horn formulae.

H.-J. Hoehnke
[a] On partial algebras. Proceedings of the Colloquium held in Esztergom, 1977. Colloquia Mathematica Societatis János Bolyai. North-Holland Publishing Co., Amsterdam.

H. Höft
[1972] Operators on classes of partial algebras. Algebra Universalis 2, 118–127.
[1973] A characterization of strong homomorphisms. Colloq. Math. 28, 189–193.
[1973 a] Weak and strong equations in partial algebras. Algebra Universalis 3, 203–215.
[1974] A normal form for some semigroups generated by idempotents. Fund. Math. 84, no. 1, 75–78.
[1974 a] On the semilattice of extensions of a partial algebra. Colloq. Math. 30, 193–201, 317.

H. Höft and P. E. Howard
[a] Representing multi-algebras by algebras, the axiom of choice, and the axiom of dependent choice. Algebra Universalis.

W. C. Holland
[1976] The largest proper variety of lattice ordered groups. Proc. Amer. Math. Soc. 57, no. 1, 25–28.
(*See also* A. M. W. Glass, W. C. Holland and S. H. McCleary)

W. Holsztyński
(*See* S. Fajtlowicz, W. Holsztyński, J. Mycielski and B. Węglorz)

A. Horn
[1969] Free *L*-algebras. J. Symbolic Logic 34, 475–480.
(*See also* G. A. Fraser and A. Horn; F. Galvin and A. Horn)

A. HORN AND N. KIMURA
[1971] The category of semilattices. Algebra Universalis 1, 26–38.

P. E. HOWARD
(*See* H. HÖFT AND P. E. HOWARD)

J. M. HOWIE
[1962] Embedding theorems with amalgamation for semigroups. Proc. London Math. Soc. (3) 12, 511–534.

J. M. HOWIE AND J. R. ISBELL
[1967] Epimorphisms and dominions. II. J. Algebra 6, 7–21.

TAH-KAI HU
[1968] Residually simple and characteristically simple universal algebras. Math. Nachr. 36, 333–344.
[1969] On the fundamental subdirect factorization theorems of primal algebra theory. Math. Z. 112, 154–162.
[1969 a] Stone duality for primal algebra theory. Math. Z. 110, 180–198.
[1969 b] Weak products of simple universal algebras. Math. Nachr. 42, 157–171.
[1970] On equational classes of algebras in which congruences on finite products are induced by congruences on their factors. Manuscript.
[1971] On the topological duality for primal algebra theory. Algebra Universalis 1, 152–154.
[1971 a] Characterization of algebraic functions in equational classes generated by independent primal algebras. Algebra Universalis 1, 187–191.
[1973] Locally equational classes of universal algebras. Chinese J. Math. 1, no. 2, 143–165.

TAH-KAI HU AND P. KELENSON
[1971] Independence and direct factorization of universal algebras. Math. Nachr. 51, 83–99.

A. P. HUHN
[1977] n-distributivity and some questions of the equational theory of lattices. Contributions to universal algebra. Proceedings of the Colloquium held in Szeged, 1975. Colloquia Mathematica Societatis János Bolyai, Vol. 17. North-Holland Publishing Co., Amsterdam, pp. 167–178.
[a] Weakly distributive lattices. Acta Fac. Rerum Natur. Univ. Comenian. Math. (1978).
[b] Congruence varieties associated with reducts of Abelian group varieties. Algebra Universalis 9 (1979), 133–134.

A. HULANICKI AND E. MARCZEWSKI
[1966] Exchange of independent sets in abstract algebras. I. Colloq. Math. 14, 203–215.

H. HULE
[1969] Polynome über universalen Algebren. Monatsh. Math. 73, 329–340.
[1970] Algebraische Gleichungen über universalen Algebren. Monatsh. Math. 74, 50–55.
[1977] Polynomial normal forms and the embedding of polynomial algebras. Contributions to universal algebra. Proceedings of the Colloquium held in Szeged, 1975. Colloquia Mathematica Societatis János Bolyai, Vol. 17. North-Holland Publishing Co., Amsterdam, pp. 179–187.

H. HULE AND W. B. MÜLLER
[1976] On the compatibility of algebraic equations with extensions. J. Austral. Math. Soc. (A)21, 381–383.

G. HUTCHINSON
(*See* G. CZÉDLI AND G. HUTCHINSON)

J. HYMAN
[1974] Automorphisms of 1-unary algebras. Algebra Universalis 4, 61–77.

J. HYMAN AND J. B. NATION
[1974] Automorphisms of 1-unary algebras. II. Algebra Universalis 4, 127–131.

V. I. Igošin

[1971] Characterizable varieties of lattices. (Russian) Ordered sets and lattices, No. 1 (Russian) pp. 22–30. Izdat. Saratov. Univ., Saratov.

[1973] h-characterizable classes of algebraic systems. (Russian) Studies in algebra, No. 3 (Russian), pp. 14–19, 79. Izdat. Saratov. Univ., Saratov.

[1974] Quasivarieties of lattices. (Russian) Mat. Zametki 16, 49–56.

[1974 a] Remarks on characterizable classes of algebras with examples from the theory of semigroups. (Russian) Theory of Semigroups and Its Applications (Russian), Saratov, 3, pp. 30–39.

[1974 b] Characterizable classes of algebraic systems. (Russian) Studies in Algebra, No. 4 (Russian), pp. 27–42. Izdat. Saratov. Univ., Saratov.

B. Imreh

[1968] On a theorem of G. Birkhoff. Publ. Math. Debrecen 15, 147–148.

K. Indermark

(*See* W. Damm, E. Fehr and K. Indermark)

J. R. Isbell

[1968] Small adequate subcategories. J. London Math. Soc. 43, 242–246.

[1973] Epimorphisms and dominions, V. Algebra Universalis 3, 318–320.

[1973 a] Functorial implicit operations. Israel J. Math. 15, 185–188.

[1974] Parametrizable algebras. J. London Math. Soc. (2) 8, 750–752.

[1976] Compatibility and extensions of algebraic theories. Algebra Universalis 6, 37–51.

(*See also* J. M. Howie and J. R. Isbell)

J. R. Isbell, M. I. Klun and S. H. Shanuel

[1977] Affine parts of algebraic theories, I. J. Algebra 44, 1–8.

[a] Affine parts of algebraic theories, II. Canad. J. Math.

A. A. Iskander

[1971] Subalgebra systems of powers of partial universal algebras. Pacific J. Math. 38, 457–463.

[1972] Algebraic functions on p-rings. Colloq. Math. 25, 37–41.

[1972 a] On Boolean extensions of primal algebras. Math. Z. 124, 203–207.

[1972 b] On subalgebra lattices of universal algebras. Proc. Amer. Math. Soc. 32, 32–36.

[1977] Coverings in the lattice of varieties. Contributions to universal algebra. Proceedings of the Colloquium held in Szeged, 1975. Colloquia Mathematica Societatis János Bolyai, Vol. 17. North-Holland Publishing Co., Amsterdam, pp. 189–203.

M. Istinger and H. K. Kaiser

[a] A characterization of polynomially complete algebras.

M. Istinger, H. K. Kaiser and A. F. Pixley

[a] Interpolation in congruence permutable algebras. Colloq. Math.

A. Iwanik

[1972] Remarks on infinite complete algebras. Bull. Acad. Polon. Sci. Sér. Sci. Math. Astronom. Phys. 20, 909–910.

[1974] A remark on a paper of H. Höft (Fund. Math. 84 (1974), no. 1, 75–78). Fund. Math. 84, no. 1, 79–80.

[1974 a] On infinite complete algebras. Colloq. Math. 29, 195–199, 307.

(*See also* K. Głazek and A. Iwanik)

E. Jacobs and R. Schwabauer

[1964] The lattice of equational classes of algebras with one unary operation. Amer. Math. Monthly 71, 151–155.

H. H. James

[1976] Cardinalities of endomorphism monoids of direct powers of universal algebras. (Abstract) Notices Amer. Math. Soc. 23, A-573.

(*See also* M. Gould and H. H. James)

T. M. V. Janssen

(*See* P. van Emde Boas and T. M. V. Janssen)

G. JARFE
(*See* W. FELSCHER AND G. JARFE)

J. JEŽEK
[1968] On the equivalence between primitive classes of universal algebras. Z. Math. Logik Grundlagen Math. 14, 309–320.
[1968 a] Reduced dimension of primitive classes of universal algebras. Comment. Math. Univ. Carolinae 9, 103–108.
[1968 b] Principal dual ideals in lattices of primitive classes. Comment. Math. Univ. Carolinae 9, 533–545.
[1969] Primitive classes of algebras with unary and nullary operations. Colloq. Math. 20, 159–179.
[1970] On atoms in lattices of primitive classes. Comment. Math. Univ. Carolinae 11, 515–532.
[1970 a] On categories of structures and classes of algebras. Dissertationes Math. Rozprawy Mat. 75, 33 pp.
[1971] The existence of upper semicomplements in lattices of primitive classes. Comment. Math. Univ. Carolinae 12, 519–532.
[1971 a] Upper semicomplements and a definable element in the lattice of groupoid varieties. Comment. Math. Univ. Carolinae 12, 565–586.
[1972] Algebraicity of endomorphisms of some relational structures. Acta Univ. Carolinae—Math. et Phys. 13, no. 2, 43–52.
[1974] Realization of small concrete categories by algebras and injective homomorphisms. Colloq. Math. 29, 61–69.
[1975] Algebraicity of endomorphisms of some structures. Collection of papers on the theory of ordered sets and general algebra. Acta Fac. Rerum Natur. Univ. Comenian. Math., Special No., 17–18.
[1976] EDZ-varieties: the Schreier property and epimorphisms onto. Comment. Math. Univ. Carolinae 17, no. 2, 281–290.
[1976 a] Intervals in the lattice of varieties. Algebra Universalis 6, 147–158.
[1977] Endomorphism semigroups and subgroupoid lattices. Contributions to universal algebra. Proceedings of the Colloquium held in Szeged, 1975. Colloquia Mathematica Societatis János Bolyai, Vol. 17. North-Holland Publishing Co., Amsterdam, pp. 209–212.
[a] Varieties of algebras with equationally definable zeros. Czechoslovak Math. J.
[b] A note on complex groupoids. Proceedings of the Colloquium held in Esztergom, 1977. Colloquia Mathematica Societatis János Bolyai. North-Holland Publishing Co., Amsterdam.

J. JEŽEK AND T. KEPKA
[1975] Extensive varieties. Acta Univ. Carolinae—Math. et Phys. 16, no. 2, 79–87.

R. JOHN
[1976] On classes of algebras defined by regular equations. Colloq. Math. 36, 17–21.
[1977] A note on implicational subcategories. Contributions to universal algebra. Proceedings of the Colloquium held in Szeged, 1975. Colloquia Mathematica Societatis János Bolyai, Vol. 17. North-Holland Publishing Co., Amsterdam, pp. 213–222.
[1978] Gültigkeitsbegriffe für Gleichungen in partiellen Algebren. Math. Z. 159, 25–35.

J. S. JOHNSON
[1969] Marczewski independence in mono-unary algebras. Colloq. Math. 20, 7–11.
[1971] Congruences and strong congruences of partial algebras. Manuscript.
[1972] Triples for classes which are not varieties. Algebra Universalis 2, 3–6.
(*See also* S. D. COMER AND J. S. JOHNSON)

J. S. JOHNSON AND E. G. MANES
[1970] On modules over a semiring. J. Algebra 15, 57–67.

B. JÓNSSON
[1953] On the representation of lattices. Math. Scand. 1, 193–206.

[1954] Modular lattices and Desargues' theorem. Math. Scand. 2, 295–314.
[1959] Representations of modular lattices and relation algebras. Trans. Amer. Math. Soc. 92, 449–464.
[1967] Algebras whose congruence lattices are distributive. Math. Scand. 21, 110–121.
[1968] Algebraic structures with prescribed automorphism groups. Colloq. Math. 19, 1–4.
[1972] Topics in universal algebra. Lecture Notes in Mathematics, Vol. 250. Springer-Verlag, Berlin-New York.
[1972 a] The class of Arguesian lattices is selfdual. Algebra Universalis 2, 396.
[1974] Some recent trends in general algebra. Proceedings of the Tarski Symposium (Proc. Sympos. Pure Math., Vol. XXV, Univ. of California, Berkeley, Calif., 1971), pp. 1–19. Amer. Math. Soc., Providence, R.I.
[1975] Varieties of algebras and their congruence varieties. Proc. International Congress of Mathematicians, Vancouver, B.C., 1974, Canadian Math. Congress, 315–320.
[1976] Sums of finitely based lattice varieties. Advances in Math. 14, 454–468.
[1976 a] Identities in congruence varieties. Lattice theory. Proceedings of the Colloquium held in Szeged, 1974. Colloquia Mathematica Societatis János Bolyai, Vol. 14. North-Holland Publishing Co., Amsterdam, pp. 195–205.
[a] A short proof of Baker's finite basis theorem.
(*See also* R. Freese and B. Jónsson; G. Grätzer, B. Jónsson and H. Lakser)

B. Jónsson, G. McNulty and R. W. Quackenbush
[1975] The ascending and descending varietal chains of a variety. Canad. J. Math. 27, 25–31.

B. Jónsson and G. S. Monk
[1969] Representations of primary Arguesian lattices. Pacific J. Math. 30, 95–139.

B. Jónsson and E. Nelson
[1974] Relatively free products in regular varieties. Algebra Universalis 4, 14–19.

B. Jónsson and T. P. Whaley
[1974] Congruence relations and multiplicity types of algebras. Pacific J. Math. 50, 505–520.

H. K. Kaiser
[1974] A class of locally complete universal algebras. J. London Math. Soc. (2) 9, 5–8.
[1975] On a problem in the theory of primal algebras. Algebra Universalis 5, 307–311.
[1975 a] Über lokal polynomvollständige universale Algebren. Abh. Math. Sem. Univ. Hamburg 43, 158–165.
[a] Contributions to the theory of polynomially complete algebras. An. Acad. Brasil. Ci.
[b] Über kompatible Funktionen in universalen Algebren. Acta Math. Acad. Sci. Hungar.
[c] A note on simple universal algebras. Proceedings of the Colloquium held in Esztergom, 1977. Colloquia Mathematica Societatis János Bolyai. North-Holland Publishing Co., Amsterdam.
(*See also* M. Istinger and H. K. Kaiser; M. Istinger, H. K. Kaiser and A. F. Pixley)

H. K. Kaiser and R. Lidl
[1975] Erweiterungs- und Rédeipolynomvollständigkeit universaler Algebren. Acta Math. Acad. Sci. Hungar. 26, no. 3–4, 251–257.

K. Kaiser
(*See* M. Armbrust and K. Kaiser)

I. Kaplansky
[1954] Infinite Abelian groups. The University of Michigan Press, Ann Arbor.

[1969] Infinite Abelian groups (revised edition). The University of Michigan Press, Ann Arbor.

A. B. KARATAY

[1974] A generalization of the characterization theorem of Boolean extensions. Algebra Universalis 4, 185–189.

O. KEANE

[1975] Abstract Horn theories. Model theory and topoi, pp. 15–50. Lecture Notes in Math., Vol. 445, Springer, Berlin.

K. KEIMEL

[1970] Darstellung von Halbgruppen und universellen Algebren durch Schnitte in Garben; bireguläre Halbgruppen. Math. Nach. 45, 81–96.

(*See also* S. BULMAN-FLEMING, I. FLEISCHER AND K. KEIMEL)

K. KEIMEL AND H. WERNER

[1974] Stone duality for varieties generated by quasi-primal algebras. Recent advances in the representation theory of rings and C^*-algebras by continuous sections (Sem., Tulane Univ., New Orleans, La., 1973), pp. 59–85. Mem. Amer. Math. Soc., No. 148, Amer. Math. Soc., Providence, R.I.

H. J. KEISLER

(*See* C. C. CHANG AND H. J. KEISLER)

P. KELENSON

[1972] Regular (normal) Schreier varieties of universal algebras. (Abstracts) Notices Amer. Math. Soc. 19, A-435, A-567.

[1973] Generalized semi-categorical algebras. Math. Nachr. 55, 21–32.

[1973 a] Generalized structure theorems of primal algebra theory. Math. Nachr. 55, 1–19.

(*See also* TAH-KAI HU AND P. KELENSON)

D. KELLY

[1972] A note on equationally compact lattices. Algebra Universalis 2, 80–84.

[1973] Basic equations: Word problems and Mal'cev conditions. (Abstract) Notices Amer. Math. Soc. 20, A-54.

(*See also* G. GRÄTZER AND D. KELLY)

R. T. KEL'TENOVA

[1975] Equational compactness of algebras of unary operations. (Russian) Izv. Akad. Nauk Kazah. SSR Ser. Fiz.-Mat., 82–84.

A. B. KEMPE

[1890] On the relation between the logical theory of classes and the geometrical theory of points. Proc. London Math. Soc. (1) 21, 147–182.

T. KEPKA

[1977] Extensive groupoid varieties. Contributions to universal algebra. Proceedings of the Colloquium held in Szeged, 1975. Colloquia Mathematica Societatis János Bolyai, Vol. 17. North-Holland Publishing Co., Amsterdam, pp. 259–285.

[1977 a] Epimorphisms in some groupoid varieties. Comment. Math. Univ. Carolinae 18, 265–279.

(*See* J. JEŽEK AND T. KEPKA)

R. KERKHOFF

[1965] Eine Konstruktion absolut freier Algebren. Math. Ann. 158, 109–112.

[1970] Gleichungsdefinierbare Klassen partieller Algebren. Math. Ann. 185, 112–133.

N. KIMURA

[1957] On semigroups. Ph.D. Thesis, Tulane.

(*See also* A. HORN AND N. KIMURA)

A. KISIELEWICZ

[a] On the number of polynomials of idempotent algebras. Algebra Universalis.

E. KISS

(*See* B. BIRÓ, E. KISS AND P. P. PÁLFY)

L. KLUKOVITS

[1973] On commutative universal algebras. Acta Sci. Math. (Szeged) 34, 171–174.

[1975] Hamiltonian varieties of universal algebras. Acat Sci. Math. (Szeged) 37, 11–15.

M. I. KLUN

(*See* J. R. ISBELL, M. I. KLUN AND S. H. SHANUEL)

R. A. KNOEBEL

[1970] Primal extensions of universal algebras. Math. Z. 115, 53–57.

[1970 a] Simplicity vis-à-vis functional completeness. Math. Ann. 189, 299–307.

[1973] Products of independent algebras with finitely generated identities. Algebra Universalis 3, 147–151.

[1974] A simplification of the functional completeness proofs of Quackenbush and Sierpinski. Algebra Universalis 4, 133–134.

[1974 a] Congruence-preserving functions in quasiprimal varieties. Algebra Universalis 4, 287–288.

D. E. KNUTH AND P. B. BENDIX

[1970] Simple word problems in universal algebras. Computational Problems in Abstract Algebra (Proc. Conf., Oxford, 1967), pp. 263–297. Pergamon, Oxford.

S. R. KOGALOVSKIĬ

[1961] On categorical characterizations of axiomatizable classes. (Russian) Uspehi Mat. Nauk 16, 209.

[1961 a] On multiplicative semigroups of rings. (Russian) Dokl. Akad. Nauk SSSR 140, 1005–1007.

[1963] On quasi-projective classes of models. (Russian) Dokl. Akad. Nauk SSSR, 148, 505–507.

[1964] A remark on ultraproducts. (Russian) Works of Young Scientists (Russian), pp. 52–54. Izdat. Saratov. Univ., Saratov.

[1965] On multiplicative semigroups of rings. (Russian) Theory of Semigroups and Applications I. (Russian), pp. 251–261. Izdat. Saratov. Univ., Saratov.

[1965 a] On classes of structures of higher degrees. (Russian) VIIth All-Union Algebra Colloquium, General Algebra (Russian), Kišinev.

[1966] Some simple remarks on the undecidability. (Russian) XXIInd Scientific-Technical Conference (Russian), Ivanovo Textile Inst., pp. 3–5.

[1966 a] On semantics of the theory of types. (Russian) Izv. Vysš. Učebn. Zaved. Matematika 1, 89–98.

[1966 b] On the logic of higher degrees. (Russian) Dokl. Akad. Nauk SSSR 171, 1272–1274.

[1967] On algebraic constructions preserving compactness. (Russian) Sibirsk. Mat. Ž. 8, 1202–1205.

[1968] A remark on compact classes of algebraic systems. (Russian) Dokl. Akad. Nauk SSSR 180, 1029–1032.

[1974] Certain simple consequences of the axiom of constructibility. (Russian) Fund. Math. 82, 245–267.

[1974 a] Reductions in the logic of the second order. (Russian) In Philosophy and Logic (Russian), Izdat. Nauka, Moscow, pp. 363–397.

[1975] Logic of predicates of higher orders. (Russian) In Encyclopedia of Cybernetics (Russian), Kiev, 1, 538–540.

[1976] On local properties. (Russian) Dokl. Akad. Nauk SSSR 230, 1275–1278.

(*See also* B. A. ČEPURNOV AND S. R. KOGALOVSKIĬ)

S. R. KOGALOVSKIĬ AND M. A. ROBER

[1974] On the definability of the concept of definability. (Russian) Učen. Zap. Ivanov. Gos. Ped. Inst. 125, 46–72.

S. R. KOGALOVSKIĬ AND V. V. SOLDATOVA

[1977] On the definability by contents. (Russian) In Methods of Logical Analysis (Russian), Moscow, Izdat. Nauka, pp. 41–45.

P. Köhler

[a] The semigroup of varieties of Brouwerian semilattices. Trans. Amer. Math. Soc. 241 (1978), 331–342.
(*See also* W. J. Blok and P. Köhler; J. Berman, A. J. Burger and P. Köhler)

M. Kolibiar

[1977] Primitive subsets of algebras. Contributions to universal algebra. Proceedings of the Colloquium held in Szeged, 1975. Colloquia Mathematica Societatis János Bolyai, Vol. 17. North-Holland Publishing Co., Amsterdam, pp. 287–293.

J. Kollár

[a] The category of unary algebras containing a given subalgebra. Acta. Math. Acad. Sci. Hung.
[b] The category of idempotent 2-unary algebras containing a given subalgebra. Proceedings of the Colloquium held in Esztergom, 1977. Colloquia Mathematica Societatis János Bolyai. North-Holland Publishing Co., Amsterdam.
[c] Some subcategories of integral domains. J. Algebra.
[d] Automorphisms of integral domains containing a given subdomain. Periodica Math. Hung.
[e] Automorphism group of subalgebras, a concrete characterization. Acta. Sci. Math. (Szeged).
[f] Interpolation property in semigroups. Semigroup Forum.
(*See also* E. Fried and J. Kollár)

J. G. Korotenkov

[1975] Homotopy of universal algebras. (Russian) Mat. Issled. 10, no. 1 (35), 165–184, 299.

A. I. Kostrikin

(*See* O. N. Golovin, A. I. Kostrikin, L. A. Skornjakov and A. L. Smel'kin)

V. Koubek and V. Trnková

[a] Isomorphisms of sums of Boolean algebras. Proceedings of the Colloquium held in Szeged, 1976. Colloquia Mathematica Societatis János Bolyai. North-Holland Publishing Co., Amsterdam.

M. Kozák

[1976] Finiteness conditions in EDZ-varieties. Comment. Math. Univ. Carolinae 17, 461–472.

P. Kratochvil

[1969] On the theorem of M. Gould and G. Grätzer. Summer Session. Theory of Ordered Sets and General Algebra, Cikhaj, Brno, 44–48.
[1971] Note on a representation of universal algebras as subdirect powers. Colloq. Math. 24, 11–14.

P. H. Krauss

[1972] Extending congruence relations. Proc. Amer. Math. Soc. 31, 517–520.
[1972 a] On primal algebras. Algebra Universalis 2, 62–67.
[1973] On quasi-primal algebras. Math. Z. 134, 85–89.
(*See also* D. M. Clark and P. H. Krauss)

P. H. Krauss and D. M. Clark

[a] Global subdirect products. Memoirs Amer. Math. Soc.

P. Křivka

[1972] On representations of monoids as monoids of polynomials. Comment. Math. Univ. Carolinae 13, 121–136.

R. L. Kruse

[1973] Identities satisfied by a finite ring. J. Algebra 26, 298–318.

L. Kučera

[1975] On universal concrete categories. Algebra Universalis 5, 149–151.

V. T. KULIK

[1969] $\sigma \circ \rho^{-1}$-stable and $\rho \circ \sigma$-stable classes of universal algebras. (Russian) Works of young scientists: mathematics and mechanics, No. 2 (Russian), pp. 53–58. Izdat. Saratov. Univ., Saratov.

[1970] Sentences preserved under homomorphisms that are one-to-one at zero. (Russian) Izv. Vysš. Učebn. Zaved. Matematika, no. 4 (95), 56–63.

[1971] The compact elements of the lattice of all strong congruence relations of a partial universal algebra. (Russian) Ordered sets and lattices, No. 1 (Russian), pp. 43–48. Izdat. Saratov. Univ., Saratov.

R. KUMMER

[1971] Über Schreiersche Erweiterungen von Universalen Algebren, I. Beiträge Alg. Geom. 1, 201–219.

A. G. KUROŠ

[1970] General algebra. (Russian) Lectures of the 1969–70 academic year, Moscow State University.

[1971] Free sums of multiple-operator linear near-algebras. Mat. Issled. 6, 83–87.

A. H. LACHLAN

(*See* J. T. BALDWIN AND A. H. LACHLAN)

H. LAKSER

[1973] Injective completeness of varieties of unary algebras: a remark on a paper of Higgs. Algebra Universalis 3, 129–130.

[1977] Semisimplicial algebras that satisfy the Kan extension condition. (Abstract) Notices Amer. Math. Soc. 24, A-526.

[a] The lattice of quasivarieties of pseudocomplemented distributive lattices is not modular. (Abstract) Notices Amer. Math. Soc. 25 (1978), A–228.

(*See also* G. BRUNS AND H. LAKSER; G. GRÄTZER AND H. LAKSER; G. GRÄTZER, B. JÓNSSON AND H. LAKSER; G. GRÄTZER, H. LAKSER AND J. PŁONKA)

H. LAKSER, R. PADMANABHAN AND C. R. PLATT

[1972] Subdirect decomposition of Płonka sums. Duke Math. J. 39, 485–488.

J. LAMBEK

(*See* Z. HEDRLÍN AND J. LAMBEK)

W. A. LAMPE

[1969] On related structures of a universal algebra. Ph.D. Thesis, Pennsylvania State University.

[1972] Notes on related structures of a universal algebra. Pacific J. Math. 43, 189–205.

[1972 a] The independence of certain related structures of a universal algebra. I–IV. Algebra Universalis 2, 99–112, 270–283, 286–295, 296–302.

[1973] On the congruence lattice characterization theorem. Trans. Amer. Math. Soc. 182, 43–60.

[1973 a] Representations of lattices as congruence lattices. Proceedings of the University of Houston Lattice Theory Conference (Houston, Tex., 1973), pp. 1–4. Dept. Math., Univ. Houston, Houston, Tex.

[1974] Subalgebra lattices of unary algebras and an axiom of choice. Colloq. Math. 30, 41–55.

[1977] A note on algebras of fixed similarity type. (Abstract) Notices Amer. Math. Soc. 24, A-371.

[a] Notes on congruence lattices of groupoids.

(*See also* G. GRÄTZER AND W. A. LAMPE)

P. J. LANDIN

(*See* R. M. BURSTALL AND P. J. LANDIN)

R. LANEKAN

[1970] Das verallgemeinerte freie Produkt in primitiven Klassen universeller Algebren. I. Publ. Math. Debrecen 17, 321–332.

H. LÄUCHLI
[1971] Concerning infinite graphs and the Boolean prime ideal theorem. Israel J. Math. 4, 422–429.

H. LAUSCH AND W. NÖBAUER
[1973] Algebra of polynomials. North-Holland Mathematical Library, Vol. 5. North-Holland Publishing Co., Amsterdam-London; American Elsevier Publishing Co., Inc., New York.

J. LAWRENCE
(*See* S. BURRIS AND J. LAWRENCE)

D. J. LEHMAN AND M. B. SMYTH
[1977] Data types. University of Warwick, Department of Computer Science, Report 19.

J. J. LE TOURNEAU
(*See* S. D. COMER AND J. J. LE TOURNEAU)

F. W. LEVI AND B. L. VAN DER WAERDEN
[1933] Über eine besondere Klasse von Gruppen. Abh. Math. Sem. Hamburg 9, 154–158.

J. LEWIN
[1968] On Schreier varieties of linear algebras. Trans. Amer. Math. Soc. 132, 553–562.

S. A. LIBER
[1973] Free compact algebras. (Russian) Mat. Sb. (N.S.) 91 (133), 109–133, 144.
[1973 a] Replica-complete classes of topological algebras. (Russian) Studies in algebra, No. 3 (Russian), pp. 34–36, 81. Izdat. Saratov. Univ., Saratov.
[1973 b] Topological ringoids over universal algebras. (Russian) Izv. Vysš. Učebn. Zaved. Matematika, no. 2 (129), 58–63.
[1974] On free algebras of normal closures of varieties. (Russian) Ordered sets and lattices, No. 2 (Russian), pp. 51–53. Izdat. Saratov. Univ., Saratov.

J. LIBICHER
[1972] Some properties of universal algebras preserved under direct products. (Czech) Sb. Prać. Ped. Fak. Ostravé Ser. A7, no. 29, 5–14.

R. LIDL
(*See* H. K. KAISER AND R. LIDL)

C. C. LINDNER
[1971] Identities preserved by the singular direct product. Algebra Universalis 1, 86–89.
[1972] Identities preserved by the singular direct product II. Algebra Universalis 2, 113–117.
(*See also* T. EVANS AND C. C. LINDNER)

J. D. LIPSON
(*See* G. BIRKHOFF AND J. D. LIPSON)

J. ŁOŚ
[1957] Abelian groups that are direct summands of every Abelian group which contains them as pure subgroups. Fund. Math. 44, 84–90.
[1959] Linear equations and pure subgroups. Bull. Acad. Polon. Sci. Sér. Sci. Math. Astr. Phys. 7, 13–18.
[1959 a] Generalized limits in algebraically compact groups. Bull. Acad. Polon. Sci. Sér. Sci. Math. Astr. Phys. 7, 19–21.
[1966] Direct sums in general algebra. Colloq. Math. 14, 33–38.

L. LOVÁSZ
[1967] Operations with structures. Acta Math. Acad. Sci. Hungar. 18, 321–328.
[1971] On the cancellation law among finite relational structures. Period. Math. Hungar. 1, no. 2, 145–156.
[1972] Direct products in locally finite categories. Acta. Sci. Math. 33, 319–322.
(*See also* R. R. APPLESON AND L. LOVÁSZ)

H. F. J. Lowig

[1968] On the definition of an absolutely free algebra. Czechoslovak Math. J. 18, 396–399.

[1974] On algebras generatable by a given set of algebras. Math. Japonicae 19, 83–91.

H. Lugowski

[1976] Grundzüge der Universellen Algebra. Teubner Texte zur Mathematik, Vol. 235, Leipzig B. G. Teubner.

[a] On completing the algebra of flow diagrams. Proceedings of the Colloquium held in Esztergom, 1977. Colloquia Mathematica Societatis János Bolyai. North-Holland Publishing Co., Amsterdam.

I. V. Lvov

[1973] Varieties of associative rings. I, II. (Russian) Algebra i Logika 12, 269–297, 363; ibid. 12, 667–688, 735.

P. Macchi

[1970] Sull'operatore di pseudo prodotto. Ann. Univ. Ferrara Sez. VII (N.S.) 15, 103–112.

J. L. MacDonald

[1974] Conditions for a universal mapping of algebras to be a monomorphism. Bull. Amer. Math. Soc. 80, 888–892.

S. O. MacDonald

[1973] Various varieties. J. Austral. Math. Soc. 16, 363–367.

[1974] Varieties generated by finite algebras. Proceedings of the Second International Conference on the Theory of Groups (Australian Nat. Univ., Canberra, 1973), pp. 446–447. Lecture Notes in Math., Vol. 372, Springer, Berlin.

(*See also* S. Oates)

C. J. Machado

[1964] *N*-ality for rings. (Portuguese) Univ. Lisboa Revista Fac. Ci. A (2) 11, 187–196.

R. Magari

[1967] Su una classe equazionale de algebre. Ann. Mat. Pura Appl. (4) 75, 277–311.

[1967 a] Sulla varietà generata da un'algebra funzionalmente completa di cardinalità infinita. Ann. Mat. Pura Appl. (4) 76, 305–324.

[1969] Una dimostrazione del fatto che ogni varietà ammette algebre semplici. Ann. Univ. Ferrara Sez. VII (N.S.) 14, 1–4.

[1969 a] Varietà a quozienti filtrali. Ann. Univ. Ferrara Sez. VII (N.S.) 14, 5–20.

[1969 b] Costruzione di classi filtrali. Ann. Univ. Ferrara Sez. VII (N.S.) 14, 35–52.

[1969 c] Un'osservazione sulle classi metafiltrali. Ann. Univ. Ferrara Sez. VII (N.S.) 14, 145–147.

[1970] Varietà a congruenze ideali (congruenze ideali II). Ann. Univ. Ferrara Sez. VII (N.S.) 15, 113–129.

[1970 a] Classi metaideali di algebre simili (congruenze ideali III). Ann. Univ. Ferrara Sez. VII (N.S.) 15, 131–143.

[1971] Sulle varietà generate da classi ideali. Boll. Un. Mat. Ital. (4) 4, 1007–1009.

[1971 a] Algebre a congruenze speciali. Symposia Mathematica, Vol. V (INDAM, Rome, 1969/70), pp. 83–111. Academic Press, London.

[1973] The classification of idealizable varieties. Congruenze ideali. IV. J. Algebra 26, 152–165.

[1973 a] Classi e schemi ideali. Ann. Scuola Norm. Sup. Pisa (3) 27, 687–706.

A. A. Mahmudov

(*See* R. A. Baĭramov, R. B. Feĭzullaev and A. A. Mahmudov)

M. Makkai

[1969] Preservation theorems. J. Symbolic Logic 34, 437–459.

[1973] A proof of Baker's finite-base theorem on equational classes generated by finite elements of congruence distributive varieties. Algebra Universalis 3, 174–181.

M. Makkai and G. E. Reyes

[1977] First order categorical logic. Lecture Notes in Mathematics, Vol. 611. Springer-Verlag, Berlin-New York.

A. I. Mal'cev

[1967] A strengthening of the theorems of Slupecki and Jablonskiĭ. (Russian) Algebra i Logika Sem. 6, no. 3, 61–75.

[1967 a] Multiplication of classes of algebraic systems. (Russian) Sibirsk. Mat. Ž. 8, 346–365.

[1968] Some borderline problems of algebra and logic. (Russian) Proc. Internat. Congr. Math. (Moscow, 1966), pp. 217–231. Izdat. Mir, Moscow.

[1970] Algebraic systems. (Russian) Izdat. Nauka, Moscow.

[1971] The metamathematics of algebraic systems. Collected papers. 1936–1967. Studies in Logic and the Foundations of Mathematics, Vol. 66. North-Holland Publishing Co., Amsterdam-London.

[1973] Algebraic systems. Die Grundlehren der mathematischen Wissenschaften, Band 192. Springer-Verlag, New York-Heidelberg.

[1976] Iterative Post algebras. (Russian) Lecture notes; a posthumous edition of the Novosibirsk State University.

W. G. Malcolm

[1973] Application of higher-order ultraproducts to the theory of local properties in universal algebras and relational systems. Proc. London Math. Soc. (3) 27, 617–637.

K. I. Mandelberg

(*See* T. Evans, K. I. Mandelberg and M. F. Neff)

E. G. Manes

[1975] Algebraic theories. Graduate Texts in Mathematics, Vol. 26. Springer-Verlag, Berlin-New York.

(*See also* J. S. Johnson and E. G. Manes)

P. Mangani

[1968] Calcoli generali con connettivi. Matematiche (Catania) 23, 83–106.

A. Marcja

[1972] Una generalizzazione del concetto di prodotto diretto. Symposia Mathematica, Vol. VIII (Note varie, INDAM, Roma, 1972), pp. 363–377. Academic Press, London.

E. Marczewski

[1969] Independence with respect to a family of mappings. Colloq. Math. 20, 17–21.

[1970] A remark on independence in abstract algebras. Comment. Math. Prace Mat. 14, 15–17.

(*See also* S. Fajtlowicz and E. Marczewski; A. Hulanicki and E. Marczewski)

L. Márki

[1975] Über einige Vollständigkeitsbegriffe in Halbgruppen und Gruppen. Acta Math. Acad. Sci. Hung. 26, 343–346.

G. Markowsky

[1976] Chain-complete posets and directed sets with applications. Algebra Universalis 6, 53–68.

[1977] A motivation and generalization of Scott's notion of a continuous lattice. IBM Research Report, RC. 6617.

[1977 a] Categories of chain-complete posets. Theor. Comput. Sci. 4, 125–135.

C. F. Martin

[1973] Equational theories of natural numbers and transfinite ordinals. Ph.D. Thesis, Berkeley.

L. M. MARTYNOV
(*See* L. N. ŠEVRIN AND L. M. MARTYNOV)

G. MATTHIESSEN
[1976] Theorie der heterogenen Algebren. Mathematik—Arbeitspapiere Nr. 3, Universität Bremen Teil A.
[a] Regular and strongly finitary structures over strongly algebroidal categories. Canad. J. Math. 30 (1978), 250–261.

V. P. MATUS
[1970] Free decompositions in an intersection of varieties of universal algebras. (Russian) Mat. Zametki 7, 551–562.
[1972] Subalgebras of free products of algebras of the variety $\mathfrak{H}_{m,n}$. (Russian) Mat. Zametki 12, 303–311.

G. I. MAURER AND M. SZILÁGYI
[1971] Sur une équation de type Fredholm définie dans des algèbres universelles topologiques. Rend. Ist. Mat. Univ. Trieste 3, 200–206.

G. MAZZANTI
[1972] Classi localmente filtrali. Ann. Univ. Ferrara Sez. VII (N.S.) 17, 143–147.
[1972 a] Classi filtrali e distributività delle congruenze. Ann. Univ. Ferrara Sez. VII (N.S.) 17, 149–157.
[1974] Classi ideali e distributività delle congruenze. Ann. Univ. Ferrara Sez. VII (N.S.) 19, 145–156.

S. H. MCCLEARY
(*See* A. M. W. GLASS, W. C. HOLLAND AND S. H. MCCLEARY)

R. MCKENZIE
[1965] On the unique factorization problem for finite commutative semigroups. (Abstract) Notices Amer. Math. Soc. 12, 315.
[1968] On finite groupoids and $\mathscr{K}$-prime algebras. Trans. Amer. Math. Soc. 133, 115–129.
[1970] Equational bases for lattice theories. Math. Scand. 27, 24–38.
[1971] Cardinal multiplication of structures with a reflexive relation. Fund. Math. 70, no. 1, 59–101.
[1971 a] Definability in lattices of equational theories. Ann. Math. Logic 3, no. 2, 197–237.
[1971 b] $\aleph_1$-incompactness of Z. Colloq. Math. 23, 199–202.
[1972] A method for obtaining refinement theorems, with an application to direct products of semigroups. Algebra Universalis 2, 324–338.
[1972 a] Equational bases and non-modular lattice varieties. Trans. Amer. Math. Soc. 174, 1–43.
[1973] Some unsolved problems between lattice theory and equational logic. Proceedings of the University of Houston Lattice Theory Conference (Houston, Tex., 1973), pp. 564–573. Dept. Math., Univ. Houston, Houston, Tex.
[1975] On spectra, and the negative solution of the decision problem for identities having a finite nontrivial model. J. Symbolic Logic 40, 186–196.
[1978] A finite algebra A with SP(A) not elementary. Algebra Universalis 8, 5–7.
[1978 a] Para-primal varieties: A study of finite axiomatizability and definable principal congruences in locally finite varieties. Algebra Universalis 8, 336–348.
[a] On minimal locally finite varieties with permuting congruence relations.
(*See also* G. GRÄTZER AND R. MCKENZIE)

R. MCKENZIE AND S. SHELAH
[1974] The cardinals of simple models for universal theories. Proceedings of the Tarski Symposium (Proc. Sympos. Pure Math., Vol. XXV, Univ. of California, Berkeley, Calif., 1971), pp. 53–74. Amer. Math. Soc., Providence, R.I.

G. McNulty

[1976] The decision problem for equational bases of algebras. Ann. Math. Logic 10, 193–259.

[1976 a] Undecidable properties of finite sets of equations. J. Symbolic Logic 41, 589–604.

[a] Fragments of first order logic. I: Universal Horn logic. J. Symbolic Logic 42 (1977), 221–237.

[b] Structural diversity in the lattice of equational classes. (Abstract) Notices Amer. Math. Soc. 23, A-401.

(*See also* B. Jónsson, G. McNulty and R. W. Quackenbush)

G. McNulty and W. Taylor

[1975] Combinatory interpolation theorems. Discrete Math. 12, 193–200.

P. Mederly

[1975] Three Mal'cev type theorems and their application. Mat. Časopis Sloven. Akad. Vied 25, no. 1, 83–95.

L. Megyesi

(*See* B. Csákány and L. Megyesi)

I. I. Mel'nik

[1969] Varieties of Ω-algebras. (Russian) Studies in Algebra (Russian), pp. 32–40. Izdat. Saratov. Univ., Saratov.

[1971] Normal closures of perfect varieties of universal algebras. (Russian) Ordered sets and lattices, No. 1 (Russian), pp. 56–65. Izdat. Saratov. Univ., Saratov.

[1973] Nilpotent shifts of varieties. (Russian) Mat. Zametki 14, 703–712.

[1974] Majorant shifts of varieties. (Russian) Studies in Algebra, No. 4 (Russian), pp. 70–78. Izdat. Saratov Univ., Saratov.

[1974 a] Majorant translations and lattices of varieties. (Russian) Ordered sets and lattices, No. 4 (Russian), pp. 85–98. Izdat. Saratov Univ., Saratov.

E. Mendelsohn

(*See* Z. Hedrlín and E. Mendelsohn)

N. S. Mendelsohn

[1975] Algebraic construction of combinatorial designs. Proc. Conf. on Algebraic Aspects of Combinatorics, University of Toronto, 157–168.

[1977] The spectrum of idempotent varieties of algebras with binary operators based on two variable identities. Aequationes Mathematicae.

C. A. Meredith and A. N. Prior

[1968] Equational logic. Notre Dame J. Formal Logic 9, 212–226.

S. Meskin

[1969] On some Schreier varieties of universal algebras. J. Austral. Math. Soc. 10, 442–444.

J. Michalski

(*See* K. Głazek and J. Michalski)

G. Michler and R. Wille

[1970] Die primitiven Klassen arithmetischer Ringe. Math. Z. 113, 369–372.

R. Milner and R. Weyrauch

[1972] Proving compiler correctness in a mechanized logic. Machine Intelligence 7, 51–72.

A. Mitschke

[1971] Implication algebras are 3-permutable and 3-distributive. Algebra Universalis 1, 182–186.

[1973] On a representation of groupoids as sums of directed systems. Colloq. Math. 28, 11–18.

[1978] Near unanimity identities and congruence distributivity in equational classes. Algebra Universalis 8, 29–32.

(*See also* J. Hagemann and A. Mitschke)

R. Mlitz

[1977] Jacobson's density theorem in universal algebra. Contributions to universal algebra. Proceedings of the Colloquium held in Szeged, 1975. Colloquia Mathematica Societatis János Bolyai, Vol. 17. North-Holland Publishing Co., Amsterdam, pp. 331–340.

[1978] Cyclic radicals in universal algebra. Algebra Universalis 8, 33–44.

G. S. Monk

(*See* B. Jónsson and G. S. Monk)

J. D. Monk

[1965] General Algebra. Lecture Notes, University of Colorado.

[1975] Some cardinal functions on algebras. Algebra Universalis 5, 76–81.

[1975 a] Some cardinal functions on algebras. II. Algebra Universalis 5, 361–366.

(*See also* L. Henkin, J. D. Monk and A. Tarski)

F. Montagna

[1974] Sulle classi quasi ideali. Boll. Un. Mat. Ital. (4) 10, 85–97.

H. G. Moore

[1970] Free algebra structure: Categorical algebras. Bull. Austral. Math. Soc. 3, 207–215.

H. G. Moore and A. Yaqub

[1968] An existence theorem for semi-primal algebras. Ann. Scuola Norm. Sup. Pisa (3) 22, 559–570.

[1969] On the structure of certain free algebras. Math. Japon. 14, 105–110.

R. Moors

[1974] A propos des théories de Galois finies et infinies. Colloq. Math. 31, 161–164.

A. C. Morel

[1973] Algebras with a maximal *C*-dependence property. Algebra Universalis 3, 160–173.

J. M. Movsisjan

[1974] Biprimitive classes of second degree algebras. (Russian) Mat. Issled. 9, vyp. 1 (31), 70–82.

W. B. Müller

(*See* H. Hule and W. B. Müller)

V. L. Murskiĭ

[1975] The existence of a finite basis of identities, and other properties of "almost all" finite algebras. (Russian) Problemy Kibernet. No. 30, 43–56.

J. Mycielski

(*See also* A. Ehrenfeucht, S. Fajtlowicz and J. Mycielski; S. Fajtlowicz, W. Holsztyński, J. Mycielski and B. Węglorz; S. Fajtlowicz and J. Mycielski)

J. Mycielski and W. Taylor

[1976] A compactification of the algebra of terms. Algebra Universalis 6, 159–163.

D. Myers

[1974] The back-and-forth isomorphism construction. Pacific J. Math. 55, 521–529.

Y. Nakano

[1971] An application of A. Robinson's proof of the completeness theorem. Proc. Japan Acad. 47, suppl. II, 929–931.

(*See also* T. Fujiwara and Y. Nakano)

T. Nakayama

(*See* N. Funayama and T. Nakayama)

J. B. Nation

[1974] Varieties whose congruences satisfy certain lattice identities. Algebra Universalis 4, 78–88.

[1974 a] Congruence lattices of relatively free unary algebras. Algebra Universalis 4, 132.

(*See also* R. Freese and J. B. Nation; J. Hyman and J. B. Nation)

M. F. NEFF
(*See* A. B. CRUSE AND M. F. NEFF; T. EVANS, K. I. MANDELBERG AND M. F. NEFF)

E. NELSON
[1967] Finiteness of semigroups of operators in universal algebra. Canad. J. Math. 19, 764–768.
[1974] Infinitary equational compactness. Algebra Universalis 4, 1–13.
[1974 a] Not every equational class of infinitary algebras contains a simple algebra. Colloq. Math. 30, 27–30.
[1974 b] The independence of the subalgebra lattice, congruence lattice, and automorphism group of an infinitary algebra. Manuscript.
[1975] Some functorial aspects of atomic compactness. Algebra Universalis 5, 367–378.
[1975 a] On the adjointness between operations and relations and its impact on atomic compactness. Colloq. Math. 33, 33–40.
[1975 b] Semilattices do not have equationally compact hulls. Colloq. Math. 34, 1–5.
[1977] Classes defined by implications. Algebra Universalis 7, 405–407.
[1978] Filtered products of congruences. Algebra Universalis 8, 266–268.
(*See also* B. BANASCHEWSKI AND E. NELSON; S. BURRIS AND E. NELSON; B. JÓNSSON AND E. NELSON)

I. NÉMETI
(*See* H. ANDRÉKA AND I. NÉMETI; H. ANDRÉKA, B. DAHN AND I. NÉMETI; H. ANDRÉKA, T. GERGELY AND I. NÉMETI)

I. NÉMETI AND I. SAIN
[a] Cone-implicational subcategories and some Birkhoff-type theorems. Proceedings of the Colloquium held in Esztergom, 1977. Colloquia Mathematica Societatis János Bolyai. North-Holland Publishing Co., Amsterdam.

B. H. NEUMANN
[1952] A note on algebraically closed groups. J. London Math. Soc. 27, 247–249.
[1970] Properties of countable character. Actes Congrès Intern. Math., Vol. 1, 293–296.
(*See also* G. HIGMAN AND B. H. NEUMANN).

B. H. NEUMANN AND E. C. WIEGOLD
[1966] A semigroup representation of varieties of algebras. Colloq. Math. 14, 111–114.

H. NEUMANN
[1967] Varieties of groups. Ergebnisse der Mathematik und ihrer Grenzgebiete, Band 37. Springer-Verlag, Berlin-New York.

P. M. NEUMANN
[1970] The inequality of SQPS and QSP as operators on classes of groups. Bull. Amer. Math. Soc. 76, 1067–1069.
(*See also* A. BLASS AND P. M. NEUMANN)

W. D. NEUMANN
[1969] On cardinalities of free algebras and ranks of operations. Arch. Math. (Basel) 20, 132–133.
[1970] Representing varieties of algebras by algebras. J. Austral. Math. Soc. 11, 1–8.
[1974] On Mal'cev conditions. J. Austral. Math. Soc. 17, 376–384.

M. H. A. NEWMAN
[1941] A characterization of Boolean lattices and rings. J. London Math. Soc. 16, 256–272.

M. NIVAT
(*See* A. ARNOLD AND M. NIVAT; B. COURCELLE, I. GUESSARIAN AND M. NIVAT; B. COURCELLE AND M. NIVAT)

W. Nöbauer
[1973] Polynome und algebraische Gleichungen über universalen Algebren. Jahresbericht. Deutsch. Math.-Verein. 75, 101–113.
[1976] Über die affin vollständigen, endlich erzeugbaren Moduln. Monatsh. Math. 82, 187–198.
(*See also* H. Lausch and W. Nöbauer)

L. Nolin
[1957] Sur les classes d'algèbres équationnelles et les théorèmes de représentation. C. R. Acad. Sci. Paris 244, 1862–1863.

H. Numakura
[1952] On bicompact semigroups. Math. J. Okayama Univ. 1, 99–108.

S. Oates
(*See* S. O. MacDonald)

S. Oates and M. B. Powell
[1964] Identical relations in finite groups. J. Algebra 1, 11–39.

T. Ohkuma
[1966] Ultrapowers in categories. Yokohama Math. J. 14, 17–37.

A. Yu. Ol'shanskiĭ
[1974] Conditional identities in finite groups. (Russian) Sibirsk. Mat. Ž. 15, 1409–1413.

L. Pacholski and B. Węglorz
[1968] Topologically compact structures and positive formulas. Colloq. Math. 19, 37–42.

R. Padmanabhan
[1969] Regular identities in lattices. (Abstracts) Notices Amer. Math. Soc. 16, 652, 969.
[1972] Equational theory of idempotent algebras. Algebra Universalis 2, 57–61.
[1976] Equational theory of algebras with a majority polynomial. (Abstract) Notices Amer. Math. Soc. 23, A-392.
(*See also* H. Lakser, R. Padmanabhan and C. R. Platt)

R. Padmanabhan and R. W. Quackenbush
[1973] Equational theories of algebras with distributive congruences. Proc. Amer. Math. Soc. 41, 373–377.

P. Pagli
[1969] Sui cloni di operazioni definite su un insieme di due elementi. Boll. Un. Mat. Ital. (4) 2, 632–638.

P. P. Pálfy
[a] On certain congruence lattices of finite unary algebras. Comment. Math. Univ. Carolinae 19 (1978), 89–95.
(*See also* B. Biró, E. Kiss and P. P. Pálfy)

P. P. Pálfy and P. Pudlák
[a] Congruence lattices of finite algebras and intervals in subgroup lattices of finite groups. Algebra Universalis.

E. A. Paljutin
[1972] Complete quasivarieties. (Russian) Algebra i Logika 11, 689–693.
[1973] Categorical quasivarieties of arbitrary signature. (Russian) Sibirsk. Mat. Ž. 14, 1285–1303, 1367.
[1975] The description of categorical quasivarieties. (Russian) Algebra i Logika 14, 145–185.
(*See also* A. I. Abakumov, E. A. Paljutin, M. A. Taĭclin and Ju. E. Šišmarev)

B. Pareigis
[1972] Algebraische Kategorien. Überblicke Mathematik, Band 5, pp. 111–144. Bibliographisches Inst., Mannheim.

R. E. PARK
[1976] Equational classes of non-associative ordered algebras. Ph.D. Thesis, UCLA.

V. V. PAŠENKOV
[1974] Duality of topological models. (Russian) Dokl. Akad. Nauk SSSR 218, 291–294.

A. PASINI
[1971] Sull'isomorfismo delle serie normali di un'algebra finitaria (soluzione di un problema di G. Grätzer). Boll. Un. Mat. Ital. (4) 4, 388–394.
[1971 a] Sul reticolo di congruenze forti in algebre parziali finitarie. Boll. Un. Mat. Ital. (4) 4, 630–634.
[1971 b] Sul reticolo delle classi di congruenza di un'algebra finitaria. Matematiche (Catania) 26, 274–290.
[1972] Osservazioni sul reticolo delle classi di congruenza di un'algebra finitaria. Matematiche (Catania) 27, 383–397.
[1973] Sulla sottoalgebra di Frattini di un'algebra. Matematiche (Catania) 28, 161–173.
[1974] Sulle algebre commutative con zero. Boll. Un. Mat. Ital. (4) 10, 683–688.

F. PASTIJN
(*See* L. BABAI AND F. PASTIJN)

P. PERKINS
[1969] Bases for equational theories of semigroups. J. Algebra 11, 298–314.

A. PETRESCU
[1977] Certain questions of the theory of homotopy of universal algebras. Contributions to universal algebra. Proceedings of the Colloquium held in Szeged, 1975. Colloquia Mathematica Societatis János Bolyai, Vol. 17. North-Holland Publishing Co., Amsterdam, pp. 341–355.

J. PFANZAGL
[1967] Homomorphisms for distributive operations in partial algebras. Applications to linear operators and measure theory. Math. Z. 99, 270–278.

E. PICHAT
[1969] Algorithms for finding the maximal elements of a finite universal algebra. (With discussion) Information Processing 68 (Proc. IFIP Congress, Edinburgh, 1968), Vol. 1: Mathematics, Software, pp. 214–218. North-Holland, Amsterdam.

H. E. PICKETT
[1967] Homomorphisms and subalgebras of multialgebras. Pacific J. Math. 21, 327–342.

G. PIEGARI
[1972] On subalgebra lattices of unary algebras. (Abstract) Notices Amer. Math. Soc. 19, A-434.
[1975] Automorphism and subalgebra structure in universal algebra. Ph.D. Thesis, Vanderbilt University.

R. S. PIERCE
[1968] Introduction to the theory of abstract algebras. Holt, Rinehart and Winston, New York.

D. PIGOZZI
[1970] Amalgamation, congruence-extension, and interpolation properties in cylindric algebras. Ph.D. Thesis, Berkeley.
[1971] Amalgamation, congruence-extension, and interpolation properties in algebras. Algebra Universalis 1, 269–349.
[1972] On some operations on classes of algebras. Algebra Universalis 2, 346–353.
[1974] The join of equational theories. Colloq. Math. 30, 15–25.
[1976] The universality of the variety of quasigroups. J. Austral. Math. Soc. Ser. A 21, no. 2, 194–219.

[1976 a] Base-undecidable properties of universal varieties. Algebra Universalis 6, 193–223.
[a] On the structure of equationally complete varieties. I and II.
[b] Universal equational theories and varieties of algebras.
[c] Minimal, locally-finite varieties that are not finitely axiomatizable. Algebra Universalis.

A. I. PILATOVSKAJA
[1968] Free objects in categories. (Russian) Sakharth. SSR Mecn. Akad. Moambe 51, 541–544.
[1968 a] The theory of free decompositions in categories. (Russian). Sakharth. SSR Mecn. Akad. Moambe 52, 31–34.
[1970] On the isomorphic continuation theorem. (Russian) Mat. Zametki 7, 683–686.

A. F. PIXLEY
[1963] Distributivity and permutability of congruence relations in equational classes of algebras. Proc. Amer. Math. Soc. 14, 105–109.
[1970] Functionally complete algebras generating distributive and permutable classes. Math. Z. 114, 361–372.
[1971] The ternary discriminator function in universal algebra. Math. Ann. 191, 167–180.
[1972] A note on hemi-primal algebras. Math. Z. 124, 213–214.
[1972 a] Completeness in arithmetical algebras. Algebra Universalis 2, 179–196.
[1972 b] Local Malcev conditions. Canad. Math. Bull. 15, 559–568.
[1972 c] Local weak independence and primal algebra theory. Boll. Un. Mat. Ital. (4) 5, 381–399.
[1974] Equationally semi-complete varieties. Algebra Universalis 4, 323–327.
[a] Characterizations of arithmetical varieties. Algebra Universalis 9 (1979), 87–98.
[b] Equational properties of congruence lattices.
[c] A survey of interpolation in universal algebra. Proceedings of the Colloquium held in Esztergom, 1977. Colloquia Mathematica Societatis János Bolyai. North-Holland Publishing Co., Amsterdam.
(*See also* K. A. BAKER AND A. F. PIXLEY; A. L. FOSTER AND A. F. PIXLEY; E. FRIED AND A. F. PIXLEY; M. ISTINGER, H. K. KAISER AND A. F. PIXLEY)

C. R. PLATT
[1970] A note on endomorphism semigroups. Canad. Math. Bull. 13, 47–48.
[1971] One-to-one and onto in algebraic categories. Algebra Universalis 1, 117–124.
[1971 a] Iterated limits of universal algebras. Algebra Universalis 1, 167–181.
[1971 b] Example of a concrete category with no embedding into sets under which every epimorphism is represented by a surjection. Algebra Universalis 1, 128.
[1974] Iterated limits of lattices. Canad. J. Math. 26, 1301–1320.
[1977] Finite transferable lattices are sharply transferable. (Abstract) Notices Amer. Math. Soc. 24, A-527.
(*See also* H. GASKILL, G. GRÄTZER AND C. R. PLATT; M. GOULD AND C. R. PLATT; G. GRÄTZER AND C. R. PLATT; G. GRÄTZER, C. R. PLATT AND B. SANDS; H. LAKSER, R. PADMANABHAN AND C. R. PLATT)

E. PŁONKA
[1968] On a problem of Bjarni Jónsson concerning automorphisms of a general algebra. Colloq. Math. 19, 5–8.
[1970] The automorphism groups of symmetric algebras. Colloq. Math. 22, 1–7.
[1971] On weak automorphisms of algebras having a basis. Colloq. Math. 24, 7–10.
(*See also* J. DUDEK AND E. PŁONKA)

J. PŁONKA
[1967] On a method of construction of abstract algebras. Fund. Math. 61, 183–189.
[1967 a] A representation theorem for idempotent medial algebras. Fund. Math. 61, 191–198.

[1967 b] On the methods of representation of some algebras. Colloq. Math. 16, 249–251.

[1967 c] Sums of direct systems of abstract algebras. Bull. Acad. Polon. Sci. Sér. Sci. Math. Astronom. Phys. 15, 133–135.

[1967 d] On some properties and applications of the notion of the sum of a direct system of abstract algebras. Bull. Acad. Polon. Sci. Sér. Sci. Math. Astronom. Phys. 15, 681–682.

[1968] On the number of independent elements in finite abstract algebras with two binary symmetrical operations. Colloq. Math. 19, 9–21.

[1968 a] Some remarks on sums of direct systems of algebras. Fund. Math. 62, 301–308.

[1969] On equational classes of abstract algebras defined by regular equations. Fund. Math. 64, 241–247.

[1969 a] On sums of direct systems of Boolean algebras. Colloq. Math. 20, 209–214.

[1971] A note on the join and subdirect product of equational classes. Algebra Universalis 1, 163–164.

[1971 a] On free algebras and algebraic decompositions of algebras from some equational classes defined by regular equations. Algebra Universalis 1, 261–264.

[1971 b] On the number of polynomials of a universal algebra. III. Colloq. Math. 22, 177–180.

[1971 c] On algebras with n distinct essentially n-ary operations. Algebra Universalis 1, 73–79.

[1971 d] On algebras with at most n distinct essentially n-ary operations. Algebra Universalis 1, 80–85.

[1972] On the number of polynomials of a universal algebra. IV. Colloq. Math. 25, 11–14.

[1973] On binary reducts of idempotent algebras. Algebra Universalis 3, 330–334.

[1973 a] On the minimal extension of the sequence $\langle 0, 0, 2, 4\rangle$. Algebra Universalis 3, 335–340.

[1973 b] On the sum of a direct system of relational systems. Bull. Acad. Polon. Sci. Sér. Sci. Math. Astronom. Phys. 21, 595–597.

[1973 c] On splitting-automorphisms of algebras. Bull. Soc. Roy. Sci. Liège 42, 302–306.

[1973 d] Note on the direct products of some equational classes of algebras. Bull. Soc. Roy. Sci. Liège 42, 561–562.

[1973 e] On the join of equational classes of idempotent algebras and algebras with constants. Colloq. Math. 27, 193–195.

[1974] On groups in which idempotent reducts form a chain. Colloq. Math. 29, 87–91.

[1974 a] On the subdirect product of some equational classes of algebras. Math. Nachr. 63, 303–305.

[1974 b] On connections between the decomposition of an algebra into sums of direct systems of subalgebras. Fund. Math. 84, no. 3, 237–244.

[1975] On relations between decompositions of an algebra into the sums of direct systems of subalgebras. Collection of papers on the theory of ordered sets and general algebra. Acta Fac. Rerum Natur. Univ. Comenian. Math., Special No., 35–37.

[1975 a] Remark on direct products and the sums of direct systems of algebras. Bull. Acad. Polon. Sci. Sér. Sci. Math. Astronom. Phys. 23, no. 5, 515–518. (*See also* B. Ganter, J. Płonka and H. Werner; G. Grätzer, H. Lakser and J. Płonka; G. Grätzer and J. Płonka; G. Grätzer, J. Płonka and A. Sekaninová)

B. I. Plotkin

[1971] The varieties and quasivarieties that are connected with group representations. (Russian) Dokl. Akad. Nauk SSSR 196, 527–530.

[1972] Groups of automorphisms of algebraic systems. Noordhoff International Publishing.

G. D. Plotkin

[1976] A powerdomain construction. SIAM J. Comput. 5, 452–487.

L. Polák

[1977] On varieties of non-indexed algebras. Arch. Math. (Brno) 13, 169–174.

S. V. Polin

[1969] Subalgebras of free algebras of certain varieties of multiple operator algebras. (Russian) Uspehi Mat. Nauk. 24, no. 1 (145), 17–26.

[1973] Functor categories in varieties of universal algebras. (Russian) Mat. Issled. 8, vyp. 1 (27), 130–140, 239.

[1976] On the identities of finite algebras. (Russian) Sib. Math. J. 17, 1356–1366.

[1977] On identities in congruence lattices of universal algebras. (Russian) Mat. Zametki 22, 443–451.

M. Polinová

[1974] Representation of lattices by equivalence relations. Mat. Časopis Sloven. Akad. Vied. 24, 3–6.

D. H. Potts

[1965] Axioms for semi-lattices. Canad. Math. Bull. 8, 519.

M. B. Powell

(*See* S. Oates and M. B. Powell)

V. S. Poythress

[1973] Partial morphisms on partial algebras. Algebra Universalis 3, 182–202.

A. Preller

[1968] On the relationship between the classical and the categorical direct products of algebras. Indag. Math. 30, 512–516.

K. Prikry

(*See* F. Galvin and K. Prikry)

A. N. Prior

(*See* C. A. Meredith and A. N. Prior)

P. Pudlák

[1976] A new proof of the congruence lattice representation theorem. Algebra Universalis 6, 269–275.

[1977] Distributivity of strongly representable lattices. Algebra Universalis 7, 85–92.

(*See also* P. P. Pálfy and P. Pudlák)

P. Pudlák and J. Tůma

[1976] Yeast graphs and fermentation of algebraic lattices. Lattice theory. Proceedings of the Colloquium held in Szeged, 1974. Colloquia Mathematica Societatis János Bolyai, Vol. 14. North-Holland Publishing Co., Amsterdam, pp. 301–341.

[a] Every finite lattice can be embedded in the lattice of all equivalences over a finite set. Algebra Universalis.

N. K. Pukharev

[1966] Construction of A_n^k-algebras. (Russian) Siberian Math. J. 7, 577–579.

A. Pultr

[1964] Concerning universal categories. Comment. Math. Univ. Carolinae 5, 227–239.

[1968] Eine Bemerkung über volle Einbettungen von Kategorien von Algebren. Math. Ann. 178, 78–82.

(*See also* Z. Hedrlín and A. Pultr)

A. Pultr and J. Sichler

[1969] Primitive classes of algebras with two unary idempotent operations, containing all algebraic categories as full subcategories. Comment. Math. Univ. Carolinae 10, 425–445.

A. PULTR AND V. TRNKOVÁ

[1972] Strong embeddings into categories of algebras. Illinois J. Math. 16, 183–195.

[a] On combinatorial, algebraic and topological representations of groups, semigroups and categories. North-Holland, Amsterdam.

R. W. QUACKENBUSH

[1971] Demi-semi-primal algebras and Mal'cev-type conditions. Math. Z. 122, 166–176.

[1971 a] On the composition of idempotent functions. Algebra Universalis 1, 7-12.

[1971 b] Equational classes generated by finite algebras. Algebra Universalis 1, 265–266.

[1972] The triangle is functionally complete. Algebra Universalis 2, 128.

[1972 a] Some remarks on categorical algebras. Algebra Universalis 2, 246.

[1974] Semi-simple equational classes with distributive congruence lattices. Ann. Univ. Sci. Budapest. Eötvös Sect. Math. 17, 15–19.

[1974 a] Some classes of idempotent functions and their compositions. Colloq. Math. 29, 71–81.

[1974 b] Structure theory for equational classes generated by quasi-primal algebras. Trans. Amer. Math. Soc. 187, 127–145.

[1974 c] Near-boolean algebras I: Combinatorial aspects. Discrete Math. 10, 301–308.

[1975] Near-vector spaces over GF(q) and (v, $q+1$, 1)-BIBDS. Linear Algebra and its Applications 10, 259–266.

[1976] Varieties of Steiner loops and Steiner quasigroups. Canad. J. Math. 28, 1187–1198.

[1977] Linear spaces with prime power block sizes. Arch. Math. (Basel) 28, 381–386.

[1977 a] A note on a problem of Goralčík. Contributions to universal algebra. Proceedings of the Colloquium held in Szeged, 1975. Colloquia Mathematica Societatis János Bolyai, Vol. 17. North-Holland Publishing Co., Amsterdam, pp. 363–364.

[a] Algebras with minimal spectrum. Algebra Universalis 9 (1979).

(*See also* E. FRIED, G. GRÄTZER AND R. W. QUACKENBUSH; J. FROEMKE AND R. W. QUACKENBUSH; B. JÓNSSON, G. MCNULTY AND R. W. QUACKENBUSH; R. PADMANABHAN AND R. W. QUACKENBUSH)

R. W. QUACKENBUSH AND B. WOLK

[1971] Strong representations of congruence lattices. Algebra Universalis 1, 165–166.

R. RADO

[1966] Abstract linear dependence. Colloq. Math. 14, 257–264.

(*See also* P. ERDÖS AND R. RADO)

M. D. RADOJČIĆ

[1968] On the embedding of universal algebras in groupoids holding the law $XY * ZU ** = XZ * YU**$. Mat. Vesnik 5 (20), 353–356.

J. C. RAOULT

(*See* B. COURCELLE AND J. C. RAOULT)

JU. K. REBANE

[1967] Primitive classes of single-type algebras. (Russian) Eesti NSV Tead. Akad. Toimetised Füüs.-Mat. 16, 143–145.

[1967 a] On primitive classes of a similarity type. Izvestia Akad. Nauk. Estonia 16, 141–145.

[1968] The representation of universal algebras in semigroups with two-sided cancellation and in commutative semigroups with cancellation. (Russian) Eesti NSV Tead. Akad. Toimetised Füüs.-Mat. 17, 375–378.

[1969] The representation of universal algebras in nilpotent semigroups. (Russian) Sibirsk. Mat. Ž. 10, 945–949.

E. REDI

[1971] Representation of Menger systems by multiplace endomorphisms. (Russian) Tartu Riikl. Ül. Toimetised Vih. 277, 47–51.

J. Reiterman
(*See* J. Adámek and J. Reiterman)

B. Rendi
[1970] Théorie des idéaux et des congruences pour les algèbres universelles. I. An. Univ. Timişoara Ser. Şti. Fiz.-Chim. 8, 199–205.
[1971] Théorie des idéaux et des congruences pour les algèbres universelles. II. An. Univ. Timişoara Ser. Şti. Fiz.-Chim. 9, 109–114.
[1971 a] Théorie des idéaux et des congruences pour les algèbres universelles. III. An. Univ. Timişoara Ser. Şti. Fiz.-Chim. 9, 201–203.

G. E. Reyes
(*See* M. Makkai and G. E. Reyes)

Ju. M. Rjabuhin
[1967] Certain varieties of universal algebras. (Russian) Bul. Akad. Štiince RSS Moldoven. no. 8, 25–53.

M. A. Rorer
(*See* S. R. Kogalovskiĭ and M. A. Rorer)

I. Rosenberg
[1965] La structure des fonctions de plusiers variables sur un ensemble fini. C. R. Acad. Sci. Paris Vie Académique 260, 3817–3819.
[1966] Zu einigen Fragen der Superpositionen von Funktionen mehrerer Veränderlichen. Bul. Inst. Politehn. Iaşi 12(13), Fasc. 1–2, 7–15.
[1966 a] Abbildung abgeschlossener Algebren. Bull. Math. Soc. Sci. Math. R. S. Roumanie 10 (58), 329–334.
[1969] Maximal clones on algebras A and A^r. Rend. Circ. Mat. Palermo (2) 18, 329–333.
[1970] Complete sets for finite algebras. Math. Nachr. 44, 253–258.
[1970 a] Algebren und Relationen. Elektronische Informationsverarbeitung und Kybernetik 6, 115–124.
[1970 b] Über die funktionale Vollständigkeit in den mehrwertigen Logiken. Academia, Praha.
[1971] A classification of universal algebras by infinitary relations. Algebra Universalis 1, 350–354.
[1973] Strongly rigid relations. Rocky Mountain J. Math. 3, no. 4, 631–639.
[1974] Une correspondence de Galois entre les algèbres universelles et les relations dans le même univers. C. R. Acad. Sci. Paris Sér. A 279, 581–582.
[1974 a] Some maximal closed classes of operations on infinite sets. Math. Ann. 212, 157–164.
[1974 b] Universal algebras with all operations of bounded range. Colloq. Math. 30, 177–185.
[1975] Une correspondence de Galois entre les algèbres universelles et les relations dans le même univers. C. R. Acad. Sci. Paris Sér. A-B 280, A615-A616.
[1977] Completeness properties of multiple valued logic algebras. Computer science and multiple valued logic—theory and applications. North-Holland, Amsterdam.
[1978] The subalgebra systems of direct powers. Algebra Universalis 8, 221–227.
(*See also* I. Fleischer and I. Rosenberg)

B. Rotman
[1971] Remarks on some theorems of Rado on universal graphs. J. London Math Soc. (2) 4, 123–126.

G. Rousseau
[1967] Completeness in finite algebras with a single operation. Proc. Amer. Math. Soc. 18, 1009–1013.

D. Rowlands-Hughes
(*See* P. D. Bacsich and D. Rowlands-Hughes)

V. V. ROZEN

[1966] Application of the general theory of algebraic operations to the theory of ordered sets. (Russian) Certain Appl. Theory Binary Relations (Russian), pp. 3–9. Izdat. Saratov. Univ., Saratov.

[1966 a] Isomorphisms of lattices associated with ordered sets. (Russian) Interuniv. Sci. Sympos. General Algebra (Russian), pp. 114–116. Tartu. Gos. Univ., Tartu.

[1969] Imbeddings of partial operatives. (Russian) Works of young scientists: mathematics and mechanics, No. 2 (Russian), pp. 118–123. Izdat. Saratov. Univ., Saratov.

[1973] Partial operations in ordered sets. (Russian) Izdat. Saratov. Univ., Saratov.

G. RUBANOVIČ

[1971] Ordered unary algebras. (Russian) Učen. Zap. Tartu. Gos. Univ. Trudi Mat. Meh. 11, 34–48.

C. RYLL-NARDZEWSKI AND B. WĘGLORZ

[1967] Compactness and homomorphisms of algebraic structures. Colloq. Math. 18, 233–237.

M. SAADE

[1969] A comment on a paper of Evans. Z. Math. Logik Grundlagen 15, 97–100.

G. SABBAGH

[1971] On properties of countable character. Bull. Austral. Math. Soc. 4, 183–192.

G. SABIDUSSI

[1975] Subdirect representations of graphs. Infinite and finite sets (Colloq., Keszthely, 1973; dedicated to P. Erdös on his 60th birthday), Vol. III, pp. 1199–1226. Colloq. Math. Soc. János Bolyai, Vol. 10, North-Holland, Amsterdam.

B. M. ŠAĬN (SCHEIN)

[1965] Relation algebras. Bull. Acad. Polon. Sci. Sér. Sci. Math. Astronom. Phys. 13, 1–5.

[1966] Theory of semigroups as a theory of superpositions of many-place functions. (Russian) Interuniv. Sci. Sympos. General Algebra (Russian), pp. 169–190. Tartu. Gos. Univ., Tartu.

[1970] Relation algebras and function semigroups. Semigroup Forum 1, 1–62.

[1972] An example of two non-isomorphic finite subdirectly irreducible bands generating the same variety of bands. Semigroup Forum 4, 365–366.

(*See also* V. S. GARVACKIĬ AND B. M. ŠAĬN)

B. M. ŠAĬN AND V. S. TROHIMENKO

[1978] Algebras of multiplace functions. Semigroup Forum 16.

I. SAIN

(*See* I. NÉMETI AND I. SAIN)

V. N. SALIĬ

[1967] Semigroups of $\mathfrak{L}$-sets. (Russian) Sibirsk. Mat. Ž. 8, 659–668.

[1969] Equationally normal varieties of universal algebras. (Russian) Works of young scientists: mathematics and mechanics, No. 2 (Russian), pp. 124–130. Izdat. Saratov. Univ., Saratov.

[1973] Boolean-valued algebras. (Russian) Mat. Sb. (N.S.) 92 (134), 550–563, 647.

A. SALOMAA

[1965] On the heights of closed sets of operations in finite algebras. Ann. Acad. Sci. Fenn. Ser. A I No. 363, 12 pp.

D. J. SAMUELSON

[1970] Semi-primal clusters. Ann. Scuola Norm. Sup. Pisa (3) 24, 689–701

[1971] On the conversion of binary algebras into semi-primal algebras. Ann. Scuola Norm. Sup. Pisa (3) 25, 249–267.

[1972] Independent classes of semiprimal algebras. Rocky Mountain J. Math. 2, no. 4, 631–639.

M. D. SANDIK

[1965] On units in n-loops. (Russian) Studies in Algebra and Math. Anal. (Russian), pp. 140–146. Izdat. Karta Moldovenjaske, Kishinev.

[1965 a] Uniqueness of the representation of n-quasigroups. (Russian) Studies in General Algebra (Sem.) (Russian), pp. 123–135. Akad. Nauk Moldav. SSR, Kishinev.

[1977] Invertible multi-operations and permutations. (Russian) Acta Sci. Math. (Szeged) 39, 153–161.

B. SANDS

(See G. GRÄTZER, C. R. PLATT AND B. SANDS)

A. L. A. SANGALLI

[1972] Une approche transformationnelle à l'algèbre universelle. Manuscripta Math. 6, 177–205.

H. P. SANKAPPANAVAR

(See S. BURRIS AND H. P. SANKAPPANAVAR)

N. SAUER AND M. G. STONE

[1977] The algebraic closure of a semigroup of functions. Algebra Universalis 7, 219–233.

[1977 a] The algebraic closure of a function. Algebra Universalis 7, 295–306.

[a] Endomorphism and subalgebra structure; a concrete characterization. Acta Sci. Math. (Szeged) 40.

[b] A Galois correspondence between algebras and endomorphisms. Proceedings of the Colloquium held in Esztergom, 1977. Colloquia Mathematica Societatis János Bolyai. North-Holland Publishing Co., Amsterdam.

B. M. SCHEIN

(See B. M. ŠAĬN)

B. SCHEPULL

[1976] Über Quasivarietäten von partiellen Algebren. Ph.D. Thesis, Akademie der Wissenschaften der DDR.

J. SCHMERL

[1978] On $\aleph_0$-categoricity of filtered Boolean extensions. Algebra Universalis 8, 159–161.

E. T. SCHMIDT

[1969] Kongruenzrelationen algebraischer Strukturen. Mathematische Forschungsberichte, XXV. VEB Deutscher Verlag der Wissenschaften, Berlin.

[1970] Über reguläre Mannigfaltigkeiten. Acta Sci. Math. (Szeged) 31, 197–201.

[1972] On n-permutable equational classes. Acta Sci. Math. (Szeged) 33, 29–30.

(See also B. CSÁKÁNY AND E. T. SCHMIDT)

J. SCHMIDT

[1968] Direct sums of partial algebras and final algebraic structures. Canad. J. Math. 20, 872–887.

[1970] A homomorphism theorem for partial algebras. Colloq. Math. 21, 5–21.

(See also P. BURMEISTER AND J. SCHMIDT)

H. SCHNEIDER

(See M. N. BLEICHER, H. SCHNEIDER AND R. L. WILSON)

K.-H. SCHRIEVER

[1974] Morphismen von Relationen- und Verknüpfungsalgebren. Ph.D. Thesis, Universität Fridericiana Karlsruhe (Technische Hochschule), Karlsruhe.

P. E. SCHUPP

[a] Varieties and algebraically closed algebras. Algebra Universalis.

M. P. SCHÜTZENBERGER

[1945] Sur certaines axiomes de la theorie des structures. C. R. Acad. Sci. Paris 221, 218–220.

(See also S. EILENBERG AND M. P. SCHÜTZENBERGER)

R. SCHWABAUER
(*See* E. JACOBS AND R. SCHWABAUER)

D. SCOTT
[1976] Data types as lattices. SIAM J. Comput. 5, 522–587.

W. SCOTT
[1951] Algebraically closed groups. Proc. Amer. Math. Soc. 2, 118–121.

L. SEDLÁČEK
[1966] An application of the generalized Jordan-Hölder theorem in the theory of direct products of sets with operators. (Czech) Sb. Prací Přírodověd. Fak. Univ. Palackého v Olomouci 21, 45–57.

R. J. SEIFERT
[1971] On prime binary relational structures. Fund. Math. 70, no. 2, 187–203.

M. SEKANINA
[1971] Number of polynomials in ordered algebras. Colloq. Math. 22, 181–192.
[1972] Algebraische Operationen in universellen Algebren. Studien zur Algebra und ihre Anwendungen (mit Anwendungen in der Mathematik, Physik und Rechentechnik) pp. 83–86. Schr. Zentralinst. Math. Mech. Akad. Wissensch. DDR, Heft 16. Akademie Verlag, Berlin.
[1972 a] Realisation of ordered sets by means of universal algebras, especially by semigroups. Theory of sets and topology (in honour of Felix Hausdorff, 1868–1942), pp. 455–466. VEB Deutsch. Verlag Wissensch., Berlin.
(*See also* A. SEKANINOVÁ AND M. SEKANINA)

A. SEKANINOVÁ
[1973] On algebras having at most two algebraic operations depending on n variables. Časopis Pěst. Mat. 98, 113–121, 212.
(*See also* G. GRÄTZER, J. PŁONKA AND A. SEKANINOVÁ)

A. SEKANINOVÁ AND M. SEKANINA
[1971] On the number of polynomials in ordered algebra. Czechoslovak Math. J. 21 (96), 391–398.

A. SELMAN
[1972] Completeness of calculi for axiomatically defined classes of algebras. Algebra Universalis 2, 20–32.

J. R. SENFT
[1970] On weak automorphisms of universal algebras. Dissertationes Math. Rozprawy Mat. 74.
[1970 a] Endomorphism semigroups of free algebras. (Abstract) Notices Amer. Math. Soc. 17, 562.
(*See also* A. GOETZ AND J. R. SENFT)

M. SERVI
[1967] Sulle sottalgebre "normali" di una W-algebra. Riv. Mat. Univ. Parma (2) 8, 243–250.

L. N. ŠEVRIN
[1972] Densely imbedded ideals of algebras. (Russian) Mat. Sb. (N.S.) 88 (130), 218–228.

L. N. ŠEVRIN AND L. M. MARTYNOV
[1971] The attainable classes of algebras. (Russian) Sibirsk. Mat. Ž. 12, 1363–1381.

A. SHAFAAT
[1968] Clusters of algebras. Algebra i Logika 7, no. 5, 109–116.
[1969] On implicationally defined classes of algebras. J. London Math. Soc. 44, 137–140.
[1970] Characterizations of some universal classes of algebras. J. London Math. Soc. (2) 2, 385–388.
[1970 a] A note on quasiprimitive classes of algebras. J. London Math. Soc. (2) 2, 489–492.

[1970 b] Subcartesian products of finitely many finite algebras. Proc. Amer. Math. Soc. 26, 401–404.
[1971] Lattices of subsemivarieties of certain varieties. J. Austral. Math. Soc. 12, 15–20.
[1973] On products of relational structures. Acta Math. Acad. Sci. Hungar. 24, 13–19.
[1973 a] Remarks on quasivarieties of algebras. Arch. Math. (Brno) 9, 67–71.
[1974] A note on Mal'cevian varieties. Canad. Math. Bull. 17, no. 4, 609.
[1974 a] On implicational completeness. Canad. J. Math. 26, 761–768.
[1974 b] On varieties closed under the construction of power algebras. Bull. Austral. Math. Soc. 11, 213–218.
[1975] The number of proper minimal quasivarieties of groupoids. Proc. Amer. Math. Soc. 49, 54–58.
(*See also* W. S. HATCHER AND A. SHAFAAT)

S. H. SHANUEL
(*See* J. R. ISBELL, M. I. KLUN AND S. H. SHANUEL)

S. SHELAH
[1971] Every two elementarily equivalent models have isomorphic ultrapowers. Israel J. Math. 10, 224–233.
[1975] A compactness theorem for singular cardinals, free algebras, Whitehead problem and transversals. Israel J. Math. 21, no. 4, 319–349.
(*See also* R. MCKENZIE AND S. SHELAH)

J. SICHLER
[1967] Concerning endomorphisms of finite algebra. Comment. Math. Univ. Carolinae 8, 405–414.
[1967 a] Category of commutative groupoids is binding. Comment. Math. Univ. Carolinae 8, 753–755.
[1968] $\mathfrak{A}(1, 1)$ can be strongly embedded into category of semigroups. Comment. Math. Univ. Carolinae 9, 257–262.
[1968 a] Concerning minimal primitive classes of algebras containing any category of algebras as full subcategory. Comment. Math. Univ. Carolinae 9, 627–635.
[1971] One-to-one and onto homomorphisms of bounded lattices. Algebra Universalis 1, 267.
[1973] Weak automorphisms of universal algebras. Algebra Universalis 3, 1–7.
[1973 a] Testing categories and strong universality. Canad. J. Math. 25, 370–385.
[a] Group-universal unary varieties. Algebra Universalis.
(*See also* M. E. ADAMS AND J. SICHLER; E. FRIED AND J. SICHLER; G. GRÄTZER AND J. SICHLER; Z. HEDRLÍN AND J. SICHLER; A. PULTR AND J. SICHLER)

H. SIMMONS
[1972] Existentially closed structures. J. Symbolic Logic 37, 293–310.
[1974] The use of injective-like structures in model theory. Compositio Math. 28, 113–142.

F. M. SIOSON
[1964] Equational bases of Boolean algebras. J. Symbolic Logic 29, 115–124.
[1967] A note on congruences. Proc. Japan Acad. 43, 103–107.

JU. E. ŠIŠMAREV
(*See* A. I. ABAKUMOV, E. A. PALJUTIN, M. A. TAĬCLIN AND JU. E. ŠIŠMAREV)

H. L. SKALA
[1971] Trellis theory. Algebra Universalis 1, 218–233.

L. A. SKORNJAKOV
[1971] Complements in a lattice of congruences. (Russian) Mat. Sb. (N.S.) 88 (130), 148–181.
[1973] Almost direct products. (Russian) Trudy Moskov. Mat. Obšč. 29, 215–222.
[1973 a] Radicals of Ω-rings. (Russian) Selected questions of algebra and logic (A collection dedicated to the memory of A. I. Mal'cev) (Russian), pp. 283–299. Izdat. Nauka Sibirsk. Otdel., Novosibirsk.
(*See also* O. N. GOLOVIN, A. I. KOSTRIKIN, L. A. SKORNJAKOV AND A. L. SMEL'KIN)

J. SŁOMIŃSKI
[1966] On mappings between quasi-algebras. Colloq. Math. 15, 25–44.
[1967] On systems of mappings between models. Colloq. Math. 17, 247–268.
[1968] Peano-algebras and quasi-algebras. Dissertationes Math. Rozprawy Mat. 57, 60 pp.
[1974] On the greatest congruence relation contained in an equivalence relation and its applications to the algebraic theory of machines. Colloq. Math. 29, 31–43, 159.

A. L. SMEL'KIN
(*See* O. N. GOLOVIN, A. I. KOSTRIKIN, L. A. SKORNJAKOV AND A. L. SMEL'KIN)

D. M. SMIRNOV
[1969] Lattices of varieties, and free algebras. (Russian) Sibirsk. Mat. Ž. 10, 1144–1160.
[1971] Cantor algebras with one generator. I, II. (Russian) Algebra i Logika 10, 61–75, 658–667.
[a] Varieties and quasivarieties of algebras. Proceedings of the Colloquium held in Esztergom, 1977. Colloquia Mathematica Societatis János Bolyai. North-Holland Publishing Co., Amsterdam.
(*See also* A. A. AKATAEV AND D. M. SMIRNOV)

D. B. SMITH
[1971] Universal homogeneous algebras. Algebra Universalis 1, 254–260.

J. D. H. SMITH
[1976] Mal'cev varieties. Lecture Notes in Mathematics, Vol. 554. Springer-Verlag, Berlin-New York.

R. S. SMITH
(*See* O. FRINK AND R. S. SMITH)

M. B. SMYTH
(*See* D. J. LEHMAN AND M. B. SMYTH)

T. V. SOKOLOVSKAJA
[1967] Multioperation groups as universal algebras with a single operation subordinate to a single identity. (Russian) Sibirsk. Mat. Ž. 8, 853–858.
[1971] Representations of finite universal algebras in finite semigroups. (Russian) Mat. Zametki 9, 285–290.

V. V. SOLDATOVA
(*See* S. R. KOGALOVSKIĬ AND V. V. SOLDATOVA)

S. K. STEIN
[1963] Finite models of identities. Proc. Amer. Math. Soc. 14, 216–222.

M. STEINBY
[1977] On algebras as tree automata. Contributions to universal algebra. Proceedings of the Colloquium held in Szeged, 1975. Colloquia Mathematica Societatis János Bolyai, Vol. 17. North-Holland Publishing Co., Amsterdam, pp. 441–455.

P. N. STEWART
(*See* W. H. CORNISH AND P. N. STEWART)

M. G. STONE
[1969] On endomorphism semigroup structure in universal algebras. Ph.D. Thesis, University of Colorado.
[1971] Proper congruences do not imply a modular congruence lattice. Colloq. Math. 23, 25–27.
[1972] Subalgebra and automorphism structure in universal algebras; a concrete characterization. Acta Sci. Math. (Szeged) 33, 45–48.
[1975] On endomorphism structure for algebras over a fixed set. Colloq. Math. 33, no. 1, 41–45.
(*See also* N. SAUER AND M. G. STONE)

A. R. STRALKA

[1971] The CEP for compact topological lattices. Pacific J. Math. 38, 795–802.

H. SUBRAMANIAN

[1969] On theory of x-ideals. J. Algebra 12, 134–142.

R. SUSZKO

[1974] Equational logic and theories in sentential languages. Colloq. Math. 29, 19–23.

D. SUTER

[1972] Charakter von Mengensystemen. Math. Ann. 199, 37–43.

[1973] n-Hüllenoperatoren und algebraische Operationen. Indag. Math. 35, 108–112.

[1974] Relative Algebraisierbarkeit von Untervollverbänden eines Unteralgebrensystems. Algebra Universalis 4, 229–234.

U. M. SWAMY

[1974] Representation of universal algebras by sheaves. Proc. Amer. Math. Soc. 45, 55–58.

L. SZABÓ

[1975] Endomorphism monoids and clones. Acta Math. Acad. Sci. Hungar. 26, no. 3–4, 279–280.

[1975 a] Characterization of some related semigroups of universal algebras. Acta Sci. Math. (Szeged) 37, 143–147.

[1978] Concrete representation of related structures of universal algebras. I. Acta Sci. Math. (Szeged) 40, 175–184.

F. SZÁSZ

[1973] On Hashimotoian universal algebras with some properties of Hopf. Math. Japon. 18, 229–234.

S. SZÉKELY

(*See* F. GÉCSEG AND S. SZÉKELY)

A. SZENDREI

[1976] The operation ISKP on classes of algebras. Algebra Universalis 6, 349–353.

[1976 a] Idempotent reducts of abelian groups. Acta Sci. Math. (Szeged) 38, 171–182.

[1977] On affine modules. Contributions to universal algebra. Proceedings of the Colloquium held in Szeged, 1975. Colloquia Mathematica Societatis János Bolyai, Vol. 17; North-Holland Publishing Co., Amsterdam, pp. 457–464.

[a] On the idempotent reducts of modules. I and II. Colloquia Mathematica Societatis János Bolyai.

[b] On the arity of affine modules. Colloq. Math.

[c] On modules in which idempotent reducts form a chain. Colloq. Math.

J. SZÉP

[1968] Sulle algebre universali. I. Rend. Mat. (6) 1, 363–370.

M. SZILÁGYI

[1972] Fixed point theorem in the category of universal algebras. An. Univ. Timişoara Ser. Şti. Mat. 10, 215–220.

(*See also* G. I. MAURER AND M. SZILÁGYI)

H. TABATA

[1969] Free structures and universal Horn sentences. Math. Japon. 14, 101–104.

[1971] A generalized free structure and several properties of universal Horn classes. Math. Japon. 16, 91–102.

M. A. TAĬCLIN

(*See* A. I. ABAKUMOV, E. A. PALJUTIN, M. A. TAĬCLIN AND JU. E. ŠIŠMAREV; O. V. BELEGRADEK AND M. A. TAĬCLIN)

A. D. TAĬMANOV

[1974] On the elementary theory of topological algebras. (Russian) Fund. Math. 81, 331–342.

D. TAMARI
[1974] Formulae for well formed formulae and their enumeration. J. Austral. Math. Soc. 17, 154–162.

T. TAMURA
[1966] Attainability of systems of identities on semigroups. J. Algebra 3, 261–276.
[1968] Maximal or greatest homomorphic images of given type. Canad. J. Math. 20, 264–271.

T. TAMURA AND F. M. YAQUB
[1965] Examples related to attainability of identities on lattices and rings. Math. Japon. 10, 35–39.

S. K. TAN
[1970] A note on the universal isomorphism theorem of universal algebras. Nanta Math. 4, 117–122.

A. TARSKI
[1955] A lattice-theoretical fixpoint theorem and its applications. Pacific J. Math. 5, 285–309.
[1956] Equationally complete rings and relation algebras. Indag. Math. 18, 39–46.
[1968] Equational logic and equational theories of algebras. Contributions to Math. Logic (Colloquium, Hannover, 1966), pp. 275–288. North-Holland, Amsterdam.
[1975] An interpolation theorem for irredundant bases of closure structures. Discrete Math. 12, 185–192.
(*See also* J. DONER AND A. TARSKI; T. C. GREEN AND A. TARSKI; L. HENKIN, J. D. MONK AND A. TARSKI)

M. A. TAYLOR
[1975] *R*- and *T*-groupoids: a generalization of groups. Aequationes Math. 12, no. 2/3, 242–248.

W. TAYLOR
[1969] Atomic compactness and graph theory. Fund. Math. 65, 139–145.
[1970] Compactness and chromatic number. Fund. Math. 67, 147–153.
[1970 a] Convergence in relational structures. Math. Ann. 186, 215–227.
[1971] Atomic compactness and elementary equivalence. Fund. Math. 71, no. 2, 103–112.
[1971 a] Some constructions of compact algebras. Ann. Math. Logic 3, no. 4, 395–435.
[1972] Residually small varieties. Algebra Universalis 2, 33–53.
[1972 a] Fixed points of endomorphisms. Algebra Universalis 2, 74–76.
[1972 b] Note on pure-essential extensions. Algebra Universalis 2, 234–237.
[1972 c] On equationally compact semigroups. Semigroup Forum 5, 81–88.
[1973] Characterizing Mal'cev conditions. Algebra Universalis 3, 351–397.
[1973 a] Products of absolute retracts. Algebra Universalis 3, 400–401.
[1974] Pure-irreducible mono-unary algebras. Algebra Universalis 4, 235–243.
[1974 a] Uniformity of congruences. Algebra Universalis 4, 342–360.
[1975] The fine spectrum of a variety. Algebra Universalis 5, 263–303.
[1975 a] Review of seventeen papers on equational compactness. J. Symbolic Logic 40, 88–92.
[1975 b] Continuum many Mal'cev conditions. Algebra Universalis 5, 333–335.
[1976] Pure compactifications in quasi-primal varieties. Canad. J. Math. 28, no. 1, 50–62.
[1976 a] Remarks on compactifying semigroups. Semigroup Forum 12, 215–219.
[1977] Equational logic. Contributions to universal algebra. Proceedings of the Colloquium held in Szeged, 1975. Colloquia Mathematica Societatis János Bolyai, Vol. 17. North-Holland Publishing Co., Amsterdam, pp. 465–501.
[1977 a] Congruence representation of modular lattices. (Abstract) Notices Amer. Math. Soc. 24, A-420, 421.

[1978] Baker's finite basis theorem. Algebra Universalis 8, 191–196.
[a] Varieties of topological algebras. J. Austral. Math. Soc. Ser. A 23 (1977), 207–241.
[b] Varieties obeying homotopy laws. Canad. J. Math. 29 (1977), 498–527.
[c] Topological algebras in non-trivial idempotent varieties.
[d] Primal topological algebras. Algebra Universalis.
(*See also* S. Bulman-Fleming, A. Day and W. Taylor; S. Bulman-Fleming and W. Taylor; G. Fuhrken and W. Taylor; J. Mycielski and W. Taylor; G. McNulty and W. Taylor)

H. H. Teh
[1968] A universal isomorphism theorem of universal algebras. Nanta Math. 3, 1–12.
[1968 a] Generalizations of Zassenhaus' lemma for universal algebras. Nanta Math. 3, 13–19.

H. H. Teh and C. C. Chen
[1964] Some contributions to the study of universal algebras. Bull. Math. Soc. Nanyang Univ., 1–62.
[1965] Direct limit of universal algebras. Bull. Math. Soc. Nanyang Univ. 48–58.
[1968] Extensions of partial orders on a class of universal algebras. Nanta Math. 2, 54–67.

A. A. Terehov
[1958] On free products and permutable congruence relations in primitive classes of algebras. (Russian) Uspehi Mat. Nauk 13, 232.

J. W. Thatcher
(*See* J. A. Goguen, J. W. Thatcher, E. G. Wagner and J. B. Wright)

J. W. Thatcher, E. G. Wagner and J. B. Wright
[1977] Free continuous theories. IBM Research Report, RC 6906.

T. Tichý and J. Vinárek
[1972] On the algebraic characterization of systems of 1-1 partial mappings. Comment. Math. Univ. Carolinae 13, 711–720.

J. Timm
[1969] Produkttreue Klassen universeller Algebren. Arch. Math. (Basel) 20, 485–490.
[1977] On regular algebras. Contributions to universal algebra. Proceedings of the Colloquium held in Szeged, 1975. Colloquia Mathematica Societatis János Bolyai, Vol. 17. North-Holland Publishing Co., Amsterdam, pp. 503–514.

J. Tiuryn
[a] Fixed points and algebras with infinitely long expressions, Parts I and II. Fund. Informat.

V. V. Topentcharov
[1967] Sur les algèbres universelles et la définition des catégories. Bul. Inst. Politehn. Iaşi (N.S.) 13 (17), fasc. 3–4, 5–9.

L. Toti-Rigatelli
(*See* R. Franci and L. Toti-Rigatelli)

T. Traczyk
[1971] On automorphisms in an algebra with a basis. Bull. Acad. Polon. Sci. Sér. Sci. Math. Astronom. Phys. 19, 701–703.

A. N. Trahtman
[1974] Covering elements in the lattice of varieties of algebras. (Russian) Mat. Zametki 15, 307–312.

Tran Duc Mai
[1974] Partitions and congruences in algebras. I. Basic properties. Arch. Math. (Brno) 10, no. 2, 111–122.

V. Trnková
[1966] Universal category with limits of finite diagrams. Comment. Math. Univ. Carolinae 7, 447–456.
[1968] Strong embedding of category of all groupoids into category of semigroups. Comment. Math. Univ. Carolinae 9, 251–256.
[1969] On products in generalized algebraic categories. Comment. Math. Univ. Carolinae 10, 49–89.
[1975] On a representation of commutative semigroups. Semigroup Forum 10, 203–214.
[1975 a] Representation of semigroups by products in a category.
[1976] On products of binary relational structures. Comment. Math. Univ. Carolinae 17, 513–521.
[a] Isomorphism of products and representation of commutative semigroups. Contributions to universal algebra. Proceedings of the Colloquium held in Szeged, 1976. Colloquia Mathematica Societatis János Bolyai, North-Holland Publishing Co., Amsterdam.
(*See also* J. Adámek and V. Trnková; V. Koubek and V. Trnková; A. Pultr and V. Trnková)

V. S. Trohimenko
(*See* B. M. Schein and V. S. Trohimenko)

K. Truöl
[1969] Mengentheoretische Eigenschaften von Systemen unabhängiger Mengen. Gesellschaft für Mathematik und Datenverarbeitung, Bonn.

S. Tulipani
[1971] Cardinali misurabili e algebre semplici in una varietà generata da un'algebra infinitaria m-funzionalmonte completa. Boll. Un. Mat. Ital. (4) 4, 882–887.
[1972] Proprietà metamatematiche di alcune classi di algebre. Rend. Sem. Mat. Univ. Padova 47, 177–186.
[1972 a] Sulla completezza e sulla categoricità della teoria delle W-algebre semplici. Ann. Univ. Ferrara Sez. VII (N.S.) 17, 1–11.
[1973] Sull'aggiunto del funtore dimenticante tra due classi equazionali. Atti Accad. Naz. Lincei Rend. Cl. Sci. Fis. Mat. Natur. (8) 54, 503–508.

J. Tůma
(*See* P. Pudlák and J. Tůma)

K. Urbanik
[1968] On some numerical constants associated with abstract algebras. II. Fund. Math. 62, 191–210.
[1969] Remarks on congruence relations and weak automorphisms in abstract algebras. Colloq. Math. 20, 1–5.
[1969 a] A remark on v^*-algebras. Colloq. Math. 20, 197–202.

A. Ursini
[1972] Sulle varietà di algebre con una buona teoria degli ideali. Boll. Un. Mat. Ital. (4) 6, 90–95.
[1973] Osservazioni sulle varietà BIT. Boll. Un. Mat. Ital. (4) 7, 205–211.

I. I. Valucè
[1968] Ideals of the endomorphism algebra of a free universal algebra. (Russian) Mat. Issled. 3, vyp. 2 (8), 104–112.
[1976] Mappings. (Russian) Izdat. Stiinca, Kishinev.

R. M. Vancko
[1969] Local independence in finite universal algebras. Ph.D. Thesis, Pennsylvania State University.
[1971] The spectrum of some classes of free universal algebras. Algebra Universalis 1, 46–53.
[1972] The family of locally independent sets in finite algebras. Algebra Universalis 2, 68–73.

[1974] The class of algebras in which weak independence is equivalent to direct sums independence. Colloq. Math. 30, 187–191.

J. C. Varlet

[1970] Endomorphisms and fully invariant congruences in unary algebras $\langle A; \Gamma\rangle$. Bull. Soc. Roy. Sci. Liège 39, 575–589.

[1977] Remarks on fully invariant congruences. Contributions to universal algebra. Proceedings of the Colloquium held in Szeged, 1975. Colloquia Mathematica Societatis János Bolyai, Vol. 17. North-Holland Publishing Co., Amsterdam, pp. 515–554.

M. R. Vaughan-Lee

[a] Laws in finite loops. Algebra Universalis.

J. Vinárek

(*See* T. Tichý and J. Vinárek)

J. Vincze and M. Vincze

[1967] Sur l'approximabilité des structures algébriques. C. R. Acad. Sci. Paris Sér. A-B 265, A167–A168.

M. Vincze

(*See* J. Vincze and M. Vincze)

A. A. Vinogradov

[1965] On the decomposability of algebras in a certain quasivariety. (Russian) Algebra i Logika Sem. 4, no. 5, 47–53.

[1965 a] Maximum primitive classes in certain quasivarieties. (Russian) Algebra i Logika Sem. 4, no. 5, 55–65.

[1965 b] On the decomposability of algebras of a certain class into a direct product of simple algebras. (Russian) Dokl. Akad. Nauk SSSR 163, 14–17.

H.-J. Vollrath

[1972] Zur Charakterisierung von Kongruenzrelationen durch Abbildungen. Elem. Math. 27, 133–134.

J. Vonden Steinen

[1967] Homomorphismen bei Algebren mit mehrdeutigen Operationen. Math. Z. 99, 182–192.

S. J. R. Vorster

(*See* R. J. Wille and S. J. R. Vorster)

B. L. van der Waerden

(*See* F. W. Levi and B. L. van der Waerden)

E. G. Wagner

(*See* J. W. Thatcher, E. G. Wagner and J. B. Wright; J. A. Goguen, J. W. Thatcher, E. G. Wagner and J. B. Wright)

Shih-chiang Wang

[1964] A remark on the automorphism groups of algebraic systems. (Chinese) Shuxue Jinzhan 7, 213–218.

R. B. Warfield

[1969] Purity and algebraic compactness for modules. Pacific J. Math. 28, 699–719.

A. G. Waterman

[1965] The free lattice with 3 generators over N_5. Portugal. Math. 26, 285–288.

D. Webb

[1936] Definition of Post's generalized negative and maximum in terms of one binary operation. Amer. J. Math. 58, 173–194.

B. Węglorz

[1965] Compactness of algebraic systems. Bull. Acad. Polon. Sci. Sér. Math. Astronom. Phys. 13, 705–706.

(*See also* J. Anusiak and B. Węglorz; S. Fajtlowicz, W. Holsztyński, J. Mycielski and B. Węglorz; L. Pacholski and B. Węglorz; C. Ryll-Nardzewski and B. Węglorz)

B. Węglorz and A. Wojciechowska
[1968] Summability of pure extensions of relational structures. Colloq. Math. 19, 27–35.

G. H. Wenzel
[1966] On Marczewski's sixtuple of constants in finite universal algebras. Ph.D. Thesis, Pennsylvania State University.
[1969] Automorphism groups of unary algebras on groups. Canad. J. Math. 21, 1165–1171.
[1970] Extensions of congruence relations on infinitary partial algebras. A problem of G. Grätzer. Fund. Math. 67, 163–169.
[1970 a] Subdirect irreducibility and equational compactness in unary algebras $\langle A; f \rangle$. Arch. Math. (Basel) 21, 256–264.
[1970 b] Relative solvability of polynomial equations in universal algebras. Habilitationsschrift, Univ. Mannheim.
[1971] On $(\mathfrak{S}, \mathfrak{A}, \mathfrak{m})$-atomic compact relational systems. Math. Ann. 194, 12–18.
[1973] Eine Charakterisierung gleichungskompakter universeller Algebren. Z. Math. Logik Grundlagen Math. 19, 283–287.

H. Werner
[1970] Eine Charakterisierung funktional vollständiger Algebren. Arch. Math. (Basel) 21, 381–385.
[1971] Produkte von Kongruenzklassengeometrien universeller Algebren. Math. Z. 121, 111–140.
[1973] A Mal'cev condition for admissible relations. Algebra Universalis 3, 263.
[1974] Congruences on products of algebras and functionally complete algebras. Algebra Universalis 4, 99–105.
[1974 a] Diagonal-products and affine completeness. Algebra Universalis 4, 269–270.
[1975] Algebraic representation and model theoretic properties of algebras with the ternary discriminator. Habilitationsschrift. Technische Hochschule Darmstadt.
[1976] Which partition lattices are congruence lattices? Lattice theory. Proceedings of the Colloquium held in Szeged, 1974. Colloquia Mathematica Societatis János Bolyai, Vol. 14. North-Holland Publishing Co., Amsterdam, pp. 433–453.
[1977] Varieties generated by quasi-primal algebras have decidable theories. Contributions to universal algebra. Proceedings of the Colloquium held in Szeged, 1975. Colloquia Mathematica Societatis János Bolyai, Vol. 17. North-Holland Publishing Co., Amsterdam, pp. 555–575.
(*See also* S. Bulman-Fleming and H. Werner; S. Burris and H. Werner; B. A. Davey and H. Werner; B. Ganter, J. Płonka and H. Werner; B. Ganter and H. Werner; K. Keimel and H. Werner)

H. Werner and R. Wille
[1970] Charakterisierungen der primitiven Klassen arithmetischer Ringe. Math. Z. 115, 197–200.

R. Weyrauch
(*See* R. Milner and R. Weyrauch)

T. P. Whaley
[1969] Algebras satisfying the descending chain condition for subalgebras. Pacific J. Math. 28, 217–223.
[1969 a] Endomorphisms of partial algebras. (Abstract) Notices Amer. Math. Soc. 16, 817–818.
[1971] Multiplicity type and congruence relations in universal algebras. Pacific J. Math. 39, 261–268.
(*See also* B. Jónsson and T. P. Whaley)

A. Whiteman
[1937] Postulates for Boolean algebra in terms of ternary rejection. Bull. Amer. Math. Soc. 43, 293–298.

S. WHITNEY
(*See* G. GRÄTZER AND S. WHITNEY)

E. C. WIEGOLD
(*See* B. H. NEUMANN AND E. C. WIEGOLD)

J. WIESENBAUER
[1977] On the polynomial completeness defect of universal algebras. Contributions to universal algebra. Proceedings of the Colloquium held in Szeged, 1975. Colloquia Mathematica Societatis János Bolyai, Vol. 17. North-Holland Publishing Co., Amsterdam, pp. 577–579.
[a] Some categorical aspects of polynomial algebras. Proceedings of the Colloquium held in Esztergom, 1977. Colloquia Mathematica Societatis János Bolyai. North-Holland Publishing Co., Amsterdam.

R. WILLE
[1969] Subdirekte Produkte und konjunkte Summen. J. Reine Angew. Math. 239/240, 333–338.
[1969 a] On a problem of G. Grätzer. (Abstract) Notices Amer. Math. Soc. 16, A-40.
[1970] Kongruenzklassengeometrien. Lecture Notes in Mathematics, Vol. 113. Springer-Verlag, Berlin-New York.
[1976] Allgemeine Algebra—zwischen Grundlagenforschung und Anwendbarkeit. In Der Mathematikunterricht, Jahrg. 22, Heft 2, pp. 40–64. Ernst Klett Verlag, Stuttgart.
(*See also* G. MICHLER AND R. WILLE; H. WERNER AND R. WILLE)

R. J. WILLE AND S. J. R. VORSTER
[1970] Tensor products in primitive categories. Indag. Math. 32, 110–115.

R. L. WILSON
(*See* M. N. BLEICHER, H. SCHNEIDER AND R. L. WILSON)

A. WOJCIECHOWSKA
[1969] Generalized limit powers. Bull. Acad. Polon. Sci. 17, 121–122.
[1969 a] Limit reduced powers. Coll. Math. 20, 203–208.
(*See also* B. WĘGLORZ AND A. WOJCIECHOWSKA)

B. WOJDYLO
[1970] On some problems of J. Slomiński concerning equations in quasi-algebras. Colloq. Math. 21, 1–4.
[1973] Remarks on lattices of congruence relations of quasi-algebras. Colloq. Math. 27, 187–191.
[1975] On equationally definable classes of quasi-algebras. Collection of papers on the theory of ordered sets and general algebra. Acta Fac. Rerum Natur. Univ. Comenian. Math., Special No., 55–57.

A. WOLF
[1974] Sheaf-representations of arithmetical algebras. Recent advances in the representation theory of rings and C^*-algebras by continuous sections. (Sem., Tulane Univ., New Orleans, La., 1973), pp. 87–93. Mem. Amer. Math. Soc., No. 148, Amer. Math. Soc., Providence, R.I.

B. WOLK
(*See* J. BERMAN AND B. WOLK; R. W. QUACKENBUSH AND B. WOLK)

G. C. WRAITH
[1970] Algebraic theories. Aarhus University Lecture Notes Series, 22.

J. B. WRIGHT
(*See* S. EILENBERG AND J. B. WRIGHT; J. A. GOGUEN, J. W. THATCHER, E. G. WAGNER AND J. B. WRIGHT; J. W. THATCHER, E. G. WAGNER AND J. B. WRIGHT)

A. WROŃSKI
(*See* E. GRACZYŃSKA AND A. WROŃSKI)

A. Yaqub
[1967] On the equational interdefinability of certain operations. Portugal. Math. 26, 125–131.
[1967 a] Primal clusters and local binary algebras. Ann. Scuola Norm. Sup. Pisa (3) 21, 111–119.
(*See also* H. G. Moore and A. Yaqub)

F. M. Yaqub
(*See* T. Tamura and F. M. Yaqub)

M. Yasuhara
[1974] The amalgamation property, the universal-homogeneous models and the generic models. Math. Scand. 34, 5–36.

B. Zelinka
[1970] Tolerance in algebraic structures. Czechoslovak Math. J. 20, 179–183.
[1975] Relations between graph theory and abstract algebra. (Czech) Knižnice Odborn. Věd. Spisů Vysoké. Učení Tech. v Brně B 56, 109–114.

I. Žembery
[1973] Chains of decompositions and n-ary relations. Mat. Casopis Sloven. Akad. Vied 23, 297–300.
[1974] Characterization of operations in algebras by means of partially ordered sets. (Russian) Mat. Časopis Sloven. Akad. Vied 24, 277–282.
[1977] Proper and improper free algebras. Contributions to universal algebra. Proceedings of the Colloquium held in Szeged, 1975. Colloquia Mathematica Societatis János Bolyai, Vol. 17. North-Holland Publishing Co., Amsterdam, pp. 595–601.

G. I. Žitomirskiĭ
[1970] Stable binary relations on universal algebras. (Russian) Mat. Sb. (N.S.) 82 (124), 163–174.

P. Zlatoš
[a] Reduced and μ-bounded Boolean extensions of first order structures.

INDEX

EPILOGUE

I met the young man of about twenty-eight at the Polo Park shopping mall in Winnipeg.[1] He walked much faster than I, so I was looking at his back as he passed by. He looked very familiar. Rather thin, with a lot of brown hair, obviously in a hurry. I caught up with him when he paused in front of a shop window. He turned around half-way; he immediately knew who I was. I cannot say that he was happy to see me.

"I did not do so badly," I stammered.

"Really," he responded. "Just compare. When I wrote *Universal Algebra,* I knew it all. Remember? At Penn State, we spent three weeks in the seminar to decide *not* to include an article in the book. I knew most everything that was published. Can you say the same?"

"No, I cannot," I replied.

"And remember your undertaking: Even though you started on *General Lattice Theory* after completing *Universal Algebra,* you resolved to keep your work evenly balanced between the two fields," he called me to account.

"True, but the numbers were against me. Since I finished *Universal Algebra* in 1966, more than 5,000 papers were published in this field and over 13,000 in lattice theory. I would have had to average two papers a day (including more than a hundred books!), just to keep up," I replied.

The young man was mad at me, and with good reason. For about ten years after I finished *Universal Algebra,* I concentrated on lattices. There was so much to do. You cannot write a book on lattices without free products and uniquely complemented lattices, and so much else. And so little was known.... Indeed, fewer than 20% of my papers after 1966 were written on universal algebraic topics and most of them were written before 1980.

* * *

[1]After F. Karinthy, Atheneum, 1913.

A lot has happened in Universal Algebra in forty years. I cannot attempt here to survey the 5,000 papers and dozens of books. But I would like to point out that many of the important papers in these forty years are in—or utilize the results of—five main developments.

1. Theory of quasivarieties. The oldest of these five fields is the theory of quasivarieties started by A. I. Maltsev but seriously explored by V. Gorbunov, his students, and colleagues in Siberia. More recently, some East European and some American mathematicians have also made important contributions. A part of this theory was covered in the book V. A. Gorbunov, *Algebraic theory of quasivarieties.*[2] In fact, he considers universal Horn classes, which include quasivarieties and anti-varieties. Thus, the results apply to a wide variety of algebraic systems that may have relational symbols in their language: graphs, ordered structures, some geometrical models, formal languages, and others. Gorbunov's book contains more than 300 references.

The Birkhoff-Maltsev problem has been one of the main driving forces in this field for the last 25 years. It asks for a characterization of lattices that can be represented as the lattice of all subquasivarieties of a quasivariety. The problem is still open even for finite lattices.

A very thorough coverage of this field was published in a special double issue of Studia Logica **78** (2004). A survey article by M. E. Adams, K. V. Adaricheva, W. Dziobiak, and A. V. Kravchenko and eighteen research articles in almost 400 pages survey the various aspects of the field, including a listing of open problems.

2. Commutator theory. Originally introduced by J. D. H. Smith for permutable varieties in 1976 and extended to modular varieties by C. Herrmann and J. Hagemann in 1979, commutators were fully developed in the book, *Commutator theory for congruence modular varieties* by R. Freese and R. N. McKenzie.[3] This book shows that a natural commutator operation can be defined on the congruence lattice of an algebra in a variety with modular congruence lattices.

As S. Oates-Williams wrote: "It is quite remarkable that anything resembling a commutator can be defined in a general algebra; that it should have such nice properties when restricted to algebras lying in a congruence modular variety is almost too much to expect."

Freese and McKenzie develop the theory and use it to prove deep results. For example, the subdirectly irreducible algebras in a finitely generated congruence modular variety either have a finite bound on their cardinality or no cardinal bound at all. Their concepts and results found many applications in later papers.

---

[2]Translated from the Russian. Siberian School of Algebra and Logic. Consultants Bureau, New York, 1998.

[3]London Mathematical Society Lecture Note Series, 125. Cambridge University Press, Cambridge, 1987.

3. Tame congruence theory. Apart from a paper of P. P. Pálfy, this theory—like Athena, born fully grown from the forehead of Zeus—burst onto the scene fully developed in D. Hobby and R. McKenzie, *The structure of finite algebras.*[4] An excellent write up of this book, is J. Berman's review in the Mathematical Reviews. Three important papers by McKenzie motivated by tame congruence theory to obtain deep results are also jointly reviewed by J. Berman.[5] They include the solution, in the negative, of A. Tarski's famous problem from the early 1960's: Does there exist an algorithm which, when given an effective description of a finite algebra **A**, determines whether or not **A** has a finite basis for its equational theory?

Tame congruence theory has been effectively applied in a large number of papers to solve various problems related to finite algebras and locally finite varieties.

4. *The shape of congruence lattices.* The forthcoming book by K. A. Kearnes and E. W. Kiss presents a beautiful theory. Here is one example: *The congruence lattices of a variety* **V** *satisfy* $SD_\vee$ *iff they satisfy* $SD_\wedge$ *and* **V** *satisfies a nontrivial congruence identity.* In particular, $SD_\wedge$ *and* $SD_\vee$ *are equivalent for varieties satisfying a nontrivial congruence identity.* This was known for locally finite varieties (using tame congruence theory) but that is far more restrictive than the full result.

5. Natural duality theory. The foundations of this theory were laid in 1980 by B. A. Davey and H. Werner. They showed that there is a common universal-algebraic framework for various classical topological dualities, including the dualities for abelian groups (L. S. Pontryagin, 1934), Boolean algebras (M. H. Stone, 1936), and distributive lattices (H. A. Priestley, 1970). They developed methods for finding many topological dualities, in addition to the classical ones. For example, if a finite algebra has a near-unanimity term—and in particular, if it is lattice-based—then there is a natural duality for the quasivariety it generates.

The book by D. M. Clark and B. A. Davey, *Natural dualities for the working algebraist*[6] covers the first eighteen years of the theory of natural dualities. More recent results, including work on full dualities and strong dualities, is covered in the book by J. G. Pitkethly and B. A. Davey, *Dualisability: unary algebras and beyond.*[7]

* * *

And what has happened with the problems I proposed? In Appendix 2, I report on (partial) solutions to a large number of problems. I would say that roughly half of the problems remain unresolved. For instance, the problem:

---

[4]Contemporary Mathematics. **76**. American Mathematical Society, Providence, RI, 1988.
[5]MR1371732-4 (97e:08002a-c)
[6]Studies in Advanced Mathematics. **57**. Cambridge University Press, Cambridge, 1998.
[7]Advances in Mathematics. **9**. Springer, New York, 2005.

Is every finite lattice the congruence lattice of a finite algebra?

now seems very difficult, because a positive solution to the "much easier problem" to embed every finite lattice into a finite partition lattice turned out to be discouragingly hard.

Some problems shifted in focus. In 1979, it seemed likely that every distributive algebraic lattice could be represented as the congruence lattice of a lattice. Now that F. Wehrung has provided a counterexample, we ask:

1. Can every distributive algebraic lattice can be represented as the congruence lattice of an algebra in a congruence distributive variety?

or even stronger:

2. Is there a congruence distributive variety **V** *such that every distributive algebraic lattice can be represented as the congruence lattice of an algebra in* **V***?*

* * *

So what does the excited young man say in this book that is still relevant after all these profound developments? A lot, I think. A vast superstructure has been built up, but the foundation is still basically the same. Despite the numerous books on specialized topics—such as the few mentioned above—it may still be the best introduction to Universal Algebra to learn the basic concepts as presented here and to work out some of the 750 exercises. Then one could proceed to the specialized books. This is a good way to get started. And remember, these specialized topics are all interconnected. Even if you cannot become a researcher in them all, you must have a passing knowledge of all five fields to become successful. Maybe, I am prejudiced. But if I were a young man, this is how I would proceed.

Acknowledgement: I would like to thank K. Adaricheva, J. Berman, G.M. Bergman, B. Davey, R. Freese, K.A. Kearnes, R.N. McKenzie, J.B. Nation, W. Taylor, and F. Wehrung for their help and encouragement.

George Gratzer